Photoperiodism

Photoperiodism

The Biological Calendar

Edited by

Randy J. Nelson

David L. Denlinger

David E. Somers

OXFORD
UNIVERSITY PRESS

2010

OXFORD
UNIVERSITY PRESS

Oxford University Press, Inc., publishes works that further
Oxford University's objective of excellence
in research, scholarship, and education.

Oxford   New York
Auckland   Cape Town   Dar es Salaam   Hong Kong   Karachi
Kuala Lumpur   Madrid   Melbourne   Mexico City   Nairobi
New Delhi   Shanghai   Taipei   Toronto

With offices in
Argentina   Austria   Brazil   Chile   Czech Republic   France   Greece
Guatemala   Hungary   Italy   Japan   Poland   Portugal   Singapore
South Korea   Switzerland   Thailand   Turkey   Ukraine   Vietnam

Published by Oxford University Press, Inc.
198 Madison Avenue, New York, New York 10016
www.oup.com

Oxford is a registered trademark of Oxford University Press

Library of Congress Cataloging-in-Publication Data
Photoperiodism : the biological calendar / edited by Randy J. Nelson,
David L. Denlinger, and David E. Somers.
p. cm.
Includes bibliographical references.
ISBN 978-0-19-533590-3
1. Photoperiodism. I. Nelson, Randy Joe. II. Denlinger, David L. III.
Somers, David E., 1954–
QP82.P45 2009
612'.022—dc22   2008041919

1 3 5 7 9 8 6 4 2
Printed in the United States of America
on acid-free paper

Dedicated to our families

Part II. Photoperiodism in Invertebrates

Part III. Photoperiodism in Vertebrates

Contributors

George E. Bentley
Department of Integrative Biology
Helen Wills Neuroscience Institute
University of California
Berkeley, CA 94720 USA

Eric L. Bittman
Department of Biology
Program in Neuroscience and Behavior, and
Center for Neuroendocrine Studies
University of Massachusetts
Amherst, MA 01003 USA

Bertil Borg
Department of Zoology
Stockholm University
SE-10691
Stockholm, Sweden

David Crews
Section for Integrative Biology
Division of Biological Sciences
University of Texas
Austin, TX 78712 USA

Gregory E. Demas
Department of Biology
Center for the Integrative Study of Animal
Behavior
Program in Neuroscience
Indiana University
Bloomington, IN 47405 USA

David L. Denlinger
Department of Entomology
The Ohio State University
Columbus, OH 43210 USA

Manfred Gödel
Institute for Medical Psychology
University of Munich
Goethestrasse 31 80336
Munich, Germany

Shin S. Goto
Graduate School of Science
Osaka City University
Osaka 558-8585, Japan

Jim Hardie
Department of Life Sciences
Imperial College London
Silwood Park Campus
Ascot, Berks SL5 7PY UK

Ryosuke Hayama
Max Planck Institute for Plant Breeding
Research
Carl-von-Linne Weg 10
Cologne, Germany D-50829

David Hazlerigg
Institute of Biological and Environmental
Sciences
University of Aberdeen,
Aberdeen, Scotland
AB24 2TZ, UK

Pekka Heino
Department of Biological and Environmental
Sciences
University of Helsinki
FIN-00014
Helsinki, Finland

Kumiko Ito-Miwa
Nagoya University, Graduate School
of Science
Division of Biological Science
Furo-cho, Chikusa-ku, Nagoya, Aichi 464-8602
Japan

Takeshi Izawa
National Institute of Agrobiological Sciences
2-1-2 Kannondai
Tsukuba, Ibaraki
305-8602, Japan

Lance J. Kriegsfeld

Department of Psychology
Helen Wills Neuroscience Institute
University of California
Berkeley, CA 94720 USA

Nancy H. Marcus

Department of Oceanography
Florida State University
Tallahassee, FL 32306 USA

Martha Merrow

Department of Chronobiology
Biological Centre
University of Groningen
Postbus 14, 9750AA Haren
The Netherlands

Scott Michaels

Department of Biology and Molecular
Biology Institute
Indiana University
Bloomington, IN 47405

Randy J. Nelson

Departments of Psychology and
Neuroscience
The Ohio State University
Columbus, OH 43210 USA

H. Frederik Nijhout

Department of Biology
Duke University
Durham, NC 27708 USA

Ove Nilsson

Umeå Plant Science Center
Department of Forest Genetics and Plant
Physiology
Swedish University of Agricultural Sciences
S-91083, Umeå, Sweden

Hideharu Numata

Graduate School of Science
Osaka City University
Osaka 558-8585, Japan

Tokitaka Oyama

Nagoya University, Graduate School of
Science
Division of Biological Science
Furo-cho, Chikusa-ku, Nagoya,
Aichi 464-8602, Japan

Tapio Palva

Department of Biological and Environmental
Sciences
University of Helsinki
FIN-00014
Helsinki, Finland

Joanna Putterill

School of Biological Sciences
University of Auckland
Private Bag 92019
Auckland, New Zealand

Tanja Radic

Institute for Medical Psychology
University of Munich
Goethestrasse 31 80336
Munich, Germany

Till Roenneberg

Institute for Medical Psychology
University of Munich
Goethestrasse 31 80336
Munich, Germany

David S. Saunders

Division of Biological Sciences
Institute of Cell, Animal and Population
Biology
Ashworth Laboratories
University of Edinburgh
West Mains Road
Edinburgh EH0 3JT, UK

Lindsay P. Scheef

Department of Oceanography
Florida State University
Tallahassee, FL 32306 USA

Paul S. Schmidt
Department of Biology
University of Pennsylvania
Philadelphia, PA 19104-6018 USA

Peter J. Sharp
Division of Genetics and Genomics
The Roslin Institute
University of Edinburgh
Roslin, Midlothian, EH25 9PS, UK

Sakiko Shiga
Graduate School of Science
Osaka City University
Osaka 558-8585, Japan

Marla B. Sokolowski
Biology Department
University of Toronto, Mississauga
Mississauga, Ontario, L5L 1C6 Canada

David E. Somers
Department of Plant Cellular and Molecular
Biology
The Ohio State University
Columbus, OH 43210 USA

Christine Stockum
School of Biological Sciences
University of Auckland
Private Bag 92019
Auckland, New Zealand

Hiroko Udaka
Graduate School of Science
Osaka City University
Osaka 558-8585, Japan

Guy Warman
Department of Anaesthesiology
University of Auckland
Private Bag 92019
Auckland, New Zealand

Zachary M. Weil
Laboratory of Neuroendocrinology
Rockefeller University
New York, NY 10021 USA

Karen D. Williams
Biology Department
University of Toronto, Mississauga
Mississauga, Ontario, L5L 1C6 Canada

Takashi Yoshimura
Laboratory of Animal Physiology and Avian
Bioscience Research Center
Graduate School of Bioagricultural
Sciences
Nagoya University, Furo-cho
Chikusa-ku, Nagoya 464-8601, Japan

Part I

Photoperiodism in Plants and Fungi

OVERVIEW

David E. Somers

> If you break open the cherry tree,
> Where are the flowers?
> But in the spring time, see how they bloom!
>
> —*Ikkyu Sojun (1394–1481)*

As organisms reliant on the sun as their primary source of energy and unable to move about the world at will, plants need to be particularly attuned to their environment. The predictably regular 24-h rotation of the earth creates a daily oscillation in the light environment available to plants. Additionally, the 23.5° tilt of the earth's axis of rotation from the perpendicular of the ecliptic adds an additional annual element to the factors contributing to the variation of light environment.

The circadian clock is one mechanism that allows plants to address the problem of daily variation in light. With light as one of the primary resetting stimuli of the clock, each dawn will reset the phase of the clock to the same point, with the particular phasing of daily clock-controlled processes following on from that. Early-morning activities, such as preparation for the day's photosynthesis, and later-day processes, such as the closing of stomata, can, in principle, all be controlled by the same circadian clock simply by adjusting how each is connected to the 24-h oscillator that constitutes the core of the clockwork.

Less obvious is a method of keeping track of the year. Seasonality would not exist were it not for that tilt of the earth's axis. But with that comes the familiar changes in day length that waxes to a maximum at the summer solstice and wanes to a minimum at the winter solstice. Importantly, these changes, at a specific latitude, occur annually as predictably as the 24-h rise and set of the sun. Additionally, the magnitude of the annual variation depends on latitude. Hence, particularly for organisms north and south of the tropics of Cancer and Capricorn, this yearly change can provide temporal information markers allowing them to position their lives within the annual progression of seasons. Also, although any specific day length will occur twice a year at any given latitude (figure PI.1), sensing whether days are increasing or decreasing in photoperiod can allow the plant to determine whether it is spring or fall.

The focus of this book is our current understanding of the mechanism of how plants, insects, and mammals use the regular changes in day length to coordinate their development and physiology to these seasonal changes. These first seven chapters are devoted to photoperiodism in plants and fungi.

The most intensively studied photoperiodic process in plants has been the transition from vegetative to reproductive growth. Not only is the onset of flowering a dramatically different stage in the life cycle of a plant, but many primary food products, such as grains, are the consequences of successful reproduction. Hence, much of the early modern research into understanding what controls flowering time was conducted in an agricultural context. The careful experiments of Garner and Allard (1920) and others clearly established that different plants respond to the same day length differently: some species are induced to flower when given periods of long days (or short days), while the same conditions will inhibit flowering in other species. Extensive studies with a wide range of species allowed researchers to classify many as "long-day" and "short-day" plants, as well as "day-neutral" plants (Thomas and Vince-Prue, 1997). Long-day plants (LDPs) will flower, or flower

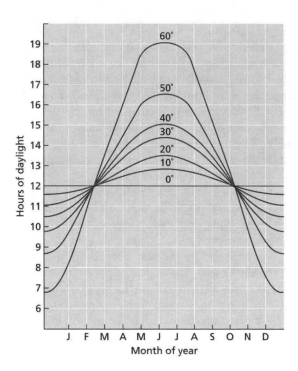

FIGURE PI.1.
The effect of latitude on day length throughout the year (reprinted with permission of Sinauer Associates from Taiz and Zeiger, *Plant Physiology*, 4th ed.).

most rapidly, only when there are more than a certain number of daylight hours in a 24-h period, whereas short-day plants (SDPs) flower when the light period is less than a certain length over a 24-h day. Day-neutral plants (DNPs) flower independent of day length. The period of light necessary for floral induction is termed the critical day length and will vary between species or ecotype even within the general LDP and SDP groupings. Additionally, some species require only a single inductive period to bring about flowering, while others may require many days of induction. Some species, such as *Arabidopsis thaliana*, are quantitative LDPs, meaning that flowering will still occur under short days, but is more rapid in long days. Other species will remain vegetative indefinitely until the appropriate critical day length is given.

Despite the emphasis on day length, numerous additional experiments indicate that it is actually the length of unbroken darkness that is the critical parameter in photoperiodic timing in plants (Thomas and Vince-Prue, 1997). Night break experiments, where a pulse or short duration of light interrupts the night, show that uninterrupted darkness is important to allow flowering in SDPs. Similar "daybreak" experiments where a period of darkness interrupts the day have no effect. In LDPs, a similar effect by a night break during a normally inhibitory long night, usually of longer duration (hours), can accelerate flowering, as well, but as a rule night breaks are not as effective in LDPs as in SDPs.

By varying the time of night at which the light pulse is given, it became clear that the effectiveness of the night break is not the same throughout the dark period. When extended dark periods (e.g., 48–72 h) are used in conjunction with a fixed photoperiod, scanning the long dark period with differently phased night breaks results in a rhythmic response in the effect on flowering time. The period between the most effective light breaks for either induction (for LDPs) or inhibition (for SDPs) is approximately 24 h (Thomas and Vince-Prue, 1997). This is very compelling evidence for the involvement of a circadian clock in the control of photoperiodic timing. A second approach involves altering the length of the dark period following a short photoperiod (e.g., 8:16, 8:20, 8:24, 8:28) and observing the effects of one or more of these altered cycles on the flowering response. These so-called resonance (or Nanda-Hamner) experiments also reveal a circadian rhythmicity to the effectiveness of certain cycle lengths. For example, cycle lengths of 24, 48, and 72 h were more effective for the suppression of flowering in the LDP *Hyoscyamus niger* and for the induction of flowering in soybean than other combinations (Hsu and Hamner, 1967).

Recent inroads into the molecular nature of the plant circadian clock have provided a basis for understanding photoperiodism in plants. In chapter 1, Putterill and coauthors outline the current molecular model of the clock, as gleaned from research with *Arabidopsis*. As this and subsequent chapters point out, a connection between the circadian clock and photoperiodism was first proposed by Bünning (1936). Prior to this, an hourglass model had been presumed, based on phytochrome's role in the night break (Hendricks, 1960), but as noted above, subsequent experiments implicated a role for a circadian timing mechanism. An evolution in thought from Bünning's first proposal to a his later elaboration and refinement (Bünning, 1960), and later still by Pittendrigh (Pittendrigh and Minis, 1964) led to the *external coincidence* model of photoperiodism. Here, the circadian clock drives a photoperiodic rhythm of sensitivity to light. When a light stimulus is received during a time in the rhythm of high light sensitivity, flowering will be inhibited in SDPs or promoted in LDPs. It is the coincidence of this external light pulse with the phase of high sensitivity that brings about the photoperiodic response. A contrasting model, also developed by Pittendrigh (1972), postulates the existence of two internal circadian oscillators that are normally out of phase, but are brought to internal coincidence by the rephasing of one or both by light occurring at the appropriate time. Light in this scheme only acts to entrain these oscillators, not to entrain and photoinduce as in the previous model. While there is little evidence for this model operating in plants, other chapters in this book address possibilities in insects and other animals.

A final consideration, further addressed in the chapters in this section, is the nature of the most effective wavelengths for the photoperiodic response in plants, and the photoreceptors involved. In SDPs, red/far-red (R/FR) reversibility experiments have unequivocally implicated phytochrome as the photoreceptor mediating night breaks. Remarkably, the effectiveness of a night break in the SDPs *Xanthium* and soybean can be predicted by knowing whether the last of a series of R-FR light pulses ends with either R or FR light. Although a 2- to 3-min R pulse will inhibit flowering, plants receiving an R-FR series will flower, as will plants receiving an R-FR-R-FR series. Plants given an R-FR-R-FR-R sequence will again be inhibited (Downs, 1956; Thomas and Vince-Prue, 1997). Chapters 2 and 3, in the discussion of the SDPs rice and Japanese morning glory (*Pharbitis* = *Ipomoea*), respectively, update the night break and photoreceptor studies in these two species at the molecular level.

Similar R/FR experiments are hard to come by for LDPs, in part because when night breaks are effective, much longer periods of light are necessary, and reversibility experiments become ambiguous. However, end-of-day FR light extensions into a normally long dark period, which accelerates flowering in LDPs, do indicate a role for phytochrome. Additionally, when short-day-grown LDPs are given FR pulses on top of R or fluorescent light extensions, there is a circadian rhythmicity to the effectiveness of the FR pulse to promote flowering (Deitzer et al., 1982). These kinds of experiments demonstrate a role both for the clock and for phytochrome for floral induction in LDPs. In chapter 1, Joanna Putterill and coauthors take this discussion further, examining studies linking the role of phytochrome A and the blue light photoreceptor cryptochrome (CRY) in the flowering of *Arabidopsis thaliana* to the expression level of flowering genes such as the floral integrator FT (Yanovsky and Kay, 2002).

Although the change from the vegetative to floral state occurs in the apical bud, the perception of the photoperiodic stimulus occurs in the leaves. This necessitates the movement of a signal from leaf to bud. The search for this elusive "florigen" has occupied countless researchers for many years, with no definitive result until very recently. Putterill and coauthors (chapter 1), Takeshi Izawa (chapter 2), Ryosuke Hayama (chapter 3), and Pekka Heino and coauthors (chapter 5) integrate recent findings in annual LDPs (*Arabidopsis*) and SDPs (rice and morning glory), and in woody plants (*Populus*), respectively, to put the answer to the florigen question in a molecular perspective.

Response to photoperiod is only one mechanism used by plants to determine the transition to flowering. An extended low-temperature treatment prior to exposure to the appropriate photoperiod is often necessary to elicit the most rapid flowering in a species, a process known as vernalization. As outlined in chapter 6, the molecular relationship between photoperiodic timing and vernalization has recently become clearer, and Scott Michaels puts these findings in context.

Although photoperiodic control of flowering is the best studied, other developmental processes are also responsive to day length (Thomas and Vince-Prue, 1997). The onset of bud dormancy in woody trees and shrubs is clearly a process that could benefit from photoperiodic control. Continued vegetative growth too far into the fall could result in frost damage, while too early cessation could result in unnecessarily diminished growth that could put the plant at a competitive disadvantage. Hence, it comes as no surprise that dormancy onset in buds is day length responsive, and Pekka Heino and coauthors (chapter 5) discuss the most recent results.

Most molecular research on photoperiodism has focused on specific SDPs and LDPs of very different species, making it difficult to make direct comparisons of findings across these large genomic distances. Kumiko Ito-Miwa and Tokitaka Oyama (chapter 4) summarize the historical experiments, and recent molecular approaches, conducted in two species of the small, floating duckweed (*Lemna* spp.), one of which is an SDP and the other an LDP.

Finally, Till Roenneberg and coauthors (chapter 7) summarize the past and present perspectives on seasonality and photoperiodism in fungi, with *Neurospora*, a well-known model for circadian research providing the molecular underpinnings for recent studies.

Many of the early experiments alluded to above were conducted in the "premolecular" era of plant physiology, when little was known of the genes involved in flowering or the circadian clock, and the concept of "model organism" had not yet been well developed among plant researchers. Hence, it would be very fruitful to return to many of these species, orthologs of genes from *Arabidopsis* and rice in hand, and perform molecular and biochemical analyses similar to those described in the following chapters. Such future work, together with the current compilation, would address the fundamental question of whether there is a common molecular mechanism of photoperiodic timing in plants.

References

Bünning E. (1960). Circadian rhythms and the time measurement in photoperiodism. Cold Spring Harbor Symp Quant Biol 25: 249–256.

Deitzer GF, Hayes RG, and Jabben M. (1982). Phase-shift in circadian rhythm of floral promotion by far-red energy in *Hordeum vulgare*. Plant Physiol 69: 597–601.

Downs RJ. (1956). Photoreversibility of Flower Initiation. Plant Physiol 31: 279–284.

Garner WW, and Allard HA. (1920). Effect of the relative length of day and night and other factors of the environment on growth and reproduction in plants. J Agric Res 18: 553–606.

Hendricks SB. (1960). Rates of change of phytochrome as an essential factor determining photoperiodism in plants. Cold Spring Harb Symp Quant Biol 25: 245–248.

Hsu JC, and Hamner KC. (1967). Studies of the involvement of an endogenous rhythm in the photoperiodic response of *Hyoscyamus niger*. Plant Physiol 42: 725–730.

Pittendrigh CS. (1972). Circadian surfaces and the diversity of possible roles of circadian organization in photoperiodic induction. Proc Natl Acad Sci USA 69: 2734–2737.

Pittendrigh CS, and Minis DH. (1964). The entrainment of circadian oscillations by light and their role as photoperiodic clocks. Am Nat 98: 261–294.

Thomas B, and Vince-Prue D. (1997). *Photoperiodism in Plants*. London: Academic Press.

Yanovsky MJ, and Kay SA. (2002). Molecular basis of seasonal time measurement in *Arabidopsis*. Nature 419: 308–312.

1

Photoperiodic Flowering in the Long-Day Plant *Arabidopsis thaliana*

Joanna Putterill, Christine Stockum, and Guy Warman

The study of biological rhythms and the clocks that control them (chronobiology) has its roots firmly in the plant world. The French astronomer J.-J. Ortons de Mairan was the first scientist to suggest that organisms possess innate biological clocks when he found, presumably to his surprise, that *Mimosa* plants maintained in constant conditions show the same daily rhythms in leaf movement that they show during the course of the normal day-night cycle. This finding is as well known to the chronobiology community as the flower clock that Linnaeus designed in 1745, which enabled him to tell the time of day based on which flowers were open and which were closed.

Less well known, perhaps, is that all of the major advances in the field of chronobiology over the next 200 years were made using plants. An eloquent and extensive review of this period in chronobiology history is provided by the German botanist Erwin Bünning in his opening address at the 1960 Cold Spring Harbor Symposium on Quantitative Biology. In short, Duhamel and Zinn confirmed de Mairan's findings with *Mimosa* in 1758 and 1759. De Candolle, Pfeffer, and Darwin recognized the endogenous nature of plant rhythms and debated their inheritance, or the lack thereof, using *Mimosa*, *Phaseolus*, and *Calendula*. Darwin assumed that plant clocks were of no selective advantage because he could not imagine the advantage of daily leaf movements, but Sachs recognized that the overt rhythms in leaf movement may present a marker of more physiologically important plant rhythms, or "the hands of the clock." It was not until the 1930s that animal chronobiology caught up with the advances that had been made in the plant world (Bünning, 1960).

In the 1930s, Bünning, generally recognized as one of the founding fathers of modern chronobiology, made enormous advances in our understanding of how circadian clocks work by demonstrating that the endogenous period (or tau) of the clock was strain specific in the bean *Phaseolus* (Bünning, 1935) and that tau was heritable. In addition, Bünning and Kleinhoonte showed the influence of light on the plant clock, and the ability of specifically timed light exposure to advance and delay the phase of the clock.

The decade following the end of World War II saw vast expansion in the field of chronobiology, which culminated in the Cold Spring Harbor Symposium on Quantitative Biology in 1960, which marks an important milestone in the formal study of chronobiology and our understanding of clock function in plants and animals.

At this stage, the concept of the circadian clock as an endogenous self-sustaining oscillator with a period that is seldom exactly 24 h, but that is entrained on a daily basis to 24 h by a forcing geophysical oscillation (or *zeitgeber*), was proven to hold for a wide range of species of plants, animals, and fungi. The location and "physical reality" of clocks was a topic of investigation together with the ways in which clocks aid in navigation and the timing of seasonal activities such as flowering, migration, and hibernation.

The discovery of photoperiodism (or day-length responses) in soy bean and tobacco plants, and subsequently numerous other plant species, by Garner and Allard in 1920 paved the way for the investigation of the mechanisms controlling seasonal rhythms in plants and animals such as flowering, growth, and reproduction. In an interesting excerpt from their 1920 paper they state: "As to animal life, nothing definite can be said, but it may be found eventually that the animal organism is capable of responding to the stimulus of certain day lengths. It has occurred to the writers that possibly the migration of birds furnishes an interesting illustration of this response" (Garner and Allard, 1920, p. 604).

It was Bünning who proposed the first formal model to explain the mechanism underlying photoperiodic responses in plants (and later animals) with his 1936 model for the execution of the photoperiodic time measurement by a circadian oscillation (Bünning, 1936). By then, plant physiologists had determined that the site of detection of photoperiod in plants was in the leaves, and the Russian scientist Chailakhyan had proposed the "florigen hypothesis"—that the floral stimulus was a flowering hormone, florigen, produced in the leaves and translocated to the shoot apex (growing tip of the plant) (Kobayashi and Weigel, 2007), leading to a shift from vegetative to reproductive development and production of flowers. Later, key physiological experiments involving grafting indicated that the floral stimulus was graft transmissible, required intact vasculature, and moved at about the same rate toward the shoot apex as did assimilates in the phloem, suggesting that it was translocated via the phloem. Between-species grafting indicated that the floral stimulus might be universally effective in plants (Thomas and Vince-Prue, 1997).

From the 1990s onward, the intensive study of the molecular biology and genetics of flowering, mainly in the Brassica *Arabidopsis thaliana* (*Arabidopsis*), but later also in rice and other cereals, has provided evidence for the molecules that underpin flowering responses. This chapter provides an overview of the theoretical models proposed to explain the mechanisms by which photoperiodic time measurement is achieved and our current understanding of the molecular basis of photoperiodic flowering in *Arabidopsis*. It also highlights some of the recent discoveries about the biochemical activities and function of the genes, including the exciting recent identification of FT protein as a florigen.

DAY LENGTH: HOW LONG IS A PIECE OF STRING?

Photoperiodic regulation of seasonal responses is common in many organisms from many different phyla. In insects alone photoperiodic control of seasonality has been described in more than 500 species (Nishizuka et al., 1988). In plants, many aspects of physiology and behavior are photoperiodically controlled, from the induction of bud dormancy and the formation of storage organs to the regulation of the transition to flowering.

All photoperiodic responses share the common feature of a highly accurate measurement of day length (or night length) with a critical day length that, when reached or exceeded, elicits a rapid change in seasonal response over a relatively small change in photoperiod (the period of the 24-h day occupied by light; figure 1.1). Organisms classified as "long-day species" exhibit a response such as flowering in photoperiods that exceed the critical day length. Conversely, short-day species exhibit their response when the length of the day falls below the critical day length. The photoperiodic system does not merely rely on single day-length exposures to determine season but can also accumulate successive short or long days in a photoperiodic counter system to an internal threshold at which the photoperiodic response is elicited (Saunders, 1977).

In order to use day length as an indicator of seasonal change, the first challenge is how to measure the length of a day to determine whether it is short or long (or more important, whether it is getting longer or shorter). The theoretical mechanisms for day-length measurement fall into two different camps: those involving a circadian clock-based system in which day length is measured by reference to the clock, and those based on a nonoscillatory hourglass system that is set in motion each day by the transition from light to dark, or dark to light.

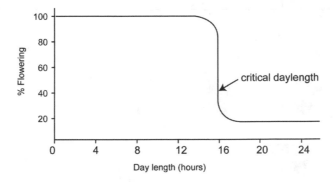

FIGURE 1.1.
Hypothetical photoperiodic phase response curve for a short-day flowering plant with a critical day length of 16 h. Note that the change from high-percentage flowering to low-percentage flowering occurs over a very narrow change in day length.

BÜNNING'S HYPOTHESIS

A description of the development of these models, and the support for and against the models, is perhaps best dealt with in a chronological manner. In 1936, Bünning described how photoperiodic time measurement could utilize a circadian oscillation. In his paper titled "Die endogene Tagesrhythmik als Grundlage der photoperiodis-chen Reaktion" (The endogenous day rhythm as a basis of photoperiodic reactions), Bünning proposed that the 24-h day was composed of two 12-h half cycles differing in light intensity: the photophil (or light-requiring phase) and the scotophil (or dark-requiring phase). Short-day effects were seen to result when light was restricted to the photophil, while long-day effects were produced when light encroached into scotophil (figure 1.2).

In this model, light has two roles: entrainment of the circadian clock to 24 h (because most clocks free-run with a non-24-h period) and illumination of the light-sensitive phase that results in the long-day response. Night-interruption (or Bünsow) experiments in which a long night (16+ h) following a short day (8 h) is systemati-cally interrupted by short bursts of light, supported Bünning's model by showing that light falling during the scotophil reversed the short-day flowering response elicited by *Kalanchoe blossfeldiana*. Support for the involvement of the circadian clock was provided by Bünsow experiments in *Kalanchoe*, the soybean *Glycine max*, and numerous animal species. The ability of short pulses of light to elicit long-day responses (or prevent short-day responses) in an artificially long night (60–70 h)

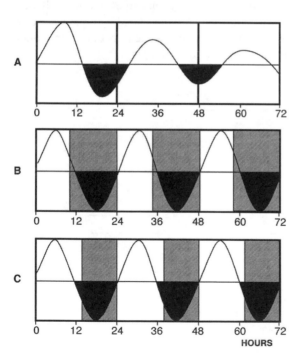

FIGURE 1.2.
Bünning's hypothesis for the clock control of photoperiodism. The area of the sine wave above the threshold indicates the photophil, while that below the threshold (in black) indicates the scotophil. The free-running oscillator is shown to damp in constant darkness (A). In short days (B), light is restricted to the photophil. In long days (C), light encroaches on the scotophil, thus eliciting long-day responses. Gray areas indicate night and white areas the day (redrawn from Bünning, 1960).

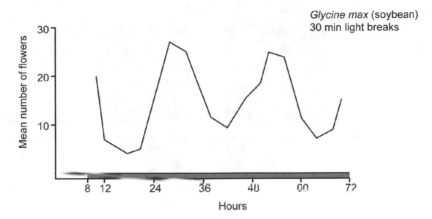

FIGURE 1.3.
Night interruption (or Büsnow) experiments in the soybean *Glycine* max, a short-day plant: a short day of 8 h is followed by a long night of 64 h. Thirty-minute night interruptions are conducted throughout the extended night with different plants. Peaks in the induction of flowering occur with a circadian rhythmicity (redrawn from Saunders, 1977).

was shown to exhibit a circadian periodicity with peak sensitivities falling approximately 24 h apart (figure 1.3).

Further support for the involvement of circadian rhythmicity came from the so-called Nanda-Hamner experiments (or T-experiments) in which peaks in flowering of *Glycine max* were observed after an initial short day was followed by a night of lengths varying from 12 to 60 h. Peaks in flowering occurred when the total day–night length was 24, 48, and 72 h. The responses would have been the same after different extended night lengths if a circadian clock was not involved (see figure 1.5).

While Bünning's hypothesis received wide support (apart from the notable exception of Lees, discussed below), one essential problem with this model was that it accounted for only critical day lengths of 12 h. Many organisms show critical day lengths of greater or less than 12 h. Furthermore, critical day lengths can differ within the same species depending on latitude, with more extreme latitudes having longer critical day lengths (Danilevskii, 1965; Pittendrigh and Takamura, 1989). Bünning's 1936 model was thus developed, or made more explicit by, Pittendrigh and Minis, who in 1964 first proposed the external coincidence model (Pittendrigh and Minis, 1964), and then later the internal coincidence model (Pittendrigh, 1972).

THE EXTERNAL COINCIDENCE MODEL

This model, originally titled the coincidence model, but later named the external coincidence model to distinguish it from the internal coincidence model discussed

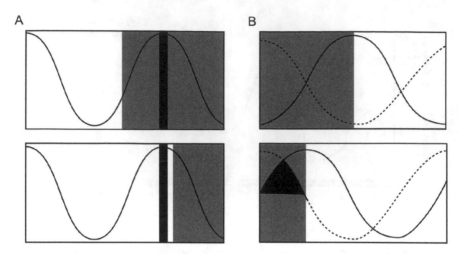

FIGURE 1.4.
Outline of the external and internal coincidence models. Day is represented by white, and night by gray. The upper panels represent short days (LD 12:12), and lower panels long days (LD 18:6). (A) The external coincidence model in which light has the dual role of entraining the circadian clock and hitting or missing the photoinducible phase (represented here in black). (B) Internal coincidence model. Here light has the single role of entraining subpopulations of dawn and dusk oscillators. In short days, there is no internal coincidence between dawn and dusk oscillators, but in long days the phase of the two is closer together. As a result, there is internal coincidence (indicated in black), which elicits long-day effects.

below, was developed from data on the eclosion (emergence of the adult fly from the puparium) of *Drosophila pseudoobscura* by Pittendrigh and Minis (1964), but has subsequently been applied to a large range of plant and animal species. In this, model light is proposed to have a dual role of (1) entraining the circadian system such that a short period of high sensitivity to light (a photoinducible phase φ_i) is maintained at the same time each night and (2) inducing the photoperiodic response by hitting (or missing) the photoinducible phase (figure 1.4A). A large amount of literature describes the evidence for external coincidence models (an extensive review can be found a 2001 special issue of *Journal of Biological Rhythms* Goldman (2001)). Recent advances into the molecular basis of photoperiodism have provided proof of the external coincidence model in animals (Yasuo et al., 2003) and in plants (Hayama and Coupland, 2004).

THE INTERNAL COINCIDENCE MODEL

Pittendrigh wrote in his 1972 paper, "Circadian Surfaces and the Diversity of Possible Roles of Circadian Organization in Photoperiodic Induction," that there are "some facts (especially in birds) that perhaps can only be explained by external coincidence" but that "there are other facts that external coincidence cannot explain" (p. 2734). The internal coincidence model was proposed based on the concept that the circadian clock is composed of a population of circadian oscillators whose mutual phase relationships affect physiology, and that a change in photoperiod "would inevitably change the entrained steady-state of the multioscillator system" (Pittendrigh, 1972, p. 2735). Under some photoperiods, internal coincidence occurs between dawn- and dusk-phased oscillators, and at other photoperiods no internal coincidence occurs (figure 1.4B). In this model, light has the single role of entraining the circadian clock, and day-length changes are detected by the change in phase relationship between oscillator populations entrained to dawn or to dusk (figure 1.4B). Current data do not support this scheme operating in plants.

THE HOURGLASS MODEL

The alternative model to those invoking the use of a clock is the hourglass model, which was proposed by Anthony D. Lees of Imperial College. In his model to explain the photoperiodic induction of sexual dimorphism in the aphid *Megoura viciae*, Lees proposed that measurement of night length was achieved by an hourglass that is set in motion in each cycle by the transition from light to dark (Lees, 1973). Night length was thus measured by whether dawn occurred before or after the critical night length (or when the hourglass ran out) 9.5 h after dusk. Lees found support for his hourglass model from Nanda-Hamner experiments with *M. viciae* (such as shown in figure 1.5) in which extension of the night length above the critical

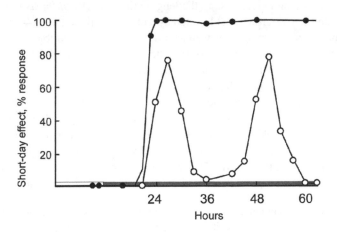

FIGURE 1.5.
Hypothetical representation of the results from Nanda-Hamner experiments in a short-day species showing the difference between an hourglass response (solid circles) and a clock-based response (open circles). A single day (12 h) is followed by a night length of 12–72 h.

day length results in a nonoscillatory plateau rather than an oscillation indicative of the involvement of the circadian system.

OVERVIEW OF THE GENETICS OF FLOWERING TIME REGULATION IN *ARABIDOPSIS*

The timing of flowering to seasons of the year that favor the production of seeds and fruits is critical for successful sexual reproduction. Flowering regulation is also needed to ensure synchronous flowering of outcrossing species. As might be expected for such an important developmental step in plants, the decision to flower is made in response to not just one signal, but to an array of signals such as the seasonal cues of photoperiod or extended winter cold (vernalization), as well as internal developmental signals such as the age and developmental stage of the plant. Flowering can also be regulated by local changes in light quality and intensity as well as nutrient and water availability.

The model laboratory plant *Arabidopsis* is found mainly in temperate regions in the northern hemisphere, and the transition to flowering is strongly promoted by long day lengths experienced in spring and summer. However, it is a facultative long-day plant and will flower eventually in short days, even those as short as 5 h light (Martinez-Zapater et al., 1994). Many varieties (ecotypes) of *Arabidopsis* come from regions with cold winters and have a vernalization requirement to ensure that flowering occurs in the spring, rather than in the autumn. These plants have adopted a winter annual lifestyle: they germinate and grow in the summer, overwinter vegetatively where they are exposed to 1–3 months of cold temperatures, and then flower in spring in response to the lengthening day. *Arabidopsis* flowering can be accelerated by overcrowding (detected by a change in light quality—a decrease in the ratio of red to far red light) as well as modest increases in ambient temperature.

Arabidopsis is an ideal model to study flowering time control. There is considerable natural variation in flowering in different ecotypes, and a large range of flowering time mutants are available. Mapping and map based cloning, as well as insertional mutagenesis, can be used to isolate genes involved in flowering time control.

Figure 1.6 shows an *Arabidopsis* wild-type plant (30 days old) and the *gigantea* (*gi*) mutant (77 days old). The latter flowers significantly later than the wild type, resulting in a "gigantic" plant at the time of flowering. Studies of this mutant led to the discovery that the *GIGANTEA* gene is a promoter of flowering in long day lengths. Other advantages of the model plant *Arabidopsis* are as follows:

- It lends itself to genetic studies because it is a small plant that produces thousands of seeds, needs little space, and is easy and inexpensive to grow. Commonly used lab strains have a rapid life cycle of 6–8 weeks.
- It has a compact and fully sequenced genome that contains at least 28,000 genes.

FIGURE 1.6.
Arabidopsis wild-type plant
(30 days old) and the *gigantea*
(*gi*) mutant (77 days old).

- More than 200,000 insertion lines are available that facilitate gene cloning and functional analysis.
- Transgenic plants are easily generated using the natural genetic engineer Agrobacterium.

Genetic, physiological, and molecular analyses in *Arabidopsis* have revealed distinct and overlapping signaling pathways that regulate flowering in response to external and internal cues. The genetics of flowering time regulation began to be investigated by studying natural variation in flowering time between different accessions and by the isolation of mutants with altered flowering time (Martinez-Zapater et al., 1994). These were mostly late-flowering mutants and included the first identification of the *co* and *gi* mutants by Redei in 1962 (figure 1.6; see table 1.1 for full gene names). An important paper by the Koornneef lab in 1991 reported the first systematic genetic study of a series of these late-flowering mutants and their

TABLE 1.1. *Arabidopsis* genes involved in photoperiodic flowering.

Gene abbreviation	Gene name	Predicted gene product	Pathway/function
API	*APETALA 1*	MADS transcription factor	Floral meristem identity gene
CCA1	*CIRCADIAN CLOCK ASSOCIATED 1*	MYB transcription factor	Circadian clock
CDF1	*CYCLING DOF FACTOR 1*	Dof transcription factor	Repressor of *CO*, represses flowering
CO	*CONSTANS*	Zinc finger transcription factor	Promotes flowering
COP1	*CONSTITUTIVELY PHOTOMORPHOGENIC 1*	E3 ubiquitin ligase	Negative regulator of light signaling
CRY 1 and 2	*CRYPTOCHROME 1 and 2*	UV/blue-light photoreceptors	Light perception, promote flowering
ELF 3	*EARLY FLOWERING 3*	Novel nuclear protein	Circadian clock
ELF 4	*EARLY FLOWERING 4*	Novel protein	Circadian clock
FD	*FLOWERING LOCUS D*	bZIP transcription factor	Promotes flowering
FKF1	*FLAVIN-BINDING, KELCH REPEAT, F-BOX*	Blue light photoreceptor	Promotes flowering
FT	*FLOWERING LOCUS T*	Protein similar to RAF kinase inhibitor protein	Floral integrator gene, promotes flowering
GI	*GIGANTEA*	Novel protein	Circadian clock, promotes flowering
LFY	*LEAFY*	Novel transcription factor	Floral integrator gene, floral meristem identity gene
LHY	*LATE ELONGATED HYPOCOTYL*	MYB transcription factor	Circadian clock
LKP2	*LOV KELCH PROTEIN 2*	Putative blue light photoreceptor	Circadian clock
LUX	*LUX ARRHYTHMO*	MYB transcription factor	Circadian clock
PHY A	*PHYTOCHROME A*	Red/far-red and blue light photoreceptor	Light perception, promotes flowering
PHY B, D, and E	*PHYTOCHROME B, D, and E*	Red/far-red light photoreceptors	Light perception, repress flowering
PIF3	*PHYTOCHROME INTERACTING FACTOR 3*	Basic/helix-loop-helix transcription factor	Light signaling

(Continued)

TABLE 1.1. (*Continued*)

Gene abbreviation	Gene name	Predicted gene product	Pathway/function
PRR5, -7, and -9	*PSEUDO-RESPONSE REGULATOR 5, 7,* and *9*	Putative transcription factors with PRR domain	Circadian clock
SOC1	*SUPPRESSOR OF OVEREXPRESSION OF CONSTANS 1*	MADS transcription factor	Floral integrator gene, promotes flowering
SPA1	*SUPPRESSOR OF PHYA-105*	Novel protein with COP1-like WD-repeat, coiled coil and kinase-like domain	Represses flowering
TOC1	*TIMING OF CHLOROPHYLL A/B BINDING PROTEIN 1*	Putative transcription factor with PRR domain	Circadian clock
ZTL	*ZEITLUPE*	Blue light photoreceptor	Circadian clock

response to different environmental conditions such as photoperiod (Koornneef et al., 1991).

The *gi* and *co* mutations were placed in the same group because both mutants appeared to have lost the ability to respond to the long-day signal. They flowered late in long days compared to wild type, but at a similar time to wild type in short days. In addition, they could be placed in the same epistatic group because plants carrying both mutations flowered no later than the latest parent. From these analyses, flowering control pathways began to be defined, including a pathway that promotes flowering in long days that became known later as the long-day pathway or the photoperiod pathway. These were elaborated upon with the identification of new mutants, including early-flowering mutants, and by gene isolation and subsequent molecular analyses that allowed a transcriptional hierarchy to be established among the genes (Putterill et al., 2004).

Thus, the study of the regulation of *Arabidopsis* flowering time has ultimately uncovered a genetic network of activators and repressors of flowering in different signaling pathways with cross talk at several points in the network (figure 1.7). No single mutation prevents the flowering of *Arabidopsis*, although there are many that significantly delay flowering. To block flowering in almost all conditions, mutants in three separate pathways are required (Reeves and Coupland, 2001). These pathways converge upon flowering time integrator genes, the most important being *FT*, *SOC1*, and *LFY*, where the molecular "decision" is made—to flower or not to flower (Putterill et al., 2004; Yoo et al., 2005). Once these genes are activated, they in turn promote the expression of floral meristem identity genes such as *AP1* and the

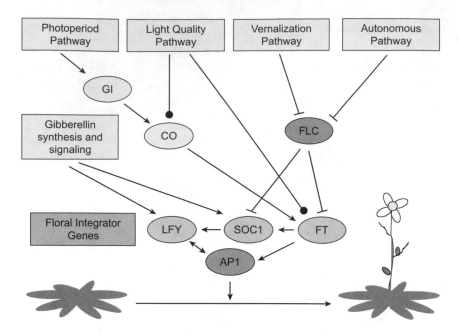

FIGURE 1.7.
Outline of the five major floral promotion pathways in *Arabidopsis*. Exogenous signals, such as day length (photoperiod) and winter cold (vernalization), and endogenous signals are perceived via different pathways and integrated at the level of the floral integrators *SOC1*, *FT* and *LFY*. Up-regulation of the floral integrators leads to further promotion of *LFY* and other floral identity genes such as *AP1*, which trigger flowering. CO is a key integrator of light and clock signals in the photoperiod pathway. *FLOWERING LOCUS C* (*FLC*) is a major repressor of flowering is targeted by the autonomous and vernalization pathways. The gibberellin hormone pathway is important for promoting flowering in short day lengths. Lines with arrows are promotive, lines with bars indicate repression, and lines with round heads indicate either activation or repression flowering depending on light quality.

transition from vegetative to reproductive development ensues with formation of flowers on the flanks of the shoot apical meristem instead of leaves.

THE PHOTOPERIOD PATHWAY

The intense molecular and genetic analyses in *Arabidopsis* have provided an excellent opportunity to identify the molecules and molecular processes that were first predicted by more classical work for how perception of day length might be transduced into responses such as flowering. Many genes have been identified that contribute to photoperiodic regulation of flowering (table 1.1). As first predicted by Bünning and then by Pittendrigh with the external coincidence model, the components of the photoperiod pathway comprise photoreceptors, the circadian clock, and clock- and light-regulated genes that regulate flowering (figure 1.8).

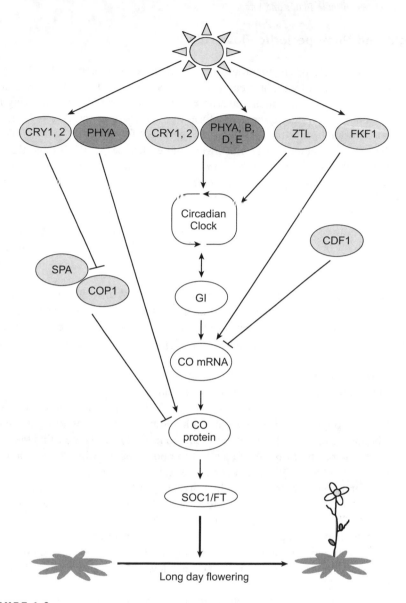

FIGURE 1.8.

Outline of the *Arabidopsis* photoperiod pathway. The three main components, the photoreceptors, the circadian clock and the circadian-regulated flowering output pathway, are shown. The photoreceptors entrain the circadian clock and have direct effects on the transcription and stability of the clock-output flowering pathway components such as CO leading to promotion of flowering in long days. The SPA proteins target CO protein for proteasomal degradation via their interaction with COP1. Lines with arrows indicate up-regulation of gene expression/activation of proteins; lines with bars indicate repression. Red and far-red light receptors (PHY) are dark-shaded, and blue light receptors (CRY, ZTL, FKF1) are light-shaded.

Light and Photoperiodic Flowering

The important role of phytochrome photoreceptors in photoperiodic flowering was first highlighted by night-break experiments carried out in short-day plants. Phytochromes exist in two interconvertible forms, with the P_r form perceiving red light and being converted to the active form of phytochrome (the P_{fr} type). The P_{fr} form is converted back to the P_r form by perception of far-red light. For example, as described in Carré et al. (2005), Borthwick and colleagues in 1952 found that the flowering of short-day plants such as *Xanthium strumarium*, which require nights of 8.5 h or longer, was inhibited by a 2-min red light flash during the night period. This effect was reversed by subsequent illumination with 3 min of far red light. The interpretation was that the flash of red light would convert the inactive form of phytochrome (P_r) to the P_{fr} active form that would inhibit flowering, while the second flash (of far red light) would convert the P_{fr} to P_r and release the inhibition. They proposed that that the slow decay of the active form P_{fr} during the night period would provide a means of measuring how long the dark period was. Once the level of P_{fr} had dropped below a critical threshold (after 8.5 h), then flowering would be induced. However, subsequent experiments showed that many short-day plants have a rhythmic response to night breaks given at different times of the night period, which was consistent not with this hourglass type model but with the involvement of the circadian clock.

In *Arabidopsis*, it has been known for many years that the spectral quality of light has a major effect on flowering time (Martinez-Zapater et al., 1994), with far red light promoting flowering and red light inhibiting it. *Arabidopsis* flowering is also strongly promoted by blue light. Similar responses to light quality were obtained with night break experiments to induce flowering of this long-day plant, with far red night breaks given in the middle of the night being the most effective (Goto et al., 1991). Light is perceived in *Arabidopsis* by approximately 12 different photoreceptors, including five phytochromes that detect red/far red light, two cryptochromes that detect blue/UVA light, three novel blue light receptors (ZTL, FKF1, and LKP2), and two phototropins that detect blue light. These photoreceptors regulate processes that occur through the life time of the plant from seed germination, photomorphogenesis, phototropism, photosynthesis, shade avoidance, plant architecture, and flowering time (Chen et al., 2004; Franklin et al., 2005).

Analysis of *Arabidopsis* photoreceptor mutants showed that some of these photoreceptors have major effects on flowering time, and these are consistent with the effect of different light qualities on flowering. They are the main far red light receptor PHYA, the major red light receptor PHYB, and the blue light receptors CRY2 and FKF1, with others having more minor or redundant roles. *cry2* and *fkf1* mutants, for example, are late flowerers in long-day conditions (Koornneef et al., 1991; Nelson et al., 2000), and *phyA* is late flowering in long-day extensions (Johnson et al., 1994), suggesting that all of these photoreceptors promote flowering in long days. In contrast, the *phyB* mutant flowers early in long and short-day conditions, indicating that it is an inhibitor of flowering (Goto et al., 1991). There is cross talk between the pathways, and *CRY2* has been proposed to function by antagonizing PHYB activity, because *cry2* mutations do not cause late flowering in a *phyB* mutant background (Guo et al., 1998; Mockler et al., 1999).

Molecular genetic analyses suggest that light has a dual role in the *Arabidopsis* photoperiod pathway. First, it signals to the circadian clock to synchronize clock activity with the daily onset of light. Four phytochromes (A, B, D, and E) have been shown to contribute to the light resetting of the clock (Somers et al., 1998; Yanovsky et al., 2000). Although unlike some animals, cryptochromes are not key components of the plant circadian clockwork, *CRY1* and *CRY2* do mediate blue light input to the clock, as shown by the period length defects seen in single and double *cry* mutants (Somers et al., 1998; Devlin and Kay, 2000; Yanovsky et al., 2000). Second, the photoreceptors also have direct effects on the activity of key flowering time genes in the photoperiod pathway such as *CO* (see section below on external coincidence) at both the transcriptional and posttranscriptional level.

The structures of the photoreceptors involved in photoperiodic flowering are shown in figure 1.9. Their structure and function have been comprehensively reviewed (Franklin et al., 2005; Wang, 2005) and is briefly outlined below. The red/far red photoreceptors PHYA and PHYB are soluble proteins that form dimers and perceive light via a covalently bound tetrapyrrole chromophore. They have a domain similar to histidine kinases and are subject to autophosphorylation, as

Phytochromes A to E

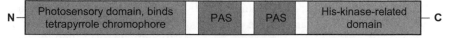

Cryptochromes 1 and 2

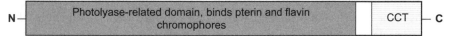

ZTL/FKF1/LKP2

FIGURE 1.9.
Diagram of the structure of members of the three photoreceptor families involved in photoperiodic flowering in *Arabidopsis*. PAS, Per-ARNT-Sim, signaling domain identified in *Drosophila* period circadian protein, aryl hydrocarbon receptor nuclear translocator protein, and *Drosophila* single-minded protein; CCT, cryptochrome c-terminus; LOV, a subform of PAS that is activated by light, oxygen, or voltage; FMN, flavin mononucleotide.

well as phosphorylating another target protein that negatively regulates phytochrome signaling. Phytochromes accumulate in the cytoplasm but, once activated by light, can move to the nucleus, where they can interact directly with transcription factors such as PIF3 to regulate gene expression. COP1 is a negative regulator of photomorphogenesis and an ubiquitin E3 ligase that targets proteins for degradation by the proteasome. Phytochromes and cryptochromes interact physically with COP1 in a light-dependent manner and may transmit light signals by inhibiting its activity. The cryptochrome CRY2 perceives blue/UVA via linked flavin and pterin chromophores. CRY2 is localized in the nucleus and dimerizes, and its levels are influenced by blue-light–dependent auto phosphorylation, which also affects its conformation, intermolecular interactions, and physiological activities. FKF1 perceives blue light via a noncovalently linked flavin mononucleotide (FMN) chromophore, localizes in the nucleus (Sawa et al., 2007), and has an F-box domain implicating it is part of an SCF complex that selects proteins for degradation by the proteasome (Han et al., 2004).

The Circadian Clock and Photoperiodic Flowering

As many as 6% of *Arabidopsis* genes are rhythmically expressed under the control of the circadian clock (Harmer et al., 2000). These include genes for flowering time; photosynthesis; photoreceptors; cold protection; carbon, nitrogen, and sulfur pathways; and starch mobilization (Harmer et al., 2000). *Arabidopsis* mutants with defects in circadian rhythmicity, such as the *lhy*, *cca1*, and *elf3* mutants, also have altered flowering time (Hicks et al., 1996; McWatters et al., 2000; Covington et al., 2001; Alabadi et al., 2002; Mizoguchi et al., 2002), and circadian-clock–regulated genes, including *GI*, *CO* and *FT*, lie at the heart of the molecular mechanism of day-length perception and response (Fowler et al., 1999; Park et al., 1999; Suarez-Lopez et al., 2001).

The molecular mechanism of the circadian clock has been extensively studied in many model systems, including *Arabidopsis*. A common feature of all clocks seems to be that the rhythms are generated by the interactions of rhythmically expressed genes that form positive and negative feedback loops (Dunlap, 1999; McWatters et al., 2001). Some of the components of this autoregulatory feedback loop, in which clock proteins cyclically regulate their own expression, have been identified in *Arabidopsis* (figure 1.10; McClung, 2006; McWatters et al., 2007).

The simplest *Arabidopsis* model has a central clock loop, with the two single-domain MYB transcription factors LHY and CCA1 and a pseudoresponse regulator 1 (PRR1) named TOC1 (Alabadi et al., 2001). The loop starts in the morning, when *LHY* and *CCA1* expression are peaking as these two transcription factors are up-regulated by TOC1. At the same time, light promotes the expression of *LHY* and *CCA1* in this early part of the day (Martinez-Garcia et al., 2000; Kim et al., 2003). These high levels of LHY and CCA1 repress the expression of *TOC1* through binding to its promoter. As TOC1 levels decrease, they no longer promote *LHY* and

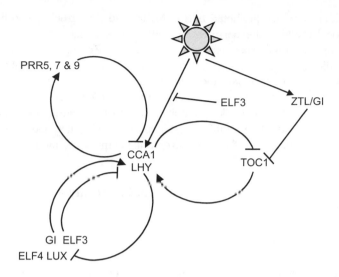

FIGURE 1.10.
Outline of the *Arabidopsis* circadian clock. There is a central circadian autoregulatory negative feedback loop with the genes *CCA1*, *LHY*, and *TOC1*. This intersects with additional interlocking circadian loops consisting of circadian regulated factors that feed back on the central loop components. These include genes expressed in the evening, such as *ELF3*, *ELF4*, *GI*, and *LUX*, and genes expressed during the day, including the *PRR5*, *-7*, and *-9*. Blue light stimulates ZTL and GI proteins to interact and ensure robust cycling of TOC1 protein. ELF3 helps to synchronize the clock to long days and gates light input to the clock at dusk.

CCA1 expression, and the corresponding protein levels decrease during the night. This decrease in the repressors of *TOC1* allows expression of these two genes, and TOC1 protein levels are elevated. The resulting peak in protein levels is proposed to up-regulate *LHY* and *CCA1* in the morning, and the cycle begins again.

This model is a very simplified representation of the molecular interactions that form the circadian oscillator in *Arabidopsis*. For example, it is known that posttranscriptional regulation and interactions with other components such as ZTL and GI affect the robustness of clock rhythms. The *ztl* mutant has a long circadian period phenotype (Somers et al., 2000), and the ZTL protein has been found to be part of an SCF E3 ubiquitin ligase complex and to mediate the degradation of the clock proteins TOC1 and PRR5 (Somers et al., 2000; Más et al., 2003; Han et al., 2004; Kiba et al., 2007; Fujiwara et al., 2008). The *gi* mutant has altered accumulation of the TOC1 protein, and this is mediated by a physical interaction between GI and ZTL, which stabilizes ZTL (Kim et al., 2007). In addition, many factors, including GI and the PRR proteins 5, 7, and 9, are transcriptionally regulated by the circadian clock and feed back into the clock to form a series of interlocking loops (figure 1.9). GI regulates circadian rhythms in both light and dark and temperature compensation in the clock (Mizoguchi et al., 2005; Gould et al., 2006).

Multiple loops are also predicted by the Millar and Kay groups, who have used mathematical analyses of experimentally gathered data related to clock associated genes and their expression patterns (Locke et al., 2006; Zeilinger et al., 2006). Their results predict that the clock consists of at least two interlocking feedback loops, which both receive light signals and contain yet unidentified components.

Taken together, the data collected about the *Arabidopsis* circadian clock show how its general mechanism works and that it is capable of creating rhythmic expression patterns as required for establishing a photoinducible phase and maintaining it at the correct phase, in accordance to the external coincidence model.

The Clock-Regulated Flowering Output Pathway and the External Coincidence Model

As the relationships and regulation of genes in the flowering network were being explored, it became clear that the floral integrator gene *FT* was a direct target of the photoperiod pathway and specifically of *CO*. *CO* encodes a nuclear zinc finger protein, and *FT* a small protein with homology to the Raf kinase inhibitor protein (Putterill et al., 1995; Kardailsky et al., 1999; Kobayashi et al., 1999).

FT mRNA is strongly up-regulated in floral inductive long days compared to short days in wild-type plants, but not in *co* mutants (Harmer et al., 2000; Suarez-Lopez et al., 2001; Wigge et al., 2005). The use of an inducible transgenic version of the *CO* gene identified *FT* as an early target of CO, very possibly a direct target, as *FT* could be up-regulated even in the presence of a protein translation inhibitor (Samach et al., 2000). In later global expression experiments using microarrays, *FT* appears as the major target of CO (Wigge et al., 2005). Finally, a strong mutant allele of *ft* was able to completely block the action of a floral promoting action of a CO transgene, reinforcing the idea that *FT* is the major downstream target of CO (Yoo et al., 2005). CO has not been shown to have DNA binding activity on its own, but recent evidence suggests that CO is part of a modified HAP protein transcriptional complex that binds CCAAT boxes, which provides a biochemical mechanism by which it could effect changes to downstream gene expression (Ben-Naim et al., 2006; Wenkel et al., 2006; Cai et al., 2007).

How though does *CO* up-regulate *FT* in long days, but not in short days? *CO* mRNA levels are regulated by the circadian clock with overall transcript abundance similar in long and short days; it peaks in the night and drops during the day to begin to rise again in the later part of the day. A slightly earlier shoulder of expression occurs in the afternoon in long days that is not seen in short days (Suarez-Lopez et al., 2001) (figure 1.11). However, because of the longer light period and a slightly earlier peak in the afternoon, *CO* transcript coincides with light in long but not short days. The overlap between *CO* expression and light and up-regulation of *FT* in long days led the Coupland group to make the hugely influential proposition that the coincidence of CO expression in the light was critical for the *Arabidopsis* day-length response (Suarez-Lopez et al., 2001).

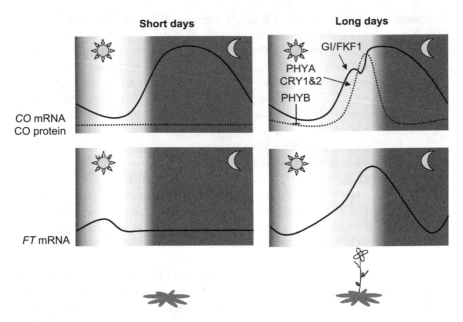

FIGURE 1.11.
Promotion of *Arabidopsis* flowering by long days is consistent with the external coincidence model. There is experimental evidence for stabilization of CO protein in the light, but degradation via the proteasome in the dark. *CO* transcription and CO protein stability are promoted in long-day conditions, but not in short days. This leads to up-regulation of the strong floral promoter *FT* in long days and activation of the transition to flowering. The main photoreceptors and proteins involved in long days are shown. GI and FKF1 promote CO transcription, and CRY2 and PHYA stabilize CO protein, while PHYB leads to its degradation.

This important idea was then tested by the Kay group using a short period circadian clock mutant *toc1-2* which showed early flowering in short days (Yanovsky and Kay, 2002). This unusual flowering phenotype correlated with a shift of CO expression into the light period even in short days. However, when the day–night cycle (the *T* cycle) was adjusted to 21 h to match the short period rhythm of this mutant, *CO* expression changed so that peak expression occurred in the night in short days and the plants flowered normally in long days and short days. These and other experiments with different *T* cycles, which led to a displacement of *CO* expression into light in wild-type plants (Roden et al., 2002), highlighted the importance of coincidence of *CO* expression with light for promoting flowering.

Why? One reason is that CO protein is degraded rapidly in the dark by being targeted to the proteasome (Valverde et al., 2004). This occurs by CO association with COP1, a negative regulator of photomorphogenesis that is an E3 ubiquitin ligase that targets proteins for destruction by the proteasome, and the four SPA proteins (Laubinger et al., 2004, 2006; Jang et al., 2008; Liu et al., 2008). In the light, blue and far red light perceived by the photoreceptors CRY2 and the PHYA, respectively, were shown to be needed for *CO* up-regulation of *FT* (Yanovsky and Kay, 2002). These photoreceptors and CRY1 were then shown to promote CO protein accumulation (Valverde et al., 2004) (figure 1.11). CRY1 and CRY2 may stabilize CO protein by physically interacting with and repressing COP1 (Wang et al., 2001). On the other hand, PHYB, the main photoreceptor responsible for responses to red light and the shade avoidance response, promotes degradation of the CO protein in the morning (Valverde et al., 2004).

These effects on CO protein abundance are consistent with the other genetic and physiological experiments (discussed above) that showed that PHYB and red light have an inhibitory role in flowering in *Arabidopsis*, while blue and far red light and the photoreceptors CRY1, CRY2, and PHYA that perceive these wavelengths promote flowering. These lines of evidence support the idea that *Arabidopsis* uses the external coincidence mechanism to detect and respond to long day lengths, with CO protein being the circadian-regulated light-sensitive molecule predicted by the model.

Indeed, multiple coincidences may be the rule in long-day promotion of flowering in *Arabidopsis*. For example, the *FKF1* gene is responsible for the earlier shoulder of *CO* expression in the afternoon (Imaizumi et al., 2003). It is cyclically expressed, and its protein activity relies on coincidence with light (blue) in the afternoon in long days. At this time of the day, FKF1 mediates proteasome degradation of CDF1 protein, which is a transcription factor that directly represses *CO* expression (Imaizumi et al., 2005). Exciting recent results using chromatin immunoprecipitation (ChIP) analysis in *Arabidopsis* suggest that FKF1, CDF1, and GI proteins interact with *CO* chromatin in the late afternoon to regulate *CO* expression (Sawa et al., 2007). FKF1 activity is dependent on GI, and they associate in a blue-light–dependent manner (Sawa et al., 2007). *GI* transcript abundance itself is regulated by the circadian clock and modulated by photoperiod, with a peak in the late afternoon in long and short days and an earlier peak of expression just after dawn (Fowler et al., 1999; Park et al., 1999). GI protein levels cycle with a major peak in the late afternoon, and there is evidence that GI protein is degraded in the dark via the proteasome in a mechanism that involves ELF3 and COP1 (David et al., 2006; Yu et al., 2008).

GI is a classical gene in the *Arabidopsis* photoperiod pathway and known to promote flowering largely via regulating *CO* expression. *CO* transcript abundance is lower at all time points in a *gi* mutant, while overexpression of *CO* rescues the late flowering of *gi* mutants (Suarez-Lopez et al., 2001), but it was not known how directly *GI* regulated *CO*. The demonstration of the physical interaction between two positive regulators of *CO* expression, GI and FKF1, with *CO* chromatin is very intriguing because it indicates that CO is directly downstream of GI. It also casts some light on the biochemical role of GI that was previously unknown: *GI* encodes a large plant-specific large protein with no known domains (Fowler et al., 1999; Park et al., 1999).

A further puzzle has been the molecular basis of the pleiotropy exhibited by *gi* mutants. For example, plants carrying mutations in the *gi* gene can show changes to the robustness of circadian rhythms (Park et al., 1999; Mizoguchi et al., 2005; Martin-Tryon et al., 2007) and temperature compensation in the clock (Gould et al., 2006), hypocotyl elongation (Huq et al., 2000; Martin-Tryon et al., 2007; Oliverio et al., 2007), and flowering time. However, molecular genetic analyses indicated that *GI*'s involvement in flowering control appears to be separate from its function in the circadian clock (Mizoguchi et al., 2005; Martin-Tryon et al., 2007). The recent discovery that GI interacts with the two related proteins, with FKF1 to promote flowering (Sawa et al., 2007) and with ZTL to modulate circadian rhythms (Kim et al., 2007), begins to provide an explanation of how GI can have distinct functions.

THE CO-FT MODULE AND FLORIGEN

Classical models show day-length measurement occurring in the leaves, followed by production of a substance that can be transmitted by grafting, most likely through the phloem, which leads to floral evocation at the shoot apex. The sections above have detailed how central the CO and FT proteins are to day-length detection and flowering in *Arabidopsis*. But how does the *CO-FT* gene module map to the predicted spatial distribution of the pathway? Might *CO* or *FT* encode florigen, the long-distance signaling molecule first predicted in the 1930s?

Initial gene expression analyses using reporter gene fusions indicated that both genes were expressed in the vasculature of the leaves and, in the case of *CO*, other tissues, as well (An et al., 2004; Takada and Goto, 2003). In 2004, the Coupland and Turgeon groups reported excellent progress on the question of where in the plant *CO* and *FT* are required to promote flowering and which of them might be florigen. They used tissue-specific promoters to drive CO expression in transgenic plants. *CO* expressed in the leaves or the phloem companion cells (of the major veins or the minor veins) was able to rescue the late flowering of *co* mutants (An et al., 2004; Ayre and Turgeon, 2004). However, *CO* expressed directly in the shoot apical meristem was not functional. Thus, if CO was florigen, it might need to be expressed in the phloem in order to be modified appropriately. Alternatively, *CO* might lie upstream of florigen and be involved in generating the florigenic signal.

In a similar approach, expressing the blue light photoreceptor CRY2 fused to the green fluorescent protein (GFP) in the phloem companion cells rescued the late flowering of *cry2* mutants and restored *FT* expression to wild-type levels, suggesting that CRY2 might modify CO protein stability/activity in a cell autonomous fashion (Endo et al., 2007). This requirement for CRY2 activity in the phloem companion cells may help to explain why expression of *CO* specifically in the shoot apex does not promote flowering (An et al., 2004). Interestingly, expression of the red/far red photoreceptor PHYB in the mesophyll cells, but not the vasculature, inhibited flowering and FT expression (Endo et al., 2005).

Grafting studies showed that wild-type plants were able to partly rescue *co* mutants when they were grafted onto them, indicating that *CO* did indeed control the expression of a graft-transmissible floral promoting substance (An et al., 2004;

Ayre and Turgeon, 2004). This also implicated CO as a leaf/phloem-expressed factor that might be florigen or at least be involved in generating a florigenic signal. However, no physical movement into the shoot apex of a functional CO-GFP reporter fusion from the phloem could be seen, which counted against CO as the mobile protein signal (An et al., 2004).

At the same time, the Coupland group also used a similar batch of promoters to express *FT* in different parts of the plant; *FT* expressed in the shoot apex or in the leaf or phloem was able to rescue the late flowering of *ft* mutants, suggesting that FT mRNA or protein had the capacity to be florigen because it could function at the shoot apex (An et al., 2004). These authors also showed that *FT* was expressed in the phloem in response to *CO* and that the promotion of flowering by *CO* in the phloem was at least in part dependent on *FT*. Later it was shown, using artificial microRNAs targeted against *FT* in the companion cells of the phloem, that *FT* expression there is vital for its function (Mathieu et al., 2007).

FT is likely to act as a transcriptional cofactor of a bZIP protein FD to directly activate target genes such as *AP1* and the evocation of flowers (Abe et al., 2005; Wigge et al., 2005). Yet, FD, a protein partner of FT, is expressed only at the shoot apex, indicating that FT protein would need to be present in the shoot apex (perhaps by moving there from the phloem) for the interaction to occur and promote floral evocation via up-regulation of floral integrator genes such as *SOC1* and later-acting floral meristem identity genes such as *AP1* (figure 1.12).

Thus, attention turned to *FT* and whether *FT* mRNA or FT protein might be a mobile florigen. *FT* encodes a small protein of 20 kDa, making it an attractive candidate for being a mobile florigen. A rice *FT* gene (*Hd3A*) had already been shown genetically to be involved in regulating photoperiodic flowering in rice (Izawa et al., 2002; Kojima et al., 2002), which fitted well with the idea of a universally acting substance. This idea received a big boost with the publication in 2006 of a paper showing that in another family of plants, the Solanaceae, overexpressed tomato *FT* mRNA in a transgenic stock could rescue a scion of a late-flowering *ft* mutant (*sft, single flower truss*) grafted to it (Lifschitz et al., 2006). Excitingly, this stock also produced graft-transmissible signals that could promote flowering in another solanaceous plant—Maryland Mammoth—studied by Garner and Allard in their early controlled experiments on photoperiodic flowering in the 1920s. However, no *FT* mRNA could be detected moving across the graft junction, pointing to FT protein as a substance that was moving, or perhaps a relay system where FT did not move but a signal was relayed through the phloem to finally turn on *FT* in the shoot apex.

A series of papers in 2007 working in different plants have brought more clarity to this issue, and these have been comprehensively reviewed (Kobayashi and Weigel, 2007). Of particular importance, they have demonstrated that FT protein can unload from the phloem and move into the shoot apex, and that it can move across graft junctions. For example, the Coupland group expressed FT-GFP fusion proteins specifically in the phloem companion cells in *Arabidopsis* and showed that they unload from the phloem in the shoot apex and promote flowering and move

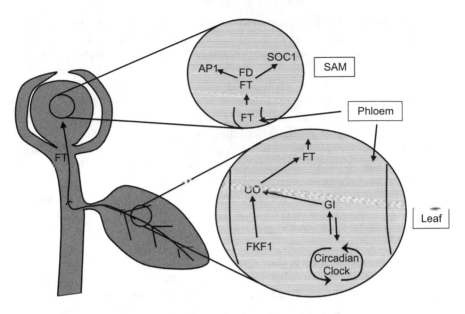

FIGURE 1.12.
Long-distance signaling of photoperiod with FT protein as florigen. In long days, CO is up-regulated in the vasculature of the leaves (phloem companion cells) by GI and FKF1. This stimulates *FT* transcription and production of FT protein, which translocates via the phloem and unloads in the shoot apex. FT interacts with the FD transcription factor in the shoot apical meristem (SAM), which triggers the development of flowers by activation of *SOC1* and *AP1*.

across graft junctions (Corbesier et al., 2007). Rice FT-GFP fusion proteins have also been observed to move from the leaf to the apex and induce flowering (Tamaki et al., 2007). In another approach, mass spectrometry on grafted cucurbits with different responses to photoperiod detected graft transmissibility of FT proteins, finding FT protein typical of the stock in the scion (Lin et al., 2007). Other studies support these results by showing that FT does not function to promote flowering if targeted only to the nucleus, which prevents it from trafficking to adjacent cells (Jaeger and Wigge, 2007), or if tethered in the phloem (Mathieu et al., 2007).

PERSPECTIVES

Possible systems for measurement of day length include circadian-based or hourglass systems (although the two are by no means mutually exclusive if an hourglass timer is considered in fact to be a damped oscillator). Of the clock-based models, there are the possibilities of external or internal coincidence that find support in different species. The determination of which system an organism uses ultimately relies on the molecular dissection of the elements that are essential for photoperiodic time measurement. *Arabidopsis* has provided an excellent model for this, and great progress has been made. This work has had the additional benefit of clearly defining a clock output pathway, the photoperiod flowering pathway, which has yet to be elucidated in many animal species.

Within the *Arabidopsis* photoperiod pathway, many specific aspects of the molecular and biochemical mechanisms remain to be understood (Kobayashi and Weigel, 2007). These include, for example, how FT protein is loaded and translocated in the phloem and unloaded from the phloem, and how it moves in the shoot apex to the cells in which its partner FD is expressed. In addition, up-regulation of *FT* transcript in the leaves is needed only for a short period of time to stably induce flowering (Corbesier et al., 2007). How the plant "remembers" this for the remainder of its life so that it does not revert to the vegetative state is not known. *CO* and *FT* mRNA are, however, more widely expressed in other tissues, and *CO* is expressed at night, but the functional consequences for plant development of these expression patterns are not understood.

The flowering gene network is made up of distinct yet overlapping pathways with cross talk between the pathways. From the discussions above, the light-quality pathway and the photoperiod pathway clearly share light perception and signaling components. In addition, there are important interactions between the photoperiod and vernalization pathways. These are particularly important for winter annual *Arabidopsis* which over winter vegetatively and may only flower in the lengthening days of spring once they have experienced an extended period of cold temperatures. The vernalization pathway mediates repression of the repressor *FLOWERING LOCUS C* to enable flowering in the spring.

How well conserved are the mechanisms of photoperiodic flowering in plants? As described above, there is great support for FT protein functioning as a mobile flowering signal in a range of diverse plants, and many of the flowering time genes are highly conserved in seed plants and have been shown to regulate flowering in a range of plants. However, the interactions in regulatory hierarchy can differ significantly. Analysis of the molecular genetics of photoperiodic flowering in rice suggests that many of the same components play a role, including rice CO, GI, FT and phytochromes, despite rice having the opposite response to photoperiod than *Arabidopsis*. However, a key difference is that CO represses *FT* and flowering in long day lengths in rice, but promotes it via *FT* in short day lengths. In addition, in a recent report of control of photoperiodic flowering in *Pharbitis nil*, a short-day plant, it appears that CO does not regulate *FT*, but rather *FT* is up-regulated in response to long nights and is under the control of an oscillator that is set by lights off at dusk (Hayama et al., 2007).

Thus, further studies of photoperiodic flowering in a range of plants, between and within families, with different photoperiodic responses, is likely to yield fundamental information on flowering regulation. This knowledge should greatly expand the tools available to customize flowering to particular regions and climates and yield novel varieties with altered flowering time.

References

Abe M, Kobayashi Y, Yamamoto S, Daimon Y, Yamaguchi A, Ikeda Y, Ichinoki H, Notaguchi M, Goto K, and Araki T. (2005). FD, a bZIP protein mediating signals from the floral pathway integrator FT at the shoot apex. Science 309: 1052–1056.

Alabadi D, Oyama T, Yanovsky MJ, Harmon FG, Mas P, and Kay SA. (2001). Reciprocal regulation between *TOC1* and *LHY/CCA1* within the *Arabidopsis* circadian clock. Science 293: 880–883.

Alabadi D, Yanovsky MJ, Mas P, Harmer SL, and Kay SA. (2002). Critical role for *CCA1* and *LHY* in maintaining circadian rhythmicity in *Arabidopsis*. Curr Biol 12: 757–761.

An HL, Roussot C, Suarez-Lopez P, Corbesler L, Vincent C, Pineiro M, Hepworth S, Mouradov A, Justin J, Turnbull C, and Coupland G. (2004). CONSTANS acts in the phloem to regulate a systemic signal that induces photoperiodic flowering of *Arabidopsis*. Development 131: 3615–3626.

Ayre BG and Turgeon R. (2004). Graft transmission of a floral stimulant derived from CONSTANS. Plant Physiol 135: 2271–2278.

Ben-Naim O, Eshed R, Parnis A, Teper-Bamnolker P, Shalit A, Coupland G, Samach A, and Lifschitz E. (2006). The CCAAT binding factor can mediate interactions between CONSTANS-like proteins and DNA. Plant J 46: 462–476.

Bünning E. (1935). Zur Kenntnis der erblichen Tagesperiodizität bei den Primärblättern von *Phaseolus multiforus*. Jahr Wiss Bot 81: 411–418.

Bünning E. (1936). Die endogene Tagesrhythmik als Grundlage der photoperiodischen Reaktion. Ber Dtsch Bot Ges 54: 590–607.

Bünning E. (1960). Opening Address: Biological clocks. Cold Spring Harbor Symp Quant Biol 25: 1–9.

Cai XN, Ballif J, Endo S, Davis E, Liang MX, Chen D, DeWald D, Kreps J, Zhu T, and Wu YJ. (2007). A putative CCAAT-binding transcription factor is a regulator of flowering timing in *Arabidopsis*. Plant Physiol 145: 98–105.

Carré I, Coupland G, and Putterill J. (2005). Photoperiodic responses and the regulation of flowering. In *Endogenous Plant Rhythms* (A Hall and H McWatters, eds). Oxford: Blackwell, pp. 167–190.

Chen M, Chory J, and Fankhauser C. (2004). Light signal transduction in higher plants. Annu Rev Genet 38: 87–117.

Corbesier L, Vincent C, Jang SH, Fornara F, Fan QZ, Searle I, Giakountis A, Farrona S, Gissot L, Turnbull C, and Coupland G. (2007). FT protein movement contributes to long-distance signaling in floral induction of *Arabidopsis*. Science 316: 1030–1033.

Covington MF, Panda S, Liu XL, Strayer CA, Wagner DR, and Kay SA. (2001). *ELF3* modulates resetting of the circadian clock in *Arabidopsis*. Plant Cell 13: 1305–1315.

Danilevskii AS. (1965). *Photoperiodism and Seasonal Development of Insects*. London: Oliver and Boyd.

David KM, Armbruster U, Tama N, and Putterill J. (2006). *Arabidopsis* GIGANTEA protein is post-transcriptionally regulated by light and dark. FEBS Lett 580: 1193–1197.

Devlin PF and Kay SA. (2000). Cryptochromes are required for phytochrome signaling to the circadian clock but not for rhythmicity. Plant Cell 12: 2499–2509.

Dunlap J. (1999). Molecular bases for circadian clocks. Cell 96: 271–290.

Endo M, Mochizuki N, Suzuki T, and Nagatani A. (2007). CRYPTOCHROME2 in vascular bundles regulates flowering in *Arabidopsis*. Plant Cell 19: 84–93.

Endo M, Nakamura S, Araki T, Mochizuki N, and Nagatani A. (2005). Phytochrome B in the mesophyll delays flowering by suppressing *FLOWERING LOCUS T* expression in *Arabidopsis* vascular bundles. Plant Cell 17: 1941–1952.

Fowler S, Lee K, Onouchi H, Samach A, Richardson K, Coupland G, and Putterill J. (1999). *GIGANTEA*: A circadian clock-controlled gene that regulates photoperiodic flowering in

Arabidopsis and encodes a protein with several possible membrane-spanning domains. EMBO J 18: 4679–4688.

Franklin KA, Larner VS, and Whitelam GC. (2005). The signal transducing photoreceptors of plants. Int J Dev Biol 49: 653–664.

Fujiwara S, Wang L, Han L, Suh SS, Salome PA, McClung CR, and Somers DE. (2008). Post-translational regulation of the *Arabidopsis* circadian clock through selective proteolysis and phosphorylation of pseudo-response regulator proteins. J Biol Chem 283: 23073–23083.

Garner WW and Allard HA. (1920). Effect of the relative length of day and night and other factors of the environment on growth and reproduction in plants. J Agric Res 18: 553–606.

Goldman BD. (2001). Mammalian photoperiodic system: formal properties and neuroendocrine mechanisms of photoperiodic time measurement. J Biol Rhythms 16: 283–301.

Goto N, Kumagai T, and Kornneef M. (1991). Flowering responses to light-breaks in photomorphogenic mutants of *Arabidopsis thaliana* a long day plant. Physiol Plant 83: 209–215.

Gould P, Locke J, Larue C, Southern M, Davis S, Hanano S, Moyle R, Milich R, Putterill J, Millar A, and Hall A. (2006). The molecular basis of temperature compensation in the *Arabidopsis* circadian clock. Plant Cell 18: 1177–1187.

Guo HW, Yang WY, Mockler TC, and Lin CT. (1998). Regulation of flowering time by *Arabidopsis* photoreceptors. Science 279: 1360–1363.

Han LQ, Mason M, Risseeuw EP, Crosby WL, and Somers DE. (2004). Formation of an SCF ZTL complex is required for proper regulation of circadian timing. Plant J 40: 291–301.

Harmer SL, Hogenesch LB, Straume M, Chang HS, Han B, Zhu T, Wang X, Kreps JA, and Kay SA. (2000). Orchestrated transcription of key pathways in *Arabidopsis* by the circadian clock. Science 290: 2110–2113.

Hayama R, Agashe B, Luley E, King R, and Coupland G. (2007). A circadian rhythm set by dusk determines the expression of *FT* homologs and the short-day photoperiodic flowering response in *Pharbitis*. Plant Cell 19: 2988–3000.

Hayama R and Coupland G. (2004). The molecular basis of diversity in the photoperiodic flowering responses of *Arabidopsis* and rice. Plant Physiol 135: 677–684.

Hicks KA, Millar AJ, Carre IA, Somers DE, Straume M, Meeks Wagner DR, and Kay SA. (1996). Conditional circadian dysfunction of the *Arabidopsis early flowering 3* mutant. Science 274: 790–792.

Huq E, Tepperman JM, and Quail PH. (2000). GIGANTEA is a nuclear protein involved in phytochrome signaling in *Arabidopsis*. Proc Natl Acad Sci USA 97: 9789–9794.

Imaizumi T, Schultz T, Harmon F, Ho L, and Kay S. (2005). FKF1 F-box protein mediates cyclic degradation of a repressor of *CONSTANS* in *Arabidopsis*. Science 309: 293–297.

Imaizumi T, Tran HG, Swartz TE, Briggs WR, and Kay SA. (2003). FKF1 is essential for photoperiodic-specific light signalling in *Arabidopsis*. Nature 426: 302–306.

Izawa T, Oikawa T, Sugiyama N, Tanisaka T, Yano M, and Shimamoto K. (2002). Phytochrome mediates the external light signal to repress *FT* orthologs in photoperiodic flowering of rice. Genes Dev 16: 2006–2020.

Jaeger KE and Wigge PA. (2007). FT protein acts as a long-range signal in *Arabidopsis*. Curr Biol 17: 1050–1054.

Jang S, Marchal V, Panigrahi KC, Wenkel S, Soppe W, Deng XW, Valverde F, and Coupland G. (2008). *Arabidopsis* COP1 shapes the temporal pattern of CO accumulation conferring a photoperiodic flowering response. EMBO J 27: 1277–1288.

Johnson E, Bradley M, Harberd NP, and Whitelam GC. (1994). Photoresponses of light-grown *phyA* mutants of *Arabidopsis*: Phytochrome A is required for the perception of daylength extensions. Plant Physiol 105: 141–149.

Kardailsky I, Shukla VK, Ahn JH, Dagenais N, Christensen SK, Nguyen JT, Chory J, Harrison MJ, and Weigel D. (1999). Activation tagging of the floral inducer *FT*. Science 286: 1962–1965.

Kiba T, Henriques R, Sakakibara H, Chua NH. (2007). Targeted degradation of PSEUDO-RESPONSE REGULATOR5 by an SCFZTL complex regulates clock function and photomorphogenesis in *Arabidopsis thaliana*. Plant Cell 19: 2516–2530.

Kim JY, Song HR, Taylor BL, and Carre IA. (2003). Light-regulated translation mediates gated induction of the *Arabidopsis* clock protein LHY. EMBO J 22: 935–944.

Kim WY, Fujiwara S, Suh SS, Kim J, Kim Y, Han LQ, David K, Putterill J, Nam HG, and Somers DE. (2007). ZEITLUPE is a circadian photoreceptor stabilized by GIGANTEA in blue light. Nature 449: 356–360.

Kobayashi Y, Kaya H, Goto K, Iwabuchi M, and Araki T. (1999). A pair of related genes with antagonistic roles in mediating flowering signals. Science 286: 1960–1962.

Kobayashi Y and Weigel D. (2007). Move on up, it's time for change—mobile signals controlling photoperiod-dependent flowering. Genes Dev 21: 2371–2384.

Kojima S, Takahashi Y, Kobayashi Y, Monna L, Sasaki T, Araki T, and Yano M. (2002). *Hd3a*, a rice ortholog of the *Arabidopsis FT* gene, promotes transition to flowering downstream of *Hd1* under short-day conditions. Plant Cell Physiol 43: 1096–1105.

Koornneef M, Hanhart CJ, and Van der Veen JH. (1991). A genetic and physiological analysis of late flowering mutants in *Arabidopsis thaliana*. Mol Gen Genet 229: 57–66.

Laubinger S, Fittinghoff K, and Hoecker U. (2004). The SPA quartet: A family of WD-repeat proteins with a central role in suppression of photomorphogenesis in *Arabidopsis*. Plant Cell 16: 2293–2306.

Laubinger S, Marchal V, Le Gourrierec J, Wenkel S, Adrian J, Jang S, Kulajta C, Braun H, Coupland G, and Hoecker U. (2006). *Arabidopsis* SPA proteins regulate photoperiodic flowering and interact with the floral inducer CONSTANS to regulate its stability. Development 133: 3213–3222.

Lees A. (1973). Photoperiodic time measurement in the aphid *Megoura viciae*. J Insect Physiol 19: 2279–2316.

Lifschitz E, Eviatar T, Rozman A, Shalit A, Goldshmidt A, Amsellem Z, Alvarez JP, and Eshed Y. (2006). The tomato *FT* ortholog triggers systemic signals that regulate growth and flowering and substitute for diverse environmental stimuli. Proc Natl Acad Sci USA 103: 6398–6403.

Lin MK, Belanger H, Lee YJ, Varkonyi-Gasic E, Taoka KI, Miura E, Xoconostle-Cazares B, Gendler K, Jorgensene RA, Phinney B, Lough TJ, and Lucas WJ. (2007). FLOWERING LOCUS T protein may act as the long-distance florigenic signal in the cucurbits. Plant Cell 19: 1488–1506.

Liu LJ, Zhang YC, Li QH, Sang Y, Mao J, Lian HL, Wang L, and Yang HQ. (2008). COP1-mediated ubiquitination of CONSTANS is implicated in cryptochrome regulation of flowering in *Arabidopsis*. Plant Cell 20: 292–306.

Locke JC, Kozma-Bognar L, Gould PD, Feher B, Kevei E, Nagy F, Turner MS, Hall A, and Millar AJ. (2006). Experimental validation of a predicted feedback loop in the multi-oscillator clock of *Arabidopsis* thaliana. Mol Syst Biol 2: 59.

Martin-Tryon E, Kreps J, and Harmer S. (2007). *GIGANTEA* acts in blue light signaling and has biochemically separable roles in circadian clock and flowering time regulation. Plant Physiol 143: 473–486.

Martinez-Garcia JF, Huq E, and Quail PH. (2000). Direct targeting of light signals to a promoter element-bound transcription factor. Science 288: 859–863.

Martinez-Zapater JM, Coupland G, Dean C, and Koornneef M. (1994). The transition to flowering in *Arabidopsis*. In *Arabidopsis* (EM Meyerowitz and CR Somerville, eds). Plainview, NY: Cold Spring Harbor Laboratory Press, pp. 403–434.

Más P, Kim W, Somers D, and Kay S. (2003). Targeted degradation of TOC1 by ZTL modulates circadian function in *Arabidopsis thaliana*. Nature 426: 567–570.

Mathieu J, Warthmann N, Kuttner F, and Schmid M. (2007). Export of FT protein from phloem companion cells is sufficient for floral induction in *Arabidopsis*. Curr Biol 17: 1055–1060.

McClung CR. (2006). Plant circadian rhythms. Plant Cell 18: 792–803.

McWatters HG, Bastow RM, Hall A, and Millar AJ. (2000). The *ELF3* zeitnehmer regulates light signalling to the circadian clock. Nature 408: 716–720.

McWatters HG, Kolmos E, Hall A, Doyle MR, Amasino RM, Gyula P, Nagy F, Millar AJ, and Davis SJ. (2007). ELF4 is required for oscillatory properties of the circadian clock. Plant Physiol 144: 391–401.

McWatters H, Roden L, and Staiger D. (2001). Picking out parallels: Plant circadian clocks in context. Philos Trans Roy Soc Lond Ser B Biol Sci 356: 1735–1743.

Mizoguchi T, Wheatley K, Hanzawa Y, Wright L, Mizoguchi M, Song HR, Carre IA, and Coupland G. (2002). *LHY* and *CCA1* are partially redundant genes required to maintain circadian rhythms in *Arabidopsis*. Dev Cell 2: 629–641.

Mizoguchi T, Wright L, Fujiwara S, Cremer F, Lee K, Onouchi H, Mouradov A, Fowler S, Kamada H, Putterill J, and Coupland G. (2005). Distinct roles of GIGANTEA in promoting flowering and plant circadian rhythms in *Arabidopsis*. Plant Cell 17: 2255–2270.

Mockler TC, Guo HW, Yang HY, Duong H, and Lin CT. (1999). Antagonistic actions of *Arabidopsis* cryptochromes and phytochrome B in the regulation of floral induction. Development 126: 2073–2082.

Nelson DC, Lasswell J, Rogg LE, Cohen MA, and Bartel B. (2000). FKF1, a clock-controlled gene that regulates the transition to flowering in *Arabidopsis*. Cell 101: 331–340.

Nishizuka M, Azuma A, and Kostal V. (1988). Diapause response to photoperiod and temperature in *Lepisma saccharina* Linneaus (Thysanura: Lepismatidae). Entomol Sci 1: 7–14.

Oliverio KA, Crepy M, Martin-Tryon EL, Milich R, Harmer SL, Putterill J, Yanovsky MJ, and Casal JJ. (2007). GIGANTEA regulates phytochrome A-mediated photomorphogenesis independently of its role in the circadian clock. Plant Physiol 144: 495–502.

Park DH, Somers DE, Kim YS, Choy YH, Lim HK, Soh MS, Kim HJ, Kay SA, and Nam HG. (1999). Control of circadian rhythms and photoperiodic flowering by the *Arabidopsis GIGANTEA* gene. Science 285: 1579–1582.

Pittendrigh C. (1972). Circadian surfaces and the diversity of possible roles of circadian organization in photoperiodic induction. Proc Natl Acad Sci USA 69: 2734–2737.

Pittendrigh C and Minis D. (1964). The entrainment of circadian oscillators by light and their role as photoperiodic clocks. Am Nat 98: 261–294.

Pittendrigh C and Takamura T. (1989). Latitudinal clines in the properties of a circadian pacemaker. J Biol Rhythms 4: 217–235.

Putterill J, Laurie R, and Macknight R. (2004). It's time to flower: The genetic control of flowering time. BioEssays 26: 363–373.

Putterill J, Robson F, Lee K, Simon R, and Coupland G. (1995). The *CONSTANS* gene of *Arabidopsis* promotes flowering and encodes a protein showing similarities to zinc finger transcription factors. Cell 80: 847–857.

Redei GP. (1962). Supervital mutants in *Arabidopsis*. Genetics 47: 443–460.

Reeves PH and Coupland G. (2001). Analysis of flowering time control in *Arabidopsis* by comparison of double and triple mutants. Plant Physiol 126: 1085–1091.

Roden LC, Song H-R, Jackson SD, Morris K, and Carre IA. (2002). Floral responses to photoperiod are correlated with the timing of rhythmic expression relative to dawn and dusk in *Arabidopsis*. Proc Natl Acad Sci USA 99: 13313–13318.

Samach A, Onouchi H, Gold SE, Ditta GS, Schwarz-Sommer Z, Yanofsky MF, and Coupland G. (2000). Distinct roles of *CONSTANS* target genes in reproductive development of *Arabidopsis*. Science 288: 1613–1616.

Saunders DS. (1977). *An Introduction to Biological Rhythms*. London: Blackie.

Sawa M, Nusinow DA, Kay SA, and Imaizumi T. (2007). FKF1 and GIGANTEA complex formation is required for day-length measurement in *Arabidopsis*. Science 318: 261–265.

Somers DE, Devlin PF, and Kay SA. (1998). Phytochromes and cryptochromes in the entrainment of the *Arabidopsis* circadian clock. Science 282: 1488–1490.

Somers DE, Schultz TF, Milnamow M, and Kay SA. (2000). ZEITLUPE encodes a novel clock-associated PAS protein from *Arabidopsis*. Cell 101: 319–329.

Suarez-Lopez P, Wheatley K, Robson F, Onouchi H, Valverde F, and Coupland G. (2001). *CONSTANS* mediates between the circadian clock and the control of flowering in *Arabidopsis*. Nature 410: 1116–1120.

Takada S and Goto K. (2003). TERMINAL FLOWER2, a HETEROCHROMATIN PROTEIN1-Like Protein of *Arabidopsis*, counteracts the activation of *FLOWERING LOCUS T* by CONSTANS in the vascular tissues of leaves to regulate flowering time. Plant Cell 15: 2856–2865.

Tamaki S, Matsuo S, Wong HL, Yokoi S, and Shimamoto K. (2007). Hd3a protein is a mobile flowering signal in rice. Science 316: 1033–1036.

Thomas B and Vince-Prue D. (1997). *Photoperiodism in Plants*. San Diego: Academic Press.

Valverde F, Mouradov A, Soppe W, Ravenscroft D, Samach A, and Coupland G. (2004). Photoreceptor regulation of CONSTANS protein in photoperiodic flowering. Science 303: 1003–1006.

Wang HY. (2005). Signaling mechanisms of higher plant photoreceptors: A structure-function perspective. Curr Top Dev Biol 68: 227–261.

Wang HY, Ma LG, Li JM, Zhao HY, and Deng XW. (2001). Direct interaction of *Arabidopsis* cryptochromes with COP1 in light control development. Science 294: 154–158.

Wenkel S, Turck F, Singer K, Gissot L, Le Gourrierec J, Samach A, and Coupland G. (2006). CONSTANS and the CCAAT box binding complex share a functionally important domain and interact to regulate flowering of *Arabidopsis*. Plant Cell 18: 2971–2984.

Wigge PA, Kim MC, Jaeger KE, Busch W, Schmid M, Lohmann JU, and Weigel D. (2005). Integration of spatial and temporal information during floral induction in *Arabidopsis*. Science 309: 1056–1059.

Yanovsky MJ and Kay SA. (2002). Molecular basis of seasonal time measurement in *Arabidopsis*. Nature 419: 308–312.

Yanovsky MJ, Mazzella MA, and Casal JJ. (2000). A quadruple photoreceptor mutant still keeps track of time. Curr Biol 10: 1013–1015.

Yasuo S, Watanabe M, Okabayashi N, Ebihara S, and Yoshimura T. (2003). Circadian clock genes and photoperiodism: Comprehensive analysis of clock gene expression in the mediobasal hypothalamus, the suprachiasmatic nucleus, and the pineal gland of Japanese quail under various light schedules. Endocrinology 144: 3742.

Yoo SK, Chung KS, Kim J, Lee JH, Hong SM, Yoo SJ, Yoo SY, Lee JS, and Ahn JH. (2005). CONSTANS activates *SUPPRESSOR OF OVEREXPRESSION OF CONSTANS 1* through *FLOWERING LOCUS T* to promote flowering in *Arabidopsis*. Plant Physiol 139: 770–778.

Yu J-W, Rubio V, Lee N-Y, Bai S, Lee S-Y, Kim S-S, Liu L, Zhang Y, Irigoyen ML, Sullivan JA, Zhang Y, Lee I, Xie Q, Paek N-C, and Deng XW (2008). COP1 and ELF3 control circadian function and photoperiodic flowering by regulating GI stability. Molec. Cell 32: 617–630.

Zeilinger MN, Farre EM, Taylor SR, Kay SA, and Doyle FJ III. (2006). A novel computational model of the circadian clock in *Arabidopsis* that incorporates PRR7 and PRR9. Mol Syst Biol 2: 58.

2

Photoperiodic Control of Flowering in the Short-Day Plant *Oryza sativa*

Takeshi Izawa

Garner and Allard (1920) first reported that many higher plants can recognize the seasonal changes of day length to determine flowering time, such that members of a population will flower synchronously. Day-length recognition has also been observed in insects, fish, birds, and mammals (Lumsden and Millar, 1998). This characteristic is termed photoperiodism (Thomas, 1998). Higher plants can be categorized into two major groups: short-day plants, which flower when day length becomes shorter than a certain threshold, and long-day plants, which flower when day length becomes longer than a certain threshold. Numerous physiological and biochemical studies have investigated the mechanisms underlying timekeeping in plants, and several promising models have been proposed to explain timekeeping (for a review, see Thomas and Vince-Prue, 1997). However, it is difficult to test these models using only physiological and biochemical approaches. With the advent of molecular genetics, model plants such as *Arabidopsis thaliana*, a typical long-day plant, and rice (*Oryza sativa*), a typical short-day plant, can be used to investigate how plants measure day length and regulate their floral transitions (Henderson et al., 2003; Izawa et al., 2003; Putterill et al., 2004; Sung and Amasino, 2005; Corbesier and Coupland, 2006; Imaizumi and Kay, 2006; Jaeger et al., 2006; Izawa, 2007a, 2007b; Kobayashi and Weigel, 2007). This chapter summarizes the progress made in understanding photoperiodic flowering in rice.

PHOTOPERIODIC CONTROL OF FLOWERING IN RICE

Rice plants produce tiny, inconspicuous flowers (figure 2.1). As shown in figure 2.2, rice is a typical short-day plant, requiring a certain number of hours of darkness in each 24-h period (a short day length) before floral development can begin. Rice generally has a critical day length around 14 h (i.e., 10 h of darkness). When day length exceeds 14 h, it takes typical rice cultivars dramatically longer to flower. For instance, under constant light, rice plants do not flower at all, even at 1 year after sowing.

Oryza sativa is the major crop species of rice and arose by the domestication of a wild ancestor (Garris et al., 2005; Doebley et al., 2006; Londo et al., 2006).

FIGURE 2.1.
Tiny rice flowers open with extruding anthers.

Its closest wild relative, *Oryza rufipogon*, is a native of Asia and grows in tropical areas, with a northern limit of approximately 25°N latitude. In contrast, cultivated *O. sativa* can grow in more northerly regions that have much longer day lengths in summer, with a northern limit of approximately 45°N (figure 2.3). The movement northward of the region of rice cultivation resulted from domestication and breeding, which relied on genetic changes related to flowering-time regulation (Izawa, 2007b). Therefore, let us first consider how rice is grown under natural seasonal conditions.

Adaptation of rice to northern environments strictly limited the appropriate flowering time, because flower development, meiosis, fertilization, and embryogenesis are very sensitive to cold. Rice cultivars bred for northern Japan (40–45°N) start the floral transition at the end of June, flower at the end of July, and are harvested at the beginning of September. In order to flower with the appropriate timing for autumn harvest, these northern cultivars must flower extremely early after

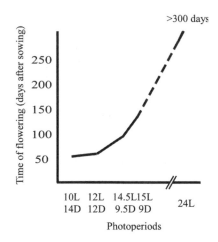

FIGURE 2.2.
The flowering response of a typical rice cultivar, Nipponbare, grown under distinct photoperiods.

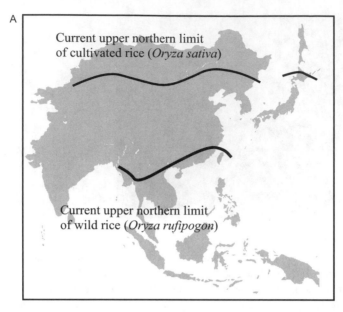

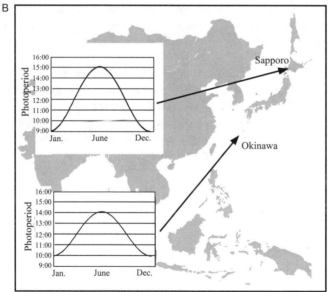

FIGURE 2.3.
(A) A map of Asia showing the northern limits of *Oryza rufipogon* (gray) and *O. sativa* (black).
(B) Changes in natural photoperiod over the year in the northern (43°N, top) and southern (25°N, bottom) regions.

sowing and be insensitive to changes in the day length. Similarly, to suit the crop-ping season that is currently very popular in Japan, many rice cultivars start the flo-ral transition at the end of June or the beginning of July and flower at the beginning of August, even in temperate regions (around 35°N). The rice cultivars grown in temperate regions, however, are still sensitive to photoperiodic changes. Therefore, many rice cultivars grown in Japan start the floral transition just after the summer solstice, and in temperate and cold areas the plants are grown under long-day condi-tions (approximately 14- to 16-h day length). Thus, how can rice, a short-day plant, start the floral transition under these long-day conditions? Molecular genetics stud-ies have revealed some answers to this question.

QUANTITATIVE TRAIL LOCUS ANALYSIS TO ELUCIDATE FLOWERING-TIME REGULATION IN RICE

In 1991, the Rice Genome Project in Japan began producing the complete genetic map of rice based on restriction-fragment-length polymorphism markers between Nipponbare (an *O. sativa* ssp. *japonica* cultivar) and Kasalath (a ssp. *indica* culti-var; Kurata et al., 1994; Harushima et al., 1998). Masahiro Yano at the National Institute of Agrobiological Sciences and his colleagues next conducted quantitative trait locus (QTL) analysis of many traits, including those related to flowering time, seed shattering, and yield, using F_2 and advanced back-crosses between Nipponbare and Kasalath (Lin et al., 1997, 1998, 2002, 2003; Yano and Sasaki, 1997; Yano et al., 1997, 2001; Yamamoto et al., 1998). Using F_2 populations, Yano et al. (1997) found several QTLs for flowering time, including *Hd1*, *Hd2*, and *Hd3*. Based on studies of further backcrossed inbred lines, more than 10 QTLs for flowering time have been discovered in rice (Yano et al., 2001).

These fundamental works contributed much to the cloning of flowering-time genes in rice, including *Hd1*, *Hd6*, and *Hd3a* (Takahashi et al., 2001; Yano et al., 2000; Kojima et al., 2002; table 2.1). Several flowering-time QTLs are often detected among parent pairs, especially in *japonica* × *indica* crosses. The identification of flowering-time QTLs among cultivars is significant because flowering time is an important agronomic trait for locally adapted cultivars, and the responsible QTL may be the result of domestication or breeding. QTL analysis is also critical when we are mapping induced mutations, including those in flowering-time genes. The results of these rice QTL studies provide a platform for elucidating the molecular mechanisms of photoperiodic control of flowering in rice.

MOLECULAR GENETIC STUDIES OF FLOWERING-TIME CONTROL IN RICE

The first clues to the molecular mechanisms of flowering-time control in rice came from the analysis of an early-flowering mutant, *se5* (photoperiod <u>se</u>nsitivity <u>5</u>)

TABLE 2.1. The genes that control flowering time in rice.

Gene	Protein	Mutant phenotype	Mutations	Primary publications
SE5 (*Photosensitivity 5*)	Heme oxygenase	Early in SD, very early in LD	Induced	Izawa et al. (2000, 2002)
Hd1 (*Heading date 1*)	Transcription factor	Late in SD, early in LD	Natural variation	Yano et al. (2000)
Hd6	CK2a	Early in LD	Natural variation	Takahashi et al. (2001)
Hd3a	Florigen	Late in SD	Natural variation	Kojima et al. (2002)
RFT1	Florigen	No phenotype in SD	RNAi	Komiya et al. (2008)
Ehd1 (*Early heading date 1*)	Type B response regulator	Late in SD, late in LD	Natural variation	Doi et al. (2004)
OsMADS50	Type II MADS-box factor	Late in LD	T-DNA	Lee et al. (2004)
Phytochrome B/C	R/FR photoreceptors	Early in LD	Tos17 or Induced	Takano et al. (2005)
OsMADS51	Type I MADS-box factor	Late in SD	T-DNA	S.L. Kim et al. (2004)
Lhd4 (*Late heading date 4*) (or Ghd7)	CCT motif protein	Early in SD, very early in LD	Natural variation	Kitagawa et al., unpublished data, Xue et al. 2008
OsGI	F-box interacting protein	Late in SD	Induced in se5	Izawa et al., unpublished data

LD, long-day conditions; SD, short-day conditions.

(Yokoo and Okuno, 1993; Izawa et al., 2000; table 2.1). The *se5* mutant plants are insensitive to photoperiod and have physiologically and biochemically undetectable levels of phytochromes. Our studies demonstrated that the *SE5* gene encodes a protein homologous to that of the *HY1* gene (Muramoto et al., 1999) in *Arabidopsis thaliana* (Izawa et al., 2000). HY1 protein exhibits heme oxygenase activity, which is required for phytochrome-chromophore biosynthesis. The loss of photoperiod sensitivity in *se5* mutants revealed that phytochromes are the photoreceptors in rice that allow for recognition of the photoperiod. In addition, Yano and his colleagues cloned the *Hd1* gene and found that it is homologous to *CO* (*CONSTANS*; Putterill et al., 1995) in *A. thaliana* (Yano et al., 2000). The completion of the rice genome map enabled us to compare sequences between these plants and to conclude that *Hd1* is an ortholog of *CO* (Izawa et al., 2003). *Hd1* has a dual function: floral promotion under short-day conditions and floral suppression under long-day conditions (figure 2.4). (The molecular basis for the opposite actions in *Hd1* and *CO* is discussed further below.) The *Hd3* QTL was divided into two neighboring loci, *Hd3a* and *Hd3b* (Monna et al., 2002), and additional work revealed that *Hd3a*

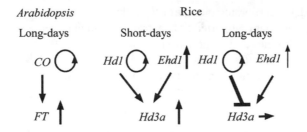

FIGURE 2.4.

A schematic comparison of key regulators involved in the photoperiodic control of flowering in *Arabidopsis* and rice. A circle with an arrowhead indicates diurnal expression controlled by circadian clocks; these mRNA levels are not changed by photoperiodic conditions. In contrast, genes such as *FT* and *Hd3a* with arrows act like molecular switches.

encodes an *FT* ortholog in rice (Kojima et al., 2002; see chapter 1 for a discussion of *FT*).

Further investigations looked at the interaction between phytochrome signaling and circadian clock signaling in the control of flowering time in rice. Izawa et al. (2002) demonstrated that *Hd1* mRNA transcription is regulated by the circadian clock in rice, as is that of *CO* in *A. thaliana* (Suarez-Lopez et al., 2001), and its floral suppression activity is further modulated by phytochrome signaling. This finding is consistent with the external coincidence model (figure 2.5), which was proposed by Pittendrigh and Minis (1964) on the basis of numerous physiological studies of plants and *Drosophila*. This model is a refined version of Bünning's hypothesis proposed in the 1930s (for review, see Bünning, 1960). The major difference between the external coincidence model and Bünning's hypothesis is

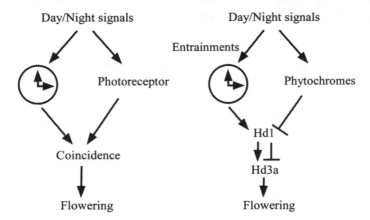

FIGURE 2.5.

A schematic comparison between the physiological external coincidence model (left) and the relationship of functional networks of key flowering-time genes in rice (right).

that the model incorporates the nature of light entrainment in circadian clocks. Molecular genetic data supported light entrainment in plant circadian clocks mediated by phytochromes and cryptochromes (Somers et al., 1998). Similarly, studies by Yanovsky and Kay (2002, 2003) revealed that the external coincidence model fits with the molecular mechanisms for photoperiodic flowering in *A. thaliana* (see chapter 1).

HD6 ENCODES A CONSERVED PROTEIN KINASE

Yano and his colleagues also cloned the QTL *Hd6* (Takahashi et al., 2001; table 2.1), which encodes a $CK2\alpha$ subunit. The Nipponbare allele of *Hd6* contains a premature stop codon and does not encode a functional CK2 protein. This research demonstrated that *Hd6* functions as a floral suppressor only under long-day conditions. In *A. thaliana*, CK2 phosphorylates the central circadian clock component CCA1 (Sugano et al., 1998, 1999; Daniel et al., 2004). Therefore, in collaboration with Yano's group we examined flowering time and circadian clock–regulated gene expression in both *hd6*-deficient and *Hd6*-overexpressing rice (E. Ogiso et al., unpublished data). We found that *Hd6* did not control the expression of genes for any critical components in the circadian clock and *Hd1* mRNA, but it controls *Hd3a* mRNA transcription and flowering time only in the presence of a functional *Hd1* allele. Further biochemical analysis strongly suggested that *Hd6* does not phosphorylate the Hd1 protein itself. Because *Hd6* mRNA was not rhythmically expressed under constant light and dark conditions, we concluded that *Hd6* controls flowering time of rice not as a component of the circadian clock, but through some modulation of the Hd1 suppressor activity due to phosphorylation. CK2 is a well-known member of circadian clocks in fungi, insects, mammals, and some other plants (Mizoguchi et al., 2006), so further work is warranted to investigate why *Hd6* is not a component of the circadian clock in rice.

UNIQUE MOLECULES IN THE PHOTOPERIODIC CONTROL OF FLOWERING OF RICE

In collaboration with Kazuyuki Doi (Kyushu University), our group cloned the QTL *Ehd1* (*Eearly heading date 1*; Doi et al., 2004). The cloning of *Ehd1* revealed that the late-flowering phenotype was conferred by a mutation in the DNA-binding domain of *Ehd1* in one of the parent cultivars, Taichung 65 (T65). Our results clearly indicated that *Ehd1* is a floral promoter under both long-day and short-day conditions (figure 2.4). The promotion activity by *Ehd1* under short-day conditions was stronger than that under long-day conditions, which is consistent with the fact that *Ehd1* mRNA is induced under short-day conditions. *Ehd1* encodes a B-type response regulator for which we could not find an orthologous gene in the *A. thaliana* genome. The B-type response regulator is believed to be a transcription factor that contains a

DNA-binding GARP domain with a typical receiver domain (Heyl and Schmülling, 2003; Ishida et al., 2008). In *A. thaliana*, more than 10 B-type response regulator genes have been discovered, and all tested ones are somehow involved in cytokinin signaling. In contrast, *Ehd1* can control flowering-time genes such as *Hd3a* just after the short-day induction of *Ehd1* expression (Doi et al., 2004). Therefore, *Ehd1* is a genuine flowering-time gene and is unique to the control of flowering time in rice; *Ehd1* may be a direct transcription factor of *Hd3a* in rice.

In the QTL analysis culminating in the cloning of *Ehd1*, another QTL was discovered around the *Hd1* region (Doi et al., 2004). This led us to examine the *Hd1* genomic sequence in T65, revealing that T65 also has a loss-of-function *Hd1* allele due to the integration of a retro-element. When the functional *Hd1* allele of Nipponbare was introduced into T65, under long day conditions (14.5 L : 9.5 D) the T65 plant with *Hd1* did not flower until 180 days after sowing (Doi et al., 2004). Thus, *Ehd1* function is required for flowering under long-day conditions (figure 2.4). The distinction between the biological functions of *Hd1* and *Ehd1* are discussed further below. The important point here, however, is that both *Hd1* and *Ehd1* are involved in the photoperiodic control of flowering in rice, and both regulate *Hd3a* expression.

In *A. thaliana*, *FT* is sometimes called a floral integrator (Imaizumi and Kay, 2006). Several distinct controls of flowering, such as photoperiodic and vernalization-related signals, are integrated in the control of diurnal *FT* mRNA expression patterns. The long-winter signal can desuppress *FT* expression through *VIN3* and *VRN2* epigenetically, and then the long-day signal in spring may induce *FT* expression though the transcriptional activation by CO protein (Sung and Amasino, 2006). This heterochronic regulation of vernalization and photoperiodism is a core system for flowering regulation in plants with a winter-annual ecotype, such as *A. thaliana*. In rice, competition between *Hd1* suppression and *Ehd1* promotion may be a core system for flowering regulation in temperate and colder areas, in which rice plants should start floral transition under the long-day conditions (Izawa, 2007b). In addition, the balance between *Hd1* suppression and *Ehd1* promotion may be involved in local adaptation. That is, the synchronized competition between *Hd1* and *Ehd1* may contribute to the ability of rice cultivars to be grown across widely diverse regions (figures 2.3, 2.4).

Our group together with Yano's group recently cloned a novel QTL, termed *Lhd4* (Kitazawa et al., unpublished data), which encodes a CCT-motif protein homologous to wheat *VRN2* (Yan et al., 2004). Lhd4 and VRN2 belong to a monocot-specific clade in the phylogenetic tree of CCT-motif proteins. Although both *Lhd4* and *VRN2* function as strong floral suppressors, their nonsyntenic chromosome positions suggest that they are not true orthologs. We also found that *Lhd4* mRNA is induced only under long-day conditions at the seedling stage, and the long-day induction of *Lhd4* mRNA requires phytochrome signals (figure 2.6). These findings clarify the relationship of *Lhd4* with the Hd1 suppressor function modulated by phytochromes observed in *se5* and *hd1* mutants (Izawa et al., 2002). If Lhd4 formed a complex with Hd1, that would easily explain how Hd1 functions as a strong suppressor only under

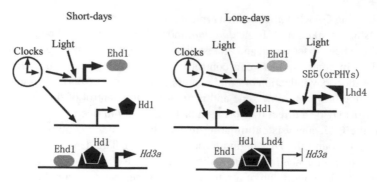

FIGURE 2.6.
A molecular model to explain the photoperiodic control of flowering in rice. Lhd4 may form a strong floral suppressor complex with Hdl, although biochemical evidence is lacking. Ehdl is transcribed more under short-day conditions (thick arrow) than under long-day conditions.

long-day conditions. Indeed, with a nonfunctional *Lhd4* allele, *Hd1* can promote *Hd3a* mRNA expression even under long-day conditions. This result is consistent with our current models (figure 2.6). Another group in China cloned the same gene, termed *Ghd7*, using different cultivars as parents (Xue et al., 2008), indicating that some variations in this gene occur naturally, thus facilitating the control of flowering time during the domestication and breeding of rice.

The identification of the unique genes *Ehd1* and *Lhd4 (Ghd7)* involved in the photoperiodic control of flowering in rice suggests a wealth of still-hidden factors in the regulation of flowering among the higher plants. For instance, some bamboo species flower once every 67 years and more. Both bamboo and rice are related monocots and are more closely related to one another than are rice and *A. thaliana*. Surveys using high-speed sequencing of the genomes of bamboo and other monocot cereals may provide more clues to the molecular diversity in plant flowering-time regulation.

INVOLVEMENT OF MADS-BOX GENES IN THE CONTROL OF FLOWERING IN RICE

S. L. Kim et al. (2007) reported that the MADS-box transcription factor *OsMADS51* functions upstream of the *Ehd1* gene to control flowering time in rice (table 2.1). *OsMADS51* is required for expression of *Ehd1* and those genes downstream of *Ehd1*, such as *Hd3a* and *OsMADS14* (a FUL-clade MADS-box gene), especially under short-day conditions. Rice plants overexpressing *OsMADS51* cDNA flowered early under short-day conditions, because this overexpression resulted in *Ehd1* induction only under short-day conditions. This finding implies that some unknown floral suppression mechanisms that work only under long-day conditions control *Ehd1* in rice.

OsMADS51 is a type I MADS-box gene. In contrast, many *A. thaliana* MADS-box genes involved in floral transition and flower formation, including most ABC model genes, are type II (or MIKC type) MADS-box genes that contain a K-box domain in addition to the SRF DNA-binding domain (Rijpkema et al., 2007). Three type II MADS-box floral-switch genes controlling the floral transition in *A. thaliana*, *SOC*, *SVP*, and *FLC*, are well characterized (Hartmann et al., 2000; Onouchi et al., 2000; Michaels et al., 2003). *OsMADS50*, a MADS-box of rice that is orthologous to *SOC1* of *A. thaliana*, was shown to activate floral transition in rice (Lee et al., 2004; table 2.1). Although their role in the photoperiodic control of flowering is not clear, both *SOC1* and *OsMADS50* function as floral promoters. Lee et al. (2008) reported that *SVP* orthologs in rice were involved in plant architecture regulated by brassinosteroid signaling, but not in floral transition. *FLC*, which functions as a strong floral suppressor, is a critical regulator of the vernalization response in *A. thaliana* (Sung and Amasino, 2005). However, no type II MADS-box genes belonging to the same clade as *FLC* have been identified in the rice genome (Izawa et al., 2003). Furthermore, the *VRN1* gene, which is involved in the floral transition in wheat through vernalization treatment, encodes a type II MADS-box gene (Yan et al., 2003), but it remains unclear whether *OsMADS14*, the true rice ortholog of *VRN1*, is involved in the floral transition in rice. Although some orthologous molecular mechanisms function similarly in rice and *A. thaliana*, such as those of *Hd1-CO* and *Hd3a-FT*, these examples indicate that an orthologous relationship does not guarantee concordant biological functions in evolutionarily distant plant species.

ROLES OF CIRCADIAN CLOCK COMPONENTS IN THE PHOTOPERIODIC FLOWERING OF RICE

Direct light signals mediated through phytochromes are critical for floral suppression in rice under long-day conditions. Because phytochrome signals may entrain components of the circadian clock (Somers et al., 1998), we examined the behavior of the circadian clock under various conditions in *se5*, the phytochrome-deficient mutant of rice (Izawa et al., 2002), but we did not observe any defects in the circadian clocks of *se5*. Because phytochromes are sensitive to light in the red to far-red region of the visible spectrum, blue light signals may compensate for the deficiencies in *se5* mutants.

Circadian clocks play critical roles in the regulation of *Hd1* and *Lhd4* expression, as indicated by their mRNA expression patterns. Therefore, to identify novel circadian clock mutants in rice, we screened suppressor mutants for flowering-time phenotypes in the *se5* background. Compared with *se5*, the *se5* suppressor mutants clearly exhibited late-flowering phenotypes under both long-day and short-day conditions. Cloning of these mutants revealed that a gene that suppresses *se5* encodes an ortholog of *Arabidopsis GIGANTEA* (*GI*; Fowler et al., 1999; Parks et al., 1999) in rice (i.e., *OsGI*; Hayama et al., 2002; Izawa et al., unpublished data; table 2.1).

GI controls both the circadian clock and *CO* transcription in *A. thaliana* (Suarez-Lopez et al., 2001; Mizoguchi et al., 2005). Recent studies have shown that this regulation can be controlled by the interaction of *GI* with *ZTL* and *FKF1* (W. Y. Kim et al., 2007; Sawa et al., 2007), both of which encode an F-box protein with an LOV domain (i.e., a blue-light perception domain). *FKF1* and *ZTL* control degradation of a *CO* suppressor, CDF1, and a circadian clock component, TOC1, respectively (Imaizumi et al., 2003, 2005; Más et al., 2003). Therefore, *OsGI* may play a similar role in the regulation of rice flowering.

The phenotypes of a single *osgi* mutant, however, revealed differences in regulation between rice and *A. thaliana*. The mutant flowered later than the wild-type under short-day conditions, but at the same time as the wild-type under long-day conditions. The *hd1* mutation caused significantly early flowering under long-day conditions, indicating that *OsGI* is not required for the floral suppression by *Hd1* under long-day conditions. In addition, it has been proposed that, using computational modeling analysis, *GI* is a member in a genetically redundant loop of the *A. thaliana* circadian clock system (Locke et al., 2005, 2006), because a single *gi* mutation did not cause any drastic changes in the gene expression of other circadian clock components, such as *TOC1* (Gould et al., 2006). In contrast, the *se5* suppressor mutation in rice caused dramatic up-regulation of *OsPRR1*, an ortholog of *TOC1/PRR1* in *A. thaliana*. The suppressor mutants also showed a clear down-regulation of *OsLHY*, another putative core component in rice circadian clock. These findings highlight the importance of *OsGI* in the circadian clock of rice. As described above, *OsGI* also plays an important role in the control of photoperiodic flowering in rice, although the network of circadian clock components and flowering-time regulation associated with *OsGI* is clearly distinct from that in *A. thaliana*.

The typical behavior of circadian clock components in rice could be monitored using *cab1R::luc* as a reporter gene (Sugiyama et al., 2001), indicating that circadian clock regulation of *cab* expression is basically conserved in rice. Furthermore, orthologs corresponding to major circadian clock components in *Arabidopsis* were identified in the rice genome (Izawa et al., 2003). Five pseudo-response regulator (PRR) genes in rice were expressed in a circadian-clock–dependent manner (Murakami et al., 2003). Murakami et al. (2005) identified a putative functional nucleotide polymorphism in a PRR gene, *OsPRR37*, which is a possible circadian clock gene in rice. The QTL *Hd2* lies near *OsPRR37*, although it remains unclear whether *Hd2* encodes *OsPRR37*. The *OsPRR37* ortholog in barley, a long-day plant that also is a member of Poaceae, was recently reported as a flowering-time gene (Turner et al., 2005).

HD3A IS A FLORIGEN

For many years, researchers have sought to identify florigens, phytohormones that function as long-distance signals transmitted from the grafts to induce floral transition at the apex in many plant species (Corbesier and Coupland, 2006). Molecular genetic research in rice led to the first breakthrough in this search, when Shimamoto and his colleagues (Tamaki et al., 2007) reported that Hd3a protein is a florigen in rice.

Many previous studies led to this discovery. After the cloning of *FT* in *A. thaliana* by Kardailsky et al. (1999) and Kobayashi et al. (1999), it took some time to find where *FT* mRNA is transcribed owing to the low levels of *FT* mRNA and a long genomic fragment of the *FT* gene required to complement the mutant phenotype completely. Several critical studies revealed that *FT* is expressed in the sieve element cells in the vascular tissues, but not in the meristematic region at the shoot apex in *A. thaliana* (Takada and Goto, 2003; An et al., 2004). Abe et al. (2005) found that *fd* mutations suppress the early-flowering phenotypes induced by the ectopic expression of *FT* cDNA. *FD* encodes a bZIP-type transcription factor that is specifically expressed in the meristematic region in *A. thaliana*. In addition, FD can interact physically with FT (Abe et al. 2005; Wigge et al., 2005). These results strongly suggest that *FT* mRNA or FT protein is a long-distance mobile signal, that is, a florigen. Soon after publication of these works, Huang et al. (2005) reported that *FT* mRNA was translocated into the meristematic region in *A. thaliana* after artificial induction of *FT* transcription driven by a heat-shock promoter. However, this means that it is difficult to determine whether FT protein is movable if *FT* mRNA is moving.

Shimamoto's group had begun to work on *Hd3a*, an *FT* ortholog in rice, to investigate the activity of an Hd3a-GFP (green fluorescent protein) fusion protein. The fusion protein could induce early flowering in rice when ectopically expressed (Tamaki et al., 2007); they were able to detect the GFP signal (possibly from the fusion protein) in the meristematic region in rice, but could not detect *Hd3a* mRNA in the region. Corbesier et al. (2007) took a similar approach using *A. thaliana* and demonstrated that FT activity was graft-transmissible, as would be expected of a florigen. Thus, the work of numerous research groups shows that the evolutionarily conserved peptides Hd3a and FT are florigens. (Note that the paper by Huang et al. [2005] was retracted later by the corresponding author.)

NIGHT-BREAK EXPERIMENTS IN RICE

Night-break experiments are classic physiological experiments used to distinguish the photoperiod effect from the light quantity effect (Hamner and Bonner, 1938). If a light pulse in the night affected the flowering response, one could conclude that the phenomenon is a photoperiodic response rather than a response related to light quantity, such as a photosynthesis response. Night-break experiments were used to answer the questions of how plants recognize the change of day length and which qualities of light are involved in the phenomenon (Parker et al., 1946; Lumsden and Furuya, 1986). These experiments clearly demonstrated that a single or daily short light pulse during the dark period significantly delayed flowering in many short-day plants. When these works were published, the results were considered to show the importance of the period of darkness in the recognition of day length. Currently, some modification of these explanations is required to incorporate the interaction between circadian clocks and direct light signals.

In addition, night-break experiments were used to investigate which wavelengths of light were effective at the inhibition of flowering. Red light was effective in many cases, indicating that phytochrome-mediated light signals were involved (Thomas and Vince-Prue, 1997). In rare cases, far-red light immediately following red-light exposure could cancel the inhibitory effect of red-light pulses, which also supported the involvement of phytochromes in flowering-time regulation. Since the long-day plants including *A. thaliana* are not so sensitive to this kind of treatment this, molecular genetics using *A. thaliana* was not applicable to the night-break experiments.

Recently, molecular genetic approaches have been combined with night-break experiments. Ishikawa et al. (2005) reported that phyB signaling, but neither phyA nor phyC signaling, was required for night-break inhibition of flowering in rice. The effects of night-break treatments were cumulative, and a single red light pulse inhibited *Hd3a* mRNA induction only in the morning following the light pulse. In addition, the ability of a single pulse to inhibit flowering depended on the developmental stage: rice plants that were 3 or more weeks old were more sensitive to the single light pulse than were 2-week-old seedlings (Ishikawa et al., 2005). Repetition of night-break treatments caused an additive delay in flowering. For instance, an additional 1-week night-break treatment delayed flowering by an additional week. These findings help to clarify how the florigen signals induce the floral transition in rice. Consecutive daily expression of *Hd3a* mRNA under short-day conditions may have a sort of synergistic effect on floral induction; a single night-break treatment inhibits *Hd3a* mRNA on the following day, so it may be that repeated night-break treatments restrict florigen accumulation to levels that are too low to reach the critical threshold needed for a signal to travel to the apex.

Another interesting finding of Ishikawa et al. (2005) was that there is a particularly sensitive phase during the night for floral inhibition. Plants exposed to a pulse at 7 h after lights off under short-day conditions showed the maximum inhibition of *Hd3a* mRNA the following morning, suggesting that this night-break effect may be distinct from the photoperiodic responses in rice flowering. Entrainments of circadian clocks can be seen in plants treated with light pulses just after dusk or just before dawn, and these types of pulses may mimic long-day conditions. Further molecular analyses are required to clarify the relationship between long-day inhibition and night-break inhibition of flowering in rice.

Under field conditions in Japan (i.e. long-day conditions), rice *phyA* mutants flowered like the wild-type plants, while *phyB* and *phyC* single mutants flowered early (Takano et al., 2005). In addition, the double mutants *phyAphyB* and *phyA-phyC*, but not *phyBphyC*, flowered very early like *se5*, indicating that phyA may have a redundant role in the control of flowering time under natural conditions. In contrast, under short-day conditions only *phyB* single mutants, but not *phyA* or *phyC* mutants, flowered early. The double mutant *phyBphyC* was comparable to *phyB* under short-day conditions, whereas *phyAphyC* and *phyAphyB* flowered slightly late under short-day conditions. Therefore, the phyB involvement in night-break treatments was a part of floral regulation since contributions mediated by each phytochrome are complicated to control flowering time in rice under laboratory short-day, long-day, and the natural conditions.

EPIGENETIC CONTROL OF FLOWERING TIME IN RICE

Epigenetic control of gene expression is a key mechanism in the control of the strong floral suppressor FLC, which is involved in the vernalization responses of *A. thaliana* (Sung and Amasino, 2005). Epigenetic control helps to explain how plants can "memorize" the history of environmental conditions to regulate processes such as floral development. Gene suppression mediated by chromatin remodeling underlies this "memory" in *A. thaliana*. In short-day plants, it appears that such regulation is not required since short-day plants need not survive during the long-winter seasons. A recent study by Komiya et al. (2008), however, suggests the epigenetic regulation of floral induction in rice. *Hd3a* is part of a small gene family, and the closest paralog is *RFT1*, a neighboring gene in the rice genome (Izawa et al., 2002, Kojima et al., 2002). Komiya et al. (2008) produced RNAi (RNA interference; in transgenic plants the specific target gene can be repressed artificially) plants suppressing only *Hd3a* or *RFT1* and examined the effects of both genes on flowering time. *RFT1* expression was induced only when *Hd3a* was suppressed by RNAi. This desuppression of *RFT1* caused a compensation of late flowering in *Hd3a* RNAi plants since rice plants did not flower under the tested short-day conditions when both genes were suppressed. Furthermore, histone-3–lysine-9 was significantly acetylated in the promoter region of *RFT1* in the *Hd3a* RNAi plants, indicating that the desuppression was an epigenetic effect. Thus, *RFT1* serves as a sort of insurance in case *Hd3a* function is lost, but *RFT1* is not involved in developmental regulation such as vernalization responses. This result could be a general epigenetic response for such duplicated genes since *Hd3a* and *RFT1* were likely to be duplicated recently in the evolution of rice. Even so, this epigenetic control of gene expression for the real rice florigen genes may imply a potential that rice may have such epigenetic control in the photoperiodic control of flowering.

CHANGES IN FLOWERING DURING THE DOMESTICATION AND BREEDING OF RICE

Molecular genetic and archeological findings suggest that rice was domesticated more than 10,000 years ago (Doebley et al., 2006; Izawa, 2007b; Kovach et al., 2007; Sang and Ge, 2007). Currently, the most closely related wild rice species, *O. rufipogon*, grows naturally only in tropical areas in Asia (around 20–25°N), including Indochina, whereas *O. sativa* can be grown even in the temperate and cold regions of Japan, Korea, and China. Rice cultivation was introduced into Japan about 3,000 years ago. Therefore, during the domestication and breeding process, rice cultivation has moved north and northeast over time (Izawa, 2007b; figure 2.3). How was it possible for rice to have adapted across such a broad range of climatic regions?

Of course, natural traits for cold tolerance must have been improved. In addition to this, however, fine floral regulation was necessary to limit the flowering time of landraces and cultivars adapted to particular areas. For instance, to obtain a sufficient harvest in northern Japan (40–45°N), the appropriate flowering time for rice cultivars was strictly limited to one week around the end of July. In this particular case, loss of *Lhd4* function was a critical step in the adaptation of rice cultivars to such cold regions. The *lhd4* mutations resulted in early flowering due to desuppression of *Hd3a* mRNA expression in rice seedlings, regardless of the photoperiod (Kitazawa et al., unpublished data).

As described above, many photoperiod-sensitive cultivars can start the floral transition under long-day conditions, soon after the summer solstice (Izawa, 2007b). To investigate the long-day transition of cultivars under natural conditions, we examined flowering time of the photoperiod-sensitive cultivar Nipponbare and its nearly isogenic line (NIL) with a null allele of *Hd1*, with seedlings transplanted at various dates in Tsukuba, Japan (around 33°N; figure 2.7). Because farmers usually transplant rice seedlings at the beginning of May in Tsukuba, the plants start the floral transition under long-day conditions in terms of the Hd1 dual function. When seedlings were transplanted before July 6, the long-day floral suppression by *Hd1* was clearly observed. When seedlings were transplanted on July 6, both Nipponbare and the *hd1*-deficient NIL plants flowered around September 1.

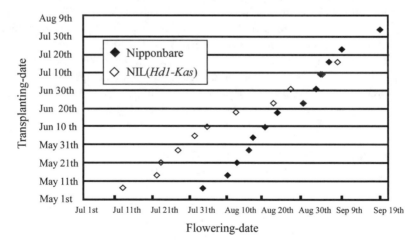

FIGURE 2.7.
Flowering time in rice is critically affected by the date of transplantation. Hd1-dependent late flowering was clearly observed when seedlings were transplanted before July 6.

Therefore, the floral transition was estimated to have occurred at the end of July, at which time *Hd1* would not affect flowering time because the day length was neutral with regard to Hd1 recognition. Because many *japonica* cultivars are closely related to Nipponbare, these findings suggest that many rice cultivars can start the floral transition in the field under long-day conditions in temperate and cold areas.

Apparently, the floral transition under long-day conditions has become possible because *Ehd1* promotes flowering under long-day conditions. Without the *Ehd1* promotion activity, rice plants did not flower under long-day conditions in the laboratory (Doi et al., 2004). During the process of domestication and modern breeding, natural variations allowed for a change in the balance between Ehd1 promotion and Hd1 suppression activities. In fact, Yano et al. (2000) reported natural variations in *Hd1* and the subsequent control of flowering time in rice cultivars.

CONCLUSION

Over the last decade, major questions regarding the photoperiodic control of flowering in plants have been addressed. First, the external coincidence model was supported at the molecular level in both long-day and short-day plants (Izawa et al., 2002; Yanovsky and Kay, 2002). Next, studies identified the major genetic components that differ between long-day and short-day plants (Doi et al., 2004; Imaizumi and Kay, 2006; Izawa, 2007a). Their opposite responses to photoperiods were attributed to the distinct transcriptional mechanisms of florigen genes, such as *FT* and *Hd3a* (Izawa et al., 2002). Finally, the molecular nature of these florigens was revealed (Corbesier et al., 2007; Tamaki et al., 2007).

In addition to these achievements, biochemical studies were undertaken to investigate the molecular mechanisms underlying the interaction between light signals and circadian clocks to explain how *A. thaliana* can recognize day length and consequently express *FT* (Imaizumi et al., 2003, 2004; Valverde et al., 2004; Sawa et al., 2007). FKF1, an F-box domain with an LOV domain that perceives blue light, is expressed at dusk by circadian clock components, and interacts with CDF1, a suppressor of *CO*. Only under long-day conditions would light signals at dusk degrade CDF1 protein through light-activated FKF1 and induce *CO* transcription (Imaizumi et al., 2005). Further fine regulations of CO activity by light signals mediated by phytochromes and cryptochromes have been reported to explain the log-day induction of FT transcription (Valverde et al., 2004). In rice, phytochrome signals can induce *Lhd4* in the morning only under long-day conditions, although the underlying molecular mechanisms remain unclear. Further work is also need to elucidate the molecular mechanisms conferring short-day induction of *Ehd1* expression (figure 2.4). In principle, a simple interaction between light signals and circadian clocks in rice may be enough for plants to recognize day length and induce *Hd3a* expression only under short-day conditions. However, more refined molecular mechanisms are likely involved in the photoperiodic control of flowering in rice.

References

Abe M, Kobayashi Y, Yamamoto S, Daimon Y, Yamaguchi A, Ikeda Y, Ichinoki H, Notaguchi M, Goto K, and Araki T. (2005). FD, a bZIP protein mediating signals from the floral pathway integrator FT at the shoot apex. Science 309: 1052–1056.

An H, Roussot C, Suarez-Lopez P, Corbesier L, Vincent C, Pineiro M, Hepworth S, Mourdov A, Justin S, Turnbull C, and Coupland G. (2004). CONSTANS acts in the phloem to regulate a systemic signal that induces photoperiodic flowering of *Arabidopsis*. Development 131: 3615–3626.

Bünning E. (1960). Circadian rhythms and the time measurement in photoperiodism. Cold Spring Harbor Symp Quant Biol 25: 249–256.

Corbesier L and Coupland G. (2006). The quest for florigen: A review of recent progress. J Exp Bot 57: 3395–3403.

Corbesier L Vincent C, Jang S, Formara F, Fan F, Searle I, Giakountis A, Farrona S, Turnbull C, and Coupland G. (2007). FT protein movement contributes to long-distance signaling in floral induction of *Arabidopsis*. Science 316: 1030–1033.

Daniel X, Sugano S, and Tobin EM. (2004). CK2 phosphorylation of CCA1 is necessary for its circadian oscillator function in *Arabidopsis*. Proc Natl Acad Sci USA 101: 3292–3297.

Doebley JF, Gaut BS, and Smith BD. (2006). The molecular genetics of crop domestication. Cell 127: 1309–1321.

Doi K, Izawa T, Fuse T, Yamanouchi U, Kubo T, Shimatani Z, Yano M, and Yoshimura A. (2004). Ehd1, a B-type response regulator in rice, confers short-day promotion of flowering and controls *FT*-like gene expression independently of *Hd1*. Genes Dev 18: 926–936.

Fowler S, Lee K, Onouchi H, Samach A, Richardson K, Morris B, Coupland G, and Putrill J. (1999). GIGANTEA: A circadian clock-controlled gene that regulates photoperiodic flowering in *Arabidopsis* and encodes a protein with several possible membrane-spanning domains. EMBO J 18: 4679–4688.

Garner WW and Allard HA. (1920). Effect of the relative length of day and night and other factors of the environment on growth and reproduction in plants. J Agrobiol Res 18: 553–606.

Garris AJ, Tai TH, Coburn J, Kresovich S, and McCouch S. (2005) Genetic structure and diversity in *Oryza sativa* L. Genetics 169: 1631–1638.

Gould PD, Locke JC, Larue C, Southern MM, Davis SJ, Hanano S, Moyle R, Milich R, Putterill J, Millar AJ, and Hall A. (2006). The molecular basis of temperature compensation in the *Arabidopsis* circadian clock. Plant Cell 18: 1177–1187.

Hamner KC and Bonner J. (1938). Photoperiodism in relation to hormones as factors in floral initiation and development. Bot Gaz 100: 388–431.

Hartmann U, Höhmann S, Nettesheim K, Wisman E, Saedler H, and Huijser P. (2000). Molecular cloning of SVP: A negative regulator of the floral transition in *Arabidopsis*. Plant J 21: 351–360.

Harushima Y, Yano M, Shomura A, Sato M, Shimano T, Kuboki Y, Yamamoto T, Lin SY, Antonio BA, Parco A, Kajiya H, Huang N, Yamamoto K, Nagamura Y, Kurata N, Khush GS, and Sasaki T. (1998). A high-density rice genetic linkage map with 2275 markers using a single F2 population. Genetics 148: 479–494.

Hayama R, Izawa T, and Shimamoto K. (2002). Isolation of rice genes possibly involved in the photoperiodic control of flowering by a differential display method. Plant Cell Physiol 43: 494–504.

Henderson IR, Shindo C, and Dean C. (2003). The need for winter in the switch to flowering. Annual Rev Genet 37: 371–392.

Heyl A and Schmülling T. (2003). Cytokinin signal perception and transduction. Curr Opin Plant Biol 6: 480–488.

Huang T, Bohlenius H, Eriksson S, Parcy F, and Nilsson O. (2005). The mRNA of the *Arabidopsis* gene *FT* moves from leaf to shoot apex and induces flowering. Science 309: 1694–1696.

Imaizumi T and Kay SA. (2006). Photoperiodic control of flowering: Not only by coincidence. Trends Plant Sci 11: 550–558.

Imaizumi T, Schultz TF, Harmon FG, Ho LA, and Kay SA. (2005). FKF1 F-box protein mediates cyclic degradation of a repressor of CONSTANS in *Arabidopsis*. Science 309: 293–297.

Imaizumi T, Tran HG, Swartz TE, Briggs WR, and Kay SA. (2003). *FKF1* is essential for photoperiodic-specific light signalling in *Arabidopsis*. Nature 426: 302–306.

Ishida K, Yamashino T, Yokoyama A, and Mizuno T. (2008). Three type-B response regulators, ARR1, ARR10 and ARR12, play essential but redundant roles in cytokinin signal transduction throughout the life cycle of *Arabidopsis thaliana*. Plant Cell Physiol 49: 47–57.

Ishikawa R, Tamaki S, Yokoi S, Inagaki N, Shinomura T, Takano M, and Shimamoto K. (2005). Suppression of the floral activator *Hd3a* is the principle cause of the night break effect in rice. Plant Cell 17: 3326–3336.

Izawa T. (2007a). Daylength measurements by rice plants in photoperiodic short-day flowering. Int Rev Cytol 256: 191–222.

Izawa T. (2007b). Adaptation of flowering-time by natural and artificial selection in *Arabidopsis* and rice. J Exp Bot 58. 3091–3097.

Izawa T, Oikawa T, Tokutomi S, Okuno K, and Shimamoto K. (2000). Phytochromes confer the photoperiodic control of flowering in rice (a short-day plant). Plant J 22: 391–399.

Izawa T, Oikawa T, Sugiyama N, Tanisaka T, Yano M, and Shimamoto K. (2002). Phytochromes mediate the external light signal to repress *FT* orthologs in photoperiodic flowering of rice. Genes Dev 16: 2003–2020.

Izawa T, Takahashi Y, and Yano M. (2003). Comparative biology has come to bloom: Genomic and genetic comparison of flowering pathways in rice and *Arabidopsis*. Curr Opin Plant Biol 6: 113–120.

Jaeger KE, Graf A, and Wigge PA. (2006). The control of flowering in time and space. J Exp Bot 57: 3415–3418.

Kardailsky I, Shukla VK, Ahn JH, Dagenais N, Christensen SK, Nguyen JT, Chory J, Harrison MJ, and Weigel D. (1999). Activation tagging of the floral inducer *FT*. Science 286: 1962–1965.

Kim SL, Lee S, Kim HJ, Nam HG, and An G. (2007). OsMADS51 is a short-day flowering promoter that functions upstream of *Ehd1*, *OsMADS14*, and *Hd3a*. Plant Physiol 145: 1484–1494.

Kim WY, Fujiwara S, Suh SS, Kim J, Kim Y, Han L, David K, Putterill J, Nam HG, and Somers DE. (2007). ZEITLUPE is a circadian photoreceptor stabilized by GIGANTEA in blue light. Nature 449: 356–360.

Kobayashi Y, Kaya H, Goto K, Iwabuchi M, and Araki T. (1999). A pair of related genes with antagonistic roles in mediating flowering signals. Science 286: 1960–1962.

Kobayashi Y and Weigel D. (2007) Move on up, it's time for change – mobile signal controlling photodependent flowering. Genes & Development 21: 2371–2384.

Kojima S, Takahashi Y, Kobayashi Y, Monna L, Sasaki T, Araki T, and Yano M. (2002). *Hd3a*, a rice ortholog of the *Arabidopsis FT* gene, promotes transition to flowering downstream of *Hd1* under short-day conditions. Plant Cell Physiol 43: 1096–1105.

Komiya R, Ikegami A, Tamaki S, Yokoi S, and Shimamoto K. (2008). *Hd3a* and *RFT1* are essential for flowering in rice. Development 135: 767–774.

Kovach MJ, Sweeney MT, and McCouch SR. (2007). New insights into the history of rice domestication. Trends Genet 23: 578–587.

Kurata N, Nagamura Y, Yamamoto K, Harushima Y, Sue N, Wu J, Antonio BA, Shomura A, Shimizu T, Lin SY, et al. (1994). A 300 kilobase interval genetic map of rice. Nat Genet 8: 365–372.

Lee S, Choi SCC, and An G. (2008). Rice SVP-group MADS-box proteins, OsMADS22 and OsMADS44, are negative regulators of brassinosteroid responses. Plant J 54: 93–105.

Lee S, Kim J, Han JJ, Han JM, and An G. (2004). Functional analyses of the flowering time gene *OsMADS50*, the putative *SUPPRESSOR OF OVEREXPRESSION OF CO1/AGAMOUS-LIKE 20 (SOC1/AGL20)* ortholog in rice. Plant J 38: 754–764.

Lin HX, Yamamoto T, Sasaki T, and Yano M. (1998). Characterization and detection of epistatic interactions of 3QTLs, *Hd1*, *Hd2*, and *Hd3*, controlling heading date in rice using nearly isogenic lines. Theor Appl Genet 101: 1021–1028.

Lin HX, Yamamoto T, Sasaki T, and Yano M. (2002). Identification and characterization of a quantitative trait locus, *Hd9*, controlling heading date in rice. Breed Sci 52: 35–41.

Lin HX, Liang ZW, Sasaki T, and Yano M. (2003). Fine mapping and characterization of quantitative trait loci *Hd4* and *Hd5* controlling heading date in rice. Breed Sci 53: 51–59.

Lin SY, Sasaki T, and Yano M. (1997). Mapping quantitative trait loci controlling seed dormancy and heading date in rice, *Oryza sativa* L., using backcross inbred lines. *Theor. Applied Genetics* 96, 997–1003.

Locke JC, Southern MM, Kozma-Bognar L, Hibberd V, Brown PE, Turner MS, and Millar AJ. (2005). Extension of a genetic network model by iterative experimentation and mathematical analysis. Mol Syst Biol 1: E1–E9.

Locke JC, Kozma-Bognár L, Gould PD, Fehér B, Kevei E, Nagy F, Turner MS, Hall A, and Millar AJ. (2006). Experimental validation of a predicted feedback loop in the multi-oscillator clock of *Arabidopsis thaliana*. Mol Syst Biol 2: 59.

Londo JP, Chiang YC, Hung KH, Chiang TY, and Schaal BA. (2006). Phylogeography of Asian wild rice, *Oryza rufipogon*, reveals multiple independent domestications of cultivated rice, *Oryza sativa*. Proc Natl Acad Sci USA 103: 9578–9583.

Lumsden PJ and Furuya M. (1986). Evidence for two actions of light in the photoperiodic induction of flowering in *Pharbitis nil*. Plant Cell Physiol 27: 1541–1551.

Lumsden PJ and Millar AJ. (1998). *Biological Rhythms and Photoperiodism in Plants*. Oxford: Bios Scientific.

Más P, Kim WY, Somers DE, and Kay SA. (2003). Targeted degradation of TOC1 by ZTL modulates circadian function in *Arabidopsis thaliana*. Nature 426: 567–570.

Michaels SD, He Y, Scortecci KC, and Amasino RM. (2003). Attenuation of *FLOWERING LOCUS C* activity as a mechanism for the evolution of summer-annual flowering behavior in *Arabidopsis*. Proc Natl Acad Sci USA 100: 10102–10107.

Mizoguchi T, Putterill J, and Ohkoshi Y. (2006). Kinase and phosphatase: The cog and spring of the circadian clock. Int Rev Cytol 250: 47–72.

Mizoguchi T, Wright L, Fujiwara S, Cremer F, Lee K, Onouchi H, Mouradov A, Fowler S, Kamada H, Putterill J, and Coupland G. (2005). Distinct roles of GIGANTEA in promoting flowering and regulating circadian rhythms in *Arabidopsis*. Plant Cell 17: 2255–2270.

Monna L, Lin X, Kojima S, Sasaki T, and Yano M. (2002). Genetic dissection of a genomic region for a quantitative trait locus, *Hd3*, into two loci, *Hd3a* and *Hd3b*, controlling heading date in rice. Theor Appl Genet 104: 772–778.

Murakami M, Ashikari M, Miura K, Yamashino T, and Mizuno T. (2003). The evolutionarily conserved OsPRR quintet: Rice pseudo-response regulators implicated in circadian rhythm. Plant Cell Physiol 44: 1229–1236.

Murakami M, Matsushika A, Ashikari M, Yamashino T, and Mizuno T. (2005). Circadian-associated rice pseudo response regulators (OsPRRs): Insight into the control of flowering-time. Biosci Biotechnol Biochem 69: 410–414.

Muramoto T, Kohchi T, Yokota A, Hwang I, and Goodman HM. (1999). The *Arabidopsis* photomorphogenic mutant *hy1* is deficient in phytochrome chromophore biosynthesis as a result of a mutation in a plastid heme oxygenase. Plant Cell 11: 335–348.

Onouchi H, Igeño MI, Périlleux C, Graves K, and Coupland G. (2000). Mutagenesis of plants overexpressing CONSTANS demonstrates novel interactions among *Arabidopsis* flowering-time genes. Plant Cell 12: 885–900.

Pittendrigh CS and Minis DH. (1964). The entrainment of circadian oscillations by light and their role as photoperiodic clocks. Am Nat 108: 261–293.

Parker MW, Hendricks SB, Borthwick HA, and Scully NJ. (1946). Action spectra for the photoperiodic control of floral initiation of short-day plants. Bot Gaz 108: 1–26.

Parks DH, Somers DE, Kim YS, Choy YH, Lim HK, Soh MS, Kim HJ, Kay SA, and Nam HG. (1999). Control of circadian rhythms and photoperiodic flowering by the *Arabidopsis* GIGANTEA gene. Science 285: 1579–1582.

Putterill J, Laurie R, and Macknight R. (2004). It's time to flower: The genetic control of flowering time. Bioessays 26: 363–373.

Putterill J, Robson F, Lee K, Simon R, and Coupland G. (1995). The *CONSTANS* gene of *Arabidopsis* promotes flowering and encodes a protein showing similarities to zinc finger transcription factors. Cell 80: 847–857.

Rijpkema AS, Gerats T, and Vandenbussche M. (2007). Evolutionary complexity of MADS complexes. Curr Opin Plant Biol 10: 32–38.

Sang T and Ge S. (2007). Genetics and phylogenetics of rice domestication. Curr Opin Genet Dev 17: 533–538.

Sawa M, Nusinow DA, Kay SA, and Imaizumi T. (2007). FKF1 and GIGANTEA complex formation is required for day-length measurement in *Arabidopsis*. Science 318: 261–265.

Somers DE, Devlin PF, and Kay SA. (1998). Phytochromes and cryptochromes in the entrainment of the *Arabidopsis* circadian clock. Science 282: 1488–1490.

Suaréz-Lupez P, Wheatley K, Robson F, Onouchi H, Valverde F, and Coupland G. (2001). CONSTANS mediates between the circadian clock and the control of flowering in *Arabidopsis*. Nature 410. 1116–1120.

Sugano S, Andronis C, Green RM, Wang ZY, and Tobin EM. (1998). Protein kinase CK2 interacts with and phosphorylates the *Arabidopsis* circadian clock-associated 1 protein. Proc Natl Acad Sci USA 95: 11020–11025.

Sugano S, Andronis C, Ong MS, Green RM, and Tobin EM. (1999). The protein kinase CK2 is involved in regulation of circadian rhythms in *Arabidopsis*. Proc Natl Acad Sci USA 96: 12362–12366.

Sugiyama N, Izawa T, Oikawa T, and Shimamoto K. (2001). Light regulation of circadian clock-controlled gene expression in rice. Plant J 26: 607–615.

Sung S and Amasino RM. (2005). Remembering winter: Toward a molecular understanding of vernalization. Annu Rev Plant Biol 56: 491–508.

Takahashi Y, Shomura A, Sasaki T, and Yano M. (2001). Hd6, a rice quantitative trait locus involved in photoperiod sensitivity, encodes the alpha subunit of protein kinase CK2. Proc Natl Acad Sci USA 98: 7922–7927.

Takada S and Goto K. (2003). Terminal flower2, an *Arabidopsis* homolog of heterochromatin protein1, counteracts the activation of flowering locus T by CONSTANS in the vascular tissues of leaves to regulate flowering time. Plant Cell 15: 2856–2865.

Takano M, Inagaki N, Xie X, Yuzurihara N, Hihara F, Ishizuka T, Yano M, Nishimura M, Miyao A, Hirochika H, and Shinomura T. (2005). Distinct and cooperative functions of phytochromes A, B, and C in the control of deetiolation and flowering in rice. Plant Cell 17: 3311–3325.

Tamaki S, Matsuo S, Wong H-L, Yokoi S, and Shimamoto K. (2007). Hd3a protein is a mobile flowering signal in rice. Science 316: 1033–1036.

Thomas B. (1998). Photoperiodism: An overview. In *Biological Rhythms and Photoperiodism in Plants* (PJ Lumsden and AJ Millar, eds). Oxford: Bios Scientific, pp. 151–181.

Thomas B and Vince-Prue D. (1997). *Photoperiodism in Plants*. London: Academic Press.

Turner A, Beales J, Faure S, Dunford RP, and Laurie DA. (2005). The pseudo-response regulator Ppd-H1 provides adaptation to photoperiod in barley. Science 310: 1031–1034.

Valverde F, Mouradov A, Soppe W, Ravenscroft D, Samach A, and Coupland G. (2004). Photoreceptor regulation of CONSTANS protein in photoperiodic flowering. Science 303: 1003–1006.

Wigge PA, Kim MC, Jaeger KE, Busch W, Schmidt M, Lohman JU, and Weigel D. (2005). Integration of spatial and temporal information during floral induction in *Arabidopsis*. Science 309: 1056–1059.

Xue W, Xing Y, Weng X, Zhao Y, Tang W, Wang L, Zhou H, Yu S, Xu C, Li X, and Zhang Q. (2008) Natural variation in Ghd7 is an important regulator of heading date and yield potential in rice. Nat Genet 40: 761–767.

Yamamoto T, Kuboki Y, Lin SY, Sasaki T, and Yano M. (1998). Fine mapping of quantitative trait loci Hd1, Hd2, and Hd3, controlling heading date of rice, as single Mendelian factors. Theor Appl Genet 97: 37–44.

Yan L, Loukoianov A, Blechl A, Tranquilli G, Ramakrishna W, SanMiguel P, Bennetzen JL, Echenique V, and Dubcovsky J. (2004). The wheat VRN2 gene is a flowering repressor down-regulated by vernalization. Science 303: 1640–1644.

Yan L, Loukoianov A, Tranquilli G, Helguera M, Fahima T, and Dubcovsky J. (2003). Positional cloning of the wheat vernalization gene *VRN1*. Proc Natl. Acad. Sci USA 100: 6263–6268.

Yano M. (2001). Genetic and molecular dissection of naturally occurring variation. Curr Opin Plant Biol 4: 130–135.

Yano M, Harushima Y, Nagamura Y, Kurata N, Minobe Y, and Sasaki T. (1997). Identification of quantitative trait loci controlling heading date in rice using a high-density linkage map. Theor Appl Genet 95: 1025–1032.

Yano M, Katayose Y, Ashikari M, Yamanouchi U, Monna L, Fuse T, Baba T, Yamamoto K, Umehara Y, Nagamura Y, and Sasaki T. (2000). *Hd1*, a major photoperiod sensitivity quantitative trait locus in rice, is closely related to the *Arabidopsis* flowering time gene CONSTANS. Plant Cell 12: 2473–2483.

Yano M, Kojima S, Takahashi Y, Lin H, and Sasaki T. (2001). Genetic control of flowering time in rice, a short-day plant. Plant Physiol 127: 1425–1429.

Yano M and Sasaki T. (1997). Genetic and molecular dissection of quantitative traits in rice. Plant Mol Biol 35: 145–153.

Yanovsky MJ and Kay SA. (2002). Molecular basis of seasonal time measurement in *Arabidopsis*. Nature 419: 308–312.

Yanovsky MJ and Kay SA. (2003). Living by the calendar: How plants know when to flower. Nat Rev Mol Cell Biol 4: 265–275.

Yokoo M and Okuno K. (1993). Genetic analysis of earliness mutations induced in the rice cultivar Norin 8. Jpn J Breed 43: 1–11.

3

The Photoperiodic Flowering Response in *Pharbitis nil*

Ryosuke Hayama

Throughout the seasonal changes that occur during the year, many plant species recognize day length as a determinant in the timing of developmental changes, such as initiation of flowering. In plants the flowering response to a shift in day length is sometimes drastic and extremely precise (figure 3.1), with some species showing a very accurate flowering response to changes in day length (Thomas and Vince-Prue, 1997). The photoperiodic flowering response is accomplished by a mechanism that enables plants to measure day length precisely, and from physiological and molecular genetic studies of this flowering response, this mechanism is proposed to involve the interaction between environmental signals such as light and the timekeeping mechanism that is associated with the circadian clock (Thomas and Vince-Prue, 1997).

The flowering response to day length has diverged widely among plant species, as a consequence of plants having evolved a mechanism to determine the timing of proper floral initiation in their own growth environment. Plants that flower in response to day length longer than a certain minimal length are called long-day plants (LDPs), and plants that flower in response to day length shorter than a certain maximal length are called short-day plants (SDPs). A characteristic feature of these plants is that they show completely opposite flowering responses to a given day length: exposure to long days (LDs) promotes flowering in LDPs but inhibits flowering in SDPs, and exposure to short days (SD) inhibits flowering in LDPs but promoting flowering in SDPs. Recent molecular and genetic studies into the control of flowering time in *Arabidopsis* (LDP) and rice (SDP) have defined similar photoperiodic flowering pathways, which at their core consist of *GIGANTEA (GI)*, *CONSTANS (CO)* and *FLOWERING LOCUS T (FT)*, and these results have led to the proposal that *CO* acts oppositely in controlling *FT* transcription in those pathways to generate LD and SD responses in *Arabidopsis* and rice, respectively (see chapters 1 and 2; Hayama and Coupland, 2004; Imaizumi and Kay, 2006). In *Arabidopsis*, transcription of the clock-controlled gene *CO* is regulated by *FKF1*, *GI*, and *CDF1* and exhibits a diurnal pattern with its expression accumulating at a certain time after dawn both under LD and SD conditions (Suarez-Lopez et al., 2001; Imaizumi et al., 2003, 2005; Sawa et al., 2007). Light posttranscriptionally activates *CO*, stabilizing CO protein and leading to CO protein accumulation

FIGURE 3.1.
Photoperiodic flowering response in *Pharbitis nil* cv. Violet. The left and right plants show *Pharbitis* grown under long days (LDs) and short days (SDs), respectively.

when the light period is sufficiently long (Valverde et al., 2004). This accumulation induces transcription of *FT* and promotes flowering only under LD (figure 3.2A). In rice, *Hd1*, a homolog of *CO* in *Arabidopsis*, is also expressed with a diurnal pattern similar to that of *CO* under LD and SD and is also controlled by light at the post-transcriptional level (Yano et al., 2000; Izawa et al., 2002; Hayama et al., 2003). In the light, Hd1 protein forms a suppressor of transcription of *Hd3a*, a homolog of *Arabidopsis FT* that functions to promote flowering, whereas in the dark it forms as an activator of *Hd3a* transcription (Izawa et al., 2002; Kojima et al., 2002). Thus, *Hd1* suppresses *Hd3a* expression if the duration of the photoperiod is sufficiently long, and flowering is inhibited under LD, whereas if the photoperiod is short *Hd1* induces *Hd3a* expression in the dark, contributing to *Hd3a* induction under SD (figure 3.2B) (Izawa et al., 2002).

A common feature of the timekeeping mechanisms proposed in both species is that these mechanisms are associated with a circadian rhythm that functions to measure the length of the day, in which clock-controlled genes, such as *CO* and *Hd1*, are expressed after a certain time from dawn under LD and SD and are activated

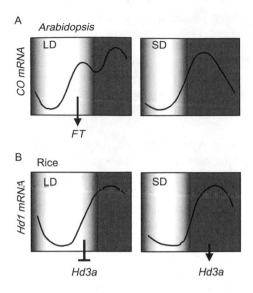

FIGURE 3.2.
A model for the timekeeping mechanism in *Arabidopsis* and rice. (A) In *Arabidopsis*,
transcript levels of *CO* exhibit a diurnal pattern of rising and falling at certain times of
the day under LD and SD. Light posttranscriptionally stabilizes CO protein, so under LD
CO mRNA is expressed at high levels during the photoperiod, allowing CO protein to
accumulate and induce *FT* expression. Under SD, high *CO* mRNA levels do not coincide
with the photoperiod, CO protein does not accumulate, and *FT* is not induced. (B) In
rice, transcripts of *Hd1* exhibit a similar pattern to that of *CO* under SD and LD. Light
posttranscriptionally controls Hd1, but here in the light Hd1 forms a suppressor of *Hd3a*
transcription, whereas in the dark Hd1 activates *Hd3a* transcription. Under LD, *Hd1*
mRNA is expressed while plants are exposed to light, leading to the accumulation of
a suppressor form of Hd1. Under SD, the suppressor form of Hd1 does not accumulate
because *Hd1* mRNA is not expressed while plants are exposed to light. Instead, an
activator form of Hd1 is produced, contributing to the induction of *Hd3a* transcription.

by light at the posttranscriptional level. However, a series of physiological studies
especially with several SDPs such as *Pharbitis* and *Xanthium* have suggested that
these plants retain a timekeeping mechanism that functions to measure the length
of the night rather than the length of the day, indicating that the mechanism for
day-length measurement in plants may have diverged in the course of evolution
(Thomas and Vince-Prue, 1997).

Here, we introduce a mechanism for controlling the photoperiodic flowering
response in an SDP, *Pharbitis* (*Ipomoea*) *nil* (Japanese morning glory). During the
last half century, *Pharbitis* has widely been used as a model plant for studying the
photoperiodic response of flowering in plants at the physiological level, and fundamental physiological information of the timekeeping mechanism has been accrued
(Thomas and Vince-Prue, 1997). Recently, molecular genetic studies have begun
to uncover the molecular mechanism that controls the photoperiodic flowering

response in this plant. We discuss the possible mechanism for the photoperiodic response in *Pharbitis* and compare it with that proposed for *Arabidopsis* and rice.

THE CIRCADIAN SYSTEM CONTROLLING THE PHOTOPERIODIC RESPONSE OF FLOWERING IN *PHARBITIS*

The circadian clock is an autonomous system that generates free running rhythms with a period length of approximately 24 h in constant conditions. This allows organisms to prepare to follow changes in the environment such as light–dark or dark–light transition or temperature changes over 24-h cycles by generating a wide range of rhythms at the physiological and gene expression level (Dunlap, 1999; Schultz and Kay, 2003). The circadian clock is entrained to daily changes in environmental conditions such as light and temperature, which ensures oscillations with an exact period of 24 h. One of studies that implicated the involvement of the circadian clock in controlling flowering time is a series of night-break experiments, in which plants exhibit a rhythmic change in the levels of floral inhibition (or promotion) in response to a light pulse (night break) given at various times during a long night (Thomas and Vince-Prue, 1997). Flowering of *Pharbitis* is inhibited by such a night break. Also, *Pharbitis* exhibits a rhythmic sensitivity in the level of inhibition of flowering when plants are exposed to a night break at different times during a long dark period (Takimoto and Hamner, 1965). The precise mechanism for determining a plant's sensitivity to a night break has been unknown until recently. However, new studies at the molecular level in SD rice have revealed that a light pulse can inhibit flowering without affecting the state of the circadian clock, perhaps by affecting the activity of a clock-controlled gene, *Hd1*, at a posttranscriptional level (see chapter 2; Ishikawa et al., 2005). This idea is based on the observation that a light pulse given during the night period under SD effectively inhibits the expression of *Hd3a* without affecting the diurnal pattern of *Hd1* expression (Ishikawa et al., 2005).

Using the technique of the night break, an experiment was carried out to understand the entrainment of the circadian clock in *Pharbitis*. Dark-grown plants were transferred to light periods of various durations, and the effect of the length of the light period on the phase of the sensitivity to night breaks was examined (Lumsden et al., 1982). When the light period was greater than 6 h, the time of maximum sensitivity to night breaks (i.e., the time at which a night break inhibits flowering most efficiently) is constant from the beginning of when plants were transferred to the dark period (figure 3.3A) (Lumsden et al., 1982). This indicates that the phase of the circadian clock generating the sensitivity to a night break is set by lights off at dusk if the light period is more than a certain duration. For light periods shorter than 6 h, the time of maximum sensitivity to night breaks was constant as measured from the beginning of the light period (figure 3.3A). Under these conditions, the circadian clock seems to enter the night period with its phase not being affected by lights off

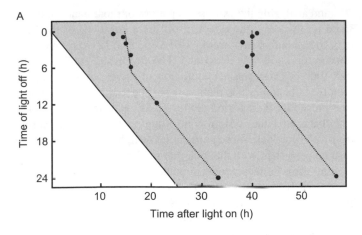

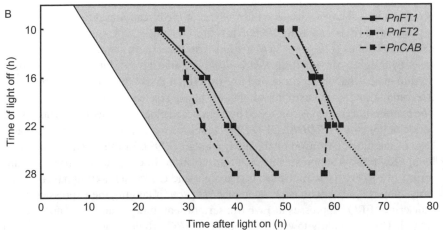

FIGURE 3.3.

Effect of the length of the light period on the circadian phase of a photoperiodic rhythm in *Pharbitis*. (A) Each set of dark-grown plants received a light period of a specific but different duration and were then transferred to continuous darkness. Plants then received a night break at specific times during the continuous darkness. The maximum sensitivity to night breaks (as measured by flowering time) is plotted. After Lumsden et al. (1982). (B) Sets of plants grown in light–dark cycles each received a light period of a specific duration and were then transferred to continuous darkness. Peak expression time of the circadian rhythm of *PnFT1, PnFT2,* and *CAB* mRNA is plotted. After Hayama et al. (2007).

(Lumsden et al., 1982). Similar observations were found in a study of *PnFT1* and *PnFT2* expression, *Pharbitis* genes that encode a protein similar to *Arabidopsis FT*. Transcripts of these genes are induced specifically under SD, and overexpression of *PnFT1* causes strong early flowering in both *Arabidopsis* and *Pharbitis* (Hayama et al., 2007). Thus, these two genes may play roles in the photoperiodic flowering response in *Pharbitis*, and analyzing the expression pattern of those genes in detail may be fruitful. Based on this idea, the effect of different durations of the light period on the phase of the circadian rhythm of *PnFT1* and *PnFT2* expression during a subsequent constant darkness was examined (Hayama et al., 2007). Results showed that the peak time of transcript accumulation of these genes during the dark period is constant from the time when plants are shifted to the dark, regardless of the duration of the preceding light period (Figure 3.3B; Hayama et al., 2007). These results indicate that a circadian system controlling photoperiodic flowering response in *Pharbitis* involves a clock whose phase is set by lights off at dusk. These observations may also be consistent with the idea that the length of the night is an important environmental factor for its photoperiodic flowering response.

The mechanism for the lights-off setting of the circadian clock in *Pharbitis* is unknown. In fact, molecular observations related to the circadian system in *Arabidopsis* have shown that contrary to the physiological and molecular data for photoperiodic flowering response in *Pharbitis*, phase resetting in *Arabidopsis* occurs mainly at dawn, based on the analysis of the circadian rhythm of *CAB* gene expression (Millar and Kay, 1996; McWatters et al., 2000). On the other hand, molecular studies of the clock in *Neurospora* implicate a mechanism whereby phase is set by a lights-off signal. In *Neurospora*, conidiation exhibits a clear circadian rhythm in continuous darkness, and like the circadian rhythm of *PnFT* expression, the conidiation rhythm exhibits a lights-off phase resetting (Aronson et al., 1994). The conidiation rhythm is controlled by a circadian clock that involves as one of the core components *FREQUENCY* (*FRQ*), which forms part of a negative-feedback loop of the clock (Aronson et al., 1994; Dunlap, 1999). In this negative-feedback loop, FRQ protein suppresses its own transcription. In continuous darkness, transcripts of *FRQ* rise late in the subjective night and exhibit a maximum peak in the early subjective day, with expression of *FRQ* protein lagging the expression of its transcripts. FRQ suppresses its own transcription at midday, and *FRQ* mRNA levels fall. This eventually lowers the amount of *FRQ* protein, and thus transcription of its mRNA starts again (Aronson et al., 1994; Garceau et al., 1997). Transcription of *FRQ* is strongly induced by light, and this mechanism has been proposed to be involved in light resetting of the clock (Crosthwaite et al., 1995). Consistent with this notion is the finding that strong light induction of *FRQ* transcription occurs at any time during the night despite the normal FRQ-mediated suppression of its own transcription. Additionally, transcripts of *FRQ* exhibit constitutively high levels in continuous light. This suggests that *Neurospora* retains a light-sensitive clock, which suspends activity at a certain phase in the light by sustaining constitutively high levels of *FRQ* and restarts oscillations with FRQ protein suppressing *FRQ* transcription after the beginning of the night period (Crosthwaite et al., 1995). This system could serve as a model for how a circadian clock controls *PnFT* expression and flowering time in *Pharbitis*. Such a clock could be one that is reset by exposure to light and afterward suspends or stops at a certain phase in the light, only to restart oscillation when the night period begins (figure 3.4).

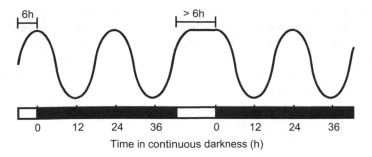

FIGURE 3.4.

Possible mechanism for lights-off phase resetting of the circadian clock in *Pharbitis*. Light resets the circadian clock at dawn. If the duration of the light period is less than 6 h, the circadian clock continues to oscillate in the dark. If the light period is longer than 6 h, the circadian clock suspends at a certain phase or a state until the end of the photoperiod, whereupon it restarts after entering the dark period.

In contrast to the normal phase resetting of the circadian clock by lights on at dawn in *Arabidopsis*, it has been shown that the *Arabidopsis early flowering 3* (*elf3*) mutant exhibits a lights-off resetting of the circadian rhythm of *CAB* gene expression, with its peak appearing at the same time in continuous darkness irrespective of the duration of the preceding light period (McWatters et al., 2000). Also, the normal restriction ("gating") by the clock of *CAB* gene responsiveness to light pulses to the subjective night in wild type is lost in *elf3* mutants, and acute photoinduction of gene expression occurs during both the subjective day and subjective night. Based on these results, it has been suggested that the circadian system in *Arabidopsis* involves a mechanism in which clock-regulated *ELF3* restricts or "gates" light input to the clock mainly at dusk, so the clock can be reset mainly at dawn (McWatters et al., 2000). Thus, a possible mechanism underlying a circadian system controlling flowering time in *Pharbitis* may be one in which a light input pathway does not involve *ELF3* or a gene with a function similar to *ELF3*, resulting in a clock that suspends during the photoperiod and restarts at the beginning of the dark period.

Contrary to the phase resetting of the circadian rhythm of *PnFT* expression, results for different clock-output genes such as *CAB* in *Pharbitis* show a different phase resetting mainly set by lights on, similar to that observed for *CAB* in *Arabidopsis* (figure 3.3B) (Hayama et al., 2007; Hayama and Coupland, unpublished data). This indicates that *PnFT* expression and the photoperiodic flowering response are controlled by a circadian clock that is distinct from the clock system regulating the circadian rhythm of *CAB* gene expression. Studies of the circadian clock in *Arabidopsis* have suggested the possibility of cell-type–specific circadian systems, where cell-type distinct clock-output genes exhibit rhythms with slightly different circadian period lengths (Thain et al., 2000, 2002). Also, clock-controlled genes regulating flowering time such as *CO*, *CDF1*, and *FT* are expressed in the vascular tissue in *Arabidopsis* (Takada and Goto, 2003; An et al., 2004; Imaizumi et al., 2005). Thus, one possibility is that a clock whose phase is set by dusk and regulates *PnFT* gene expression may function specifically in the vascular tissue in *Pharbitis*, whereas a circadian clock regulating other clock-controlled genes, such as *CAB* gene expression, could be expressed in other tissues such as the leaf mesophyll. Recently, *FT* has been proposed to encode a mobile protein that moves from the vascular tissue to the meristem to promote flowering. A distinct clock system that evolved for measuring the length of the night might exist in the vascular tissue and control *PnFT* expression and flowering time in *Pharbitis*.

MECHANISM FOR DARK-PERIOD MEASUREMENT AND ACTION OF LIGHT IN *PHARBITIS*

Molecular genetic studies of the photoperiodic flowering response in *Arabidopsis* and rice suggest that these species retain a mechanism for measuring the length of the day, in which the phase of certain clock-controlled genes, such as *CO* and *Hd1*, begin to accumulate from dawn and are posttranscriptionally activated by light to generate their activities under *LD*. These activated proteins then induce or suppress the expression of *FT* or *Hd3a* in *Arabidopsis* and rice, respectively (see chapters 1 and 2; Hayama and Coupland, 2004). In both species, light is postulated to have two independent actions, one to set the phase of the circadian clock and the other to regulate directly the activity of a clock-controlled protein. The observation that a night break strongly suppresses *Hd3a* expression under SD without affecting the diurnal pattern of *Hd1* is consistent with the idea of light having two independent actions in controlling photoperiodic flowering response (Ishikawa et al., 2005).

Physiological studies have shown that *Pharbitis* retains a mechanism for measuring the length of the night. One direct piece of evidence is that the critical night length required for initiation of flowering is always constant even when preceded by a light period with more than 8 h of various durations (figure 3.5) (Saji et al., 1984). This indicates that the critical night length is not affected by the preceding

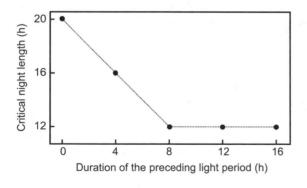

FIGURE 3.5.
Effect of a preceding light period on critical night length. Dark-grown plants received a light period of various durations and were transferred to a dark period of various durations to determine the critical night length (for flowering) for each light period. After Saji et al. (1984).

light period, if it is more than a certain duration, and that a timekeeping mechanism specifically measures the length of the dark period. When the photoperiod is less than 8 h, the critical night length was shortened as the photoperiod was lengthened, but the time from the beginning of the photoperiod until the end of the dark period of the critical night was a constant 20 h (i.e., the sum of the duration of light period and critical night length; figure 3.5) (Saji et al., 1984). Under these conditions, the circadian clock may still oscillate during the light period and enter the night period with its phase not being affected. On the other hand, when the photoperiod is longer than 8 h, activity of the circadian clock may suspend during this time and restart to run after entering the dark period. As noted above, a recent molecular study describing a detailed expression analysis of *PnFT* has revealed the involvement of a dark-induced activity rhythm in controlling photoperiodic flowering in *Pharbitis* (Hayama et al., 2007). Expression analyses of *PnFT* showed that a weak circadian rhythm of *PnFT* was sustained when plants were transferred from noninductive LD to continuous darkness, whereas expression remained extremely low when plants were shifted to continuous light (figure 3.6). As well, the circadian oscillation of *PnFT* expression in continuous darkness preceded by an inductive SD dampened to a low level when plants were transferred to continuous light, but remained robust in continuous darkness (figure 3.6) (Hayama et al., 2007). Together with the observation that the phase of *PnFT* expression is set by a lights-off signal (figure 3.3B), these data support that notion that *PnFT* induction under SD occurs through a mechanism for night-length measurement, in which an unknown clock-controlled gene(s) whose phase is set by dusk generates a dusk-set activity rhythm that controls *PnFT* expression during the night (Hayama et al., 2007).

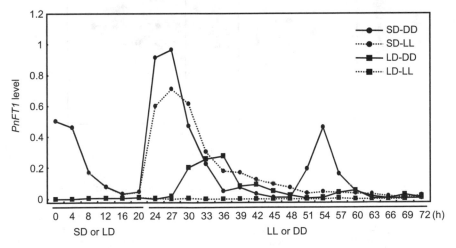

FIGURE 3.6.
Effect of light on the circadian rhythm of *PnFT* expression. Plants were grown either under LDs or SDs and transferred to continuous light (LL) or continuous darkness (DD). *PnFTI* mRNA expression is shown for each of these four treatments. Based on Hayama et al. (2007).

In this model, light is proposed to at least set the phase of the circadian clock regulating the clock-controlled gene, although this clock differs from that in *Arabidopsis* and rice, since the action of light on this *Pharbitis* clock is to set the phase at lights off. To date, molecular analysis has not implicated a direct action of light on the activity of a specific clock-controlled output gene in *Pharbitis*. However, studies have suggested that light also acts downstream of the clock to control flowering time. In this experiment, a night break of various light intensities was given during the first dark period of the long night, and following each case, the phase of the subsequent rhythm was "scanned" by giving another light flash at various times (Lumsden and Furuya, 1986). Results showed that when the night break is less than a certain light intensity, it can inhibit flowering without affecting the time of maximum sensitivity to night breaks in the following dark period (figure 3.7) (Lumsden and Furuya, 1986). This suggests that light can directly inhibit flowering without affecting the phase of the circadian clock. Thus, a possible mechanism for measuring the length of the night in *Pharbitis* is that a circadian clock whose phase is set by lights off acts to determine the phase of a clock-controlled gene. This clock-controlled gene is activated in the dark (or is

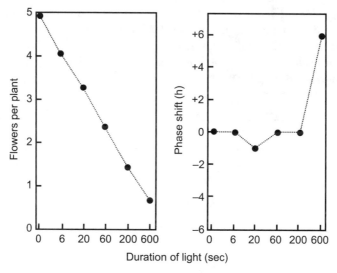

Duration of light (sec)

FIGURE 3.7.
Dose-dependent effects of a night break on floral inhibition and phase shifting of a photoperiodic rhythm. (Left) Plants received a night break with various durations of red light 8 h after the beginning of a 72-h dark period, and the magnitude of floral inhibition for each night break was examined. (Right) For each case, plants received another night break at different times during the following night period, and the time of maximum sensitivity to those night breaks was examined. Effect of the duration of the first night break on phase shifting was examined based on the comparison of the time of maximum sensitivity after the first night break with that without a night break. After Lumsden and Furuya (1986).

suppressed by light) and eventually drives a dusk-set activity rhythm in the dark, which, if the duration of the night period is sufficiently long, induces *PnFT* expression (figure 3.8).

A clock-controlled gene regulating flowering time in *Pharbitis* could be *CO*-related gene, and in fact, genes that encode a protein similar to CO have been isolated. Transcripts of those genes, *PnCO* and *PnCOL*, respectively, show a circadian rhythm and also exhibit higher levels under SD, consistent with the idea that these genes are involved in control of flowering in *Pharbitis* (Liu et al., 2001; S.J. Kim et al., 2003). Moreover, overexpression of *PnCO* causes early flowering in *Arabidopsis*, suggesting a function similar to *CO* in *Arabidopsis* (Liu et al., 2001). However, expression of *PnCO* shows a circadian rhythm whose phase is set by lights on, not lights off, which differs from that observed for the circadian rhythm of *PnFT* expression (Hayama et al., 2007). This result indicates that *PnCO* is not the factor determining the dusk-set rhythm of *PnFT* expression. Together with the lack of the functional data for this gene in *Pharbitis*, it is still unknown whether *PnCO* is involved in controlling flowering time in *Pharbitis*. To date, a clock-controlled gene whose phase is set by lights off and specifically regulates *PnFT* expression remains unknown.

There have been a number of attempts to isolate genes involved in the photoperiodic flowering response in *Pharbitis*. Because physiological observations have suggested the involvement of circadian oscillations in the dark in the control of flowering, genes whose expression cycles in continuous darkness or are induced under SD were isolated using differential hybridization of a *Pharbitis* cDNA library or differential display (O'Neill et al., 1994; Sage-Ono et al., 1998; K.-C. Kim et al., 2003). Among them, expressions of *PnFL1* and *PnC401*, both of which encode

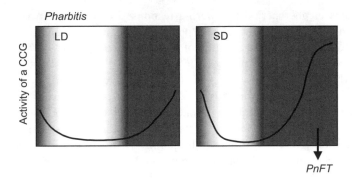

FIGURE 3.8.
A possible timekeeping mechanism in *Pharbitis*. The circadian clock whose phase is set by lights off determines the phase of a clock-controlled gene (CCG) or genes, and this drives a dusk-set activity rhythm during the night, which controls *PnFT* expression. Under LD, *PnFT* is not induced because the activity of the CCG is terminated by light before it reaches its peak level, whereas under SD *PnFT* is induced because the activity of CCG can reach its maximum level during the longer night.

unknown proteins, were found to exhibit a circadian rhythm in continuous darkness and be induced under SD and were also suppressed by a night break (Sage-Ono et al., 1998; K.-C. Kim et al., 2003). Additionally, a germin-like protein was isolated as one whose levels are increased during a flower-inductive dark period in *Pharbitis* cotyledons, through the analysis of *in vivo* labeled proteins with two-dimensional gel electrophoresis (Ono et al., 1996). Expression pattern analysis of this gene was similar to that observed for *PnFL1* and *PnC401* (Ono et al., 1996). Whether the circadian rhythm of these genes shows a lights-off phase resetting is not known, however, and it remains unclear whether these candidates are involved in the photoperiodic response in *Pharbitis*.

LIGHT QUALITY AND THE CONTROL OF PHOTOPERIODIC FLOWERING RESPONSE IN *PHARBITIS*

It is evident that red, far-red, and blue light are involved in entraining the circadian clock in *Arabidopsis* (Somers et al., 1998; Salome and McClung, 2005). Physiological studies in *Pharbitis* also suggest the involvement of the same wavelengths in clock entrainment. One experiment showed that the time of maximum sensitivity to a night break is not altered regardless of whether red, far-red, or blue light preceded the dark period (Vince-Prue, 1981). Moreover, molecular studies with *PnFT* show that the time of the peak expression of *PnFT* rhythm in continuous darkness is the same following lights off, regardless of whether red, far-red, or blue light precede the dark period (Hayama and Coupland, unpublished data). Those results indicate that these wavelengths can all generate a lights-off signal and phase resetting of the circadian clock that controls flowering time in *Pharbitis*.

Studies of photoperiodic flowering response in *Arabidopsis* have revealed that blue and far-red light activate *CO* posttranscriptionally, which ensures activation of CO under LD if the duration of light is sufficiently long (Valverde et al., 2004). Red light has been proposed to suppress CO posttranscriptionally especially early in the day (Valverde et al., 2004). On the other hand, the light quality responsible for regulating the activity of the downstream process of the circadian clock in *Pharbitis* is still largely unknown. Studies using *PnFT* show that red, far-red, or blue light can all suppress *PnFT* expression when used in a light–dark cycle, indicating that all of these light qualities are able to inhibit flowering in *Pharbitis* (Hayama and Coupland, unpublished data). Currently, it is unknown whether this light-dependent suppression of *PnFT* expression occurs through an effect on state of the circadian clock, through a direct effect on the activity of a clock-controlled gene, or through both mechanisms. However, as noted above, a red-light night break can inhibit flowering without phase shifting the clock (Lumsden and Furuya, 1986), suggesting that at least red light may have a role in directly regulating flowering activity downstream of the circadian clock in *Pharbitis* (figure 3.7).

CONCLUSION

In this chapter I have attempted to demonstrate the mechanism of photoperiodic flowering response in *Pharbitis*, drawing on the physiological and molecular genetic studies carried out to date. In *Pharbitis*, a circadian system whose phase is set by lights off through the action of red, far-red, and/or blue light during the light period controls the photoperiodic flowering response. This circadian system may be distinct from that which controls other circadian rhythms, such as *CAB* gene expression. This circadian clock determines the phase of one or more clock-controlled genes, which generate a dusk-set activity rhythm in the dark and induces *PnFT* expression if the dark period is sufficiently long. In this model, one or more clock-controlled genes may be activated specifically in the dark, while red light may directly inhibit such activity during the light period. This mechanism is in contrast to that proposed for *Arabidopsis* and rice, both of which possess a mechanism for measuring the length of the day in which a clock-controlled gene, whose phase is mainly set by lights on at dawn is directly activated by light posttranscriptionally.

That a single family in the plant kingdom involves both SDPs and LDPs indicates that the mechanism controlling photoperiodic flowering response in plants has diverged rapidly in the evolution. While molecular genetic studies in *Arabidopsis* and rice suggest a model for a molecular mechanism to measure the length of the day, the molecular nature of night-length measurement in plants is still not known. Recent molecular and genetic studies, together with the establishment of a stable transformation method (Ono et al., 2000), have expanded the potential for *Pharbitis* to become a model plant to study the mechanism of night-length measurement. Advances in understanding the photoperiodic flowering response in *Pharbitis* may allow us to increase our general understanding of the mechanisms for day-length measurement underlying the plant kingdom and may enable us to begin to study how the diversity of photoperiodic mechanism in plants came about.

References

An H, Roussot C, Suarez-Lopez P, Corbesier L, Vincent C, Pineiro M, Hepworth S, Mouradov A, Justin S, Turnbull C, and Coupland G. (2004). CONSTANS acts in the phloem to regulate a systemic signal that induces photoperiodic flowering of *Arabidopsis*. Development 131: 3615–3626.

Aronson BD, Johnson KA, Loros JJ, and Dunlap JC. (1994). Negative feedback defining a circadian clock: Autoregulation of the clock gene *frequency*. Science 263: 1578–1584.

Crosthwaite SK, Loros JJ, and Dunlap JC. (1995). Light-induced resetting of a circadian clock is mediated by a rapid increase in *frequency* transcript. Cell 81: 1003–1012.

Dunlap JC. (1999). Molecular bases for circadian clocks. Cell 96: 271–290.

Garceau NY, Liu Y, Loros JJ, and Dunlap JC. (1997). Alternative initiation of translation and time-specific phosphorylation yield multiple forms of the essential clock protein FREQUENCY. Cell 89: 469–476.

Hayama R, Agashe B, Luley E, King R, and Coupland G. (2007). A circadian rhythm set by dusk determines the expression of *FT* homologs and the short-day photoperiodic flowering response in *Pharbitis*. Plant Cell 19: 2988–3000.

Hayama R and Coupland G. (2004). The molecular basis of diversity in the photoperiodic flowering responses of *Arabidopsis* and rice. Plant Physiol 135: 677–684.

Hayama R, Yokoi S, Tamaki S, Yano M, and Shimamoto K. (2003). Adaptation of photoperiodic control pathways produces short-day flowering in rice. Nature 422: 719–722.

Imaizumi T and Kay SA. (2006). Photoperiodic control of flowering: Not only by coincidence. Trends Plant Sci 11: 550–558.

Imaizumi T, Schultz TF, Harmon FG, Ho LA, and Kay SA. (2005). FKF1 F-box protein mediates cyclic degradation of a repressor of *CONSTANS* in *Arabidopsis*. Science 309: 293–297.

Imaizumi T, Tran HG, Swartz TE, Briggs WR, and Kay SA. (2003). FKF1 is essential for photoperiodic-specific light signalling in *Arabidopsis*. Nature 426: 302–306.

Ishikawa R, Tamaki S, Yokoi S, Inagaki N, Shinomura T, Takano M, and Shimamoto K. (2005). Suppression of the floral activator *Hd3a* is the principal cause of the night break effect in rice. Plant Cell 17: 3326–3336.

Izawa T, Oikawa T, Sugiyama N, Tanisaka T, Yano M, and Shimamoto K. (2002). Phytochrome mediates the external light signal to repress *FT* orthologs in photoperiodic flowering of rice. Genes Dev 16: 2006–2020.

Kim K-C, Hur Y, and Maeng J. (2003). Isolation of a gene, *PnFL-1*, expressed in *Pharbitis* cotyledons during floral Induction. Mol Cell 16: 54–59.

Kim SJ, Moon J, Lee I, Maeng J, and Kim SR. (2003). Molecular cloning and expression analysis of a *CONSTANS* homologue, *PnCOL1*, from *Pharbitis nil*. J Exp Bot 54: 1879–1887.

Kojima S, Takahashi Y, Kobayashi Y, Monna L, Sasaki T, Araki T, and Yano M. (2002). *Hd3a*, a rice ortholog of the *Arabidopsis FT* gene, promotes transition to flowering downstream of *Hd1* under short-day conditions. Plant Cell Physiol 43: 1096–1105.

Liu J, Yu J, McIntosh L, Kende H, and Zeevaart JA. (2001). Isolation of a *CONSTANS* ortholog from *Pharbitis nil* and its role in flowering. Plant Physiol 125: 1821–1830.

Lumsden PJ and Furuya M. (1986). Evidence for two actions of light in the photoperiodic induction of flowering in *Pharbitis nil*. Plant Cell Physiol 27: 1541–1551.

Lumsden P, Thomas B, and Vince-Prue D. (1982). Photoperiodic control of flowering in dark-grown seedlings of *Pharbitis nil* Choisy: The effect of skeleton and continuous light photoperiods. Plant Physiol 70: 277–282.

McWatters HG, Bastow RM, Hall A, and Millar AJ. (2000). The *ELF3 zeitnehmer* regulates light signalling to the circadian clock. Nature 408: 716–720.

Millar AJ and Kay SA. (1996). Integration of circadian and phototransduction pathways in the network controlling CAB gene transcription in *Arabidopsis*. Proc Natl Acad Sci USA 93: 15491–15496.

O'Neill SD, Zhang XS, and Zheng CC. (1994). Dark and circadian regulation of mRNA accumulation in the short-day plant *Pharbitis nil*. Plant Physiol 104: 569–580.

Ono M, Sage-Ono K, Inoue M, Kamada H, and Harada H. (1996). Transient increase in the level of mRNA for a germin-like protein in leaves of the short-day plant *Pharbitis nil* during the photoperiodic induction of flowering. Plant Cell Physiol 37: 855–861.

Ono M, Sage-Ono K, Kawakami M, Hasebe M, Ueda K, Masuda K, Inoue M, and Kamada H. (2000). Agrobacterium-mediated transformation and regeneration of *Pharbitis nil*. Plant Biotechnol 17: 211–216.

Sage-Ono K, Ono M, Harada H, and Kamada H. (1998). Accumulation of a clock-regulated transcript during flower-inductive darkness in *Pharbitis nil*. Plant Physiol 116: 1479–1485.

Saji H, Furuya M, and Takimoto A. (1984). Role of the photoperiod preceding a flower-inductive dark period in dark-grown seedlings of *Pharbitis nil* Choisy. Plant Cell Physiol 25: 715–720.

Salome PA and McClung CR. (2005). What makes the *Arabidopsis* clock tick on time? Light and temperature entrainment of the circadian clock. Plant Cell Environ 28: 21–38.

Sawa M, Nusinow DA, Kay SA, and Imaizumi T. (2007). FKF1 and GIGANTEA complex formation is required for day-length measurement in *Arabidopsis*. Science 318: 261–265.

Schultz TF and Kay SA. (2003). Circadian clocks in daily and seasonal control of development. Science 301: 326–328.

Somers DE, Devlin PF, and Kay SA. (1998). Phytochromes and cryptochromes in the entrainment of the *Arabidopsis* circadian clock. Science 282: 1488–1490.

Suarez-Lopez P, Wheatley K, Robson F, Onouchi H, Valverde F, and Coupland G. (2001). CONSTANS mediates between the circadian clock and the control of flowering in *Arabidopsis*. Nature 410: 1116–1120.

Takada S and Goto K. (2003). TERMINAL FLOWER2, an *Arabidopsis* homolog of HETERO-CHROMACHIN PROTEIN1, counteracts the activation of *FLOWERING LOCUS T* by CONSTANS in the vascular tissues of leaves to regulate flowering time. Plant Cell 15: 2856–2865.

Takimoto A and Hamner KC (1965). Studies on red light interruption in relation to timing mechanisms involved in the photoperiodic response of *Pharbitis nil*. Plant Physiol 40: 852–854.

Thain SC, Hall A, and Millar AJ. (2000). Functional independence of circadian clocks that regulate plant gene expression. Curr Biol 10: 951–956.

Thain SC, Murtas G, Lynn JR, McGrath RB, and Millar AJ. (2002). The circadian clock that controls gene expression in *Arabidopsis* is tissue specific. Plant Physiol 130: 102–110.

Thomas B and Vince-Prue D. (1997). *Photoperiodism in Plants*. San Diego, CA: Academic Press.

Valverde F, Mouradov A, Soppe W, Ravenscroft D, Samach A, and Coupland G. (2004). Photoreceptor regulation of CONSTANS protein in photoperiodic flowering. Science 303: 1003–1006.

Vince-Prue D. (1981). Daylight and photoperiodism. In *Plants and the Daylight Spectrum* (H Smith, ed). London: Academic Press, pp. 223–242.

Yano M, Katayose Y, Ashikari M, Yamanouchi U, Monna L, Fuse T, Baba T, Yamamoto K, Umehara Y, Nagamura Y, and Sasaki T. (2000). *Hd1*, a major photoperiod sensitivity quantitative trait locus in rice, is closely related to the *Arabidopsis* flowering time gene *CONSTANS*. Plant Cell 12: 2473–2484.

4

Photoperiodic Control of Flowering in *Lemna*

Kumiko Ito-Miwa and Tokitaka Oyama

The duckweeds (Lemnaceae, monocotyledon)—the smallest flowering plants—inhabit freshwater ponds and pools. Geographically, they are widely distributed at high and low latitudes, including tropical zones (Urbanska-Worytkiewicz, 1975). Within this family, the *Lemna* genus includes plants with leaflike bodies called fronds and a root. Each frond has two pockets from which new fronds arise during the vegetative growth phase (figure 4.1). When flowering is induced, a flower with a pistil and two stamina emerges from one of the pockets of a frond. The floating, tiny plant bodies and rapid growth under aseptic culture conditions make *Lemna* plants attractive for experimental investigations.

Photoperiodic flowering has been examined in *L. paucicostata* 6746 (also called *L. perpusilla* 6746 until about 1976) and *L. gibba* G3 (figure 4.1; Kandeler, 1955; Kandeler and Hügel, 1974; Hillman, 1959, 1961). These two species show distinct responses to day length; *L. paucicostata* 6746 is a short-day plant, whereas *L. gibba* G3 is a long-day plant. The close phylogenetic relationship between these two species is advantageous when comparing the underlying mechanisms driving their photoperiodic flowering. Under certain experimental conditions, these plants show obligatory and sensitive responses to the day length and can be used for quantitative flowering experiments that last a week. To evaluate flowering, the numbers of vegetative fronds and flowering fronds are counted under a dissecting microscope, and the flowering percentage (FL%) is determined as the flowering frond number divided by the total frond number. A single long night for *L. paucicostata* 6746 and three long days for *L. gibba* G3 cause the subsequent production of flowering fronds

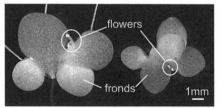

L. gibba G3   L. paucicostata 6746

FIGURE 4.1.
The short-day plant *L. paucicostata* 6746 (right) and the long-day plant *L. gibba* G3 (left). Flowers and fronds are indicated.

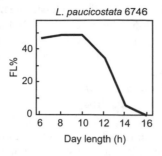 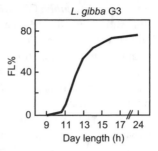

FIGURE 4.2.
Effects of day length on the flowering of *L. paucicostata* 6746 and *L. gibba* G3. *L. paucicostata* 6746 cultured under continuous light and *L. gibba* G3 cultured under short day conditions (9 h light/15 h dark) were subjected to the indicated photoperiods for four and nine days, respectively. After treatment, plants were grown under continuous light for three more days, and the FL% values were determined (adapted from Hillman, 1959; Cleland and Briggs, 1967).

(Hillman, 1959; Umemura et al., 1963). The critical day length in both *L. paucicostata* 6746 and *L. gibba* G3 has been reported to be 12–13 h (figure 4.2; Hillman, 1959; Umemura et al., 1963; Cleland and Briggs, 1967). Because they can grow in carbon-source–containing media under photoperiods with very short exposures to light, many photoperiodic schedules can be tested for flower induction without interference from the effects of photosynthesis.

PHYSIOLOGY OF FLOWERING OF *LEMNA* PLANTS

L. paucicostata 6746 can be induced to flower by a single dark period, with periods of 12 h or longer inducing similar levels of flowering (Hillman, 1959). Shibata and Takimoto examined the flowering response of *L. paucicostata* 6746 to light interruption during a single prolonged dark period (Shibata and Takimoto, 1975). The plants were subjected to a single dark period (60 or 72 h) with a 10-min pulse of white light, and their flowering was evaluated. In each condition, light interruptions given approximately 6 h after the onset of darkness, and approximately 12 h before the end of the dark period strongly inhibited flowering, suggesting the involvement of a time measurement mechanism in photoperiodic flowering and the presence of an inhibitory signal that overrode the inductive signal. A light pulse near the end of the dark period appeared to reset a circadian clock, suggesting the plants detected the onset of the light period.

The effects of light quality on photoperiodic control in the flowering of *L. paucicostata* 6746 and *L. gibba* G3 have also been investigated. Red light strongly inhibits the flowering of *L. paucicostata* 6746, whereas blue and far-red light have little effect (Hillman, 1965; Esashi and Oda, 1966). This observation suggested that phytochrome signaling was responsible for the inhibition. On the other hand, red light induced the flowering of *L. gibba* G3, indicating that the phytochrome had a different effect in this strain (Esashi and Oda, 1966; Ishiguri et al., 1975). Furthermore, blue/violet light strongly induced flowering in *L. gibba* G3, suggesting that receptors for blue light, such as cryptochrome, were involved in the signaling pathway (Ishiguri et al., 1975).

The photoperiodic responses of *L. paucicostata* have been ecologically analyzed. Twenty-two strains of *L. paucicostata* from throughout the Japanese archipelago (from 24°N to 44°N, including subtropic and subfrigid zones) were examined for two photoperiodic behaviors: sensitivity to the inducing photoperiods and the critical night length (Yukawa and Takimoto, 1976; Beppu and Takimoto, 1981a, 1981b). Seventeen strains showed photoperiodic responses that characterized them as short-day plants; the critical dark lengths varied from 9.5 h to 13.5 h. Strains collected from northern areas tended to require shorter dark periods for flower induction than those from southern regions. Four strains from southern areas did not flower, whereas a strain flowered irrespective of the day length. This variation in the characteristics of photoperiodic flowering provided an opportunity for analysis of the ecological significance of the photoperiodic behaviors. Because *Lemna* plants vegetatively grow eternally, strains collected from natural environments can be maintained in a laboratory and used reproducibly in experiments with maintenance of their genetic backgrounds.

The most famous experiments examining photoperiodic flowering of *Lemna* plants were performed by Hillman, who analyzed the response of *L. paucicostata* 6746 to "skeleton photoperiods" (Hillman, 1964). These were the first experiments that clearly demonstrated the three critical components of the photoperiodic response for flower induction: environmental photoperiods, the endogenous circadian clock, and the recognition of day length.

BÜNNING'S HYPOTHESIS OF PHOTOPERIODIC TIME MEASUREMENTS

In 1920, Garner and Allard reported photoperiodic effects on the flowering of soybean and tobacco plants. Since this discovery, how organisms measure the length of the day or night has been the focus of numerous studies. The scientific literature on circadian rhythms can be traced back approximately 200 years before this discovery when de Mairan (1729) reported that the daily leaf movement of a sensitive heliotrope plant (probably *Mimosa pudica*) persisted in constant darkness, demonstrating an endogenous origin for this rhythmic behavior. Through the nineteenth and early twentieth centuries, the endogenous rhythms of the leaf, tendril, and shoot movements of many plants were described by eminent plant biologists, including Darwin, Dutrochet, Hofmeister, Sachs, and Pfeffer (Sweeney, 1987). In particular, Darwin devised a system for recording plant organ movements, which he used to confirm the widespread occurrence of sleep movements (Darwin and Darwin, 1880). After these early studies, Bünning established the concept of a circadian clock. Beginning in the 1930s, he detailed the leaf movement of *Phaseolus* plants (common bean) and realized that leaf movements were manifestations of a fundamental ability of the plant to tell time, that is, a circadian clock. Moreover, he suggested that circadian rhythms were involved in photoperiodic measurements of day length (Bünning, 1936). Bünning suggested the foundation of endogenous cellular activity rhythms consisted of two half-cycles based on a "light-loving" (or "photophil") and a "dark-loving" (or "scotophil") process (figure 4.3A). He postulated that light detected by short-day plants during the scotophil phase inhibits flowering. On the other hand, in long-day plants, exposure to light during the scotophil phase would induce flowering.

His hypothesis was verified by examining the effects of short exposures to light during the night (night breaks). Hamner and Bonner (1938; Hamner, 1940) showed that the flowering of short-day plants (the cocklebur *Xanthium pennsylvanicum* and Biloxi soybean, *Glycine max*) was partially or completely inhibited if the inductive dark period was interrupted by light. Maximum inhibition resulted when the light interruption occurred in the middle of the dark period. This result suggested that

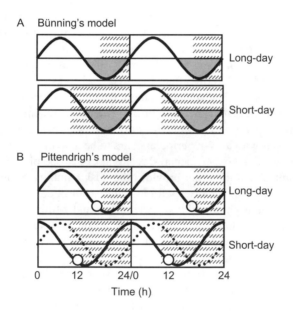

FIGURE 4.3.
External coincidence models. A circadian rhythm that changes the light sensitivity of a photoperiodic response is shown. White and striped areas represent light and dark periods, respectively. (A) Bünning's model. The gray half cycle indicates the "scotophil" phase, and the other half cycle represents the "photophil" phase. Light stimuli in the scotophil phase cause the long-day reaction, whereas darkness produces the short-day reaction. (B) Pittendrigh's model. The phases of the rhythms are different under the long-day and short-day conditions. Open circles indicate the "inducible" phase. Recognition of day length depends on illumination during in this phase. The dotted line in B traces the rhythm under long-day conditions (adapted from Bünning, 1936; Pittendrigh and Minis, 1964).

an endogenous clock contributed to the sensitivity to the light interruption, and supported Bünning's hypothesis that circadian rhythms were involved in the measurement of day length for photoperiodism.

ENTRAINMENT OF THE CIRCADIAN RHYTHM BY SKELETON PHOTOPERIODS

In 1964, Pittendrigh and Minis (1964; Pittendrigh, 1972) proposed an external coincidence model that modified Bünning's hypothesis (figure 4.3B). Bünning assumed that the phase of the endogenous rhythm was locked based on the timing of dawn and independent of the day–night schedule (figure 4.3A). The phase of circadian systems, however, is dependent on the photoperiod. Pittendrigh and Minis studied the entrainment of the circadian rhythm by assaying the timing of adult eclosion in *Drosophila pseudoobscura* (Pittendrigh and Minis, 1964; Pittendrigh, 1966). Light and dark signals serve as environmental synchronizers (*zeitgebers*) to cue the circadian clock. The phase of the *Drosophila pseudoobscura* eclosion rhythm in constant darkness was shifted by a single light pulse, and the magnitude of the shift depended on the time of treatment. A plot of the magnitude and direction of the phase shifts versus the time of treatment is referred to as a phase response curve (PRC) (figure 4.4A). Generally, light pulses delivered during the subjective day do not markedly affect the phase, whereas those during the early subjective night or in the late subjective night delay or advance the phase, respectively. PRCs have suggested that a light pulse detected during the early or late subjective night is perceived as a cue for dusk or dawn, respectively. Moreover, a pair of light pulses with a 24-h period appears to be able to reset the phase of the circadian rhythm to the experimental day–night cycle; one of the pulses reflects the dawn stimulus, and the other provides the dusk stimulus. Schedules that are given using two short light pulses with a 24-h period in constant darkness have been termed skeleton photoperiods (Pittendrigh and Minis, 1964; Pittendrigh, 1966). In the case of the eclosion rhythm, when the interval between the light pulses was >13.7 h (or <10.3 h), the smaller time interval was recognized as daytime after the phase of the eclosion rhythm was stably entrained (figure 4.4B). On the other hand, when the time interval between the light pulses was between 10.3 h and 13.7 h, either the longer or shorter time interval could be recognized as daytime, a phenomenon referred to as "bistablity." Determination of which interval was recognized as daytime depended on (1) the phase of the clock subjected to the first pulse and (2) the first interval of the skeleton schedule. Interestingly, the phase determination in skeleton photoperiods was theoretically predicted based on the PRC.

In the external coincidence model, Pittendrigh hypothesized that the "inducible phase" was particularly sensitive to light (figure 4.3B). When light illuminates an organism during the "inducible phase," it detects the schedule as a long day; otherwise the schedule is detected as a short day. Thus, the environmental

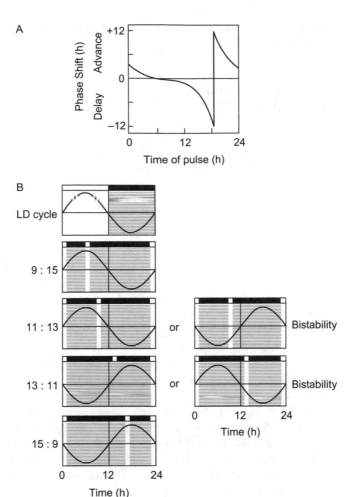

FIGURE 4.4.
PRC and the phase of the rhythm under skeleton photoperiods. (A) A schematic representation of the PRC for the *Drosophila pseudoobscura* eclosion rhythm in response to single light pulses in constant darkness. On the x-axis, 0 h/24 h and 12 h represent the subjective times at dawn and dusk, respectively. Phase shifts versus the timing of the light pulse are plotted. The direction of the shifts is denoted by "Delay" or "Advance." (B) An ideal representative skeleton photoperiod experiment based on the PRC in A. The circadian rhythm shows a peak in the daytime under complete light/dark conditions (LD cycle). Two short (15 min) light pulses divide the darkness into segments, for example, 9 h and 15 h in the 9:15 skeleton schedule. The stable phase relationship after multiple skeleton photoperiods is shown. The 9:15 and 15:9 skeletons entrain the rhythm to have a peak in the shorter skeleton period (9-h period), irrespective of the phase at the time of the first light pulse. The 11:13 and 13:11 skeletons entrain the rhythm with a peak phase in either the shorter or longer period. The phase relationship is dependent on the phase at the time of the first pulse. This phenomenon is called "bistability" and is predicted by the PRC (adapted from Pittendrigh and Minis, 1964; Pittendrigh, 1966).

light–dark cycle plays two roles: determining the phase of the circadian clock and providing the signal for coincidence during the inducible phase.

THE FLOWERING OF *L. PAUCICOSTATA* 6746 IN RESPONSE TO SKELETON PHOTOPERIODS

Drosophila pseudoobscura eclosion provided a good marker to monitor phase shifts of a circadian clock, but the eclosion rate itself is not affected by the day length. Therefore, this process could not be used to confirm the external coincidence model of photoperiodism. Around the same time as Pittendrigh's study, Hillman solved this issue using *Lemna paucicostata* 6746. He examined the effects of skeleton photoperiods on photoperiodic flowering (Hillman, 1964; Oda, 1969). As shown in figure 4.2, *L. paucicostata* 6746 plants subjected to a cycle of 11-h light/13-h dark showed increased flowering compared with those in 13-h light/11-h dark cycles. Plants grown under constant light were subjected to cycles with various dark periods to reset the circadian rhythm (figure 4.5A). Then, they were placed under 11:13 skeleton schedules (15 min of light/10.5 h of darkness/15 min of light/13 h of darkness) or 13:11 skeleton schedules (15 min of light/13 h of darkness/15 min of light/10.5 h darkness) for six to seven days. The FL% elicited by each schedule showed a 24-h periodicity and was mediated by the duration of the dark period before the skeleton schedules. For the plants under the 11:13 skeleton schedule, 0 and 24 h of darkness resulted in low flowering levels, whereas 15 h of darkness produced a high flowering level. On the other hand, for those plants subjected to the 13:11 skeleton schedules, 0 and 24 h of darkness resulted in high flowering levels, whereas flowering was reduced with 10 h of darkness (figure 4.5B). Interestingly, the FL% curve for the 13:11 skeleton schedule was nearly the mirror image of that observed with the 11:13 schedule. Moreover, the rhythmic change in FL% persisted for two cycles (Hillman, 1964). Thus, the determination of which interval was recognized as daytime depended on both the length of the dark period before the first pulse and the first interval of the skeleton schedule.

This conclusion was the same as that obtained from Pittendrigh's skeleton photoperiod experiments using the eclosion rhythm of *Drosophila pseudoobscura* (Pittendrigh, 1966). Therefore, the results of Hillman's skeleton experiments support the bistability of the circadian rhythm and the external coincidence model. In the skeleton experiments, the circadian rhythm of the *Lemna* plants was reset based on dusk when they were transferred from constant light to darkness. The next light stimulus would be a dawn or dusk stimulus depending on the phase of the clock at the time. For example, the first dark period of 24 h would make the first light pulse a dusk stimulus, and if the next pulse was provided 13 h later (13:11 skeleton schedule), the 13-h interval was recognized as night. Of note, this 13:11 schedule represents the short-day conditions that yielded a high flowering level. If an 11:13 schedule was applied instead, the 11-h interval was recognized as night, that is, long-day conditions. According to the theory of bistability, the light pulses in the following 13:11 or 11:13 skeleton schedule would shift the phases of the circadian

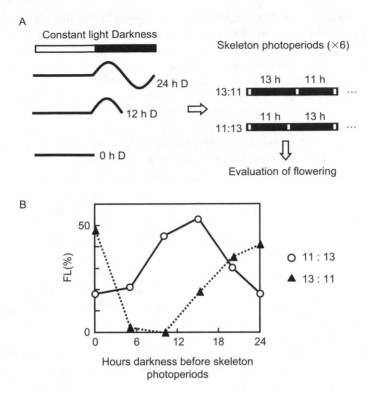

FIGURE 4.5.

Hillman's skeleton photoperiod experiment using *L. paucicostata* 6746. (A) A schematic of the experimental procedures. Plants that were grown under continuous light were subjected to various periods of darkness. Pretreated plants were then transferred into the 11:13 or 13:11 skeleton schedule. They finished the skeleton schedule six to seven times and then FL% values were determined. (B) The flowering of *L. paucicostata* 6746 treated with various dark periods before the repeated 11:13 (open circles) or 13:11 (solid triangles) skeleton schedule (adapted from Hillman, 1964).

rhythm, although the shift would not be sufficient to reverse the subjective daytime/ night profile. Thus, Hillman's experiments using *L. paucicostata* 6746 clearly provided evidence that measurement of the day length is associated with the phase of a circadian rhythm, and photoperiodic flowering can be differentially controlled by daily environmental cycles with the same light–dark ratios.

DETECTING THE BISTABILITY OF THE *L. PAUCICOSTATA* 6746 CIRCADIAN CLOCK USING A BIOLUMINESCENT REPORTER SYSTEM

Although Hillman's skeleton photoperiod experiments were theoretically successful, the circadian rhythm of *Lemna* plants had never been measured directly, because there was no way to verify the bistability of the rhythm at that time. Hillman measured the circadian rhythm of *L. paucicostata* 6746 using CO_2 output, although the detected circadian rhythm was not robust (Hillman, 1970). The CO_2 output of plants under skeleton photoperiods was also monitored, resulting in traces that appeared different between plants subjected to 13:11 and 11:13 schedules after the same initial dark periods. The phases of the rhythm and the courses of entrainment, however, were not consistent enough to draw definitive conclusions.

More than three decades after Hillman's experiments on circadian rhythms, we were able to directly observe the bistability phenomenon. Bioluminescent reporter systems using firefly luciferase have been successfully used in plants, animals, and cyanobacteria to detect circadian rhythms in living cells (Welsh and Kay, 2005). Using a particle bombardment method, we employed this system in *Lemna* plants and monitored the circadian rhythms (figure 4.6; Miwa et al., 2006). The firefly luciferase gene driven by the *Arabidopsis CCA1* rhythmic promoter (an *Arabidopsis* clock gene that is expressed around dawn) was introduced into *Lemna* plants, and a luminescence dish monitoring system automatically monitored the bioluminescence from sample dishes (figure 4.6). This reporter showed robust bioluminescence rhythms in *L. paucicostata* 6746 plants under both long-day (15 h light/9 h dark) and short-day (9 h light/15 h dark) conditions (figure 4.6C). The peak bioluminescence signals occurred in the morning, similar to observations in *Arabidopsis* (Nakamichi et al., 2004). When compared with the phases observed under the long-day and short-day conditions, the long-day phase was delayed by approximately 5 h, suggesting that it had been reset in the manner predicted by Pittendrigh (figures 4.3, 4.4).

We then monitored the bioluminescence rhythm of *L. paucicostata* 6746 under skeleton photoperiods. Plants that had been grown under constant light schedules were subjected to particle bombardment and transferred to the dark (0 h or 10 h). Then, the bioluminescence rhythms were monitored under a 16:8, 13:11, or 11:13 skeleton schedule (figure 4.7). Under the 16:8 schedule, the peak phase was entrained to the light pulse at the beginning of an 8-h dark interval irrespective of the initial period in the dark. Because the peak occurred at the beginning of the light period in the light–dark cycle, the 8-h interval was always recognized as daytime. Under the 13:11 schedule, plants that did not receive an initial dark treatment showed a bioluminescence rhythm with a peak at the beginning of the 11-h dark interval. Therefore, the shorter dark interval was recognized as daytime. Plants that were initially subjected to a 10-h dark period, however, produced peaks at the beginning of the 13-h dark interval, suggesting the longer interval was recognized as daytime. Together, these results demonstrate the bistability of the circadian clock in the 13:11 skeleton photoperiod. Results from the 11:13 skeleton schedule resulted in the same conclusion. Thus, recognition of the day length in the skeleton schedule was consistently observed in Hillman's skeleton photoperiod experiments and our bioluminescence rhythm assays, supporting an external coincidence model of photoperiodism in the short-day plant *L. paucicostata* 6746.

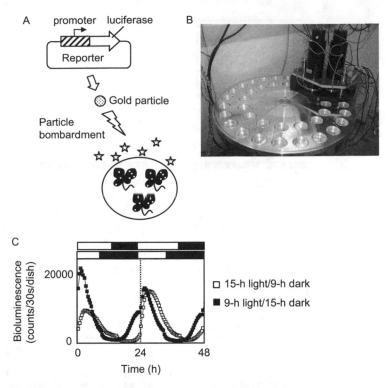

FIGURE 4.6.
The bioluminescent reporter system for monitoring rhythms in *Lemna* plants. (A) A
schematic of the experimental procedures. The bioluminescence reporter gene
firefly luciferase was introduced into plants using a particle bombardment method.
(B) Bioluminescence was measured with a luminescence dish monitoring system with
the photomultiplier tubes. (C) Bioluminescence rhythms from luciferase driven by the
Arabidopsis CCA1 promoter under long-day (15 h light/9 h dark; open squares) or short-day
(9 h light/15 h dark; solid squares) conditions (adapted from Miwa et al., 2006).

RECENT MOLECULAR STUDIES
AND CONCLUDING REMARKS

In this chapter, we have summarized the photoperiodic induction of flowering in
L. paucicostata 6746. Although this description conforms to the external coincidence model, it does not rule out the involvement of time measurement systems
based on other mechanisms, such as an internal coincidence model (Pittendrigh,
1960; Tyshchenko, 1966). In this model, photoperiodic induction depends on the
phase relationship between two circadian rhythms that are subject to different
entrainment signals. In natural settings that include temperature and light-quality
changes, measurements of the day length may be performed using multiple mechanisms. In fact, it was reported that several strains of *L. paucicostata* under culture

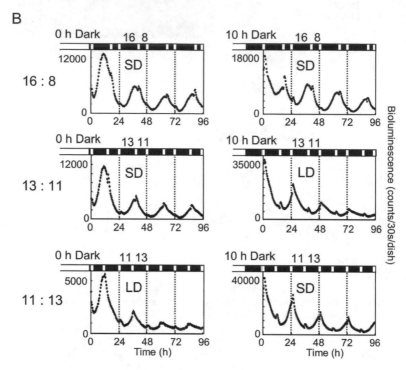

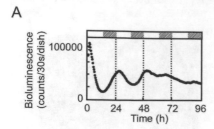

FIGURE 4.7.

Bistability in the skeleton photoperiod experiment. (A) A bioluminescence circadian rhythm of luciferase driven by the *L. gibba LHY* promoter under continuous light. This promoter produced a similar rhythm as that observed with the *Arabidopsis CCA1* promoter (figure 4.6A), but its expression level is high enough to monitor the rhythm in the dark. Hatched areas denote the periods of subjective night. Note that the peak occurs around dawn. (B) Bioluminescence rhythms of plants under skeleton photoperiods. Plants that were grown under continuous light were transferred directly to skeleton photoperiods (16:8, 13:11, or 11:13; left) or to a 10-h dark period before the skeleton photoperiods (right). Bioluminescence was measured from the time of the first pulse of each skeleton schedule. Recognition of day length was estimated by examining the time of the peak in the second day relative to the reference phase shown in A. For example, in the 16:8 skeleton without an initial dark period, the peak was around the time of the light pulse at the beginning of the shorter period (8 h), suggesting that this pulse was interpreted as dawn. Therefore, the plants recognized this photoperiod as a short day. The results for each skeleton schedule have been interpreted as SD (short day) or LD (long day) in each graph (Miwa et al., unpublished data).

conditions that mimic natural environments showed flowering behaviors with different photoperiodic properties than those observed under more traditional laboratory light–dark cycles (Yukawa and Takimoto, 1976; Beppu and Takimoto, 1981b).

We have recently initiated further molecular analyses of the machinery of photoperiodic flowering in *L. paucicostata* 6746 and *L. gibba* G3 (Miwa et al., 2006; Serikawa et al., 2008). Homologs of genes that have been shown to be involved in photoperiodism and circadian oscillation in *Arabidopsis* (a long-day plant; dicotyledon) and rice (a short-day plant; monocotyledon) were isolated from *Lemna* plants and analyzed. All of the *Lemna* homologs to the three clock-related genes (*LHY*, *GI*, and *ELF3*) showed expression profiles that were comparable with those of the corresponding genes in *Arabidopsis*, suggesting that the *Lemna* plants use a similar circadian clock machinery. The results also suggest that a similar circadian system is used by long-day and short-day *Lemna* plants, as the pattern of expression was similar in the two species.

As shown in figures 4.6 and 4.7, a bioluminescent reporter system that employs particle bombardment is an easy and powerful way to directly detail the dynamics of circadian behaviors. The reporter activity can be sustained for more than one week, long enough to allow examinations of circadian rhythms and the entire course of floral induction, which can be visually observed within a week after photoperiodic transitions (Hillman, 1959; Umemura et al., 1963). Examining *Lemna* plants using these molecular tools should assist in analyses of the processes that drive the recognition of day length to flower induction and the determination of critical day lengths, as well as evolutionary divergence in day-length responses.

As noted in chapter 2, one way in which the long-day and short-day flowering can differ is due to the opposite effect of CO protein (Hd1 in rice) on *FT* gene expression in *Arabidopsis* and rice. In both species, the clock controls the circadian expression of *CO* (*Hd1*), which is critical for day-length measurement. Since the mechanisms regulating circadian gene expression are likely to be highly conserved, the difference between long-day and short-day flowering might have arisen via either through the functional modification of common flowering-related genes or through the recruitment of additional genes with flowering-related functions. The example of *CO* and *Hd1* fits with the former hypothesis. On the other hand, *Ehd1*, an apparent rice-specific photoperiodic flowering gene, may fit with the latter idea (Doi et al., 2004). Since the genetic distance between *L. gibba* and *L. paucicostata* is much smaller than that between *Arabidopsis* and rice, one or both of the above mechanisms may be in force here. The puzzle of how opposite photoperiodic responses are realized in these closely related species is a particularly intriguing question that is now addressable at the molecular level.

References

Beppu T and Takimoto A. (1981a). Geographical distribution and cytological variation of *Lemna paucicostata* Hegelm. in Japan. Bot Mag Tokyo 94: 11–20.

Beppu T and Takimoto A. (1981b). Further studies on the flowering of *Lemna paucicostata* in Japan. Bot Mag Tokyo 94: 69–76.

Bünning E. (1936). Die endogene Tagesrhythmik als Grundlage der photoperiodischen Reaktion. Ber Dtsch Bot Ges 54: 590–607.

Cleland CF and Briggs WR. (1967). Flowering responses of the long-day plant *Lemna gibba* G3. Plant Physiol 42: 1553–1561.

Darwin C and Darwin F. (1880). *The Power of Movement in Plants.* London: J. Murray.

de Mairan J. (1729). Observation botanique. Hist Acad Roy Sci 35–36.

Doi K, Izawa T, Fuse T, Yamanouchi U, Kubo T, Shimatani Z, Yano M, and Yoshimura A(2004). *Ehd1*, a B-type response regulator in rice, confers short-day promotion of flowering and controls *FT*-like gene expression independently of *Hd1*. Genes Dev 18: 926–936.

Esashi Y and Oda Y. (1966). Two light reactions in the photoperiodic control of flowering of *Lemna perpusilla* and *L. gibba*. Plant Cell Physiol 7: 59–74.

Garner WW and Allard HA. (1920). Effect of the relative length of day and night and other factors of the environment on growth and reproduction in plants. J Agric Res 18: 533–606.

Hamner KC. (1940). Interrelation of light and darkness in photoperiodic induction. Bot Gaz 101: 658–687.

Hamner KC and Bonner J. (1938). Photoperiodism in relation to hormones as factors in floral initiation and development Bot Gaz 100: 388–431.

Hillman WS. (1959). Experimental control of flowering in *Lemna*. I. General methods. Photoperiodism in *L. perpusilla* 6746. Am J Bot 46: 466–473.

Hillman WS. (1961). Experimental control of flowering in *Lemna* III. A relationship between medium composition and the opposite photoperiodic responses of *L. perpusilla* 6746 and *L. gibba* G3. Am J Bot 48: 413–419.

Hillman WS. (1964). Endogenous circadian rhythms and the response of *Lemna perpusilla* to skeleton photoperiods. Am Nat 98: 323–328.

Hillman WS. (1965). Red light, blue light, and copper ion in the photoperiodic control of flowering in *Lemna perpusilla* 6746. Plant Cell Physiol 6: 499–506.

Hillman WS. (1970). Carbon dioxide output as an index of circadian timing in *Lemna* photoperiodism. Plant Physiol 45: 273–279.

Ishiguri Y, Oda Y, and Inada K. (1975). Spectral dependences of flowering in *Lemna perpusilla* and *L. gibba*. Plant Cell Physiol 16: 521–523.

Kandeler R. (1955). Über die Blütenbildung bei *Lemna gibba* L. I. Kulturbedingungen und Tageslängenabhängigkeit. Z Bot 43: 61–71.

Kandeler R and Hügel B. (1974). Wiederentdeckung der echten *Lemna perpusilla* Torr. und Vergleich mit *L. paucicostata* Hegelm. Plant Syst Evol 123: 83–96.

Miwa K, Serikawa M, Suzuki S, Kondo T, and Oyama T. (2006). Conserved expression profiles of circadian clock-related genes in two *Lemna* species showing long-day and short-day photoperiodic flowering responses. Plant Cell Physiol 47: 601–612.

Nakamichi N, Ito S, Oyama T, Yamashino T, Kondo T, and Mizuno T. (2004). Characterization of plant circadian rhythms by employing *Arabidopsis* cultured cells with bioluminescence reporters. Plant Cell Physiol 45: 57–67.

Oda Y. (1969). The action of skeleton photoperiods on flowering in *Lemna perpusilla*. Plant Cell Physiol 10: 399–409.

Pittendrigh CS. (1960). Circadian rhythms and the circadian organization of living systems. Cold Spring Harbor Symp Quant Biol 25: 159–184.

Pittendrigh CS. (1966). The circadian oscillation in *Drosophila pseudoobscura* pupae: A model for the photoperiodic clock. Z Pflanzenphysiol 54: 275–307.

Pittendrigh CS. (1972). Circadian surfaces and the diversity of possible roles of circadian organization in photoperiodic induction. Proc Natl Acad Sci USA 69: 2734–2737.

Pittendrigh CS and Minis DH. (1964). The entrainment of circadian oscillations by light and their role as photoperiodic clocks. Am Nat 98: 261–294.

Serikawa M, Miwa K, Kondo T, and Oyama T. (2008). Functional conservation of clock-related genes in flowering plants: Overexpression and RNA interference analyses of the circadian rhythm in the monocotyledon *Lemna gibba*. Plant Physiol 146: 1952–1963.

Shibata O and Takimoto A. (1975). Flowering response of *Lemna perpusilla* 6746 to a single dark period. Plant Cell Physiol 16: 513–519.

Sweeney BM. (1987). *Rhythmic Phenomena in Plants*, 2nd ed. San Diego: Academic Press.

Tyshchenko VP. (1966). Two-oscillatory model of the physiological mechanism of the photoperiodic reaction of insects. Zh Obshch Biol 27: 209–222.

Umemura K, Inokuchi H, and Oota Y. (1963). Flowering in *Lemna gibba* G3. Plant Cell Physiol 4: 289–292.

Urbanska-Worytkiewicz K. (1975). Cytological variation within *Lemna* L. Aquat Bot 1: 377–394.

Welsh DK and Kay SA. (2005). Bioluminescence imaging in living organisms. Curr Opin Biotechnol 16. 73–78.

Yukawa I and Takimoto A. (1976). Flowering response of *Lemna paucicostata* in Japan. Bot Mag Tokyo 89: 241–250.

5

Photoperiodic Control of Dormancy and Flowering in Trees

Pekka Heino, Ove Nilsson, and Tapio Palva

Plants as sessile and poikilothermic organisms have to grow, reproduce, and survive in a wide variety of environmental conditions. Consequently, plants have evolved an array of adaptive strategies, both physiological and developmental, that allow them to recognize different environmental variables, which then trigger the responses required for optimal growth and survival. Perennial plants, including trees, are recurrently exposed to unfavorable conditions during their life time. A common strategy to cope with such conditions evolved in perennials is the ability to suspend and reactivate growth in response to environmental conditions, including temperature, light, and water availability, by using annual cycles with altering states of active growth and vegetative dormancy that are closely timed with seasonal changes in climate. These seasonal changes are particularly prominent in trees growing in temperate and boreal zones that have the ability to anticipate the coming winter and initiate processes that lead to development of dormancy and cold hardiness. This ability is genetically determined and the response is governed by distinct environmental cues, of which the most important are shortening of the photoperiod and decrease in temperature. Such alterations are often seasonally determined and manifest in late summer, when the shortening of the day length acts as an input signal to trigger growth cessation and formation of the apical bud and to initiate the development of dormancy in bud and cambial meristems. Subsequent exposure to low temperature has a dual function in the annual cycle: for development of the extreme freezing tolerance found in trees during midwinter, and for subsequent release of dormancy. In the dormant state, the meristematic tissues are not able to react to growth-promoting substances, even under favorable conditions, and thus the dormancy release is essential for bud burst and regrowth in the following spring. The release of dormancy in most woody plants requires a period of chilling and depends on the accumulated thermal time during winter (chill-unit accumulation; Arora et al., 2003). The timing of growth cessation and bud set in autumn and bud burst and initiation of growth in spring constitute the limits for the active growth period of a tree. Therefore, late bud set and early bud burst lead to longer growth periods, thus increasing the competitiveness of the tree. On the other hand, both late bud set and dormancy development and early bud burst increase the risk that the tree will be damaged by frost during autumn and spring, respectively.

Therefore, the exact timing of the annual growth cycle can be seen as a compromise between growth and avoidance of damage.

DORMANCY IN TREES

Dormancy can be broadly defined as reversible suppression of visible growth in any plant tissue containing meristem (Lang et al., 1987). Lang et al. (1987) have introduced the division of dormancy into three types: (1) ecodormancy, which is regulated by environmental factors and prevents growth in unfavorable conditions; (2) paradormancy, regulated by physical factors outside the dormant organ, best exemplified with apical dominance, where auxin produced in the apical meristem, together with a carotenoid derived signaling molecule, inhibits the growth of axillary buds (Booker et al., 2004; Schwartz et al., 2004; Mouchel and Leyser, 2007); and (3) endodormancy, which is regulated by the factors inside the dormant organ itself and triggered by environmental cues, light being the most important (Horvath et al., 2003). Characteristic to endodormant tissues is their lack of the ability to resume growth, even under favorable conditions. Growth can only occur after dormancy has been released (Rohde and Bhalerao, 2007). During the annual growth cycle of trees endodormancy develops in buds and cambial meristems during autumn and is released during winter, after which the trees are in an ecodormant state and can resume active growth when the conditions become favorable.

Development of Dormancy

The first step in establishment of dormancy is growth cessation, which is required for subsequent development of endodormancy. Growth cessation is known to be triggered by shortening of the photoperiod, the light signal being recognized by the phytochrome photoreceptors (Nitsch, 1957; Olsen et al., 1997; Ingvarsson et al., 2006; see below). Simultaneously with growth cessation, the development of cold hardiness and drought tolerance is initiated (Weiser, 1970), and these processes are continuing throughout the subsequent development of dormancy (Ruttink et al., 2007). After growth cessation, dormancy develops gradually, but the specific molecular events leading to dormancy have been somewhat difficult to analyze, because development of bud dormancy is intimately connected to bud formation and parallels acclimation to cold and drought. Especially in natural environments, the development of dormancy and cold acclimation are closely connected. However, under experimental conditions, it has been possible to separate these processes to some extent (Welling et al., 2002; Mølmann et al., 2005; Ruttink et al., 2007). Additionally, ectopic expression of the *Arabidopsis ETR1*, encoding a dominant negative form of the ethylene receptor, in birch and overexpression of *ABI3* in *Populus* have been shown to inhibit bud formation but not to prevent dormancy development, indicating that bud formation is not a prerequisite for dormancy development

(Rohde et al., 2002; Ruonala et al., 2006). In wild-type trees, however, development of buds is tightly connected to dormancy development, and bud set precedes dormancy development. During apical bud development, the primordia that were present before the short-day signal was perceived will still develop into leaves, and the primordia initiated after the perception of the short days will develop into bud scales and embryonic leaves forming inside the developing bud (Rohde et al., 2002, 2007; Ruttink et al., 2007).

Cell Cycle

One of the first events characteristic to dormancy development is inhibition of cell division in meristematic tissues. Espinosa-Ruiz et al. (2004) have analyzed the alterations in the molecular components of the cell cycle machinery during dormancy development in cambial meristem of poplar. Cambium provides a good model for dissection of dormancy-related processes because the developmental alterations that complicate the analysis in buds do not take place in the cambium (Espinosa-Ruiz et al., 2004; Schrader et al., 2004). The induction of dormancy in poplar cambium could be temporally divided into two phases. After four weeks of short-day treatment, the tissues were in an ecodormant state, which was followed by development of endodormancy during the following two weeks (Espinosa-Ruiz et al., 2004). Development of ecodormancy was correlating with a decline in the activities of cyclin-dependent kinases CDKA and CDKB, which in plants are the key regulators of G1–S and G2–M transitions, respectively (Inze and de Veylder, 2006). However, the amounts of mRNAs encoding CDKA or CDKB or the corresponding protein levels were not affected (Espinosa-Ruiz et al., 2004). This is in agreement with the fact that when the tissues are in ecodormant state, the cell cycle machinery is still able to react to growth-promoting signals. Prolonged short-day treatment and development of endodormancy were needed for decline in the levels of *CDKB* mRNA and protein and the level of CDKA protein (Espinosa-Ruiz et al., 2004). The endodormant cambial cells are arrested in the G1 phase (Zhong et al., 1995), which suggests that the G1–S transition is inhibited during dormancy development. In accordance with this, genes encoding D-type cyclins, key regulators of the G1–S transition, have been shown to be down-regulated during cell cycle arrest in poplar cambium (Druart et al., 2007). Analogously to animal cells the G1–S transition in plants is regulated by the EF2-Rb (adenovirus E2 promoter binding factor-retinoblastoma) pathway (Shen, 2002; Inze and de Veylder, 2006). In this pathway, the EF2 transcription factor activates genes that commit the cells to DNA replication, and Rb is a repressor of E2F. The affinity of Rb to E2F depends on the phosphorylation status, and it binds E2F in a dephosphorylated form. In late G1, Rb is phosphorylated by CDKA, which leads to reduced affinity, and therefore activation of E2F (Shen, 2002; Inze and de Veylder, 2006). During endodormancy development in poplar cambium, the Rb phosphorylation activity of CDKA was reduced, indicating that the E2F-mediated transcriptional activation was inhibited, leading to cell cycle arrest at G1 phase (Espinosa-Ruiz et al., 2004). Furthermore, during the transition from eco- to endodormancy, there was a transient increase in the E2F phosphorylation activity of CDKA (Espinosa-Ruiz et al., 2004). E2F phosphoryla-tion mediates the inactivation and subsequent degradation of E2F via the ubiquitin-proteasome pathway in both mammalian and plant cells (Ohta and Xiong, 2001; del Pozo et al., 2002), suggesting that the arrest of cell cycle during development of endodormancy includes both activation of the Rb repressor function and reduction in the amount of its target protein E2F (Espinosa-Ruiz et al., 2004).

Hormonal Regulation of Dormancy Development

The hormones that have traditionally been connected to regulation of growth and dormancy are gibberellins (GA), GA_1 being the active form, and abscisic acid (ABA) (Eagles and Warcing, 1964; Olsen et al., 1995). The level of GA is higher during long than during short days, and exposure of plants to short day length appears to block GA_1 biosynthesis (Jackson and Thomas, 1997; Olsen et al., 1997). However, the role of GA in growth cessation appears not to be restricted to the decrease in level of active GA but also affects the GA response. Ruttink et al. (2007) have shown that two genes encoding the GA response inhibitor GAI are highly up-regulated when poplar trees are exposed to short-day photoperiod. In addition, *PttRGA1*, a gene with high similarity to *RGA1*, encoding another repressor of gibberellin response has been shown to be induced in the dormant cambium of poplar (Schrader et al., 2004; Druart et al., 2007).

The reports describing the role of ABA in bud set and dormancy develop-ment have been controversial (Arora et al., 2003). ABA levels have been shown to fluctuate in several tree species, the levels being highest during late summer and early autumn and then declining during winter (Harrison and Saunders, 1975; Rinne et al., 1994). However, no unambiguous correlation between ABA level and dormancy cycle have been reported, mainly due to the inconsistencies in the mea-sured levels of endogenous ABA during dormancy induction (Lenton et al., 1972; Rinne et al., 1994). In addition, exogenous ABA acts as growth retardant but alone is not able to induce growth cessation and dormancy in long days (Johansen et al., 1986; Welling et al., 1997; Li et al., 2003). However, exogenous ABA is able to promote growth cessation and dormancy development in short days (Welling et al., 1997; Li et al., 2003). It therefore appears that even if ABA is not directly trigger-ing dormancy development, it still may have a role establishing dormancy as well as in dormancy-related processes, especially as a regulator for acclimation to cold and drought (Welling et al., 1997; Rinne et al., 1998; Arora et al., 2003). ABA also appears to control bud development. Rohde et al. (2002) analyzed the involve-ment of the poplar homolog of the *Arabidopsis ABI3* (Parcy et al., 1994) in bud set and dormancy development. They demonstrated that the *PtABI3* is expressed in buds at the time of growth arrest and, by fusing the *PtABI3* to the GUS reporter gene, that the expression mainly takes place in young embryonic leaves in buds during short-day–induced bud set concomitantly with the increase in the endog-enous ABA level in buds (Rohde et al., 2002). Furthermore, transgenic poplars that overexpressed *PtABI3* developed aberrant buds with larger embryonic leaves and smaller bud scales than did the corresponding wild type (Rohde et al., 2002). On the other hand, transgenic trees in which the corresponding gene had been silenced by antisense strategy showed pronounced bud scale development and smaller leaves than seen in the wild type (Rohde et al., 2002). This indicates that ABA/*ABI3* module is essential for growth and differentiation of embryonic leaves prior to growth cessation. In accordance with this, it was recently shown that genes encoding proteins involved in ABA biosynthesis and signal transduction are transiently induced in poplar buds during bud development but before cessation of meristematic activity (Ruttink et al., 2007).

Ethylene signaling has been shown to be essential for proper bud development. By expressing the dominant negative form of the ethylene receptor, ETR1, from *Arabidopsis* in transgenic birch (*Betula pendula*), Ruonala et al. (2006) demonstrated that ethylene signaling was not required for short-day–induced growth cessation or development of dormancy, but it was required for bud formation. Genes involved in ethylene biosynthesis and signaling have been shown to be transiently activated after two weeks of short-day treatment in poplar (Ruttink et al., 2007). The activation of ethylene signaling after two weeks in short days indicates that this activation is not a direct response to short days, and it has been suggested that depletion of sugars due to extended nights is acting as a trigger for ethylene biosynthesis and action (Ruttink et al., 2007) as has been shown to be the case in *Arabidopsis* (Thimm et al., 2004). The timing of ethylene signaling, before growth cessation and bud development, suggests that ethylene acts as a signal for proper progression of the bud development (Ruttink et al., 2007). Interestingly, Ruonala et al. (2006) have shown that ethylene signaling is also needed for the increase in endogenous ABA in the bud during short-day–induced growth cessation in birch, indicating that ethylene promotes ABA accumulation in these conditions and that both ethylene and ABA are involved in terminal bud formation in trees.

Acclimation to Cold

Simultaneously with short-day–induced growth cessation and bud development, the tolerance to cold and drought also increases. Short-day photoperiod is sufficient to trigger cold acclimation and subsequently a moderate increase in cold hardiness in birch (Welling et al., 1997; Li et al., 2003). However, exposure to low temperature is needed for full cold hardiness (Weiser, 1970; Rinne et al., 2001; Puhakainen et al., 2004; Li et al., 2005; Welling and Palva, 2006). Short-day–derived signals alone are sufficient to induce the expression of tree homologs of diagnostic low-temperature response genes in herbaceous plants (Puhakainen et al., 2004; Druart et al., 2007; Ruttink et al., 2007). Many of these genes have previously been assigned as members of the CBF regulon activated in response to low temperature in *Arabidopsis* (Fowler and Thomashow, 2002). However, no short-day induction was seen for *CBF1–3* or *ICE1*, the upstream components of this regulon, indicating that short-day photoperiod and low temperature activate cold-responsive genes through different signaling pathways (Druart et al., 2007; Ruttink et al., 2007). Druart et al.

(2007) characterized the expression patterns of cold-acclimation–related genes in poplar cambium during the growth–dormancy cycle. They were able to divide the genes into five different classes according to their expression patterns. Classes I–III were associated with transition from growth to dormancy, and classes IV–V, with reactivation of growth during spring. In classes I and II, the activation of expression took place in response to short days, prior to the decrease in temperature, and in class III significantly later, when the trees have been exposed to low temperature (Druart et al., 2007). They suggested that the activation of class III by low temperature might be modulated by short-day signals. Short day length has previously been shown to potentiate the low-temperature induction of gene expression in birch (Pulukainen et al., 2004). It appears that the two-phase induction of cold hardiness in trees is at least partially mediated by an overlapping set of genes, and the level of hardiness obtained depends on the level of their expression. The early triggering of the genes in classes I and II temporally overlaps with growth cessation, suggesting that their activation is involving a signal that derives from shortening of the photoperiod and also acts in growth cessation (Druart et al., 2007).

Release of Dormancy

When the endodormancy has been fully developed in autumn, the tissues are not able to react to growth promoting substances, even in favorable conditions, and to be able to reactivate growth in spring, the dormancy needs to be released (Rohde and Bhalerao, 2007). In most tree species the release of dormancy requires a period of low temperature (Arora et al., 2003). Even if the required thermal time for dormancy release is to a large extent species specific, some trees are also able to modify the requirement according to environmental cues. For example, for black alder, birch, and Norway spruce, increased autumn temperatures lead to increased chilling requirement for bud burst (Heide, 2003; Junttila et al., 2003; Søgaard et al., 2008), indicating that the temperature during bud set and dormancy development affects the timing of dormancy release. It is noteworthy that the meristematic cells themselves need to experience low temperature, and the chilling signal cannot be exported from other parts of the tree. During the dormant state, the meristems are unable to react to growth-promoting substances (Rohde and Bhalerao, 2007). It has been suggested that this lack of response to growth-promoting substances is at least partially due to the symplasmic isolation of the meristematic cells in dormant tissues. In actively growing meristems all the cells are connected by plasmodesmata, allowing cell-to-cell communication and passage of signaling molecules and hormones (Rinne and van der Schoot, 1998). During dormancy development, this symplasmic connection is prevented by formation of 1,3-β-d-glucan–containing sphincters on plasmodesmata through local activation of 1,3-β-d-glucan synthase complexes (Rinne and van der Schoot, 1998; Rinne et al., 2001; Ruonala et al., 2008). This process is reversed during chilling, by both enhancing the production of 1,3-β-d-glucanase and relocating the preexisting glucanases in the vicinity of plasmodesmata (Rinne et al., 2001). Therefore, the chilling requirement for dormancy release could include a gradual restoration of the symplastic connections between the cells during chilling (Rinne et al., 2001). After the endodormancy has been released, the tissues are in ecodormant state and can resume growth when the environmental conditions again will be favorable.

PHOTOPERIODIC REGULATION OF DORMANCY IN TREES

Light is the main environmental cue controlling plant life and, in addition to being the main energy source, acts as a signal in a wide variety of developmental and adaptive responses during the plant life cycle. Plants measure the levels of ambient light through several distinct photoreceptors acting on different wavelengths. Blue and UVA light are perceived by the cryptochromes (CRY1 and CRY2) (Lin and Shalitin, 2003); blue light also by phototropins (Briggs and Christie, 2002), red/far red by phytochromes (Quail, 2002), and UVB light through currently uncharacterized photoreceptors.

The daily fluctuation in light quality and quantity is governed by the regular day/night cycles derived from the rotation of the earth. These cycles form the photoperiod, which is generally defined as the number of hours of light in a 24-h period. The sensing of photoperiod is used by plants to adjust their physiology and development to both diurnal and annual/seasonal changes. The interactions of light signals with intrinsic circadian rhythms are forming the basis of day-length measurement.

The plant circadian clock is the internal chronometer by which plants measure time and anticipate the environmental changes that are related to the photoperiodism (Webb, 2003). The clock generates the circadian rhythms, which manifest in periodic alterations in molecular composition of the cells and physiology of the plants (Harmer et al., 2000; Webb, 2003). The circadian clock is characteristically described to contain three parts. First, a central oscillator creates a near 24-h periodicity. Input pathways to the oscillator synchronize the oscillation to photoperiod and temperature cycles, and outputs from the oscillator lead to altered patterns of gene expression and physiology (Dunlap, 1999). The circadian clock is composed of a network of transcription factors arranged in interlocking positive and negative feedback loops. In *Arabidopsis*, the central oscillator of the clock is formed, in part, from the MYB- transcription factors CIRCADIAN AND CLOCK ASSOCIATED 1 (CCA1) and LATE ELONGATED HYPOCOTYL (LHY), which have been proposed to repress the expression of timing of CAB EXPRESSION 1 (TOC1), encoding a pseudo-response regulator with still unknown function. TOC1 then activates CCA1 and LHY, thus generating a transcriptional feedback loop (Alabadi et al., 2001). Recent results indicate that a second and third feedback loop, with GIGANTEA (GI) and and other pseudo-response regulators (PRR7 and PRR9) as likely components, may act in concert with the first loop (Locke et al., 2006; Zeilinger et al., 2006). Characteristic of the circadian rhythm is that it is maintained in a wide range of temperatures, and in *Arabidopsis* this robustness has been shown to be mediated by GI (Gould et al., 2006).

Photoperiod in nature is not static, but changes throughout the year. This leads to shortening and lengthening of the day length during autumn and spring months, respectively. This alteration in day length, usually in combination with the circadian clock, is utilized in plants to trigger the day-length–specific processes. In herbaceous plants, the photoperiodic regulation manifests mainly in flowering, and many plants can be classified according to their flowering response: long-day plants, which flower under a long photoperiod, or short-day plants, which flower under a short photoperiod. However, some plants are day-neutral and do not respond to changes in photoperiod. In trees, the photoperiod has traditionally been connected to growth cessation and development of dormancy and winter hardiness (Nitsch, 1957). However, recent studies have demonstrated the possible involvement of photoperiod in control of flowering also in trees (Böhlenius et al., 2006; see below).

The first step in establishment of dormancy in trees is growth cessation (see above), which is triggered by shortening of the day length (Nitsch, 1957). When the photoperiod is shorter than the critical day length, defined as the least amount of light that plants need to recognize days as long days, growth cessation is induced, trees start to prepare for overwintering, and development of dormancy is induced. Depending on the latitude of their growth habit, trees have different critical day lengths, and therefore, populations of trees growing in different latitudes form so-called photoperiodic ecotypes or provenances. Photoperiodic ecotypes have been described for several temperate tree species (Vaartaja, 1957), including *Populus*, *Salix*, and *Betula* (Junttila, 1980; Howe et al., 1995; Li et al., 2003, 2005). The timing of growth cessation and bud set, and thus the sensing of the critical day length, is under genetic control, indicating that different provenances have adapted to the local climate conditions (Bradshaw and Stettler, 1995; Chen et al., 2002). To avoid damage caused by early frosts to the growing tissues, the northern provenances have longer critical day lengths, so their growth cessation and bud set take place earlier than in the southern provenances of the same species. Alterations in the photoperiod are perceived by leaves, and the signal is subsequently transmitted to the apex, where growth is inhibited (Wareing, 1956). The signal transmission has long been known to be mediated by translocatable compound(s) because it is well known that leaf extracts or grafts from short-day–grown plants can evoke growth cessation in long-day–grown plants (Wareing, 1956; Eagles and Wareing, 1964). However, Ruonala et al. (2008) have recently suggested that the rib meristem in the apex also perceives photoperiodic signals and that this plays a role in growth cessation.

Phytochromes are the primary receptors for photoperiodically controlled light signaling in plants (Smith, 1995). In *Populus*, PhyA is the major photoreceptor for short-day signaling (Olsen et al., 1997), although also one of the two PhyB-encoding genes (*PhyB2*) appears to be involved in this process (Ingvarsson et al., 2006). Olsen et al. (1997) overexpressed the oat *PhyA* gene in hybrid aspen (*Populus tremula* × *Populus tremuloides*) and demonstrated that the *PhyA* overexpressors had altered critical day-length requirement and responded only to significantly shorter days (from 16 h in the wild type up to 6 h in transgenic trees) for growth cessation. They also showed that the *PhyA*-overexpressing trees had lower levels of indole acetic acid (IAA) and gibberellins, which correlated with the stunted growth of the transgenic trees. The reduced sensitivity to the short-day signal was associated with impaired down-regulation of the levels of active gibberellins and suggested that reduction in GA level is part of the growth cessation process, as described also for *Salix pentandra* (Olsen et al., 2005).

Recently, work by Böhlenius et al. (2006) has identified some of the molecular components in the short-day signal transduction leading to growth cessation. They demonstrated that, analogous to their involvement in photoperiodic control of flowering, the *CONSTANS* and *FT* genes also act as regulators for short-day induced growth cessation in *Populus*. Normally, growth cessation is induced within four weeks after the trees have been transferred to short-day photoperiod. However, when the *Populus* homolog of *FT* was overexpressed in *Populus tremula* × *tremuloides*, the transgenic trees did not exhibit growth cessation but continued to grow under a short-day photoperiod (Böhlenius et al., 2006). In accordance, silencing of the endogenous *FT* gene in *Populus* by RNA interference caused highly accelerated cessation of growth and bud formation in the transgenic tree lines, regardless of the day length (Böhlenius et al., 2006). The expression of the FT thus appears to efficiently suppress the short-day–induced growth cessation. In long-day–grown *Populus*, the *FT* gene is under diurnal regulation, with the expression peak early during the night. The shift to short days caused rapid decrease of the *FT* mRNA levels, followed by the loss of diurnal variation after a week in short days (Böhlenius et al., 2006). The down-regulation of the *FT* expression thus serves as an early marker for short-day–induced growth cessation. The *FT* gene is regulated by *CO*, and as expected, silencing of the *Populus CO* gene also led to accelerated growth cessation (Böhlenius et al., 2006). *CO* exhibits diurnal regulation with the expression peak late in the light period when trees are grown in long days. Because *CO* is needed for *FT* expression, and the CO protein (being labile in darkness) is present only when the gene is expressed during the light period, this regulation system nicely explains the critical day-length model, where the critical day length is the amount of hours of light that is sufficient for CO to accumulate and induce *FT*, which then sustains growth. When the day length is shorter than the critical day length, *CO* expression takes place in darkness, no CO protein is present, and thus there is no *FT* expression, which leads to growth cessation and bud set (Böhlenius et al., 2006). Furthermore, according to this model, the different provenances within a species, which have different critical photoperiods, should also have different diurnal regulation patterns of the *CO* gene. This also appears to be so. Böhlenius et al. (2006) analyzed the expression patterns of the *CO* gene in the different provenances of European aspen (*Populus tremula*) and demonstrated that the expression of *CO* was diurnally regulated but triggered at different times in different provenances, and the timing of expression actually formed a latitudinal gradient: the more southern the growth habit, the earlier the expression was triggered. This indicates that in the southern provenances shorter day length is required for the situation when *CO* expression is triggered in darkness and no *FT* expression takes place, therefore extending the growth period of trees by the end of the summer (Böhlenius et al., 2006).

Interestingly, *CENTRORADIALIS-LIKE1* (*CENL1*), the poplar homolog of the *Arabidopsis TERMINAL FLOWER1* (*TFL1*), has been shown to be somewhat up-regulated in poplar apex during the first three weeks of short-day treatment, after which expression declines simultaneously with growth cessation (Ruonala et al., 2008). In *Arabidopsis*, TFL1 counteracts the effect of FT in flowering, and therefore, it might be that the CENL1 suppresses the FT activity after the initial perception of the short-day signal and then is down-regulated when FT no longer is present. *TFL1/CENL1* in poplar has also been shown to be highly expressed in poplar buds after dormancy release, prior the time of bud flush, and overexpression of *TFL1/CENL1* has been shown to delay bud flush in transgenic poplars, indicating that TFL/CENL1 is needed for maintenance of bud dormancy (Mohamed, 2006).

In peach, a mutation called *evergrowing* prevents dormancy development (Bielenberg et al., 2004). The mutation has been shown to consist of a large deletion containing a group of tandemly arranged MADS-box genes (Bielenberg et al., 2004). The sequences of these *DORMANCY ASSOCIATED MADSBOX* (*DAM*) genes are related to the *Arabidopsis SVP* and *AGL24*. In *Arabidopsis*, SVP has been shown to bind the *FT* promoter and regulate the expression of *FT* in response to ambient temperature (Lee et al., 2007). *DAM* genes have been shown to be up-regulated during dormancy development and maintenance (Druart et al., 2007; Ruttink et al., 2007; Horvath et al., 2008). Horvath et al. (2008) characterized the changes in transcriptome of leafy spurge (*Euphorbia esula*) during the annual dormancy cycle and found that two *DAM* genes were induced, albeit with different temporal pattern during endodormancy and through the subsequent period of ecodormancy (from October to February). The expression of the *DAM* genes correlates with the down-regulation of *FT* expression, and they suggested that *DAM* genes might regulate dormancy transitions through interactions with *FT* (Horvath et al., 2008).

PHOTOPERIODIC INDUCTION OF FLOWERING IN TREES

Although the photoperiodic regulation of flowering has been recognized and studied in many plant species for almost 100 years (Kobayashi and Weigel, 2007), the molecular basis for the photoperiodic regulation of plant growth and development has mostly been studied in relation to the regulation of flowering in *Arabidopsis* and rice (Kobayashi and Weigel, 2007). *Arabidopsis* is a facultative long-day plant, meaning that it flowers much earlier under long-day conditions than under short days. The length of the day is measured by the CONSTANS (CO) protein (Valverde et al., 2004). The transcription of the *CO* gene is diurnally regulated, with a peak in mRNA accumulation that under short-day conditions occurs in darkness in the beginning of the night, and in long days occurs in light, at the end of day (Suarez-Lopez et al., 2001; Yanovsky and Kay, 2002). Since the CO protein is extremely labile in darkness, there is no accumulation of CO protein in short days (Valverde et al., 2004). However, after a shift to long-day conditions, the CO protein is stabilized and can accumulate in order to activate downstream targets (Valverde et al., 2004). The most important CO target for the activation of flowering is the gene *FT* (Kardailsky et al., 1999; Kobayashi et al., 1999; Abe et al., 2005; Wigge et al., 2005). *FT* is a powerful activator of *Arabidopsis* flowering when overexpressed (Kardailsky et al., 1999; Kobayashi et al., 1999), and the *ft* mutant is late flowering in long days, making it essentially day-length insensitive (Koornneef et al., 1991). The CO protein activates *FT* transcription in the phloem companion cells of source leaves (Takada and Goto, 2003; An et al., 2004). The FT protein is then thought to move from the leaf to the shoot apex, where it is believed to induce downstream targets important for floral initiation, including the flower meristem-identity gene *APETALA1* (*AP1*) (Abe et al., 2005; Wigge et al., 2005; Corbesier et al., 2007).

In the short-day plant rice, the same general *CO*/*FT* pathway for sensing of photoperiod and the induction of flowering seems to be conserved (Yano et al., 2000; Izawa et al., 2002; Kojima et al., 2002; Hayama et al., 2003); this makes it likely that this type of regulation is conserved in at least all annual angiosperms.

What then is known about photoperiodic regulation of flowering in trees? Unfortunately, the answer is, almost nothing, mainly because in trees growing in temperate regions, this question is very difficult to address experimentally. First of all, trees in temperate regions usually induce the production of dormant flower buds the year before these buds flush, giving rise to the inflorescence and flower development that we usually think of as "tree flowering." This means that it is often very hard to pinpoint the exact time of floral initiation since the early stages of floral bud production is very inconspicuous and hard to separate from the development of a vegetative bud. Furthermore, since most trees display an extended juvenile period that can last from years to decades, this usually means that when the trees become reproductively mature, they have also gained such a considerable height that it is very difficult to perform experiments involving controlled changes in day length. The little we know about photoperiodic regulation of tree flowering comes from correlative data from tropical trees, where it is easier to couple the flowering response to small changes in day length since these trees do not go through a period of dormancy. By combining field observations and analysis of herbarium collections, Rivera and Borchert (2001) established the existence of highly synchronous flowering periods for more than 25 tropical tree species. By systematic examination of a set of criteria, they concluded that the most plausible explanation for the synchronous flowering was a decline in photoperiod of less than 30 min. Later, they also showed that close to the equator, where there is no difference in day length over the year, synchronous flowering could be correlated to small variations in the timing of sunrise and sunset over the year, caused by the tilting of the earth's axis (Borchert et al., 2005). However, no experimental data exist to couple these correlations to differential activities of genes such as *CO* and *FT*.

In terms of floral initiation and the determination of flower meristem identity, it has been known for quite some time that these processes seem to be conserved among all angiosperms. First, it was shown that the flower-meristem-identity gene *LEAFY (LFY)* is functionally conserved between *Arabidopsis* and hybrid aspen trees, since *Arabidopsis LEAFY* overexpression in hybrid aspen led to a dramatic shortening of the juvenile phase and the induction of ectopic flowers (Weigel and Nilsson, 1995). Also the flower meristem-identity gene *AP1* could induce early flowering when expressed in transgenic citrange trees (Peña et al., 2001). Even in conifers, *LFY* homologs appear to have a role in the initiation of reproductive structures, based on expression patterns (Carlsbecker et al., 2004; Vázquez-Lobo et al., 2007). Also, conifer homologs of MADS-box genes known to be involved in the determination of floral organ identity in angiosperms are specifically expressed during various stages of the development of the conifer reproductive structures (Rutledge et al., 1998; Tandre et al., 1998, Mourudov et al., 1999; Sundström et al., 1999; Sundström and Engström, 2002).

However, flower meristem-identity genes are not really involved in determining the timing of flowering, but rather in determining the identity of the developing flower primordium. In *Arabidopsis*, the environmental effect on the time to flowering is determined by the flowering time genes, such as *CO* and *FT*, that are involved in the long-day induction of flowering and the autonomous pathway genes, such as *FLC*, *FCA*, *FY* and *FVE*, that are involved in the temperature regulation of flowering and the response to vernalization (Boss et al., 2004; Putterill et al., 2004; Simpson, 2004; Parcy, 2005; Imaizumi and Kay, 2006). However, it is by no means obvious that the function of the flowering-time genes would also be conserved between annual plants and trees. This is because of the long juvenile phase of trees when they simply do not respond to any environmental signals that could be involved in triggering flowering. It could be that a completely different mechanism controls the timing of flowering in trees, a mechanism that could be more related to the age or size of the tree.

However, recent studies have shown that trees also possess orthologs of *CO* and *FT* genes and that overexpression of *Populus FT* orthologs can trigger extremely early flowering in transgenic *Populus* trees (Böhlenius et al., 2006; Hsu et al., 2006). *FT* expression also displays a gradual increase as the trees grow older, suggesting that the trees have to reach a critical threshold level of *FT* expression in order to induce flowering (Böhlenius et al., 2006; Hsu et al., 2006). The function of the tree *FT* gene was also conserved when expressed in transgenic *Arabidopsis* (Böhlenius et al., 2006; Hsu et al., 2006). Together, these data suggest that the difference between the flowering time of annual plants and trees lies in the regulation of *FT* expression—in essence, time point *FT* at which expression reaches a critical level to be able to induce flowering.

It was also shown that the function of the *CO/FT* regulon is conserved between *Arabidopsis* and *Populus* trees in the sense that the *Populus CO* ortholog displays the same kind of diurnal expression pattern as the *Arabidopsis CO* and appears to regulate *FT* expression in exactly the same way in response to the day-length information (Böhlenius et al., 2006). *Populus CO* expression is also necessary for expression of the *Populus FT* ortholog (Böhlenius et al., 2006). This would confirm that *Populus* trees behave as long-day plants in terms of the regulation of the *CO/FT* regulon, which is consistent with the *Populus* floral initiation occurring in late spring or early summer. Is this not a good proof for a photoperiodic regulation of flowering in trees?

This question is complicated by the fact that in *Populus* trees the *CO/FT* regulon is involved in a completely different process, the regulation of short-day–induced growth cessation and bud set occurring in the fall (Böhlenius et al., 2006). This means that when the day length drops below a critical length, *FT* expression is dramatically down-regulated, and this appears to be the signal to cease growth and induce bud set (Böhlenius et al., 2006). In spring, the signal to release the buds from dormancy and induce bud flush is not the increasing day length, but rather the temperature sum experienced after the tree has experienced a period of chilling temperatures. Accordingly, bud flush can be induced simply by transferring trees to warmer temperatures, but still in short-day conditions. Under those conditions the bud will flush, but after a period of shoot elongation, the shoot will again go into growth cessation and bud set.

In *Populus* trees *FT* expression is induced in spring, at a time that roughly coincides with the expected time for floral initiation (Böhlenius et al., 2006; Hsu et al., 2006). The expression is not induced in all leaves but seems to be, as in *Arabidopsis*, confined to the older, more developed source leaves (Hsu et al., 2006). Although the timing of this induction may coincide with a day length that is longer than the critical day length for growth cessation, the induction of *FT* expression could just as well be under developmental regulation, when the leaf reaches a certain size. If the trees are induced to flush under short-day conditions and then go back to growth cessation, it is possible that the leaves never get the time to reach the appropriate size, and this is the reason for a putative lack of *FT* induction. However, this has not yet been experimentally tested, and it will be of great interest to see if the *CO/FT* regulon has a role during bud flush.

However, given the conservation of the function of the *CO/FT* regulon between species, the most straightforward hypothesis is that in trees growing in temperate regions, CO induces *FT* expression in late spring/early summer when the critical day length is reached. This induction then serves two purposes. First, it is necessary to maintain vegetative growth and prevent the tree from going into growth cessation and bud set. Second, as the tree grows older, the *Populus FT* gene gets more and more responsive to the day-length–induced *CO* activity. This leads to a gradual increase in the peak value of *FT* expression during late spring and summer. When the peak expression then reaches a certain threshold, at a certain age, FT induces flowering.

An important question is to what extent the function of the *CO/FT* regulon is conserved between trees belonging to different systematic groups. Interestingly, it was recently shown that even in the conifer tree Norway spruce, a family of PEBP-type genes, where *FT* and the very closely related repressor of flowering *TERMINAL FLOWER 1* (*TFL1*) belong, exist (Gyllenstrand et al., 2007). Furthermore, one of these genes, called *PaFT4*, showed a differential expression pattern, tightly coupled to the induction of growth cessation and the bud flush, suggesting that it might have a role in these processes (Gyllenstrand et al., 2007). Since no functional data exist for this gene, it will be of great interest to see if its function is more related to that of *FT*, or if it is rather involved in a *TFL1*-like activity. A *Populus* homolog of *TFL1*, called *PtCENL1*, was recently suggested to have a role in growth cessation and transition to dormancy in *Populus* (Ruonala et al., 2008).

More research is required to understand the details of how the *CO/FT* regulon can control two such completely different developmental processes as flowering and growth cessation, and to what extent these processes are controlled by similar downstream FT targets.

References

Abe M, Kobayashi Y, Yamamoto S, Daimon Y, Yamaguchi A, Ikeda Y, Ichinoki H, Notaguchi M, Goto K, and Araki T. (2005). FD, a bZIP protein mediating signals from the floral pathway integrator FT at the shoot apex. Science 309: 1052–1056.

Alabadi D, Oyama T, Yanovsky MJ, Harmon FG, Mas P, and Kay SA. (2001). Reciprocal regulation between TOC1 and LHY/CCA1. Science 293: 880–883.

An H, Roussot C, Suarez-Lopez, P, Corbesier L, Vincent C, Pineiro M, Hepworth S, Mouradov A, Justin S, Turnbull C, and Coupland G. (2004). CONSTANS acts in the phloem to regulate a systemic signal that induces photoperiodic flowering of *Arabidopsis*. Development 131: 3615–3626.

Arora R, Rowland LJ, and Tanino K. (2003). Induction and release of bud dormancy in woody perennials: A science comes of age. HortScience 38: 911–921.

Bielenberg DG, Wang Y, Fan S, Reighard GL, Scorza R, and Abbott AG. (2004). A deletion affecting several gene candidates is present in the *evergrowing* peach mutant. J Hered 95: 436–444.

Böhlenius H, Huang T, Charbonnel-Campaa L, Brunner AM, Jansson S, Strauss SH, and Nilsson O. (2006). CO/FT regulatory module controls timing of flowering and seasonal growth cessation in trees. Science 312: 1040–1043.

Booker J, Auldridge M, Wills S, McCarty D, Klee H, and Leyser O. (2004). MAX3/CCD7 is a carotenoid cleavage dioxygenase required for synthesis of a novel plant signaling molecule. Curr Biol 27: 1232–1238.

Borchert R, Renner SS, Calle Z, Navarrete D, Tye A, Gautier L, Spichiger R, and von Hildebrand P. (2005). Photoperiodic induction of synchronous flowering near the Equator. Nature 433: 627–629.

Boss PK, Bastow RM, Mylne JS, and Dean C. (2004). Multiple pathways in the decision to flower: Enabling, promoting, and resetting. Plant Cell 16: S18–S31.

Bradshaw HD and Stettler RF. (1995). Molecular genetics of growth and development in Populus. IV. Mapping QTLs with large effects on growth, form, and phenology traits in a forest tree. Genetics 139: 963–973.

Briggs WR and Christie JM. (2002). Phototropins 1 and 2: Versatile plant blue-light receptors. Trends Plant Sci 7: 204–210.

Carlsbecker A, Tandre K, Johanson U, Englund M, and Engström P. (2004). The MADS-box gene *DAL1* is a potential mediator of the juvenile-to-adult transition in Norway spruce (*Picea abies*). Plant J 40: 546–557.

Chen THH, Howe GT, and Bradshaw HD. (2002). Molecular genetic analysis of dormancy-related traits in poplars. Weed Sci 50: 232–240.

Corbesier L, Vincent C, Jang S, Fornara F, Fan Q, Searle I, Giakountis A, Farrona S, Gissot L, Turnbull C, and Coupland G. (2007). FT protein movement contributes to long-distance signaling in floral induction of *Arabidopsis*. Science 316: 1030–1033.

del Pozo JC, Boniotti MB, and Guitiérrez C. (2002). *Arabidopsis* E2Fc functions in cell division and is degraded by the ubuquitin-SCFAtSKP2 pathway in response to light. Plant Cell 14: 3057–3071.

Druart N, Johansson A, Baba K, Schrader J, Sjödin A, Bhalerao RR, Resman L, Trygg J, Moritz T, and Bhalerao RP. (2007). Environmental and hormonal regulation of the activity-dormancy cycle in the cambial meristem involves stage-specific modulation of transcriptional and meta-bolic networks. Plant J 50: 557–573.

Dunlap JC. (1999). Molecular bases for circadian clocks. Cell 96: 271–290.

Eagles CF and Wareing PF. (1964). The role of growth substances in the regulation of bud dormancy. Physiol Plant 17: 697–709.

Espinosa-Ruiz A, Saxena S, Schmidt J, Mellerowitcz E, Miscolczi P, Bako L, and Bhalerao RP. (2004). Differential stage-specific regulation of cyclin-dependent kinases during cambial dor-mancy in hybrid aspen. Plant J 38: 603–615.

Fowler S and Thomashow MF. (2002). *Arabidopsis* transcriptome profiling indicates that multiple regulatory pathways are activated during cold acclimation in addition to the CBF cold response pathway. Plant Cell 14: 1675–1690.

Gould PD, Locke JCW, Larue C, Southern MM, Davis SJ, Hanano S, Moyle R, Milich R, Putterill J, Millar AJ, and Hall A. (2006). The molecular basis of temperature compensation in the *Arabidopsis* circadian clock. Plant Cell 18: 1177–1187.

Gyllenstrand N, Clapham D, Källman T, and Lagercrantz U. (2007). A Norway spruce *FLOWERING LOCUS T* homolog is implicated in control of growth rhythm in conifers. Plant Physiol 144: 248–257.

Harmer SL, Hogenesch JB, Straume M, Chang HS, Han B, Zhu T, Wang X, Kreps JA, and Kay SA. (2000). Orchestrated transcription of key pathways in *Arabidopsis* by the circadian clock. Science 290: 2110–2113.

Harrison MA and Saunders PF. (1975). The abscisic acid content of dormant birch buds. Planta 123: 291–298.

Hayama R, Yokoi S, Tamaki S, Yano M, and Shimamoto K. (2003). Adaptation of photoperiodic control pathways produces short-day flowering in rice. Nature 422: 719–722.

Heide OM. (2003). High autumn temperature delays spring bud burst in boreal trees, counterbalanc-ing the effect of climatic warming. Tree Physiol 23: 931–936.

Horvath DH, Anderson JV, Chao WS, and Foley ME. (2003). Knowing when to grow: Signals regu-lating bud dormancy. Trends Plant Sci 8: 534–540.

Horvath DP, Chao WS, Suttle JC, Thimmapuram J, and Anderson JV. (2008). Transcriptome analy-sis identifies novel responses and potential regulatory genes involved in seasonal dormancy transitions of leafy spurge (Euphorbia esula L.). BMC Genomics 9:536.

Howe GT, Hackett WP, Furnier GR, and Klevorn RE. (1995). Photoperiodic responses of a northern and southern ecotype of black cottonwood. Physiol Palnt 93: 695–708.

Hsu CY, Liu Y, Luthe DS, and Yuceer C. (2006). Poplar *FT2* shortens the juvenile phase and pro-motes seasonal flowering. Plant Cell 18: 1846–1861.

Imaizumi T and Kay SA. (2006). Photoperiodic control of flowering: Not only by coincidence. Trends Plant Sci 11: 550–558.

Ingvarsson PK, Garcia MV, Hall D, Luquez V, and Jansson S. (2006). Clinal variation in phyB2, a candidate gene for day-length-induced growth cessation and bud set, across a latitudinal gradi-ent in European aspen (*Populus tremula*). Genetics 172: 1845–1853.

Inze D and de Veylder L. (2006). Cell cycle regulation in plant development. Annu Rev Genet 40: 77–105.

Izawa T, Oikawa T, Sugiyama N, Tanisaka T, Yano M, and Shimamoto K. (2002). Phytochrome mediates the external light signal to repress *FT* orthologs in photoperiodic flowering of rice. Genes Dev 16: 2006–2020.

Jackson S and Thomas B. (1997). Photoreceptors and signals in the photoperiodic control of develop-ment. Plant Cell Environ 20: 790–795.

Johansen LG, Odén P-C, and Junttila O. (1986). Abscisic acid and cessation of apical growth in *Salix pentandra*. Physiol Plant 66: 409–412.

Junttila O. (1980). Effect of photoperiod and temperature on apical growth cessation in two ecotypes of *Salix* and *Betula*. Physiol Plant 48: 347–352.

Junttila O, Nilsen J, and Igeland B. (2003). Effect of temperature on the induction of bud dormancy in ecotypes of *Betula pubescens* and *Betula pendula*. Scand J For Res 18: 208–217.

Kardailsky I, Shukla VK, Ahn JH, Dagenais N, Christensen SK, Nguyen JT, Chory J, Harrison MJ, and Weigel D. (1999). Activation tagging of the floral inducer *FT*. Science 286: 1962–1965.

Kobayashi Y, Kaya H, Goto K, Iwabuchi M, and Araki T. (1999). A pair of related genes with antagonistic roles in mediating flowering signals. Science 286: 1960–1962.

Kobayashi Y and Weigel D. (2007). Move on up, it's time for change—mobile signals controlling photoperiod-dependent flowering. Genes Dev 21: 2371–2384.

Kojima S, Takahashi Y, Kobayashi Y, Monna L, Sasaki T, Araki T, and Yano M. (2002). *Hd3a*, a rice ortholog of the *Arabidopsis FT* gene, promotes transition to flowering downstream of *Hd1* under short-day conditions. Plant Cell Physiol 43: 1096–1105.

Koornneef M, Hanhart CJ, and van der Veen JH. (1991). A genetic and physiological analysis of late flowering mutants in *Arabidopsis thaliana*. Mol Gen Genet 229: 57–66.

Lang GA, Early JD, Martin GC, and Darnell RL. (1987). Endo-, para- and eco-dormancy: Physiological terminology and classification for dormancy research. Hortic Sci 22: 371–377.

Lee JH, Yoo SJ, Park SH, Hwang I, Lee JS, and Ahn JH. (2007). Role of *SVP* in the control of flowering time by ambient temperature in *Arabidopsis*. Genes Dev 21: 397–402.

Lenton JR, Perry VM, and Saunders PF. (1972). Endogenous abscisic acid in relation to photoperiodically induced bud dormancy. Planta 106: 13–22.

Li C, Junttila O, Ernstsen A, Heino P, and Palva ET. (2003). Photoperiodic control of of growth, cold acclimation and dormancy development in silver birch (*Betula pendula*) ecotypes. Physiol Plant 117: 206–212.

Li C, Welling A, Puhakainen T, Viherä-Aarnio A, Ernstsen A, Heino P, and Palva ET. (2005). Differential responses of silver birch (*Betula Pendula*) ecotypes to short-day photoperiod and low temperature. Tree Physiol 25: 1563–1569.

Lin C and Shalitin D. (2003). Cryptochrome structure and signal transduction. Annu Rev Plant Biol 54: 469–496.

Locke JC, Kozma-Bognar L, Gould PD, Feher B, Kevei E, Nagy F, et al. (2006). Experimental validation of a predicted feedback loop in the multi-oscillator clock of Arabidopsis thaliana. Mol Syst Biol 2: 59.

Mohamed R. (2006). Expression and Function of Populus Homologs to *TERMINAL FLOWER1* Genes: Roles in Onsetting of Flowering and Shoot Phenology. PhD dissertation, Oregon State University, Corvallis.

Mølmann JA, Asante DKA, Jensen JB, Krane MN, Ernstsen A, Junttila O, and Olsen JE. (2005). Low night temperature and inhibition of gibberellin biosynthesis override phytochrome action and induce bud set and cold acclimation, but not dormancy in PHYA overexpressors and wild-type of hybrid aspen. Plant Cell Environ 28: 1579–1588.

Mouchel CF and Leyser O. (2007). Novel phytohormones involved in long-range signaling. Curr Opin Plant Biol 10: 473–476.

Mouradov A, Hamdorf B, Teasdale RD, Kim JT, Winter KU, and Theissen G. (1999). A *DEF/GLO*-like MADS-box gene from a gymnosperm: *Pinus radiata* contains an ortholog of angiosperm B class floral homeotic genes. Dev Genet 25: 245–252.

Nitsch JP. (1957). Photoperiodism in woody plants. Proc Am Soc Hort Sci 70: 526–544.

Ohta T and Xiong Y. (2001). Phosphorylation- and Skp1-independent in vitro ubiquitination of E2F1 by multiple ROC-cullin ligases. Cancer Res 61: 1347–1353.

Olsen JE, Junttila O, and Moritz T. (1995). A localized decrease of GA_1 in shoot tips of *Salix pentandra* seedlings precedes cessation of shoot elongation under short photoperiod. Physiol Plant 95: 627–632.

Olsen JE, Junttila O, Nilsen J, Eriksson ME, Martinussen I, Olsson O, Sandberg G, and Moritz T. (1997). Ectopic expression of oat phytochrome A in hybrid aspen changes critical day length for growth and prevents cold acclimatization. Plant J 12: 1339–1350.

Parcy F. (2005). Flowering: A time for integration. Int J Dev Biol 49: 585–593.

Parcy F, Valon C, Raynal M, Gaubier-Gomella P, Delsney M, and Giraudad J. (1994). Regulation of gene expression programs during *Arabidopsis* seed development. Roles of the ABI3 locus and of endogenous abscisic acid. Plant Cell 6: 1567–1582.

Peña L, Martín-Trillo M, Juárez J, Pina JA, Navarro L, and Martínez-Zapater JM. (2001). Constitutive expression of *Arabidopsis LEAFY* or *APETALA1* genes in citrus reduces their generation time. Nat Biotechnol 19: 263–267.

Puhakainen T, Li C, Boije-Malm M, Kangasjärvi J, Heino P, and Palva ET. (2004). Short-day potentiation of low temperature-induced gene expression of a C-repeat binding factor-controlled gene during cold acclimation in silver birch. Plant Physiol 136: 4299–4307.

Putterill J, Laurie R, and Macknight R. (2004). It's time to flower: The genetic control of flowering time. Bioessays 26: 363–373.

Quail PH. (2002). Photosensory perception and signalling in plant cells: New paradigms? Curr Opin Cell Biol 14: 180–188.

Rinne PLH, Kaikuranta PM, and van der Schoot C. (2001). The shoot apical meristem restores its symplasmic organization during chilling-induced release from dormancy. Plant J 26: 249–264.

Rinne P, Tuominen H, and Junttila O. (1994). Seasonal changes in bud dormancy in relation to bud morphology, water and starch content, and abscisic acid concentration in adult trees of *Betula pubescens*. Tree Physiol 14: 549–561.

Rinne PLH and van der Schoot C. (1998). Symplasmic fields in the tunica of the shoot apical meristem coordinate morphogenetic events. Development 125: 1477–1485.

Rivera G and Borchert R. (2001). Induction of flowering in tropical trees by a 30-min reduction in photoperiod: Evidence from field observations and herbarium specimens. Tree Physiol 21: 201–212.

Rohde A and Bhalerao RP. (2007). Plant dormancy in perennial context. Trends Plant Sci 12: 217–223.

Rohde A, Prinsen E, De Rycke D, Engler G, Van Montagu M, and Boerjan W. (2002). PtABI3 impinges on the growth and differentiation of embryonic leaves during bud set in poplar. Plant Cell 14: 1885–1901.

Rohde A, Ruttink T, Hostyn V, Sterck L, Van Driessche K, and Boerjan W. (2007). Gene expresssion during the induction, maintenance and release of dormancy in apical buds of poplar. J Exp Bot 58: 4047–4060.

Ruonala R, Rinne PLH, Baghour M, Moritz T, Tuominen H, and Kangasjärvi J. (2006). Transitions in the functioning of the shoot apical meristem in birch (*Betula pendula*) involve ethylene. Plant J 46: 628–640.

Ruonala R, Rinne PL, Kangasjärvi J, and van der Schoot C. (2008). *CENL1* expression in the rib meristem affects stem elongation and the transition to dormancy in *Populus*. Plant Cell 20: 59–74.

Rutledge R, Regan S, Nicolas O, Fobert P, Côté C, Bosnich W, Kauffeldt C, Sunohara G, Séguin A, and Stewart D. (1998). Characterization of an *AGAMOUS* homologue from the conifer black spruce (*Picea mariana*) that produces floral homeotic conversions when expressed in *Arabidopsis*. Plant J 15: 625–634.

Ruttink T, Arend M, Morreel K, Storme V, Rombauts S, Fromm J, Bhalerao RP, Boerjan W, and Rohde A. (2007). A molecular timetable for apical bud formation and dormancy induction in poplar. Plant Cell 19: 2370–2390.

Schrader J, Moyle R, Bhalerao R, Hertzberg M, Lundeberg J, Nilsson P, and Bhalerao RP. (2004). Cambial meristem dormancy in trees involves extensive remodeling of the transcriptome. Plant J 40: 173–187.

Schwartz S, Qin X, and Loewen M. (2004). The biochemical characterization of two carotenoid cleavage enzymes from *Arabidopsis* indicates that a carotenoid-derived compound inhibits lateral branching. J Biol Chem 279: 46940–46945.

Shen W-H. (2002). The plant E2F-Rb pathway and epigenetic control. Trends Plant Sci 11: 505–511.

Simpson GG. (2004). The autonomous pathway: Epigenetic and post-transcriptional gene regulation in the control of *Arabidopsis* flowering time. Curr Opin Plant Biol 7: 570–574.

Smith H. (1995). Physiological and ecological function within the phytochrome family. Annu Rev Plant Physiol Plant Mol Biol 46: 289–315.

Søgaard G, Johnsen Ø, Nilsen J, and Junttila O. (2008). Climatic control of bud burst in young seedlings of nine provenances of Norway spruce. Tree Physiol 28: 311–320.

Suarez-Lopez P, Wheatley K, Robson F, Onouchi H, Valverde F, and Coupland G. (2001). CONSTANS mediates between the circadian clock and the control of flowering in *Arabidopsis*. Nature 410: 1116–1120.

Sundström J, Carlsbecker A, Svensson ME, Svenson M, Johanson U, Theissen G, and Engström P. (1999). MADS-box genes active in developing pollen cones of Norway spruce (*Picea abies*) are homologous to the B-class floral homeotic genes in angiosperms. Dev Genet 25: 253–266.

Sundström J and Engström P. (2002). Conifer reproductive development involves B-type MADS-box genes with distinct and different activities in male organ primordia. Plant J 31: 161–169.

Takada S and Goto K. (2003). TERMINAL FLOWER2, an *Arabidopsis* homolog of HETERO-CHROMATIN PROTEIN1, counteracts the activation of *FLOWERING LOCUS T* by CONSTANS in the vascular tissues of leaves to regulate flowering time. Plant Cell 15: 2856–2865.

Tandre K, Svenson M, Svensson ME, and Engström P. (1998). Conservation of gene structure and activity in the regulation of reproductive organ development of conifers and angiosperms. Plant J 15: 615–623.

Thimm O, Blasing O, Gibon Y, Nagel A, Meyer S, Kruger P, et al. (2004) MAPMAN: a user-driven tool to display genomics data sets onto diagrams of metabolic pathways and other biological processes. Plant J 37: 914–939.

Vaartaja O. (1959). Evidence of photoperiodic ecotypes in trees. Ecol Monogr 29(2): 91–111.

Valverde F, Mouradov A, Soppe W, Ravenscroft D, Samach A, and Coupland G. (2004). Photo-receptor regulation of CONSTANS protein in photoperiodic flowering. Science 303: 1003–1006.

Vázquez-Lobo A, Carlsbecker A, Vergara-Silva F, Alvarez-Buylla ER, Piñero D, and Engström P. (2007). Characterization of the expression patterns of *LEAFY/FLORICAULA* and *NEEDLY* orthologs in female and male cones of the conifer genera *Picea*, *Podocarpus*, and *Taxus*: Implications for current evo-devo hypotheses for gymnosperms. Evol Dev 9: 446–459.

Wareing PF. (1956). Photoperiodism in woody plants. Annu Rev Plant Physiol 7: 191–214.

Webb AAR. (2003). The physiology of circadian rhythms in plants. New Phytol 160: 281–303.

Weigel D and Nilsson O. (1995). A developmental switch sufficient for flower initiation in diverse plants. Nature 377: 495–500.

Weiser C. (1970). Cold resistance and injury in woody plants. Science 169: 1269–1278.

Welling A, Kaikuranta P, and Rinne P. (1997). Photoperiodic induction of dormancy and freezing tolerance in *Betula pubescens*. Involvement of ABA and dehydrins. Physiol Plant 100: 119–125.

Welling A, Moritz T, Palva ET, and Junttila O. (2002). Independent activation of cold acclimation by low temperature and short photoperiod in hybrid aspen. Plant Physiol 129: 1633–1641.

Welling A and Palva ET. (2006). Molecular control of cold acclimation in trees. Physiol Plant 127: 167–181.

Wigge PA, Kim MC, Jaeger KE, Busch W, Schmid M, Lohmann JU, and Weigel D. (2005). Integration of spatial and temporal information during floral induction in *Arabidopsis*. Science 309: 1056–1059.

Yano M, Katayose Y, Ashikari M, Yamanouchi U, Monna L, Fuse T, Baba T, Yamamoto K, Umehara Y, Nagamura Y, and Sasaki T. (2000). *Hd1*, a major photoperiod sensitivity Quantitative Trait Locus in rice, is closely related to the *Arabidopsis* flowering-time gene *CONSTANS*. Plant Cell 12: 2473–2483.

Yanovsky MJ and Kay SA. (2002). Molecular basis of seasonal time measurement in *Arabidopsis*. Nature 419: 308–312.

Zeilinger MN, Farre EM, Taylor SR, Kay SA, and Doyle FJ, III (2006). A novel computational model of the circadian clock in Arabidopsis that incorporates PRR7 and PRR9. Mol Syst Biol 2:58.

Zhong Y, Mellerowich EJ, Llyod AD, Leinhos V, Riding RT, and Little CHA. (1995). Seasonal variation in the nuclear genome size of ray cells in the vascular cambium of *Fraxinus americana*. Physiol Plant 2: 302–311.

6

Integration of Photoperiodic Timing and Vernalization in *Arabidopsis*

Scott D. Michaels

What is the meaning of life? Myriad suggestions have been brought forth by mankind in an effort to answer this question. In a biological sense, however, one can make a strong case that the meaning, or at least the purpose, of life is to pass on one's genetic material to the next generation. Therefore, it is not surprising that organisms have evolved complex behaviors in an effort to maximize their chances of successful reproduction. Plants, of course, are no exception. One of the defining characteristics of plants is their sessile nature. Because plants remain anchored in one place from germination to eventual senescence, plants have evolved a number of pathways that allow them to take in information about their environment and respond by customizing their growth and development to maximize their chances of survival. Such environmental cues include light quality (wavelength), intensity, and duration (photoperiod), as well as temperature. Of these cues, photoperiod and temperature play the most major roles in the regulation of reproductive growth in many plant species. By monitoring annual fluctuations in day length and average temperature, plants can coordinate reproductive development with the seasons and ensure that reproduction takes place at a favorable time of year.

Because the regulation of flowering time by photoperiod is discussed in detail elsewhere in this volume (chapters 1–5), this chapter focuses on pathways that regulate flowering in response to temperature and ultimately how signals from temperature- and photoperiod-sensing pathways are integrated.

VERNALIZATION AND MERISTEM COMPETENCE

The life cycle of most plants begins with a phase of vegetative growth. During this phase the shoot apical meristem (SAM), a population of stem cells in the growing tip of the plant, produces primordia that will form the vegetative parts of the plant. At some point later in development, however, the SAM switches to produce the reproductive structures (i.e., flowers). In most plants, the timing of the transition from reproductive development to flowering is not predetermined. Rather, the induction of flowering can be accelerated, delayed, or entirely prevented depending on the day-length and temperature regimes that the plants experience. Many plants that have a day-length response for flowering can be divided into long-day plants, in which flowering is promoted by day lengths longer than some critical day length, and short day plants, in which flowering is promoted by day lengths shorter than some critical day length. Depending on the species, the proper photoperiod can be an obligate requirement for flowering (plants will not flower without inductive day lengths); alternatively, inductive photoperiods can have facultative effect (plants will flower faster under inductive day lengths, but flowering will eventually occur regardless of photoperiod). In addition to day length, the prolonged cold temperatures of winter provide a seasonal landmark for many plant species in temperate climates.

Flowering in many species of plants found in temperate climates is accelerated by a prolonged period of cold exposure, known as vernalization. Chouard (1960) defined vernalization as "acquisition or acceleration of the ability to flower by a chilling treatment" (p. 193). Thus, vernalization does not typically induce flowering in and of itself, but rather creates the capacity for subsequent flowering. Plants with an obligate vernalization requirement cannot flower in the absence of cold treatment, whereas species with facultative vernalization responses flower more rapidly after cold treatment, but will eventually flower without cold (figure 6.1). The terms biennial and winter annual are often used to describe plants with obligate and facultative vernalization responses, respectively.

FIGURE 6.1.
Sugar beet is a biennial plant and therefore cannot flower unless vernalized. This specimen has been grown in a greenhouse for 2.5 years. Because it has never been cold treated, it continues to grow vegetatively.

Typical conditions for vernalization treatment is generally 1–7°C for a duration of 1–3 months (Chouard, 1960; Lang, 1965; Bernier et al., 1981). Optimum conditions, however, are somewhat species specific. Some cereals can be vernalized at temperatures as low as –6°C (Bernier et al., 1981), whereas temperatures as high as 13°C are effective in some plants native to warmer climates, such as species of olive (Hackett and Hartmann, 1967). Cold treatments as short as 8 days can cause a significant acceleration of flowering in celery; however, more than a month of cold is required for maximal response (Thompson, 1944). Vernalization takes place only in metabolically active plants; thus, desiccated seeds cannot be vernalized. Many species can be vernalized as imbibed seeds or seedlings, but some cannot and must reach a critical age or stage in development before vernalization can occur. *Hyoscyamus niger*, for example, cannot be vernalized before 10 days after germination (Lang, 1986).

Localized cooling and grafting experiments have demonstrated that the SAM is the site of cold perception (Lang, 1965). Despite the fact that vernalization and the transition from vegetative to reproductive development both take place in the SAM, it is important to note that vernalization does not induce flowering per se. Rather, vernalization makes the SAM competent to respond to other pathways that regulate flowering (e.g., photoperiod). This point is illustrated nicely in experiments performed in *H. niger*, which has obligate requirements for both vernalization and long days before flowering can occur (Lang, 1986). In the absence of vernalization, *H. niger* will not flower under long or short days. Following vernalization, plants grown under long days flower rapidly, whereas those grown under short days still do not flower. Thus, vernalization does not induce flowering, but renders the plants competent to respond to inductive photoperiods.

PLANTS HAVE A MITOTICALLY STABLE MEMORY OF VERNALIZATION

One of the most interesting properties of vernalization is that cold-treated plants retain a relatively permanent memory of vernalization. This memory has been demonstrated in *H. niger*, which requires both vernalization and long days for flowering to occur (described above). Interestingly, short-day–grown vernalized plants will flower rapidly when transferred to long days even after >300 days of growth in short days (Lang, 1986), demonstrating that the meristem competence conferred by vernalization has been maintained. Perhaps an even more dramatic illustration of the permanent memory of vernalization comes from the biennial *Lunaria biennis*. Working with cuttings of *Lunaria*, Wellensiek (1962) showed that cuttings taken from cold-treated plants would regenerate into flowering plants, whereas cuttings from non-cold-treated plants yielded only vegetative plants after regeneration. Thus, the memory of vernalization was stable even through the regeneration of plants from tissue culture. Although cells have a mitotically stable memory of vernalization, the vernalized state is not passed on to the next generation.

Wellensiek's experiments in *Lunaria* also shed light on the site of perception of vernalization. In intact plants, the site of vernalization is the SAM (Lang, 1965). Wellensiek (1964) found, however, that cuttings taken from any portion of the plant that was mitotically active during the cold treatment would regenerate into flowering plants (even roots), suggesting that any actively dividing cells are capable of being vernalized.

THE VERNALIZATION REQUIREMENT IN *ARABIDOPSIS*

The biennial or winter annual growth habit can be thought of as consisting of two parts, the creation of a vernalization-responsive block to flowering and the subsequent removal of this block by vernalization. In *Arabidopsis*, the vernalization-responsive block to flowering is created by *FLOWERING LOCUS C* (*FLC*). *FLC* encodes a MADS-domain–containing transcription factor that acts to delay flowering (Michaels and Amasino, 1999; Sheldon et al., 1999). *FLC* is expressed most highly in the shoot and root apex and constitutive expression of *FLC* is sufficient to block flowering (Michaels and Amasino, 1999, 2000). *FLC* shows a complex pattern of regulation by both endogenous and environmental signals; the details of this regulation are discussed below.

FLC Is Repressed by the Autonomous Floral-Promotion Pathway in Rapid-Cycling *Arabidopsis*

Most laboratory strains and many naturally occurring accessions of *Arabidopsis* flower rapidly even in the absence of vernalization. Mutagenesis of these rapid-cycling strains, however, has yielded a number of mutants that show delayed flowering (Redei, 1962; Koornneef et al., 1991). Flowering time is typically expressed as the number of leaves formed by the SAM prior to flowering (i.e., the longer the plant is in the vegetative stage of development, the larger the number of leaves produced). Counting the number of leaves formed, rather than the number of days to flowering, also allows for the identification of mutants that flower late as a result of overall slower growth. Such slow-growing mutants are not typically considered to be flowering time mutants. Late-flowering mutants obtained from the mutagenesis of rapid-cycling strains can be divided into two classes based on their response to vernalization and photoperiod. *Arabidopsis* is a facultative long-day plant, meaning that it will flower more rapidly in long days but will still eventually flower under short days. One group of mutants flowers similarly under long days and short days (flowering is similar to wild type under short days, but is delayed under long days; table 6.1). The genes identified by these mutations compose the photoperiod pathway and are required the promotion of flowering in response to inductive photoperiods. A second group of mutants remains sensitive to photoperiod, but flowering is delayed under both long days and short days (table 6.1). Because the photoperiod

TABLE 6.1. Approximate leaf number at flowering of *Arabidopsis* grown under the indicated conditions.

Plant type	Short days (noninductive)	Long days (inductive)	Long days + vernalization
Rapid-cycling wild type	50	15	12
Photoperiod pathway mutant	50	50	40
Autonomous pathway mutant	>100	60	12
FRI-containing accessions	>100	60	12

response of these mutants is unaffected, these mutants are often referred to as the "autonomous" floral-promotion pathway (AP). Thus, the genes of the AP act to constitutively promote flowering under both long days and short days. Another distinction between photoperiod and AP mutants is their response to vernalization. The late-flowering phenotype of AP mutants is eliminated by vernalization (typically imbibed seeds or seedlings are cold treated at 30–40 days at 4°C prior to planting), whereas photoperiod pathway mutants show a much smaller acceleration of flowering in response to vernalization (table 6.1, figure 6.2).

Given their late-flowering vernalization-responsive phenotype, AP mutants behave as winter annuals and make an attractive model for the study of vernalization in *Arabidopsis*. The AP promotes flowering by repressing *FLC* expression (figure 6.3). Therefore, AP mutants have elevated levels of *FLC* expression and are late flowering. In addition, the late-flowering phenotype of AP mutants is eliminated by loss-of-function mutations in *FLC* (Michaels and Amasino, 2001). There are seven "classic" AP genes: *FCA, FLOWERING LOCUS D (FLD), FLOWERING LOCUS K (FLK), FPA, FVE, FY,* and *LUMINIDEPENDENS (LD)*. Although the many of the molecular details of how these genes interact to repress *FLC* expression are still unknown, an intriguing picture is beginning to emerge that links RNA metabolism with chromatin structure and transcription.

Genes of the AP

Of the seven AP genes, it is interesting to note that, based on their sequences, four are predicted to have RNA-related activities. FCA, FPA, and FLK all contain RNA-binding motifs. FLK contains three K homology (KH) domains (Lim et al., 2004; Mockler et al., 2004), whereas FCA and FPA contain two and three RRM-type RNA-binding domains, respectively (Macknight et al., 1997; Schomburg et al., 2001). Although FCA and FPA both contain RRM domains, they do not otherwise share significant sequence similarity. FY is related to a group of eukaryotic proteins that play a role in 3'-end processing of RNA transcripts (Simpson et al., 2003). The fact that mutations in any of these four genes leads to elevated *FLC* steady-state RNA levels suggests the possibility that FCA, FLK, FPA, and FY may regulate *FLC* posttranscriptionally. Although no direct evidence has been reported

FIGURE 6.2.
The vernalization response in *Arabidopsis*. (A) Rapid-cycling accessions flower early in the absence of vernalization. AP mutants (B and D) or FRI-containing strains (C and E) are late flowering in the absence of vernalization (B and C), but flower rapidly if cold treated for 40 days as young seedlings (D and E).

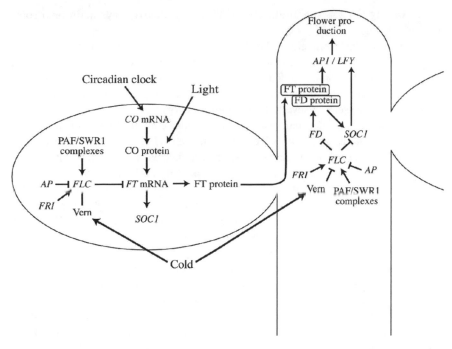

FIGURE 6.3.
A model for the integration of flowering signals from the vernalization and photoperiod pathways.

that indicates that *FLC* is posttranscriptionally regulated, evidence does exist for posttranscriptional regulation within the AP.

Four alternatively spliced forms of the *FCA* transcript have been identified (alpha, beta, delta, and gamma) (Macknight et al., 1997). Of these splice forms, however, only the gamma form appears to encode a functional protein; when each of these splice forms is expressed in an *fca* mutant, only the gamma form is able to rescue the mutant phenotype (Macknight et al., 2002). Normally, the gamma transcript comprises approximately 35% of the *FCA* transcript pool (Macknight et al., 1997). Interestingly, FCA is able to negatively regulate the splicing of its own transcript (Quesada et al., 2003). Overexpression of FCA leads to reduced production of the gamma transcript and increased production of the inactive beta splice form. The effect of FCA on transcript splicing and flowering time is dependent on a physical interaction with FY (Simpson et al., 2003). The WW protein interaction domain of FCA interacts with the PPLP domain of FY to promote formation of the beta splice form. Thus, FCA and FY physically interact to regulate the splicing of the *FCA* transcript. It is interesting to note that, although loss-of-function mutations in most AP genes result only in late flowering, null alleles of *FY* are embryo lethal and hypomorphic alleles are late flowering (Henderson et al., 2005). Thus, *FY* clearly plays additional roles in development.

The remaining members of the AP are likely to play roles in regulating transcription and/or chromatin structure. LD possesses several motifs that suggest that it acts as a transcription factor. The amino-terminal portion of LD contains a divergent homeodomain and the carboxy terminus contains a glutamine-rich region that is similar to those seen in several transcription factors (Lee et al., 1994; Aukerman and Amasino, 1996; Aukerman et al., 1999). LD also contains two bipartite nuclear localization domains and is localized to the nucleus (Lee et al., 1994; Aukerman et al., 1999). Although its domains structure and nuclear localization suggest that LD acts as a transcription factor, sites of LD binding in the genome have not yet been identified. In addition, it should be noted that there are examples of homeodomain-containing proteins that bind RNA (Dubnau and Struhl, 1996; Rivera-Pomar et al., 1996). Thus, one cannot rule out the possibility that, like FCA, FLK, and FPA, LD may also function as an RNA-binding protein.

FLD and *FVE* encode proteins that are predicted to act as part of a chromatin-remodeling complex. FLD is homologous to the human protein *LYSINE SPECIFIC DEMETHYLASE* (*LSD1*), which has been shown to act as a histone demethylase (Shi et al., 2004). Whereas, FVE contains six WD repeats and is similar to mammalian retinoblastoma-associated proteins that are present in complexes involved in chromatin assembly and modification. In addition, homologs of both FLD and FVE (KIAA0601 and RbAp48/46, respectively) have been shown to be associated with a histone deacetylase (HDAC) complex in humans (He et al., 2003; Ausin et al., 2004). Histone hypoacetylation is associated with transcriptional repression in many species (Iizuka and Smith, 2003). The observation that *FLC* is repressed by FLD and FVE (i.e., *FLC* expression is elevated in *fld* and *fve* mutants) suggests that FLD and FVE may play a direct role in *FLC* repression, and indeed, histones at the *FLC* locus are hyperacetylated in *fld* and *fve* mutants (He et al., 2003). It is interesting to note, however, that although *FLC* expression is up-regulated in all AP mutants, histone acetylation at *FLC* is not increased in all mutants. In contrast to *fld* and *fve*, mutations in *fca, fpa,* and *ld* increase *FLC* expression without increased histone acetylation (He et al., 2003). This suggests that there are likely to be multiple mechanisms by which the AP acts to repress *FLC* expression.

Creation of the Vernalization Requirement in Naturally Occurring *Arabidopsis*

In contrast to rapid-cycling laboratory strains, many, if not most, naturally occurring accessions of *Arabidopsis* are late flowering and vernalization responsive. These accessions, therefore, behave as winter annuals, and their flowering behavior is virtually indistinguishable from AP mutants. In most cases, crosses between winter-annual and rapid-cycling accessions reveal that the winter-annual phenotype is conferred by a single dominant gene, *FRIGIDA* (*FRI*) (Napp-Zinn, 1979;

Burn et al., 1993; Lee et al., 1993; Clarke and Dean, 1994). *FRI* encodes coiled-coil-domain–containing gene of unknown biochemical function (Johanson et al., 2000). Although the biochemical function of *FRI* is unclear, functional alleles of *FRI* lead to increased expression of *FLC* (Michaels and Amasino, 1999; Sheldon et al., 1999). Thus, the fact that both AP mutations and *FRI* result in increased *FLC* expression provides an explanation for their similar late-flowering phenotypes (figure 6.2). It is important to note, however, that *FRI* and the AP have opposing functions in the regulation of *FLC*. Functional alleles of *FRI* up-regulate the expression of *FLC*, whereas the normal function of the AP is to repress *FLC* (figure 6.3). In rapid-cycling accessions, which lack *FRI*, the AP acts to repress *FLC* and promote flowering. If functional *FRI* alleles are introduced in to a rapid-cycling background, however, *FRI* acts to up regulate *FLC* and confer a late-flowering phenotype. Thus, *FRI* is epistatic to the AP.

The late-flowering habit of naturally occurring winter annuals requires the synergistic interaction of both *FRI* and *FLC*. Mutagenesis of winter annuals has shown that loss-of-function mutations in either gene causes rapid flowering (Michaels and Amasino, 1999; Johanson et al., 2000); in the absence of *FRI*, *FLC* is expressed at very low levels, and in the absence of *FLC*, *FRI* has no effect on flowering time. Sequence analysis of the *FRI* and *FLC* genes from many winter-annual and rapid-cycling accessions has revealed that winter annuals contain functional alleles of both genes, whereas rapid-cycling accessions contain mutations in either *FRI* or *FLC* (Johanson et al., 2000; Gazzani et al., 2003; Shindo et al., 2005). This suggests that the ancestral form of *Arabidopsis* was likely a winter annual and that over the course of evolution *FRI* and *FLC* activity has been attenuated, leading to a rapid-cycling flowering habit. The vast majority of rapid-cycling accessions contain loss-of-function mutations in *FRI* (Johanson et al., 2000; Shindo et al., 2005). Interestingly, greater than 20 different *FRI* mutations have been observed in rapid-cycling accessions, suggesting that *FRI* activity has been lost many times over the course of evolution (Johanson et al., 2000; Gazzani et al., 2003; Shindo et al., 2005).

Although most rapid-cycling accessions contain loss-of-function mutations in *FRI*, some have been identified that contain mutations in *FLC*. In contrast to naturally occurring *FRI* mutants, the mutations that have been identified in *FLC* do not appear to be null mutations (Gazzani et al., 2003; Michaels et al., 2003). The accessions Shakhdara and Da (1)-12, for example, contain functional alleles of *FRI* but flower early due to atypical weak *FLC* alleles (transformation of these accessions with a functional copy of *FLC* results in late flowering) (Michaels et al., 2003). In addition, the laboratory strain Landsberg *erecta* (L*er*) contains both a null allele of *FRI* and a weak allele of *FLC*. The *FLC* alleles in these accessions do not contain missense or nonsense mutations but are expressed at much lower levels than strong alleles. Thus, these *FLC* alleles appear to act as weak alleles due to changes in regulation rather than mutations that would alter the FLC protein. In the case of L*er* and Da (1)-12, the reduction in *FLC* expression appears to be due to the insertions in the first intron of *FLC* (Michaels et al., 2003), which is normally

3.5 kb in length. The L*er* and Da (1)-12 *FLC* alleles contain additional 1.2-kb and 4.2-kb insertions, respectively. Although the sites of integration are different and the inserted sequences show no homology, both insertions appear to be due to ret-rotransposons. Shakhdara, in contrast, does not contain any significant insertions. The observation that rapid-cycling accessions that have lost *FRI* function typically contain null mutations, whereas accessions that are early flowering due to *FLC* contain weak alleles, suggests that the role of *FRI* may be more dispensable than that of *FLC*. This notion is supported by evidence that *FLC* may play roles outside of flowering, such as in regulation of the circadian clock (Salathia et al., 2006).

Genes Required for High Levels of *FLC* Expression

Mutant screens in rapid-cycling backgrounds have been quite successful in iden-tifying repressors of *FLC*. *FLC* is normally expressed at very low levels in these early-flowering strains and mutations in *FLC* repressors (i.e., the AP) lead to up-regulation of *FLC* and delayed flowering. Because *FLC* levels are already quite low, mutations in positive regulators (which would cause a further reduction in *FLC* expression) would be predicted to have little effect on flowering time. Indeed, loss-of-function mutations in *FLC* have little effect on flowering time in rapid-cycling backgrounds (Michaels and Amasino, 1999). In order to identify positive regulators of *FLC*, a number of laboratories have conducted genetic screens in *FRI*-containing winter-annual backgrounds. Because these lines are late flowering due to high lev-els of *FLC* expression, mutations in positive regulators of *FLC* result in reduced *FLC* expression and an early-flowering phenotype. These screens have been highly successful and have resulted in the identification of a large number of genes that are required for high levels of *FLC* expression (table 6.2). Although much work remains in determining the function of all of these genes in the regulation of *FLC*, it is possible to place many of these genes into groups based on their phenotypes and predicted functions.

Although *FRI* and AP mutations both lead to elevated levels of *FLC* and late flowering, they are likely to do so through pathways that are at least par-tially independent. This model is supported by the observation that some muta-tions that suppress *FLC* expression in a *FRI*-containing background have little or no affect on *FLC* expression in an AP-mutant background. Thus, these positive regulators are required for the up-regulation of *FLC* by *FRI*, but not for the up-regulation of *FLC* by AP mutations. Interestingly, these *FRI*-specific regulators of *FLC* include two *FRI*-related genes, *FRI-LIKE 1* (*FRL1*) and *FRI-LIKE 2* (*FRL2*) (Michaels et al., 2004). The *Arabidopsis* genome contains six genes with significant similarity to *FRI*. FRL1 and FRL2 are approximately 54% identical to each other and 25% identical to FRI. Although FRI and FRL1 are related, it should be noted that FRI contains additional N-terminal and C-terminal domains, and there are no large blocks of primary sequence identity that might suggest a shared functional domain between FRI and FRL1/FRL2. Despite sequence

TABLE 6.2. Genes affecting the expression of *FLC*.

Gene class/ name	Gene number	Domains	Probable function	Effect on FLC
FLC and related genes				
FLC	*At5g10140*	MADS	Transcription factor	—
FLM/MAF1	*At1g77080*	MADS	Transcription factor	None reported
MAF2	*At5g65050*	MADS	Transcription factor	None reported
MAF3	*At5g65060*	MADS	Transcription factor	None reported
MAF4	*At5g65070*	MADS	Transcription factor	None reported
MAF5	*At5g65080*	MADS	Transcription factor	None reported
Autonomous pathway genes				
FCA	*At4g16280*	RRM domains	RNA binding	Repress
FLD	*At3g10390*	Histone demethylase	Histone demethylase	Repress
FLK	*At3g04610*	KH domains	RNA binding	Repress
FPA	*At2g43410*	RRM domains	RNA binding	Repress
FVE	*At2g19520*	WD-40 repeat	Chromatin remodeling complex	Repress
FY	*At5g13480*	Polyadenylation factor	RNA processing	Repress
LD	*At4g02560*	Homeodomain	Transcription factor	Repress
REF6	*At3g48430*	Jumonji, zinc finger	Histone demethylase	Repress
ELF6	*At5g04240*	Jumonji, zinc finger	Histone demethylase	Repress
FRI-specific FLC activators				
FRI	*At4g00650*	Coiled coil	Unknown	Activate
FRL1	*At5g16320*	Coiled coil	Unknown	Activate
FRL2	*At1g31814*	Coiled coil	Unknown	Activate
FES1	*At2g33835*	CCCH zinc finger	Nuclear protein	Activate
SUF4	*At1g30970*	BED, C_2H_2 zinc fingers	Transcription factor	Activate
FLX	*At2g30120*	Leucine zipper	Unknown	Activate
PAF-like complex				
EFS	*At1g77300*	SET domain	Histone methyltransferase	Activate
ELF7	*At1g79730*	Yeast PAF1 homolog	PAF1-like complex	Activate
ELF8/VIP6	*At2g06210*	Yeast ScCTR9 homolog	PAF1-like complex	Activate
VIP4	*At5g61150*	Yeast ScLEO1 homolog	PAF1-like complex	Activate
VIP5	*At1g61040*	Yeast Rtf1 homolog	PAF1-like complex	Activate

(*Continued*)

TABLE 6.2. *(Continued)*

Gene class/ name	Gene number	Domains	Probable function	Effect on FLC
SWR1-like complex				
SEF/SWC6	At5g37055	HIT-Zn finger	Chromatin remodeling complex	Activate
SUF3/ARP6/ ESD1	At3g33520	Actin-related protein 6	Chromatin remodeling complex	Activate
PIE1	At3g12810	ISWI family	Chromatin remodeling complex	Activate
Other FLC activators				
ABH	At2g13540	Cap binding protein 80	mRNA cap-binding complex	Activate
ELF5	At5g62640	Proline-rich domains	Nuclear protein	Activate
HUA2	At5g23150	RPR domain	Pre-mRNA processing	Activate
VIP3	At4g29830	WD motifs	Unknown	Activate
Vernalization- associated genes				
VRN1	At3g18990	B3 DNA binding	Chromatin remodeling	Repress
VRN2	At4g16845	Zinc finger	Chromatin remodeling	Repress
VIN3	At5g57380	PHD	Chromatin remodeling	Repress
VIL1/VRN5	At3g24440	PHD	Chromatin remodeling	Repress
LHP1	At5g17690	chromodomain	Chromatin remodeling	Repress

similarity, genetic evidence suggests that FRL1 has a biochemical activity that is distinct from FRI; overexpression of *FRL1* cannot compensate for a loss of *FRI* activity, and vice versa (Michaels et al., 2004). Although the majority of natural variation in flowering time can be explained by mutations at *FRI* and *FLC*, there is evidence that variation at *FRL1* and *FRL2* exists, as well. In the Columbia background, *frl1* mutants strongly suppress the late-flowering pheno-type of *FRI*, whereas *frl2* mutations have a relatively minor effect. This suggests that, in Columbia, *FRL1* plays the primary role in *FLC* activation. In another genetic background, however, the roles appear to be reversed. The *Ler* allele of

FRL1 contains a naturally occurring nonsense mutation and functional evidence suggests that *FRL2* is responsible for the up-regulation of *FLC* by *FRI* in this accession (Schlappi, 2006).

In addition to *FRL1* and *FRL2*, three other genes are specifically required for the up-regulation of *FLC* by *FRI*. *FLC EXPRESSOR X* (*FLX*) (Andersson et al., 2007) encodes a protein of unknown function and *FRI-ESSENTIAL1* (*FES1*) (Schmitz et al., 2005) and *SUPPRESSOR OF FRI4* (*SUF4*) (Kim et al., 2006; Kim and Michaels, 2006) encode proteins with one CCCH-type and two C_2H_2-type zinc fingers, respectively. Like *FRL1* and *FRL2*, mutations in these genes eliminate the up-regulation of *FLC* by *FRI* but have little effect on *FLC* expression in AP-mutant backgrounds. In all cases where it has been examined, overexpression of one of these genes cannot up-regulate *FLC* if one of the others is mutant. This suggests that these genes do not act in a linear pathway but may function as a complex. In support of this model SUF4 has been shown to physically interact with FRI and FRL1 in yeast and when transiently expressed in tobacco (Kim et al., 2006). Furthermore, SUF4 has been shown to localize to the *FLC* promoter region, suggesting that SUF4 and associated proteins may act to regulate *FLC* directly (Kim et al., 2006).

As described above, *FRL1, FRL2, FLX, FES1*, and *SUF4* appear to function specifically in the up-regulation of *FLC* by *FRI*. A second group of genes have been identified, however, that appear to function more broadly (table 6.2). Mutations in these genes suppress *FLC* expression in *FRI* or AP-mutant backgrounds; furthermore this group of mutants also displays various non-flowering-time phenotypes. Thus, the function of these genes includes, but is not limited to, flowering time. The molecular mechanism by which all of these genes affect *FLC* expression has not been elucidated. Many of the genes, however, appear to be homologs of genes involved in two chromatin-remodeling complexes in yeast.

In yeast, the RNA polymerase II-associated factor 1 (PAF1) complex promotes gene expression in part by recruiting a histone H3 lysine 4 (H3K4) methyltransferase–containing complex to target-gene chromatin (Krogan et al., 2003; Ng et al., 2003). Increased levels of histone H3K4 trimethylation is often associated with actively transcribed genes (Krogan et al., 2003; Ng et al., 2003). The PAF1 complex in yeast is a five-member complex consisting of PAF1, CTR9, LEO1, RTF1, and CDC73 (Krogan et al., 2002; Squazzo et al., 2002). Mutations in the *Arabidopsis* homologs of PAF1 (*EARLY FLOWERING 7, ELF7*), CTR9 (*EARLY FLOWERING 8, ELF8*), LEO1 (*VERNALIZATION INDEPENDENCE 4, VIP4*), and RTF1 (*VERNALIZATION INDEPENDENCE 5, VIP5*) lead to reduced H3K4 trimethylation in *FLC* chromatin and reduced *FLC* expression (Zhang and van Nocker, 2002; He et al., 2004; Oh et al., 2004). The PAF-like complex in *Arabidopsis* may recruit the histone methyltransferase *EARLY FLOWERING IN SHORT DAYS* (*EFS*) to the *FLC* locus. Mutations in *efs* have been shown to reduce levels of H3K4 and H3K36 methylation in *FLC* chromatin (Soppe et al., 1999; Kim et al., 2005; Zhao et al., 2005).

The SWR1 complex of yeast also plays an important role in regulation of gene expression through changes in chromatin structure. The SWR1 complex is responsible for catalyzing the replacement of histone 2A for the histone variant histone 2A.Z, which is required for full activation of some genes in yeast (Raisner and Madhani, 2006). Putative *Arabidopsis* orthologs of SWR1 complex components include *EARLY IN SHORT DAYS 1 (ESD1)/SUPPRESSOR OF FRI 3 (SUF3)/ACTIN RELATED PROTEIN 6 (ARP6)* (Choi et al., 2005; Deal et al., 2005; Martin-Trillo et al., 2006), *PHOTOPERIOD INDEPENDENCE 1 (PIE1)* (Noh and Amasino, 2003), and *SERRATED AND EARLY FLOWERING (SEF)/AtSWC6* (March-Diaz et al., 2007; Lazaro et al., 2008). In support of the model that these proteins form a SWR1-like complex in *Arabidopsis*, PIE1 and AtSWC6 have been shown to interact with H2A.Z, and H2A.Z has been identified in *FLC* chromatin (Deal et al., 2005; Choi et al., 2007). Furthermore, in *arp6* and *pie1* mutants, H2A.Z is lost from *FLC* chromatin and *FLC* expression is suppressed (Deal et al., 2005).

VERNALIZATION IN *ARABIDOPSIS*

Vernalization Is Genetically Separable from Other Cold Responses

In *Arabidopsis*, as with most vernalization-responsive species, vernalization requires long periods of cold exposure. Winter-annual or AP-mutant strains show some acceleration of flowering with cold treatments as short as five to seven days; maximum promotion of flowering, however, requires approximately 40 days. This is not to say that plants cannot react quickly to cold temperature. Processes such as stratification (Wareing, 1971) (the breaking of seed dormancy by cold) and cold acclimation (Thomashow, 1999) (which helps to protect against frost damage) occur after hours or several days of cold exposure. In addition to the differences in response kinetics, vernalization and cold acclimation appear to involve distinct molecular mechanisms. During cold acclimation the transcription factor *C-REPEAT/DRE-BINDING FACTOR1 (CBF1)* activates *COLD-REGULATED (COR)* genes (Thomashow, 1999). Constitutive expression of *CBF1* is sufficient to induce the expression of *COR* genes in the absence of cold exposure; however, *CBF1* overexpression has no effect on vernalization (Liu et al., 2002).

Vernalization Results in an Epigenetic Repression of *FLC*

As described in the preceding section, *FLC* expression is key to the creation of the vernalization requirement in *Arabidopsis* (i.e., blocking flowering prior to vernalization). Research has also shown that the promotion of flowering by vernalization is also accomplished through *FLC*. Vernalization promotes flowering in FRI-containing or AP-mutant backgrounds by inducing a mitotically stable repression of *FLC* expression; *FLC* expression is repressed during cold treatment and remains repressed even following transfer to warm conditions (Michaels and Amasino, 1999; Sheldon et al., 1999). This epigenetic repression of *FLC* provides a molecular basis for earlier results that have demonstrated that many vernalized plants have a mitotically stable memory of cold exposure. *FLC* levels are reset to high levels in the next generation, however; thus, each generation of plants must be vernalized (Michaels and Amasino, 1999; Sheldon et al., 1999, 2008). Therefore, winter-annual seedlings that germinate in summer or fall will have high levels of *FLC* that will act to block flowering prior to winter. Over winter, vernalization causes a stable repression of *FLC* that will allow plants to flower the following spring.

Vernalization Results in Repressive Modifications in *FLC* Chromatin

Posttranslational modifications of histones play key roles in eukaryotic gene regulation and are often associated with epigenetic changes in gene expression. Prior to vernalization, *FLC* is highly expressed and *FLC* chromatin carries histone modifications that are hallmarks of actively transcribed genes, such as acetylation of H3K9 and H3K14 and methylation of H3K4 (Margueron et al., 2005). During vernalization, the levels of these activating modifications are reduced and are replaced by repressive histone modifications, in particular H3K9 and H3K27 methylation (Bastow et al., 2004; Sung and Amasino, 2004; Mylne et al., 2006; Sung et al., 2006a). In other eukaryotic systems H3K9 and H3K27 methylation leads to a mitotically stable suppression of gene expression through the recruitment of additional repressor complexes (Margueron et al., 2005).

Although many of the molecular details of the chromatin remodeling and stable repression of *FLC* in response to vernalization have yet to be determined, forward and reverse genetics has identified several important components. Mutations in *VERNALIZATION 1* (*VRN1*), *VRN2*, *VERNALIZATION INSENSITIVE 3* (*VIN3*), and *LIKE HETEROCHROMATIN 1* (*LHP1*) all result in a reduced vernalization response. An analysis of *FLC* expression in these mutants has shown than the stable repression of *FLC* by vernalization requires both an initial repression of *FLC* during the cold treatment and maintenance of repression after return to warm temperatures (figure 6.3). In *vin3* mutants, *FLC* expression is not reduced during cold treatment, whereas *FLC* is repressed normally during cold treatment in *vrn1*, *vrn2*, and *lhp1* mutants, but this repression is not maintained in subsequent warm temperatures (figure 6.4) (Gendall et al., 2001; Levy et al., 2002; Sung and Amasino, 2004). VIN3 contains a plant homeodomain (PHD) domain; other PHD domains have been shown to bind to histones with specific posttranslational modifications (Sung and Amasino, 2004; Soliman and Riabowol, 2007). *VIN3* itself is regulated by vernalization, and its transcription is maximally induced only after several weeks of cold treatment (Sung and Amasino, 2004). This finding is significant in that it shows that the mechanism by which plants sense the duration of cold exposure lies

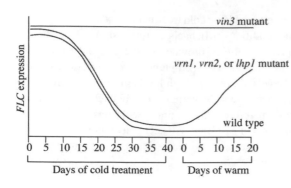

FIGURE 6.4.
Expression of *FLC* during cold treatment and after recovery in warm temperatures in *FRI*-containing wild type and vernalization mutants.

upstream of *VIN3*. Consistent with the result that *FLC* levels are not suppressed by vernalization in *vin3* mutants, histone acetylation does not decrease and methylation at H3K9 and H3K27 does not accumulate in *vin3* mutants following vernalization. Interestingly, VIN3 has been shown to interact with a VIN3-related protein, VIN3-LIKE1 (VIL1)/VERNALIZATION5 (VRN5), in yeast (Sung et al., 2006b). Consistent with this interaction, *vil1* mutants show a similar vernalization-insensitive phenotype to *vin3* (Sung et al., 2006b; Greb et al., 2007). Thus, it appears that VIN3-related proteins are required for the vernalization response.

VRN2 encodes a homolog of suppressor of Zeste 12 and has been shown to be present in a polycomb repression complex 2-like complex that includes VIN3, FERTILIZATION INDEPENDENT ENDOSPERM, CURLY LEAF, and SWINGER (Gendall et al., 2001; Wood et al., 2006). In *vrn2* mutants, histone acetylation at *FLC* drops during the cold treatment but is restored to high levels after return to warm temperatures (figure 6.4). Also, the repressive H3K9 and H3K27 marks fail to accumulate (Bastow et al., 2004; Sung and Amasino, 2004). Like *vrn2*, *vrn1* mutants do not accumulate H3K9 methylation in response to vernalization; however, the molecular phenotype of *vrn1* mutant is distinct from that of *vrn2* mutants in that increased H3K27 methylation does occur (Bastow et al., 2004; Sung and Amasino, 2004). This result indicates that both H3K27 and H3K9 methylation are required for *FLC* suppression. In animals, the *LHP1* homolog *HETEROCHROMATIN 1* is physically associated with heterochromatic regions of the genome. LHP1 association with *FLC* chromatin increases with vernalization and thus may be directly involved in the stable repression of *FLC* following vernalization (Mylne et al., 2006; Sung et al., 2006a). Interestingly, H3K9 methylation accumulates at *FLC* during cold treatment in *lhp1* mutants but is not maintained in warm temperatures. This indicates that *LHP1* is required for maintenance, but not initiation, of H3K9 methylation.

FLC Is the Primary, but Not Only, Target of Vernalization

FLC is the major determinate of the winter annual habit. Consistent with this model, the late-flowering phenotype of *FRI* or AP mutants is also eliminated by loss-of-function mutations in *FLC*. If vernalization promotes flowering exclusively through the repression of *FLC*, one would predict that *flc* null mutants would be completely insensitive to vernalization. However, this is not the case. If *flc* mutants are grown under noninductive short days, vernalized plants flower significantly earlier than nonvernalized plants (Michaels and Amasino, 2001). Several *FLC*-related genes have been implicated in *FLC*-independent vernalization responses. *Arabidopsis* contains >100 MADS-domain-containing transcription factors (Parenicova et al., 2003), many of which have been shown to play a role in the regulation of flowering time. *FLC* belongs to a clade of MADS genes with five other members, *FLOWERING LOCUS M (FLM)/MADS AFFECTING FLOWERING 1 (MAF1)*, *MAF2*, *MAF3*, *MAF4*, and *MAF5* (Ratcliffe et al., 2001, 2003, Scortecci et al., 2001, 2003). Several of these genes have been shown to affect flowering time and are regulated by vernalization and mutations that suppress *FLC* expression (Ratcliffe et al., 2001, 2003; Scortecci et al., 2001, 2003; He et al., 2004; Oh et al., 2004; Deal et al., 2005; Doyle et al., 2005; Kim et al., 2005; Martin-Trillo et al., 2006; Sung et al., 2006b; Andersson et al., 2007; Lazaro et al., 2008). Detailed genetic analysis of these *FLC*-related genes is complicated by the fact that *MAF2*, *MAF3*, *MAF4*, and *MAF5* are arranged in tandem on chromosome V. In addition to the genes of the *FLC* clade, another MADS-domain-containing transcription factor, *AGAMOUS-LIKE19 (AGL19)*, has also been implicated in the *FLC*-independent vernalization response. Unlike *FLC*, *AGL19* acts as a floral promoter and its expression is up-regulated by vernalization. Interestingly, like the repression of *FLC* by vernalization, the activation of *AGL19* also requires the activity of *VRN2* and *VIN3* (Schonrock et al., 2006). This suggests that *FLC*-dependent and *FLC*-independent vernalization responses may both be controlled by the same upstream components.

COMMON TARGETS OF *FLC* AND THE PHOTOPERIOD PATHWAY

The integration of signals from the vernalization and photoperiod pathways occurs primary through the antagonistic regulation of a group of floral promoters referred to as floral integrators. The expression of the floral integrators is repressed by *FLC* and promoted by the photoperiod pathway through the floral promoter *CONSTANS (CO)*. Because the photoperiod pathway is discussed in greater detail in chapter 1, the regulation of *CO* by photoperiod is only briefly summarized here.

CO Promotes Flowering under Inductive Long-Day Conditions

Arabidopsis is a facultative long-day plant; thus, plants flower more rapidly under long days than short days. Plants containing mutations in *CO* flower normally under short days, but are late under long days (Redei, 1962; Koornneef et al., 1991).

Therefore, *CO* is required for the promotion of flowering in response to inductive photoperiods. The long-day–specific promotion of flowering by CO is brought about through the regulation of both *CO* mRNA and protein levels. *CO* mRNA levels are regulated by the circadian clock and peak expression occurs late in the day in long days, but after dusk in short days (Suarez-Lopez et al., 2001; Yanovsky and Kay, 2002). CO protein is in turn stabilized by light and degraded in darkness (Valverde et al., 2004). Thus, under long days *CO* mRNA levels are high during the light, allowing CO protein to be stabilized; under short days, however, *CO* transcript is high only during the dark, where CO is degraded. As a result CO protein only accumulates under long days.

The Floral Integrators

The genes *FT*, *TWIN SISTER OF FT* (*TSF*), and *SUPPRESSOR OF OVER-EXPRESSION OF CO 1* (*SOC1*)/*AGAMOUS LIKE 20* (*AGL20*) are considered to act as floral integrators because their expression is regulated by both vernalization (through *FLC*) and photoperiod (through *CO*). These three genes all act as strong promoters of flowering (Kardailsky et al., 1999; Kobayashi et al., 1999; Borner et al., 2000; Lee et al., 2000; Samach et al., 2000). Because the downstream targets of the floral integrators are genes critical to flower formation, such as *LEAFY* (*LFY*) and *AGAMOUS* (*AG*), it appears that these genes act at or near the end of the pathways that regulate the timing of flowering (Kobayashi and Weigel, 2007).

    FT encodes a small protein with limited similarity to Raf kinase inhibitor proteins from animals (Kardailsky et al., 1999; Kobayashi et al., 1999). The significance of this homology is unclear; however, genetic and biochemical analysis of FT indicates that it is an essential cofactor for the bZIP transcription factor FD (Abe et al., 2005; Wigge et al., 2005). The FT-FD complex is proposed to activate the expression of floral meristem identity genes and *SOC1* (Abe et al., 2005; Michaels et al., 2005; Wigge et al., 2005; Yamaguchi et al., 2005). FT is a member of a small group of related genes in *Arabidopsis*. TSF is approximately 82% identical to FT, and, similar to FT, overexpression of TSF strongly accelerates flowering (Kobayashi et al., 1999; Michaels et al., 2005; Yamaguchi et al., 2005). Although their similar sequence and overexpression phenotypes suggest that FT and TSF have similar biochemical functions, genetic experiments suggest that FT plays a larger role in the regulation of flowering time. *ft* mutants have a strong late-flowering phenotype, whereas *tsf* mutants show little change in flowering time (Michaels et al., 2005; Yamaguchi et al., 2005). Given the dominant role of *FT* in the regulation of flowering time, more effort has been placed on its characterization, and the role of *TSF* has been somewhat less well characterized. SOC1 is a MADS-domain-containing transcription factor that, like FT and TSF, acts as a strong promoter of flowering. *soc1* mutants are late flowering and overexpression of *SOC1* causes extreme early flowering (Borner et al., 2000; Lee et al., 2000; Samach et al., 2000). Mutations in *FT* lead to reduced expression of *SOC1*, indicating that *SOC1* is genetically downstream of *FT* (Searle et al., 2006). Consistent with this result, overexpression of *FT* also leads to increased *SOC1* expression (Michaels et al., 2005; Yamaguchi et al., 2005; Yoo et al., 2005).

Regulation of the Floral Integrators by *CO* and *FLC*

In nonvernalized winter annuals or AP mutants, *FLC* is highly expressed and acts to repress the expression of *FT*, *SOC1*, and possibly *TSF* (Borner et al., 2000; Lee et al., 2000; Samach et al., 2000; Hepworth et al., 2002; Michaels et al., 2005). In addition to genetic and transgenic experiments that have shown the negative regulation of *FT* and *SOC1* by *FLC*, several papers have provided evidence that this repression is direct. *SOC1* contains a putative CArG box approximately 400 bp from the transcriptional start site. CArG boxes have been shown to be the binding sites of MADS-domain transcription factors (Shore and Sharrocks, 1995). Promoter deletion analysis has shown that this region is important for the suppression of *SOC1* expression by *FLC* (Hepworth et al., 2002). Furthermore, FLC has been shown to bind to this region of the *SOC1* promoter in vitro and in vivo (Hepworth et al., 2002; Helliwell et al., 2006). *FT* contains two putative CArG boxes, one in the promoter and one in the first intron (Helliwell et al., 2006). Interestingly, chromatin immunoprecipitation experiments detected FLC binding only in intron I. This suggests that, like *SOC1*, *FT* is directly regulated by FLC.

Unlike FLC, CO has not been demonstrated to bind DNA directly. Experiments using an inducible form of CO protein in conjunction with cyclohexamide, however, indicate that the up-regulation of *FT* and *SOC1* by CO does not require the synthesis of additional proteins and therefore may be direct (Samach et al., 2000). A current model for the biochemical activity of CO is that it acts as part of a heme activator protein (HAP)-like complex (Ben-Naim et al., 2006; Wenkel et al., 2006; Cai et al., 2007). In yeast, the HAP complex is composed of HAP2, HAP3, and HAP5 subunits and binds to CCAAT boxes, which are present in the promoters of approximately 25% of eukaryotic promoters (Gelinas et al., 1985). CO contains domains that exhibit similarity to HAP2, suggesting that CO might replace HAP2 in a HAP complex. This model is supported by the findings that CO interacts with HAP3 and HAP5 in yeast and in plants, and that alteration of HAP expression levels does affect flowering time (Ben-Naim et al., 2006; Wenkel et al., 2006; Cai et al., 2007). A complicating factor in the investigation of the role of the CO/HAP3/HAP5 complex is that each of the HAP subunits is encoded by a family of 10–13 genes in *Arabidopsis* (Gusmaroli et al., 2002).

Spatial Aspects of the Integration of Vernalization and Photoperiod

A large body of work from a variety of species has shown that the sites of photoperiod and cold perception are physically separated (Zeevaart, 1962; Lang, 1965).

Experiments involving grafting and localized application of cold or photoperiod have demonstrated that photoperiod is perceived primarily in the leaves, whereas vernalization occurs in the shoot apex. The observation that the sites of photoperiod perception and floral initiation (at the apex) are separate led to the hypothesis that a floral-promoting substance (florigen) must be produced in the leaves and translocated to the SAM in response to inductive photoperiods (Chailakhyan, 1936). Thus, to understand how flowering time is regulated, it is necessary to examine the events that take place in the leaves and the shoot apex and ultimately to determine how signals from both locations are integrated into the flowering decision.

A key component in the promotion of flowering by long days in *Arabidopsis* is the up-regulation of *FT* by *CO*. Consistent with this regulation being direct, both genes are expressed most strongly in the phloem of leaves (Takada and Goto, 2003; An et al., 2004). The expression of these genes in the vasculature suggested that they might be involved in the production of a mobile signal that would be transported to the SAM via the phloem. Grafting and experiments using tissue-specific promoters showed that *CO* was effective in promoting flowering when expressed in the phloem but not when expressed in the SAM (An et al., 2004; Ayre and Turgeon, 2004). This result suggests that CO itself is not the mobile signal, and indeed, in grafting experiments using CO::GFP fusions, CO movement was not detected. In contrast, *FT* expression from either phloem- or SAM-specific promoters is effective in promoting flowering, indicating that FT can directly promote flowering in the SAM (An et al., 2004). In support of the model that FT acts in the meristem, FT has been shown to physically interact with FD, which is expressed in the SAM (Abe et al., 2005). Recently, a number of papers have explored the role of FT as a mobile flowering signal in *Arabidopsis*, rice, and tomato (Corbesier et al., 2007; Jaeger and Wigge, 2007; Lin et al., 2007; Mathieu et al., 2007; Tamaki et al., 2007). Despite a lack of direct evidence (e.g., detection of the movement of endogenous FT protein), a strong case can now be made that FT protein is a mobile flowering signal (i.e., florigen).

The site of vernalization is the SAM and consistent with *FLC* being a major target of the vernalization pathway, *FLC* is expressed most strongly in the shoot apex (Michaels and Amasino, 2000). It should be noted, however, that the expression of *FLC* is not restricted to the SAM only; young leaves also show significant *FLC* expression (Michaels and Amasino, 2000). It is likely that the expression of *FLC* in both the apex and the leaves plays important roles in the repression floral promoters; the expression patterns of *FD* (apex) and *FT* (leaves) do not overlap, yet both genes are repressed by *FLC* (Takada and Goto, 2003; Abe et al., 2005; Wigge et al., 2005). *SOC1* is also a target of *FLC* and has a similar expression pattern to *FLC*, showing highest levels of expression in the SAM and young leaves (Borner et al., 2000; Lee et al., 2000; Samach et al., 2000). Because vernalization represses *FLC* expression in both the apex and leaves, vernalized plants contain higher levels of *FD* and *SOC1* expression in the apex, as well as increased *FT* and *SOC1* levels in the leaves.

Summary: The Reproductive Cycle of Winter-Annual *Arabidopsis*

Winter-annual *Arabidopsis* have proven a valuable model in understanding how both photoperiod and vernalization pathways contribute to the coordination of flowering with seasonal changes in day length and temperature. *FRI*-containing winter annuals typically germinate in late summer or fall (Korves et al., 2007). Even if the days are sufficiently long to induce CO protein accumulation, high levels of *FLC* act to block the expression of *FT, FD*, and *SOC1*, thereby preventing flowering. The plants then overwinter as a vegetative rosette. During this time, vernalization takes place, and the action of *VIN3, VRN1, VRN2, VIL1*, and *LHP1* leads to repressive histone modifications in the chromatin of *FLC*. This creates an epigenetic repression of *FLC* that is stable even after temperatures warm in the spring. It is important to note, however, that the suppression of *FLC* is not sufficient to directly induce flowering. *FT* and *SOC1* are not highly expressed until the days of spring become sufficiently long to induce CO protein accumulation. At that point, with the repression of *FLC* removed by vernalization, CO can induce the expression of *FT*. FT protein then moves to the apex where it interacts with FD to activate the expression of *SOC1* and the floral meristem-associated genes *AP1* and *LFY*, to trigger the production of flowers. In the embryos formed from these flowers, *FLC* expression is reset to high levels, restoring the vernalization requirement in the next generation.

Acknowledgments Preparation of this chapter was supported by grants from the National Science Foundation (grant no. IOB-0447583) and the National Institutes of Health (grant no. 1R01GM075060).

References

Abe M, Kobayashi Y, Yamamoto S, Daimon Y, Yamaguchi A, Ikeda Y, Ichinoki H, Notaguchi M, Goto K, and Araki T. (2005). FD, a bZIP protein mediating signals from the floral pathway integrator FT at the shoot apex. Science 309: 1052–1056.

An H, Roussot C, Suarez-Lopez P, Corbesier L, Vincent C, Pineiro M, Hepworth S, Mouradov A, Justin S, Turnbull C, et al. (2004). CONSTANS acts in the phloem to regulate a systemic signal that induces photoperiodic flowering of *Arabidopsis*. Development 131: 3615–3626.

Andersson CR, Helliwell CA, Bagnall DJ, Hughes TP, Finnegan EJ, Peacock WJ, and Dennis ES. (2008). The FLX gene of *Arabidopsis* is required for FRI-dependent activation of FLC expression. Plant Cell Physiol 49(2): 191–200.

Aukerman M and Amasino RM. (1996). Molecular genetic analysis of flowering time in *Arabidopsis*. *Semin Dev Biol* 7: 427–434.

Aukerman MJ, Lee I, Weigel D, and Amasino RM. (1999). The *Arabidopsis* flowering-time gene LUMINIDEPENDENS is expressed primarily in regions of cell proliferation and encodes a nuclear protein that regulates LEAFY expression. Plant J 18: 195–203.

Ausin I, Alonso-Blanco C, Jarillo JA, Ruiz-Garcia L, and Martinez-Zapater JM. (2004). Regulation of flowering time by FVE, a retinoblastoma-associated protein. Nat Genet 36: 162–166.

Ayre BG and Turgeon R. (2004). Graft transmission of a floral stimulant derived from CONSTANS. Plant Physiol 135: 2271–2278.

Bastow R, Mylne JS, Lister C, Lippman Z, Martienssen RA, and Dean C. (2004). Vernalization requires epigenetic silencing of FLC by histone methylation. Nature 427: 164–167.

Ben-Naim O, Eshed R, Parnis A, Teper-Bamnolker P, Shalit A, Coupland G, Samach A, and Lifschitz E. (2006). The CCAAT binding factor can mediate interactions between CONSTANS-like proteins and DNA. Plant J 46: 462–476.

Bernier G, Kinet J-M, and Sachs RM. (1981). *The Physiology of Flowering.* Boca Raton, FL: CRC Press.

Borner R, Kampmann G, Chandler J, Gleissner R, Wisman E, Apel K, and Melzer S. (2000). A MADS domain gene involved in the transition to flowering in *Arabidopsis.* Plant J 24: 591–599.

Burn JE, Smyth DR, Peacock WJ, and Dennis ES. (1993). Genes Conferring Late Flowering in *Arabidopsis thaliana.* Genetica 90: 147–155.

Cai X, Ballif J, Endo S, Davis E, Liang M, Chen D, DeWald D, Kreps J, Zhu T, and Wu Y. (2007). A putative CCAAT-binding transcription factor is a regulator of flowering timing in *Arabidopsis.* Plant Physiol 145: 98–105.

Chailakhyan MK. (1936). New facts in support of the hormonal theory of plant development. C R (Dokl) Acad Sci USSR 13: 79–83.

Choi K, Kim S, Kim SY, Kim M, Hyun Y, Lee H, Choe S, Kim SG, Michaels S, and Lee I. (2005). SUPPRESSOR OF FRIGIDA3 encodes a nuclear ACTIN-RELATED PROTEIN6 required for floral repression in *Arabidopsis.* Plant Cell 17: 2647–2660.

Choi K, Park C, Lee J, Oh M, Noh B, and Lee I. (2007). *Arabidopsis* homologs of components of the SWR1 complex regulate flowering and plant development. Development 134: 1931–1941.

Chouard P. (1960). Vernalization and its relations to dormancy. Annu Rev Plant Physiol 11: 191–238.

Clarke JH and Dean C. (1994). Mapping *FRI,* a locus controlling flowering time and vernalization response in *Arabidopsis thaliana.* Mol Gen Genet 242: 81–89.

Corbesier L, Vincent C, Jang S, Fornara F, Fan Q, Searle I, Giakountis A, Farrona S, Gissot L, Turnbull C, et al. (2007). FT protein movement contributes to long-distance signaling in floral induction of *Arabidopsis.* Science 316: 1030–1033.

Deal RB, Kandasamy MK, McKinney EC, and Meagher RB. (2005). The nuclear actin-related protein ARP6 is a pleiotropic developmental regulator required for the maintenance of FLOWERING LOCUS C expression and repression of flowering in *Arabidopsis.* Plant Cell 17: 2633–2646.

Doyle MR, Bizzell CM, Keller MR, Michaels SD, Song J, Noh YS, and Amasino RM. (2005). HUA2 is required for the expression of floral repressors in *Arabidopsis* thaliana. Plant J 41: 376–385.

Dubnau J and Struhl G. (1996). RNA recognition and translational regulation by a homeodomain protein. Nature 379: 694–699.

Gazzani S, Gendall AR, Lister C, and Dean C. (2003). Analysis of the molecular basis of flowering time variation in *Arabidopsis* accessions. Plant Physiol 132: 1107–1114.

Gelinas R, Endlich B, Pfeiffer C, Yagi M, and Stamatoyannopoulos G. (1985). G to A substitution in the distal CCAAT box of the A gamma-globin gene in Greek hereditary persistence of fetal haemoglobin. Nature 313: 323–325.

Gendall AR, Levy YY, Wilson A, and Dean C. (2001). The VERNALIZATION 2 gene mediates the epigenetic regulation of vernalization in *Arabidopsis.* Cell 107: 525–535.

Greb T, Mylne JS, Crevillen P, Geraldo N, An H, Gendall AR, and Dean C. (2007). The PHD finger protein VRN5 functions in the epigenetic silencing of *Arabidopsis* FLC. Curr Biol 17: 73–78.

Gusmaroli G, Tonelli C, and Mantovani R. (2002). Regulation of novel members of the *Arabidopsis* thaliana CCAAT-binding nuclear factor Y subunits. Gene 283: 41–48.

Hackett WP and Hartmann HT. (1967). The influence of temperature on floral initiation in the olive. Physiol Plant 20: 430–436.

He Y, Doyle MR, and Amasino RM. (2004). PAF1-complex-mediated histone methylation of FLOWERING LOCUS C chromatin is required for the vernalization-responsive, winter-annual habit in *Arabidopsis.* Genes Dev 18: 2774–2784.

He Y, Michaels SD, and Amasino RM. (2003). Regulation of flowering time by histone acetylation in *Arabidopsis.* Science 302: 1751–1754.

Helliwell CA, Wood CC, Robertson M, James Peacock W, and Dennis ES. (2006). The *Arabidopsis* FLC protein interacts directly in vivo with SOC1 and FT chromatin and is part of a high-molecular-weight protein complex. Plant J 46: 183–192.

Henderson IR, Liu F, Drea S, Simpson GG, and Dean C. (2005). An allelic series reveals essential roles for FY in plant development in addition to flowering-time control. Development 132: 3597–3607.

Hepworth SR, Valverde F, Ravenscroft D, Mouradov A, and Coupland G. (2002). Antagonistic regulation of flowering-time gene SOC1 by CONSTANS and FLC via separate promoter motifs. EMBO J 21: 4327–4337.

Iizuka M and Smith MM. (2003). Functional consequences of histone modifications. Curr Opin Genet Dev 13: 154–160.

Jaeger KE and Wigge PA. (2007). FT protein acts as a long-range signal in *Arabidopsis*. Curr Biol 17: 1050–1054.

Johanson U, West J, Lister C, Michaels S, Amasino R, and Dean C. (2000). Molecular analysis of *FRIGIDA*, a major determinant of natural variation in *Arabidopsis* flowering time. Science 290: 344–347.

Kardailsky I, Shukla VK, Ahn JH, Dagenais N, Christensen SK, Nguyen JT, Chory J, Harrison MJ, and Weigel D. (1999). Activation Tagging of the Floral Inducer FT. Science 286: 1962–1965.

Kim S, Choi K, Park C, Hwang HJ, and Lee I. (2006). SUPPRESSOR OF FRIGIDA4, encoding a C_2H_2-type zinc finger protein, represses flowering by transcriptional activation of *Arabidopsis* FLOWERING LOCUS C. Plant Cell 18: 2985–2998.

Kim SY, He Y, Jacob Y, Noh YS, Michaels S, and Amasino R. (2005). Establishment of the vernalization-responsive, winter-annual habit in *Arabidopsis* requires a putative histone H3 methyl transferase. Plant Cell 17: 3301–3310.

Kim SY and Michaels SD. (2006). SUPPRESSOR OF FRI 4 encodes a nuclear-localized protein that is required for delayed flowering in winter-annual *Arabidopsis*. Development 133: 4699–4707.

Kobayashi Y, Kaya H, Goto K, Iwabuchi M, and Araki T. (1999). A pair of related genes with antagonistic roles in mediating flowering signals. Science 286: 1960–1962.

Kobayashi Y and Weigel D. (2007). Move on up, it's time for change—mobile signals controlling photoperiod-dependent flowering. Genes Dev 21: 2371–2384.

Koornneef M, Hanhart CJ, and van der Veen JH. (1991). A genetic and physiological analysis of late flowering mutants in *Arabidopsis thaliana*. Mol Gen Genet 229: 57–66.

Korves TM, Schmid KJ, Caicedo AL, Mays C, Stinchcombe JR, Purugganan MD, and Schmitt J. (2007). Fitness effects associated with the major flowering time gene FRIGIDA in *Arabidopsis* thaliana in the field. Am Nat 169: E141–E157.

Krogan NJ, Dover J, Wood A, Schneider J, Heidt J, Boateng MA, Dean K, Ryan OW, Golshani A, Johnston M, et al. (2003). The Paf1 complex is required for histone H3 methylation by COMPASS and Dot1p: Linking transcriptional elongation to histone methylation. Mol Cell 11: 721–729.

Krogan NJ, Kim M, Ahn SH, Zhong G, Kobor MS, Cagney G, Emili A, Shilatifard A, Buratowski S, and Greenblatt JF. (2002). RNA polymerase II elongation factors of Saccharomyces cerevisiae: A targeted proteomics approach. Mol Cell Biol 22: 6979–6992.

Lang A. (1965). Physiology of flower initiation. In *Encyclopedia of Plant Physiology*, vol 15, pt 1 (W Ruhland, ed). Berlin: Springer-Verlag, pp. 1371–1536.

Lang A. (1986). *Hyoscyamus niger*. In *CRC Handbook of Flowering*, vol 5 (AH Halevy, ed). Boca Raton, FL: CRC Press, pp. 144–186.

Lazaro A, Gomez-Zambrano A, Lopez-Gonzalez L, Pineiro M, and Jarillo JA. (2008). Mutations in the *Arabidopsis* SWC6 gene, encoding a component of the SWR1 chromatin remodelling complex, accelerate flowering time and alter leaf and flower development. J Exp Bot 59: 653–666.

Lee H, Suh SS, Park E, Cho E, Ahn JH, Kim SG, Lee JS, Kwon YM, and Lee I. (2000). The AGAMOUS-LIKE 20 MADS domain protein integrates floral inductive pathways in *Arabidopsis*. Genes Dev 14: 2366–2376.

Lee I, Aukerman MJ, Gore SL, Lohman KN, Michaels SD, Weaver LM, John MC, Feldmann KA, and Amasino RM. (1994). Isolation of *LUMINIDEPENDENS*—a gene involved in the control of flowering time in *Arabidopsis*. Plant Cell 6: 75–83.

Lee I, Bleecker A, and Amasino R. (1993). Analysis of naturally occurring late flowering in *Arabidopsis-thaliana*. Mol Gen Genet 237: 171–176.

Levy YY, Mesnage S, Mylne JS, Gendall AR, and Dean C. (2002). Multiple roles of *Arabidopsis* VRN1 in vernalization and flowering time control. Science 297: 243–246.

Lim MH, Kim J, Kim YS, Chung KS, Seo YH, Lee I, Hong CB, Kim HJ, and Park CM. (2004). A new *Arabidopsis* gene, FLK, encodes an RNA binding protein with K homology motifs and regulates flowering time via FLOWERING LOCUS C. Plant Cell 16: 731–740.

Lin MK, Belanger H, Lee YJ, Varkonyi-Gasic E, Taoka K, Miura E, Xoconostle-Cazares B, Gendler K, Jorgensen RA, Phinney B, et al. (2007). FLOWERING LOCUS T protein may act as the long-distance florigenic signal in the cucurbits. Plant Cell 19: 1488–1506.

Liu J, Gilmour SJ, Thomashow MF, and Van Nocker S. (2002). Cold signalling associated with vernalization in *Arabidopsis* thaliana does not involve CBF1 or abscisic acid. Physiol Plant 114: 125–134.

Macknight R, Bancroft I, Page T, Lister C, Schmidt R, Love L, Westphal L, Murphy G, Sherson S, Cobbett C, et al. (1997). FCA, a gene controlling flowering time in *Arabidopsis*, encodes a protein containing RNA-binding domains. Cell 89: 737–745.

Macknight R, Duroux M, Laurie R, Dijkwel P, Simpson G, and Dean C. (2002). Functional significance of the alternative transcript processing of the *Arabidopsis* floral promoter FCA. Plant Cell 14: 877–888.

March-Diaz R, Garcia-Dominguez M, Florencio FJ, and Reyes JC. (2007). SEF, a new protein required for flowering repression in *Arabidopsis*, interacts with PIE1 and ARP6. Plant Physiol 143: 893–901.

Margueron R, Trojer P, and Reinberg D. (2005). The key to development: Interpreting the histone code? Curr Opin Genet Dev 15: 163–176.

Martin-Trillo M, Lazaro A, Poethig RS, Gomez-Mena C, Pineiro MA, Martinez-Zapater JM, and Jarillo JA. (2006). EARLY IN SHORT DAYS 1 (ESD1) encodes ACTIN-RELATED PROTEIN 6 (AtARP6), a putative component of chromatin remodelling complexes that positively regulates FLC accumulation in *Arabidopsis*. Development 133: 1241–1252.

Mathieu J, Warthmann N, Kuttner F, and Schmid M. (2007). Export of FT protein from phloem companion cells is sufficient for floral induction in *Arabidopsis*. Curr Biol 17: 1055–1060.

Michaels S and Amasino R. (1999). FLOWERING LOCUS C encodes a novel MADS domain protein that acts as a repressor of flowering. Plant Cell 11: 949–956.

Michaels S and Amasino R. (2000). Memories of winter: Vernalization and the competence to flower. Plant Cell Environ 23: 1145–1154.

Michaels SD and Amasino RM. (2001). Loss of FLOWERING LOCUS C activity eliminates the late-flowering phenotype of FRIGIDA and autonomous pathway mutations but not responsiveness to vernalization. Plant Cell 13: 935–942.

Michaels SD, Bezerra IC, and Amasino RM. (2004). FRIGIDA-related genes are required for the winter-annual habit in *Arabidopsis*. Proc Natl Acad Sci USA 101: 3281–3285.

Michaels SD, He Y, Scortecci KC, and Amasino RM. (2003). Attenuation of FLOWERING LOCUS C activity as a mechanism for the evolution of summer-annual flowering behavior in *Arabidopsis*. Proc Natl Acad Sci USA 100: 10102–10107.

Michaels SD, Himelblau E, Kim SY, Schomburg FM, and Amasino RM. (2005). Integration of flowering signals in winter-annual *Arabidopsis*. Plant Physiol 137: 149–156.

Mockler TC, Yu X, Shalitin D, Parikh D, Michael TP, Liou J, Huang J, Smith Z, Alonso JM, Ecker JR, et al. (2004). Regulation of flowering time in *Arabidopsis* by K homology domain proteins. Proc Natl Acad Sci USA 101: 12759–12764.

Mylne JS, Barrett L, Tessadori F, Mesnage S, Johnson L, Bernatavichute YV, Jacobsen SE, Fransz P, and Dean C. (2006). LHP1, the *Arabidopsis* homologue of HETEROCHROMATIN PROTEIN1, is required for epigenetic silencing of FLC. Proc Natl Acad Sci USA 103: 5012–5017.

Napp-Zinn K. (1979). On the genetical basis of vernalization requirement in *Arabidopsis thaliana* (L.) Heynh. In *La Physiologie de la Floraison* (P Champagnat and R. Jaques, eds). Paris: Collège International CNRS, pp. 217–220.

Ng HH, Robert F, Young RA, and Struhl K. (2003). Targeted recruitment of Set1 histone methylase by elongating Pol II provides a localized mark and memory of recent transcriptional activity. Mol Cell 11: 709–719.

Noh YS and Amasino RM. (2003). PIE1, an ISWI family gene, is required for FLC activation and floral repression in *Arabidopsis*. Plant Cell 15: 1671–1682.

Oh S, Zhang H, Ludwig P, and van Nocker S. (2004). A mechanism related to the yeast transcriptional regulator Paf1c is required for expression of the *Arabidopsis* FLC/MAF MADS box gene family. Plant Cell 16: 2940–2953.

Parenicova L, de Folter S, Kieffer M, Horner DS, Favalli C, Busscher J, Cook HE, Ingram RM, Kater MM, Davies B, et al. (2003). Molecular and phylogenetic analyses of the complete MADS box transcription factor family in *Arabidopsis*: New openings to the MADS world. Plant Cell 15: 1538–1551.

Quesada V, Macknight R, Dean C, and Simpson GG. (2003). Autoregulation of FCA pre-mRNA processing controls *Arabidopsis* flowering time. EMBO J 22: 3142–3152.

Raisner RM and Madhani HD. (2006). Patterning chromatin: Form and function for H2A.Z variant nucleosomes. Curr Opin Genet Dev 16: 119–124.

Ratcliffe OJ, Kumimoto RW, Wong BJ, and Riechmann JL. (2003). Analysis of the *Arabidopsis* MADS AFFECTING FLOWERING gene family: MAF2 prevents vernalization by short periods of cold. Plant Cell 15: 1159–1169.

Ratcliffe OJ, Nadzan GC, Reuber TL, and Riechmann JL. (2001). Regulation of flowering in *Arabidopsis* by an FLC homologue. Plant Physiol 126: 122–132.

Redei GP. (1962). Supervital mutants in *Arabidopsis*. Genetics 47: 443–460.

Rivera-Pomar R, Niessing D, Schmidt-Ott U, Gehring WJ, and Jackle H. (1996). RNA binding and translational suppression by bicoid. Nature 379: 746–749.

Salathia N, Davis SJ, Lynn JR, Michaels SD, Amasino RM, and Millar AJ. (2006). FLOWERING LOCUS C-dependent and -independent regulation of the circadian clock by the autonomous and vernalization pathways. BMC Plant Biol 6: 10.

Samach A, Onouchi H, Gold SE, Ditta GS, Schwarz-Sommer Z, Yanofsky MF, and Coupland G. (2000). Distinct roles of CONSTANS target genes in reproductive development of *Arabidopsis*. Science 288: 1613–1616.

Schlappi MR. (2006). FRIGIDA LIKE 2 is a functional allele in Landsberg erecta and compensates for a nonsense allele of FRIGIDA LIKE 1. Plant Physiol 142: 1728–1738.

Schmitz RJ, Hong L, Michaels S, and Amasino RM. (2005). FRIGIDA-ESSENTIAL 1 interacts genetically with FRIGIDA and FRIGIDA-LIKE 1 to promote the winter-annual habit of *Arabidopsis* thaliana. Development 132: 5471–5478.

Schomburg FM, Patton DA, Meinke DW, and Amasino RM. (2001). FPA, a gene involved in floral induction in *Arabidopsis*, encodes a protein containing RNA-recognition motifs. Plant Cell 13: 1427–1436.

Schonrock N, Bouveret R, Leroy O, Borghi L, Kohler C, Gruissem W, and Hennig L. (2006). Polycomb-group proteins repress the floral activator AGL19 in the FLC-independent vernalization pathway. Genes Dev 20: 1667–1678.

Scortecci KC, Michaels SD, and Amasino RM. (2001). Identification of a MADS-box gene, FLOWERING LOCUS M, that represses flowering. Plant J 26: 229–236.

Scortecci K, Michaels SD, and Amasino RM. (2003). Genetic interactions between FLM and other flowering-time genes in *Arabidopsis* thaliana. Plant Mol Biol 52: 915–922.

Searle I, He Y, Turck F, Vincent C, Fornara F, Krober S, Amasino RA, and Coupland G. (2006). The transcription factor FLC confers a flowering response to vernalization by repressing meristem competence and systemic signaling in *Arabidopsis*. Genes Dev 20: 898–912.

Sheldon CC, Burn JE, Perez PP, Metzger J, Edwards JA, Peacock WJ, and Dennis ES. (1999). The FLF MADS box gene. A repressor of flowering in arabidopsis regulated by vernalization and methylation. Plant Cell 11: 445–458.

Sheldon CC, Hills MJ, Lister C, Dean C, Dennis ES, and Peacock WJ. (2008). Resetting of FLOWERING LOCUS C expression after epigenetic repression by vernalization. Proc Natl Acad Sci USA 105: 2214–2219.

Shi Y, Lan F, Matson C, Mulligan P, Whetstine JR, Cole PA, Casero RA, and Shi Y. (2004). Histone demethylation mediated by the nuclear amine oxidase homolog LSD1. Cell 119: 941–953.

Shindo C, Aranzana MJ, Lister C, Baxter C, Nicholls C, Nordborg M, and Dean C. (2005). Role of FRIGIDA and FLOWERING LOCUS C in determining variation in flowering time of *Arabidopsis*. Plant Physiol 138: 1163–1173.

Shore P and Sharrocks AD. (1995). The MADS-box family of transcription factors. Eur J Biochem 229: 1–13.

Simpson GG, Dijkwel PP, Quesada V, Henderson I, and Dean C. (2003). FY is an RNA 3' end-processing factor that interacts with FCA to control the *Arabidopsis* floral transition. Cell 113: 777–787.

Soliman MA and Riabowol K. (2007). After a decade of study-ING, a PHD for a versatile family of proteins. Trends Biochem Sci 32: 509–519.

Soppe WJ, Bentsink L, and Koornneef M. (1999). The early-flowering mutant efs is involved in the autonomous promotion pathway of *Arabidopsis* thaliana. Development 126: 4763–4770.

Squazzo SL, Costa PJ, Lindstrom DL, Kumer KE, Simic R, Jennings JL, Link AJ, Arndt KM, and Hartzog GA. (2002). The Paf1 complex physically and functionally associates with transcription elongation factors in vivo. EMBO J 21: 1764–1774.

Suarez-Lopez P, Wheatley K, Robson F, Onouchi H, Valverde F, and Coupland G. (2001). CONSTANS mediates between the circadian clock and the control of flowering in *Arabidopsis*. Nature 410: 1116–1120.

Sung S and Amasino RM. (2004). Vernalization in *Arabidopsis* thaliana is mediated by the PHD finger protein VIN3. Nature 427: 159–164.

Sung S, He Y, Eshoo TW, Tamada Y, Johnson L, Nakahigashi K, Goto K, Jacobsen SE, and Amasino RM. (2006a). Epigenetic maintenance of the vernalized state in *Arabidopsis* thaliana requires LIKE HETEROCHROMATIN PROTEIN 1. Nat Genet 38: 706–710.

Sung S, Schmitz RJ, and Amasino RM. (2006b). A PHD finger protein involved in both the vernalization and photoperiod pathways in *Arabidopsis*. Genes Dev 20: 3244–3248.

Takada S and Goto K. (2003). Terminal flower2, an *Arabidopsis* homolog of heterochromatin protein1, counteracts the activation of flowering locus T by constans in the vascular tissues of leaves to regulate flowering time. Plant Cell 15: 2856–2865.

Tamaki S, Matsuo S, Wong HL, Yokoi S, and Shimamoto K. (2007). Hd3a protein is a mobile flowering signal in rice. Science 316: 1033–1036.

Thomashow MF. (1999). PLANT COLD ACCLIMATION: Freezing Tolerance Genes and Regulatory Mechanisms. Annu Rev Plant Physiol Plant Mol Biol 50: 571–599.

Thompson HC. (1944). Further studies on effect of temperature on initiation of flowering in celery. Proc Am Soc Horic Sci 35: 425–430.

Valverde F, Mouradov A, Soppe W, Ravenscroft D, Samach A, and Coupland G. (2004). Photoreceptor regulation of CONSTANS protein in photoperiodic flowering. Science 303: 1003–1006.

Wareing PF. (1971). The control of seed dormancy. Biochem J 124: 1P–2P.

Wellensiek SJ. (1962). Dividing cells as the locus for vernalization. Nature 195: 307–308.

Wellensiek SJ. (1964). Dividing cells as the prerequisite for vernalization. Plant Physiol 39: 832–835.

Wenkel S, Turck F, Singer K, Gissot L, Le Gourrierec J, Samach A, and Coupland G. (2006). CONSTANS and the CCAAT box binding complex share a functionally important domain and interact to regulate flowering of *Arabidopsis*. Plant Cell 18: 2971–2984.

Wigge PA, Kim MC, Jaeger KE, Busch W, Schmid M, Lohmann JU, and Weigel D. (2005). Integration of spatial and temporal information during floral induction in *Arabidopsis*. Science 309: 1056–1059.

Wood CC, Robertson M, Tanner G, Peacock WJ, Dennis ES, and Helliwell CA. (2006). The *Arabidopsis thaliana* vernalization response requires a polycomb-like protein complex that also includes VERNALIZATION INSENSITIVE 3. Proc Natl Acad Sci USA 103: 14631–14636.

Yamaguchi A, Kobayashi Y, Goto K, Abe M, and Araki T. (2005). TWIN SISTER OF FT (TSF) acts as a floral pathway integrator redundantly with FT. Plant Cell Physiol 46: 1175–1189.

Yanovsky MJ and Kay SA. (2002). Molecular basis of seasonal time measurement in *Arabidopsis*. Nature 419: 308–312.

Yoo SK, Chung KS, Kim J, Lee JH, Hong SM, Yoo SJ, Yoo SY, Lee JS, and Ahn JH. (2005). CONSTANS activates SUPPRESSOR OF OVEREXPRESSION OF CONSTANS 1 through FLOWERING LOCUS T to promote flowering in *Arabidopsis*. Plant Physiol 139: 770–778.

Zeevaart JAD. (1962). Physiology of flowering. Science 58: 531–542.

Zhang H and van Nocker S. (2002). The VERNALIZATION INDEPENDENCE 4 gene encodes a novel regulator of FLOWERING LOCUS C. Plant J 31: 663–673.

Zhao Z, Yu Y, Meyer D, Wu C, and Shen WH. (2005). Prevention of early flowering by expression of FLOWERING LOCUS C requires methylation of histone H3 K36. Nat Cell Biol 7: 1156–1160.

7

Seasonality and Photoperiodism in Fungi

Till Roenneberg, Tanja Radic, Manfred Gödel, and Martha Merrow

Chronobiology investigates the endogenous mechanisms that allow organisms to adapt and anticipate the predictable temporal "spaces" of their environment. These include the regular alternations of tides, night and day, lunar phases, and seasons. Although self-sustained endogenous rhythmic processes, commonly called "clocks" or oscillators, are important mechanistic solutions to the problems of internal timing and anticipation, presumably several other mechanisms participate in an optimal adaptation to regular environmental time structures. For example, circadian clocks that run in synchrony with the earth's rotation can theoretically be either entrained or merely masked. While entrainment involves the active process of repetitively shifting the phase of an oscillator, eventually resulting in a specific phase angle between the endogenous oscillator and the rhythmic environment (stable phase of entrainment or chronotype), masking is often viewed as a direct response of the system triggered by environmental stimuli. Experiments have shown, however, that the phenomenon of masking involves far more than a simple, direct "stimulus response." For example, masking responses often display oscillator properties themselves (for references, see Honma et al., 1983). By analogy, adaptation to seasons can utilize a circannual clock (Gwinner, 1986) but can also be triggered by an environmental signal. Seasonal responses can theoretically be triggered by many features of the environment (temperature, humidity, rainfall, food availability, the presence of other organisms, and many more), but the seasonal trigger that seems the most important in seasonal adaptation, and that we know most about, is day or night length, that is, photo- or scotoperiod (Nunes and Saunders, 1999). The main difference between adaptation processes based on oscillators and those that are triggered is that the former continues to function even in constant conditions while the latter damps without an environmental input. As in the case of masking in the daily domain, photoperiodism in the annual domain is far more than a simple, triggered response. It has been proposed that the anticipation of seasons based on photoperiodism is a result of at least two interrelated processes (Nunes and Saunders, 1999): a timer that measures night length (rarely day length; see Saunders, 1982, 1987) and a memory that can integrate seasonal changes over time.

Compared to the overwhelming evidence of active seasonal adaptation in plants and animals, we know fairly little about seasonality and photoperiodism in fungi.

To approach this topic in any organism, several questions have to be answered: (1) Is there evidence for seasonality? (2) Is there evidence for photobiology—a prerequisite for detecting changing lengths of day and night? (3) Is there evidence for a circadian system that could be used as internal reference to make sense of the changes in day or night length?

In this chapter, we summarize what is known about the questions enumerated above. In addition, we describe the first results of investigating seasonality and photoperiodism in *Neurospora crassa*, a classic model organism in circadian research. As in other organisms, the *Neurospora* photoperiodic responses rely on a functional circadian clock that involves determination of night length.

PREREQUISITES FOR SEASONALITY AND PHOTOPERIODISM IN FUNGI

Seasonal adaptation can be demonstrated by many diverse physiologies: foliage in trees, weight control or hibernation in many animals, or more species-specific functions such as molt or migration in birds and the shedding of antlers in deer. However, the most prominent seasonal adaptation is often apparent in reproduction. We therefore begin with a review of reproduction in fungi, with special emphasis on *Neurospora*, and discuss the possible adaptive benefits of controlling fungal reproduction according to season.

Reproduction in Fungi

Generation cycles in fungi are among the most complex reproductive systems in biology. In addition to the various forms of asexual reproduction via conidia, split hyphae, and budding, there are very diverse sexual reproduction strategies even within genera. While *Candida albicans* exists predominantly in a diploid form with meiosis before gamete formation, *C. glabrata* grows mainly in a haploid state by going through meiosis shortly after zygote formation (similar to *Neurospora crassa*). *Saccharomyces* switches between haploid and diploid life cycles, existing primarily as a diploid (or polyploid) in the wild.

Fungi that inhabit environments characterized by changes in day length, and thus also ambient temperature, may have strategies to seasonally control their reproduction. Growth rates will be altered as temperatures control fungal metabolic rates, and the entire ecosystem (i.e., food sources and predators) will similarly adjust. Any mushrooms collector knows that the appearance of fungal fruiting bodies is restricted to certain times of the year. The concentration of airborne fungal spores (most frequently assessed by air filtration) also changes drastically over the course of a year (see examples in figure 7.1). These annual rhythms are thought to be correlated directly to environmental conditions such as available nutrients, humidity, wind speed, or temperature (Ingold, 1971).

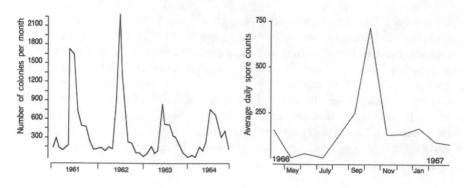

FIGURE 7.1.
Annual distributions of (sexual) ascospores in the air. Examples shown for *Cladosporium* collected in Wales, with maximum concentrations in June (left, redrawn from Harvey, 1967) and for *Hypoxylon rubiginosum* with a maximum in October (right, redrawn from Hodgkiss and Harvey, 1969).

Sexual Reproduction

Sexual reproduction requires nuclei of two mating types; in some cases, they are inherently part of the same thallus (homothallic, e.g., *Sordaria* or *Neurospora tetrasperma*). Alternatively, they derive from mixing opposite mating types from distinct thalli or colonies (heterothallic, e.g., *Ustilago* or *Neurospora crassa*). Again, different environmental signals may trigger sexual reproduction. In *Neurospora*, for example, mating is initiated via shortage of carbon or nitrogen. Commingling of nuclei is achieved by first developing a specialized structure (a protoperithecium), initially by forming as a tangle of hyphae. This pear-shaped container sends out a single hypha called a trichogyne, which grows chemotactically in the direction of attracting pheromones of conidia or tissues of the opposite mating type. The trichogyne can undergo plasmogamy with micro/macroconidia or hyphae and transfer the received nuclei to a recipient ascogonial cell in the perithecium, where karyogamy takes place. Meiotic divisions produce eight haploid ascospores in an elongated spore container (ascus). Mature perithecia hold a lot of asci and appear in a dark color. Approximately 10 days following fertilization, they start to shoot the spores out with great force. *Neurospora*'s close cousin, *Sordaria*, also shoots sexual spores from perithecia, and a detailed analysis of this process (including modeling the forces required; see Ingold and Hadland, 1959) documents how many spores are shot in a given release (0 to 8) as well as how far they travel (1 cm for single spores, 6–8 cm for groups of eight). One centimeter is about 75,000 times the size of a single spore. Ascospores can survive for years until germination.

Exhaustive genetic evaluation of wild-type isolates indicates that *Neurospora* in nature relies extensively on sexual reproduction (Perkins and Turner, 1988). Spores serve as a stable repository of the genome until favorable conditions reappear. Their germination often occurs only under special conditions. In *Neurospora*, for example, this is achieved by intense heat. *Neurospora* isolates are often harvested following burns (sugar cane fields, forest fires, etc.; see Jacobson et al., 2006), where they appear as pioneer species. It is unlikely that *Neurospora* has evolved a specialized sexual reproduction exclusively in response to a sporadic fire. Given that *Neurospora* is frequently found in hot climates, it is likely that sexual reproduction can also be triggered when the sun heats the surface soil or the bark of a tree or topsoil to temperatures promoting the germination process (in the laboratory, germination is routinely triggered by heating the spores by 60°C for 30 min). Alternatively, germination is also promoted by chemicals released into the soil from decaying compost, a process that also involves heat production or, alternatively, may signal fresh nutritional sources for development of a haploid colony.

Asexual Reproduction

Neurospora's asexual developmental program (conidiation) is regulated primarily by light and/or the circadian clock, but also by nutrition. Furthermore, field studies demonstrate that this asexual reproduction is temporally distinct from sexual development, by as much as several months (as observed near Bangalore, 13°N, by Pandit and Maheshwari, 1994). The first morphological change associated with conidiation in *Neurospora* is the formation of aerial hyphae. At some point the hyphae will form microscopically visible indentations (minor constrictions, becoming major constrictions). Eventually, each point of constriction forms an independent structure, an individual macroconidium (often containing several nuclei, in contrast to a uninucleate microconidium). Conidia essentially segregate mitotic nuclei. As they mature, they are liberated simply by movement of air, and germination of conidia to conidial germ tube can occur in a matter of hours.

As with most plants, fungi are predominantly sedentary and often spread (in search for nutrition, sexual partners, etc.) simply by growing (e.g., the huge subterranean *Armillaria*; see Smith et al., 1992). Some fungal species are also capable of moving physically, as the amoeboid slime molds (e.g., *Dictyostelium*). A substantial problem for any sedentary organism is exposure to sunlight, to extreme temperatures, and to varying access to water and nutrients. The fungi produce an impressive collection of pigmentation, which may help to shield them from the effects of UV light, thus reducing UV-induced mutagenesis. In the case of *Neurospora* this includes carotenoids in mycelia and asexual spores and melanin in sexual spores.

Photoperiodism in Fungi

Many photoperiodic plants require light induction followed by a dark period to initiate their seasonal reproduction (Cumming, 1971). This is also the case in some fungi, namely, in the "diurnal sporulators" (Durand, 1982; Leach, 1967). Like in short-day plants, their maturation process is exquisitely light sensitive and can be inhibited when a short light pulse (in the range of seconds) interrupts the required dark period (Durand, 1982). As in plants, the timing of the light pulse within the obligatory dark period is crucial for its inhibitory effect. Durand (1982) found that the timing of the pulse also depends on the respective night length (scotoperiod), with maximal inhibition when the pulse is applied two-thirds into the dark period.

Like in other poikilotherms, fungal seasonality will be mediated both by photoperiod and temperature. The interplay between these factors has been extensively studied in insects (Pittendrigh et al., 1991). However, while there is no question about seasonality in fungal biology, investigations into photoperiodic mechanisms and/or photoperiodic memory are very sparse. Many of the investigations into the influence of photoperiod on fungal development and reproduction concern host-infecting pathogens. In these cases, it is often difficult to determine whether a photoperiodic response is due to the fungus, to the host, or to both. In some cases, however, the pathogen is grown on artificial media or on plant debris, making a distinction possible. For example, when the entomopathogenic (insect-infecting) fungus *Metarhizium anisopliae* is grown in different temperatures (25°C, 28°C, and 30°C) and photoperiods (24, 16, 12, and 8 h), its colony size and the number of spores produced are maximal at 28°C and a photoperiod of 16 h (Alves et al., 1984). Linear mycelial growth in *Colletotrichum manihotis* (a plant pathogen from the Congo) is greater in constant darkness (DD) than in constant light (LL) (Makambila, 1984). The inhibitory effect of light depends, however, systematically on photoperiod and is lower in light:dark (LD) cycles of 12 h:12 h (12:12) compared to both longer and shorter photoperiods. The amplitude of this photoperiodic response increases with light fluence and temperature and is absent at lower temperatures (20°C). The formation of chlamydospores (thick-walled, asexual resting spores) of the corn pathogen *Exserohilum turcicum* is also affected by photoperiod and temperature (Levy, 1995). In addition to photoperiodic effects on growth and asexual spore formation, day or night length can also modify the development of sexual spores. The formation of perithecia in *Mycosphaerella pinodes*, for example, favors temperatures around 20°C and a 16-h photoperiod when it is grown under controlled laboratory conditions (Roger and Tivoli, 1996).

This incomplete list of the effects of photoperiod on fungal reproduction concerns many species that are either difficult to study in the laboratory or that have not been characterized genetically and are not readily transformable. A thorough characterization of photoperiodic responses, proof for the involvement of the circadian system or the existence of a photoperiodic memory will involve a wide range of experiments, which include a good characterization of the organism's circadian system and its photobiology, photoperiodic response curves, and night-interruption and Nanda-Hamner–type experiments (Nanda and Hamner, 1958). Thus, an organism that has good genetics and transformation possibilities would be extremely helpful.

A good candidate for such an approach is *Neurospora crassa*. It is amenable to biochemical, molecular, genetic, and physiological experiments and can easily be grown in large quantities. The sequence of its (haploid) genome is completed, and microarrays are being produced. Many mutants are available and have been characterized. In addition, it constitutes one of the pioneer model systems, both for the circadian clock and for photobiology.

Light Reception in Fungi

Without light reception, there can be no photoperiodic response. The fungi, as a kingdom, have been utilized extensively to characterize photobiology (Corrochano, 2007). Light regulates reproduction (e.g., promotion of development or inhibition germination; see Ingold and Nawaz, 1967; Brook, 1969; Calpouzos and Chang, 1971; Durand, 1982; Degli-Innocenti and Russo, 1984), carotenoid or other pigment formation, and phototropism (Delbruck et al., 1976; Linden et al., 1999). Action spectra and fluence response curves have been determined for photoinhibition of spore maturation (Leach, 1968), carotenoid induction in several fungi (see list in Berjarano et al., 1990) including *Neurospora* (De Fabo et al., 1976), and phase shifting of circadian rhythms (Sargent et al., 1956; Dharmananda, 1980; Crosthwaite et al., 1995).

Most of the photo responses in fungi have a maximum in the short end of the light spectrum (sometimes including the UV region, e.g., reported for circadian phase shifting; see West, 1976). There are, however, also reports of red or far-red light responses in the fungi. In the apple scab, *Venturia inaequalis*, ascospore release is greatly enhanced by light in the range of 710–730 nm, resembling phytochrome-mediated responses in plants (Brook, 1969). In *Sphaerobolus*, sporophores mature more efficiently with red than with shorter wavelength light, despite the requirement of blue light for development overall (Ingold and Nawaz, 1967).

Are all fungi responsive to light? Notably absent from the catalog of photoresponsive fungi is budding yeast *Saccharomyces cerevisiae*. Although decreased growth rate has been observed with increasing illumination in yeast, the phenotype is generally weak and apparent only at low temperatures (Edmunds et al., 1978, 1979b). In addition, cell division and amino acid transport in *S. cerevisiae* can be synchronized in the circadian range to LD cycles (Edmunds et al., 1979a). This has also been reported for the fission yeast, *Schizosaccharomyces pombe* (Kippert et al., 1991b), which separated in evolution from *S. cerevisiae* approximately 400 million years ago (Sipiczki, 2000). Considering that wild yeasts from nature (via fermentation vats or grapes) are heterogeneous in their appearance relative to the smooth, white, standard lab strains, it could be that photoresponsiveness was selected out. A similar situation exists in *Aspergillus nidulans*, where a mutant strain was adopted as the lab strain, based on its uniform, smooth colonies that constantly conidiate. Characterization of this mutant strain indicates that it has lost light regulation of conidiation (Mooney and Yager, 1990). The issue of yeast photobiology is still open, but modern techniques will increase the chances of it being resolved, for example, by identifying light-inducible genes with the help of chromatin immunoprecipitation analysis of wild yeasts grown in darkness and given a light pulse.

Although numerous light-regulated or light-dependent physiologies have been described for the fungi, until recently only flavin-mediated blue light receptors were known (Sano et al., 2007). Pharmacological experiments or experiments with mutants have suggested cytochrome b (Kippert et al., 1991a, 1991b) and molybdenum cofactors (Ninnemann, 1991) participating in light reception of fission yeast and *Neurospora*, respectively. Genetic approaches have used *Neurospora crassa* to isolate light reception-impaired mutants, leading to identification of photoreceptor genes and proteins. Via mutagenesis, two loci, *white-collar-1* and *white-collar-2* (*wc-1* and *wc-2*), have been identified (Ballario et al., 1996; Linden and Macino, 1997), apparently deficient in most (H. Ninnemann, personal communication; Dragovic et al., 2002), if not all (Degli-Innocenti and Russo, 1984; Russo, 1988), photoresponses (details provided below). *wc-1* is homologous to the plant *NPH1* gene (Huala et al., 1997). An additional blue-light photoreceptor was identified based on abnormally high carotenoid production in *Neurospora*, namely, VIVID (VVD). Loss of VVD function results in abnormal (deficient) photoadaptation.

Genome sequencing projects have revealed that some classes of fungi (including *Aspergillus*, *Ustilago*, and *Neurospora*) have phytochromes that sense red light (Blumenstein et al., 2005). In *Aspergillus*, these proteins regulate asexual and sexual development, but in opposite ways. Opsin-like proteins have also been identified, apparently with sensitivity to green light (Bieszke et al., 1999). Their role in photoreception has only been demonstrated in heterologous expression systems, and there is as yet no known green-light–regulated *Neurospora* biology.

Daily Rhythms in Fungi

A second requirement for photoperiodism—the assessment of the light or the dark period—entails a timer, which in plants and animals is somehow linked to the circadian system. In *Neurospora*, a distinction between light reception and the circadian system is difficult since the two seem to constitute a fused system—the *white collar* genes are also integral for the circadian clock, and *frequency* (*frq*), previously considered only essential for the circadian system, is also required for light-regulated conidiation (Chang and Nakashima, 1997; Merrow et al., 1999; Lakin-Thomas and Brody, 2000). Further, many light-induced responses in *Neurospora* (even those that are not rhythmic in constant conditions) are modulated by the circadian system (Merrow et al., 2001).

The endogenous versus exogenous dispute in early circadian research has also been carried out on the back of fungi. Specifically, the dung-loving fungus *Pilobolus* has interested many scientists inside (Bruce et al., 1960) and outside (Delbruck et al., 1976) of the circadian field because of its spectacular rhythm of shooting asexual spores as far as 2 m! Although Pfeffer (1915) and Bünning (1932) had published unequivocal proof of the endogenous nature of the leaf movement rhythm in plants, the discussion whether the sporulation rhythm in *Pilobolus* is caused by exogenous or endogenous factors was not resolved until the 1950s. The first scientific investigation of this rhythm did not challenge the phenomenon in constant conditions and thus attributed its cause to the LD cycle (Coemans, 1861; Klein, 1872; Brefeld, 1881). Two independent reinvestigations in the 1940s included experiments in constant light. The author of the first study (McVickar, 1942) also excluded endogenous processes because rhythmicity ceased in LL, while the author of the other study

(Klein, 1948) did not commit himself because only one or at most two additional peaks persisted. Klein also submitted *Pilobolus* to various symmetrical non-24-h LD cycles and found that the spore-shooting rhythm was absent in LD 4:4, 8:8, 24:24 cycles but showed a large amplitude in LD 16:16 cycles, even larger than in any 24-h cycles with varying photoperiods. This series of experiments probably represents the first systematic investigation of the influence of photoperiod on the sporulation rhythm: both short (LD 4:20) and long (LD 20:4) photoperiods appear to suppress rhythmicity, while it is present with a low amplitude in LD 9:15 and 15:9 and with a higher amplitude in LD 12:12.

The first study to prove the endogenous nature of this rhythmicity included experiments both in LL and DD (Schmidle, 1951). When *Pilobolus* is kept in an LD cycle of 12:12 and released to constant light, the rhythm rapidly dampens, whereas it continues when released to DD. Rhythmicity in DD also persists after release from several days in LL. Although spore shooting is arrhythmic in LL, the amount of spores shot is 20-fold compared to DD. Additional evidence for circadian qualities in *Pilobolus* comes from the remarkable thesis work of Esther-Ruth Uebelmesser (1954). Many of her experiments anticipated circadian protocols, frequently used in later years (different T-cycles and photoperiods, reciprocity, night-interruption experiments, entrainment by temperature cycles, etc.). Uebelmesser was a true pioneer of our field and has certainly inspired Colin Pittendrigh to use *Pilobolus* as a circadian model system (Bruce et al., 1960).

Uebelmesser's work confirmed the endogenous basis of the spore-shooting rhythm and showed that the phase angle of the rhythm in LD cycles is species specific. While *P. sphaerosphorus* peaks at external time ExT 09, *P. crystallinus* reaches its discharge maximum at ExT 20 (for definition of ExT, see Daan et al., 2002). She showed that this phase angle is specific for different photoperiods and is not strictly coupled to dawn or dusk (figure 7.2).

Sordaria fimicola and *Daldinia concentrica* (both also *Ascomycetes*, like *Neurospora*) show circadian rhythms in sexual spore shooting in DD, which are entrainable in LD (Austin, 1968; Ingold and Cox, 1955). In *Daldinia*, circadian rhythmicity persists in LL, though it damps after several days; in *Sordaria*, LL appears to suppress rhythmic release (Austin, 1968). Interestingly, *S. fimicola* shares a functional *frq* ortholog, relative to *N. crassa* (Merrow and Dunlap, 1994), suggesting mechanistic similarities in the photobiology and in the circadian programs of these species. Like *Daldinia*, spore release of the basidiomycete *Pellicularia filamentosa* is circadian in either LL or DD and entrainable by LD cycles (Carpenter, 1949). Although not tested under constant conditions, many fungi, mainly of the *Ascomycota* and the *Deuteromycota* but also of the *Zygomycota* (we could not find reports from the lichens), are rhythmic when measured in nature or under laboratory LD cycles either as diurnal or as nocturnal sporulators (see excellent review in Ingold, 1971). Among them are, however, also species that show rhythmicity in LD but neither in LL nor DD (e.g., *Pyricularia*; see Barksdale and Asai, 1961), indicating that not all of the rhythmic sporulators are controlled by a circadian clock or that the appropriate laboratory conditions have not been identified (we favor the latter

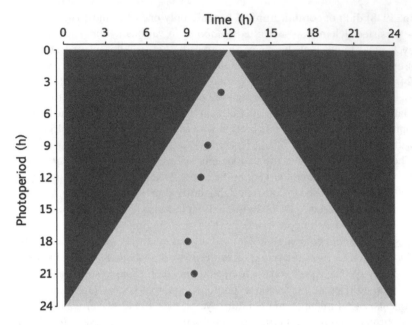

FIGURE 7.2.
Phase angle of the sporulation maximum of *Pilobolus spaerosphorus* in different photoperiods (drawn after Uebelmesser, 1954).

explanation). Among the fungal organisms, reports for rhythmic spore discharge have only been published for the Oomycota (e.g., *Phytophthora*, see Hirst, 1953). Considering that fungi are in many cases plant or animal pathogens, the observations that their reproductive success depends on the light regime are of paramount importance (e.g., the oat pathogen *Erysiphe graminis* produces many fewer spore-forming structures in constant light, relative to DD or LD; see Carver et al., 1994).

In view of the obvious phenotype of spore production and release, relatively few nonreproductive rhythms have been evaluated for the fungi. In *Neurospora* numerous metabolic enzymes are rhythmically expressed as determined by function (Hochberg and Sargent, 1974) or gene expression (Bell-Pedersen et al., 1996; Nowrousian et al., 2003). Rhythmic CO_2 production has been reported, though the data do not appear in print (Lakin-Thomas et al., 1990). An additional factor that must now be considered for our interpretation of clock-regulated functions in fungi based on *Neurospora* as a model system is that the lab strain used to monitor conidiation (and for most other clock work) is itself a mutant, namely, in the *ras* gene (Belden et al., 2007). So far, we know of no major discrepancies between the fungal clock in a wild type versus a *ras*-mutant background, but relatively little has been investigated until now. Light responses at the RNA level are exaggerated in the mutant background, and considering clock regulation by light, there may still be some surprises to come.

EVIDENCE FOR SEASONAL ADAPTATION IN *NEUROSPORA*

Most organisms integrate the information from both LD and temperature cycles for an optimal adaptation to seasons. This is evidently the case in poikilotherms but has even been suggested for seasonal control in humans (Roenneberg and Aschoff, 1990a, 1990b). Temperature will surely play a role in seasonal adaptation within the fungal kingdom, even if these effects will merely act via phototransduction.

Effects of LD and Temperature Cycles on *Neurospora*

Uebelmesser's entrainment experiments on *Pilobolus* included two types of LD cycles—those with the length of the LD cycle was changed, with light and darkness always filling half of the cycle (for reference, see Roenneberg and Merrow, 2001a), and those with only the photoperiod altered in the context of a 24-h cycle (figure 7.2). While, the rhythm appears to be driven in the former type of experiments (so that spore shooting always peaked 28 h after light onset), phase angles in the latter type of experiments were locked neither to dawn nor to dusk (figure 7.2). These results are analogous to our findings in *Neurospora crassa*. In symmetrical LD T-cycles, bands always begin to form after a constant, strain-specific delay after the onset of darkness (approximately one-third of strain's circadian period in DD, i.e., 7 h in the *frq*[+] strain; see figure 7.3A), indicating a nonentrained, masking response. And, as in *Pilobolus*, phases of entrainment are independent of dawn or dusk, when different photoperiods are presented in the context of a 24-h cycle (figure 7.3B).

The striking difference between the distribution of phase angles in the two conditions either stress the complexity of the mechanisms underlying the adaptive processes in different LD conditions or reflect our limited understanding of the system. Several factors may be involved: development and maturation, refractoriness to light, length and timing of the light and dark periods, and circadian control. The constant lag of 6–7 h in the condition shown in figure 7.3A indicates that the developmental program underlying the formation of bands cannot be compressed to less than the observed lag. The orientation to midnight under the conditions shown in figure 7.3B suggests seasonal adaptation in conjunction with circadian entrainment. As described further below (see "Understanding the Molecular Mechanisms Underlying Photoperiodism"), the observed midnight constancy is not restricted to the apparent output of the system, that is, the formation of conidial bands, but is also present at the molecular level and concerning clock components.

The *Neurospora* circadian system (Merrow et al., 1999), its responsiveness to light pulses (Nakashima and Feldman, 1980; Gooch et al., 1994), and the levels of *frq* RNA and FRQ protein are strongly affected by temperature (Liu et al., 1998; Roenneberg and Merrow, 2001b) despite the period of the *Neurospora* clock being as temperature compensated as in other systems (Feldman and Hoyle, 1973; Loros and Feldman, 1986). We have shown that *Neurospora* can be entrained in the circadian range to temperature cycles even without the expression of FRQ

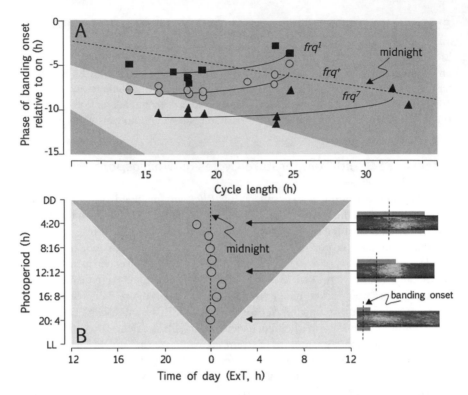

FIGURE 7.3.

(A) Entrainment of *Neurospora's* conidiation rhythm in symmetrical LD cycles of different period length (T). The phase angle relative to the light:dark transition is specific for the different *frq* strains (circles: *frq+*, period = 21.5 h; squares: *frq¹*, period = 16 h; triangles: *frq⁷*, period = 29 h) and is independent of T, except in longer cycles. Conidiation starts after approximately one-third of the circadian period, as determined in DD (redrawn after Merrow et al., 1999). (B) Unlike in the symmetrical LD conditions shown in A, the phases of entrainment are independent of the LD changes when different photoperiods are presented in the context of a 24-h cycle, as exemplified by the banding rhythm in the *frq+* strain (redrawn after Tan et al., 2004a). Examples of the banding onsets relative to the dark phase are shown on the right.

protein (Merrow et al., 1999; Roenneberg et al., 2005), indicating that circadian qualities exist in *Neurospora* outside of the *frq*/FRQ feedback loop. In addition, the circadian system is light-blind in FRQ-less strains (Chang and Nakashima, 1997; Merrow et al., 1999, 2001; Lakin-Thomas and Brody, 2000) while other light responses remain intact. It will be interesting to see what role FRQ plays in measuring night length and transducing this information to reproductive development and how temperature (as level and/or as cycle) affects the qualities of the photoperiodic timing.

Photoperiodism in *Neurospora*

To investigate photoperiodism in *Neurospora*, we chose three different physiologies as readouts: sexual reproduction (by measuring protoperithecial production), asexual propagation (by measuring conidia production), and pigmentation (by measuring carotenoid production). To test the involvement of the circadian clock in photoperiodic measurement, we compared the responses in the commonly used laboratory wild-type strain (*bd*) with those of a strain deficient in the *frequency* gene (*bd frq*^{10}) in each of these assays.

The production of protoperithecia is the first stage in *Neurospora*'s sexual development. While protoperithecial production by the wild-type strain is 5- to 8-fold in a 14 h photoperiod, compared to LL or DD (solid circles in figure 7.4), it does not change systematically between photoperiods in the mutant (open circles in figure 7.4). Simple irradiance cannot account for the photoperiodic response in the wild type, since LL yields numbers equivalent to DD conditions.

As with protoperithecia, the photoperiod-dependent production of asexual spores requires a functional FRQ oscillator (figure 7.5). Again, simple irradiance cannot account for this response, since LL yields numbers equivalent to DD conditions. These results indicate that *Neurospora* uses photoperiodic information to control both asexual and sexual development and that this mechanism relies on an intact circadian system. The absence of a functional FRQ oscillator results in the

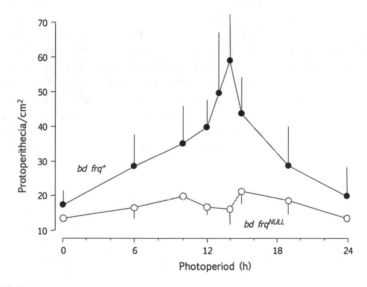

FIGURE 7.4.
Protoperithecial production in different photoperiods (redrawn after Tan et al., 2004b). The photoperiod-dependent development of the presexual structures, protoperithecia, is compared in a laboratory wild-type strain (bd frq$^+$; solid circles) and a mutant deficient in the FRQ oscillator (bd frq^{10}; open circles).

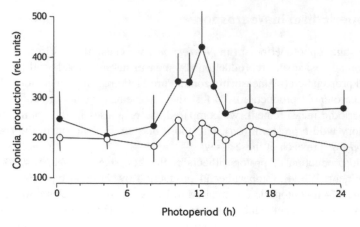

FIGURE 7.5.
Conidia production in different photoperiods (redrawn after Tan et al., 2004b). As for protoperithecial production (figure 7.4), the photoperiod-dependent production of asexual spores requires a functional FRQ oscillator. Symbols as in figure 7.4.

loss of several clock properties, including light-regulated conidial development in LD cycles (Aronson et al., 1994a; Merrow et al., 1999).

N. crassa has long been a model system in photobiology. Numerous well-characterized light-regulated functions have been described that can now be investigated for regulation by photoperiod. We chose to test the induction of carotenoids in mycelia, which is controlled by a light transduction pathway physiologically and genetically distinct from (though coupled to) that regulating the formation of conidial bands (Dragovic et al., 2002). For this assay, we used the aconidial *fluffy* (*fl*) mutant, to eliminate any background from constitutive production of carotenoids by conidia. Mycelial carotenogenesis is triggered by light, and 2 h of light per day induces an approximate 10-fold increase in production compared to mycelia incubated in DD (see figure 7.7). Between LD 2:22 and 10:14, carotenoid levels remain constant, despite a 5-fold increase in daily irradiance (figure 7.6). From LD 10:14 to 14:10, however, carotenoid levels climb, remaining high until LD 20:4. In even longer photoperiods, carotenoid levels decrease to amounts lower than those found with 2 h of light per day. Hence, a measurement of day or night length is evident also in this nonreproductive output.

LD cycles fail to consolidate conidial bands in *frq*[10], but light does enhance mycelial carotenoid production in this mutant, reaching levels approximately 50% compared to wild type (Merrow et al., 2001). Thus, the *frq* knockout strain can be used to distinguish the different light transduction pathways leading to conidiation and carotenoid production. The aconidial, arrhythmic *fl frq*[10] double mutant was therefore incubated in varying photoperiods and then assayed for carotenoid levels. The results show that FRQ is also essential for the photoperiodic response of this nonreproductive output (open circles in figure 7.6). However in this case, the

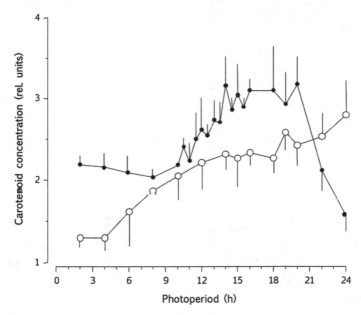

FIGURE 7.6.
Carotenoid production depends on day length, and this photoperiodic regulation requires a functional FRQ oscillator (redrawn after Tan et al., 2004b). Symbols as in figure 7.4.

insensitivity to different photoperiods is entirely different from those in figures 7.2 and 7.3: carotenoid production increases at an almost constant rate from LD 2:22 to LL.

Neurospora Measures the Length of the Night

Most photoperiodic response mechanisms measure night rather than day length (Saunders, 1982). For this, some timer must measure the length of darkness. A likely candidate for measuring scotoperiod in *Neurospora* is FRQ oscillator acting as a night timer. Light levels as low as moonlight prevent the decay of *frq* RNA or FRQ protein, perhaps through continued light-induced production (Crosthwaite et al., 1995; Merrow et al., 1997), while these molecules readily decay in darkness (Merrow et al., 1997, 1999; Liu et al., 2000). Thus, different forms of *frq*/FRQ affect not only the period in DD but also the phase of synchronization in symmetrical LD cycles. Furthermore, the *frq*-less mutant is insensitive to photoperiods in any of the three assays (figures 7.4–7.6), indicating FRQ as a critical component in measuring night length in *Neurospora*. The regulation of both *frq* RNA and FRQ protein has been extensively investigated both in DD and LL (Aronson et al., 1994b; Crosthwaite et al., 1995; Garceau et al., 1997; Collett et al., 2002), but rarely in LD cycles. *frq* RNA starts to rise 8–12 h after release into DD, while LL conditions maintain noisy

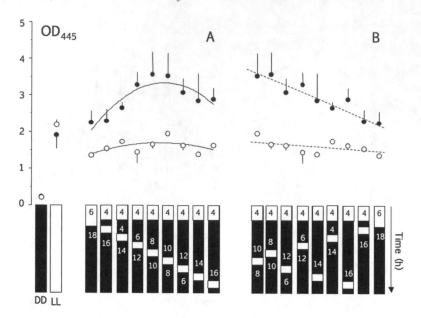

FIGURE 7.7.
Neurospora measures night length (redrawn after Tan et al., 2004b). Light-induced mycelial carotenogenesis (as measured by OD445) was measured in different night break experiments using the *frq*-sufficient and the *frq*-null strain (solid and open circles, respectively). The different LD regimes are shown below the graphs (with numbers indicating the length of the different subportions in h). The results shown in A are rearranged in B according to the longest uninterrupted portion of darkness. Except for the cultures kept in constant darkness or constant light (graphed on the far left of A), all samples received a total of 6 h of light, either as one or two blocks (4 plus 2 h).

but constitutively high *frq* levels. One hypothesis would therefore predict that *frq* (or rather FRQ) levels could be used as an hourglass timer for night length—reset by each dusk. We performed night break experiments based on a traditional protocol for investigating scotoperiodism (Bünning, 1960) in which the same amount of light in the form of two 2-h pulses is administering with varying durations of intermittent darkness (figure 7.7A). The readout for these experiments was light-induced carotenogenesis based on an LD cycle of 6:18 that does not stimulate carotenoid production in consolidated photoperiods (see figure 7.6). In the different conditions, the longest (of the two possible) dark periods varied between 18 and 10 h, in which carotenoid production showed a negative correlation (figure 7.7B). Hence, measuring night length appears to be at least one of the critical mechanisms for photoperiodism in *Neurospora*, as it is in many other organisms (Saunders, 1982). The *frq*[10] mutant showed almost no response to the different night interruptions, consistent with the lack of a photoperiodic response in this genetic background.

UNDERSTANDING THE MOLECULAR MECHANISMS UNDERLYING PHOTOPERIODISM

One of the advantages of *Neurospora* as a model system for photoperiod research is the detailed molecular description of both the light input pathway and the circadian clock. *Neurospora* can be used to define genes and proteins that participate in photoperiodic mechanisms. Good candidates would be, of course, genes that participate in daily timing, since a timer that measures and compared day and night lengths on a daily basis is required for photoperiodic responses. Experiments in *Drosophila* indicated, however, that the contribution of clock genes to photoperiodism is not simple and straightforward: the *per* gene has systematic sequence differences according to latitude, corresponding functionally to temperature compensation of the circadian system (Sawyer et al., 1997), yet photoperiodic response curves and Nanda-Hamner–type experiments determined that photoperiodic responses remain intact in *per⁰* flies (Saunders, 1990).

The *Neurospora* Circadian System

Figure 7.8 summarizes the components of the *Neurospora* circadian transcriptional-translational feedback loop. The products of the two *white collar* genes (*wc-1* and *wc-2*) form a protein complex (WCC) that also interacts with the FRQ protein (Merrow et al., 2001). *frq* transcription is activated by WCC and feeds back negatively onto its own expression (Aronson et al., 1994b; Lee et al., 2000). An oscillator (*frq*-less oscillator, FLO), entrainable by temperature cycles in the circadian

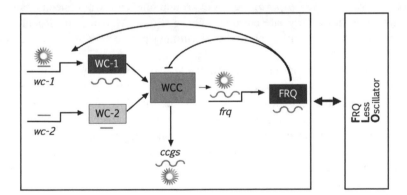

FIGURE 7.8.
The components of the *Neurospora* circadian transcriptional-translational feedback loop (see text for details). Rhythmic or constitutive expressions are indicated by wiggle or straight lines, respectively, and light inducibility, by the sun symbol (redrawn after Merrow and Roenneberg, 2001).

range, is present in FRQ-less strains and interacts with the components in the transcriptional-translational feedback loop (Merrow et al., 1999; Roenneberg and Merrow, 2001b). The *white collar* genes were identified through screens looking for mutations in light perception (Harding and Turner, 1981; Degli-Innocenti and Russo, 1984; Ballario et al., 1996; Linden and Macino, 1997). They were demonstrated as integral for clock function some years later (Russo, 1988; Crosthwaite et al., 1997). The *white collar* gene products are transcription factors, as judged by sequence analysis and DNA binding by WC-1 (Froehlich et al., 2002). WC-1 is photoreceptive, via its FAD molecule (He et al., 2002), and WC-2 is an essential cofactor, in that mutations in each gene lead to more or less identical phenotypes. WC-1 mediates transcription as part of the normal response to light in *N. crassa*. In a wild-type strain, some rapidly induced RNA species appear within minutes (Sommer et al., 1989). Interestingly, it has been noted that perithecial development, even in darkness, is poor in *wc* mutants (i.e., in the absence of light responsiveness). This suggests a role for the clock in the reproductive success of *N. crassa*, something that has been alluded to in the past (Lakin-Thomas et al., 1990). Given that perithecial development is stimulated by light (Degli-Innocenti and Russo, 1984), that light input pathway components are inseparable from the clock, and that they are themselves under circadian control (Lee et al., 2000; Merrow et al., 2001), perhaps the endogenous rhythmicity contributes an ersatz light stimulus in DD. Regardless, the involvement of photoperiodism in the reproductive success of *Neurospora* must be considered.

A precedence exists in plants, where many of the phytochrome and cryptochrome photoreceptors (they are multigene families) have effects on free-running rhythms and photoperiodism (Somers et al., 1998; Martinez-Garcia et al., 2000). Furthermore, there may be several more photoreceptors that could contribute to photoperiodism in *Neurospora*: the VVD protein binds a FMN molecule, and the mutant strain has a very late entrained phase (Boesl, Merrow, and Roenneberg, unpublished data), despite a normal free running period (Heintzen et al., 2001; Shrode et al., 2001).

Circadian oscillations are seen from the level of molecules to complex outputs such as behavior. A set of "clock genes" that forms an autoregulatory, transcription/ translation negative feedback loop has been described as the basis of circadian rhythmicity in all model organisms. Research into the molecular mechanisms of entrainment has primarily focused on how clock components are regulated following introduction of light. The first demonstration therein showed a dramatic increase in clock gene RNA levels in *Neurospora* (Crosthwaite et al., 1995). Subsequently, an alternative mechanism, namely, light-dependent degradation of a clock protein (TIMELESS, in *Drosophila*), was described (Hunter-Ensor et al., 1996; Lee et al., 1996; Myers et al., 1996). These mechanisms can be used to explain phase shifts and have been recruited to describe entrainment (Crosthwaite et al., 1995; Young, 1998; Liu, 2003). Yet how they mediate entrainment to full days and nights and how they contribute to the phenomenon of photoperiodism are far from understood.

The *Neurospora* Molecular Clock in Different Photoperiods

In contrast to our knowledge of molecular components in entrained conditions, there is a wealth of information concerning regulation of *frq* RNA and FRQ protein in constant darkness (DD) and constant light (LL). In LL, *frq* RNA and protein levels are noisy but constitutively high (Crosthwaite et al., 1995; Collett et al., 2002). After transfer to DD, *frq* RNA first decreases and then starts to rise in approximately 8 h. FRQ protein accumulates some 4 h later (Garceau et al., 1997). If darkness is maintained, the circadian clock takes over the control of *frq* and FRQ, repeating a wave of expression about once every 22 h (Aronson et al., 1994b; Garceau et al., 1997). Thus, judged by the kinetics of *frq* RNA and protein in constant conditions, the clock is stalled during light, while it clearly progresses during dark. Given the evidence for involvement of the *frequency* gene in photoperiodism, as shown on the behavioral level (figures 7.4–7.6; Tan et al., 2004b), FRQ could be an excellent timer for night length—a straightforward hypothesis that can be easily tested by determining a molecular profiles of *frq* and FRQ in a series of different photoperiods (figure 7.9).

In all photoperiods tested, *frq* RNA rapidly increases approximately 10-fold with the onset of light (figure 7.9, left panels). Levels then adapt to about half-maximal levels, staying constant throughout the light period while they rapidly decrease upon transition to darkness. In scotoperiods longer than 10 h, *frq* levels start to rise approximately 8 h into the night (presumably under the control of the circadian system).

Thus, except for a small bump in RNA levels in longer scotoperiods, RNA levels of *frq* are driven by light. This driven response is not *frq* specific but can be observed for other light-regulated genes (Tan et al., 2004a) whether they are coregulated by light *and* the circadian clock, such as *vivid* (*vvd*; Heintzen et al., 2001) and *conidiation-10* (*con-10*; Lauter and Yanofsky, 1993), or only by light, such as *white-collar-1* (*wc-1*; Lee et al., 2000; Merrow et al., 2001) and *albino-1* (*alb-1*; Merrow et al., 2001). *vvd* encodes a protein that contributes to photoadaptation in *Neurospora* (Heintzen et al., 2001; Schwerdtfeger and Linden, 2001; Shrode et al., 2001) and thus could regulate light signaling pathways that contribute to entrainment. As with *frq*, expression of *vvd* and *con-10* (a gene involved in forming conidia) RNA is driven by light, although in the case of *con-10*, RNA levels adapt down to dark levels in long days. As in the case of *frq*, *vvd* and *con-10* RNAs spontaneously increase in longer nights with a constant delay of about 8 h.

Among the light-induced RNAs that are not controlled by the circadian clock in DD, *wc-1* shows the acute response to lights on in all photoperiods, immediately adapting to baseline levels, where it remains until the next experimental dawn. The adaptation kinetics of *al-1* expression are similar to those of *frq* and *vvd*.

Thus, all five RNA profiles mirror the light environment with a rapid response to lights on, adaptation, and (when their levels do not drop down to baseline levels) a rapid decrease in response to lights off. This drivenness at the RNA level contradicts the concept of a transcriptional-translational feedback loop as the molecular

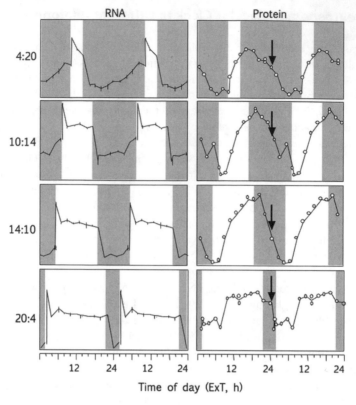

FIGURE 7.9.

frq RNA and FRQ protein expression (left and right panels, respectively) in different photoperiods (LD cycles indicated on the left). Gray areas correspond to the dark portions of the cycles. Arrows in the right panels indicate when FRQ reach half-maximal levels that correspond to middark in each of the different photo- (scoot-) periods (redrawn from Tan et al., 2004a).

basis for circadian rhythmicity—especially since it is safe to presume that the molecular mechanisms responsible for circadian rhythmicity were never selected for in evolution under constant conditions (Roenneberg and Merrow, 2002). They also predict that the phase angle between internal and external events would stay constant in *zeitgeber* cycles of different lengths (T-cycles). On the behavioral level, however, *Neurospora* does establish systematic phase angles in different T-cycles when temperature is used as the *zeitgeber* (Merrow et al., 1999), and it also shows independence relative to dawn and dusk in different photoperiods when the cycle length is fixed at 24 h (see figure 7.3B; Tan et al., 2004a). If one presumes adaptive entrainment to be the signature of a functioning circadian system, these results indicate that clock function is independent of RNA kinetics in light cycles and that the circadian kinetics of the RNAs in temperature cycles is rather a consequence (output) of the oscillator than a prerequisite for its rhythmicity. In light of recent work elaborating translational regulation of *frq*/FRQ as a key mechanism for adjusting clock function according to temperature, this is an attractive interpretation (Diernfellner et al., 2007). The *frq*/FRQ loop may be important for sustaining the oscillation in DD, but *frq* fails to oscillate in light cycles and is thus not imperative to maintain the *Neurospora* circadian clock under entrained conditions, where it obviously also functions adequately as an adaptive timer. Since the *frq* gene has, however, been shown to have drastic effects on the circadian clock in *Neurospora*, one has to presume that its key contribution must be restricted to the level of proteins.

Unlike the kinetics of *frq* RNA levels in different photoperiods, FRQ protein levels are far more consistent with molecular mechanisms underlying adaptive entrainment. In contrast to their kinetics in DD, where the timing (or phase) of the FRQ oscillation is strictly determined by the last transfer from light, the phase of the FRQ oscillation in LD cycles is not keyed to these transitions. After long nights (figure 7.9, right panels), the protein surges up to high levels within 2 h after lights on, consistent with the traditional observations made in experiments where tissue is grown in DD and transferred to LL (Collett et al., 2002). However, the response to dawn after shorter nights contrasts all prior observations. A significant increase in FRQ levels can be delayed by as much as 8 h into the light phase. Thus, in long nights, the surge of FRQ coincides with dawn, while it is progressively delayed in shorter nights (longer days). The rate of FRQ decline also depends on photoperiod, such that it reaches approximately half-maximal levels shortly after midnight in all photoperiods tested (arrows in figure 7.9). In LD 20:4, RNA and protein actually decline coincidentally, challenging the traditional view of negative feedback, the function of which has always been associated with a predictable delay between the oscillations of RNA and protein (Aronson et al., 1994b).

Unlike *frq* RNA, FRQ protein levels are as independent of dawn or dusk as is the observed behavior (figure 7.3B), where the onset of conidiation is around midnight in all photoperiods. Although FRQ—even as a constitutively expressed component—is essential for light-regulated conidiation (Merrow et al., 2001), the biochemical association of FRQ to the conidiation pathway is not known. These data suggest that high FRQ levels suppress conidiation, while FRQ decline supports it.

Temporal programs such as the circadian clock integrate environmental history and use this *memory* to endogenously generate functional niches within the daily or seasonal cycle. The experiments under entrained conditions in different photoperiods show that expression of FRQ protein, unlike *frq* RNA, is also history dependent. Because *frq* transcription is acutely induced by light regardless of FRQ concentration, it was hypothesized that light overrides the negative feedback by FRQ (Crosthwaite et al., 1995). The entrainment results shown here indicate a more complex regulation, beyond the transcriptional feedback loop, involving posttranscriptional regulation of the *Neurospora* circadian clock. Depending on the structure of the LD cycle, for example, the declines of *frq* RNA and protein in darkness can either occur with a lag or can coincide. Candidates for the components responsible for this posttranscriptional regulation have been detected in microarray experiments (the expression of translational and posttranslational regulators show circadian oscillation) and determinants of clock protein production and destruction are already suggested by genetics and biochemistry (Liu et al., 2000; Akhtar et al., 2001; Görl et al., 2001; Grima et al., 2002; Ko et al., 2002; Panda et al., 2002). The photoperiod-specific turnover kinetics of FRQ and regulation of conidial band formation also show that the circadian clock in *Neurospora* must be running in LD cycles and not only during darkness, as has been suggested by most prior observations (Liu, 2003). Without this feature, one of the central qualities of a circadian system, namely, the ability to systematically vary the phase angle of entrainment depending on *zeitgeber* conditions, that is, in different seasons (Roden et al., 2002), would be lost.

Given that *Neurospora* is considered a model system for higher eukaryotes, will similar mechanisms be described in mammals? In hamsters and mice, the expression of selected clock gene RNAs and proteins in the clock pacemaker in the brain (the suprachiasmatic nucleus, SCN) mirrors the light cycle in long and short days; for other genes (e.g., Cry-1), it can take several days to resynchronize with the new LD cycle (Nuesslein-Hildesheim et al., 2000; Reddy et al., 2002). However, as shown for the mammalian clock gene, *Perl*, entrainment of gene expression can be dissociated from entrainment of SCN neurophysiology and behavior (Vansteensel et al., 2003). This is comparable to dissociation of clock gene RNA versus protein levels and behavior in *Neurospora*.

CONCLUSIONS

Neurospora Is Found Worldwide

In spite of the fact that the first reports of the bread mold *Neurospora* came from France (Perkins, 1991), *Neurospora* is generally associated with geographical regions that do not exhibit large photoperiodic changes over the course of the year—most strains were collected in the southern United States or India. However, *Neurospora crassa* strains have been found over a large range of southern and northern latitudes (figure 7.10). *N. intermedia* and other *Neurospora* species have been found as far north as Alaska and as far south as New Zealand, so the genus *Neurospora* is found at latitudes corresponding to regions where nights and days can be as long as 16 h and as short as 8 h. *Neurospora*, therefore, amends itself to investigations similar to the ones performed by Pittendrigh on *Drosophila* (Pittendrigh et al., 1991; Pittendrigh, 1993).

The distribution of collected samples shown in figure 7.10 reflects only the locations where scientists have actively looked for *N. crassa* or where they found it by chance, rather than *Neurospora*'s actual geographical distribution. The large number of samples collected in Louisiana, Florida, and India, however, might indicate *Neurospora*'s preferred habitat, concerning both nutrition and climate. *Neurospora* is a pioneer organism after fires have destroyed most of the vegetation, and it can specifically utilize those carbon sources that remain after fires (e.g., quinic acid). It

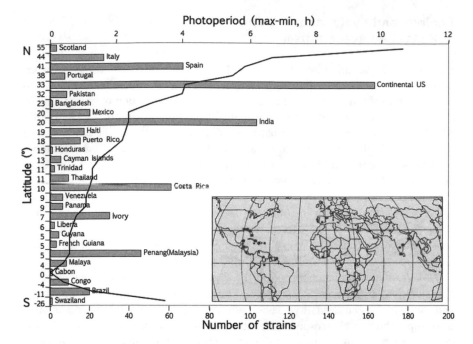

FIGURE 7.10.
Latitudinal and geographical distribution of the wild-type strains of *Neurospora crassa* deposited in the Fungal Genetic Stock Center, Kansas (www.fgsc.net; R. Metzenberg, D. Perkins, and D. Jacobson personal communication; Turner et al., 2001; Jacobson et al., 2006). The respective photoperiodic differences between longest and shortest day of the year are drawn as a line (upper horizontal axis). The global map inserted at the bottom right shows the locations where *Neurospora crassa* was found.

thrives under moist and warm conditions. The strain used by practically all laboratories that study *Neurospora* was first isolated by Shear and Dodge (1927) from sugar cane bagasse in Louisiana. In these regions, humidity remains relatively constant over the course of the year (yet, with large daily changes), while the annual changes in photoperiod amount to 4 h even in this southern location (30°N). Annual temperature changes can range from 8°C to 34°C. These changes would certainly make it advantageous for *Neurospora* to anticipate seasons with the help of photoperiodic and/or thermoperiodic mechanisms, especially as these changes correlate with large differences in rainfall. It should be noted that the largest amplitudes in annual human reproduction occur at latitudes around 30° (Roenneberg and Aschoff, 1990a, 1990b).

Based on the evidence that there is a distinct physiological behavior of *Neurospora crassa* in response to different photo- (scoto-) periods, future initiatives should address the ecological and evolutionary adaptiveness of photoperiodism in fungi, its molecular basis as well as the relationship between the circadian and seasonal timing system.

Ecology and Evolutionary Adaptiveness of Fungal Photoperiodism

Experimental data on seasonal events in the life cycle of *Neurospora crassa* are rare and will therefore have to be investigated under controlled laboratory conditions. Such experiments should make use of the large collection of more than 160 *N. crassa* strains isolated from more than 80 locations from all over the globe, which represent many diverse climatic, seasonal, and photoperiodic environments (figure 7.10). Light conditions vary dramatically between these different locations both in photoperiod and in the sun's inclination. While the latter peaks once per year at latitudes equal to or greater than 23.5°, it peaks every six months closer to the equator. Both day length and inclination affect the annual changes in temperature via the intensity of irradiation. The sun's inclination also affects the spectral composition of sunlight at different times of day. The flat slope of the sun's daily course during a polar summer exposes organisms to red-enriched light for a much longer time of the day compared to the tropics.

Modern culture chambers permit the simulation of a variety of these seasonally changing environmental factors (e.g., temperature, photoperiod, light fluence, and the light's spectral quality). Thus, different outputs, such as the production of (proto-)perithecia and ascospores, asexual spores (macroconidia), or different carotenoids, can be compared under different conditions in the numerous *Neurospora* isolates. The comparisons concern the immediate effects of different environmental simulations on the different stains as well as their long-term effects on propagation and reproduction. Experiments probing the latter would provide us with insights on whether and how adaptation to a given seasonal environment translates to fitness. Fitness can be assessed by measuring propagation (both asexual and sexual), by assessing how seasonal adaptation (e.g., via carotenoid production) might protect different strains from accumulating UV-light dependent mutations, or how it might affect the efficiency to exploit different substrates (e.g., by determining mycelial growth rates).

Molecular Biology of Photoperiodism

The identification of photoperiodism in a simple organism such as the fungus *Neurospora crassa* offers excellent possibilities to discover the molecular mechanisms underlying this ubiquitous biological function in addition to its relationships to the circadian system on the level of genes and proteins. The development of a luciferase reporter construct (Loros et al., 2007; Gooch et al., 2008) will allow high throughput monitoring of photoperiodic responses of many genes at the transcriptional and the translational level (by using fusion constructs). Expression patterns can be related to different photoperiodic profiles recorded at the functional and behavioral levels in strains from different latitudes in combination with simulations of seasonally changing environmental factors (see above). Photoperiodic adaptation can vary extensively across latitudes. While a fungus occupying niches at the polar circle is submitted to changes in day length of, on average, 8 min per day, those living near the equator experience minute changes in photoperiod over the course of the year and might use different seasonal cues to adjust their biological functions to appropriate times of year. All these measurements will most certainly involve some aspects of the circadian machinery as an internal reference for seasonal changes. Concerning the involvement of temperature in photo- and thermoperiodic responses, the role of temperature-dependent gene regulation, such as differential gene splicing, as has been shown for *frq* (Diernfellner et al., 2007), has to be taken into account.

During some parts of the year, organisms living at high latitudes are confronted with either constant light or constant darkness. It has been shown for reindeer that the circadian system—at least on the physiological and behavioral level—does not run free under these conditions (van Oort et al., 2005). One might therefore expect that the capacity to produce a self-sustained oscillation might vary in *Neurospora* strains collected from different latitudes. These and many other questions can now be addressed with the help of reporter genes and a large collection of strains collected at different locations on the globe.

Acknowledgments   This chapter is dedicated to David D. Perkins. *Neurospora* scientists will miss his experience, knowledge, and generosity. Our work is supported by the Deutsche Forschungsgemeinschaft, the Nederlandse Organisatsie voor Wetenschappelijk Onderzoek, and the University of Groningen Rosalind Franklin Research Fellowship Program.

References

Akhtar RA, Reddy AB, Maywood ES, Clayton JD, King VM, Smoth AG, Gant TW, Hastings MH, and Kyriacou CP. (2001). Circadian cycling of the mouse liver transcriptome, as revealed by cDNA microarray, is driven by the suprachiasmatic nucleus. Curr Biol 12: 540–550.

Alves SB, Risco SH, and Almeida LC. (1984). Influence of photoperiod and temperature on the development and sporulation of *Metarhizium anisopilae* (Metsch,) Sorok. Z ang Ent 97: 127–129.

Aronson BD, Johnson KA, and Dunlap JC. (1994a). The circadian clock locus *frequency*: A single ORF defines period length and temperature compensation. Proc Natl Acad Sci USA 91: 7683–7687.

Aronson BD, Johnson KA, Loros JJ, and Dunlap JC. (1994b). Negative feedback defining a circadian clock: Autoregulation of the clock gene *frequency*. Science 263: 1578–1584.

Austin B. (1968). An endogenous rhythm of spore discharge in *Sordaria fimicola*. Ann Bot 32: 261–278.

Ballario P, Vittorioso P, Magrelli A, Talora C, Cabibbo A, and Macino G. (1996). *White collar-1*, a central regulator of blue light responses in *Neurospora*, is a zinc finger protein. EMBO J 15: 1650–1657.

Barksdale TH and Asai GN. (1961). Diurnal spore release of *Piricularia oryzea* from rice leaves. Phytopathology 51: 313–317.

Belden WJ, Larrondo LF, Froehlich AC, Shi M, Chen CH, Loros JJ, and Dunlap JC. (2007). The band mutation in *Neurospora crassa* is a dominant allele of ras-1 implicating RAS signaling in circadian output. Genes Dev 21: 1494–1505.

Bell-Pedersen D, Shinohara ML, Loros J, and Dunlap JC. (1996). Circadian clock-controlled genes isolated form *Neurospora crassa* are late night- to early morning-specific. Proc Natl Acad Sci USA 93: 13096–13101.

Berjarano ER, Avalos J, Lipson ED, and Cerda-Olmeda E. (1990). Photoinduced accumulation of carotene in *Phycomyces*. Planta 183: 1–9.

Bieszke JA, Spudich EN, Scott KL, Borkovich KA, and Spudich JL. (1999). A eukaryotic protein, NOP-1, binds retinal to form an archaeal rhodopsin-like photochemically reactive pigment. Biochemistry 38: 14138–14144.

Blumenstein A, Vienken K, Tasler R, Purschwitz J, Veith D, Frankenberg-Dinkel N, and Fischer R. (2005). The *Aspergillus nidulans* phytochrome FphA represses sexual development in red light. Curr Biol 15: 1833–1838.

Brefeld O. (1881). *Botanische Untersuchuchungen über Schimmelpilze*. Berlin: Springer.

Brook PJ. (1969). Stimulation of ascospore release in *Venturia inaequalis* by far red light. Nature 222: 390–392.

Bruce VC, Weight F, and Pittendrigh CS. (1960). Resetting the sporulation rhythm in *Pilobolus* with short light flashes of high intensity. Science 131: 728–730.

Bünning E. (1960). Circadian Rhythms and the Time Measurement in Photoperiodism. Cold Spring Harbor Symp Quant Biol 25: 249–256.

Bünning E. (1932). Über die Erblichkeit der Tagesperiodizität bei den *Phaseolus* Blättern. Jb wiss Bot 81: 411–418.

Calpouzos L and Chang H. (1971). Fungus spore germination inhibited by blue and far red radiation. Plant Physiol 47: 729–730.

Carpenter JB. (1949). Production and discharge of basidiospores of *Pellicularia filamentosa* (Pat.) Rogers on *Hevea* rubber. Phytopathology 39: 980–985.

Carver TL, Ingerson-Morris SM, Thomas BJ, and Gay AP. (1994). Light-mediated delay of primary haustorium formation by *Erysiphe graminis* f. sp avenae. Physiol Mol Plant Pathol 45: 59–79.

Chang B and Nakashima H. (1997). Effects of light-dark cycles on the circadian conidiation rhythm in *Neurospora crassa*. J Plant Res 110: 449–453.

Coemans E. (1861). Monographie du genre *Pilobolus*. Mém cour et des sav étrang Acad, roy de Belge 30: 1–10.

Collett MA, Garceau N, Dunlap JC, and Loros JJ. (2002). Light and clock expression of the *Neurospora* clock gene frequency is differentially driven by but dependent on WHITE COLLAR-2. Genetics 160: 149–158.

Corrochano LM. (2007). Fungal photoreceptors: Sensory molecules for fungal development and behaviour. Photochem Photobiol Sci 6(7):725–736.

Crosthwaite SK, Dunlap JC, and Loros JJ. (1997). *Neurospora wc-1* and *wc-2*: Transcription, photoresponses, and the origin of circadian rhythmicity. Science 276: 763–769.

Crosthwaite SK, Loros JJ, and Dunlap JC. (1995). Light-induced resetting of a circadian clock is mediated by a rapid increase in *frequency* transcript. Cell 81: 1003–1012.

Cummings BG. (1971) Endogenous rhythms and photoperiodism in plants. In *Biochronometry* (M Menaker, ed). Washington, DC: National Academy of Sciences, pp. 281–291.

Daan S, Merrow M, and Roenneberg T. (2002). External time—internal time. J Biol Rhythms 17: 107–109.

De Fabo EC, Harding RW, and Shropshire W Jr. (1976). Action spectrum between 260 and 800 nanometers for the photoinduction of carotenoid biosynthesis in *Neurospora crassa*. Plant Physiol 57: 440–445.

Degli-Innocenti F and Russo VE. (1984). Isolation of new *white collar* mutants of *Neurospora crassa* and studies on their behavior in the blue light-induced formation of protoperithecia. J Bacteriol 159: 757–761.

Delbruck M, Katzir A, and Presti D. (1976). Responses of *Phycomyces* indicating optical excitation of the lowest triplet state of riboflavin. Proc Natl Acad Sci USA 73: 1969–1973.

Dharmananda S. (1980). Studies of the circadian clock of *Neurospora crassa*: Light-induced phase shifting. Ph.D. thesis, University of California at Santa Cruz.

Diernfellner A, Colot HV, Dintsis O, Loros JJ, Dunlap JC, and Brunner M. (2007). Long and short isoforms of *Neurospora* clock protein FRQ support temperature-compensated circadian rhythms. FEBS Lett 581: 5759–5764.

Dragovic Z, Tan Y, Görl M, Roenneberg T, and Merrow M. (2002). Light reception and circadian behavior in "blind" and "clock-less" mutants of *Neurospora crassa*. EMBO J 21: 3643–3651.

Durand R. (1982). Photoperiodic response of *Coprinus congregatus*: Effects of light breaks on fruiting. Physiol Plant 55: 226–230.

Edmunds LN Jr, Apter RI, Rosenthal PJ, Shen W-K, and Woodward JR. (1979a). Light effects in yeast: Persisting oscillations in cell division activity and amino acid transport in cultures of *Saccharomyces cerevisiae* entrained by light.dark cycles. Photochem Photobiol 30: 595–601.

Edmunds LN Jr, Ulaszewski S, Mamouneas T, Shen W-K, Rosenthal PJ, Woodward JR, and Cirillo VP. (1979b). Light effects in yeast: Evidence for participation of cytochromes in photoinhibition of growth and transport in *Saccharomyces cerevisiae* cultured at low temperatures. J Bact 138: 523–529.

Edmunds LN Jr, Woodward JR, and Cirillo VP. (1978). Light effects in yeast: Inhibition by visible light of growth and transport in *Saccharomyces cerevisiae* grown at low temperatures. J Bact 133: 692–698.

Feldman JF and Hoyle MN. (1973). Isolation of circadian clock mutants of *Neurospora crassa*. Genetics 75: 605–613.

Froehlich AC, Liu Y, Loros JJ, and Dunlap JC. (2002). White Collar-1, a circadian blue light photoreceptor, binding to the *frequency* promoter. Science 297: 815–819.

Garceau NY, Liu Y, Loros JJ, and Dunlap J. (1997). Alternative initiation of translation and time specific phosphorylation yield multiple forms of essential clock protein FREQUENCY. Cell 89: 469–476.

Gooch VD, Mehra A, Larrondo LF, Fox J, Touroutoutoudis M, Loros JJ, and Dunlap JC. (2008). Fully codon-optimized luciferase uncovers novel temperature characteristics of the *Neurospora* clock. Eukaryot Cell 7: 28–37.

Gooch VD, Wehseler RA, and Gross CG. (1994). Temperature effects on the resetting of the phase of the *Neurospora* circadian rhythm. J Biol Rhythms 9: 83–94.

Görl M, Merrow M, Huttner B, Johnson J, Roenneberg T, and Brunner M. (2001). A PEST-like element in FREQUENCY determines the length of the circadian period in *Neurospora crassa*. EMBO J 20: 7074–7084.

Grima B, Lamouroux A, Chelot E, Papin C, Limbourg-Bouchon B, and Rouyer F. (2002). The F-box protein Slimb controls levels of clock proteins Period and Timeless. Nature 420: 178–182.

Gwinner E. (1986). *Circannual Rhythms*. Berlin: Springer Verlag.

Harding RW and Turner RV. (1981). Photoregulation of the carotenoid biosynthetic pathway in *albino* and *white collar* mutants of *Neurospora crassa*. Plant Physiol 68: 745–749.

Harvey R. (1967). Air-spora studies at Cardiff I. *Cladosporium*. Trans Br Mycol Soc 50: 479–495.

He Q, Cheng P, Yang Y, Wang L, Gardner KH, and Liu Y. (2002). White Collar-1, a DNA binding transcription factor and light sensor. Science 297: 840–843.

Heintzen C, Loros JJ, and Dunlap JC. (2001). The PAS protein VIVID defines a clock-associated feedback loop that represses light input, modulates gating, and regulates clock resetting. Cell 104: 453–464.

Hirst JM. (1953). Changes in atmospheric spore content: Diurnal periodicity and the efects of weather. Trans Br Mycol Soc 36: 375–393.

Hochberg ML and Sargent ML. (1974). Rhythms of enzyme activity associated with circadian conidiation in *Neurospora crassa*. J Bacteriol 120: 1164–1175.

Hodgkiss IJ and Harvey R. (1969). Spore discharge rhythms in *Pyrenomycetes*. VI The effect of climatic factors on seasonal and diurnal periodicities. Trans Brit Mycol Soc 52: 355–363.

Honma K, von Goetz C, and Aschoff J. (1983). Effects of restricted daily feeding on free running circadian rhythms in rats. Physiol Behav 30: 905–913.

Huala E, Oeller PW, Liscum E, Han I-S, Larsen E, and Briggs WR. (1997). *Arabidopsis* NPH1: A protein kinase with a putative redox-sensing domain. Science 278: 2120–2123.

Hunter-Ensor M, Ousley A, and Sehgal A. (1996). Regulation of the *Drosophila* protein Timeless suggests a mechanism for resetting the circadian clock by light. Cell 84: 677–685.

Ingold CT. (1971). Periodicity. In *Fungal Spores*. Oxford: Oxford University Press, pp. 214–238.

Ingold CT and Cox VJ. (1955). Periodicity of spore discharge in *Daldinia*. Ann Bot 29: 201–209.

Ingold CT and Hadland SA. (1959). The ballistics of Sodaria. New Phytol 58: 46–57.

Ingold CT and Nawaz M. (1967). Sporophore development in *Sphaerobolus*: Effect of blue and red light. Ann Bot 31: 469–477.

Jacobson DJ, Dettman JR, Adams R, Boesl C, Sultana S, Roenneberg T, Merrow M, Duarte M, Marques I, Ushakova A, Carneiro P, Videira A, Navarro-Sampedro L, Olmedo M, Corrochano LM, and Taylor JW. (2006). New findings of *Neurospora* in Europe and comparisons of diversity in temperate climates on continental scales. Mycologia 98 (4):550–559.

Kippert F, Engelmann W, and Ninnemann H. (1991a). "Blind" cytochrome b mutants of yeast can still be entrained to 24 hour temperature cycles. J Interdiscip Cycle Res 22: 137.

Kippert F, Ninnemann H, and Engelmann W. (1991b). Photosynchronization of the circadian clock of *Schizosaccharomyces pombe*: Mitochondrial cytochrome b in an essential component. Curr Genet 19: 103–107.

Klein D. (1948). Influence of varying periods of light and dark on asexual reproduction of *Pilobolus kleinii*. Bot Gaz 110: 139.

Klein J. (1872). Zur Kenntnis des *Pilobolus*. Jahrb F Wissencsch Bot 8: 305–308.

Ko HW, Jiang J, and Edery I. (2002). Role for Slimb in the degradation of *Drosophila* Period protein phosphorylated by Doubletime. Nature 420: 673–678.

Lakin-Thomas PL, and Brody S. (2000). Circadian rhythms in *Neurospora crassa*: Lipid deficiencies restore robust rhythmicity to null *frequency* and *white-collar* mutants. Proc Natl Acad Sci USA 97: 256–261.

Lakin-Thomas PL, Coté GG, and Brody S. (1990). Circadian rhythms in *Neurospora crassa*: Biochemistry and genetics. Crit Rev Microbiol 17: 365–416.

Lauter F-R and Yanofsky C. (1993). Day/night and circadian rhythm control of con gene expression in *Neurospora*. Proc Natl Acad Sci USA 90: 8249–8253.

Leach CM. (1967). Interaction of near-ultraviolet and temperature on sporulation of the fungi *Alternaria*, *Cercosporella*, *Fusarium*, *Helminthosporium*, and *Stemphylium*. Can J Bot 45: 1999–2016.

Leach CM. (1968). An action spectrum for light inhibition of the "terminal phase" of photosporogenesis in the fungus *Stemphylium bortyosum*. Mycolcogia 60: 532–546.

Lee C, Parikh V, Itsukaichi T, Bea K, and Edery I. (1996). Resetting the *Drosophila* clock by photic regulation of PER and PER-TIM complex. Science 271: 1740–1744.

Lee K, Loros JJ, and Dunlap JC. (2000). Interconnected feedback loops in the *Neurospora* circadian system. Science 289: 107–110.

Levy Y. (1995). Inoculum survival of *Exserohilum turcicum* on corn between and during growing periods. Can J Plant Pathol 17: 144–146.

Linden H, Ballario P, Arpaia G, and Macino G. (1999). Seeing the light: News in *Neurospora* blue light signal transduction. Adv Genet 41: 35–54.

Linden H and Macino G. (1997). *White collar 2*, a partner in blue-light signal transduction controlling expression of light-regulated genes in *Neurospora crassa*. EMBO J 16: 98–109.

Liu Y. (2003). Molecular mechanisms of entrainment in the *Neurospora* circadian clock. J Biol Rhythms 18: 195–205.

Liu Y, Loros J, and Dunlap JC. (2000). Phosphorylation of the *Neurospora* clock protein FREQUENCY determines its degradation rate and strongly influences the period length of the circadian clock. Proc Natl Acad Sci USA 97: 234–239.

Liu Y, Merrow M, Loros JL, and Dunlap JC. (1998). How temperature changes reset a circadian oscillator. Science 281: 825–829.

Loros JJ, Dunlap JC, Larrondo LF, Shi M, Belden WJ, Gooch VD, Chen CH, Baker CL, Mehra A, Colot HV, Schwerdtfeger C, Lambreghts R, Collopy PD, Gamsby JJ, and Hong CI. (2007). Circadian output, input, and intracellular oscillators: Insights into the circadian systems of single cells. Cold Spring Harb Symp Quant Biol 72: 201–214.

Loros JJ and Feldman JF. (1986). Loss of temperature compensation of circadian period length in the *frq-9* mutant of *Neurospora crassa*. J Biol Rhythms 1: 187–198.

Makambila C. (1984). Étude de quelques interactions de la lumière et de la temperature sur la croissance linére et la sporulation chez *Colletotrichum manihotus* Henn. FYTON 1984: 1–7.

Martinez-Garcia JF, Huq E, and Quail PH. (2000). Direct targeting of light signals to a promoter element-bound transcription factor. Science 288: 859–863.

McVickar DL. (1942). The light controlled diurnal rhythm of asexual reproduction in *Pilobolus*. Am J Bot 29: 372–375.

Merrow M, Brunner M, and Roenneberg T. (1999). Assignment of circadian function for the *Neurospora* clock gene *frequency*. Nature 399: 584–586.

Merrow MW and Dunlap JC. (1994). Intergeneric complementation of a circadian rhythmicity defect: Phylogenetic conservation of structure and function of the clock gene *frequency*. EMBO J 13: 2257–2266.

Merrow M, Franchi L, Dragovic Z, Görl M, Johnson J, Brunner M, Macino G, and Roenneberg T. (2001). Circadian regulation of the light input pathway in *Neurospora crassa*. EMBO J 20: 307–315.

Merrow M, Garceau N, and Dunlap J. (1997). Dissection of a circadian oscillation into discrete domains. Proc Natl Acad Sci USA 94: 3877–3882.

Merrow M and Roenneberg T. (2001). The circadian cycle: Is the whole greater than the sum of its parts? Trends Genet 17: 4–7.

Mooney JL and Yager LN. (1990). Light is required for conidiation in *Aspergillus nidulans*. Genes Dev 4: 1473–1482.

Myers M, Wagersmith K, Rothenfluhhilfiker A, and Young M. (1996). Light induced degeneration of *timeless* and entrainment of the *Drosophila* circadian clock. Science 271: 1736–1740.

Nakashima H and Feldman J. (1980). Temperature sensitivity of light-induced phase shifting of the circadian clock of *Neurospora*. Photochem Photobiol 32: 247–251.

Nanda KK and Hamner KC. (1958). Studies on the nature of the endogenous rhythm affecting photoperiodic response of Biloxi soybean. Bot Gaz 120: 14–25.

Ninnemann H. (1991). Participation of the molybdenum cofactor of nitrate reductase from *Neurospora crassa* in light promoted conidiation. J Plant Physiol 137: 677–682.

Nowrousian M, Duffield GE, Loros JJ, and Dunlap JC. (2003). The frequency gene is required for temperature-dependent regulation of many clock-controlled genes in *Neurospora crassa*. Genetics 164: 923–933.

Nuesslein-Hildesheim B, O'Brien JA, Ebling FJP, Maywood ES, and Hastings MH. (2000). The circadian cycle of mPER clock gene products in the suprachiasmatic nucleus of the Siberian hamster encodes both daily and seasonal time. Eur J Neurosci 12: 2856–2864.

Nunes MV and Saunders DS. (1999). Photoperiodic time measurement in insects: A review of clock models. J Biol Rhythms 14: 84–104.

Panda S, Antoch MP, Millar BH, Su AI, Schook AB, Straume M, Schultz PG, Kay SA, Takahashi JS, and Hogenesch JB. (2002). Coordinated transcription of key pathways in the mouse by the circadian clock. Cell 109: 307–320.

Pandit A and Maheshwari R. (1994). Sexual reproduction by *Neurospora* in nature. Fung Genet Newslett 41: 67–68.

Perkins DD. (1991). The first published scientific study of *Neurospora*, including a description of photoinduction of carotenoids. Fung Genet Newslett 38: 64–65.

Perkins DD and Turner BC. (1988). *Neurospra* from natural populations: Toward the population biology of a hapoid eukaryote. Exp Mycol 12: 91–131.

Pfeffer W. (1915). Beiträge zur Kenntnis der Entstehung der Schlafbewegungen. Abh sächs Akad Wiss Math-phys Kl 34: 1–154.

Pittendrigh, CS. (1993). Temporal organization: reflections of a Darwinian clock-watcher. Annu Rev. Physiol 55: 17–54.

Pittendrigh CS, Kyner WT, and Takamura T. (1991). The amplitude of circadian oscillators: Temperature dependence, latitudinal clines, and the photoperiodic time measurement. J Biol Rhythms 6: 299–314.

Reddy AB, Fields MD, Maywood ES, and Hastings MH. (2002). Differential resynchronisation of circadian clock gene expression within the suprachiasmatic nuclei of mice subjected to experimental jet lag. J Neurosci 22: 7326–7330.

Roden L, Song H, Jackson S, Morris K, and Carre IA. (2002). Floral responses to photoperiod are correlated with timing of rhythmic expression relative to dawn and dusk in *Arabidopsis*. Proc Natl Acad Sci USA 99: 13313–13318.

Roenneberg T and Aschoff J. (1990a). Annual rhythm of human reproduction: I. Biology, sociology, or both? J Biol Rhythms 5: 195–216.

Roenneberg T and Aschoff J. (1990b). Annual rhythm of human reproduction: II. Environmental correlations. J Biol Rhythms 5: 217–240.

Roenneberg T, Dragovic Z, and Merrow M. (2005). Demasking biological oscillators: Properties and principles of entrainment exemplified by the *Neurospora* circadian clock. Proc Natl Acad Sci USA 102: 7742–7747.

Roenneberg T and Merrow M. (2001a). Seasonality and photoperiodism in fungi. J Biol Rhythms 16: 403–414.

Roenneberg T and Merrow M. (2001b). The role of feedbacks in circadian systems. In *Zeitgebers, Entrainment and Masking of the Circadian System* (K Honma and S Honma, eds). Sapporo: Hokkaido University Press, pp. 113–129.

Roenneberg T and Merrow M. (2002). Life before the clock—modeling circadian evolution. J Biol Rhythms 17: 495–505.

Roger C and Tivoli B. (1996). Effect of culture medium, light and temperature on sexual and asexual reproduction of four strains of *Mycosphaerella pinodes*. Mycol Res 100: 304–306.

Russo VE. (1988). Blue light induces circadian rhythms in the *bd* mutant of *Neurospora*: Double mutants *bd, wc-1* and *bd, wc-2* are blind. Photochem Photobiol 2: 59–65.

Sano H, Narikiyo T, Kaneko S, Yamazaki T, and Shishido K. (2007). Sequence analysis and expression of a blue-light photoreceptor gene, Le.phrA from the basidiomycetous mushroom *Lentinula edodes*. Biosci Biotechnol Biochem 71: 2206–2213.

Sargent ML, Briggs WR, and Woodward DO. (1956). Circadian nature of a rhythm expressed by an invertaseless strain of *Neurospora crassa*. Plant Physiol 41: 1343–1349.

Saunders DS, ed. (1982). *Insect Clocks*. Oxford: Pergamon.

Saunders DS. (1987). Insect photoperiodism: The linden bug, *Phyrrhocoris apterus*, a species that measures day length rather than night length. Experientia 43: 935–937.

Saunders DS. (1990). The circadian basis of ovarian diapause regulation in *Drosophila melanogaster*: Is the period gene causally involved in photoperiodic time measurement? *J Biol Rhythms* 5: 315–332.

Sawyer LA, Hennessy JM, Peixoto AA, Rosato E, Parkinson H, Costa R, and Kyriacou CP. (1997). Natural variation in a *Drosophila* clock gene and temperatur compensation. Science 278: 2117–2120.

Schmidle A. (1951). Die Tagesperiodizität der asexuellen Reproduktion von *Pilobolus spaerosporus*. Arch Mikrobiol 16: 80–100.

Schwerdtfeger C and Linden H. (2001). Blue light adaptation and desensitization of light signal transduction in *Neurospora crassa*. Mol Microbiol 39: 1080–1087.

Shear CL and Dodge BO. (1927). Life histories and heterothallism of the red bread-mold fungi of the Monilia sitophila group. J Agric Res 34: 1019–1042.

Shrode LB, Lewis ZA, White LD, Bell-Pedersen D, and Ebbole DJ. (2001). *vvd* is required for light adaptation of conidiation-specific genes of *Neurospora crassa*, but not circadian conidiation. Fung Gen Biol 32: 169–181.

Sipiczki M. (2000). Where does fission yeast sit on the tree of life? *Genome Biology* 1: 1011.1011–1011.1014.

Smith ML, Bruhn JN, and Anderson JB. (1992). The fungus *Armillaria bulbosa* is among the largest and oldest living organisms. Nature 356: 428–431.

Somers DE, Devlin PF, and Kay SA. (1998). Phytochromes and cryptochromes in the entrainment of the *Arabidopsis* circadian clock. Science 282: 1488–1490.

Sommer T, Chambers JA, Eberle J, Lauter FR, and Russo VE. (1989). Fast light-regulated genes of *Neurospora crassa*. Nucl Acid Res 17: 5713–5723.

Tan Y, Dragovic Z, Roenneberg T, and Merrow M. (2004a). Entrainment of the circadian clock: Translational and post-translational control as key elements. Curr Biol 14: 433–438.

Tan Y, Merrow M, and Roenneberg T. (2004b). Photoperiodism in *Neurospora crassa*. J Biol Rhythms 19: 135–143.

Turner BC, Perkins DD, and Farifield A. (2001). *Neurrospora* from natural populations: A global study. Fung Gen Biol 32: 67–92.

Uebelmesser E-R. (1954). Über den endogenen Tagesrhythmus der Sporangienbildung von *Pilobolus*. Arch Mikrobiol 20: 1–33.

van Oort BE, Tyler NJ, Gerkema MP, Folkow L, Blix AS, and Stokkan KA. (2005). Circadian organization in reindeer. Nature 438: 1095–1096.

Vansteensel MJ, Yamazaki S, Albus H, Deboer T, Block GD, and Meijer JH. (2003). Dissociation between circadian Per1 and neuronal and behavioral rhythms following a shifted environmental cycle. Curr Biol 13: 1538–1542.

West DJ. (1976). Phase shifting of circadian rhythm of conidiation in response to ultraviolet light. *Neurospora* Newslett 23: 17–18.

Young MW. (1998). The molecular control of circadian behavioral rhythms and their entrainment in *Drosophila*. Annu Rev Biochem 67: 135–152.

Part II

Photoperiodism in Invertebrates

OVERVIEW

David L. Denlinger

The little caterpillar creeps
Awhile before in silk it sleeps.
It sleeps awhile before it flies,
And flies away before it dies,
And that's the end of three good tries.

—*David McCord*

The daily rotation of the earth on its axis and the annual rotation of the earth around the sun establish two fundamental rhythms that dictate the common physiological cycles of plants and animals. The earth's rotation on its axis, establishing the daily cycle of light and darkness, is responsible for the rhythms that occur with a frequency of approximately 24 h, hence the term circadian ("circa" meaning approximately and "dian" referring to the 24-h day). By contrast, the annual cycle dictated by the rotation of the earth around the sun generates our seasons and the seasonal changes in day length that are so dramatic at high latitudes. It is this second type of rhythm that we refer to as photoperiodism. While circadian patterns orchestrate regular periods of sleep and activity, and all the finely tuned daily cycles of metabolism that are essential for our well-being, it is photoperiodism that dictates large-scale changes in activity, development, and reproduction that are essential for coping with a seasonally changing environment.

The pioneering work on photoperiodism in plants was followed in a few years by classic demonstrations of photoperiodism in aphids (Marcovitch, 1923) and silk moths (Kogure, 1933). It is now evident that photoperiodism features prominently in the seasonal adaptations of most invertebrates. As poikilotherms, invertebrates are unable to continue active development throughout the year. They are particularly vulnerable to the stress of winter. Not only are temperatures low at that time of year, but food supplies are frequently nonexistent, thus requiring these cold-blooded animals not only to escape the challenge of low temperature but to also sequester sufficient food reserves to circumvent the inimical seasons. Frequently, 9–10 months of the year are spent in a form of arrested development referred to as diapause. Alerting the animal that an unfavorable season is approaching is, of course, the purview of photoperiodism. The short day lengths of late summer and early autumn provide an incredibly reliable indicator of the advent of winter, and most invertebrates indeed exploit this powerful signaling system as a token stimulus signaling the change of seasons. Photoperiodism, of course, does not provide adequate seasonal cues on the equator where day length is constant throughout the year, but beyond 5° north or south of the equator, photoperiodism has been well documented (Denlinger, 1986), even though the longest days at 5°N or 5°S are only about 30 min longer than the shortest days. This does not mean that seasonal phenotypic changes and interruptions of development, features we commonly associate with photoperiodism, do not occur in equatorial regions, but rather that these seasonal events are governed by other sorts of environmental input such as temperature changes and cues provided by wet and dry seasons.

Though the entry into diapause is by far the most common end result of photoperiodism in invertebrates, it is not the only response dictated by seasonal changes in day length. A few insects, such as the monarch butterfly, respond to the short day lengths of late summer by stocking up on energy reserves and migrating thousands of miles southward from Canada and the northern United States to a highland spot in central Mexico (Brower et al., 1977). Once they arrive in Mexico, the adult monarchs enter a classic adult diapause characterized by reduced activity and cessation of reproduction; thus, the prediapause flight is actually a component of the overwintering diapause syndrome. Many other insects also migrate to protected sites before entering into diapause, but in most cases the prediapause migration is not nearly as dramatic as the lengthy flight of the monarch. Color patterns of certain butterfly species are dictated by day length. Frequently, this results in color patterns that are coordinated with seasonal changes in background vegetation. Whether an insect, such as an aphid, develops wings is sometimes a developmental decision based on photoperiodism, as demonstrated in pioneering work by A.D. Lees (1966). Wingless forms prevail during the long days of early summer, but in response to short day lengths, late-summer generations of many aphid species develop wings, thus facilitating dispersal of the fall generation.

Two features are essential for the timekeeping mechanism of photoperiodism. First, the organism must be capable of measuring the length of the day (or night). This requires a photoreceptor capable of detecting the length of light exposure during the day. Usually, very low levels of light intensity are sufficient for an animal to distinguish the photophase from the scotophase. Even a codling moth larva in the core of an apple receives sufficient light during the daytime to distinguish day from night. It is a common perception that invertebrates respond to the decrease in day length as a meaningful signal, but that is seldom the case. In most cases, invertebrates respond to the absolute day length as a threshold character; that is, day lengths are interpreted as being either short or long, not decreasing or increasing. For most photoperiodic responses such as diapause, one can easily calculate a critical photoperiod, the photoperiod that elicits a 50% response. For example, flesh flies reared at 13.5 h of light per day yield a 50% diapause response; longer day lengths avert diapause, while shorter day lengths elicit a full diapause response. The transition between diapause and nondiapause is quite abrupt, and within a half hour, between day lengths of 13.75 and 13.25 h of light/day, the whole population shifts from nondiapause to diapause. Shorter days do not increase the diapause incidence, nor does an incremental shortening of the days to mimic the natural decrease in day length alter the response. Flesh flies, like most other invertebrates, respond to day length as a threshold.

The second requirement for a successful photoperiodic response is that the animal be able to "count" the number of short days it has received. In most cases, invoking a photoperiodic response requires a certain number of short days that must be received within a particular developmental window, known as the photosensitive stage. This photosensitive stage usually occurs well in advance of the photoperiodic response, for example, diapause, thus enabling the animal to garner additional energy reserves and seek a well-protected site in preparation for the long upcoming winter. Thus, most photoperiodic responses are anticipated responses elicited by signals of day length that occur well in advance of the actual event. This implies that the organism must thus be capable of distinguishing short days from long days and be able to add the number of short days (or long days) it has experienced during its photosensitive stage to determine its response. One simple model proposed by Saunders (1971) suggests that an insect may have a fixed number of required short days needed to elicit diapause. When temperature is lowered, the insect moves through its sensitive period more slowly, thus accumulating more short days. This model nicely explains why insects usually have a higher diapause incidence when reared at lower temperatures.

Because the overt response, such as diapause, does not usually occur immediately, the diapause program must be stored within the brain for deployment at a later stage of development. Brain transplant experiments, beginning with the classic experiments of Carroll Williams (1946, 1952), elegantly demonstrated the brain to be the repository of this information. The brain not only holds the program determining whether the insect should enter diapause but also determines how long the insect should remain in diapause. Based on the number of short days received as embryos and larvae, pupae of the tobacco hornworm enter diapauses of different lengths. A few short days, as would be received by larvae in early summer, dictate a pupal diapause of long duration, while short-day exposure throughout development, typical of late summer development, yields a diapause of shorter duration. This program determining diapause length also resides within the brain, as can be demonstrated by transplanting brains from one type of pupa to another and thereby also transferring the program controlling diapause duration (Denlinger and Bradfield, 1981).

We are particularly indebted to insects for the discovery of the major clock genes, work that had its foundation in pioneering experiments by Konopka and Benzer (1971) leading to characterization of the first clock gene, *period*. In the intervening years, a whole suite of clock genes have been described, and we now know quite a bit about the timekeeping mechanism, the photoreceptor pigment, the interactions of the key clock genes, and the tissue and cellular localization of the proteins encoded by these genes (Hall, 2003). Most of this literature, however, is based on experiments focusing on circadian rhythmicity, and we actually know relatively little about how these clock genes function in photoperiodism. Although it is tempting to assume that the timekeeping mechanism that coordinates circadian rhythmicity would also function in photoperiodism, we still have little empirical evidence to support this contention. Several lines of evidence, however, do point to *timeless* as a potentially important gene involved in photoperiodism (Sandrelli et al., 2007; Tauber et al., 2007).

The developmental arrests that result from the action of the clock genes elicit patterns of gene expression that are sometimes shared by diverse taxa. For example, a suite of heat-shock proteins is up-regulated during dormancy in a wide variety of insect taxa, as well as in the dauer stage of the roundworm *Caenorhabditis elegans* and in diapausing embryos of the brine shrimp *Artemia franciscana* (Rinehart et al., 2007). Another pathway that is likely to be common to dormancies of many types is the insulin signaling pathway. This pathway has been implicated in the dormancies of nematodes (Lee et al., 2001) and insects (Tatar et al., 2001). The dormancies of *C. elegans* and *A. franciscana*, however, are not programmed by photoperiod, so it is evident that the common downstream pathways noted here can be accessed by diverse environmental signals.

Central to the photoperiodic responses in invertebrates is the adaptive advantage they offer in enabling the animal to finely coordinate its development and reproduction with favorable seasons. This is really what photoperiodism is all about, coordinating development and reproduction with seasons of food abundance, being present when other members of your species are around, and preparing for upcoming unfavorable seasons. The link to photoperiodism allows the fine-tuning needed to precisely coordinate development with the appropriate season. There is usually sufficient variability in photoperiodic responses to provide the grist for selection, thus enabling an organism to modify its photoperiodic response as it expands into new ranges, yet the current rate of accelerated global warming could easily desynchronize the appearance of plant food sources with the photoperiodic responses of insects and organisms in other trophic levels that depend on that food source. Synchronizing the photoperiodic responses of organisms at different trophic levels is a problem of practical significance for the biological control industry, as well. Commercially reared parasitic wasps and predatory insects are used widely to control pest insects and weeds. Usually this involves searching for potential biocontrol agents in foreign lands and then importing them to control a pest population, a species that itself was often accidentally introduced. However, for this strategy to work, the photoperiodic responses of the introduced biocontrol species must indeed match the photoperiodic response of the species to be controlled. Not only must the two be present at the same time, but the biocontrol agent must be equipped with a photoperiodic response that gives it the capacity to enter and survive an overwintering diapause in its new environment. Many control efforts have failed as a result of this sort of environmental incompatibility.

Photoperiodic responses can sometimes also be used to determine the origin of a pest species. For example, when the Asian tiger mosquito, *Aedes albopictus*, first appeared in the southern United States nearly 20 years ago, a check of its photoperiodic response revealed that the newly arrived population most likely came from Japan (Hawley et al., 1987). This mosquito ranges throughout Asia from the tropics to northern Japan. Tropical species lack a photoperiodic diapause, while the more northern ones enter an embryonic diapause elicited by short day length. The new arrivals in the United States displayed a photoperiodic response curve characteristic of north-central Japan, thus linking the arrival of this new pest species to a shipment of used tires sent from Japan to Houston.

Photoperiodism in invertebrates thus has many dimensions. Invertebrates, especially the insects, have been examined extensively to understand how development is coordinated with the seasonal environment. The insect work has been especially valuable for probing the molecular mechanisms involved in animal photoperiodism and has indeed revealed the key timekeeping clock genes. And, the economic importance of insects has provided a special impetus for understanding how photoperiodism can be used to predict seasonal distributions of pest species as well as the parasitic and predatory species used for their control. Sources discussing the ecological context for insect dormancy and other seasonal adaptations include excellent reviews by Tauber et al. (1986) and Danks (1987). The insect clock literature is nicely reviewed by Saunders (2002). Other related insect reviews include coverage of the hormonal regulation of diapause (Denlinger et al., 2005), the molecular basis for diapause (Denlinger 2002), the dynamics of insect diapause (Kostal, 2006), and the energetics of diapause (Hahn and Denlinger, 2007). Much less has been written about photoperiodism in other invertebrates, but key papers on copepod photoperiodism include those by Hairston and Kearns (1995) and Marcus (1982). The mollusk literature includes reviews by Joosse (1984), Sokolove et al. (1984), and a more recent review by Wayne (2001). A review by Bradshaw and Holzapfel (2007) provides an excellent overview of the evolution of photoperiodism in both invertebrates and vertebrates.

Many exciting findings over the past few years have advanced our knowledge of photoperiodism in invertebrates, and part II of this book provides updates of this timely topic. Chapter 8 on snails by Hideharu Numata and Hiroko Udaka and chapter 9 on copepods by Nancy Marcus and Lindsay Scheef capture new developments on photoperiodism in these two important invertebrate groups. The content of this invertebrate section is, however, rightfully skewed toward insects because it is indeed this taxa that has been exploited most thoroughly for work on photoperiodism. Chapter 10 by David Saunders provides a comprehensive overview of the involvement of photoperiodism on insect migration and diapause. The mechanisms of light perception and the role of the clock genes in regulating insect photoperiodism are discussed in chapter 11 by Shin Goto, Sakiko Shiga, and Hideharu Numata. Chapter 12 by Karen Williams, Paul Schmidt, and Marla Sokolowski provides an update on the molecular events dictating the expression of diapause, as well as the events underpinning diapause maintenance. In chapter 13, Fred Nijhout discusses the role of photoperiodism on phenotype expression, a story based largely on the fascinating literature on butterfly wing patterns and coloration. And chapter 14 by Jim Hardie interprets the complicated role of photoperiod in regulating life history traits in aphids, a story with roots in the classic studies of A.D. Lees. Together, the chapters in part II provide a state-of-the-art review of photoperiodism in diverse taxa of invertebrates and point to exciting directions for future work.

References

Bradshaw WE and Holzapfel CM. (2007). Evolution of animal photoperiodism. Annu Rev Ecol Evol Syst 38: 1–25.

Brower LP, Calvert WH, Hedrick LE, and Christian J. (1977). Biological observations on an overwintering colony of monarch butterflies (*Danaus plexippus*, Danaidae) in Mexico. J Lepidopt Soc 31: 232–242.

Denlinger DL. (1986). Dormancy in tropical insects. Annu Rev Entomol 31: 239–264.

Denlinger DL. (2002). Regulation of diapause. Annu Rev Entomol 47: 93–122.

Denlinger DL and Bradfield JY. (1981). Duration of pupal diapause in the tobacco hornworm is determined by number of short days received by the larva. J Exp Biol 91: 331–337.

Denlinger DL, Yocum GD, and Rinehart JP. (2005). Hormonal control of diapause. In *Comprehensive Insect Molecular Science*, vol 3 (LI Gilbert, K Iatrou, and S Gill, eds). Amsterdam: Elsevier, pp. 615–650.

Danks HV. (1987). *Insect Dormancy: An Ecological Perspective*. Ottawa: Biological Survey of Canada.

Hahn DA and Denlinger DL. (2007). Meeting the energetic demands of diapause: Nutrient storage and utilization. J Insect Physiol 53: 760–773.

Hairston NG and Kearns CM. (1995). The interaction of photoperiod and temperature in diapause timing: A copepod example. Biol Bull 189: 42–48.

Hall JC. (2003). Genetics and molecular biology of rhythms in *Drosophila* and other insects. Adv Genet 48: 1–280.

Hawley WA, Reiter P, Copeland RS, Pumpuni CB, and Craig GB Jr. (1987). *Aedes albopictus* in North America: Probable introduction in used tires from northern Asia. Science 236: 1114–1116.

Joosse J. (1984). Photoperiodicity, rhythmicity and endocrinology of reproduction in the snail *Lymnaea stagnalis*. Ciba Found Symp 104: 204–220.

Kogure M. (1933). The influence of light and temperature on certain characters of the silkworm, *Bombyx mori*. J Dept Agric Kyushu Univ 4: 1–93.

Konopka R and Benzer S. (1971). Clock mutants of *Drosophila melanogaster*. Proc Natl Acad Sci USA 68: 2112–2116.

Kostal V. (2006). Eco-physiological phases of insect diapause. J Insect Physiol 52: 113–127.

Lee RY, Hench J, and Ruvkun G. (2001). Regulation of *C. elegans* DAF-16 and its human ortholog FKHRL1 by the daf-2 insulin-like signaling pathway. Curr Biol 11: 1950–1957.

Lees AD. (1966). The control of polymorphism in aphids. Adv Insect Physiol 3: 207–277.

Marcovitch S. (1923). Plant lice and light exposure. Science 58: 537–538.

Marcus NH. (1982). Photoperiodic and temperature regulation of diapause in *Labidocerca aestiva*. Biol Bull 162: 45–52.

Rinehart JP, Li A, Yocum GD, Robich RM, Hayward SAL, and Denlinger DL. (2007). Up-regulation of heat shock proteins is essential for cold survival during insect diapause. Proc Natl Acad Sci USA 104: 11130–11137.

Sandrelli F, Tauber E, Pegoraro M, Mazzotta G, Cisotto P, et al. (2007). A molecular basis for natural selection at the *timeless* locus in *Drosophila melanogaster*. Science 316: 189–1900.

Saunders DS. (1971). The temperature-compensated photoperiodic clock "programming" development and pupal diapause in the flesh-fly, *Sarcophaga argyrostoma*. J Insect Physiol 17: 801–812.

Saunders DS. (2002). *Insect Clocks*, 3rd ed. Amsterdam: Elsevier.

Sokolove PG, McCrone EJ, van Minnen J, and Duncan WC. (1984). Reproductive endocrinology and photoperiodism in a terrestrial slug. Ciba Found Symp 104: 189–203.

Tatar M, Kopelman A, Epstein D, Tu M-P, Yin C-M, et al. (2001). A mutant *Drosophila* insulin receptor homolog that extends life-span and impairs neuroendocrine function. Science 292: 107–110.

Tauber E, Zordan M, Sandrelli F, Pegoraro M, Osterwalder N, et al. (2007). Natural selection favors a newly derived *timeless* allele in *Drosophila melanogaster*. Science 316: 1895–1898.

Tauber MJ, Tauber CA, and Masaki S. (1986). *Seasonal Adaptations of Insects*. Oxford: Oxford University Press.

Wayne NL. (2001). Regulation of seasonal reproduction in mollusks. J Biol Rhythms 16: 391–402.

Williams CM. (1946). Physiology of insect diapause. I. The role of the brain in the production and termination of pupal dormancy in the giant silkworm *Platysamia cecropia*. Biol Bull 90: 234–243.

Williams CM. (1952). Physiology of insect diapause. IV. The brain and prothoracic glands as an endocrine system in the cecropia silkworm. Biol Bull 103: 120–138.

8

Photoperiodism in Mollusks

Hideharu Numata and Hiroko Udaka

Although photoperiodism was first demonstrated in plants (Garner and Allard, 1920), numerous examples have also been reported in various animals, as indicated in the chapters of this book. However, most reports are restricted to vertebrates and arthropods. Jenner (1951) reported the first example in mollusks: photoperiodic control of oviposition in the marsh pond snail *Lymnaea palustris*. Later, Stephens and Stephens (1966) reported another example in the garden snail *Helix aspersa*. The effect of photoperiod on reproduction has since been shown in some terrestrial snails and slugs as well as in freshwater snails, although these species all belong to Pulmonata, Gastropoda (table 8.1). In some of these species, moreover, photoperiod also affects growth and temperature tolerance.

In contrast, there are very few publications on photoperiodism in marine mollusks. The only reliable example of photoperiodism shown in these animals is in the Californian sea hare *Aplysia californica* (Gastropoda, Opisthobranchia), in which photoperiod shows a minor role in the regulation of reproduction, secondary to the role of temperature (Wayne and Block, 1992). The fighting conch *Strombus pugilis* (Gastropoda, Sorbeoconcha) shows a higher growth rate under continuous light than under short-day conditions or continuous darkness (Manzano et al., 1998). This is not evidence for photoperiodism in the strict sense, however, because no results were obtained under naturally occurring long-day conditions. Nevertheless, here we point out the possibility that photoperiodism is more widespread in marine species, and we argue that the following generalization is not always true: photoperiodism is not used as a seasonal signal by marine species because seawater temperature is more reliable than air temperature or freshwater temperature.

In mollusks other than gastropods, studies on photoperiodism are scarce. In the Japanese chiton *Acanthopleura japonica* (Polyplacophora), photoperiod had no effect on gametogenesis, and water temperature alone determines the reproductive season (Yoshioka, 1987). In the great scallop *Pecten maximus* (Bivalvia), Saout et al. (1999) showed that a simultaneous increase in temperature and photoperiod enhanced gonad development, although it is unclear which of the two factors is responsible for stimulation. In the common octopus *Octopus vulgaris* (Cephalopoda, Octopoda), males produced a slightly larger number of spermatophores under long-day than under short-day conditions (Laubier-Bonichon and Mangold, 1975). Females of the short-finned squid *Illex illecebrosus* (Cephalopoda, Decapoda) had larger nidamental glands (female accessory glands) when transferred to laboratory

TABLE 8.1. List of mollusks in which photoperiodism was clearly shown by experiments.

Species	Habitat	Type[a]	Photoperiod effect	Reference
Gastropoda, Pulmonata				
Lymnaea palustris	Freshwater	L	Oviposition	Jenner (1951)
Lymnaea stagnalis	Freshwater	L	Oviposition	Bohlken and Joosse (1982)
		S	Growth	Bohlken and Joosse (1982)
		S	Glycogen accumulation	Hemminga et al. (1985)
Lymnaea peregna	Freshwater	L/S	Oviposition	Lundelius and Freeman (1986)
Helix aspersa	Terrestrial	L	Oviposition, reproductive maturation	Stephens and Stephens (1966), Enée et al. (1982), Gomot et al. (1989)
		L	Growth	Gomot et al. (1982)
		S	Supercooling point	Biannic and Daguzan (1993), Ansart et al. (2001)
Helix pomatia	Terrestrial	L	Winter dormancy	Jeppesen and Nygård (1976)
		L	Oviposition	Gomot (1990)
Archachatina ventricosa	Terrestrial	L	Oviposition	Otchoumou et al. (2007)
Limicolaria flammea	Terrestrial	L	Growth, spermatogenesis	Egonmwan (1991)
Limax maximus	Terrestrial	L	Male-phase reproductive maturation	Sokolove and McCrone (1978)
		L	Growth	Sokolove and McCrone (1978)
Lehmannia valentiana	Terrestrial	L	Oviposition	Hommay et al. (2001)
		S	Reproductive maturation	Udaka and Numata (2008)
		S	Growth	Udaka and Numata (2008)
		L	High-temperature tolerance	Udaka et al. (2008)
		S	Low-temperature tolerance	Udaka et al. (2008)
Gastropoda, Opisthobranchia				
Aplysia californica	Marine	S	Oviposition	Wayne and Block (1992)

[a]L, long-day response; S, short-day response.

long-day conditions in September than did wild specimens caught in the same season (O'Dor et al., 1977). It is, however, necessary to compare statistically the results between long-day and short-day conditions at a same temperature before concluding the occurrence of photoperiodism in Cephalopoda. This chapter focuses on photoperiodism in gastropods, particularly terrestrial and freshwater species in the order Pulmonata, in which photoperiodism has been intensively examined.

REPRODUCTION

Photoperiodic Effect

Oviposition is initiated in *Lymnaea palustris* and continues for an extended period under long-day conditions of 13.5-h light and 10.5-h darkness (LD 13.5:10.5), whereas it is prevented under short-day conditions of LD 11:13, although the temperature in these experiments was not reported (Jenner, 1951). In the great pond snail *Lymnaea stagnalis*, Bohlken and Joosse (1982) showed that oviposition is controlled by photoperiod. Snails began to lay eggs earlier and laid many more egg masses under long-day conditions of LD 16:8 than under LD 12:12 or LD 8:16 at 20°C. While snails did lay more egg masses under LD 12:12 than under LD 8:16, the difference was slight, and the threshold therefore exists between LD 16:8 and LD 12:12 (Bohlken and Joosse, 1982). Even under short-day conditions, snails continued to lay eggs, but the proportion of ovipositing individuals is unknown because they were reared in groups and only the total number of egg masses in each tank was counted. Bohlken and Joosse (1982) concluded, without histological observation, that the delay of oviposition under short-day conditions reflects retarded maturation of oocytes and female accessory reproductive organs.

Lundelius and Freeman (1986) reported that different strains of the wandering snail *Lymnaea peregra* show either a long-day or a short-day response for oviposition and concluded by a crossing experiment that these responses are genetically determined by a single gene locus with the long-day allele dominant over the short-day allele.

In *Helix aspersa*, Stephens and Stephens (1966) showed that snails did not lay eggs under short-day conditions of LD 9:15 but did under long-day conditions of LD 15:9 at 17°C. Additional studies later confirmed that long-day conditions promote oviposition in this species (Enée et al., 1982; Gomot and Gomot 1985; Gomot et al., 1989). Gomot et al. (1989) observed oviposition under four combinations of two photoperiodic conditions (LD 18:6 and LD 8:16) and two temperatures (15°C and 20°C) in *H. aspersa* after winter dormancy. Although both photoperiod and temperature affected oviposition, the effect of photoperiod was greater (figure 8.1). Under long-day conditions at 20°C, all 20 snails laid eggs, and oviposition continued for the duration of the experiment (16 weeks), while under short-day conditions at 20°C, the rate of oviposition was lower. Under long-day conditions at 15°C, the onset of oviposition was delayed but the rate of oviposition was similar to that

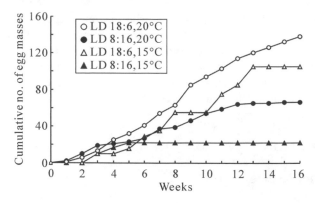

FIGURE 8.1.
Effects of photoperiod and temperature on oviposition in *Helix aspersa* (based on the data of Gomot et al., 1989). Groups of 20 snails were kept under LD 18:6 or LD 8:16 at 15°C or 20°C.

under long-day conditions at 20°C, and under short-day conditions at 15°C, only 5 of 20 snails laid eggs, and the laying period was short-lived. Moreover, Gomot et al. (1989) examined the gonad (hermaphrodite gland, or ovotestis) and albumen gland (female accessory gland) by histological observations and measurements of DNA synthesis to determine which organ or cell is responsible for the inhibition of reproduction in this species. Under short-day conditions, oocytes matured but were not released and thus degenerated. In the albumen gland, moreover, cellular multiplication and the release of secretory products decreased. However, differentiation of spermatogonia and spermatocytes were observed under all four conditions. Gomot et al. (1989) attributed the suppression of reproduction by short-day conditions to the inhibition of female reproductive functions.

In the Burgundy snail *Helix pomatia*, snails collected from the field in May continued to lay eggs under long-day conditions of LD 18:6, whereas under short-day conditions of LD 8:16 they initially laid eggs at a lower rate and eventually stopped oviposition at both 15°C and 20°C (Gomot, 1990).

The great gray slug *Limax maximus* is hermaphroditic and protandrous. Male-phase maturation precedes female-phase maturation by approximately two months under natural conditions (Sokolove and McCrone, 1978). Sokolove and McCrone (1978) examined the effect of photoperiod on reproductive maturation in *L. maximus* under long-day (LD 16:8) and short-day (LD 8:16) conditions at 15°C, but without histological examinations. Male-phase maturation, denoted by the enlargement of the gonad, was affected by photoperiod and was promoted by long-day conditions. Moreover, the long-day response was also observed in penile growth (McCrone and Sokolove, 1979). After male-phase maturation, female-phase maturation, shown by growth of the albumen gland, occurred irrespective of photoperiod and therefore is not controlled by photoperiod. Furthermore, when slugs were transferred from short-day to long-day conditions, male-phase maturation was more rapid than under continuous long-day conditions (Sokolove and McCrone, 1978). *L. maximus* thus shows a long-day response for male-phase maturation, and a change from short-day to long-day conditions is more effective than constant long-day conditions at promoting male-phase maturation.

Hommay et al. (2001) examined the effect of photoperiod on oviposition in a laboratory colony of the greenhouse slug *Lehmannia valentiana* (as *Limax valentianus*) originating from Rennes in northern France. However, they did not use constant temperatures, with the temperature differing between the photophase and scotophase. Moreover, they did not use the same temperature regime under the two photoperiodic conditions used. It is therefore difficult to interpret their results. Slugs started oviposition earlier and laid many more eggs under long-day conditions (LD 16:8) combined with a higher temperature in the photophase than under short-day conditions (LD 12:12) combined with a higher temperature in the scotophase (Hommay et al., 2001). It is probable that short-day conditions suppress oviposition in the French population of *L. valentiana*, although the unnatural combination of light and temperature conditions might have been responsible for the inhibitory effect. To demonstrate the photoperiodic effect clearly, it will be necessary to rear slugs at the same temperature under long-day and short-day conditions, as Hommay et al. (2001) themselves pointed out.

Recently, Udaka and Numata (2008) demonstrated a short-day response for reproduction in *L. valentiana* collected in Osaka, southwestern Japan. Field-collected immature slugs were reared under long-day (LD 16:8) or short-day (LD 12:12) conditions at 15°C, 20°C, or 25°C for 60 days. Gonads (hermaphrodite gland) became heavier under short-day conditions than under long-day conditions at all three temperatures (figure 8.2a). Gonad weight increased more at lower temperatures under both photoperiodic conditions.

L. valentiana is also hermaphroditic like *Limax maximus*, but in this species male- and female-phase maturation occurs almost simultaneously under natural conditions (Udaka et al., 2007; see figure 8.3). By histological examinations of field-collected slugs, Udaka et al. (2007) showed that an increase in gonad volume includes the accumulation of differentiated sperm and the enlargement of oocytes. The effects of photoperiod and temperature on oogenesis and spermatogenesis were also examined in the laboratory. Oocytes in the gonad were larger and spermatogenesis was more advanced under short-day conditions than under long-day conditions at each temperature. In addition, oocytes were smaller and spermatogenesis was less advanced under short-day conditions at 25°C than at 15°C and 20°C (figure 8.2b,c; Udaka and Numata, 2008). The Japanese population of *L. valentiana* thus shows a clear short-day response for reproductive maturation of both female and male phases, although the effect of temperature is also prominent. In this population, short-day and low-temperature conditions promoted reproductive maturation, whereas long-day and high-temperature conditions suppressed it.

Wayne and Block (1992) examined the effects of water temperature and photoperiod on oviposition in field-collected *Aplysia californica*. Animals collected in September laid eggs at 20°C but few or no eggs at 15°C. Moreover, they laid eggs

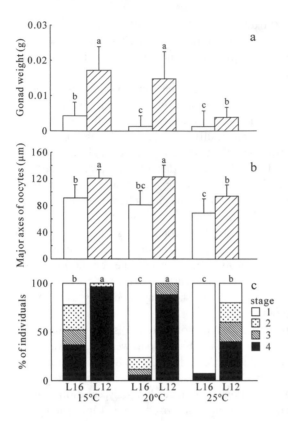

FIGURE 8.2.
Effects of temperature and photoperiod on the reproductive maturation in *Lehmannia valentiana* (modified from Udaka and Numata, 2008). Slugs were kept for 60 days under LD 16:8 (L16) or LD 12:12 (L12) at 15°C, 20°C, or 25°C. For gonad weight (a), and the major axes of the largest oocytes (b), the error bars represent SD, and the mean values with different letters are significantly different ($p < 0.05$ by Steel-Dwass test). Spermatogenesis was classified into four stages (c): stage 1, spermatogonia and spermatocytes but no spermatids; stage 2, spermatids but no sperm; stage 3, clumps of sperm but no free sperm; and stage 4, free sperm. The proportions of slugs with sperm (stages 3 and 4) were examined by Tukey's multiple comparison test for proportions, and columns with different letters are significantly different ($p < 0.05$).

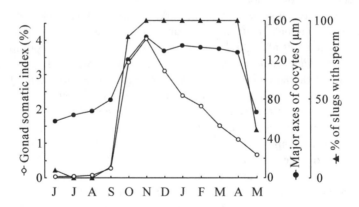

FIGURE 8.3.
Seasonal changes of the gonad somatic index, the major axes of the largest oocytes, and the proportions of slugs with mature sperm in *Lehmannia valentiana* in Osaka, Japan (based on the data of Udaka et al., 2007).

more frequently under short-day conditions (LD 8:16) than under long-day conditions (LD 16:8) at 20°C. From these results, Wayne and Block (1992) suggested that high water temperature permits oviposition, and short-day conditions further stimulate it. In sea hares collected in January and March, however, oviposition was suppressed in all conditions, including short-day and warm water conditions (Wayne and Block, 1992). In *A. californica*, therefore, there is a strong inhibition of oviposition during the winter and spring, which overrides the stimulatory effects of temperature and photoperiod. This inhibition seems to be equivalent to photorefractoriness in birds and mammals (Nicholls et al., 1988; see also chapters 17, 19, 20).

Circannual Rhythms

In the discussion of a paper on the annual periodicity in birds in the Cold Spring Harbor Symposium on Quantitative Biology on Biological Clocks in 1960, E. Segal showed that reproduction in the yellow slug *Limax flavus* is controlled by a circannual rhythm (Marshall, 1960). This was the first example of a circannual rhythm in mollusks. Oviposition continued at intervals of about 11 months under a constant photoperiod of LD 11:13 at 10°C and 20°C over three cycles. In the nearly 50 years since the symposium, however, the effect of photoperiodic changes on the phase of this rhythm has not been shown, and the *zeitgeber* for entrainment of the rhythm is still unclear.

In *Helix aspersa*, Bailey (1981) showed circannual changes not only in mating and oviposition but also in food consumption and locomotor activity under a constant photoperiod of LD 12:12 at a mean temperature of 17.4°C with little variation. During the 14-month period of his experiment, only one cycle of the rhythm was observed, and therefore there is weak evidence for a circannual rhythm using strict definition (Gwinner, 1986). Nevertheless, the results of Bailey (1981) have significance in the study of photoperiodism in mollusks because the rhythm continues with a period of one year under a photoperiod with seasonal variation, indicating that the *zeitgeber* for entrainment of the circannual rhythm is the change in photoperiod. The photoperiodic response in *H. aspersa* shown first by Stephens and Stephens (1966) is therefore not a simple long-day response, but may function to reset the phase of the circannual rhythm, as in most animals that use circannual rhythms to control their life cycles (Gwinner, 1986; Nisimura and Numata, 2003).

Ecological Significance

As mentioned above, most mollusks that rely on photoperiodism for reproduction show long-day responses as shown in insects (Beck, 1980). In the temperate zone, spring and summer are suitable for reproduction for most poikilotherms, and therefore the ecological significance of long-day responses in mollusks is obvious and has not attracted special attention. Some insects, on the other hand, have short-day responses for summer diapause, and their ecological significance has been discussed (Masaki, 1980). In mollusks, however, only a few examples of short-day responses for reproduction exist (Lundelius and Freeman, 1986; Wayne and Block, 1992; Udaka and Numata, 2008). In *Lymnaea peregra*, of which different strains show a long-day or a short-day response for oviposition, Lundelius and Freeman (1986) presumed that the long-day strains lay eggs in summer and the short-day strains do so in spring and autumn. However, because they did not observe the seasonal reproduction of this species in the field, the ecological significance of these responses is unclear.

A population of *Lehmannia valentiana* from Osaka, Japan, shows a short-day response both for male-phase and female-phase reproductive maturation (Udaka and Numata, 2008). What is the ecological significance of the short-day response? In this population, gonad development began in September, and all slugs developed mature sperm from October through to the following April. Oocyte size increased in September and reached its maximum in November. Furthermore, slugs laid eggs from early November to the following May (figure 8.3; Udaka et al., 2007). This population of *L. valentiana* thus reproduces from late autumn to spring, unlike many terrestrial and freshwater mollusks. Although the availability of food is one of the predominant factors that determine the reproductive season in many animals, it is not likely that the lack of food for young animals limits the reproductive season in omniphagous *L. valentiana*. The tolerance of eggs and juvenile slugs to various temperatures was therefore examined to determine whether summer temperature is a factor limiting the seasonal reproduction in this species. After 1 h exposure to 33°C, the survival rate for juvenile slugs was 100%, but was 0% for eggs. When eggs were kept continuously under different constant temperatures, hatching success was greater than 85% between 15°C and 20°C, yet no eggs hatched at 25°C. The daily maximum and minimum temperatures in Osaka exceed 32°C and 25°C, respectively, for approximately 20 days in July and August (Japan Meteorological Agency, 2006). Even though natural oviposition sites of *L. valentiana* might be cooler than air temperatures, oviposition from June to September seems maladaptive with respect to the survival of laid eggs. This heat susceptibility of eggs would be a key factor for restricting the reproduction of *L. valentiana* from late autumn to spring by a short-day response (Udaka et al., 2007; Udaka and Numata, 2008).

Geographic Variations

Geographic variations in photoperiodic responses have been one of the major focal points in the study of seasonal adaptations in insects, and clines in critical day length along latitudes have been shown in various insects (Danilevskii, 1965; Masaki, 1972). In mollusks, such clines have not been reported. Geographic differences in life cycles have been reported in the grey field slug *Deroceras reticulatum* (Runham and Laryea, 1968; Dmitrieva, 1969; South, 1989, 1992). However,

South (1992) attributed these differences not to geographic strains but to differences in climatic conditions, and photoperiodic responses have not been examined in this species.

Two of the four strains of *Lymnaea peregra* for which geographic origins are clear in Britain show long-day responses for oviposition and two show short-day responses (Lundelius and Freeman, 1986). There are no latitudinal clines in the photoperiodic responses, however, because the former are from the two intermediate localities and the latter are from the northernmost and the southernmost localities. It is therefore very difficult to explain these geographic differences in photoperiodic response as adaptations to the local climate.

In a French population of *Lehmannia valentiana*, it is probable that long-day conditions promote oviposition (Hommay et al., 2001), although a certain ambiguity remains (see above). In slugs collected in Japan, however, a short-day response is prominent both for spermatogenesis and for oogenesis (Udaka and Numata, 2008). Thus, the photoperiodic response is different between the populations in France and Japan, which suggests geographic variation in the photoperiodic response of *L. valentiana*. Although Hommay et al. (2001) did not observe the life cycle in the field, it is probable that *L. valentiana* in northern France reproductively matures under long-day conditions and reproduces in summer.

Endocrine Control

Thirty years ago, Sokolove and McCrone (1978) stated that in Pulmonata the endocrine control of reproduction had been intensively studied relative to the control by external factors such as photoperiod. At the 1983 Ciba Foundation Symposium on Photoperiodic Regulation of Insect and Molluscan Hormones, P.G. Sokolove and J. Joosse read papers on the endocrine regulation of reproduction in *Limax maximus* and *Lymnaea stagnalis*, respectively (Joosse, 1984; Sokolove et al., 1984). In recent years, two important reviews have been published on the regulation of reproduction in mollusks (Flari and Edwards, 2003; Wayne, 2001). Wayne (2001) reviewed the regulation of seasonal reproduction in mollusks focusing on three species: *Limax maximus*, *Lymnaea stagnalis*, and *Aplysia californica*. Flari and Edwards (2003) reviewed the endocrine system in the regulation of reproduction in terrestrial pulmonate mollusks. More recently, Lagadic et al. (2007) summarized the hormonal control of reproduction in pulmonate snails in their review on endocrine disruption in aquatic pulmonate mollusks.

Various endocrine factors have been reported in the regulation of reproduction in mollusks, mostly by classic endocrinology techniques such as extirpation and transplantation of a putative endocrine organ, and injection of organ homogenates. The optic tentacle, cerebral ganglion, dorsal body (an endocrine organ attached to the cerebral ganglion), gonad, and so forth, have been shown to be secretory organs for endocrine factors (Wayne, 2001; Flari and Edwards, 2003; Lagadic et al., 2007).

In *Limax maximus*, implantation of the cerebral ganglion from a long-day donor induced male-phase maturation in a short-day host (McCrone and Sokolove, 1979; McCrone et al., 1981). Sokolove et al. (1984) concluded that a hormone or hormones from the neurosecretory cells in a small region (area Z) of the cerebral ganglion are most important for the photoperiodic stimulation of male-phase maturation, because an extract from area Z highly induces DNA synthesis in the gonad. Melrose et al. (1983) suggested that this male gonadotropic hormone is a peptide. J. Joosse pointed out during the discussion for the Sokolove's presentation at the Ciba Foundation symposium that the dorsal body hormone (DBH) could be responsible for this male gonadotrophic activity because area Z is located close to the dorsal body (Sokolove et al., 1984; see also Flari and Edwards, 2003). No follow-up has yet been conducted on this male gonadotropic hormone from the cerebral ganglion (Wayne, 2001).

In *Lymnaea stagnalis*, Joosse (1984) summarized that both DBH, a hormone secreted by the endocrine dorsal body, and a neuroendocrine hormone secreted by the caudodorsal cells in the cerebral ganglion (the caudodorsal cell hormone, CDCH) are necessary for maturation of female reproductive functions. DBH stimulates the development of oocytes and the albumen gland, and CDCH induces ovulation and oviposition (Joosse, 1984; see also Wayne, 2001). In *L. stagnalis*, Ebberink et al. (1983) suggested that DBH is a peptide, whereas Nolte et al. (1986) proposed that it might be an ecdysteroid. The precise chemical nature of DBH remains unsolved, and it could differ among species (Flari and Edwards, 2003; Lagadic et al., 2007). On the other hand, CDCH, which stimulates ovulation in *L. stagnalis*, was purified as a peptide (CDCH-1, or ovulation hormone), and its sequence of 36 amino acids was determined (Ebberink et al., 1983). It is probable that inactivity of the dorsal body or the caudodorsal cells is responsible for the suppression of oviposition under short-day conditions, although it is unclear which plays a critical role.

In summary, our knowledge of the endocrine mechanism for photoperiodic control of reproduction remains rather poor in mollusks as compared with other invertebrates, such as adult diapause in insects (Denlinger et al., 2005).

GROWTH, ACTIVITY, AND METABOLISM

Photoperiodism in growth has been reported in several species for which reproduction was also shown to be controlled by photoperiod (table 8.1). In *Lymnaea stagnalis*, long-day conditions promote reproduction but suppress growth (Bohlken and Joosse, 1982). Moreover, short-day conditions enhance the accumulation of glycogen in the anterior mantle region and the head-foot muscles (Hemminga et al., 1985). This response is consistent with respect to energy allocation: if animals allocate more energy resources to reproduction, they would have fewer resources for growth and glycogen reserves. It is therefore assumed that this species reproduces with retarded growth under long-day conditions in spring and summer and grows without reproduction under short-day conditions in autumn (Bohlken and Joosse, 1982).

However, this seasonal allocation of growth and reproduction is not applicable for *Helix aspersa*, *Limicolaria flammea*, and *Limax maximus*, in which both reproduction and growth are promoted by long-day conditions (Stephens and Stephens, 1966; Sokolove and McCrone, 1978; Enée et al., 1982; Gomot et al. 1982, 1989; Egonmwan, 1991). In *Helix aspersa*, food consumption and locomotor activity are low in nonreproductive seasons, indicating a state of winter dormancy (hibernation) (Bailey, 1981). In *Helix pomatia*, a greater proportion of snails enter winter dormancy, producing an epiphragm when exposed to short-day conditions of LD 12:12 at 12°C, whereas no snails did so under long-day conditions of LD 18:6 (Jeppesen and Nygård, 1976). In this species, therefore, long day conditions promote growth and reproduction, whereas short-day conditions induce winter dormancy (Jeppesen and Nygård, 1976; Gomot, 1990).

A Japanese population of *Lehmannia valentiana* shows short-day responses for both reproduction and growth. Slugs reared from hatching under short-day conditions (LD 12:12) became heavier than those reared under long-day conditions (LD 16:8) at 20°C (figure 8.4; Udaka and Numata, 2008). For the three species above, the season unsuitable for reproduction might also be unsuitable for growth. Feeding activities are likely retarded in *L. maximus* and *H. aspersa* under short-day conditions and in *L. valentiana* under long-day conditions. *L. valentiana*, however, does feed and continues to gain body weight even under long-day conditions (figure 8.4; Udaka and Numata, 2008), and the state of this species under long-day conditions is therefore not summer dormancy (aestivation) as reported in some snails that withdraw into the shell and cover the opening with an epiphragm (Schmidt-Nielsen et al., 1971).

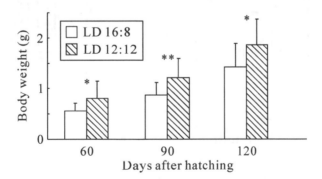

FIGURE 8.4.
Effect of photoperiod on growth in *Lehmannia valentiana* (modified from Udaka and Numata, 2008). Slugs were reared from hatching under LD 16:8 or LD 12:12 at 20°C. The error bars represent SD, and the asterisks show significant differences between the two photoperiodic conditions (**, $p < 0.01$; *, $p < 0.05$ by Student's t-test).

TEMPERATURE RELATIONS

Supercooling Point

The supercooling point is a reliable indicator of cold hardiness in freeze-intolerant animals. In *Helix aspersa*, a transfer from long-day (LD 16:8) to short-day (LD 12:12) conditions at 20°C decreases the supercooling point (Biannic and Daguzan, 1993). This effect of photoperiod on supercooling points was also found at lower temperatures of 5°C, 10°C, and 15°C when the photoperiod was gradually changed from LD 16:8 to LD 12:12 or LD 8:16 (figure 8.5; Ansart et al., 2001). Surprisingly, acclimation temperature had no effect on supercooling points in this experiment.

In *H. aspersa*, there was a positive correlation between supercooling points and water content, and a transfer to short-day conditions induced lower water content and higher osmolality (Biannic and Daguzan, 1993; Ansart et al., 2001). It is therefore probable that dehydration is involved in the decrease in supercooling points. However, even with identical water content, supercooling points were lower in snails transferred to short-day conditions than in those kept continuously under long-day conditions (Biannic and Daguzan, 1993), and individual supercooling points were not correlated with total water content (Ansart et al., 2001). Therefore, the photoperiodic effect on supercooling points cannot be explained solely by dehydration. Although Biannic and Daguzan (1993) proposed an accumulation of cryoprotectants under short-day conditions, this has not been proven. Ansart et al. (2001) pointed out that a decrease in the quantity of ice nucleating agents in the digestive tract

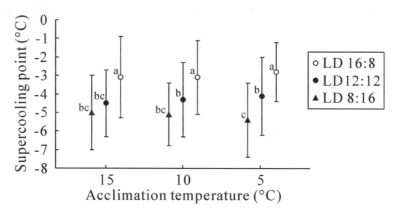

FIGURE 8.5.
Effects of temperature and photoperiod on supercooling points in *Helix aspersa* (based on the data of Ansart et al., 2001). Snails were kept continuously under LD 16:8 or gradually changed to LD 12:12 or LD 8:16 by a 2-h decrease in the photophase per week. Snails were subjected to 15°C for 1 month, then 10°C for 1 month, and finally 5°C for 1 month. The supercooling points were tested at the end of each temperature period. The error bars represent SD, and the mean values with different letters are significantly different ($p < 0.05$ by protected least significant difference Fisher test).

and the production of epiphragms may be other reasons for the lower supercooling points when *H. aspersa* enters winter dormancy under short-day conditions.

Temperature Tolerance

Recently, Udaka et al. (2008) examined the effect of temperature and photoperiod on heat and cold tolerance in a Japanese population of *Lehmannia valentiana*. Slugs were reared from hatching under short-day (LD 12:12) or long-day (LD 16:8) conditions at 20°C, and some were acclimated to 15°C or 25°C for 15 days. They were then exposed to a high or low temperature for 1 h. Cold tolerance was affected by acclimation temperature, and slugs acclimated to 15°C were more tolerant to low temperature than those that experienced 20°C and 25°C. The effect of photoperiod on cold tolerance was also notable and was prominent when slugs were acclimated to 15°C or 25°C (figure 8.6 left). Heat tolerance was enhanced by a higher acclimation temperature, and long-day conditions also increased heat tolerance, at least in slugs continuously kept at 20°C (figure 8.6 right). In *L. valentiana*, therefore, both temperature acclimation and a response to photoperiod mutually contribute to the seasonal changes in temperature tolerance.

Temperature Preference

In the ear pond snail *Lymnaea auricularia*, temperature preference had an annual oscillation, although snails kept under constant photoperiod (LD 12:12) and temperature (21°C) conditions over two years showed no oscillation. However, at a constant temperature of 22°C under a changing photoperiod in which the phase was the reverse of natural changes, that is, long days in the winter and short days in the summer, the preference also fluctuated in the reverse phase. Moreover, a decrease in temperature from 22°C to 6°C over three months had no effect on temperature preference (Rossetti et al., 1989). This is, therefore, an interesting example of photoperiod affecting the behavior of a mollusk. However, temperature preference did not differ between snails kept under LD 16:8 and LD 8:16 at 20°C for 40 days. Rossetti et al. (1989) concluded that *L. auricularia* do not respond to constant photoperiods but to a cyclic change in photoperiod.

PHOTORECEPTORS

The neural mechanisms of photoperiodism in mollusks are still unknown. The only approach that has been made toward studying these mechanisms has been on the photoreceptor for photoperiodism. Stephens and Stephens (1966) reported that after removal of the distal half of the optic tentacles, *Helix aspersa* showed a normal response to photoperiod; that is, some snails laid eggs under long-day

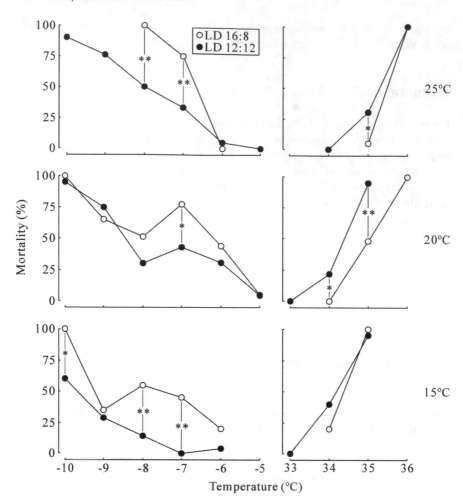

FIGURE 8.6.
Effects of temperature and photoperiod on temperature tolerance in *Lehmannia valentiana* (Udaka et al., 2008). Slugs were reared from hatching under long-day (LD 16: 8) or short-day (LD 12:12) conditions at 20°C. Forty-five days after hatching, slugs were transferred to 15°C or 25°C or kept continuously at 20°C. Sixty days after hatching, temperature tolerance was examined by exposure to a high temperature (right) or low temperature (left) for 1 h. The asterisks show significant differences between the two photoperiodic conditions (**, $p < 0.01$; *, $p < 0.05$ by Fisher's exact test).

conditions (LD 15:9), whereas none did under short-day conditions (LD 9:15) at 17°C. Because the eye is borne at the tip of the tentacle in Stylommatophora, the results clearly demonstrated that an extraocular photoreceptor plays a dominant role in photoperiodism. However, it is unclear whether the eye is involved in the photoperiodic response of intact snails. Even if the elimination of a photoreceptor

has no effect on photoperiodism, there is still a possibility that two or more receptors are involved and that after the elimination of one of the component organs the
remaining receptor(s) are enough to respond to photoperiod (Numata et al., 1997).
In fact, multiple photoreceptors for photoperiodism have been suggested in insects
(see chapter 11).

After removal of the optic tentacles, *Limax maximus* also showed a normal
response to photoperiod; that is, the gonad and penis of slugs developed after
transfer to long-day conditions (LD 16:8) with white light, whereas the gonad and
penis remained small in slugs kept continuously under short-day conditions (LD
8:16) at 15°C (figure 8.7a; McCrone and Sokolove, 1979; Sokolove et al., 1981).
The results are similar to those of Stephens and Stephens (1966) in *H. aspersa*.
Under long-day conditions with red light, however, male-phase reproductive maturation did not occur in snails with the optic tentacle removed (figure 8.7a; Sokolove
et al., 1981). In *Limax maximus*, therefore, both the eye and the extraocular photoreceptor are functional for photoperiodism, but the latter is insensitive to longer

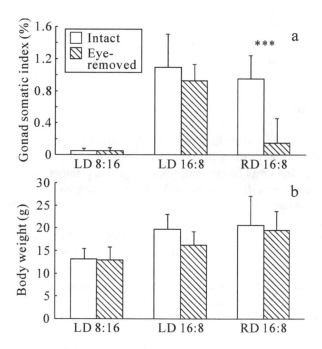

FIGURE 8.7.
Effect of bilateral eye removal on the photoperiodic responses of gonad development
(a) and growth (b) in *Limax maximus* (based on the data of Sokolove et al., 1981). Intact and
eye-removed slugs were exposed for 10 weeks to long-day (LD 16:8) or short-day (LD 8:16)
conditions with white fluorescent light, or long-day conditions with red light (wavelength
> 600 nm, RD 16:8) at 15°C. The error bars represent SD, and the asterisks show a
significant difference between intact and eye-removed slugs (***, $p < 0.001$ by Student's
t-test).

wavelengths. Interestingly, slugs grew significantly more under long-day conditions with red light than under short-day conditions even after removal of the optic tentacles (figure 8.7b; Sokolove et al., 1981). To explain these results, Sokolove et al. (1981) suggested that long-day and short-day conditions are necessary to stimulate male-phase reproductive maturation and to suppress growth, respectively. Under long-day conditions with red light, therefore, slugs were reproductively immature but grew more without long-day and short-day effects, as if in constant darkness. It is difficult to interpret the results, however, because there were no control series of short-day conditions with red light or of constant darkness. There is a possibility that two different extraocular photoreceptors are used for the control of growth and reproduction.

Contrasting results were obtained in *Aplysia californica* (Wayne and Block, 1992). Removal of the eyes eliminated the stimulation of oviposition by short-day conditions. In the control series of this experiment, however, the difference in the frequency of oviposition between long-day and short-day conditions was small but statistically significant.

To date, the photoreceptor for photoperiodism has been examined in only three species, and the results vary. However, it is not surprising that closely related animals use different photoreceptors for photoperiodism, as has been reported in insects (Numata et al., 1997; see chapter 11).

FUTURE DIRECTIONS

Jenner (1951) demonstrated a long-day response for oviposition in *Lymnaea palustris*, and interruption of the long scotophase by a light pulse showed a long-day effect. A well-defined photoperiodic response and the importance of the scotophase were thus demonstrated in mollusks more than half a century ago. However, studies on photoperiodism in mollusks are lacking compared to arthropods and vertebrates.

One possible reason for this is that photoperiod is seldom the only or the predominant controlling factor of seasonal development in mollusks, even though it is generally accepted that photoperiod is the most reliable cue for indicating seasonal information. In most species showing photoperiodism, temperature also has a significant effect (Gomot et al., 1989; Gomot, 1990; Wayne and Block, 1992; Udaka and Numata, 2008). In these species, photoperiod and temperature mutually interact to control seasonal development. Although some snails live in the desert (Schmidt-Nielsen et al., 1971), generally mollusks favor aquatic or humid terrestrial habitats. In these habitats, temperature is a more reliable seasonal cue than in dryer habitats, so many mollusks use temperature for controlling reproduction in addition to photoperiod. Moreover, nutrition and isolation/grouping also affect photoperiodic control of reproduction in mollusks (Bohlken et al., 1986, 1987). In *Helix aspersa*, in which photoperiodism has been intensively examined (.Stephens and Stephens, 1966; Enée et al., 1982; Gomot et al. 1982, 1989; Biannic and Daguzan, 1993; Ansart et al., 2001), Bailey (1981) showed the existence of a circannual rhythm.

For the entrainment of circannual rhythms, it is common that responses to photoperiod change depending on the phase in the rhythm (Gwinner, 1986; Miyazaki et al., 2005). For these reasons, photoperiodism in mollusks is quite complex and has been rather difficult to study.

Here we suggest two possibilities for future studies in mollusk photoperiodism. The first involves approaches toward understanding the ecological significance of photoperiodism. Generally, in an early phase of a research area, studies tend to focus on simpler systems. Complex systems such as mollusk photoperiodism therefore tend to be ignored. However, the evolution of complex systems is also worthy of future study with respect to comparative biology. For example, it would be interesting to clarify why *Helix pomatia* uses photoperiodism for the induction of dormancy (Jeppesen and Nygård, 1976), while a closely related species (*Helix lucorum*) enters dormancy responding directly to low humidity, with little or no effect from photoperiod (Lazaridou-Dimitriadou and Saunders, 1986). In addition to laboratory studies on photoperiodism, field observations of life cycles are also needed to answer this question. Moreover, rearing animals under natural photoperiod and temperature conditions will work as a bridge between these two study types.

Second, we suggest that the neural mechanism of photoperiodism must be clarified in mollusks. Because the neurons of mollusks are larger in size and fewer in number than those of vertebrates and arthropods, the nervous system in mollusks is well suited for identifying specific neurons in living tissues and monitoring their physiological activities (Wayne, 2001). This advantage has contributed to the studies on the neural mechanisms of learning/memory and circadian rhythm (Bailey et al., 1996; Blumenthal et al., 2001). Until now, however, studies on the neural mechanisms of photoperiodism in mollusks have not taken advantage of this.

Even in insects, photoperiodism is not always the predominant mechanism for seasonal development. In the fruit fly *Drosophila melanogaster*, for example, photoperiodism for the induction of adult diapause is shown only in a small range of temperatures, and diapause incidence varies even under the same photoperiod (Saunders et al., 1989; Saunders and Gilbert, 1990). Therefore, *D. melanogaster* poses similar difficulties to those in mollusks. Despite these difficulties, the accumulation of genetic information and the availability of various mutants in this species have contributed to studies on the molecular mechanism of photoperiodism (Saunders et al., 1989; Saunders, 1990; Sandrelli et al., 2007; Tauber et al., 2007; see chapter 11). Because the size and number of neurons make mollusks ideal for the study of neural mechanisms, the difficulties resulting from the influence of various factors such as temperature on photoperiodism may not hinder the study on the mechanism of photoperiodism. Mollusks may very well be ideal subjects for future studies on the neural mechanism of photoperiodism.

References

Ansart A, Vernon P, and Daguzan J. (2001). Photoperiod is the main cue that triggers supercooling ability in the land snail, *Helix aspersa* (Gastropoda: Helicidae). Cryobiology 42: 266–273.

Bailey CH, Bartsch D, and Kandel ER. (1996). Toward a molecular definition of long-term memory storage. Proc Natl Acad Sci USA 93: 13445–13452.

Bailey SER. (1981). Circannual and circadian rhythms in the snail *Helix aspersa* Müller and the photoperiodic control of annual activity and reproduction. J Comp Physiol A 142: 89–94.

Beck SD. (1980). *Insect Photoperiodism*, 2nd ed. New York: Academic Press.

Biannic M and Daguzan J. (1993). Cold-hardiness and freezing in the land snail *Helix aspersa* Müller (Gastropoda; Pulmonata). Comp Biochem Physiol 104A: 503–506.

Bohlken S and Joosse J. (1982). The effect of photoperiod on female reproductive activity and growth of the freshwater pulmonate snail *Lymnaea stagnalis* kept under laboratory breeding conditions. Int J Invertebr Reprod Dev 4: 213–222.

Bohlken S, Joosse J, and Geraerts WPM. (1987). Interaction of photoperiod, grouping and isolation in female reproduction of *Lymnaea stagnalis*. Int J Invertebr Reprod Dev 11: 45–58.

Bohlken S, Joosse J, van Elk R, and Geraerts WPM. (1986). Interaction of photoperiod and nutritive state in female reproduction of *Lymnaea stagnalis*. Int J Invertebr Reprod Dev 10: 151–157.

Blumenthal EM, Block GD, and Eskin A. (2001). Cellular and molecular analysis of molluscan circadian pacemakers. In *Handbook of Behavioral Neurobiology*, vol 12, *Circadian Clocks* (JS Takahashi, FW Turek, and RY Moore, eds). New York: Kluwer Academic/Plenum, pp. 371–400.

Danilevskii AS. (1965). *Photoperiodism and Seasonal Development of Insects*. London: Oliver and Boyd.

Denlinger DL, Yocum GD, and Rimehart JP. (2005). Hormonal control of diapause. In *Comprehensive Molecular Insect Science*, vol 3, *Endocrinology* (LI Gilbert, K Iatrou, and SS Gill, eds). Oxford: Elsevier, pp. 615–650.

Dmitrieva EF. (1969). Population dynamics, growth, feeding and reproduction of *Deroceras reticulatum* in the Leningrad district [in Russian, with English summary]. Zool Zh 48: 802–810.

Ebberink RHM, van Loenhout H, Geraerts WPM, and Joosse J. (1983). Purification and amino acid sequence of the ovulation neurohormone of *Lymnaea stagnalis*. Proc Natl Acad Sci USA 82: 7767–7771.

Egonmwan RI. (1991). The effects of temperature and photoperiod on growth and maturation rate of *Limicolaria flammea* Müller (Gastropoda: Pulmonata: Achatinidae). J Afr Zool 105: 69–75.

Enée J, Bonnefoy-Caudet R, and Gomot L. (1982). Effet de la photopériode artificielle sur la reproduction de l'Escargot *Helix aspersa* Müll. C R Acad Sci III 294: 357–360.

Flari VA and Edwards JP. (2003). The role of the endocrine system in the regulation of reproduction in terrestrial pulmonate gastropods. Invertebr Reprod Dev 44: 139–161.

Garner WW and Allard HA. (1920). Effect of the relative length of day and night and other factors of the environment on growth and reproduction in plants. J Agric Res 18: 553–606.

Gomot A. (1990). Photoperiod and temperature interaction in the determination of reproduction of the edible snail, *Helix pomatia*. J Reprod Fertil 90: 581–585.

Gomot L, Enée J, and Laurent J. (1982). Influence de la photopériode sur la croissance pondérale de l'Escargot *Helix aspersa* Müller en milieu contrôlé. C R Acad Sci III 294: 749–752.

Gomot P and Gomot L. (1985). Action de la photopériode sur la multiplication spermatogoniale et la reproduction de l'Escargot *Helix aspersa* Müller. Bull Soc Zool Fr Evol Zool 110: 445–459.

Gomot P, Gomot L, and Griffond B. (1989). Evidence for a light compensation of the inhibition of reproduction by low temperatures in the snail *Helix aspersa*. Ovotestis and albmen gland responsiveness to different conditions of photoperiods and temperatures. Biol Reprod 40: 1237–1245.

Gwinner E. (1986). *Circannual Rhythms*. Berlin: Springer-Verlag.

Hemminga MA, Koomen W, Maaskant JJ, and Joosse J. (1985). Effects of photoperiod and temperature on the glycogen stores in the mantle and the heat-foot muscles of the freshwater pulmonate snail *Lymnaea stagnalis*. Comp Biochem Physiol 80B: 139–143.

Hommay G, Kienlen JC, Gertz C, and Hill A. (2001). Growth and reproduction of the slug *Limax valentianus* Férussac in experimental conditions. J Mollusc Stud 67: 191–207.

Japan Meteorological Agency. (2006). Historical meteorological data search [in Japanese]. Available at: www.datajmagojp/obd/stats/etrn/indexphp.

Jenner CE. (1951). The significance of the period of darkness in animal photoperiodic responses [abstract]. Anat Rec 111: 512.

Jeppesen LL and Nygård K. (1976). The influence of photoperiod, temperature and internal factors on the hibernation of *Helix pomatia* L. (Gastropoda, Pulmonata). Videnskabelige Meddelelser fra Dansk Naturhistorisk Foreningi Kovenhavn 139: 7–20.

Joosse J. (1984). Photoperiodicity, rhythmicity and endocrinology of reproduction in the snail *Lymnaea stagnalis*. Ciba Found Symp 104: 204–219.

Lagadic L, Coutellec M-A and Caquet T. (2007). Endocrine disruption in aquatic pulmonate molluscs: Few evidences, many challenges. Ecotoxicology 16: 45–59.

Laubier-Bonichon A and Mangold K. (1975). La maturation sexuelle chez les mâles d'*Octopus vulgaris* (Cephalopoda: Octopoda), en relation avec le réflexe photo-sexuel. Mar Biol 29: 45 52.

Lazaridou-Dimitriadou M and Saunders DS. (1986). The influence of humidity, photoperiod, and temperature on the dormancy and activity of *Helix lucorum* L. J Mollusc Stud 52: 180–189.

Lundelius JW and Freeman G. (1986). A photoperiod gene regulates vitellogenesis in *Lymnaea peregra* (Mollusca: Gastropoda: Pulmonata). Int J Invertebr Reprod Dev 10: 201–226.

Manzano NB, Aranda DA, and Brulé T. (1998). Effects of photoperiod on development, growth and survival of larvae of the fighting conch *Strombus pugilis* in the laboratory. Aquaculture 167: 27–34.

Marshall AJ. (1960). Annual periodicity in the migration and reproduction of birds. Cold Spring Harbor Symp Quant Biol 25: 499–505.

Masaki S. (1972). Climatic adaptation and photoperiodic response in the band-legged ground cricket. Evolution 26: 587–600.

Masaki S. (1980). Summer diapause. Annu Rev Entomol 25: 1–25.

McCrone EJ and Sokolove PG. (1979). Brain-gonad axis and photoperiodically-stimulated sexual maturation in the slug, *Limax maximus*. J Comp Physiol A 133: 117–123.

McCrone EJ, van Minnen J, and Sokolove PG. (1981). Slug reproductive maturation hormone: In vivo evidence for long-day stimulation of secretion from brains and cerebral ganglia. J Comp Physiol A 143: 311–315.

Melrose GR, O'Neill MC, and Sokolove PG. (1983). Male gonadotropic factor in brain and blood of photoperiodically stimulated slugs. Gen Comp Endocrinol 52: 319–328.

Miyazaki Y, Nisimura T, and Numata H. (2005). A phase response curve for circannual rhythm in the varied carpet beetle *Anthrenus verbasci*. J Comp Physiol A 191: 883–887.

Nicholls TJ, Goldsmith AR, and Dawson A. (1988). Photorefractoriness in birds and comparison with mammals. Physiol Rev 68: 133–176.

Nisimura T and Numata H. (2003). Circannual control of the life cycle in the varied carpet beetle *Anthrenus verbasci*. Funct Ecol 17: 489–495.

Nolte A, Koolman J, Dorlöchter M, and Straub H. (1986). Ecdysteroids in the dorsal bodies of pulmonates (Gastropoda)—synthesis and release of ecdysone. Comp Biochem Physiol 84A: 777–782.

Numata H, Shiga S, and Morita A. (1997). Photoperiodic receptors in arthropods. Zool Sci 14: 187–197.

O'Dor RK, Durward RD, and Balch N. (1977). Maintenance and maturation of squid (*Illex illecebrosus*) in a 15 meter circular pool. Biol Bull 153: 322–335.

Otchoumou A, Dupont-Nivet M, Ocho LA, and Dosso H. (2007). Effects of photoperiod on growth and reproduction in *Archachatina ventricosa* (Gould 1850). under indoor rearing conditions. Invertebr Reprod Dev 50: 109–115.

Rossetti Y, Rossetti L, and Cabanac M. (1989). Annual oscillation of preferred temperature in the freshwater snail *Lymnaea auricularia*: Effect of light and temperature. Anim Behav 37: 897–907.

Runham NW and Laryea AA. (1968). Studies on the maturation of the reproductive system of *Agriolimax reticulatus* (Pulmonata: Limacidae). Malacologia 7: 93–108.

Sandrelli F, Tauber E, Pegoraro M, Mazzotta G, Cisotto P, Landskron J, Stanewsky R, Piccin A, Rosato E, Zordan M, Costa R, and Kyriacou CP. (2007). A molecular basis for natural selection at the *timeless* locus in *Drosophila melanogaster.* Science 316: 1898–1900.

Saout C, Quéré C, Donval A, Paulet YM, and Samain JF. (1999). An experimental study of the combined effects of temperature and photoperiod on reproductive physiology of *Pecten maximus* from the Bay of Brest (France). Aquaculture 172: 301–314.

Saunders DS. (1990). The circadian basis of ovarian diapause regulation in *Drosophila melanogaster*: Is the *period* gene causally involved in photoperiodic time measurement? J Biol Rhythms 5: 315–331.

Saunders DS and Gilbert LI. (1990). Regulation of ovarian diapause in *Drosophila melanogaster* by photoperiod and moderately low temperature. J Insect Physiol 36: 195–200.

Saunders DS, Henrich VC, and Gilbert LI. (1989). Induction of diapause in *Drosophila melanogaster*: Photoperiodic regulation and the impact of arrhythmic clock mutations on time measurement. Proc Natl Acad Sci USA 86: 3748–3752.

Schmidt-Nielsen K, Taylor CR, and Shkolnik A. (1971). Desert snails: Problems of heat, water and food. J Exp Biol 55: 385–398.

Sokolove PG, Kirgan J, and Tarr R. (1981). Red light insensitivity of the extraocular pathway for photoperiodic stimulation of reproductive development in the slug, *Limax maximus.* J Exp Zool 215: 219–223.

Sokolove PG and McCrone EJ. (1978). Reproductive maturation in the slug, *Limax maximus*, and the effects of artificial photoperiod. J Comp Physiol A 125: 317–325.

Sokolove PG, McCrone EJ, van Minnen J, and Duncan WC. (1984). Reproductive endocrinology and photoperiodism in a terrestrial slug. Ciba Found Symp 104: 189–203.

South A. (1989). A comparison of the life cycles of the slugs *Deroceras reticulatum* (Müller) and *Arion intermedius* Normand on permanent pasture. J Mollusc Stud 55: 9–22.

South A. (1992). *Terrestrial Slugs: Biology, Ecology and Control.* London: Chapman and Hall.

Stephens GJ and Stephens GC. (1966). Photoperiodic stimulation of egg laying in the land snail *Helix aspersa.* Nature 212: 1582.

Tauber E, Zordan M, Sandrelli F, Pegoraro M, Osterwalder N, Breda C, Daga A, Selmin A, Monger K, Benna C, Rosato E, Kyriacou CP, and Costa R. (2007). Natural selection favors a newly derived *timeless* allele in *Drosophila melanogaster.* Science 316: 1895–1898.

Udaka H, Goto SG, and Numata H. (2008). Effects of photoperiod and acclimation temperature on heat and cold tolerance in the terrestrial slug, Lehmannia valentiana (Pulmonata: Limacidae). Appl Entomol Zool 43: 547–551.

Udaka H, Mori M, Goto SG, and Numata H. (2007). Seasonal reproductive cycle in relation to tolerance to high temperatures in the terrestrial slug, *Lehmannia valentiana.* Invertebr Biol 126: 154–162.

Udaka H and Numata H. (2008). Short-day and lowtemperature conditions promote reproductive maturation in the terrestrial slug, *Lehmannia valentiana.* Comp Biochem Physiol A 150: 80–83.

Wayne NL. (2001). Regulation of seasonal reproduction in mollusks. J Biol Rhythms 16: 391–402.

Wayne NL and Block GD. (1992). Effects of photoperiod and temperature on egg-laying behavior in a marine mollusk, *Aplysia californica.* Biol Bull 182: 8–14.

Yoshioka E. (1987). Environmental cue to initiate gametogenesis in the chiton *Acanthopleura japonica* [in Japanese with English summary]. Venus Jpn J Malacol 46: 173–177.

9

Photoperiodism in Copepods

Nancy H. Marcus and Lindsay P. Scheef

As prominent components of most aquatic food webs in terms of numbers, biomass, and productivity, copepods play a vital role in the functioning of these systems. As such, copepods have been studied for decades, but as noted by Marcus (1986), most studies have focused on the influence of only a few physical factors (temperature, salinity, and light) and biological factors (food and predation) on the distribution and abundance of copepods. This observation is supported by a search of the electronic Web of Science (http://www.thomsonreuters.com/products_services/scientific Web_of_Science) using the term "copepod," which yielded approximately 6,000 works. Of these, 1,801 dealt with food, 864 with temperature, 804 with predation, 345 with salinity, and 298 with light. Relatively few studies, 51 and 31, respectively, considered the role of the light cycle or photoperiod on the distribution and abundance of copepods. Furthermore, initiating a search of the electronic database with the term "photoperiod" yielded approximately 12,000 studies, of which only 35 also considered copepods. Hence, despite enormous interest in the importance of photoperiodism and photoperiodic effects, the vast majority of the research effort has not focused on copepods.

With the exception of those organisms that live their entire life cycle in caves or the deep sea, all organisms on earth are exposed to light and thereby subject to the daily cycle of daylight (day) and darkness or reduced light (night). Moreover, depending on their latitudinal location, organisms also experience greater (polar) or lesser (tropical) seasonal change in the duration of day and night coincident with the annual movement of the earth around the sun. The influence of the 24-h cycle of day and night on physiological and behavioral functions has been well studied in a wide variety of organisms, including bacteria, plants, invertebrates, and vertebrates, as summarized in Dunlap et al.'s 2004 book *Chronobiology: Biological Timekeeping*. Since abiotic and biotic environmental conditions change coincident with the day–night cycle (e.g., temperature, prey and predator activity), organisms that are able to synchronize their activity to times of food availability and minimal predator activity are more likely to survive and reproduce. In some cases, the synchronization may require biochemical and physiological adjustments so those organisms that are able to anticipate the timing of changes in the environment should show enhanced survival. As noted by Dunlap et al. (2004), being able to anticipate change is far better than responding to an immediate external cue. This rationale has stimulated much of the research on biological clocks and endogenous timing mechanisms.

The results have led investigators to suggest that internal timers synchronize the many functions of organisms with the 24-h daily cycle. It is evident from these works that in most cases light serves as the entraining agent that resets the internal clock each day so that the daily timing of rhythmic functions and activities of organisms adjust to seasonal change. In some cases, it is the onset of light, and in other cases, it is actually the switch from light to darkness that serves to entrain the rhythms. Other factors may also act as entraining agents for the daily cycle. The thermocline is a unique feature of many aquatic environments. For copepods that undergo daily vertical migrations, temperature could also act as an entraining agent when the animals cross the thermocline. In addition, because water is less subject to daily variations in temperature, copepods may rely on temperature as a primary cue of impending seasonal change, as noted by Hairston and Kearns (1995).

The daily cycle of organisms is usually divided into active and quiet periods. This does not mean that an organism is constantly active during the active phase, nor does it mean that an organism remains inactive for the entire length of the quiet period. The midnight sinking (Pearre, 2003) that has been reported for some copepods that undergo daily vertical migration is consistent with this interpretation.

A critical aspect of endogenous circadian rhythms is that they are not affected by nutritional status or temperature. In other words, the basic length of the circadian rhythm is fixed. Circadian rhythms are typically demonstrated by exposing the test organisms to constant light or darkness and observing the pattern of the activity or function. The duration of the cycle should not be affected by the nutritional status or temperature within the normal range of the organism. However, the existence of an endogenous rhythm could be masked if the temperature is such that it prohibits activity (e.g., it is too warm).

While most organisms are subject to the diel cycle of daylight and darkness, the duration of the light period, the light intensity, and the spectral quality of light to which organisms are exposed is very different in aquatic compared to terrestrial systems. Light intensity is attenuated quickly in water, and even in waters devoid of particles, light quality changes with depth. Freshwater and marine organisms that live at different depths will be exposed to very different light regimes, and this may very well lead to different responses to the daily cycle of light and darkness as well as seasonal changes. For example, on a given day, organisms living at the surface may be exposed to a 12/12-h light/dark (LD) cycle, whereas organisms living at depth in the ocean may only perceive that the cycle is 11/13-h LD. Hence, experiments designed to assess a response must be conducted at conditions that are ecologically relevant to the organism. Spectral sensitivity may also be important. For these reasons, aquatic organisms may differ from terrestrial organisms in how they respond to changes in the light.

Marcus (1986) drew attention to the potential influence of photoperiodism on the population dynamics of copepods, and since that review, a number of investigators have considered the impact of light on copepods with some specifically addressing photoperiodic effects. This chapter summarizes these studies and suggests directions for additional research. The concept of photoperiodic control is credited to Bünning (1973). It refers to the use of the hours of day or night to control the timing of processes and activities.

DORMANCY

Dormancy or developmental arrest occurs in a variety of plants and animals. Comprehensive (Williams-Howze, 1997) and specialized (Grice and Marcus, 1981; Dahms, 1995; Hirche, 1996; Gyllstrom and Hansson, 2004) overviews of copepod dormancy have been published in the last two decades. In copepods, dormancy occurs during the embryonic phase (referred to as resting or diapause eggs in the literature) and copepodite phases. As reviewed by Williams-Howze (1997), studies showing that photoperiod alone or in combination with temperature influence the induction of copepodite dormancy in freshwater cyclopoid copepods date to the 1950s. However, it was not until the 1980s when Marcus (1980, 1982a, 1982b, 1984) reported the results of her studies of the marine calanoid copepod *Labidocera aestiva* that photoperiod was acknowledged to play an important role in regulating the production of diapause and nondiapause eggs in marine copepods. *L. aestiva* is generally a summer–fall species throughout its range that extends from the Cape Cod region of Massachusetts southward along the U.S. East Coast to Florida. The species also occurs in the northern Gulf of Mexico. The studies by Marcus revealed that *L. aestiva* produced a greater proportion of nondiapause eggs when reared under long day lengths (>12 h of daylight) typical of summer and relatively more diapause eggs under short day lengths (<12 h of daylight) typical of fall. Temperature modified the response, and the critical threshold response was different for populations from different latitudes. Additional study of *L. aestiva* revealed that adult females could be induced to switch the from diapause to nondiapause egg production and vice versa over a period of one to two weeks by changing the light conditions from short to long days. At the same time, Nelson Hairston and his colleagues were studying the factors that controlled embryonic dormancy in the freshwater copepod *Diaptomus sanguineus*. The results of these studies showed that this species responds to increasing day lengths and warming temperature typical of spring by switching from nondiapause to diapause egg production (Hairston and Olds, 1984, 1986, 1987; Hairston and Kearns, 1995).

Recognizing the important role that dormancy plays in affecting the distribution and abundance of copepods in aquatic systems, other researchers turned their attention to additional species. These investigations have continued to substantiate the importance of photoperiod and temperature in regulating dormancy, particularly diapause egg production in copepods (Ban, 1992; Marcus and Murray, 2001; Chinnery and Williams, 2003; Avery, 2005a, 2005b; Wu et al., 2007). Most of these studies indicate that photoperiod is the primary factor regulating diapause egg production, although other studies (Avery, 2005a, 2005b; Wu et al., 2007) concluded that temperature may play a more important role. Wu et al. (2007) determined the

type of eggs produced by the subtropical species *Centropages tenuremis* in relation to temperature and photoperiod conditions at the time of collection. They used
probit analysis to calculate a critical photoperiodic threshold of 12.95 h of daylight
and a median threshold of 13.42 h that corresponded well with the observed threshold of 13 h. They also reported that very few diapause eggs were produced when
temperatures were below 22.5°C and that this also corresponded well with predicted values of 22.8°C. Applying a binary logistic multivariate model to the data
to predict the combined effect of temperature and photoperiod, they found that temperature was a better predictor of diapause egg production than photoperiod. They
suggest that this reflects the subtropical occurrence of *C. tenuremis*. In temperate
regions, seasonal changes in photoperiod are much more dramatic than at the lower
latitudes. In waters off the coast of Xiamen, China, the location of their study site,
the change in photoperiod and temperature conditions during the time the animals
were switching the type of eggs produced was only 3.3% (<0.1 h) for day length,
but 16.4% (2°C) for temperature. They suggest that this explains the greater sensitivity to temperature. Avery (2005a, 2005b) reared batches of *Acartia hudsonica*
from Narragansett Bay, Rhode Island, and Cundy's Harbor, Maine, to elucidate
the role of photoperiod and temperature conditions in the induction of dormant
egg production. Although Avery suggested that temperature is more important than
photoperiod in regulating dormancy in *A. hudonsica*, he also acknowledged that the
experiments were not conclusive in this regard.

Understanding the role of photoperiod and other factors on the induction of
dormancy, especially critical thresholds, requires that one clarify which stages
of development are sensitive to the induction cue(s). Marcus (1982a) showed that
adults of *Labidocera aestiva* could be induced to switch the type of eggs they produced by altering the photoperiod and temperature conditions to which they were
exposed. The reversal took place over a period of one to two weeks. This appears
to also be the case for *Acartia bifilosa*. Chinnery and Williams (2003) reported for
A. bifilosa that animals brought in from the field and subjected to opposite day-
length conditions from what they were used to, switched the type of egg being
produced. On the other hand, Ban (1992) showed that the type of eggs produced by
adults was fixed, although they reported that the larval nauplii were responsive to a
change in photoperiod conditions.

Other factors have also been shown to modify the effect of photoperiod. Ban
(1992) reared the brackish species *Eurytemora affinis* in the laboratory at short and
long day lengths (10L/14D, 12L/12D, and 14L/10D) at 10°C and 15°C and observed
that diapause eggs were produced at 10L/14D and 10 and 15°C, with more diapause
eggs being produced at the colder temperature. However, Ban (1992) also found
that at 15°C and 12L/12D or 14L/10D, conditions typically resulting in nondiapause
egg production, diapause eggs were produced in cultures with high densities (>80
females per liter) of animals. He concluded that the response to photoperiod was not
only modified by temperature, but crowding as well, and suggested that unknown
chemical substances might be responsible for the effect. Subsequently, Ban and
Minoda (1994) demonstrated that rearing animals in water from crowded cultures

TABLE 9.1. Embryonic dormancy: short- and long-day–responsive marine and brackish copepod species (x indicates day length that induces dormancy).

Species	Short day	Long day	Reference
Labidocera aestiva	x		Marcus (1979)
Centropages hamatus		x	Marcus and Murray (2001)
Centropages tenuremis		x	Wu et al. (2007)
Acartia clausi		x	Uye (1985)
Acartia bifilosa		x	Chinnery and Williams (2003)
Eurytemora affinis	x		Ban (1992)

modified the effect of photoperiod and concluded that metabolites from the animals induced diapause egg production.

Many examples of freshwater copepod species that respond to photoperiod exist (see Gyllstrom and Hansson, 2004) particularly for cyclopoid copepods, and several examples of short-day species and long-day species are now known for marine and brackish water copepods (table 9.1).

Despite the findings that photoperiod plays an important role in regulating copepodite dormancy in freshwater cyclopoid copepods, evidence for the importance of photoperiod in inducing and terminating copepodite dormancy in marine copepods, specifically *Calanus* spp., remains elusive. Conclusions have generally been based on correlative evidence from field study and not causative evidence based on experimental work.

Calanus spp. undergo dormancy during the copepodite phase and for decades researchers have attempted to elucidate the factors regulating the life cycle. Progress in this regard has been slow however, largely given the difficulty in rearing the animals in the laboratory. Photoperiod has been suggested to be important in the induction of diapause based on correlative evidence. Depending on geographic location *Calanus* proceeds through one or more generations and undergoes daily vertical migrations. This pattern is interrupted at specific times during the year when the animals descend to depth, cease development, and remain in an arrested state. After some time the animals awaken, resume development and the interrupted generation is completed.

Photoperiod has been implicated in the induction of the dormancy because it is assumed that animals residing in surface waters can perceive changes in photoperiod. It has been more difficult to assume a role for photoperiod in the termination of dormancy since *Calanus* are found at depth where light levels are extremely low. Nevertheless, Miller et al. (1991) proposed that termination of the resting phase in *Calanus* is a response to the immediate photoperiodic conditions, but acknowledge that a timer could also play a role. They describe the typical pattern for *Calanus*: individuals enter dormancy in the fall after loading up the oil sac; they then descend to depth and are inactive for several months and, lastly, complete the cycle by awakening in the late winter, molting and rising to the surface to go through one or more generations before descending again.

To assess whether the animals might respond to light at depth, Miller et al. (1991) examined the depth distribution, stage of emergence from dormancy as evidenced by morphological attributes (e.g., gonad development, gnathabases [teeth]), and light regimen associated with diapausing *Calanus* from the North Atlantic Slope Water region south of New England. Distribution of the fifth stage (CV) copepodites was primarily centered around 500 m with a thickness of the layer of about 100–150 m. Based on the tooth development Miller et al. (1991) suggest that awakening occurs in February, whereas analysis of gonad development points to January. They acknowledge that in all likelihood termination of diapause is a gradual process that begins in January and proceeds into March. The daytime light intensity at the 500 m depth at which the animals remained during dormancy was approximately 10^{-5} $\mu W/cm^2$, and rapid movement of the isolume occurred at sunrise and sunset. As noted by Miller et al. (1991), the estuarine copepod *Acartia tonsa* is sensitive to a similar light level (Stearns and Forward, 1984a, 1984b), and they suggest *Calanus* should be at least as sensitive as *Acartia*. In an earlier study, Grigg and Bardwell (1982) examined the arousal process by monitoring the molting of *Calanus* that were collected in the field. They observed that molting occurred 10 days after collection and suggested that the resumption of development was due to exposure to light that came about with collection of the animals.

Johnson (2004) considered the role of photoperiod as the induction cue for diapause in *Calanus pacificus* based on an examination of tooth condition, hormone levels, and molt status. Her rationale for this approach was that if the preparatory phase for dormancy can be identified and described, then the results will point to the inducing cues, but she acknowledged that we still do not understand what those cues are. Johnson (2004) found that the onset of diapause was coincident with declining day length, but the timing did not indicate the importance of a threshold value.

REPRODUCTIVE ACTIVITY

Reproduction is essential for the survival of species, and reproductive patterns and strategies in copepods have received considerable attention (reviewed by Mauchline, 1998). Studies have considered the pattern of reproductive activity (e.g., broadcast spawning vs. the production of egg sacs, the numbers of eggs produced, the types of eggs produced, and seasonality of reproduction). Although some studies have considered the role of photoperiod in regulating life cycle timing (e.g., dormant phases as described previously), few have addressed the role of photoperiod in regulating other aspects of reproductive activity. Niehoff's thorough review of reproductive processes in calanoid copepods (Niehoff, 2006) draws attention to the importance of environmental factors such as temperature and food in this regard, but not photoperiod. Ample evidence points to food and temperature as playing a major role in affecting the number of eggs produced by copepods. However, since the impact of these factors is mediated by the actual act of feeding, if feeding is affected by photoperiod, then it is essential to consider these factors in combination to understand patterns in the field (see below section on feeding rhythms). The timing of egg release may be an important factor affecting the reproductive success of species. Where eggs are released in the water column may affect their survival if sinking into the seabed results in delayed hatching or death, or if predators in the water column consume the eggs or newly hatched nauplii. The timing of spawning has been shown to occur primarily at night in several copepod species (see Mauchline, 1998) suggesting an influence of the light–dark cycle. On the other hand the rhythm could simply be the result of a rhythm in feeding or cyclic food availability associated with vertical migration into and out of food-rich surface waters. Pagano et al. (2004) reported a diel spawning rhythm for *Acartia clausi*, and based upon observing that the timing of peak spawning activity differed between areas in an estuary but the timing of peak feeding did not, they concluded that a feeding rhythm was not likely the direct cause of the spawning rhythm. Additionally, they suggested that the differences in timing reflect genetic adaptation to different salinity regimes and the existence of distinct genetic populations. The implication, therefore, is that light affected the existence of the spawning rhythm directly.

Wu et al. (2006) examined seasonal differences in the spawning rhythm of *Centropages tenuremis*. They reported that during winter subitaneous (nondiapause) eggs were produced and mostly at night, whereas during summer mostly diapause eggs were produced, and these were released mostly during the day. They related the timing and location in the water column of the release with sinking velocity of the eggs.

Photoperiod has also been shown to affect hatching success of *Acartia tonsa*. Peck and Holste (2006) reared *A. tonsa* at photoperiods of 8, 12, 16, and 20 h of daylight. The total number of eggs produced under these conditions was not significantly different; however, egg production was much greater during the dark phase compared to the light phase. Moreover, the 48-h hatching success of the eggs did differ and was greatest for the conditions >16L (78% and 85%). The authors noted that it was not known whether the eggs that did not hatch were dormant or nonviable.

FEEDING

Many studies have reported diel patterns in copepod feeding behavior with typically greater feeding activity at night and suppression of feeding during the day (figure 9.1) (Torgersen, 2003).

Often these rhythms are associated with diel vertical migration (DVM) and movement into and out of a food-rich surface layer, but they have also been observed in cases where DVM is absent or food distribution is homogeneous (Calbet et al., 1999; Pagano et al., 2006). This evidence suggests that diel feeding patterns are not necessarily the result of migratory behavior in patchy food environments, but must have a separate adaptive advantage.

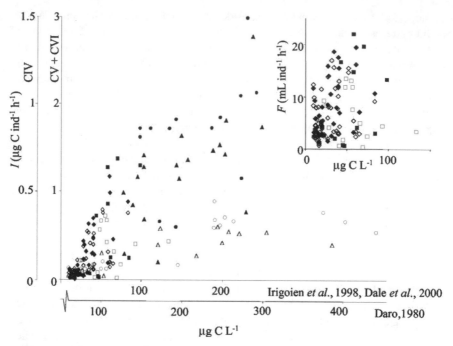

FIGURE 9.1.

Functional feeding responses from day (open symbols) and night (solid symbols) of *Calanus finmarchicus* from data in Dale and Kaartvedt (2000), Irigoien et al. (1998), and Daro (1980). These data indicate feeding activity for this copepod species is generally higher at night, with greater day/night differences at higher food concentrations. Reprinted with permission from Torgersen (2003).

It is believed this kind of rhythmic feeding behavior is the result of selection favoring the optimal trade-off between feeding and predator avoidance (vanDuren and Videler, 1996; Calbet et al., 1999; Torgersen, 2003). The swimming behavior associated with feeding can increase predator encounter probability and can make copepods more conspicuous to predators that are rheotactically or visually sensitive to motion (vanDuren and Videler, 1996; Tiselius and Jonsson, 1997; Torgersen, 2003). In species with little pigmentation, a full gut can also make feeding copepods more vulnerable to visual predation by increasing the contrast between their bodies and the surrounding water (Tsuda et al., 1998). These predation risks are reduced under low-light conditions, making more frequent nocturnal feeding bouts less risky than continuous feeding bouts.

Light is considered the proximate cue that regulates the diel feeding rhythms seen in copepods and other types of zooplankton. Because visual predation is believed to be the primary ultimate factor driving diel feeding rhythms, it is reasonable to believe that these rhythms are affected by certain intensities of light. Cieri and Stearns (1999) found that *Acartia* spp. under the threat of visual predation adjust their feeding activity at a threshold light intensity where their predator's feeding efficiency would significantly change. They observed that copepods would not feed in the presence of predators unless light levels were below that intensity. This threshold intensity occurred at twilight in the environment from which the animals were taken, which explains the higher night feeding activity of animals in the field. In the absence of predators, feeding rhythms can become weak (Calbet et al., 1999) or disappear completely (Cieri and Stearns, 1999) under changing light conditions. This suggests that while light is a cue for diel feeding rhythms, in some cases other environmental factors must be present to induce a response.

An observed feeding pattern can vary based on the cost of abstaining from feeding and the benefits of predator avoidance, and susceptibility to predators and the ability to undure periods of starvation can be dependent upon species characteristics. Therefore the expression of diel feeding rhythms can be expected to vary between species of copepods. Copepods that are more heavily pigmented or are larger in body size may be at greater risk of visual predation than smaller or more transparent species. More distinct diel feeding patterns could be expected for such vulnerable species. Some copepod species are also better able to store lipids and endure short periods of hunger without greatly compromising their fitness (Tiselius, 1998; Calbet et al., 1999). For them, restricting their feeding activity to night hours may be beneficial even if food is scarce or predation pressure is low, so they may also express more regular feeding rhythms in comparison to species that are less tolerant of short periods of starvation.

Within a species risk of predation can change with ontogenetic stage. Nauplii and copepodites may be less vulnerable to visual predators because of their smaller size. Adults, and especially females carrying eggs, may be more susceptible as their size and egg coloration can make them more conspicuous. Variable predation risk between ontogenetic stages could explain cases where diel feeding patterns are more pronounced in older stages or gravid females but less so in nauplii or copepodites (Landry et al., 1994).

Some populations of copepods have been observed to change their feeding rhythms in response to changes in food concentration or predation pressure. Some species of copepods will respond to a decrease in food concentration by feeding continuously despite light conditions. In these cases, it can be assumed the risk of starving presents a greater threat to fitness than the risk of predation. If a species is able to detect the presence of a predator by sensing the predator's kairomones, a population may change their feeding behavior based on whether such kairomones are present (Bollens and Stearns, 1992; Cieri and Sterns, 1999). The addition of predator kairomone to water may initiate nocturnal feeding behavior when continuous feeding was previously displayed in its absence, and a return to continuous feeding can be induced by removing the chemical. These observations present strong supporting evidence for feeding rhythms being the result of predator–prey interactions.

Feeding patterns that are not easily altered by manipulations of environmental factors such as predator presence, food concentration, or temperature are likely to be the result of very strong selective pressure to avoid predation (Durbin et al., 1990; Calbet et al., 1999). In populations under such high predation pressure, the light triggered diel feeding patterns of copepods can become an endogenous rhythm that will persist not only in the absence of predators, but through mistimed light cues and constant light conditions. Stearns (1986) observed that copepods in the field with a nocturnal grazing rhythm would respond to light conditions at sunset by increasing feeding activity, but would not respond to the same cue during daylight hours. In the laboratory, it was observed that nocturnal feeding rhythms would also persist under constant darkness. This study indicates that sometimes an endogenous rhythm can act in conjunction with light cues to time diel feeding responses. In these cases, light is likely to act as a *zeitgeber*, or an agent for resetting the copepods' endogenous clocks, and thereby allows for a synchronous response among individuals.

Sometimes, the pattern observed in the field is the reverse of the nocturnal feeding pattern, with more feeding occurring during the daytime. This pattern may be a response to nonvisual, tactile predators that emerge during night hours to feed, presenting a greater predation risk during darkness than during light (Torgersen, 2003). This reverse rhythm can sometimes also be related to higher food abundance during the day. Daytime feeding increases have been observed in high-nutrient, low-chlorophyll zones, on algal mats of mudflats, and in the arctic under ice sheets. In high-nutrient, low-chlorophyll zones, it is thought that feeding follows algal production, which peaks during the day (Gaudy et al., 2003). As food is grazed down after the daily peak in algal production, feeding decreases and the reverse feeding pattern emerges. Examples of this relationship between copepod feeding activity and diel patterns in algal biomass have also been observed in the benthic environment. Buffan-Dubau and Carman (2000) observed that harpacticoid copepods composing part of the meiofauna of a mudflat exhibited a midday peak in feeding activity coinciding with a peak in benthic microalgal biomass. A similar pattern plays out for zooplankton feeding under ice sheets. Daytime sunlight may warm the ice enough to slightly melt the lower layer, releasing food that was unavailable at night (Hattori and Saito, 1997). In these environments where food is extremely limiting, feeding patterns can be directly related to food abundance. Rather than a direct response to changes in light intensity, these examples suggest reverse diel feeding rhythms are indirectly related to light through algal production and slight surface temperature changes.

With increasing latitude, algal abundance, algal composition, and water temperature vary seasonally in response to changing day length. These changes have been reported to cause seasonal shifts in copepod feeding activity. Feeding activity at the community level tends to be greater during times of the year when water temperature and zooplankton biomass are higher (Froneman, 2001). However, the grazing impact of a particular species may vary seasonally as that species becomes more or less abundant depending on its temperature thresholds, food availability, and predation pressure (Landry et al., 1994). Copepods may also display seasonal shifts in diet depending on the availability of different food particles. For example, species may switch from herbivorous to predatory behavior (Adrian et al., 1999) or begin to incorporate more detritus into their diet. The incorporation of such non-photosynthetic materials into zooplankton diets must be taken into account when using gut fluorescence as an indicator of feeding activity.

The diel feeding patterns of copepods may also shift according to season, usually in association with changes in food availability. By analyzing samples a from series of cruises in the Southern California Bight, Landry et al. (1994) found that *Oncaea* spp. and two species of *Clausocalanus* showed strong diel feeding patterns with higher nighttime gut pigment content in autumn, but no pattern in spring. They concluded that in general, diel patterns in feeding activity tended to be most apparent when chlorophyll in the water column was low. However, the opposite scenario has also been observed. Dam et al. (1995) observed that the mesozooplankton community of the central Pacific exhibited diel feeding patterns in the fall when primary production was high but not in the spring when it was lower due to El Niño. Irogoien et al. (1998) found that the diel feeding patterns of *Calanus finmarchicus* in the Norwegian Sea were more pronounced during a spring bloom, with greater nighttime feeding activity than in the low-chlorophyll conditions preceding and following the bloom. This pattern follows the reasoning that when the threat of starvation exceeds the threat of predation, diurnal feeding suppression disappears as the animals compensate for low food availability (Torgersen, 2003).

Metabolic activities of copepods such as respiration and excretion are closely linked to feeding (Kiørboe et al., 1985) and may therefore be affected by diel feeding rhythms. The respiration rates of fed copepods are higher than those of starved copepods due to biosynthesis processes, and even after being starved for as little as 12 h, respiration rates can more than double with the reintroduction of food (Thor, 2003). Where copepods undergo diel feeding suppression, diel patterns in respiration rates may also be observed in association with changes in feeding activity. In vertically migrating species, diel patterns in respiration rates can also be attributed to changes in swimming activity and exposure to different temperatures and oxygen concentrations with changes in depth (Besiktepe et al., 2005). Phosphate and ammonia excretion have been measured to be higher during times of increased feeding activity in both field and laboratory studies (Macedo and Pinto-Coelho, 2000; Gaudy et al., 2004). When diel feeding patterns occur, diel patterns in excretion may also be expected. Gut evacuation rates, however, are constant regardless of feeding activity patterns (Atkinson et al., 1996). Constant evacuation rates allow for the use of gut fluorescence as a fairly reliable indicator of short-term feeding history.

SWIMMING

Although being part of the plankton implies copepod distribution within the water column is at the mercy of the currents, copepods are able to control their direction of motion to some degree, or arguably, to a large degree in the case of vertical migration. The swimming activity of copepods as part of feeding, predator evasion, and reproduction can have an immense impact on their spatial distribution (Folt and Burns, 1999), and it is therefore important to take diel variability into account.

Diel rhythms in swimming behavior can be linked to three activities: feeding, migration, and swarming.

Copepod feeding can be linked to swimming activity since copepods will actively search for food. Modern three-dimensional imaging techniques have allowed for a closer examination of copepod current generation and swimming behavior associated with feeding (Malkiel et al., 2003). The results suggest that feeding copepods will continually hop while feeding to move away from rejected particles and sample different patches of their environment. This indicates that wherever one observes diel feeding behavior, a general increase in swimming activity would accompany the increase in feeding activity.

Diel vertical migration (DVM) is a widely studied phenomenon in which copepods move between shallow and deep water layers on a daily basis. DVM occurs all over the world in both fresh and saltwater and in terms of biomass represents the largest animal migration on the planet (Hays, 2003). Three general patterns of DVM have emerged: normal, reverse, and twilight. These patterns are thought to result from trade-offs between predator avoidance, hunger and satiation, and the seeking of favorable water temperatures.

Normal vertical migration involves animals remaining in deep water during the day and swimming to surface waters at night. Often animals will feed while in the surface layer but not at depth. This indicates some diel feeding patterns may be the result of animals traveling between food-rich and food-poor parts of the water column, but the two patterns are not always linked. For example, where DVM occurs in water columns with homogeneous food distributions, feeding will still only occur at night regardless of the animals' position in the water column. It is generally accepted that the ultimate reason for DVM is predator avoidance (Ringelberg and Van Gool, 2003). At dusk, copepods come to the surface layer to feed, and at dawn, they take refuge from visual predators in deeper, darker water (figure 9.2).

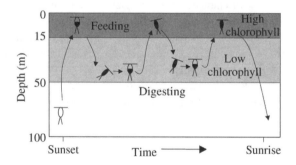

FIGURE 9.2.

Hungry copepods swim to the surface layers at sunset to feed and return to depth at sunrise. During their time at the surface, satiated copepods may cease swimming and sink as they digest but resume swimming and feeding when they become hungry again. This behavior may reduce predation and could also explain observations of uneven surface distributions and midnight sinking. Reprinted with permission from Leising et al. (2005).

In fact, the depth to which animals migrate may be determined by the ambient light levels necessary to escape predation. Frank and Widder (2002) found that several species of vertically migrating organisms adjusted their daytime vertical position in the water column during a period of increased turbidity. Some species were found more than 100 m shallower than where they were observed during clear-water conditions, indicating they were following a preferred irradiance level. Often members of a species will manifest a different daytime depth distribution according to their ontogenetic stage. Larger stages will stay deeper than nauplii and copepodites and exhibit stronger migration patterns (Osgood and Frost, 1994; Huggett and Richardson, 2000; Titelman and Fiksen, 2004). This ontogenetic variability in vertical distribution may be due to a combination of adults being more vulnerable to visual predators and smaller copepodites and nauplii having weaker swimming ability.

Further evidence for DVM being the result of predator avoidance behavior is suggested by the more pronounced migration patterns of individuals or species at greater risk of visual predation. Larger bodied animals, animals carrying eggs, animals with full guts, heavily pigmented species, and more active species have the most potential to be vulnerable to visual predators (Lampert, 1989; Bollens et al., 1993; Hays et al., 1994; Hays, 2003). Using a large data set from a continuous plankton recorder, Hays et al. (1994, 1997) were able to correlate small bodied zooplankton DVM with pigmentation and large bodied zooplankton DVM with morphology. That is, more heavily pigmented small species and less elongate large species are the ones that tend to exhibit DVM. Bollens et al. (1993) also suggest species specific mortality and fecundity could also affect predation risk at the population level, and thereby affect DVM intensity.

While predation is usually considered the ultimate cause of vertical migration, light is considered the proximate cause. That is, light is thought to be the most important factor affecting the timing of DVM behavior, while the behavior ultimately evolved as a response to predation pressure (Cohen and Forward, 2002; Ringelberg and Van Gool, 2003). Without changes in light, there would be asynchronous migration behavior due to endogenous rhythms or no migratory behavior at all. Cohen and Forward (2002) found that specific wavelengths are important for the induction of a migratory response in copepods. They tested the spectral sensitivity of four different species of estuarine copepods with differing migratory behaviors by measuring the percentage of individuals displaying positive phototaxis under different light conditions. They predicted the two normally migrating species *Centropages typicus* and *Calanopia americana* and the reverse migrating species *Anomalocera ornata* would respond to the narrow range of wavelengths experienced in their environment at twilight, and the nonmigrating, predatory species *Labidocera aestiva* would respond to a wide range of wavelengths in order to maximize their photon capture during the day. *C. typicus* and *A. ornata* both showed sensitivity to the predicted twilight wavelength range, and while *C. americana* did show responses at other wavelengths, it showed maximum responsiveness to that range. *L. aestiva* also behaved as predicted, with responses to a wide range of wavelengths. Maximum sensitivity of vertically migrating species to twilight wavelengths is indicative of their dependence on light as a cue for the initiation of diel migrations. This idea that copepods will respond to specific wavelengths as a cue is known as the absolute intensity hypothesis. Another hypothesis, known as the rate of change hypothesis, has also been used to explain the vertical movements of copepods in relation to changing light. While specific wavelengths of light are associated with twilight, this is also the time of day when light intensity is changing most rapidly. Some studies suggest it is this rapid change in light intensity that zooplankton are able to detect and respond to (Stearns and Forward, 1984; Ringelberg, 1999). Although it is the response to light that causes the diel rhythm in migratory behavior, other factors may induce or alter copepod migrations.

Food availability may affect copepod DVM behavior (Durbin et al., 1995). While studying *Calanus agulhensis* off South Africa, Huggett and Richardson (2000) found that the species did perform DVM, but this behavior was absent in circumstances where chlorophyll concentrations were low. In one case, low-chlorophyll conditions at one station improved through the day, and copepods at the surface that did not appear to be demonstrating DVM resumed the behavior under the higher chlorophyll concentrations. Adults appeared to resume migrations earlier than smaller individuals, which may be due to younger stages being more susceptible to starvation or less susceptible to predation (De Robertis et al., 2000). Dagg et al. (1997) suggest the depth to which migrating adult *Calanus pacificus* ascend is dependant on food abundance in the water column. Their study implies that in situations when surface water primary production is high and sinking cells can provide food for migrating individuals, more of the population will remain at intermediate depth rather than migrating to the surface. When surface primary production is low, a full ascent to productive surface waters from deep water would be expected.

Although it is thought that migration to surface layers at night is necessary for copepods and other migrating zooplankton to obtain food, it has been observed that DVM can still occur in cases where there is a deep-water chlorophyll maximum or vertical food distribution is homogeneous and it is not necessary to migrate to surface waters to feed. In such cases, it has been suggested that the animals may be migrating to the surface layer to seek out warmer temperatures that speed growth and reproduction, but it is difficult to determine whether such a benefit is greater than the cost of the migratory swimming (Williamson et al., 1996; Lampert et al., 2003; Winder et al., 2003). Even so, temperature does appear to have the potential to alter migration patterns. While studying seasonal variations in the DVM of *Temora longicornis* in Long Island Sound, Dam and Peterson (1993) found that although migration patterns did not seem to be affected by algal concentration, the amplitude of vertical movement decreased when surface waters became warmer, with animals ceasing to enter the surface waters once the temperature exceeded 17°C. This relationship between ascent depth and temperature was shown to be consistent over several years.

On a monthly time scale, it has been observed that moon phase can affect vertical migratory behavior. Generally, nonpredatory zooplankton will maintain lower positions in the water column during full moon but will become more abundant in surface waters during new moon (Jerling and Wooldridge, 1992; Morgado et al., 2006). This behavior varies with taxonomic group indicating a relationship to predator–prey interactions (Morgado, et al., 2006), but reproductive benefits of this behavior have also been suggested (Rios-Jara, 2005).

Although predation pressure can account for much of the benefit of DVM, currents that remove individuals from a population may also select for migratory behavior. It has been suggested that DVM can serve as a mechanism to keep a population from being flushed out of an area by currents. Advection can strongly influence the zooplankton community found at a specific location, so avoiding advective loss can be important for some populations to stay in a favorable area. In some cases, it has been shown that vertically migrating zooplankton synchronize their migrations with the tides so that tidal currents can help them maintain a specific horizontal distribution. This behavior often occurs in estuaries where tidal flushing is strong and zooplankton, especially meroplankton, need to maintain position in a specific salinity range (Kimmerer et al., 2002; Rawlinson et al., 2004). Although much tidal migration research has focused on the larvae of benthic organisms and fish, a few studies have found that copepods demonstrate tidal migration. Kimmerer et al. (2002) have monitored the migrations of zooplankton in San Francisco Bay over three years and found that the vertical migrations of several copepod species are synchronized to flood and ebb tidal currents in a way that would help retain individuals in the estuary. DVM could also serve to maintain copepods in near-shore upwelling zones. Upwelled surface water in coastal areas is highly productive, but moves offshore due to Ekman transport while deeper water moves toward shore. By spending night in the surface waters and day at depth, copepods can maintain the same relative horizontal position (Peterson, 1998; Batchelder et al., 2002).

In some cases, DVM by copepods may be the result of predatory copepods following their prey or the prey of other vertical migrators seeking to avoid their predators. Predatory copepods hunt by sensing the movements of their prey in the water around them. Tactile hunting maintains the same efficiency regardless of light conditions or may even be more efficient in darkness (Torgersen, 2003), so copepods that hunt in this manner should be able to follow their migrating prey and hunt throughout the day while possibly avoiding their own predators at depth. Many species of fish and plankton vertically migrate to avoid their predators while at the same time preying on copepods. In cases where these predators perform normal vertical migration, their copepod prey may exhibit reverse vertical migration to avoid encounters with them. Vertical migration into surface waters during the day and to depth at night is considered reverse DVM.

Twilight DVM is a third type of migration that involves multiple movements through the water column during the night, typically with an ascent to surface waters at sundown, a sinking into intermediate or deep waters around midnight known as the midnight sink, and an ascent to surface waters followed by a descent to deep water just before dawn. With no light cue at midnight to trigger a descent, the cause of this behavior has been debated. The midnight sink phenomenon was originally considered to be a sinking phase of copepods that had swum to the surface after dusk and ceased feeding and associated swimming behavior in response to becoming satiated. There have been several studies that observed higher gut pigments occurring in deep water some time after the dusk ascent of copepods, suggesting a descent of satiated individuals (Tarling et al., 2002; Pearre, 2003).

In recent years, advances in technology have allowed for a more in depth analysis of twilight DVM. The development of the acoustic Doppler current profiler (ADCP) has allowed for the measure of vertical distribution and quantity of zooplankton biomass in the water column as well as the calculation of velocity of zooplankton in a vertical plane. This data enables researchers to track population movements as well as the average swimming behavior of individuals. Tarling et al. (2002) used this technology along with net sampling to track the diel movements of *Calanus finmarchicus* through the water column of the Clyde Sea along the coast of Scotland. Where an asynchronous midnight descent of individuals was assumed due to variability in nutritive states and times to satiation, Tarling et al. (2002) found an apparent synchronized descent of *C. finmarchicus* that closely coincided with the arrival of krill, a nonvisual predator of the copepod, to the surface layer. They concluded midnight sink behavior could be attributed to predator avoidance rather than satiation and that other studies should take the vertical location of predators in the water column into account before ascribing sinking behavior to satiation. Cohen and Forward (2005) took the presence of predators and tidal currents into account during their study of twilight migration of *Calanopia americana* in the Newport River estuary in North Carolina. They could not find a clear relationship between the vertical movements of the copepods and either of those possible cues. Further investigation revealed the dawn and dusk movements to be related to light and endogenous rhythms, and the midnight sink and early morning rise to be related to endogenous rhythms or states of satiation or hunger. Pearre (2003) discusses the difficulties in examining vertical migration with midnight sink without knowing the motions of individuals. It is likely that causes of this behavior vary with environment and in many cases may act in combination.

Where seasonal variability in vertical migration patterns is observed, it can sometimes be related to seasonality in temperature, food composition, predation pressure, or a combination of these factors (Durbin et al., 1995; Falkenhaug et al., 1997). Although these environmental factors can affect migration on a seasonal basis, light is the most universal factor with the potential to influence migration patterns. Because the range of wavelengths at twilight is the cue that triggers the migrations, seasonal changes in day and night length can have a strong affect on diel migration strength and the amount of time copepods spend at the surface (figure 9.3) (Hays et al., 1996; Falkenhaug et al., 1997). These seasonal changes become greater with increasing latitude and are most dramatic at the poles. Here, clear seasonal shifts in DVM behavior can be observed. At times of midnight sun or continuous darkness at the poles, the light cue that triggers the migratory response is weak or absent, and field observations have recorded low intensity or absent vertical migration during these times of the year. During times when day and

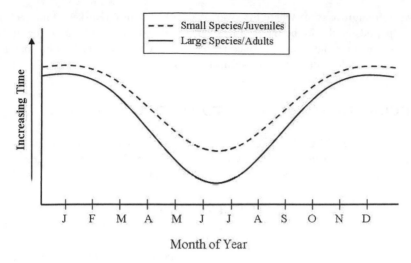

FIGURE 9.3.
The amount of time migratory copepods spend in surface waters during night hours at high northern latitudes. The migration patterns of copepod populations living at high latitudes can change drastically with seasonal changes of day and night length. This figure, based upon the work of Hays (2003) and Falkenhaug (1997), is an example of how some copepods spend less time at the surface during summer months when nights are shorter and more during winter months when nights are long. This seasonal variability may be less pronounced in smaller species and juveniles (Hays, 2003).

night hours are more balanced, migratory behavior is strongest. Falkenhaug et al. (1997) were able to find a close correlation between vertical migration strength and light variability for the copepod species *Chiridius armatus* in a Norwegian fjord. Migration amplitude was greatest during times of the year there were nearly 12 h of light and 12 h of darkness and absent during times of continuous light. They also found the amount of time some species spent at the surface varied seasonally with night length.

Like vertical migration, swarming is a type of copepod swimming behavior that involves copepods swimming to stay in a specific part of the water column during certain times of day. Possible explanations for this type of behavior may be to maintain close proximity to mates, to avoid predation, or to stay in favorable patches of the environment (Ambler, 2002). Ambler (2002) reviews the swarming behaviors of three taxonomic groups of marine copepods along with freshwater cladocerans. Oithonids and *Acartia* species both exhibit diel patterns in their swarming behavior. These copepods form swarms at dawn over specific areas of the habitat such as coral heads or sandy clearings and disperse at dusk. The swarms contain mostly older copepodite stages and adults due to the limited swimming ability of younger animals and the possible mating advantages associated with the swarming. Swarming behavior of the species *Dioithona oculata* has been induced in laboratory settings with the use of a fiber optic light to mimic a light shaft between mangrove roots where the swarms occur in nature (Buskey et al., 1995). The copepods could not be induced to swarm with exposure to light during night hours, suggesting that light cues in combination with endogenous rhythms may be responsible for the diel timing of swarming behavior.

THOUGHTS AND DIRECTIONS FOR FUTURE RESEARCH

There is ample evidence that several copepod functions manifest a pattern of rhythmicity that is synchronized to the cycle of day and night. Moreover, in regions that experience seasonal changes in the light–dark cycle, it is evident that the pattern of the rhythm and manifestation of the function may change (e.g., appearance of a dormant phase or a cessation of vertical migration). Experimental studies of many terrestrial organisms and some aquatic organisms have revealed that photoperiod acts as a signal that induces changes in biological functions. Photoperiod has received so much attention because it is viewed as the most reliable signal of impending seasonal change in the environment. This is certainly true for the terrestrial environment where temperatures may fluctuate widely on a daily basis, but in aquatic systems daily temperature change is dampened due to the heat capacity of water. As a result, it may be that temperature plays a much greater role in regulating seasonal changes in processes either alone or in concert with photoperiod. Additional opportunity for temperature acting as a daily entraining agent in aquatic systems exists for those species that vertically migrate and cross a thermocline on a daily basis.

Life Cycles

Miller et al. (1991) pointed out that life cycle issues do not receive the degree of attention given to topics relating to trophic dynamics. They suggested that the timing of life cycle adjustments should be quite responsive to environmental change and also to geographical differences in environmental conditions. For example, the onset, duration, and termination of the resting phase allow organisms to focus reproductive effort during the most favorable parts of the year.

    The factors controlling the life cycle of *Calanus* spp. have intrigued researchers for decades. First, there is the pattern of diel vertical migration by which typically animals reside at depth during the day where they experience extremely low light levels or none. How does *Calanus* know when to move up into surface waters? As noted above, some investigators have suggested that the animals follow an isolume or reside at a depth where they can still perceive sufficient light that could act as an entraining agent. If in some regions animals are dwelling below the depth of adequate light penetration, then a biological timer could enable the copepods to anticipate the onset of sunset and move upward in the water column. This type of scenario for *Calanus* seems quite similar to the situation for bats described by Dunlap et al. (2004). Many bats rest in caves or similar protected dark areas during the day, awaken several hours before sunset and initiate flying around the cave to warm up and become metabolically active. They move to the opening of the cave to assess the light conditions and then, at the appropriate threshold of light intensity, exit the cave to feed. This rhythm helps them conserve energy by being metabolically inactive during the day in cooler temperatures. Features of the dormancy pattern in *Calanus* also appear analogous to the dormancy patterns described for other animals (Dunlap et al., 2004). Squirrels are mostly dormant during their hibernation phase, but they do undergo short bouts of activity to obtain some food to sustain them. Similarly, studies of *Calanus* have reported that not all individuals in a population are dormant (i.e., some appear to feed). Those observed with food in their guts could simply represent individuals needing to "grab a bite" to sustain themselves through the long diapause phase. Likewise, bears enter caves and dwell far from any light source for much of the winter. Eventually they awaken and sample the light cycle. If conditions are not right, they reenter hibernation. Their internal timer signals them to awaken, and they begin sampling the environment. Such could be the case for *Calanus*. An internal timer could trigger an awakening from dormancy so that individuals become active and initiate some vertical movement to sample light conditions. If conditions are right, they resume development; if not, they reenter dormancy. A critical aspect of this for *Calanus* would be resumption of feeding and being prepared metabolically to deal with ingestion and digestion. Future studies should examine the coincidence of increased enzyme activity with the termination of diapause because this would help in preparation for feeding perhaps before animals begin migrating.

Environmental Change and Its Potential Impact

Much of the research on global change is focused on warming, but other features of the environment are undergoing change as a result of human activity. One needs only look at a photograph of the earth at night from space (figure 9.4) to realize how humans are altering the light conditions experienced by organisms.

In well-populated areas, light has become an ever-present condition, and this light pollution may have a negative influence on organisms (see Dunlap et al., 2004; Longcore and Rich, 2004; Navara and Nelson, 2007). Recent reports for humans indicate the negative influence of exposure to artificial light at night on melatonin production (see Navara and Nelson, 2007). If light pollution in the natural environment is indeed a problem for wild organisms, one can hypothesize that terrestrial organisms living near heavily populated and well-lit areas will show greater disruption/change in their normal behavior than those living in rural low-density regions, and that aquatic organisms living in lakes or coastal areas near heavily populated regions will be similarly more affected than those occurring in remote regions or the open ocean. In regions where artificial lighting results in nighttime light levels that never fall below natural twilight, it is reasonable to assume that some physiological, biochemical, and/or behavioral processes will be disrupted

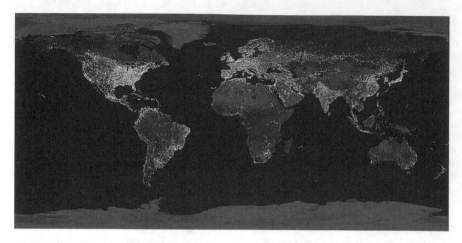

FIGURE 9.4.
Image of the earth at night depicting light pollution. The image showing the presence of man-made lighting was created by Craig Mayhew and Robert Simmon, NASA GSFC, using data from the Defense Meteorological Satellite Program (DMSP). The brightest areas result from city lights associated with heavy urbanization. Image courtesy of NASA and DMSP.

(Longcore and Rich, 2004; Navara and Nelson, 2007). As noted by Longcore and Rich (2004), the potential impact of artificial lighting depends not only on the duration and intensity of the artificial lighting, but also on the spectral quality of the lights. There are examples of animals being attracted to or avoiding artificial light at night, resulting in a disruption of normal activity, for example, nocturnal orientation of sea turtle hatchlings (Tuxbury and Salmon, 2005) and the vertical migration patterns of *Daphnia* (Moore et al., 2000). In this latter study, black and clear enclosures were used to compare the migratory patterns at night of *Daphnia* in a lake located 16 km southwest of Boston. The black enclosures eliminated 96% of the downwelling irradiance. No evidence for vertical migration by *Daphnia retrocurva* was observed in the clear enclosures or the lake itself; however, the species did vertically migrate in the black enclosures. No other cladoceran species or copepod species showed evidence of being affected by the elimination of the ambient light at night. This study draws attention to the potential disruption of normal grazing patterns by zooplankton due to reduced vertical migration and the cascading effects on ecosystems; for example, if phytoplankton are not cropped by the zooplankton, this could lead to decreased water clarity. Longcore and Rich (2004) also point to the disruption of normal predator–prey interactions noting the studies of Gliwicz (1986, 1999) in which zooplankton suffered increased predation by visual fish predators on nights illuminated by a full moon. The implication is that artificial lighting could increase the susceptibility of copepods to visual predators by illuminating water columns at night. These ideas suggest many directions for future research.

Since light penetration in aquatic systems is affected by water clarity, nutrient pollution could lead to a change in the light regimen and this could have an influence on rhythmic behaviors. Depth of light penetration, intensity, and spectral quality could be affected by changes in water clarity. Such changes are more likely to occur in heavily populated coastal regions where nutrient pollution is a much greater problem than in the open ocean. Likewise, changes in temperature due to global warming could differentially alter the development and growth of species, potentially impacting critical life cycle stages differently and hence the coincidence of stages with particular day lengths. This could disrupt life cycles by disrupting the coincidence of specific life stages with critical day lengths as shown for some plankton in a temperate eutrophic lake (Adrian et al., 2006). There could be a large scale ecosystem effect if there is a cascade of effects and the overlap period of prey and predators is affected. It would seem that there are many opportunities for research on the impact of light on copepods. Such efforts will provide a much improved understanding of these critical components of aquatic ecosystems.

References

Adrian R, Hansson S, Sandin B, De Stasio B, and Larsson U. (1999). Effects of food availability and predation on a marine zooplankton community—a study on copepods in the Baltic Sea. Int Rev Hydrobiol 84: 609–626.

Adrian R, Wilhelm S, and Gerten D. (2006). Life-history traits of lake plankton species may govern their phenological response to climate warming. Glob Change Biol 12: 652–661.

Ambler JW. (2002). Zooplankton swarms: Characteristics, proximal cues, and proposed advantages. Hydrobiologia 480: 155–164.

Atkinson A, Ward P, and Murphy EJ. (1996). Diel periodicity of subantarctic copepods: Relationships between vertical migration, gut fullness and gut evacuation rate. J Plankt Res 18: 1387–1405.

Avery DE. (2005a). Induction of embryonic dormancy in the calanoid copepod *Acartia hudsonica*: Proximal cues and variation among individuals. J Exp Mar Biol Ecol 314: 203–214.

Avery DE. (2005b). Induction of embryonic dormancy in the calanoid copepod *Acartia hudsonica*: Heritability and phenotypic plasticity in two geographically separated populations. J Exp Mar Biol Ecol 314: 215–225.

Ban S. (1992). Effects of photoperiod, temperature, and population-density on induction of diapause egg-production in *Eurytemora affinis* (Copepoda, Calanoida) in Lake Ohnuma, Hokkaido, Japan. J Crustac Biol 12: 361–367.

Ban S and Minoda T. (1994). Induction of diapause egg production in *Eurytemora affinis* by their own metabolites. Hydrobiologia 293: 185–189.

Besiktepe S, Svetlichny L, Yuneva T, Romanova Z, and Shulman G. (2005). Diurnal gut pigment rhythm and metabolic rate of *Calanus euxinus* in the Black Sea. Mar Biol 146: 1189–1198.

Batchelder HP, Edwards CA, and Powell TM. (2002). Individual-based models of copepod copulations in coastal upwelling regions: Implications of physiologically and environmentally influenced diel vertical migration on demographic success and nearshore retention. Prog Oceanogr 53: 307–333.

Bollens SM, Osgood K, Frost BW, and Watts SD. (1993). Vertical distributions and susceptibilities to vertebrate predation of the marine copepods *Metridia lucens* and *Calanus pacificus*. Limnol Oceanogr 38: 1827–1837.

Bollens SM and Stearns DE. (1992). Predator-induced changes in the diel feeding cycle of a planktonic copepod. J Exp Mar Biol Ecol 156: 179–186.

Buffan-Dubau E and Carman KR. (2000). Diel feeding behavior of meiofauna and their relationships with microalgal resources. Limnol Oceanogr 45: 381–395.

Bünning F. (1973). *The Physiological Clock, Circadian Rhythms, and Biological Chronometry.* New York: Springer-Verlag.

Buskey EJ, Peterson JO, and Ambler JW. (1995). The role of photoreception in the swarming behavior of the copepod *Diothona oculata*. Mar Freshw Behav Physiol 26: 273–285.

Calbet A, Saiz E, Irigoien X, Alcaraz M, and Trepat I. (1999). Food availability and diel feeding rhythms in the marine copepods *Acartia grani* and *Centropages typicus*. J Plankt Res 21: 1009–1015.

Chinnery FE and Williams JA. (2003). Photoperiod and temperature regulation of diapause egg production in *Acartia bifilosa* from Southampton Water. Mar Ecol Prog Ser 263: 149–157.

Cieri MD and Stearns DE. (1999). Reduction of grazing activity of two estuarine copepods in response to the exudate of a visual predator. Mar Ecol Prog Ser 177: 157–163.

Cohen JH and Forward RB. (2002). Spectral sensitivity of vertically migrating marine copepods. Biol Bull 203: 307–314.

Cohen JH and Forward RB. (2005). Diel vertical migration of the marine copepod *Calanopia americana*. I. Twilight DVM and its relationship to the diel light cycle. Mar Biol 147: 387–398.

Dagg MJ, Frost BW, and Newton JA. (1997). Vertical migration and feeding behavior of *Calanus pacificus* females during a phytoplankton bloom in Dabob Bay, US. Limnol Oceanogr 42: 974–980.

Dahms HU. (1995). Dormancy in the Copepoda—an overview. Hydrobiologia 306: 199–211.

Dale T and Kaartvedt S. (2000). Large scale distributions and population structure of *Calanus finmarchicus* in the Norwegian Sea during summer, in relation to physical environment, food, and predators. ICES J Mar Sci 57: 1800–1818.

Dam HG and Peterson WT. (1993). Seasonal contrasts in the diel vertical distribution, feeding behavior, and grazing impact of the copepod *Temora longicornis* in Long Island Sound. J Mar Res 51: 561–594.

Dam HG, Zhang XS, Butler M, and Roman MR. (1995). Mesozooplankton grazing and metabolism at the equator in the Central Pacific—implications for carbon and nitrogen fluxes. Deep-sea research part II. Top Stud Oceanogr 42: 735–756.

Daro MH. (1980). Field study of the diel feeding of a population of *Calanus finmarchicus* at the end of a phytoplankton bloom. Meteor Forsch.-Ergebnisse Reihe A 22:123–132.

De Robertis A, Jaffe JS, and Ohman MD. (2000). Size-dependent visual predation risk and the timing of vertical migration in zooplankton. Limnol Oceanogr 45: 1838–1844.

Dunlap JC, Loros JJ, and DeCoursey PJ. (2004). *Chronobiology: Biological Timekeeping*. Sunderland, MA: Sinauer Associates.

Durbin AG, Durbin EG, and Wlodarczyk E. (1990). Diel feeding-behavior in the marine copepod *Acartia tonsa* in relation to food availability. Mar Ecol Prog Ser 68: 23–45.

Durbin EG, Gilman SL, Campbell RG, and Durbin AG. (1995). Abundance, biomass, vertical migration, and estimated development rate of the copepod *Calanus finmarchicus* in the southern Gulf of Maine during late spring. Cont Shelf Res 15: 571–591.

Falkenhaug T, Tande KS, and Semenova T. (1997). Diel, seasonal, and ontogenetic variations in the vertical distributions of four marine copepods. Mar Ecol Prog Ser 149: 105–119.

Folt CL and Burns CW. (1999). Biological drivers of zooplankton patchiness. Trends Ecol Evol 14: 300–305.

Frank TM and Widder EA. (2002). Effects of a decrease in downwelling irradiance on the daytime vertical distribution patterns of zooplankton and micronekton. Mar Biol 140: 1181–1193.

Froneman PW. (2001). Seasonal changes in zooplankton biomass and grazing in a temperate estuary, South Africa. Estuar Coast Shelf Sci 52: 543–553.

Gaudy R, Champalbert G, and Le Borgne R. (2003). Feeding and metabolism of mesozooplankton in the equatorial Pacific high-nutrient, low-chlorophyll zone along 180 degrees. J Geophys Res Oceans 108.

Gaudy R, Le Borgne R, Landry MR, and Champalbert G. (2004). Biomass, feeding and metabolism of mesozooplankton in the equatorial Pacific along 180 degrees. Deep-sea research part II. Top Stud Oceanogr 51: 629–645.

Gliwicz ZM. (1986). A lunar cycle in zooplankton. Ecology 67: 883–897.

Gliwicz ZM. (1999). Predictability of seasonal and diel events in tropical and temperate lakes and reservoirs. In *Theoretical Reservoir Ecology and Its Applications* (JG Tundisi and M Straskraba, eds). Leiden: Backhuys, pp. 99–124.

Grice G and Marcus N. (1981). Egg dormancy of marine copepods. Oceanogr Mar Biol Annu Rev 19: 125–140.

Grigg H and Bardwell SJ. (1982). Seasonal observations on moulting and maturation in stage V copepodites of *Calanus finmarchicus* from the Firth of Clyde. J Mar Biol Assoc UK 62: 315–327.

Gyllstrom M and Hansson LA. (2004). Dormancy in freshwater zooplankton: Induction, termination and the importance of benthic-pelagic coupling. Aquat Sci 66: 274–295.

Hairston NG and Kearns CM. (1995). The interaction of photoperiod and temperature in diapause timing—a copepod example. Biol Bull 189: 42–48.

Hairston N and Olds EJ. (1984). Population differences in the timing of diapause: Adaptation in a spatially heterogeneous environment. Oecologia 61: 42. 48

Hairston NG Jr and Olds EJ. (1986). Partial photoperiodic control of diapause in three populations of the freshwater copepod *Diaptomus sanguineus*. Biol Bull 171: 135–142.

Hairston N and Olds EJ. (1987). Population differences in the timing of diapause: A test of hypotheses. Oecologia 71: 339–344.

Hattori H and Saito H. (1997). Diel changes in vertical distribution and feeding activity of copepods in ice-covered Resolute Passage, Canadian arctic, in spring 1992. J Mar Syst 11: 205–219.

Hays GC. (2003). A review of the adaptive significance and ecosystem consequences of zooplankton diel vertical migrations. Hydrobiologia 503: 163–170.

Hays GC, Proctor CA, John AWG, and Warner AJ. (1994). Interspecific differences in the diel vertical migration of marine copepods—the implications of size, color, and morphology. Limnol Oceanogr 39: 1621–1629.

Hays GC, Warner AJ, and Lefevre D. (1996). Long-term changes in the diel vertical migration behaviour of zooplankton. Mar Ecol Prog Ser 141: 149–159.

Hays GC, Warner AJ, and Tranter P. (1997). Why do the two most abundant copepods in the North Atlantic differ so markedly in their diel vertical migration behaviour? J Sea Res 38: 85–92.

Hirche HJ. (1996). Diapause in the marine copepod, *Calanus finmarchicus*—a review. Ophelia 44: 129–143.

Huggett JA and Richardson AJ. (2000). A review of the biology and ecology of *Calanus agulhensis* off South Africa. ICES J Mar Sci 57: 1834–1849.

Irogoien X, Head R, Klenke U, Meyer-Harris B, Harbour D, Niehoff B, Hirche HJ, and Harris R. (1998). A high frequency time series at Weathership M, Norwegian Sea, during the 1997 spring bloom: Feeding of adult female *Calanus finmarchicus*. Mar Ecol Prog Ser 172: 127–137.

Jerling HL and Wooldridge TH. (1992). Lunar influence on distribution of a calanoid copepod in the water column of a shallow, temperate estuary. Mar Biol 112: 309–312.

Johnson CL. (2004). Seasonal variation in the molt status of an oceanic copepod. Prog Oceanogr 62: 15–32.

Kimmerer WJ, Burau JR, and Bennett WA. (2002). Persistence of tidally-oriented vertical migration by zooplankton in a temperate estuary. Estuaries 25: 359–371.

Kiørboe T, Mohlenberg F, and Hamburger K. (1985). Bioenergetics of the planktonic copepod *Acartia tonsa*—relation between feeding, egg production and respiration, and composition of specific dynamic action. Mar Ecol Prog Ser 26: 85–97.

Lampert W. (1989). The adaptive significance of diel vertical migration of zooplankton. Funct Ecol 3: 21–27.

Lampert W, McCauley E, and Manly BFJ. (2003). Trade-offs in the vertical distribution of zooplankton: Ideal free distribution with costs? Proc Roy Soc Lond B Biol Sci 270: 765–773.

Landry MR, Peterson WK, and Fagerness VL. (1994). Mesozooplankton grazing in the Southern California Bight. 1. Population abundances and gut pigment contents. Mar Ecol Prog Ser 115: 55–71.

Leising AW, Pierson JJ, Cary S, and Frost BW. (2005). Copepod foraging and predation risk within the surface layer during night-time feeding forays. J Plankt Res 27: 987–1001.

Longcore T and Rich C. (2004). Ecological light pollution. Front Ecol Environ 2: 191–198.

Macedo CF and Pinto-Coelho RM. (2000). Diel variations in respiration, excretion rates, and nutritional status of zooplankton from the Pampulha Reservoir, Belo Horizonte, MG. J Exp Zool 286: 671–682.

Malkiel E, Sheng I, Katz J, and Strickler JR. (2003). The three-dimensional flow field generated by a feeding calanoid copepod measured using digital holography. J Exp Biol 206: 3657–3666.

Marcus NH. (1979). Population biology and nature of diapause of *Labidocera aestiva* (Copepoda: Calanoida). Biol Bull 157: 297–305.

Marcus N. (1980). Photoperiodic control of diapause in the marine calanoid copepod, *Labidocera aestiva*. Biol Bull 159: 311–318.

Marcus N. (1982a). The reversibility of subitaneous and diapause egg production by individual females of *Labidocera aestiva* (Copepoda: Calanoida). Biol Bull 162: 39–44.

Marcus N. (1982b). Photoperiodic and temperature regulation of diapause in *Labidocera aestiva* (Copepoda: Calanoida). Biol Bull 162: 45–52.

Marcus N. (1984). Variation in the diapause response of *Labidocera aestiva* (Copepoda:Calanoida) from different latitudes and its importance in the evolutionary process. Biol Bull 166: 127–139.

Marcus NH. (1986). Population dynamics of marine copepods: The importance of photoperiodism. Am Zool 26: 469–477.

Marcus NH and Murray M. (2001). Copepod diapause eggs: A potential source of nauplii for aquaculture. Aquaculture 201: 49–60.

Mauchline J. (1998). *Advances in Marine Biology*, vol 33, *The Biology of Calanoid Copepods*. San Diego: Academic Press.

Miller CB, Cowles TJ, Wiebe PH, Copley NJ, and Grigg H. (1991). Phenology in *Calanus finmarchicus*—Hypotheses about control mechanisms. Mar Ecol Prog Ser 72: 79–91.

Moore MV, Pierce SM, Walsh HM, Kvalvik SK, and Lim JD. (2000). Urban light pollution alters the diel vertical migration of *Daphnia*. Verhandl Int Ver Theor Angew Limnol 27: 1–4.

Morgado FM, Pastorinho MR, Quintaneiro C, and Re P. (2006). Vertical distribution and trophic structure of the macrozooplankton in a shallow temperate estuary (Ria de Aveiro, Portugal). Sci Mar 70: 177–188.

Navara KJ and Nelson RJ. (2007). The dark side of light at night: Physiological, epidemiological, and ecological consequences. J Pineal Res 43: 215–224.

Niehoff B. (2006). Life history strategies in zooplankton communities: The significance of female gonad morphology and maturation types for the reproductive biology of marine calanoid copepods. Prog Oceanogr 74: 1–47.

Osgood KE and Frost BW. (1994). Ontogenic diel vertical migration behaviors of the marine planktonic copepods *Calanus pacificus* and *Metridia lucens*. Mar Ecol Prog Ser 104: 13–25.

Pagano M, Champalbert G, Aka M, Kouassi E, Arfi R, Got P, Troussellier M, N'Dour EH, Corbin D, Bouvy M, and Ouvy B. (2006). Herbivorous and microbial grazing pathways of metazooplankton in the Senegal River Estuary (West Africa). Estuar Coast Shelf Sci 67: 369–381.

Pagano M, Kouassi E, Arfi R, Bouvy M, and Saint-Jean L. (2004). In situ spawning rate of the calanoid copepod *Acartia clausi* in a tropical lagoon (Ebrie, Cote d'Ivoire): Diel variations and effects of environmental factors. Zool Stud 43: 244–254.

Pearre S. (2003). Eat and run? The hunger/satiation hypothesis in vertical migration: History, evidence and consequences. Biol Rev 78: 1–79.

Peck MA and Holste L. (2006). Effects of salinity, photoperiod and adult stocking density on egg production and egg hatching success in *Acartia tonsa* (Calanoida: Copepoda): Optimizing intensive. Aquaculture 255: 341–350.

Peterson W. (1998). Life cycle strategies of copepods in coastal upwelling zones. J Mar Syst 15: 313–326.

Rawlinson KA, Davenport J, and Barnes DKA. (2004). Vertical migration strategies with respect to advection and stratification in a semi-enclosed lough: A comparison of mero- and holozooplankton. Mar Biol 144: 935–946.

Ringelberg J. (1999). The photobehaviour of *Daphnia* spp. as a model to explain diel vertical migration in zooplankton. Biol Rev 74: 397–423.

Ringelberg J and Van Gool E. (2003). On the combined analysis of proximate and ultimate aspects in diel vertical migration (DVM) research. Hydrobiologia 491: 85–90.

Rios-Jara E. (2005). Effects of the lunar cycle and substratum preference on zooplankton emergence in a tropical, shallow-water embayment, in southwestern Puerto Rico. Carib J Sci 41: 108–123.

Stearns DE. (1986). Copepod grazing behavior in simulated natural light and its relation to nocturnal feeding. Mar Ecol Prog Ser 30: 65–76.

Stearns DE and Forward RB. (1984a). Photosensitivity of the calanoid copepod *Acartia tonsa*. Mar Biol 82: 85–89.

Stearns D and Forward R. (1984b). Copepod photobehavior in a simulated natural light environment and its relation to nocturnal vertical migration. Mar Biol 82: 91–100.

Tarling GA, Jarvis T, Emsley SM, and Matthews JBL. (2002). Midnight sinking behaviour in *Calanus finmarchicus*: A response to satiation or krill predation? Mar Ecol Prog Ser 240: 183–194.

Thor P. (2003). Elevated respiration rates of the neritic copepod *Acartia tonsa* during recovery from starvation. J Exp Mar Biol Ecol 283: 133–143.

Tiselius P. (1998). Short term feeding responses to starvation in three species of small calanoid copepods. Mar Ecol Prog Ser 168: 119–126.

Tiselius P and Jonsson PR. (1997). Effects of copepod foraging behavior on predation risk: An experimental study of the predatory copepod *Pareuchaeta norvegica* feeding on *Acartia clausi* and *A. tonsa* (Copepoda). Limnol Oceanogr 42: 164–170.

Titelman J and Fiksen O. (2004). Ontogenetic vertical distribution vertical distribution patterns in small copepods: Field observations and model predictions. Mar Ecol Prog Ser 284: 49–63.

Torgersen T. (2003). Proximate causes for anti-predatory feeding suppression by zooplankton during the day: Reduction of contrast or motion—ingestion or clearance? J Plankt Res 25: 565–571.

Tsuda A, Saito H, and Hirose T. (1998). Effect of gut content on the vulnerability of copepods to visual predation. Limnol Oceanogr 43: 1944–1947.

Tuxbury SM and Salmon M. (2005). Competitive interactions between artificial lighting and natural cues during seafinding by hatchling marine turtles. Biol Conserv 121: 311–316.

Uye S-I. (1985). Resting egg production as a life history strategy of marine planktonic copepods. Bull Mar Sci 37: 440–449.

vanDuren LA and Videler JJ. (1996). The trade-off between feeding, mate seeking and predator avoidance in copepods: Behavioural responses to chemical cues. J Plankt Res 18: 805–818.

Williams-Howze J. (1997). Dormancy in the free-living copepod orders Cyclopoida, Calanoida, and Harpacticoida. Oceanogr Mar Biol Annu Rev 35: 257–321.

Williamson CE, Sanders RW Moeller RE, and Stutzman PL. (1996). Utilization of subsurface food resources for zooplankton reproduction: Implications for diel vertical migration theory. Limnol Oceanogr 41: 224–233.

Winder M, Boersma M, and Spaak P. (2003). On the cost of vertical migration: Are the feeding conditions really worse at greater depths? Freshw Biol 48: 383–393.

Wu LS, Wang GZ, et al. (2006). Seasonal variation in diel production pattern of *Centropages tenuiremis* in Xiamen waters, China. Hydrobiologia 559: 225–231.

Wu LS, Wang GZ, Jiang XD, and Li SJ. (2007). Seasonal reproductive biology of *Centropages tenuiremis* (Copepoda) in Xiamen waters, People's Republic of China. J Plankt Res 29: 437–466.

10

Photoperiodism in Insects: Migration and Diapause Responses

David S. Saunders

Higher latitudes are characterized by marked seasonal changes in climate, with winters becoming colder and longer to the north. Being "cold-blooded" animals, most insects find it difficult to cope with periods of cold and have evolved a number of strategies to avoid this form of stress. Broadly speaking, insects at higher latitudes may circumvent the adverse effects of winter by using two different strategies: they may move (migrate) to a more amenable climate or microclimate, or they may pass the winter in a state of dormancy (diapause). These two strategies have been regarded as alternatives (Southwood, 1962). However, in reality, there are numerous and diverse migration and diapause phenomena, and they are not always mutually exclusive. Many long-distance travelers migrate in a state of diapause, and many insects that enter diapause may do so after shorter distance movements to specific overwintering sites. Moreover, the onset of migration and the induction of diapause may be induced by the same environmental factors, most commonly day length or photoperiod. This chapter reviews these strategies, together with their physiological and behavioral characteristics and the environmental factors that induce or initiate them. We place particular emphasis on the relationship between photoperiodic time measurement and the circadian, or other "circa" systems.

MIGRATION

Long-Distance Migrants

The astonishing annual migrations of the monarch butterfly *Danaus plexippus* in North America have intrigued both biologists and the general public for many years (Brower, 1977, 1996; Reppert, 2006). Every summer, monarchs move north through the United States as far as southern Canada, producing successive genera-tions of butterflies on stands of milkweed (*Asclepias syriaca*). In the autumn, large numbers then fly south to their hibernation sites. Western populations fly down the Pacific coast to overwintering sites on just a few trees—for example, on the Monterey peninsula, south of San Francisco. Eastern and midwestern populations use a different route to a few localities in the Sierra Madre Mountains of Mexico, or to sites in Guatemala and Honduras, where they congregate in vast assemblages of up to several million insects. The southerly migrations are carried out by insects in a reproductive diapause (see below) characterized by a cessation of ovarian development and fat-body hypertrophy. The onset of ovarian diapause and the start of migration are both induced by short autumnal photoperiods. Diapause is then maintained in the overwintering sites by short days and low temperatures. Termination of diapause occurs with the onset of longer days and increasing warmth. Mating occurs in February and the mated butterflies then begin their return to more northerly breeding sites.

Migrating monarchs use a time-compensated sun compass mechanism to navigate to their overwintering grounds (Froy et al., 2003; Sauman et al., 2005; Reppert, 2006). As with foraging honey bees, which use a similar type of "continuously consulted" clock to find their way back to the hive (von Frisch, 1950; Saunders, 2002), monarchs use their internal circadian systems to compensate for the sun's apparent movement across the sky as the day proceeds. In this way, although using the sun as their principal directional cue, they can maintain a fixed compass direction. The crucial experiment to demonstrate the relationship between circadian rhythmicity and solar orientation was the observed shift of the butterflies' flight course by a predicted 90° after the circadian clock was phase advanced by 6 h. Recent work has also shown that constant bright light, which is known to disrupt circadian rhythms at both the behavioral and molecular levels, also disrupts their navigation. Ultraviolet, but not polarized, light is also important.

The migration strategies of the milkweed bug *Oncopeltus fasciatus* have been investigated by Dingle (1972, 1974) and Caldwell (1974). This species occurs from Brazil up to Canada. In the spring some insects migrate northward up the Mississippi Valley, colonizing temporary stands of milkweed. In the autumn, after producing two to four summer generations, but being unable to overwinter successfully in the northernmost parts of its distribution, it takes the return path to the south. As with the monarch butterfly, migration occurs in young postteneral adults in a state of reproductive diapause, and the migrating insects show many features characteristic of the diapause state, such as previtellogenic ovaries, fat-body hypertrophy and the suppression of vegetative functions (feeding, mating, and reproduction), in favor of sustained and unidirectional flight. Short days initiate diapause and the southerly migration; northerly migration in the spring occurs under lengthening days.

Short-Distance Migrants

Many insects that enter diapause in response to autumnal day length (see below) also undergo short movements to particular overwintering sites. Diapause-induced adults of the Colorado potato beetle *Leptinotarsa decemlineata*, for example, move from feeding to hibernation sites at the edges of fields and burrow into the soil to a depth of 10–70 cm (De Wilde, 1954). Ladybird beetles *Coleomegilla maculata* similarly move from corn fields to aggregate in leaf litter near the edges of fields

(Obrycki and Tauber, 1979). Other coccinellids such as *Semiadalia undecimnotata* and *Coccinella septempunctata* enter a similar reproductive diapause under short days and then migrate to hibernation sites in cultivated areas such as pastures or the edge of woods, often in elevated areas. Diapause-destined adults of *S. undecimnotata* migrate to winter refuges on or near the summits of hills and overwinter in large aggregations in rock crevices or at the base of plants (Hodek, 1960, 1967). In the spring, the reactivated beetles disperse in any direction.

Similar short-distance movements may be seen in species that overwinter as larvae or pupae. Flesh flies (*Sarcophaga* spp.), for example, enter diapause in the pupal instar, but diapause-destined, short-day–induced larvae show a longer postfeeding "wandering period" than do non-diapause-destined long-day larvae; this allows such larvae more time to reach suitable overwintering sites (Denlinger, 1972).

DIAPAUSE: AN OVERVIEW

Insects living in areas with marked, often seasonal, changes in climate (e.g., wet and dry, summer or winter) frequently survive adverse seasons by entering a period of dormancy. Dormancy may be of two broad types—quiescence or diapause—although these apparent alternatives are by no means mutually exclusive. Quiescence is generally regarded as a direct response to adverse environmental factors such as drought or cold, whose effects are lifted almost immediately upon rehydration or warming. Diapause, on the other hand, is a programmed developmental alternative involving changes to the neuroendocrine system (see Denlinger et al., 2005), induced by environmental factors not in themselves adverse (called "token factors") acting well in advance of the resulting dormancy, and requiring additional physiological or environmental factors to effect its termination. The principle environmental token stimuli for induction are periodic—most commonly photoperiod, but perhaps also thermoperiod.

There are many different diapause strategies to be seen in the insects (Tauber et al., 1986; Danks, 1987, 2002; Saunders, 2002; Koštál, 2006) depending on life cycle and ecological demand; here these strategies are broadly outlined by considering the seasonal responses of insects in different areas, from the tropics to the northernmost limits of insect distribution.

Diapause Induction in the Tropics

Within the tropics seasonal changes in day length are much reduced (at 20°N or 20°S to about 2 h and 25 min, including twilight) and, of course, virtually nonexistent at the equator. Tropical regions, however, frequently have clearly marked wet and dry seasons with abundant plant growth during the rains, and much of the vegetation dying back during periods of drought. Many tropical insects show marked seasonal cycles of dormancy interspersed with periods of growth, although it is not always clear from the available evidence whether such dormancies represent "true" diapause or merely quiescence.

In a review of diapause in the tropics, Denlinger (1986) described dormancies in about 73 species across six orders of insects, with developmental arrest occurring at any stage of development: egg, larva, pupa, or adult. Photoperiodically controlled diapause may be frequent at the higher latitudes within the tropics, but closer to the equator it may not be evident.

Working at Nairobi, Kenya (1°S), where annual changes in day length are only about ±7 min, Denlinger (1974) showed that pupal diapause occurred in several species of flesh flies (Sarcophagidae). Induction of diapause was shown to depend on temperature rather than photoperiod, with larval exposure to temperatures below about 18°C being the most effective stimulus. In East Africa, such cooler conditions led to the occurrence of diapause in July and August. This investigation was later extended to include sarcophagids from African localities up to 9°N and to *Sarcophaga ruficornis* from Belém, Brazil (1°S). No effect of photoperiod on diapause induction could be detected, low temperatures during larval development being the most effective. However, given that periodic environmental factors are important for diapause induction (see below) it would be of interest to investigate the possible role of temperature cycles (or thermoperiod) in this respect. The fact that daytime temperatures are more important than night time temperatures in programming diapause in the Nairobi flies (Denlinger, 1979) already suggests a role for thermoperiodism in this response.

Diapause Induction in Long-Lived Insects

Induction of diapause in some long-lived insects may be regulated by sequential changes in day length (Saunders, 2002), even within the tropics where such changes are slight. Norris (1965) for example showed that the red desert locust *Nomadacris septemfasciata* entered an intense reproductive diapause if the nymphs (hoppers) experienced long days (about 13 h/day) but the adults a short day (about 12 h). On the other hand, diapause-free development followed a transfer from short days as hoppers to long days as adults. In its natural environment—parts of tropical Africa within a few degrees of the equator—this type of response ensured that the species reproduced during the summer rains, but became dormant during the winter drought. Sequences of day length also proved to be important for regulating reproductive diapause in the fungus beetle *Stenotarsus rotundus* on Barro Colorado Island, Panama (9°N). Wolda and Denlinger (1984) showed that adults of *S. rotundus* formed dense aggregations of up to 70,000 diapausing adults on a single tree. The beetles remained in diapause for up to 10 months showing reduced metabolism, fat-body hypertrophy, resistance to desiccation and degeneration of their flight muscles. Tanaka et al. (1987a, 1987b, 1988) demonstrated that seasonal changes in photoperiod, even at this latitude, were the major environmental factors for diapause induction and termination. From June to September, some of the beetles developed their gonads under long days (13 h/day), whereas no such development occurred under a 12-h photoperiod. From October to December, beetles showed no rapid gonadal development at either photoperiod, but between January and April development of the gonads and flight muscles occurred rapidly as days lengthened. Migration from the diapausing aggregation to feeding sites began before the summer months.

Farther north, within the temperate zones, other long-lived insects show similar responses to seasonal changes in photoperiod. Two groups have been extensively investigated: dragonflies (Odonata) and dermestid beetles (Coleoptera).

The seasonal life cycles of dragonflies have been comprehensively reviewed by Corbet (1999). Life cycles of many temperate species extend over several years. Northern populations of *Aeshna juncea*, for example, may take from two to five years to complete a generation at high latitude (58–67°N) (Norling, 1984). Many species of odonates show differential responses to day length at different stages of their larval (nymphal) development, and show "cohort splitting" whereby faster developing (older) larvae may respond differently to photoperiod than slower developing (younger) larvae. In the first phase, long days of summer may induce a larval diapause that arrests further development, particularly among the faster developers. In the second phase, short days close to the autumnal equinox induce an overwintering diapause in all larvae. In the spring, rising temperatures and increasing day length prompt metamorphosis and emergence. Progressively more northern dragonfly life cycles are characterized by an increase in the time taken to complete a generation, with life cycles longer than a year comprising several overwintering diapauses in the larvae during sequentially later stadia.

In a series of papers, Blake (1958, 1959) described the seasonal development of another long-lived insect, the "carpet" beetle, *Anthrenus verbasci* (Coleoptera, Dermestidae). In its natural habitat, the larvae of this beetle feed on material of animal origin and are commonly found in house sparrows' nests. In England, the life cycle generally occupies two years, with the first winter spent in diapause as a young larva and the second as a full-grown larva, again in diapause. Some individuals, however, may take three or even four years to complete their development. After the second or last diapause the larvae pupate and the adults emerge the following spring.

Blake (1958) showed that when cultures of larvae were reared in the laboratory in constant conditions of temperature and continuous darkness a rhythm of development and diapause persisted. The period of this rhythm was found to be between 41 and 44 weeks, rather than the 52 weeks found in the natural environment. Blake recognized this 41 week interval as the rhythm's "basic periodicity". The period of the rhythm was unaffected by stationary photoperiod or by temperature (i.e., it was temperature compensated). Populations of larvae maintained in constant darkness and in different constant temperatures differed, however, in the proportions using each pupation peak. Thus, at high temperature (25°C and 22.5°C), development was rapid and all larvae pupated in the first peak, whereas at low temperature (15°C) all the larvae underwent two cycles of development and diapause, and pupated in the second peak about 41 weeks after the first. Larvae maintained at intermediate temperatures (20°C and 17.5°C) were "split," some utilizing the first and some the second pupation peak. This remarkable phenomenon was recognized as an example of a circannual rhythm, the first in an insect, and one of the very first in any animal group. It clearly demonstrated that the circannual rhythm of pupation in *A. verbasci* was a temperature-compensated and "gated" phenomenon comparable to the similar gated control of adult eclosion in *Drosophila pseudoobscura* (Pittendrigh, 1966) but on an annual rather than a daily time scale.

As with circannual rhythms in other animals (e.g., mammals) the "basic periodicity" (free-running rhythm) of about 41 weeks was corrected or entrained to an exact year by natural seasonal changes in day length (Blake, 1960, 1963). It appeared that naturally decreasing day length experienced by the larvae delayed pupation from October to January, thereby lengthening the cycle to about 52 weeks. It was later demonstrated that the first cycle, as illustrated by the development of univoltine individuals, was controlled in a different way. During the first cycle, the length of the larval period was decreased whenever larvae were reared in increasing day lengths during early larval life.

This important phenomenon has been confirmed by Numata and his colleagues using Japanese populations of *A. verbasci* reared under unchanging short days (e.g., 12/12-h light/dark cycle [LD 12:12]) and a range of temperatures from 17.5°C to 27.5°C. Under these regimes the insects showed a periodical pattern of pupation at about 40 week intervals; the rhythm was self-sustained, temperature compensated, and entrained to an exact year by changes in photoperiod (Nisimura and Numata, 2001, 2003). In later papers (Miyasaki et al., 2005, 2006) phase response curves (PRCs) were constructed by exposing cultures of larvae to four-week blocks of long days at sequentially later times during their circannual cycle under either unchanging short days (LD 12:12) or naturally changing day lengths. These PRCs resembled those for circadian rhythms—although on a different time scale—and strongly suggested that the phenomenon was underpinned by an endogenous approximately 40-week rhythm, the mechanism of which remains unknown.

The four examples reviewed here (*Nomadacris*, *Stenotarsus*, dragonflies and *Anthrenus*) have a number of features in common. All are long-lived insects with a differential sensitivity to long (lengthening) or short (shortening) photoperiods at different ages or stages, and at different times of the year. These photoperiodic influences serve to synchronize or entrain the rhythm of diapause and nondiapause development to an exact year in the two tropical examples, but to life cycles occupying two, three, or even more years in those living further north. It is possible, therefore, that all four, and others like them, represent examples of circannual rhythmicity.

Diapause Induction in Short-Lived Insects in Temperate Zones

Away from the tropics, and into the temperate zones, life cycles shorter than a year are commonplace. Many insects produce several nondiapause generations during

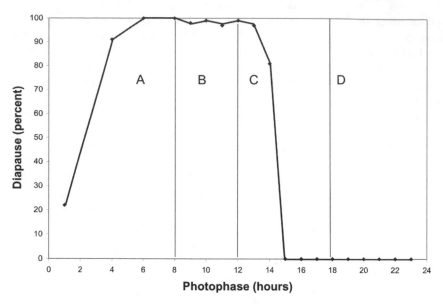

FIGURE 10.1.
Photoperiodic response curve for larval diapause induction in the blow fly, *Calliphora vicina*, a species showing maternal sensitivity to photoperiod. Adult flies were maintained at 20°C under a range of photophases; larvae were maintained under continuous darkness at 11–12°C. Zones A and D include very short and very long photophases never experienced in the natural environment; zone B includes photophases occurring during the depth of winter when the insects are in diapause and not responsive to photoperiod; zone C occurs between midsummer and autumn and is dominated by an abrupt change in the diapause response at the critical day length. Data from Saunders (1987a).

the season favorable for growth and reproduction but, in response to the appropriate photoperiod, become dormant as the winter or the dry season approaches. These facultative responses to photoperiod are often characterized by a well-defined "critical day length" (figure 10.1), which implies that the insect is able to "measure" the length of the day or the night, or perhaps both.

By analogy with plants (Garner and Allard, 1920), insects may show either short-day or long-day photoperiodic responses, defined by the season during which "actively" feeding, growing, and reproducing stages occur. In short-day insects— frequently observed in species inhabiting lower latitudes where summers may be hot and dry but winters mild and wet—the actively developing or reproducing stages may be confined to the shorter days of winter, but the hot dry summers are passed in a state of aestival diapause, often characterized by fat-body hypertrophy, reduced metabolism and a resistance to desiccation. In these short-day responses, high temperatures tend to accentuate the diapause-promoting effects of long days, whereas lower temperatures help promote nondiapause development under short days. However, aestival diapause is not restricted to areas with a hot, dry summer. It has been recorded from a wide range of habitats up to 60°N (Masaki, 1980), where it may occur in the egg, larva, pupa, or adult. The geometrid moth *Abraxas miranda*, for example, spends the summer (June to August) in a pupal diapause, but the adults emerge in September and October and the larvae of the next generation actively feed and grow during the short days of winter (Masaki, 1957). One of the best-known examples, however, is also one of the most unusual. Some strains of the commercial silk moth *Bombyx mori* have an overwintering diapause in the egg, but a short-day photoperiodic response because the stages *sensitive* to photoperiod (the eggs and young larvae of the maternal generation) occur during the preceding summer (Kogure, 1933), a full generation before the diapausing stage. In the cabbage moth *Mamestra brassicae*, there is a summer and a winter diapause in the pupal instar (Masaki and Kimura, 2001) set in response to photoperiods experienced during the egg and early larval stages. The long winter diapause is induced by photoperiods shorter than about 13 h/day; the short summer diapause by long days close to LD 16:8.

Long-day responses are typically seen in areas with a winter that is too cold for active metabolism. These species may produce a series of nondiapausing generations through the summer months, and then enter an overwintering diapause as autumnal day lengths fall below a well-marked critical value. Most of the remainder of this chapter addresses the responses of such long-day species.

Although short days may be the primary environmental stimulus inducing overwintering diapause, such day lengths occur twice in each solar year—once in the autumn and again in the spring—and insects may therefore be faced with the necessity of distinguishing between these two potentially inductive photoperiods. Some insects are known to have an overwintering diapause that may be terminated by long days, often showing critical day lengths for induction and termination that are almost "mirror images" of each other (e.g., Williams and Adkisson, 1964). In others, however, reactivation follows a period of exposure to low temperature. In the linden bug *Pyrrhocoris apterus*, for example, although long days may terminate ovarian diapause in the laboratory, such photoperiodic regulation of diapause termination may be completely overridden by 12–16 weeks at 4°C (Hodek, 1968; Saunders, 1983). Indeed, working with *P. apterus*, Hodek (1978) has shown that diapause "development" is completed in the field by the end of December, and the insects then remain dormant in a state of low temperature, postdiapause quiescence until reactivation in the spring as temperatures rise. Aspects of diapause development are discussed by Hodek (2002).

Working with the flesh fly *Sarcophaga bullata*, Henrich and Denlinger (1982) described a mechanism by which this species avoided a maladaptive diapause in the spring. It was shown that flies emerging from diapause reactivated pupae (i.e., the first spring generation) were incapable of producing diapausing progeny themselves, even under strong short-day conditions. The block to diapause induction was the result of short-day exposure of the mothers whilst they were intrauterine embryos, and normal responses to short days only reappeared after a further generation under long days. Results (Rockey et al., 1989) suggested that an unidentified "message" was passed from the brain to the ovaries sometime between the end of larval development and the third day of adult life, and that hemolymph and central nervous system factors were involved. In more northerly flesh flies (e.g., *S. argyrostoma*; Kenny et al., 1992), such a mechanism appeared to be absent, probably because the longer winter and a more prolonged pupal dormancy delayed the first postdiapause generation of flies until photoperiods exceeded the critical value. Aphids, however, may also distinguish spring from autumn by a mechanism that resembles that described above for S. *bullata*. Working with the green vetch aphid *Megoura viciae*, Lees (1960) found that the first virginoparae developing from over wintering eggs were incapable of producing sexuales (oviparae and males) under short days for at least 90 days, a time covering several generations of virginoparae. After this period, the photoperiodic response was suddenly restored and oviparae were produced under the short-day treatment, suggesting the operation of a trans-generational timer (see also chapter 14).

Insects inhabiting increasingly more northerly latitudes show features of their photoperiodic response that form well-marked south-to-north geographical clines, the best known of which is the latitudinal cline in critical day length (CDL) first recorded by Danilevskii (1965) in the knot grass moth *Acronycta rumicis*. Other important examples of this relationship are those afforded by the pitcher plant mosquito *Wyeomyia smithii* and the fruit fly *Drosophila littoralis*. In the former, 22 populations of *W. smithii* were collected from localities spanning 19° of latitude (30–49°N) in eastern North America, and the CDL was shown to lengthen systematically toward the north (Bradshaw, 1976). In the most extensive study, Lankinen (1986) recorded the photoperiodic responses of *D. littoralis* from 57 localities ranging from the Black Sea coast (41°N) to northern Finland (69°N). Critical day lengths were shown to range from about 12 h in the south to more than 19 h in the north with a high degree of correlation ($r = 0.943$). A similar relationship between latitude and CDL has been recorded in about 15 species of insects and one mite (see Saunders, 2002). The longer CDL in more northerly latitudes almost certainly affords a strong selective advantage because winters arrive earlier in the north while day lengths are still long. Conversely, a shorter CDL in the south allows insects to continue development before autumnal photoperiods curtail their activity.

In addition to these often robust south-north clines in CDL are similar clines in the number of generations per year (voltinism) and in the "depth" or "intensity" of the diapause induced. This variation is exemplified by reference to the blow fly *Calliphora vicina*, which shows a larval diapause induced by autumnal short days experienced by the female parent (Vinogradova and Zinovjeva, 1972; Saunders, 1987a, 2001). When adults were maintained under different photophases at 20°C and their larvae in continuous darkness at 11–12°C, flies originating from Barga, Italy (44°N) showed a CDL of 13.5 h/24, about 80% entering diapause under LD 12:12, and a duration of diapause of about 30 days, whereas flies from the north of Finland (65°N) showed a CDL of 15 h/24, almost all entering diapause under LD 12:12, and a diapause duration of more than 70 days. Geographical clines of diapause intensity have also been demonstrated for summer diapause in the field cricket *Teleogryllus emma* (egg diapause) and the cabbage moth *Mamestra brassicae* (pupal diapause) in the Japanese islands (Masaki, 2002), with the duration of summer diapause becoming systematically shorter at more northerly latitudes.

At the northernmost extremes of insect distribution—in arctic and subarctic areas—winters are severe and very long, and summer "growing" seasons very short (Downes, 1965). Day length during the summer also becomes extremely long, approaching that of continuous light; reliable photoperiodic cues may thus be absent. Overwintering diapause at these high latitudes may be a response to low temperature. However, periodic environmental stimuli may still be present, with low-amplitude light and temperature cycles being evident in some microhabitats. These cycles may offer important inductive cues that investigations have yet to uncover.

The Diapause Syndrome

Apart from the important *timing* aspects of photoperiodically controlled diapause— which have been reviewed above—there is a medley of associated behavioral and physiological phenomena, together referred to as the "diapause syndrome." In different species these may include, inter alia, prediapause movements or migrations, the construction of specialized hibernacula, the deposition of additional fat and protein reserves, cuticular changes to reduce desiccation, and the acquisition of cold hardiness. Cold hardiness itself may, or may not, be a part of the diapause syndrome (Denlinger, 1991). Adedokun and Denlinger (1984), for example, showed that diapausing pupae of the flesh flies *Sarcophaga crassipatpis* and *S. bullata* showed a high degree of cold tolerance (up to 25 days at –10°C), whereas nondiapausing pupae rapidly succumbed to the cold. In the beetle *Dendroides canadensis*, Horwath and Duman (1983) showed that larvae exposed to short days showed an increased synthesis of "antifreeze" proteins, whereas in those exposed to long days such synthesis was greatly reduced. The insect's circadian system was shown to be involved in the photoperiodic clock regulating the production of these proteins (Horwath and Duman, 1982).

THE PHOTOPERIODIC RESPONSE AND THE SENSITIVE PERIOD

This section will be concerned almost entirely with long-day multivoltine species, that is, those that produce a series of nondiapause generations during the summer months and then enter diapause in response to autumnal day lengths shorter than a well-marked critical value. This is a common response in many temperate species, and one that has attracted the most attention in studies concerning the mechanism(s) of photoperiodic time measurement.

Diapause Stages and Sensitive Periods

Diapause may occur at any stage of development—egg, larva, pupa, or adult, although it is generally species specific and with related species often entering diapause at the same stage (Tauber et al., 1986; Saunders, 2002). For example, aedine mosquitoes frequently enter diapause as embryos within the egg, flesh flies (Sarcophagidae) as pupae, and many drosophilids as newly eclosed adults. Diapause in these and other insects occurs as a result of an alteration to the neuroendocrine systems controlling growth, molting, or reproduction, in many cases by the retention, under environmental short days, of the cerebral neuropeptides normally regulating the synthesis of ecdysteroids and juvenile hormones. The endocrinology of insect diapause is reviewed by Denlinger et al. (2005).

Photoperiod acts during an earlier stage of development, during the so-called "sensitive period," and it is to this part of the mechanism that attention should be directed. In the example of reproductive diapause in *Drosophila* species, the sensitive period occurs in the newly eclosed adults, immediately preceding the resulting diapause (Lumme, 1978). Reproductive diapause in the linden bug *Pyrrhocoris apterus* is induced by short days experienced by the developing nymphs (Hodek, 1968; Saunders, 1983). In other species the sensitive period may occur at an earlier stage of development. Pupal diapause in *Sarcophaga* spp., for example, is regulated by photoperiods experienced by the intrauterine embryos within the maternal uterus (Denlinger, 1972) or in larvae soon after deposition. In other species, the sensitive period is truly maternal: this is the case for larval diapauses in the parasitic wasp *Nasonia vitripennis* (Saunders, 1966a) and the blow fly *Calliphora vicina* (Vinogradova and Zinovjeva, 1972; Saunders, 1987a). These delayed photoperiodic responses imply that "information" is either stored within the central nervous system or transmitted from mother to offspring via the egg.

The Photoperiodic Response Curve

Experiments in which insects are exposed to the full range of photophases, and to continuous darkness (DD) and continuous light (LL), give rise to a *photoperiodic response curve* (PPRC), an example of which is given in figure 10.1 for a typical long-day species. In such curves, regions A and D are never experienced in nature and therefore are of no ecological importance, and region B occurs only during the winter when the insect is in diapause. Section C of this curve, however, shows the responses to naturally occurring photophases and is therefore important ecologically. This part of the curve is dominated by the abrupt critical day length (CDL) separating nondiapause development under long days from the diapause-inducing short days. Although responses to ultrashort and ultralong photophases have no *ecological* significance they require explanation in physiological terms. In many species, temperature has rather little effect on the value of the CDL, but lower temperatures increase the incidence of diapause under strong short days and DD, and higher temperatures reduce it.

In some insects a rather similar response to temperature cycles (thermoperiod) has been observed. Figure 10.2 shows such a response for the parasitic wasp *Nasonia vitripennis* (Saunders, 1973a) in which adult wasps were exposed—in the complete absence of light—to daily temperature cycles of 13–23°C with different numbers of hours per day in the cooler and warmer conditions. Progeny produced by the female wasps were reared at 18°C, again in darkness, to determine the incidence of larval diapause. Results (Saunders, 1973a) showed that wasps exposed to a daily thermophase longer than about 14 h produced all their progeny as developing (nondiapause) larvae, whereas those exposed to a daily thermophase of less than 12 h produced larvae that became dormant; long- and short-day thermoperiods were separated by a sharp critical value. Similar results were obtained for the southwestern corn borer *Diatraea grandiosella* (Chippendale et al., 1976) and the cabbage butterfly *Pieris brassicae* (Dumortier and Brunnarius, 1977). In the Indian meal moth *Plodia interpuctella*, however, Masaki and Kikukawa (1981) showed that daily thermoperiod was diapause inductive under constant light as well as constant darkness. Any model for the photoperiodic response (see below)—and the molecular and biochemical explanations that flow from them—must take into account the *full* shape of the photoperiodic response curve, and the various effects of temperature.

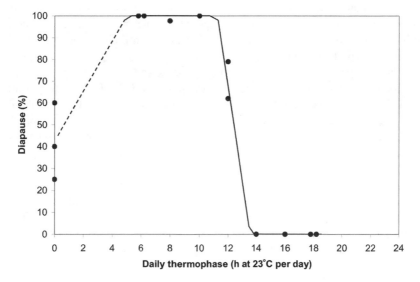

FIGURE 10.2.
Thermoperiodic response curve for the maternal induction of larval diapause in the parasitic wasp *Nasonia vitripennis*. Adult females were maintained under complete darkness in a range of square-wave daily thermoperiods changing from 23°C to 13°C. The response curve resembles that for photoperiod (see figure 10.1), with a well-defined critical thermophase at about 13°C per "day." Data from Saunders (1973a).

In a series of studies, Beck (1982, 1984 and 1987) investigated interactions between daily light and temperature cycles in the induction of larval diapause in the European corn borer *Ostrinia nubilalis* and found that the duration of the cool dark phase (= "night") was more important than the duration of the warmer light phase (= "day"). However, he concluded that the periodic nature of the input ("periodism") was of greater significance than just the light, dark or temperature per se.

The Photoperiodic "Counter"

During the sensitive period two major physiological events take place. The first of these is the "measurement" of either day length or night length, or perhaps both, by what has been dubbed the photoperiodic "clock." The second involves the accumulation or summation of photoperiodic information to some sort of internal threshold after which the insect is programmed for either the diapausing or the nondiapausing pathway. This aspect of the response has been called the photoperiodic "counter" (Saunders, 1981).

The photoperiodic counter has routinely been investigated by exposing insects, within their sensitive period, to different numbers of inductive short days (long nights) against a noninductive background, or vice versa. Such experiments, however, are greatly facilitated in species in which the sensitive period is clearly demarcated, that is, with a clearly defined beginning and end. The parasitic wasp *Nasonia vitripennis*, in which diapause occurs in the fourth larval instar but photoperiodic sensitivity is maternal, is one such case (Saunders, 1966a). In this species, the progeny developing from each day's batch of eggs act as a measure of the female wasp's physiological state, that is, whether the mother has been programmed to produce nondiapausing or diapausing larvae.

When wasps were exposed to "strong" short days (6–14 h of light per day), they produced nondiapausing offspring for the first few days of adult life and then switched over, one by one, to the production of diapausing larvae. On average this switch occurred after eight to nine short-day cycles, the inference being that females accumulated eight to nine inductive cycles until they were programmed to produce eggs destined to become diapausing larvae. This was called the required day number (RDN). Wasps exposed to "strong" long days (15.5–20 h of light per day), on the other hand, produced nearly all of their offspring as nondiapausing larvae; the switch to diapause, if it occurred, was delayed until near the end of adult life. Under intermediate day lengths (14.5–15.75 h of light per day), the RDN became gradually longer (Saunders, 1966a). Similar data suggesting the operation of a photoperiodic counter were obtained for the flesh fly *Sarcophaga argyrostoma* (Saunders, 1971) and the blow fly *Calliphora vicina* (Saunders, 1987a).

The Effects of Temperature on the Photoperiodic Counter

Females of the parasitic wasp *Nasonia vitripennis* were exposed to short days (LD 12:12) at a range of constant temperatures (15°C, 20°C, 25°C, and 30°C)

(Saunders, 1966a). All wasps reacted to LD 12:12 by switching to the production of diapausing larvae—after 8.4, 7.6, 8.4, and 6.9 days, respectively. The required day number (RDN) therefore showed a high degree of temperature compensation (Q_{10} = 1.04) (figure 10.3A). Life span and the rate of oviposition, however, showed a more normal relationship to temperature. Thus, at 30°C, the wasps showed a short life span (11.5 days) and a rapid rate of oviposition with a peak three to five days after emergence. At 15°C, on the other hand, the wasps showed a protracted life span (32.7 days) and a slower rate of oviposition with a peak about 14 days after emergence. Interaction between the temperature-compensated mechanism accumulating light–dark cycles (the RDN) and the temperature-dependent rate of oviposition (representing the sensitive period) resulted in a high incidence of larval diapause at 15°C (90.9%), somewhat lower at 20°C (71.2%) and 25°C (61.0%) and a low incidence at 30°C (27.4%). Very similar relationships between developmental rate and RDN were observed in the flesh fly *Sarcophaga argyrostoma* (Saunders, 1971) (figure 10.3B) and the blow fly *Calliphora vicina* (Saunders, 1987a). Not only do these observations provide a partial explanation for the diapause enhancing effects of lower temperature on insects exposed to short days, but they show that a temperature compensated mechanism is involved in the summation of inductive photoperiods during the sensitive period.

However, both temperature-compensated and temperature-dependent cycle accumulation has been described in three species, the aphids *Megoura viciae* (Hardie, 1990) and *Aphis fabae* (Vaz Nunes and Hardie, 1999) and the flesh fly *Sarcophaga argyrostoma* (Saunders, 1992). These studies found that summation of short days (long nights) was temperature compensated, as described above, but the accumulation of development-inducing long days (short nights) was not. This important observation suggests that diapause and nondiapause induction are regulated by somewhat different mechanisms, something we return to further below.

The Effects of Feeding on the Photoperiodic Counter

Nutrition, or the lack of it, also affects the interaction between the sensitive period and the required day number, and therefore the proportion entering diapause. In *N. vitripennis*, for example, depriving the female wasps of host pupae (from which they obtain a meal of protein) delayed ovarian development (Saunders, 1966b); the most severe conditions of starvation (7 days without hosts) delayed the start of oviposition by a week without reducing the total number of offspring. However, because the required day number was unaltered during the period of deprivation, the proportion of the progeny entering diapause rose from about 73% in the fed controls to over 99% in those experimentally deprived of hosts for 7 days. Clearly the photoperiodic counter continued to accumulate the effects of successive short-day cycles throughout the period of starvation, independently of ovarian development. A similar relationship between feeding and diapause incidence was found in the blow fly *C. vicina* (Saunders, 1987a).

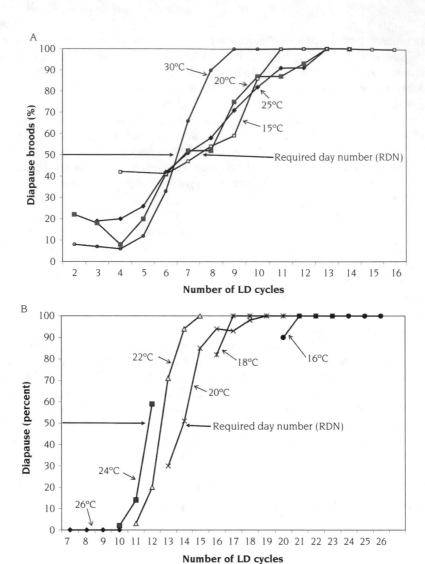

FIGURE 10.3.

The summation of diapause-inductive photocycles (LD 12:12) by adults of the parasitic wasp *Nasonia vitripennis* (A) and larvae of the flesh fly *Sarcophaga argyrostoma* (B), at constant temperatures between 15°C and 30°C. In both examples, the incidence of diapause in each day's batch of larval progeny (A) or pupae (B) rises with an increase in the number of inductive cycles of LD 12:12, to give a "family" of response curves. The number of days or cycles needed to raise the incidence of diapause to 50% (the required day number, RDN) shows a high degree of temperature compensation. Data from Saunders (1966a, 1971).

In the flesh fly *S. argyrostoma*, the relationship between the sensitive period (larval development) and the required day number was also manipulated by changing the feeding regime (Saunders, 1975a). In this species, diapause occurs in the pupal instar with sensitivity to photoperiod beginning in the intrauterine embryo and coming to an end by puparium formation. In experiments conducted at 18–20°C under a diapause-inducing photoperiod (LD 12:12), premature extraction of third instar larvae from their food led to early pupariation, a shortened sensitive period, and a lowered incidence of diapause. Temporary removal of young larvae from their food, however, led to a delay in pupariation, a lengthening of the sensitive period and a consequent increase in diapause (Saunders, 1975a, 2002). The results showed a linear relationship between the length of the sensitive period and the occurrence of pupal diapause, every day of shortening or lengthening of the sensitive period resulting in about 10% decrease or increase, respectively (figure 10.4).

The Effects of Latitude on the Photoperiodic Counter

Working with three geographically distinct populations of the rock-hole mosquito *Aedes atropalpus*, Beach (1978) showed that the required day number (RDN) varied

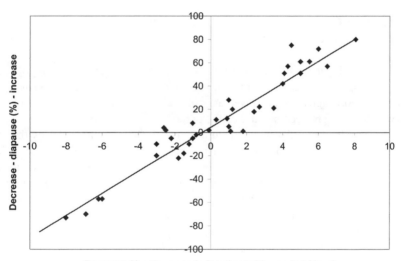

Decrease (-) or increase in larval sensitive period (days)

FIGURE 10.4.
Sarcophaga argyrostoma: manipulation of the duration of the larval sensitive period (SP) on the incidence of pupal diapause (data for 18°C and 20°C combined). The SP was shortened (minus values) by premature extraction of third instar larvae from their food; SP was lengthened (plus values) by removal of younger feeding larvae for a few days before returning them to their food. The calculated line (r2 = 0.9011; $p < 0.05$) shows that for every days shortening or lengthening of the SP the incidence of diapause fell or rose, respectively, by about 10%. Data from Saunders (1981).

with latitude. The three strains originated from Ontario, Canada (45°N); Georgia, USA (34°N); and El Salvador (14°N). The northernmost population deposited diapause eggs after only four short-day cycles, but at 34°N and 14°N the RDN number had lengthened to seven and nine, respectively. The low RDN of the Canadian strain caused it to enter diapause whenever day length dropped into the short-day range. Because of the low value of the RDN, this response was stable up to 28°C and appropriate for an area where the transition from a favorable summer to an adverse winter was rapid and closely correlated with photoperiod. The most southerly population, however, although still capable of entering diapause, rarely did so because the prevailing high temperature produced a sensitive period generally shorter than the nine required cycles. At the intermediate latitude (34°N), the mechanism was more susceptible to temperature, the insects being able to lay more nondiapause eggs at higher temperature when the sensitive period fell to less than seven days. This strategy is particularly suitable for climatic conditions at 34°N, where the onset of short days frequently coincides with a warm autumn that allows further breeding and an eventual increase in the size of the overwintering population. These results demonstrate a latitudinal cline in RDN, to join those in critical day length, the number of generations during the growing season, and diapause intensity, reviewed above.

Summary: General Evidence for Circadian Involvement in Photoperiodic Timing

Alternative diapause and nondiapause developmental programs are regulated by *periodic* aspects of the environment, principally photoperiod and thermoperiod. These periodic signals are received as trains of pulses during the insect's sensitive period (SP) and are accumulated by the photoperiodic counter mechanism.

Accumulation of diapause-inducing short-day (long night) cycles by the counter is a temperature-compensated mechanism, although the accumulation of long-day (short-night) cycles may not be. Temperature affects diapause induction in a number of ways. Apart from the direct diapause-enhancing effects of lower temperature and the diapause-inducing effects of thermoperiod, temperature affects the proportion of insects entering diapause under short days by altering the duration of the temperature-dependent SP in relation to the temperature-compensated RDN. Induction of diapause by trains of photoperiods during the sensitive period in a temperature-compensated manner suggests the operation of a circadian-based mechanism in photoperiodic timing. Evidence for this is presented in the next section.

PHOTOPERIODIC INDUCTION: RELATIONSHIP TO THE CIRCADIAN SYSTEM

Tests designed to uncover possible circadian involvement in photoperiodic time measurement are based on well-known properties of the circadian system, in particular the entrainment of circadian oscillations by complex cycles of light and temperature. The most successful of these tests have been carried out with species showing robust circadian rhythms of activity in addition to equally robust photo-periodic responses. Colin Pittendrigh called such comparisons "parallel peculiarities" and pointed out that the conclusions drawn from these experiments became more compelling as the predictions became more complex and more demanding. A full account of models for photoperiodic time measurement has been given earlier (Vaz Nunes and Saunders, 1999; Saunders, 2002, 2005; Saunders et al., 2004). This section examines the more successful of these circadian-based models for photoperiodism.

Bünning's Hypothesis: Three Predictions

In 1936, the plant physiologist Erwin Bünning suggested that photoperiodic time measurement (PPTM) was a function of endogenous daily rhythms, now called the circadian system (Bünning, 1936, 1960). The most explicit form of his model (figure 10.5) suggested that the endogenous near-24-h cycle was divided into two equal parts, a 12-h "photophil" section and a 12-h "scotophil" section; these were equivalent to the subjective day and the subjective night of modern terminology. When the rhythm was entrained to a daily light–dark cycle comprising a short day (or long night; figure 10.5B), light was restricted to the first half of the cycle, resulting in an autumnal response, but under long days (or short nights; figure 10.5C) light extended into the subjective night, producing the summer response. Light therefore had a dual role—entrainment and photoinduction with illumination or nonillumination of a part of the light-sensitive scotophil operating a seasonal switch between the summer and autumnal pathways.

Bünning made two further suggestions concerning the role of circadian rhythms in PPTM: (1) that overt behavioral rhythms—in his case the up-and-down movement of leaves—could act as "hands of the clock," enabling the experimenter to visualize the covert photoperiodic system, and (2) that the oscillation in continuous darkness (DD) was slowly dampening (Bünning, 1967) (figure 10.5A). These two important suggestions are discussed further below.

External Coincidence: The Dual Function of Light

In the 1960s, Colin Pittendrigh proposed the external coincidence model as a development of Bünning's original hypothesis. This model (figure 10.6) incorporated a more modern appreciation of the phenomenon of circadian entrainment and was more appropriate for the insects. Based on entrainment of the adult eclosion rhythm in *Drosophila pseudoobscura* (Pittendrigh and Minis, 1964; Pittendrigh, 1966), the external coincidence model retained Bünning's proposition that light had a dual role, entrainment and photoinduction. It differed from Bünning's original in that

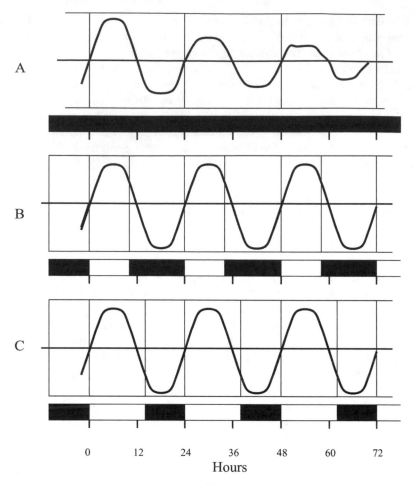

FIGURE 10.5.
Bünning's hypothesis for the role of the circadian system in photoperiodic time measurement. (A) The free-running oscillation in continuous darkness, showing slow dampening. (B) Under a short day (long night), light is restricted to the first half of the cycle (the "photophil"), giving rise to a short-day or autumnal response. (C) Under a long day (short night), light extends into the second half of the cycle (the "scotophil"), giving rise to a long-day or summer response. Schematic, after Bünning (1960).

it recognized that the insect circadian system was reset to a near constant phase (about circadian time 12) at the end of a long light phase. This property meant that any further lengthening of the light component of the cycle involved the "tracking back" of the dawn transition of the cycle to illuminate phases in the late subjective night (see figure 10.6). In insects, therefore, the photosensitive (or "photoinducible") phase was thought to lie in the latter part of the night—at the end of

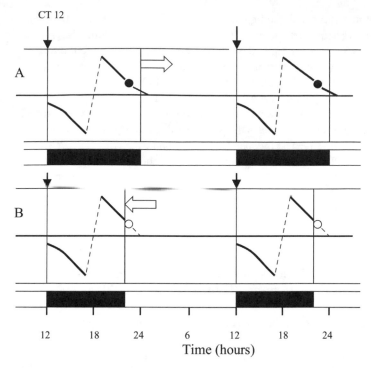

FIGURE 10.6.
The external coincidence model for photoperiodic timing in an insect (after Pittendrigh, 1972). The circadian oscillation, here shown as a phase response curve, is reset to a constant phase (circadian time, CT 12) at the end of the photophase (vertical arrow). In A (short-day cycle), the photoinducible phase falls in the dark (solid circles), leading to the autumnal pathway (diapause induction); in B (long day cycle), the photoinducible phase (open circles) is illuminated by the dawn transition of the daily photophase (open arrow) to give the summer developmental pathway (nondiapause development). Open horizontal arrows show the "movement" of the dawn transition of the daily photophase in relation to the oscillation as day length shortens in the autumn (A) or lengthens in early summer (B).

the critical night length—rather than early in the night as envisaged by Bünning for plants. Experimental evidence supporting these aspects of the model is presented in subsequent sections.

Night-Length or Day-Length Measurement?

In theory, the photoperiodic mechanism could "measure" either day length or night length (or perhaps both). In practice, however, night length is "central" to the mechanism, implying that insects generally measure the duration of the dark phase, apparently with limited reference to the duration of the light. The central importance of night length in photoperiodic time measurement has been demonstrated in experiments in which the light component of the cycle (L) and the dark component (D) are varied independently (see Saunders, 2002). In the flesh fly *Sarcophaga argyrostoma*, for example, the incidence of diapause was found to be very low in cycles containing a short night (LD 12:8 and LD 16:8) but approached 100% in cycles containing a long night (LD 12:12 and LD 16: 12), regardless of whether the accompanying photophase was "short" or "long" (Saunders, 1973b).

In some species, however, the duration of the photophase does appear to be paramount. One such example is that of the linden bug *Pyrrhocoris apterus* (Saunders, 1987b). In this species, independent variation of the dark (D = 7, 8, 9, or 12 h) and light (L = 12, 15, 16, or 17 h) components of the cycle showed that a "critical day length" of about 15.75 h/24 was more important than the reciprocal 8.25 h of darkness. In all species, however, the photophase is more than something that merely separates successive periods of darkness; these properties will become apparent below.

Parallel Peculiarities of Diapause Induction and Circadian Entrainment: "Hands of the Clock"

Bünning's suggestion that overt circadian rhythms may be used as "hands" of the covert photoperiodic mechanism has been investigated in a number of species, but with variable and sometimes rather limited utility, probably because the circadian properties of the overt and covert systems differed in certain respects (see details provided below). The most informative experiments of this type were performed with the flesh fly *Sarcophaga argyrostoma* in which the chosen overt system (the rhythm of adult eclosion) and the covert photoperiodic system had many features in common (Saunders, 1978). In this study, phase response curves (PRCs) for the eclosion rhythm exposed to light pulses between 1 and 20 h duration were constructed and these PRCs used to calculate the presumed performance of the photoperiodic system in complex trains of light pulses. This approach, which proved successful for *S. argyrostoma*, forms the basis of many of the experiments described below. The use of overt behavioral rhythms as markers of phase within the covert photoperiodic system are discussed further in chapter 11, stressing the complexity of the insect circadian system.

Transient Cycles and the Approach to Steady State

Short or "weak" light pulses cause small phase changes to a circadian oscillation and give rise to a low-amplitude (type 1) PRC. Conversely, longer or "stronger" light pulses cause larger phase shifts and produce a high-amplitude (type 0) PRC (Winfree, 1970). The magnitude of possible phase shifts is dictated by the amplitude of the PRC; that is, shifts are greater with type 0 than with type 1.

Therefore when exposed to a train of weak light pulses, with the first pulse in the train falling out of phase, the oscillation goes through a greater number of transients before it reaches steady state entrainment than it does when exposed to a train of strong pulses. This difference has its effect on diapause induction. For example, figure 10.7 shows the incidence of pupal diapause in cultures of *S. argyrostoma* exposed during their larval sensitive period to trains of weak or strong pulses of light with the first pulses commencing at either the beginning of the subjective day, CT 24/0 (initially out of phase), or at the beginning of the subjective night, CT 12 (initially in phase). With strong light pulses (10–13 h), steady state was achieved rapidly with few transients, and diapause incidence was high in insects with both starting phases. On the other hand, with weaker pulses (1–8 h) the resulting incidence of diapause was lower, especially when the initial pulse occurred at an out-of-phase position (CT 24/0) and the oscillations

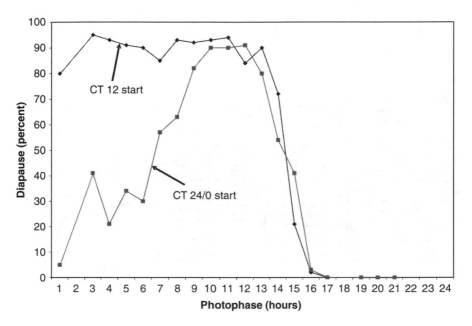

FIGURE 10.7.
Pupal diapause induction in the flesh fly *Sarcophaga argyrostoma* exposed to trains of different photophases during the larval sensitive period, and with the first pulse in the train commencing either at circadian time (CT) 12 or at CT 24/0. With an initially in-phase CT 12 start, all photophases shorter than about 13 h induce a high incidence of diapause, whereas with an out-of-phase CT 24/0 start photophases shorter than about 9 h lead to a reduced incidence of diapause. Computer modeling has shown that cultures starting initially out of phase proceed to steady state entrainment through a greater number of non-steady-state or transient cycles than those starting in phase. An increased number of transient cycles on the input (entrainment) pathway adversely affects the final incidence of diapause. Data from Saunders (2002).

had to go through a higher number of transient cycles before steady state was achieved (Saunders, 2002). This experiment provided strong evidence for a circadian mechanism on the "input pathway" to diapause regulation.

Night Interruptions: Localization of the Photoinducible Phase

In night-interruption experiments, different groups of an insect are maintained, throughout its sensitive period, in diapause-inducing light–dark cycles with the inductive long night systematically perturbed at different times by a short supplementary light pulse. Such experiments generally show two points of low diapause incidence (long day effect), one early in the night (at point A) and one late in the night (at point B) (see Saunders, 2002). Light pulses placed in the middle of the night have little effect (figure 10.8, right panel).

Night-interruption experiments of this type prompted Pittendrigh and his colleagues (Pittendrigh and Minis, 1964; Pittendrigh, 1966) to propose the external

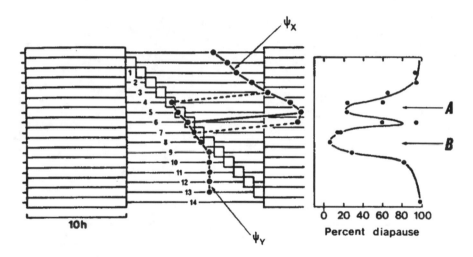

FIGURE 10.8.
Sarcophaga argyrostoma. Pupal diapause induction in cultures of larvae exposed to asymmetrical skeleton regimes consisting of a "main" photophase of 10 h and the 14 h "night" systematically interrupted by a scanning 1-h pulse of light. Solid circles show the computed phases of the photoinducible phase (at CT 21.5) in relation to these two component cycles. The right panel shows the incidence of pupal diapause in these regimes and the two points of low diapause incidence (A and B) characteristic of these night-interruption experiments. When the photoinducible phase is calculated to fall in the dark, diapause incidence is high; when it is illuminated, either by the "main" photophase or the scanning pulse, diapause incidence is low. ψ_x and ψ_y show the two characteristic phase relationships and the phase jump (solid line) between them. Dashed lines mark the earliest and latest possible phase jumps. Data from Saunders (1979).

coincidence model for photoperiodic time measurement (see figure 10.6) by comparing such results with those obtained for the entrainment of the adult eclosion rhythm (in *Drosophila pseudoobscura*) exposed to similarly complex light cycles. In short, the circadian oscillation(s) were seen to achieve steady state entrainment to the asymmetrical "skeleton" photoperiods formed from the main light phase and the scanning pulse. When the pulse fell early in the night, it was read as a terminator, or new dusk; when it fell late in the night it was read as an initiator, or new dawn. In *S. argyrostoma*, point A occurred about 9.5 h (the critical night length) before the start of the main photophase, whereas point B occurred about 9.5 h after the start of darkness (Saunders, 1978). Since the circadian system is reset to a constant phase (CT 12) at the start of the night, this suggests that the putative "photoinducible phase" occurred at point B rather than at A.

By analogy with the rhythm of adult eclosion, light pulses falling early in the night produce phase delays while those falling later in the night cause phase advances. According to the model, therefore, early pulses cause delays until the photoinducible phase (at B) coincides with lights on, producing "long-day" (diapause averting effects), whereas late pulses cause phase advances until the short scanning pulse coincides directly with the photoinducible phase (at B), again averting diapause (figure 10.8). Light thus shows the two effects predicted by Bünning. These are (1) phase-shifting (entrainment) at points A and B, but (2) the dual effects of phase-shifting *and* photic regulation of the diapause–nondiapause switch at B. In *S. argyrostoma*, such an interpretation strongly suggests that the crucial photoinducible phase, at point B, is 9.5 h after the start of the dark phase—at about circadian time (CT) 21.5.

The occurrence of the photoinducible phase late in the subjective night was also indicated in asymmetrical skeleton photoperiods using two additional designs. In the first, a supplementary 1-h pulse of light was placed 3 h after the end of a 10-h main photophase, at a point equivalent to A, and the terminal hours of darkness then systematically varied from 7 to 13 h to give overall cycle lengths ranging from 21 to 27 h (see figure 10.9, top). In the second, using an 8-h main photophase, the hours of darkness before the supplementary pulse were varied from 3 to 11 h, but the hours following it were maintained at 12 h, or longer than the critical night length (see figure 10.9, bottom). In the first design, regimes in which the terminal hours of darkness were less than about 9 h gave rise to a low incidence of diapause, but as the terminal hours of darkness increased, the incidence of diapause rose. In the second experimental design, diapause incidence was high until the supplementary pulse fell about 6–9 h after dusk, after which it rose again. Using eclosion rhythm PRCs as "hands of the clock" showed that whenever the supposed photoinducible phase (at CT 21.5) fell in the light, diapause incidence was low, but when it fell in the dark it was high. These results clearly indicate that the photoinducible phase occurs late in the subjective night as envisaged by the external coincidence model. Later experiments using the T-experimental protocol would confirm this.

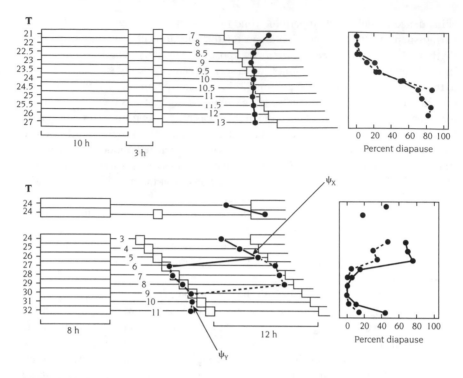

FIGURE 10.9.

Sarcophaga argyrostoma. Pupal diapause induction in night-interruption experiments in non-24-h cycles. (Top) Larvae exposed to cycles consisting of a "main" photophase of 10 h, a 1-h supplementary pulse commencing 3 h into the "night," and a terminal dark period varying from 7 to 13 h to give overall cycle lengths (*T*) between 21 and 27 h. Solid circles show the computed phases of the photoinducible phase (at CT 21.5) in relation to these two-component cycles. The right panel shows the incidence of pupal diapause (two experiments). When the photoinducible phase falls in the dark, diapause incidence is high; when it is illuminated, diapause incidence is low. (Bottom) Pupal diapause induction in experiments consisting of a "main" photophase of 8 h, a 1-h supplementary pulse falling 3–11 h into the "night," and a final dark phase longer than the critical night length (12 h), to give overall cycle lengths (*T*) between 24 and 32 h. Solid circles show the computed phases of the photoinducible phase (at CT 21.5). When the photoinducible phase falls in the dark, diapause incidence is high (two experiments); when it coincides with the scanning pulse, diapause incidence is low. Other details are as in figure 10.8. Data from Saunders (1979).

The Bünsow Protocol: The Free-Running Rhythm of Light Sensitivity

Although night-interruption experiments, as described above, provide robust evidence for circadian rhythmicity in photoperiodic timing, demonstration of a free-running rhythm—one of the canonical properties of a circadian system—can

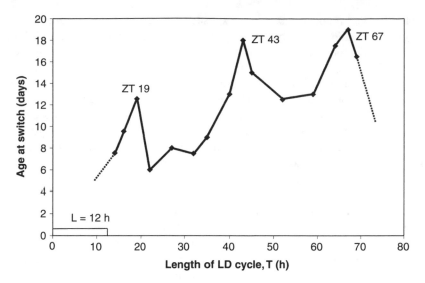

FIGURE 10.10.
Production of diapausing and nondiapausing larvae by females of the parasitic wasp
Nasonia vitripennis exposed to 72-h cycles consisting of a 12-h "main photophase, and
60 h of darkness systematically interrupted by a 2-h supplementary light pulse (Bünsow
experiment). Data show the age at the switch to diapause that is delayed in "long-day"
responses (Saunders, 1966a). Peaks of low diapause incidence (delayed switch) occur at
24-h (circadian) intervals (*zeitgeber* time, ZT 19, 43, and 67) in the extended night. Schematic
after Saunders (1970).

only be demonstrated in cycles longer than a day. The Bünsow protocol, first performed in plants (Bünsow, 1953), provides that possibility by extending the night-
interruption technique to two- or three-day light cycles with the extended "night"
systematically interrupted by a short scanning pulse. One such example is for the
parasitic wasp *Nasonia vitripennis* (Saunders, 1970) (figure 10.10) in which 2-h
light breaks introduced into the 60-h "night" phase of an LD 12:60 cycle produced
three peaks of long-day response (low diapause incidence) at roughly 24-h (circadian) intervals. The central peak was particularly informative because it was
unlikely to have arisen from a direct interaction between the pulse and the main
photophase. "Positive" Bünsow results like these have been obtained for about nine
species in about five orders of insects and one mite (see Saunders, 2002); in about
five other species, however, similar results were not obtained. The possible reasons
for these essentially "negative" responses are addressed below.

Nanda-Hamner Experiments: Multiple Ranges of Entrainment

Another type of experiment using non-24-h cycles is that known as the Nanda-
Hamner or resonance protocol (see Saunders, 2002, 2005; Saunders et al., 2004) in

which different groups of insects are exposed, throughout their sensitive periods, to regimes containing a "short" photophase (e.g., 12 h) combined with different durations of darkness (e.g., 6–72 h) to give light–dark cycles ranging across several circadian periods (i.e., from 18 to 84 h). In these experiments the light phase "comes on" at different times in the extended period of darkness and should reveal any photoinducible phase that might repeat itself with a circadian periodicity. The results frequently show alternate peaks and troughs of diapause incidence at roughly circadian intervals as the duration of darkness increases (figure 10.11A), clearly indicating a circadian component in the photoperiodic mechanism. Peaks of high diapause incidence close to 24, 48, and 72 h represent the primary, secondary and tertiary ranges of entrainment of a circadian oscillator (period close to 24 h) to the experimental light cycles. In the flesh fly *Sarcophaga argyrostoma*, application of circadian entrainment theory using phase response curves for the rhythm of adult eclosion has been used to calculate temporal interactions between the photoinducible phase and the light in such cycles; a high incidence of diapause was found to occur when the photoinducible phase fell in the dark, but low diapause incidence occurred when this phase was illuminated (Saunders, 1978).

Nanda-Hamner experiments using a very short photophase (e.g., 1 h) add further evidence for the involvement of the circadian system in the photoperiodic mechanism. In such an experiment with *S. argyrostoma* (figure 10.11B), a peak of high diapause incidence was also observed when the 1-h light pulses occurred at intervals of about 12 h (one half of the circadian period), indicating "frequency demultiplication" in the entrainment pathway. "Positive" Nanda-Hamner responses such as those shown in figure 10.11A have been reported in about 15 species from about six orders of insects and some mites (Saunders, 2002). "Negative" responses, lacking such peaks and troughs in the response, on the other hand, have been recorded in about seven other species. These cases are also discussed further below.

The T-Experiment: Isolating the Photoinducible Phase

In a special case of the Nanda-Hamner protocol, larvae of *S. argyrostoma* were exposed throughout their sensitive period to cycles containing, in different experimental subsets, a 1-h light pulse coupled with a variable duration of darkness to give cycle lengths (T values) varying from 21.5 h (i.e., LD 1:20.5) to 30.5 h (i.e., LD 1:29.5) (Saunders, 1979). This range of cycle lengths encompassed the primary range of entrainment, as judged from the 1-h pulse PRC for the rhythm of adult eclosion. This PRC was then used to calculate where the photoinducible phase fell in relation to the train of 1-h pulses. Results (figure 10.12) showed that only in a cycle of $T = 21.5$ h (LD 1:20.5) in which the light pulse came on at CT 20 and finished at CT 23.5—and therefore illuminated the photoinducible phase (at CT 21.5)—were "long-day" or low-diapause effects obtained. In all other regimes, the photoinducible phase was calculated to fall in the dark and diapause incidence was high.

Since this experiment was based on predictions arising from entrainment theory, it offered strong confirmation that photoperiodic time measurement was, in part,

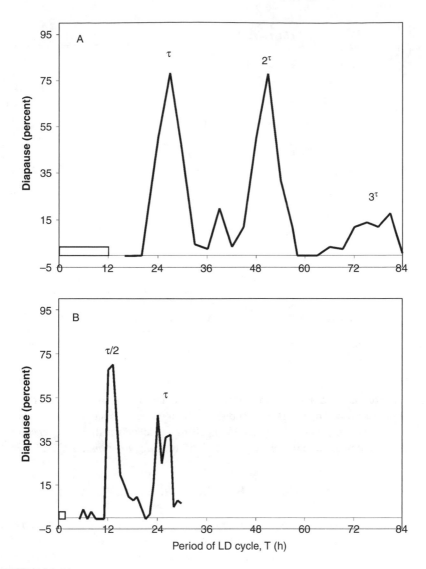

FIGURE 10.11.

Sarcophaga argyrostoma: induction of pupal diapause in cultures of larvae exposed to Nanda-Hamner experiments (see text). In A, larvae were exposed to a series of cycles comprising a 12-h photophase coupled, in different experimental subsets, with an increasing duration of darkness to give cycle lengths between 16 and 84 h. In B, larvae were exposed to cycles comprising a 1-h photophase and 4–29 h of darkness. In A, peaks of high diapause incidence are observed at circadian intervals (τ, 2τ and 3τ) as the duration of darkness is extended; in B, peaks of high diapause are observed close to 12 h ($\tau/2$) and 24 h (τ), indicating "frequency demultiplication" of the circadian period. Data from Saunders (various sources).

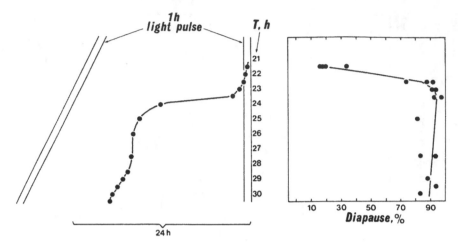

FIGURE 10.12.
Sarcophaga argyrostoma: induction of pupal diapause in cultures of larvae exposed to a
T-experiment (see text) in which different subsets of larvae were exposed to light-dark
cycles from 21.5 to 30.5 h, each containing a single 1-h pulse of light per cycle. The plotted
line shows the computed phase relationship of the photoinducible phase (at CT 21.5).
Incidence of pupal diapause (to the right) shows that diapause incidence was high when
the photoinducible phase fell in the dark; diapause incidence was reduced only when this
phase coincided with the 1-h light pulse. Data from Saunders (1979).

a function of the circadian system. Furthermore, since the postulated photoperiodic
oscillator was probed by a single short pulse of light in the absence of any other photoperiodic influence, this result offers compelling evidence for the occurrence of a
photoinducible phase late in the subjective night, in *S. argyrostoma*, at CT 21.5.

Skeleton Photoperiods: The Bistability Phenomenon

In the 1960s, Pittendrigh and his colleagues (Pittendrigh and Minis, 1964; Pittendrigh,
1966) reported that two brief pulses of light *n* hours apart within each 24-h cycle
could entrain the *Drosophila pseudoobscura* eclosion rhythm in a manner comparable to a single long duration pulse of *n* hours. These two short pulses were dubbed
a symmetrical "skeleton" photoperiod (PPs), with the first simulating light-on or
"dawn" and the second, lights off or "dusk." Such skeletons, however, differ from a
"complete" photoperiod (PPc) in one important respect: there is no continuous light
between the two pulses. For this reason a skeleton may be interpreted in two ways. A
skeleton of PPs 8:16, for example, is also a skeleton of PPs 16:8, the only difference
between the two being one of phase. When the two pulses are less than about 10 h
apart, simulation of the complete photoperiod by the skeleton is good, but when the
two pulses are initially 14 or more hours apart, the oscillation undergoes a phase-
jump to accept the shorter of the two interpretations. With skeletons close to half the
circadian period, that is, between 10 and 14 h, which of the two interpretations is

accepted—the longer or the shorter—depends on two initial conditions: (1) the first interval presented, and (2) the circadian phase of the first pulse in the train. This region around 10–14 h was called the "zone of bistability" by Pittendrigh.

Similar phenomena were observed with the adult eclosion rhythm in the flesh fly *Sarcophaga argyrostoma* and—more important in the present context—with the induction of pupal diapause (Saunders, 1978). Larvae of *S. argyrostoma* were exposed, throughout their sensitive period, to symmetrical skeleton photoperiods formed from two 1-h pulses of light, either PPs 11 (LD 1:9: 1:13) or PPs 15 (LD 1: 13: 1:9) with the first pulse in the train commencing at all circadian times. These two skeletons were chosen because they were "mirror images" of each other and, if accepted by the circadian system, would be read either as a diapause-inducing "short" day (LD 11: 13) or as a non-diapause-inducing "long" day (LD 15:9), depending on the circadian phase illuminated by the first pulse in the train. Figure 10.13 shows

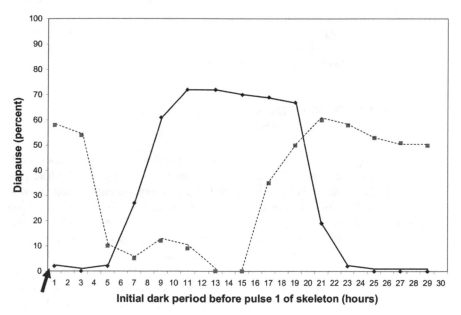

FIGURE 10.13.
Sarcophaga argyrostoma: A "positive" bistability experiment (see text) on pupal diapause induction. Cultures of larvae were exposed, at 18°C, to "symmetrical skeleton" photoperiods of LD 1:9:1:13, or LD 1:13:1:9, with the first 1-h pulse of the skeleton starting at different times after transfer of the cultures from continuous light (LL) to darkness (DD) (equivalent to circadian time, CT 12—arrow). Solid line: LD 1:9:1:13 given first. The 1:9:1 interval is interpreted as an 11-h photophase (a high incidence of pupal diapause) when the first pulse in the train commences between 7 and 21 h after the LL/DD transfer (i.e., between CT 19 and 08); in all other regimes the 1:13:1 interval is accepted as a 15-h photophase (a low incidence of diapause). Dotted line: LD 1:13:1:9 given first. The 1:13:1 interval is interpreted as a 15-h photophase (low diapause) when the first pulse commences between 5 and 15 h after the LL/DD transfer (i.e., between CT 17 and 04); in all other regimes the 1:9:1 interval is accepted as an 11-h photophase (high diapause). Data from Saunders (1975b).

that experiment matched prediction. When the 11-h interval (1:9:1) was presented first, short-day responses (a high incidence of diapause) were observed when the first pulse started between CT 19 and 08, but long-day responses (low diapause incidence) were observed when the first pulse started between CT 09 and 19. When the 15-h interval (1:13:1) was presented first, the results were a mirror image, with a high incidence of diapause between CT 04 and 17 and a low incidence of diapause between CT 17 and 04. Calculations of entrainment of the oscillation using a 1-h PRC for the eclosion rhythm showed that diapause was always high when the putative photoinducible phase (at CT 21.5) fell in the dark, but low when it fell within one of the two short pulses of light in the cycle. Other examples of a "positive" response to bistability have been recorded for the blow fly *Calliphora vicina* (Vaz Nunes et al., 1990) and the cabbage moth *Mamestra brassicae* (Kimura and Masaki, 1993), but in some others the results were negative (see below).

The positive bistability results reviewed above constitute some of the strongest and most compelling evidence that circadian rhythmicity is involved in photoperiodic time measurement, almost certainly on the photic input pathway. Such entrainment serves to position the phase relationship of the photoinducible phase (at CT 21.5) to the light cycle so that the photoperiodic switch between short and long days may be operated by seasonal changes in day length.

OSCILLATION DAMPING AND APPARENT HOURGLASS-LIKE RESPONSES

In the foregoing section, reference was made to various negative responses, principally in Nanda-Hamner and Bünsow experiments, which lacked alternate diapause peaks and troughs, but also in bistability and T-experiments, which failed to show clear evidence of circadian rhythmicity. These negative responses have been reviewed extensively in earlier publications (Lewis and Saunders, 1987; Saunders and Lewis, 1987a, 1987b; Saunders, 2002, 2005, 2007; Saunders et al., 2004) and are not dealt with here in detail. The most significant body of experimental data of this type is that for the aphid *Megoura viciae* by Lees (1973), who concluded that photoperiodic time measurement in that species was performed by a noncircadian (or hourglass-like) timer measuring night length. For example, in a Nanda-Hamner experiment with aphids exposed to an 8-h photophase in combination with a systematically increasing number of dark hours, the incidence of oviparae (the short-day response) rose abruptly as the duration of darkness exceeded the critical night length (9.5 h) but then remained as an elevated "plateau" with none of the peaks and troughs associated with a circadian response (Lees, 1973).

In a number of papers, Veerman and his colleagues also proposed a night length hourglass for the spider mite *Tetranychus urticae*, despite a mixture of negative and positive results obtained for that species. A Nanda-Hamner experiment, for example, showed a positive response with four peaks of high diapause at intervals of about 19 h (Veerman and Vaz Nunes, 1980), and a positive Bünsow experiment was also reported (Vaz Nunes and Veerman, 1984). Negative results, however, were described for a T-experiment and for bistability (Vaz Nunes and Veerman, 1997). In other species, positive and negative responses have been recorded depending on such factors as temperature (Saunders, 1973), latitude (Thiele, 1977), and even diet (Dumortier and Brunnarius, 1989). Without going into the details of these observations here, suffice it to say that it seems improbable that the photoperiodic mechanism should act as a circadian clock in one sort of experiment or under certain conditions, but as an entirely different sort of timer—an hourglass—under others. There are undoubted hourglass-like responses in some insects and under certain circumstances. The question therefore arises, what is an hourglass in the context of photoperiodic timing? The answer to this question was supplied by Bünning himself nearly 40 years ago (Bünning, 1969).

In 1969 Bünning suggested that the circadian component in photoperiodic timing was a damping oscillator (see figure 10.5). Bearing in mind that apparent dampening might be due to increasing asynchrony within a *population* of oscillators—or within a population of insects—a system that dampened in extended periods of darkness could then act as though it was an "hourglass." Using this property as a basis for a computer-based photoperiodic clock model, Lewis and Saunders (1987) further proposed that when the oscillation fell below a threshold the photoinducible phase ceased to operate, but the oscillation could be boosted above threshold following a light pulse of sufficient strength. This model has allowed computer simulations of a wide range of photoperiodic responses (Saunders and Lewis, 1987a, 1987b; Saunders et al., 2004) including those experimental results described by Lees and Veerman. The strongest hourglass-like properties of photoperiodic timing in the aphid *Megoura viciae* (Lees, 1973) could be attributed to a highly dampened system.

PARALLEL PECULIARITIES: OVERT RHYTHMS AS "HANDS OF THE CLOCK"

In the preceding section, experiments using *Sarcophaga argyrostoma* to uncover a possible circadian involvement in photoperiodic timing were accompanied by analyses of entrainment using the overt rhythm of adult eclosion. The evident success of this approach was probably because the two rhythmic systems, adult eclosion and photoperiodic timing, showed many features in common. In the flesh fly, therefore, an overt behavioral rhythm served as "hands of the photoperiodic clock" in a manner predicted by Bünning.

This approach, however, also has had its problems. Even in *S. argyrostoma*, certain differences between the circadian oscillators involved in photoperiodic induction and in eclosion soon became apparent (Saunders, 1986). Although both had endogenous periods close to 24 h and showed similar sensitivities to light perturbation, the former revealed evidence of oscillator damping, whereas the latter was fully self-sustained. In other species, these differences were more marked.

In the blow fly *Calliphora vicina*, Nanda-Hamner experiments suggested that the period of the photoperiodic oscillator was close to 24 h and dampening, whereas the period of locomotor activity was closer to 22.5 h and the rhythm fully self-sustained (Kenny and Saunders, 1991). In addition, selection for a long critical day length in *C. vicina* had no effect on the period of the locomotor activity rhythm (Saunders and Cymborowski, 2003) showing that they were physiologically separate components of the circadian system. Working with geographical strains of *Drosophila littoralis* (described above), Lankinen (1986) showed that the period of the eclosion rhythm and the critical day length for diapause induction showed rather similar latitudinal clines, but a clear cause-and-effect relationship between these two variables was not apparent. In a more recent paper (Lankinen and Forsman, 2006) generations of artificial selection led to a strain of *D. littoralis* showing extreme "southern" characteristics for diapause induction (i.e., a long critical night length) but extreme "northern" characteristics for adult eclosion (i.e., a short circadian period). The "separate" nature of the two systems was therefore made abundantly clear. In these and other examples, it became apparent that although overt behavioral rhythms and photoperiodic timing were both governed by circadian oscillators, they were separate elements of a multioscillator circadian system, differing in clock properties (period, damping coefficient, persistence) and possibly also in the nature of the light input pathways, cellular location within the central nervous system and nature of their output.

We should not be surprised by this complexity. The behavior of many insects reveals a complex of oscillators with different clock properties (Saunders, 2002). The marine midge *Clunio marinus*, for example, shows a combination of a circadian rhythm controlling adult eclosion and a circasemilunar (14-day) rhythm determining the beginning of pupation (Neumann, 1966, 1976), together with a photoperiodic regulation of overwintering larval diapause induced by autumnal short days (Neumann and Kruger, 1985). On the Pacific coast of North America the intertidal beetle *Thalassotrechus barbarae* displays nocturnal activity controlled by a circadian rhythm and a circatidal (12.4 h) rhythm that inhibits locomotor activity during periods of nocturnal high water (Evans, 1976). Different circadian oscillators regulating photoperiodism and overt behavioral rhythms in terrestrial insects become a real possibility.

Summary: Circadian Regulation of Photoperiodic Time Measurement

Tests to uncover possible circadian involvement in photoperiodic timing rely on experiments based on the canonical properties of the circadian system: free-running in darkness with a circadian frequency, temperature compensation, and entrainment to complex cycles of light and darkness in a manner similar to overt rhythms. The most useful model to emerge from these studies is that derived from Bünning's original hypothesis: the external coincidence model of Pittendrigh (1966, 1972).

This model suggests that light has a dual function, (1) entraining the constituent circadian oscillators, and (2) regulation of nondiapause/diapause responses by coinciding (or not) with a specific photoinducible phase. In most cases, night length is central to the photoperiodic mechanism. Successive (diapause-inductive) long nights are accumulated by the clock-counter system in a temperature-compensated fashion indicating a circadian oscillation. Accumulation of (non-diapause-inductive) short nights is not temperature compensated, indicating an exogenous or direct effect of light upon the photoinducible phase. Light pulses falling on initially out-of-phase positions cause the endogenous oscillator to pass through a series of non-steady-state or transient cycles; this is reflected in the incidence of diapause. Night-interruption experiments (in 24 h, or two- to three-day cycles), the Nanda-Hamner protocol, the T-experiment, and tests for bistability in the entrainment mechanism all provide, when positive, powerful evidence for the operation of the circadian system in photoperiodic time measurement. They show that photoperiodism relies on, inter alia, persistent circadian oscillations(s), temperature compensation of period, and entrainment to cycles of light and darkness over several ranges of entrainment. The bistability test, in particular, provides exceptionally strong evidence for a circadian involvement. Negative responses to these various experimental protocols, sometimes suggested as evidence for a noncircadian "hourglass-like" photoperiodic timer, may be explained by assuming that the constituent circadian oscillators are heavily dampening. Alternatively, apparent dampening may be due to a developing internal asynchrony within a population of oscillators, or population of insects. In some insects, notably *Sarcophaga argyrostoma*, an overt behavioral rhythm (adult eclosion) was used successfully as "hands of the clock." In other species, significant differences between the overt and covert systems were apparent. Entrainment pathways probably have much in common with those involved in showing different properties (period, persistence, etc.), and possibly different input pathways, different cellular substrates, and different outputs.

FUTURE DIRECTIONS

Future directions for research should uncover the molecular and genetic components of the timing mechanism, the photoreceptive pathways and pigments involved, and the way in which short days and long days regulate the retention or release of the cerebral neurohormones that regulate the diapause or nondiapause pathways. These avenues of research should be based on the formal properties of the system as revealed by earlier research and presented in this chapter. In this way, insect photoperiodism will follow that of its obvious exemplar, circadian rhythmicity, the unraveling of which has become one of the success stories of modern biology (Hall, 2003).

The dual role of light (entrainment and photoinduction), exemplified by external coincidence, should guide future research. Genes and molecules in the light input (entrainment) pathway probably have much in common with those involved in the circadian regulation of behavioral rhythms, with known "clock" genes such as *period*, *timeless*, and *cryptochrome* probably playing a significant role (see chapter 11). Details of their feedback loops, however, are probably different to allow for differences between the overt and covert systems, such as period and persistence. Photobiological events occurring when light coincides with the photoinducible phase, however, will probably involve a different set of genes including those encoding photopigments such as boceropsin (Shimizu et al., 2001). There are probably multiple photoreceptive pathways for entrainment, both within and between species, involving both compound eyes and brain (see chapter 11). Photopigments may also differ between entrainment and photoinduction; the former may involve cryptochromes (among others) whereas the latter may involve carotenoids. This difference may be reflected in differences in the respective action spectra for the two processes. Lastly, the way in which illumination or nonillumination of the photoinducible phase regulates neurohormone release is an almost unknown phenomenon.

References

Adedokun TA and Denlinger DL. (1984). Cold hardiness: A component of the diapause syndrome in pupae of the flesh flies, *Sarcophaga crassipalpis* and *S. bullata*. Physiol Entomol 9: 361–364.

Beach RF. (1978). The required day number and timely induction of diapause in geographic strains of the mosquito, *Aedes atropalpus*. J Insect Physiol 24: 449–455.

Beck SD. (1982). Thermoperiodic induction of larval diapause in the European corn borer, *Ostrinia nubilalis*. J Insect Physiol 28: 273–278.

Beck SD. (1984). Effects of temperature on thermoperiodic determination of diapause. J Insect Physiol 30: 383–386.

Beck SD. (1987). Thermoperiod-photoperiod interactions in the determination of diapause in *Ostrinia nubilalis*. J Insect Physiol 33: 707–712.

Blake GM. (1958). Diapause and the regulation of development in *Anthrenus verbasci* (L.) (Col Dermestidae). Bull Entomol Res 49: 751–775.

Blake GM. (1959). Control of diapause by an "internal clock" in *Anthrenus verbasci* (L.) (Col Dermestidae). Nature 183: 126–127.

Blake GM. (1960). Decreasing photoperiod inhibiting metamorphosis in an insect. Nature 188: 168–169.

Blake GM. (1963). Shortening of a diapause-controlled life cycle by means of increasing photoperiod. Nature 198: 462–463.

Bradshaw WE. (1976). Geography of photoperiodic response in a diapausing mosquito. Nature 262: 384–386.

Brower LP. (1977). Monarch migration. Nat Hist June/July: 41–53.

Brower LP. (1996). Monarch butterfly orientation: Missing pieces of a magnificent puzzle. J Exp Biol 199: 93–103.

Bünning E. (1936). Die endogene Tagesrhythmik als Grundlage der Photoperiodischen Reaktion. Ber Dtsch Bot Ges 54: 590–607.

Bünning E. (1960). Circadian rhythms and time measurement in photoperiodism. Cold Spring Harbor Symp Quant Biol 25: 249–256.

Bünning E. (1967). *The Physiological Clock*, 3rd ed rev. London: English Universities Press.

Bünning E. (1969). Common features of photoperiodism in plants and animals. Photochem Photobiol 9: 219–228.

Bünsow RC. (1953). Uber Tages- und Jahresrhythmische Anderungen der Photoperiodischen Lichteropfindlichkeit bei *Kalanchoe blossfeldiana* und ihre beziehungen zur endogonen Tagesrhythmik. Z Bot 41: 257–276.

Caldwell RL. (1974). A comparison of the migratory strategies of two milkweed bugs, *Oncopeltus fasciatus* and *Lygaeus kalmii*. In *Experimental Analysis of Insect Behaviour* (L Barton-Browne, ed). New York: Springer-Verlag, pp 304–316.

Chippendale GM, Reddy AS, and Catt CL. (1976). Photoperiodic and thermoperiodic interactions in the regulation of the larval diapause of *Diatraea grandiosella*. J Insect Physiol 22: 823–828.

Corbet PS. (1999). *Dragonflies: Behavior and ecology of Odonata*. Ithaca, NY: Cornell University Press.

Danilevskii AS. (1965). *Photoperiodism and Seasonal Development of Insects*, English ed. Edinburgh: Oliver and Boyd.

Danks HV (1987), *Insect Dormancy: An Ecological Perspective*. Biological Survey of Canada Monograph Series 1, Terrestrial Arthropods. Ottawa. Biological Survey of Canada. (Terrestrial Arthropods). Monograph Series 1, Ottawa.

Danks HV. (2002). The range of insect dormancy responses. Eur J Entomol 99: 127–142.

Denlinger DL. (1972). Induction and termination of pupal diapause in *Sarcophaga* (Diptera: Sarcophagidae). Biol Bull 142: 11–24.

Denlinger DL. (1974). Diapause potential in tropical flesh flies. Nature 252: 223–224.

Denlinger DL. (1979). Pupal diapause in tropical flesh flies: Environmental and endocrine regulation, metabolic rate and genetic selection. Biol Bull 156: 31–46.

Denlinger DL. (1986). Dormancy in tropical insects. Annu Rev Entomol 31: 239–264.

Denlinger DL. (1991). Relationship between cold hardiness and diapause. In *Insects at Low Temperature* (RE Lee Jr and DL Denlinger, eds). New York: Chapman and Hall, pp. 174–198.

Denlinger DL, Yochum GD, and Rinehart JP. (2005). Hormonal control of diapause. In *Comprehensive Insect Molecular Science*, vol 3 (LI Gilbert, K Iatrou, and S Gill, eds). Amsterdam: Elsevier, pp. 615–650.

de Wilde J. (1954). Aspects of diapause in adult insects, with special reference to the Colorado potato beetle, *Leptinotarsa decemlineata* Say. Arch Neerl Zool 10: 375–378.

Dingle H. (1972). Migration strategies of insects. Science 175: 1327–1335.

Dingle H. (1974). Diapause in a migrant insect, the milkweed bug *Oncopeltus fasciatus* (Dallas) (Hemiptera: Lygaeidae). Oecologia 17: 1–10.

Downes JA. (1965). Adaptations of insects in the Arctic. Annu Rev Entomol 10: 257–274.

Dumortier B and Brunnarius J. (1977). L'information thermoperiodiques et l'induction de la diapause chez *Pieris brassicae*. C R Acad Sci D 284 : 957–960.

Dumortier B and Brunnarius J. (1989). Diet-dependent switch from circadian to hourglass-like operation of an insect photoperiodic clock. J Biol Rhythms 4: 481–490.

Evans WG. (1976). Circadian and circatidal locomotory rhythms in the intertidal beetle Thalassotrechus barbarae (Horn): Carabidae. J Exp Mar Biol Ecol 22: 79–90.

Froy O, Gotter AL, Casselman AL, and Reppert SM. (2003). Illuminating the circadian clock in monarch butterfly migration. Science 300: 1303–1305.

Garner WW and Allard HA. (1920). Effects of the relative length of the day and night and other factors of the environment on growth and reproduction in plants. J Agric Res 18: 553–606.

Hall JC. (2003). Genetics and molecular biology of rhythms in Drosophila and other insects. Adv Genet 48.

Hardie J. (1990). The photoperiodic counter, quantitative day-length effects and scotophase timing in the vetch aphid *Megoura viciae*. J Insect Physiol 36: 939–949.

Henrich VC and Denlinger DL. (1982). A maternal effect that eliminates pupal diapause in progeny of the flesh fly, *Sarcophaga bullata*. J Insect Physiol 28: 881–884.

Hodek I. (1960). Hibernation bionomics in Coccinellidae. Acta Soc Czech 57: 1–20.

Hodek I. (1967). Bionomics and ecology of predaceous Coccinellidae. Annu Rev Entomol 12: 79–104.

Hodek I. (1968). Diapause in females of *Pyrrhocoris apterus* L. (Heteroptera). Acta Ent Bohemoslov 65: 422–435.

Hodek I. (2002). Controversial aspects of diapause development. Eur J Entomol 99: 163–173.

Horwath KL and Duman JG. (1982). Involvement of the circadian system in photoperiodic regulation of insect antifreeze proteins. J Exp Zool 219: 269–270.

Horwath KL and Duman JG. (1983). Photoperiodic and thermal regulation of antifreeze protein levels in the beetle *Dendroides canadensis*. J Insect Physiol 29: 907–917.

Kenny NAP, Richards DS, Bradley HK, and Saunders DS. (1992). Photoperiodic sensitivity and diapause induction during ovarian, embryonic and larval development of the flesh fly, *Sarcophaga argyrostoma*. J Biosci 17: 241–251.

Kenny NAP and Saunders DS. (1991). Adult locomotor rhythmicity as "hands" of the maternal photoperiodic clock regulating larval diapause in the blowfly, *Calliphora vicina*. J Biol Rhythms 6: 217–233.

Kimura Y and Masaki S. (1993). Hourglass and oscillator expression of photoperiodic diapause response in the cabbage moth *Mamestra brassicae*. Physiol Entomol 18: 240–246.

Kogure M. (1933). The influence of light and temperature on certain characters of the silk worm, *Bombyx mori*. J Dept Agric Kyushu Univ 4: 1–93.

Koštál V. (2006). Eco-physiological phases of insect diapause. J Insect Physiol 52: 113–127.

Lankinen P. (1986). Geographical variation in circadian eclosion rhythms and photoperiodic adult diapause in *Drosophila littoralis*. J Comp Physiol A 159: 123–142.

Lankinen P and Forsman P. (2006). Independence of genetic geographical variation between photoperiodic diapause, circadian eclosion rhythm, and Thr-Gly repeat region of the period gene in *Drosophila littoralis*. J Biol Rhythms 21: 1–10.

Lees AD. (1960). Some aspects of animal photoperiodism. Cold Spring Harbor Symp Quant Biol 25: 261–268.

Lees AD. (1973). Photoperiodic time measurement in the aphid *Megoura viciae*. J Insect Physiol 19: 2279–2316.

Lewis RD and Saunders DS. (1987). A damped circadian oscillator model of an insect photoperiodic clock. 1. Description of the model based on a feedback control system. J Theor Biol 128: 47–59.

Lumme J. (1978). Phenology and photoperiodic diapause in northern populations of *Drosophila*. In *Evolution of Insect Migration and Diapause* (H Dingle, ed). New York: Springer-Verlag, pp. 145–170.

Masaki S. (1957). Ecological significance of diapause in the seasonal cycle of *Abraxas miranda*. Bull Fac Agric Mie Univ 15: 15–24.

Masaki S. (1980). Summer diapause. Annu Rev Entomol 25: 1–25.

Masaki S. (2002). Ecophysiological consequences of variability in diapause intensity. Eur J Entomol 99: 143–154.

Masaki S and Kikakawa S. (1981). The diapause clock in a moth: Response to temperature signals. In *Biological Clocks in Seasonal Reproductive Cycles* (BK Follett and DE Follett, eds). Bristol: John Wright and Sons, pp. 101–112.

Masaki S and Kimura Y. (2001). Photoperiodic time measurement and shift of the critical photoperiod for diapause induction in a moth. In *Insect Timing: Circadian Rhythmicity to Seasonality* (DL Denlinger, JM Giebultowicz, and DS Saunders, eds). Amsterdam: Elsevier, pp. 95–112.

Miyasaki Y, Nisimura T, and Numata H. (2005). A phase response curve for circannual rhythm in the varied carpet beetle *Anthrenus verbasci*. J Comp Physiol A 191: 883–887.

Miyasaki Y, Nisimura T, and Numata H. (2006). Phase responses in the circannual rhythm of the varied carpet beetle, *Anthrenus verbasci*, under naturally changing day length. Zool Sci 23: 1031–1037.

Neumann D. (1966). Die lunare und tagliche Schlupfperiodik der Mucke *Clunio*. Steuer Abst Gez Z Vergl Physiol 53: 1–61.

Neumann D. (1976). Adaptations of chironomids to intertidal environments. Annu Rev Entomol 21: 387–414.

Neumann D and Kruger M. (1985). Combined effects of photoperiod and temperature on the diapause of an intertidal chironomid. Oecologia 67: 154–156.

Nisimura T and Numata H. (2001). Endogenous timing mechanism controlling the circannual pupation rhythm of the varied carpet beetle *Anthrenus verbasci*. J Comp Physiol A 187: 433–440.

Nisimura T and Numata H. (2003) Circannual control of the life cycle in the varied carpet beetle *Anthrenus verbasci*. Funct Ecol 17: 489–495.

Norling U. (1984). Life history patterns in the northern expansion of dragonflies. Adv Odonatol 2: 127–156.

Norris, MJ. (1965). The influence of constant and changing photoperiods on imaginal diapause in the red locust (*Nomadacris septemfasciata* Serv.). J Insect Physiol 11: 1105–1119.

Obrycki JJ and Tauber MJ. (1979). Seasonal synchrony of the parasite *Perilitus coccinellae* with its host, *Coleomegilla maculata*. Environ Entomol 8: 400–405.

Pittendrigh CS. (1966). The circadian oscillation in *Drosophila pseudoobscura* pupae: A model for the photoperiodic clock. Z Pflanzenphysiol 54: 275–307.

Pittendrigh CS. (1972). Circadian surfaces and the diversity of possible roles of circadian organization in photoperiodic induction. Proc Natl Acad Sci USA 69: 2734–2737.

Pittendrigh CS and Minis DH. (1964). The entrainment of circadian oscillations by light and their role as photoperiodic clocks. Am Nat 98: 261–294.

Reppert SM. (2006). A colourful model of the circadian clock. Cell 124: 233–236.

Rockey SJ, Miller BB, and Denlinger DL. (1989). A diapause maternal effect in the flesh fly, *Sarcophaga bullata*: Transfer of information from mother to progeny. J Insect Physiol 35: 553–558.

Sauman I, Briscoe AD, Zhu H, Shi D, Froy O, Stalleicken L, Yuan Q, Casselman A, and Reppert SM. (2005). Connecting the navigational clock to sun compass input in monarch butterfly brain. Neuron 46: 457–467.

Saunders DS. (1966a). Larval diapause of maternal origin-II. The effect of photoperiod and temperature on *Nasonia vitripennis*. J Insect Physiol 12: 569–581.

Saunders DS. (1966b). Larval diapause of maternal origin-III. The effect of host shortage on *Nasonia vitripennis*. J Insect Physiol 12: 899–908.

Saunders DS. (1970). Circadian clock in insect photoperiodism. Science 169: 601–603.

Saunders DS. (1971). The temperature-compensated photoperiodic clock "programming" development and pupal diapause in the flesh-fly, *Sarcophaga argyrastoma*. J Insect Physiol 17: 801–812.

Saunders DS. (1973a). Thermoperiodic control of diapause in an insect: Theory of internal coincidence. Science 181: 358–360.

Saunders DS. (1973b). The photoperiodic clock in the flesh fly, *Sarcophaga argyrostoma*. J Insect Physiol 19: 1941–1954.

Saunders DS. (1975a). Manipulation of the length of the sensitive period, and the induction of pupal diapause in the flesh fly, *Sarcophaga argyrostoma*. J Entomol A 50: 107–118.

Saunders DS. (1975b). "Skeleton" photoperiods and the control of diapause and development in the flesh fly, *Sarcophaga argyrostoma*. J Comp Physiol 97: 97–112.

Saunders DS. (1978). An experimental and theoretical analysis of photoperiodic induction in the flesh fly, *Sarcophaga argyrostoma*. J Comp Physiol 124: 75–95.

Saunders DS. (1979). External coincidence and the photoinducible phase in the *Sarcophaga* photoperiodic clock. J Comp Physiol 132: 179–189.

Saunders DS. (1981). Insect photoperiodism: The clock and the counter. Physiol Entomol 6: 99–116.

Saunders DS. (1983). A diapause induction-termination asymmetry in the photoperiodic responses of the linden bug, *Pyrrhocoris apterus*, and an effect of near-critical photoperiods on development. J Insect Physiol 29: 399–405.

Saunders DS. (1986). Many circadian oscillators regulate developmental and behavioral events in the flesh fly *Sarcophaga argyrostoma*. Chronobiol Int 3: 71–83.

Saunders DS. (1987a). Maternal influence on the incidence and duration of larval diapause in *Calliphora vicina*. Physiol Entomol 12: 331–338.

Saunders DS. (1987b). Insect photoperiodism: The linden bug, *Pyrrhocoris apterus*, a species that measures daylength rather than nightlength. Experientia 43: 935–937.

Saunders DS. (1992). The photoperiodic clock and "counter" in *Sarcophaga argyrostoma:* Experimental evidence consistent with "external coincidence" in insect photoperiodism. J Comp Physiol 170: 121–127.

Saunders DS. (2001). Geographical strains and selection for the diapause trait in *Calliphora vicina*. In *Insect Timing: Circadian Rhythmicity to Seasonality* (DL Denlinger, J Giebultowicz, and DS Saunders, eds). New York: Elsevier, pp. 113–121.

Saunders DS. (2002). *Insect Clocks*, 3rd ed. Amsterdam: Elsevier.

Saunders DS. (2005). Erwin Bünning and Tony Lees, two giants of chronobiology, and the problem of time measurement in insect photoperiodism. J Insect Physiol 51: 599–608.

Saunders DS. (2007). Photoperiodism in insects and other animals. In *Photobiology: The Science of Life and Light*, 2nd ed (LO Bjorn, ed). New York: Springer, pp. 389–416.

Saunders DS and Cymborowski B. (2003). Selection for high diapause incidence in blow flies (*Calliphora vicina*) maintained under long days increases the maternal critical day length: Some consequences for the photoperiodic clock. J Insect Physiol 49: 777–784.

Saunders DS and Lewis RD. (1987a). A damped circadian oscillator model of an insect photoperiodic clock. II, Simulations of the shapes of the photoperiodic response curve. J Theor Biol 128: 61–71.

Saunders DS and Lewis RD. (1987b). A damped circadian oscillator model of an insect photoperiodic clock. III. Circadian and "hourglass" responses. J Theor Biol 128: 73–85.

Saunders DS, Lewis, RD, and Warman GR. (2004). Photoperiodic induction of diapause: Opening the black box. Physiol Entomol 29: 1–15.

Shimizu I, Yamakawa Y, Shimazaki Y, and Iwasa Y. (2001). Molecular cloning of *Bombyx* cerebral opsin (Boceropsin) and cellular localization of its expression in the silkworm brain. Biochem Biophys Res Commun 287: 27–34.

Southwood TRE. (1962). Migration of terrestrial arthropods in relation to habit. Biol Rev 37: 171–214.

Tanaka S, Denlinger DL, and Wolda H. (1987a). Daylength and humidity as environmental cues for diapause termination in a tropical beetle. Physiol Entomol 12: 213–224.

Tanaka S, Denlinger DL, and Wolda H. (1988). Seasonal changes in the photoperiodic response regulating diapause in a tropical beetle, *Stenotarsus rotundus*. J Insect Physiol 34: 1135–1142.

Tanaka S, Wolda H, and Denlinger DL. (1987b). Seasonality and its physiological regulation in three neotropical insect taxa from Barro Colorado island, Panama. Insect Sci Appl 8: 507–514.

Tauber MJ, Tauber CA, and Masaki S. (1986). *Seasonal Adaptations of Insects*. Oxford: Oxford University Press.

Thiele H-D. (1977). Differences in measurement of daylength and photoperiodism in two stocks from sub-arctic and temperate climates in the carabid beetle, *Pterostichus nigrita* F. Oecologia 30: 349–365.

Vaz Nunes M and Hardie J. (1999). The effect of temperature on the photoperiodic "counters" for female morph and sex determination in two clones of the black bean aphid, *Aphis fabae*. Physiol Entomol 24: 339–345.

Vaz Nunes M, Kenny NAP, and Saunders DS. (1990). The photoperiodic clock in the blow fly *Calliphora vicina*. J Insect Physiol 36: 61–67.

Vaz Nunes M and Saunders DS. (1999). Photoperiodic time measurement in insects: A review of clock models. J Biol Rhythms 14: 84–104.

Vaz Nunes M and Veerman A. (1984). Light-break experiments and photoperiodic time measurement in the spider mite, *Tetranychus ulmi*. J Insect Physiol 30: 891–897.

Vaz Nunes M and Veerman A. (1997). "Bistability" experiments and the photoperiodic clock in the spider mite *Tetranychus urticae*. Entomol Exp Appl 84: 195–197.

Veerman A and Vaz Nunes M. (1980). Circadian rhythmicity participates in the photoperiodic determination of diapause in spider mites. Nature 287: 140–141.

Vinogradova EB and Zinovjeva KB. (1972). Maternal induction of larval diapause in the blowfly, *Calliphora vicina*. J Insect Physiol 18: 2401–2409.

von Frisch K. (1950). Die Sonne als Kompass im Lieben der Bienen. Experientia 6: 210–221.

Williams CM and Adkisson PL. (1964). Physiology of insect diapause. XIV An endocrine mechanism for the photoperiodic control of pupal diapause in the oak silkworm, *Antheraea pernyi*. Biol Bull 127: 511–525.

Winfree AT. (1970). Integrated view of resetting a circadian clock. J Theor Biol 28: 327–374.

Wolda H and Denlinger DL. (1984). Diapause in a large aggregation of a tropical beetle. Ecol Entomol 9: 217–230.

11

Photoperiodism in Insects: Perception of Light and the Role of Clock Genes

Shin G. Goto, Sakiko Shiga, and Hideharu Numata

In his book *Insect Clocks*, Saunders (2002) proposed a simple model for photoperiodic phenomena, including a "photoreceptor" to distinguish light from dark, a "clock" to measure the length of the day or night, and an "output" to regulate the seasonal response such as diapause. The output component has been investigated intensively, and today the endocrine control of insect diapause is clear even at the molecular level in some species (Denlinger et al., 2005). The characteristics or molecular mechanisms of the photoreceptor and clock for photoperiodism, however, are not well understood.

As described in chapter 10, Bünning (1936) first introduced the idea that endogenous circadian clocks for daily rhythms are involved in the photoperiodic clock for seasonal rhythms. Since then, various models for photoperiodic time measurement (measuring day or night length by photoperiodic clock) involving one or more circadian oscillators have been proposed (chapter 10; Vaz Nunez and Saunders, 1999). Because both circadian and seasonal clocks respond to environmental information from daily light–dark cycles, similar mechanisms have been assumed for the photoreceptor and clock between circadian rhythms and photoperiodism. This chapter describes photoperiodic mechanisms focusing on light perception and the roles of circadian clock genes, with an emphasis on some shared physiological mechanisms between circadian rhythms and photoperiodism.

LIGHT PERCEPTION

To receive photoperiodic information from the environment, insects must be equipped with light-sensitive cells. These can be ordinary photoreceptors, called retinal receptors for the visual system, or nonvisual cells called extraretinal photoreceptors. Photoperiodic receptors have been identified at the organ level in many species. However, neither specific cells nor molecules have been identified for light perception for photoperiodism, although some candidate molecules have been proposed. Various experiments have demonstrated that both retinal and nonretinal photoreceptors are involved in insect photoperiodism. There are no correlations between photoperiodic organs and insect phylogeny (Numata et al., 1997).

This section first describes the current understanding of photoperiodic receptors for insects at organ, cellular, and molecular levels. The photopigments or anatomical locations for photoreceptors are then compared for circadian entrainment and photoperiodism within the same species.

Organs

Several methods have been employed to identify the organs responsible for light perception for photoperiodism. External distinct organs such as the ocelli and compound eyes have been surgically removed or covered to interrupt light penetration, and responses to photoperiod were then examined. Alternatively, animals kept under short-day conditions had supplemental illumination given to certain organs so that these organs received long-day conditions. In some experiments, organs were cultured in vitro to receive photoperiodic information and then transplanted to host insects to examine whether the organ perceives and stores photoperiodic information.

Each of the above methods has its strengths and limitations. Damage caused to insects by organ elimination may interrupt photoperiodism not through the interruption of photoperiodic perception but from damage to the nervous system. Therefore, sham operations or surgical ablation of unrelated organs are necessary as controls. In addition, histological examinations should be conducted following the experiments. Although covering certain organs causes less damage compared to surgical ablation, the complete shielding of organs from light penetration is often difficult. When there is an effect on photoperiodism following the covering of organs, it suggests that these organs are responsible for light perception. However, negative results could simply be the result of incomplete shielding. The selection of appropriate paints for shielding is also important.

Supplemental illumination has been made using a light guide or luminous paints. The former uses microilluminators constructed from fine capillaries to supply supplementary illumination to different areas of the body surface (Lees, 1964). However, it is difficult to fix an animal body to a light guide for a certain period, before releasing it into darkness. To overcome this difficulty, luminous paints, such as radioluminous, phosphorescent, or chemiluminescent paints, have been used. These paints produce extra light exposure to a localized area on the body's surface. However, it can be difficult to obtain appropriate light intensity or duration. Details of this methodology are described by Numata et al. (1997). To overcome these methodological difficulties, it is preferable to combine several methods to examine photoreceptor locations.

To date, 19 species have been examined for photoperiodic receptor organs (table 11.1). Photoperiodic receptors have been localized in the head region in each of these species. No reports have shown that photoperiodic reception occurs in the thoracic or abdominal regions, nor has the involvement of the dorsal ocelli in a photoperiodic response been demonstrated.

TABLE 11.1. Insects in which the photoperiodic receptor organ was experimentally localized.

Order	Species	Sensitive stage	Photoreceptor organs[a]	References
Hemiptera	*Megoura viciae*	Adult	Br	Lees (1964), Steel and Lees (1977)
	Poecilocoris lewisi	Nymph	CE	Miyawaki et al. (2003)
	Riptortus pedestris	Adult	CE	Numata and Hidaka (1983), Numata (1985)
	Plautia stali	Adult	CE, Br[b]	Morita and Numata (1999)
	Graphosoma lineatum	Adult	CE	Nakamura and Hodkova (1998)
Orthoptera	*Anacridium aegyptium*	Adult	Br	Geldiay (1969)
	Dianemobius nigrofasciatus	Adult	CE	Shiga and Numata (1996)
	Modicogryllus siamensis	Nymph	CE	Sakamoto and Tomioka (2007)
Lepidoptera	*Pieris brassicae*	Larva	Br, St	Claret (1966), Seugé and Veith (1976)
	Pieris rapae	Larva	Br	Kono (1970, 1973)
	Cydia pomonella	Larva	Br[b]	Hayes (1971)
	Antheraea pernyi	Larva	Br[b]	Tanaka (1950)
		Pupa	Br	Williams and Adkisson (1964), Williams (1969)
	Manduca sexta	Larva	Br	Bowen et al. (1984)
	Bombyx mori	Larva	Br	Shimizu (1982), Hasegawa and Shimizu (1987), Shimizu and Hasegawa (1988)
Coleoptera	*Leptinotarsa decemlineata*	Adult	Br[b]	De Wilde et al. (1959)
	Pterostichus nigrita	Adult	CE	Ferenz (1975)
	Leptocarabus kumagaii	Larva	St	Shintani, Shiga and Numata (unpublished)
		Adult	CE	Shintani, Shiga and Numata (unpublished)
Diptera	*Calliphora vicina*	Adult	Br[b]	Saunders and Cymborowski (1996)
	Protophormia terraenovae	Adult	CE	Shiga and Numata (1997)

[a]Br, brain; CE, compound eye; St, stemmata.

[b]Photoreceptor organs are considered to be in the brain from responsiveness to photoperiod after functional elimination of the retinal eyes.

Table 11.1 includes only species in which photoperiodic receptors were determined at the organ level. There are two main types of photoperiodic receptors: retinal receptors and extraretinal receptors. In some insects, such as the brown-winged green bug *Plautia stali* (Morita and Numata, 1999), and the large cabbage white *Pieris brassicae* (Seugé and Veith, 1976), both types of receptors have been demonstrated to be involved in the photoperiodic response.

Hemiptera

Lees (1964) showed that the photoperiodic receptor in the bean and vetch aphid *Megoura viciae* is located in a region of the head other than the compound eyes, most likely in the brain. He used a fine light guide attached to the insects as a micro-illuminator for supplemental light exposure to localized regions. Photosensitivity is confined to the central region of the dorsal area of the head. Because there are no distinctive morphological features in the cuticle of this region, photoperiodic receptors are therefore suggested to be located in the underlying protocerebrum. Covering or cauterization of the compound eyes did not impair photoperiodism (Lees, 1964).

In Heteroptera, the compound eyes have been shown to be important for photoperiodic reception both at larval and adult stages. In the scutellerid bug *Poecilocoris lewisi*, the compound eyes were shown to be the principal receptor for photoperiodism in the nymphal stage. This was achieved by applying a phosphorescent paint to expose selected regions of the body surface to a longer photophase than the rest of the body (Miyawaki et al., 2003). Adults of the bean bug *Riptortus pedestris* (= *clavatus*), *P. stali*, and the striped shield bug *Graphosoma lineatum* also use compound eyes as photoperiodic receptors (Numata and Hidaka, 1983; Numata, 1985; Nakamura and Hodkova, 1998; Morita and Numata, 1999). In *R. pedestris*, supplementary illumination, and elimination experiments were performed and both methods confirmed that the compound eyes are the principal receptors for photoperiodism (Numata and Hidaka, 1983; Numata, 1985). In *P. stali*, histological examinations confirmed the success of the operations. Because a long stalk is present between the retina and the first optic neuropil (lamina), the compound eye could be removed without causing damage to the optic lobe (Morita and Numata, 1999).

The dominant role of the compound eyes for photoperiodic reception was shown in adults of *P. stali*, because removal of these eyes terminated diapause under a diapause-maintaining photoperiod. However, bilateral removal of the compound eyes did not completely prevent male adults from responding to photoperiod. For the first 10 days after the removal of the compound eyes, males discriminated long days from short days, but then they lost sensitivity (Morita and Numata, 1999). It is therefore suggested that organs other than the compound eyes, most likely the brain, are also involved in photoperiodism.

Orthoptera

In the Egyptian grasshopper *Anacridium aegyptium*, supplemental illumination showed that the central region of the head, possibly the brain, was important for photoperiodic reception (Geldiay, 1969). Supplemental illumination was given to a region of the body surface protruding from a light-proof box that prevented light from contacting the rest of the body surface. Lack of light to the compound eyes and ocelli did not prevent grasshoppers from responding to photoperiod (Geldiay, 1966, as cited in Geldiay, 1969).

Photoreception by the compound eyes for photoperiodism was demonstrated in two crickets. Adults of the band-legged cricket *Dianemobius nigrofasciatus* (= *Pteronemobius nigrofasciatus*) show a photoperiodic response controlling embryonic diapause in the progeny. When compound eyes were bilaterally removed, female adults lost sensitivity to photoperiod. Histological observation showed that this operation completely removed the compound eyes, but peripheral regions of the optic lobe in the brain were also damaged. Although it cannot be ruled out that the photoperiodic receptors reside within the optic lobe, it is more probable that the compound eyes contain the photoperiodic receptor in *D. nigrofasciatus* (Shiga and Numata, 1996).

In the cricket *Modicogryllus siamensis*, nymphal development is controlled by photoperiodism. When a compound eye was unilaterally removed on the second day following hatching, responsiveness to photoperiod was lost. This suggests that *M. siamensis* receives photoperiodic information through the compound eye, and that bilateral compound eyes are required for a complete photoperiodic response (Sakamoto and Tomioka, 2007). Interestingly, some effects on photoperiodism were also observed in *D. nigrofasciatus* after unilateral removal of a compound eye (Shiga and Numata, 1996). Integration mechanisms of bilateral components seem to be important for light-input pathways of the photoperiodic clock in crickets.

Lepidoptera

The photoperiodic receptors of a relatively large number of Lepidoptera species have been examined, and the brain has been shown as a photoperiodic receptor in each species. In two species belonging to the genus *Pieris*, of which larvae respond to photoperiod for controlling pupal diapause, the larval brain is the major photoreceptor. In *P. brassicae*, blinding of the six pairs of stemmata by coating with black enamel did not impair photoperiodism (Seugé and Veith, 1976). In addition, when the brain was transplanted into the abdomen, the abdomen became sensitive to photoperiod (Claret, 1966). After the cauterization of the stemmata, most larvae responded to a 16-h photophase of white light, whereas larvae did not respond to a photophase of 9-h white light followed by 7-h blue, green, or yellow light. Control groups responded to all photoperiodic schedules (Seugé and Veith, 1976). These results indicated that in larvae of *P. brassicae*, the predominant role of the brain in photoperiodic reception is clear, but the stemmata play a subordinate role (Seugé and Veith, 1976). In the small cabbage white *Pieris rapae*, larval brains transplanted to pupae also responded to photoperiod (Kono, 1970, 1973).

In the codling moth *Cydia pomonella* (= *Laspeyresia pomonella*), covering the larval stemmata did not interrupt photoperiodism controlling pupal diapause, suggesting that the stemmata are not directly involved in photoperiodic reception (Hayes, 1971). In the Chinese oak moth, *Antheraea pernyi*, cauterization of the larval stemmata or covering them with black paint had no effect on the photoperiodic induction of pupal diapause (Tanaka, 1950). In addition, photoperiodism for the termination of pupal diapause is obvious in *A. pernyi*, and later the photoreceptor for diapause termination was examined in pupae. The pupal brain was removed from the head and transplanted under a plastic window at the posterior tip of the abdomen, and the head and abdomen were exposed to different photoperiods. Following this transplantation of the brain, the sensitivity to photoperiod was also shifted from the head to the abdomen, suggesting that the brain is the site for photoperiodic reception (Williams and Adkisson, 1964).

Bowen et al. (1984) showed that the larval brain of the tobacco hornworm *Manduca sexta* receives photoperiodic information in vitro. Short-day photoperiod in the larval stage induces pupal diapause in *M. sexta*, although diapause can be averted if three long-day cycles intervene early in the fifth (final) instar. Brains attached with corpora cardiaca and corpora allata taken from day 1 fifth instar larvae were cultured in vitro for three days under a long-day or a short-day photoperiod, and the brains were then implanted to day 4 fifth instar larvae under short-day conditions, which would normally cause the larvae to enter diapause. Brains cultured under a long-day photoperiod reversed the diapause program in some recipient larvae, whereas short-day brains did not. The brain can therefore distinguish photoperiod in vitro, that is, when completely separated from the body.

In the silkworm *Bombyx mori*, in vitro reprogramming of photoperiodism was also successful (Hasegawa and Shimizu, 1987). A bivoltine strain of *B. mori* shows maternal induction of embryonic diapause by photoperiodism, and the ability of photoperiodic reception in the brain–subesophageal ganglion complex was examined in vitro. Complexes from newly ecdysed fifth instar females were cultured under long-day or short-day conditions for four days. After in vitro culture, a pair of complexes was implanted into the abdomen of a late fifth instar larva under short-day conditions that would normally cause larvae to become nondiapause-egg producers. The complexes cultured under short-day conditions caused the host insects to produce a large portion of diapause eggs, whereas those cultured under long-day conditions caused them to produce nondiapause eggs. These results show that photoperiodic receptors and clocks are located in the brain–subesophageal complex, possibly in the brain itself (Hasegawa and Shimizu, 1987).

These elegant in vitro experiments in *M. sexta* and *B. mori* demonstrated that photoperiodic receptors are present in the larval brain (Bowen et al., 1984; Hasegawa and Shimizu, 1987). It would be informative to culture parts of the brain rather than the whole brain to determine important regions for photoperiodic mechanisms.

Coleoptera

Three species have been examined for photoperiodic receptors. In the Colorado potato beetle *Leptinotarsa decemlineata*, covering of the compound eyes did not prevent adult beetles from responding to photoperiod for controlling reproductive diapause (De Wilde et al., 1959). After covering the compound eyes with black paint, the completeness of the covering was confirmed by phototaxis. The results suggested the presence of extraretinal photoreceptors for photoperiodism in *L. decemlineata*.

In contrast, it was demonstrated in two carabid beetles that the retinal eyes are important as photoperiodic receptors. The compound eyes are important in *Pterostichus nigrita* for controlling male gonads (Ferenz, 1975). The development of spermatozoa aggregations in the testis is promoted by short-day conditions. Under continuous light conditions, which normally suppress gonad development, gonads still developed following bilateral extirpation of the compound eyes. In another carabid beetle, *Leptocarabus kumagaii*, which shows photoperiodism in two developmental stages of larvae and adults, surgical removal of the retinal eyes showed that the stemmata and compound eyes are used for photoreception for photoperiodism in larval and adult stages, respectively (Shintani, Shiga and Numata, unpublished), indicating that different organs are used for photoperiodism in different stages within a species.

Diptera

Saunders and Cymborowski (1996) showed that the blowfly *Calliphora vicina* remains sensitive to photoperiod after the removal of the optic lobes for the maternal induction of larval diapause. This suggests that photoperiodic receptors are likely located in the cerebral lobe without the optic lobe. In contrast, in another blowfly, *Protophormia terraenovae*, it was shown that the compound eyes are the dominant photoreceptors for photoperiodic control of reproductive diapause in adult females (Shiga and Numata, 1997). Elimination of the compound eyes was performed by covering or surgical removal. Although both treatments prevented flies from responding to photoperiod, they had the opposite effect on gonadal expression. When the compound eyes were bilaterally covered by silver paint, the ovary did not develop irrespective of photoperiod, and diapause incidences were similar to those under continuous darkness. When compound eyes and a peripheral part of the optic lobe lamina were bilaterally surgically ablated, the ovary developed irrespective of photoperiodic conditions. When the antennal lobe was removed as a control, the ovary did not develop under diapause-inducing conditions. Therefore ovarian development was not due to damage or injury to the nervous tissue but to the functional elimination of the compound eyes. These experiments indicate that perception of continuous darkness by covering and the complete lack of the receptors by surgical removal have physiologically different effects on photoperiodic clocks. In addition, the blowflies *C. vicina* and *P. terraenovae* are sensitive to photoperiod in the adult stage, but the photoperiodic receptors seem to be different (Saunders and Cymborowski, 1996; Shiga and Numata, 1997).

Tissues and Cells

Among the 19 species for which photoperiodic receptor organs have been determined (table 11.1), a restricted area within the organ has been shown as the potential tissue or cell type for photoperiodic reception in only three species (figure 11.1). A specific region has been reported in the compound eye from *Riptortus pedestris* and in the brain from *Megoura viciae* and *Antheraea pernyi*.

Riptortus pedestris

In *R. pedestris*, surgical removal of bilateral compound eyes eliminated sensitivity to photoperiod (Numata, 1985; Morita and Numata, 1997). A compound eye of adult *R. pedestris* is composed of approximately 1,600 ommatidia. Do all ommatidia equally contribute to photoperiodic reception, or are there regional differences

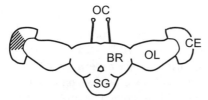

Riptortus pedestris (adult)

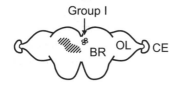

Megoura viciae (adult)

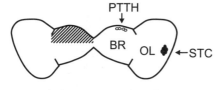

Antheraea pernyi (pupa)

FIGURE 11.1.
Location of photoperiodic receptors in the bean bug *Riptortus pedestris*, the bean and vetch aphid *Megoura viciae*, and the Chinese oak moth *Antheraea pernyi*. The photoreceptor regions (hatched area) are shown in a frontal view of the brain and eyes. BR, brain; CE, compound eye; Group I, a group of neurosecretory cells in the pars intercerebralis; PTTH, lateral neurosecretory cells producing prothoracicotropic hormone; OC, ocellus; OL, optic lobe; SG; subesophagial ganglion; STC, stemma derived cell. Based on data from Williams (1969), Steel and Lees (1977), Sauman and Reppert (1996b), Morita and Numata (1997), and Gao et al. (1999).

of receptive functions in the compound eyes? In some species, certain regions of the compound eye, called the dorsal rim area, are specified for ultraviolet perception, and spatial and spectral properties of the compound eye can be directly linked to the behavior and habitat of the animal (Stavenga, 1992). To clarify whether there are any regional differences in the compound eyes for photoperiodic perception, Morita and Numata (1997) surgically removed varying numbers of ommatidia from different regions of a compound eye under a short-day photoperiod that normally induces and maintains diapause. Since unilateral removal of a compound eye would not interrupt photoperiodic reception, all the ommatidia were eliminated from one eye while the other was used for regional ablations. When ommatidia in the anterior, posterior, dorsal, or ventral region were removed, most insects with 900 or more ommatidia left remained in diapause, whereas most insects with fewer than 800 ommatidia terminated diapause. When ommatidia in the central region were removed, however, only a few insects remained in diapause even when 900 or more ommatidia remained. In contrast, when ommatidia in the peripheral region were removed, a greater proportion of insects were in diapause even though fewer than 800 ommatidia remained. From these results, the ommatidia in the central region of the compound eye were concluded to play the principal role in photoperiodic reception in *R. pedestris* (figure 11.1).

Megoura viciae

After Lees (1964) showed that a dorsal area is necessary for photoperiodic reception in *M. viciae*, Steel and Lees (1977) surgically removed localized areas in the brain. Damaged areas were histologically examined after the experiments. Lesions confined entirely to the optic lobes invariably had no effect on photoperiodic responses. This supports the idea that the compound eyes are not involved as photoperiodic receptors because the optic lobe connects the compound eye to the central brain. In the central brain, five types of neurosecretory cells were found. Ablation experiments showed that a type called the group I cells in the pars intercerebralis was indispensable for the operation of photoperiodic mechanisms. Steel and Lees (1977) proposed that these neurosecretory cells play a role as neurosecretory effecters of the photoperiodic response. Ablation experiments also showed another important region for photoperiodism slightly lateral to the group I cells, suggesting that these areas presumably contain photoperiodic clocks or receptors (Steel and Lees, 1977).

It has been suggested that there is involvement of an opsin/retinoid-based photopigment in *M. viciae* (Lees, 1981), so opsin-containing regions were examined by immunocytochemistry. Brain slices were stained using 20 antibodies against various kinds of opsins. Seven of the 20 antibodies equally stained a crescent-shaped region in the neuropil of an anterior ventral protocerebral region, just ventral to the group I cell clusters, as well as the compound eyes (Gao et al., 1999). Steel and Lees (1977) did not surgically remove this area, so Gao et al. (1999) performed microsurgery on this area to see the effects on photoperiodism. Six of 13 aphids that had this operation failed to respond to photoperiod, whereas sham-operated aphids responded normally.

Two opsins were cloned in *M. viciae* and in situ hybridization demonstrated that the transcripts of these opsins were expressed in the retinula cells of the compound eyes (Gao et al., 2000). It would be interesting to know if these genes are expressed in the putative photoreceptor area in the brain in *M. viciae*. From these anatomical findings, together with previous localized illumination and microlesion studies (Lees, 1964; Steel and Lees, 1977), it is thought that the anterior ventral to lateral neuropil region of the protocerebrum contains the photoperiodic receptor in the aphid (figure 11.1). Because the photoreceptor region is in close proximity to the group I neurosecretory cells, it is reasonable to assume that a neural component of photoperiodic control exists in the anterior medial to lateral protocerebrum in *M. viciae*.

Antheraea pernyi

Williams (1969) localized the photoreceptor regions in the pupal brain. The entire brain was excised, and portions were then reimplanted into the pupae to examine the responses to photoperiod. When cerebral lobes without optic lobes were implanted, pupae showed photoperiodism. The pupae responded to photoperiod even when only the dorsal half of the cerebral lobe was implanted. However, when a brain with the dorsolateral area removed was implanted, the sensitivity to photoperiod was lost completely. This clearly indicates that the photoperiodic receptors reside in the dorsolateral region of the cerebral lobe (figure 11.1). Sauman and Reppert (1996b) showed that prothoracicotropic hormone (PTTH) is produced by lateral neurosecretory cells in *A. pernyi*. Because PTTH is necessary for adult development after the termination of pupal diapause, PTTH cells are thought to function as a photoperiodic effector in this species at the pupal stage. The transplantation experiment by Williams (1969) indicates the dorsolateral region in *A. pernyi* contains the three components: photoperiodic receptors, clocks, and effectors.

Molecules

For the entrainment of the circadian clocks, two photopigments are known in insects: a vitamin A–based photopigment, retinal or 3-hydroxyretinal, conjugated with a protein, opsin, and a vitamin B_2–based pigment, cryptochrome (Helfrich-Förster, 2001). Cryptochrome is known to function as a photoperiodic receptor pigment in the plant *Arabidopsis thaliana* (Guo et al., 1998). However, the roles of cryptochrome in insect photoperiodism have not yet been studied.

Most species are optimally sensitive to light in the blue-green region of the spectrum and largely insensitive to red, but in some species sensitivity extends to the red end of the spectrum (Danks, 1987; Saunders, 2002). Consequently, action spectra vary widely in different species. The pink bollworm *Pectinophora gossypiella* distinguished photoperiod between a 14/10-h light/dark cycle (LD 14:10; nondia-pause) and LD 12:12 (diapause) in which the red light (>600 nm) was used in the photophase (Pittendrigh et al., 1970). Photoperiodism in *P. gossypiella* evidently shows spectral sensitivity to red light. Because the cryptochrome identified in the fruit fly *Drosophila melanogaster* and the monarch butterfly *Danaus plexippus* undergoes a reversible absorption change upon blue light irradiation, but does not absorb red light at wavelengths above 500 nm (Berndt et al., 2007; Song et al., 2007), it is conceivable that insects sensitive to red light for photoperiodism use molecules other than cryptochrome.

Classic dietary-deficient experiments suggest that a vitamin A–based pig-ment, retinal-opsin, is more likely to be used for photoperiodic reception in insects and mites. Takeda (1978) first showed that larvae of the southwestern corn borer *Diatraea grandiosella* reared on an artificial diet deficient in carotenoids lose their photoperiodic response. In the spider mite *Tetranychus urticae*, a lowered sensitivity to photoperiod in an albino mutant, in which the uptake and oxidative metabolism of beta carotene were disturbed, suggests that carotenoid pigments are functionally involved in photoperiodic response (Veerman and Helle, 1978; Veerman, 1980). The loss of photoperiodic response by dietary deficiency of carotenoids has since been shown in many species (Saunders, 2002). It has been suggested that vitamin A or a derivative, such as retinal or 3-hydroxyretinal, is involved in photoperiodic perception, possibly conjugated with a protein to form a rhodopsin-like pigment. In insects that use retinal eyes for photoperiodism, such as *Riptortus pedestris* and *Dianemobius nigrofasciatus*, the involvement of vitamin A or a derivative in photoperiodism is quite plausible. Also, in species in which extraretinal receptors have been shown to play the principal role in the photoperiodic reception, such as *Bombyx mori* and *Pieris brassicae*, dietary-deficiency experiments of carotenoid support the importance of vitamin A or its derivatives (Claret, 1989; Claret and Volkoff, 1992). However, no direct evidence has yet been obtained.

Shimizu et al. (2001) cloned a novel opsin (boceropsin) from the larval brain in *B. mori*, in which the photoperiodic receptor might reside (Hasegawa and Shimizu, 1987). Moreover, an antibody to the boceropsin protein labeled somata in four dif-ferent brain regions: the dorsal anterior protocerebrum, ventral anterior protocere-brum, dorsal posterior protocerebrum, and lateral posterior tritocerebrum (Shimizu et al., 2001). Shimizu et al. (2001) suggested that boceropsin is a putative pho-toperiodic receptor protein in *B. mori*. The role of boceropsin in photoperiodism remains to be shown.

Comparison of Photoreceptors between Circadian Entrainment and Photoperiodism

Photoperiodic information is also used for circadian entrainment to adjust intrinsic rhythms to daily cycles of environmental changes. It has been suggested that pho-toperiodism has its basis in the circadian clock system. If the circadian clock and photoperiodism are functionally related, it is likely that circadian clock entrainment and photoperiodism use a common photoreceptor. In order to determine the functional relationships between circadian clock entrainment and photoperiodic responses, the anatomical location or photopigments for photoreceptors have been examined. Because each species has species-specific and stage-specific photoperiodic receptors as described above, it is important to compare photoreceptors between photoperiodism and circadian entrainment in the same species at the same developmental stage.

Pupae of *Antheraea pernyi* perceive photoperiod for circadian entrainment and photoperiodism. The circadian clock controls the eclosion rhythm, and photoperiodism controls the termination of diapause. Truman (1972) showed by transplantation experiments that a photoreceptor component for the entrainment of eclosion rhythm was present in the brain of *A. pernyi*. It has been shown that in the cecropia silkmoth *Hyalophora cecropia*, which is in the same Saturniinae subfamily as *A. pernyi*, the imaginal disk of the compound eye in pupae are not essential for the entrainment of the eclosion rhythm (Truman, 1972). Using the same photoperiodic chamber that had been used for photoperiodic reception by Williams and Adkisson (1964), Truman (1972) showed that when the brain was transplanted from the head to the abdomen, photosensitivity for rhythm entrainment was transplanted along with it. Because the brain transplantation experiments showed identical results for circadian entrainment and photoperiodism, photoreceptors for photoperiodism and rhythm entrainment seem to be identical in *H. cecropia* and *A. pernyi*, and are in the brain rather than the retina.

In adults of *Graphosoma lineatum*, light entrainment of the locomotor rhythms and photoperiodic regulation of diapause were examined after the removal of the compound eyes in order to localize the photoreceptor location (Nakamura and Hodkova, 1998). Bilateral removal of the eyes resulted in the loss of locomotor rhythms and of photoperiodism. Nakamura and Hodkova (1998) suggested that the compound eyes are involved in the transmission of light signals for both the entrainment of circadian rhythms and photoperiodism, and the circadian clock itself also is situated in the compound eyes. However, following the removal of the optic lobes, adult blowflies of *Calliphora vicina* responded to photoperiod to control larval diapause maternally and for the entrainment of circadian activity rhythms (Cymborowski et al., 1994; Saunders and Cymborowski, 1996), indicating that both responses used extraretinal receptors, probably elsewhere in the brain other than the optic lobes.

Although the details of receptor organs or tissues have not been identified, the same light input pathways are probable between photoperiodism and circadian entrainment in *A. pernyi*, *G. lineatum*, and *C. vicina*. However, different sets of receptors are postulated in several other species.

In *B. mori*, the effects of carotenoid depletion on the entrainment of the eclosion and hatching rhythms, and photoperiodic control of embryonic diapause were examined. Shimizu and his colleagues found that carotenoid is dispensable for the entrainment of the eclosion and hatching rhythms, suggesting that there are circadian photoreceptive molecules other than carotenoid-based pigments (Shimizu and Matsui, 1983; Sakamoto and Shimizu, 1994; Sakamoto et al., 2003). In contrast, carotenoid depletion had a critical effect on photosensitivity in photoperiodic induction of diapause (Shimizu and Kato, 1984). The light-input pathways for the two systems do not appear to be the same.

Pectinophora gossypiella senses photoperiod at the larval stage to control larval diapause. Pittendrigh et al. (1970) demonstrated that photoperiodic responses controlling larval diapause could be affected by wavelengths of 600 nm or longer, which did not entrain egg hatching, eclosion, or oviposition rhythms. Although the developmental stages examined for photoperiodism and circadian entrainment were different, the results suggest that in *P. gossypiella* the photoreceptors involved in the two systems are not identical.

In *Dianemobius nigrofasciatus* and *Protophormia terraenovae*, ablation of the compound eyes prevented adults from responding to photoperiod to control embryonic and adult diapause, respectively (Shiga and Numata, 1996, 1997). However, the same operation did not interrupt light perception for entrainment of the circadian locomotor activity rhythm in both species (Shiga et al., 1999; Hamasaka et al., 2001). Figure 11.2 summarizes these results. Under short-day conditions, most females entered reproductive diapause in *P. terraenovae* and became diapause-egg

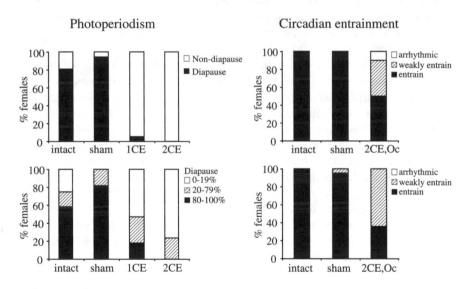

FIGURE 11.2.
Effects of removal of the compound eyes on photoperiodism and entrainment of circadian activity rhythm in the blowfly *Protophormia terraenovae* (top) and the band-legged ground cricket *Dianemobius nigrofasciatus* (bottom). (Left) Under diapause-inducing short-day conditions, unilateral (1CE) or bilateral (2CE) removal of compound eyes prevented diapause induction in both species. (Right) In contrast, removal of the compound eyes and ocelli (2CE, Oc) did not prevent adults from entraining to photoperiod. Modified from Shiga and Numata (1996, 1997), Shiga et al. (1999), and Hamasaka et al. (2001).

producers in *D. nigrofasciatus*. After uni- or bilateral removal of compound eyes, diapause incidence decreased in both species. This indicates that *P. terraenovae* and *D. nigrofasciatus* use compound eyes for the photoperiodic control of diapause. In contrast, entrainability to light–dark cycles still remains after the removal of both compound eyes and all ocelli. Extraretinal photoreceptors for circadian entrainment are therefore likely present in the brain (Shiga et al., 1999; Hamasaka et al., 2001).

However, a higher proportion of females showed weak entrainment patterns after removal of the compound eyes. This suggests that the retinal eyes are also involved in entrainment of the circadian activity rhythms. In *D. nigrofasciatus* and *P. terraenovae*, the compound eyes are used for both photoperiodism and circadian entrainment, but the photoreceptor systems are not identical. There are carotenoid-dependent photoreceptors involved solely in photoperiodism and not in circadian entrainment in *Bombyx mori*, and red-sensitive photoreceptors in *Pectinophora gossypiella*. In these species, however, the results do not rule out the possibility that other types of receptors, such as carotenoid-independent or red insensitive photoreceptors, are also involved in photoperiodism and might also be used for circadian entrainment.

In *Tetranychus urticae*, it is believed that different photopigments are used for the photoperiodic clock and the circadian clock (Veerman and Veenendaal, 2003). In this species, the Nanda-Hamner rhythm (see chapter 10) controlled by the circadian clock is necessary for the realization of photoperiodism. Using the Nanda-Hamner protocol, Veerman and Veenendaal (2003) showed that the photoperiodic clock, which discriminates long and short nights, is able to respond to orange-red light (>580 nm), whereas the circadian clock controlling the Nanda-Hamner rhythm was not. This indicates that there is a photoreceptor sensitive to orange-red light used solely for the photoperiodic time measurement. However, because output signals from the Nanda-Hamner clock are necessary for photoperiodism, light information via the orange-red light insensitive photoreceptor also affects photoperiodic responses.

In conclusion, it seems that the two light perception systems for circadian entrainment and photoperiodism share some components but are not completely identical.

ROLE OF CLOCK GENES

Circadian Clock and Photoperiodic Time Measurement

As mentioned in chapter 10, it has been assumed that the same clock circuitry is involved in both behavioral circadian rhythms and photoperiodic responses. However, the evidence is far from conclusive, and at times even contradictory (Tauber and Kyriacou, 2001; Veerman, 2001; Saunders, 2002).

The following sections describe the molecular mechanisms of the circadian clock as verified in *Drosophila melanogaster*. We then focus on clock gene expression under different photoperiodic conditions. Finally, we introduce several reports on the involvement or lack of involvement of circadian clock genes in photoperiodism.

Molecular Mechanisms of the Circadian Clock
in *Drosophila melanogaster*

Konopka and Benzar (1971) published a landmark study documenting three mutants in *D. melanogaster* with aberrant circadian rhythms. The mutations equally affected two different rhythms: the adult eclosion rhythm and the locomotor rhythm. This clearly indicated that the mutations affected a general circadian clock system rather than the observed output behaviors. Mutations were all mapped to the same gene, known as *period* (*per*). This discovery further led to the discovery of at least a half-dozen other such "clock genes" by forward genetic screens and reverse genetic methodologies. Many detailed reviews on the molecular mechanisms of the circadian clock circuits in *D. melanogaster* already exist (e.g., Williams and Sehgal, 2001; Saunders, 2002; Dunlap et al., 2003; Price, 2004), so herein we only briefly describe the mechanisms.

The *Drosophila* circadian oscillator includes interlocked positive and negative feedback loops based on transcription and translation, with a heterodimerization of PAS domain–containing proteins acting as a transcriptional activator and other proteins acting as negative elements in the feedback loop. The core interlocked feedback loops are defined by the actions of the CLOCK (CLK) and CYCLE (CYC) proteins, *per* and *timeless* (*tim*) mRNAs, and PER and TIM proteins. The clock could be reset by light input. Although opsin-mediated resetting is not fully understood, cryptochrome (CRY)-mediated resetting has been extensively investigated. CRY, a blue- and UV-sensitive photoreceptor pigment, uses flavin as a chromophore to confer light sensitivity on the clock system.

These clock components are expressed in a daily pattern. *per* and *tim* mRNAs are low during the photophase but high during the scotophase. PER and TIM also show similar patterns, but their peaks are delayed by a few hours compared with those of the mRNAs (see Price, 2004). The expression patterns of *Clk* mRNA and CLK protein are in the antiphase of the *per* and *tim* mRNAs (Bae et al., 2000). The expression of *cyc* mRNA and CYC protein does not oscillate (Rutila et al., 1998). *cry* mRNA shows a clear diel oscillation; that is, its level is higher during the photophase than the scotophase (Emery et al., 1998; Egan et al., 1999). It is of interest that CRY expression is greatly dependent on posttranscriptional or translational regulation, or possibly both. When *cry* mRNA is high during the photophase, CRY protein is low. CRY clearly increases before the mRNA rises, and peaks late in the scotophase. When *cry* mRNA peaks very early in the photophase, CRY is already greatly reduced (Emery et al., 1998).

Expression Patterns of Circadian Clock Genes
under Different Photoperiodic Conditions

The circadian clock transforms signals from daily light-dark cycles into the oscillation of clock gene expression in order to transmit the information toward circadian output systems. If the circadian clock itself also mediates seasonal cues into pho-toperiodic output, how is it performed? Does the circadian clock also transform the photoperiodic signals into the oscillation of clock genes' expression? Goto and Denlinger (2002) addressed this in the flesh fly *Sarcophaga crassipalpis*. A clear diel rhythm of *per* mRNA levels was detected: *per* mRNA was down-regulated during the photophase and up-regulated during the scotophase. *tim* mRNA revealed an obvious diel rhythm as well. Photoperiod indeed influenced the expression pat-terns of *per* and *tim* mRNA: the peak of *per* mRNA expression shifted in concert with an onset of the scotophase, whereas the shift in *tim* mRNA expression was less pronounced. The amplitude of *tim* mRNA was severely dampened under long-day conditions, but that of *per* mRNA was not affected. Similar photoperiod-dependent phase shifts in the expression of clock genes were observed in *D. melanogaster* (Majercak et al., 1999; Shafer et al., 2004) and *B. mori* (Iwai et al., 2006). In the Syrian hamster, the amplitudes and expression patterns of clock genes were also strongly affected by photoperiods (Carr et al., 2003; Tournier et al., 2003). Distinct expression patterns of *per* and *Clk* genes under short- and long-day conditions were also observed in the linden bug *Pyrrhocoris apterus* (see below). These results indicate that seasonal cues affect the expression patterns of clock genes, and such distinct expressions could be used to determine photoperiodic responses.

Geographic Variations of Photoperiodic Response and Circadian Behavior, and the Linkage between Them

If a set of genes responsible for the expression of photoperiodism is also responsi-ble for the expression of daily rhythms, adaptive modification of the photoperiodic time measurement system implies genetic modification of the circadian clock. This further implies that the two must evolve together because of the causal relation-ship between them. Only a small number of studies have focused on this issue, but they all revealed that the two characteristics could have evolved independently (see Bradshaw and Holzapfel, 2007a).

In the pitcher-plant mosquito *Wyeomyia smithii*, the critical day length for lar-val diapause increases regularly with latitude and altitude (Bradshaw and Holzapfel, 2001a, 2001b). However, neither the period nor the amplitude of the circadian rhythm is correlated with the critical photoperiod (Bradshaw et al., 2003, 2006). In addition, Mathias et al. (2006) revealed by crossing experiments that genetic mod-ification of the circadian clock has not been the basis for adaptive modification of photoperiodic time measurement. Moreover, quantitative trait loci (QTL) analyses revealed that *tim* is not causally involved in the difference in the critical photope-riod, but rather only epistatically interacts with it (Mathias et al., 2007).

Lankinen and Forsman (2006) also investigated the genetic linkage between the two timekeeping mechanisms in *Drosophila littoralis*. A northern strain shows a long critical day length for diapause, an early phase of the entrained eclo-sion rhythm under extreme short days, and a short period of the free-running

eclosion rhythm. A southern strain shows a short critical day length, a late eclosion phase, and a long free-running period. These distinct strains were crossed to
determine the linkage between photoperiodic and circadian characteristics. After
54 generations that included free recombination, artificial selection, and genetic
drift, a novel strain that had "southern" diapause and "northern" eclosion rhythm
characteristics was produced. The observed complete separation of eclosion rhythm
characteristics from photoperiodism reveals that the systems controlling the variability of the eclosion rhythm and photoperiodism are different and genetically
separable in *D. littoralis*.

We must consider the downstream cascades, however, when interpreting these
results. It is clear that downstream cascades governing circadian behaviors and photoperiodic responses are completely different. This implies that a certain type of
natural selective pressure could promote stabilizing selection for one cascade and
directional selection for the other. Thus, even if a single core oscillator is responsible for both photoperiodic responses and circadian behaviors, it is possible for the
oscillator to output different responses through distinct downstream pathways.

Photoperiodic Responses in Mutants or Variants
That Display Arrhythmic Circadian Behavior

Saunders et al. (1989) first studied the role of the circadian clock gene in photoperiodism using mutants of *D. melanogaster* and showed that a null mutant of
per and deletion strains of *per* loci were able to discriminate photoperiods for the
induction of diapause (figure 11.3). This clearly indicates that *per* is not necessary
for the photoperiodic induction of diapause in this species. However, Bradshaw and
Holzapfel (2007b) pointed out that the mutant flies lacking *per* could theoretically

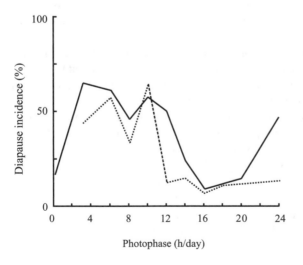

FIGURE 11.3.
Photoperiodic response
curves in wild type (Canton-S,
solid line) and *per*[ol] mutant
(broken line) *Drosophila
melanogaster* at 12°C. The
mutant was capable of
discriminating photoperiods,
but the critical day length
of the mutant was shifted
to about 2 h shorter when
compared with the wild type.
Modified from Saunders et al.
(1989).

convey information on day length through *tim*, because TIM could be degraded by light directly through CRY even in the *per* mutant (Myers et al., 1996; Zeng et al., 1996). In addition, as Saunders et al. (1989) themselves pointed out, other circadian clock genes may play crucial roles in the photoperiodic time measurement system. This idea is supported by some studies that reported the presence of a *per*-independent circadian oscillator in *D. melanogaster* (see Helfrich-Förster, 2001; Weitzel and Rensing, 1981).

Several recent studies show a causal genetic linkage between the circadian clock and the photoperiodic time measurement system. A variant named *npd* (*non-photoperiodic diapause*) in the drosophilid fly *Chymomyza costata*, first isolated from a natural population by Riihimaa and Kimura (1988), does not enter diapause even under diapause-inducing short-day conditions (figure 11.4A), although it still retains the capacity to enter diapause in response to low temperature. The variant shows an arrhythmic pattern of adult eclosion (Lankinen and Riihimaa, 1992). Distinct diurnal rhythms in *per* and *tim* mRNA abundance were noted in wild-type flies, but in the *npd* flies *per* abundance was low, and its expression was arrhythmic. In addition, *tim* mRNA was completely undetectable in the *npd* flies (figure 11.4B; Koštál and Shimada, 2001; Pavelka et al., 2003). Genetic linkage analysis revealed that the gene responsible for the nondiapause phenotype was mapped to the locus including *tim*. Moreover, suppression of *tim* mRNA expression by RNA interference (RNAi) modestly but clearly reduced the incidence of diapause (Pavelka et al., 2003). Thus, expression of the circadian clock gene *tim* is indispensable for the photoperiodic response in *C. costata*.

A variant displaying an arrhythmic adult eclosion pattern was also isolated in the flesh fly *Sarcophaga bullata* (figure 11.5A; Goto et al., 2006). Interestingly, this variant failed to enter pupal diapause even under strong diapause-inducing conditions (figure 11.5B). These results suggest that a functional circadian clock is necessary for the expression of photoperiodism in this species. One might expect that the circadian clock genes may fail to be expressed or their expression may be diminished and/or be arrhythmic in this variant, as reported in *C. costata*. This was not the case, however, as both *per* and *tim* mRNAs were present, and surprisingly, both of these clock genes were expressed higher in the nondiapause variant than in the wild-type colony (figure 11.5C). This abnormality is possibly a result of an upstream clock gene malfunction or a malfunction of the autoregulatory loop. This malfunction in the circadian clock would result in the disruption of photoperiodism.

Hodková and her colleagues also focused on circadian behavior, clock gene expression, and diapause induction in *Pyrrhocoris apterus*. A variant that failed to enter diapause even under diapause-inducing short-day conditions was isolated. The wild-type strain expressed 10-fold higher levels of *per* mRNA under diapause-inducing short-day conditions than diapause-averting long-day conditions. Interestingly, the variant exhibited low levels of *per* mRNA irrespective of photoperiodic conditions, and the levels were comparable with those observed in the wild-type strain under long-day conditions. Furthermore, low *per* mRNA levels were associated with unique long-day specific circadian characteristics (long

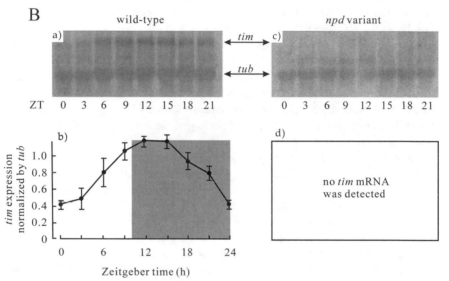

A

FIGURE 11.4.
Photoperiodic response curves and expression of *timeless* (*tim*) mRNA in *Chymomyza costata*. (A) Photoperiodic response curve at 15°C. The wild type (open circles) showed a clear photoperiodic response, but the *npd* (*non-photoperiodic diapause*) variant (solid circles) did not enter diapause irrespective of the photoperiod. Reprinted with permission from Riihimaa and Kimura (1988). (B) Northern hybridization with probes of *tim* and β-*tubulin* (*tub*) (panels *a* and *c*), and quantified radioactive signals of *tim* hybridization (normalized by *tub*) shown as mean ± standard error (*b* and *d*) of triplicate samples. The shaded area in *b* indicates scotophase. Insects were reared under short-day conditions (LD 10:14) at 18°C. RNAs were extracted from adult heads taken at 3-h intervals of *zeitgeber* time (ZT). Clear diel oscillation was observed in the wild-type flies, but *tim* mRNA was undetectable in the variant. Modified from Pavelka et al. (2003).

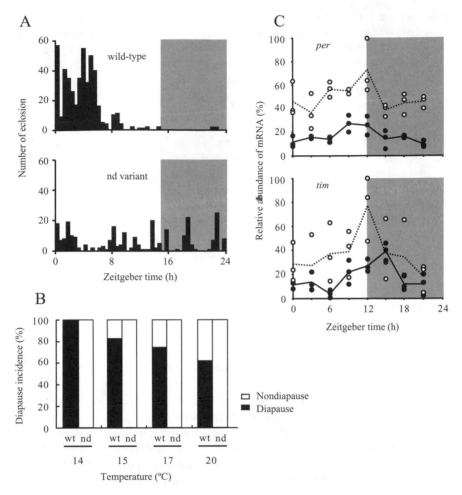

FIGURE 11.5.
Adult eclosion rhythm, diapause incidence, and expression of *per* and *tim* mRNAs
in *Sarcophaga bullata*. (A) Distribution of adult eclosion in the wild type (top) and a
nondiapause (nd) variant (bottom) under LD 15:9 at 25°C. Shaded area indicates the
scotophase. The variant showed an arrhythmic eclosion pattern. B: diapause incidence of
the wild type (wt) and the nondiapause (nd) variant under LD 12:12 at various temperatures.
The nondiapause variant did not enter diapause even under strong diapause-inducing
conditions of 14°C. (C) Relative abundance of *per* (top) and *tim* (bottom) mRNAs in the wild
type (solid circles) and the nondiapause variant (open circles) under LD 12:12 at 25°C. RNAs
were extracted from adult heads taken at 3-h intervals of *zeitgeber* time (ZT). Shaded area
indicates the scotophase. Solid and broken lines are drawn through the means of triplicate
plots for wild type and nondiapause variant, respectively. The highest level of mRNA is
determined as 100. Both of the clock genes were expressed higher in the nondiapause
variant than in the wild type. Modified from Goto et al. (2006).

free-running periods and late peaks of the activity) (Hodková et al., 2003). *Clk* also displayed similar expression patterns (Syrová et al., 2003). These results also support the idea that the circadian clock transforms photoperiodic signals into expression patterns of circadian clock genes, and the information is transmitted to both photoperiodic and circadian output. However, Doležel et al. (2007) recently showed that in this species the up-regulation of *per* is directly involved in the food-dependent pathway, which is important in diapause expression, but not in the photoperiodic time measurement system.

Despite the early works by Saunders and his colleagues (Saunders et al., 1989; Saunders 1990; Gillanders and Saunders, 1992), *D. melanogaster* has not become a model organism for the analyses of photoperiodic time measurement system. This is simply because diapause of this species is shallow and the variance of diapause incidence is often large (see Saunders et al., 1989). In addition, diapause is induced at low temperatures close to the lower developmental threshold of this species. Moreover, Tatar et al. (2001) could not confirm the presence of diapause in this species, and the authors regarded the suppression of ovarian development as quiescence. However, recent studies have revealed that some natural populations of *D. melanogaster* clearly showed diapause (Schmidt et al., 2005a, 2005b) and are still suitable for the molecular analyses of photoperiodic response (Sandrelli et al., 2007; Tauber et al., 2007).

Recent Progress in *Drosophila melanogaster*

Tauber et al. (2007) found in European *D. melanogaster* that the incidence of diapause is positively correlated with latitude; that is, the incidences in northern European populations are higher than in southern ones. In these populations, two alleles of *tim*, *s-tim* and *ls-tim*, were found. The former produces truncated TIM (S-TIM) protein and the latter produces both S-TIM and full-length TIM (LS-TIM) proteins. Molecular phylogenetic analyses revealed that one of the alleles, *s-tim*, has arisen and spread by natural selection relatively recently in Europe. A clear latitudinal cline of frequency in the *ls-tim* allele was also observed; that is, the frequency was low at high latitudes. Surprisingly, introducing one of the natural or artificial alleles of *tim* into different genetic backgrounds slightly but clearly affected the incidence of diapause, with females having *ls-tim* showing higher diapause incidence than those having *s-tim*. In addition, a null mutant of *tim*, *tim*[01], is also capable of entering diapause. These results indicate that *tim* is not causally involved in the expression of diapause but can lead to a disposition for inducing or preventing diapause. Furthermore, it is noteworthy that in *tim*[01] flies the photoperiodic response becomes ambiguous and the flies enter diapause irrespective of photoperiod. This raises the possibility that *tim* is involved in the photoperiodic time measurement system. We must analyze this possibility carefully, however, because the photoperiodic response curve of this species is less clear (see Saunders et al., 1989) when compared with other insects (e.g., the response in *Chymomyza costata*; figure 11.4A).

In addition, it is unclear why *ls-tim*, which is responsible for diapause, is more frequent in southern populations, since diapause is maladaptive for these populations. Sandrelli et al. (2007) further investigated characteristics of the *s-tim* and *ls-tim* alleles in *D. melanogaster*. The authors clarified that the ability of LS-TIM to interact with CRY is much weaker than that of S-TIM with CRY, resulting in flies that have the *ls-tim* allele showing significantly smaller phase responses in locomotor activity as compared with those having *s-tim*. This further promoted higher levels of TIM in the *ls-tim* than in the *s-tim* flies, and a higher amplitude of TIM oscillation in the *ls-tim* flies (Sandrelli et al., 2007). It is still uncertain how such a weak ability of LS-TIM to interact with CRY contributes to the photoperiodic response, but the potential role of *tim* and *cry* on photoperiodism leads to the necessity to explore the relationship between the circadian clock and the photoperiodic time measurement system in the model organism *D. melanogaster*.

CONCLUSIONS AND FUTURE DIRECTIONS

As shown in chapter 10, there is no doubt about the involvement of the circadian clock in the photoperiodic time measurement system, but it is unclear whether these two timekeeping mechanisms share some components.

The first study by Saunders et al. (1989) revealed that at least *per* is dispensable for the photoperiodic time measurement system in *D. melanogaster*. In *Wyeomyia smithii* and *Drosophila littoralis*, the mechanisms controlling the variability of circadian behavior and the photoperiodic characteristics are genetically separable. In addition, several lines of evidence have demonstrated that behavioral rhythms and photoperiodic timing are based on different clocks (see chapter 10).

On the other hand, several studies using mutants or variants revealed a possibility of the involvement of circadian clock genes in photoperiodism. The linkage between the circadian clock and photoperiodic time measurement system was particularly well demonstrated in *Chymomyza costata* (Koštál and Shimada, 2001; Pavelka et al., 2003). The linkage is also supported in recent studies on *D. melanogaster* (Sandrelli et al., 2007; Tauber et al., 2007). Furthermore, photoperiod affected the oscillation patterns of circadian clock components in some insects. This implies that clock cells transform the photoperiodic information into distinct oscillation patterns in order to transmit the information to output cascades for photoperiodic response.

Contradictory evidence therefore appears to exist for the involvement of circadian clock genes in the photoperiodic time measurement system. This may indicate that the two timekeeping circuits in insects share some components but are not identical. This idea has been proven in *Arabidopsis thaliana*. In brief, photoperiodic flowering and the regulation of the circadian clock share common elements, including photoreceptors and proteins comprising the timekeeping mechanism, yet differ in some components (Kim et al., 2007; Rubio and Deng, 2007; Sawa et al., 2007). In terms of light perception mechanisms, current knowledge also suggests that photoreceptors differ but could be partly shared between circadian entrainment and photoperiodism. In *Dianemobius nigrofasciatus* and *Protophormia terraenovae*, the compound eyes are used for both photoperiodism and circadian entrainment (Shiga and Numata, 1996, 1997; Shiga et al., 1999; Hamasaka et al., 2001). However, photoreceptors involved solely in photoperiodism and not in circadian entrainment have been demonstrated in *Bombyx mori* and *Pectinophora gossypiella* (Pittendrigh et al., 1970; Shimizu and Kato, 1984). Multiple-input pathways for photoperiodism have been demonstrated in some species (Seugé and Veith, 1976; Morita and Numata, 1999), and thus *B. mori* and *P. gossypiella* may also have other photoreceptors commonly used for circadian entrainment and photoperiodism. Multiple light-input pathways have also been shown in the circadian entrainment system in *D. melanogaster* (Helfrich-Förster et al., 2001).

Biological timing systems involved in both entrainment for circadian rhythms and photoperiodism use light perception and timekeeping clocks. It seems that cellular and molecular mechanisms underlying these two systems are not completely separable, and some shared components could be responsible for both biological timing systems.

The present chapter proposed that some circadian clock genes may also be responsible for photoperiodism. Many studies have shown the importance of the brain for photoperiodism, and thus these genes must be expressed in brain neurons. In *D. melanogaster*, different types of cells have been revealed to express circadian clock genes in the brain. Among them, ventral lateral neurons are thought to be one of the most important clock neurons that govern adult locomotor rhythms, but there are also many other PERIOD-expressing neurons in the brain (Kaneko and Hall, 2000). It is probable that in insects the cells responsible for photoperiodic time measurement are located at different sites from neurons responsible for the circadian clock in the brain.

Bradshaw and Holzapfel (2007b) emphasized that focusing on the photoperiodic timer by exhaustive studies of specific circadian clock genes may not yield much information. This may be true, but, as mentioned in this chapter, studies using mutations or variants showing circadian arrhythmicity, such as nondiapause variants of *C. costata* and *S. bullata* and circadian clock mutants in *D. melanogaster*, have certainly contributed to an understanding of photoperiodism. Focusing on clock genes will certainly aid in understanding the photoperiodic time measurement system. We must bear in mind the diversity of the clockwork in the various studies. Although the molecular circuits of the circadian clock have been precisely clarified in *D. melanogaster*, the clockwork may differ in other insects (see Sauman and Reppert, 1996a; Goto and Denlinger, 2002; Zhu et al., 2005, 2008; Rubin et al., 2006; Yuan et al., 2007). In addition, fine-scale mapping of genes by quantitative trait locus analyses, discovering genes that show differential expression under different photoperiodic conditions by subtractive hybridization, differential display, and microarrays would also be beneficial. Furthermore, looking for "good" mutants by large-scale mutagenesis, like the approach adopted in the analysis of the circadian clock by Konopka and Benzer (1971), as well as creating gain-of-function or loss-of-function transformants using transposon-mediated technologies, would also greatly contribute to the understanding of the genetic systems underlying photoperiodism. While these studies will be difficult to perform, the information obtained is too valuable to avoid.

In order to clarify the photoperiodic timekeeping system, neurobiological studies are also crucial. A clarification of the responsible genes provides details for the basic components of photoperiodism but does not explain the whole integration mechanism of photoperiodic clocks. Because it is likely that many genes have multiple tasks, it is therefore important to identify which neurons among the many clock-gene–expressing cells are responsible for photoperiodism. Since Helfrich-Förster (1995) first characterized clock-gene–expressing neurons, the understanding of circadian oscillator mechanisms composing complex behavioral rhythms has quickly progressed. To understand the mechanisms underlying photoperiodic timing systems, a combination of molecular biological and neurobiological studies is necessary.

References

Bae K, Lee C, Hardin PE, and Edery I. (2000). dCLOCK is present in limiting amounts and likely mediates daily interactions between the dCLOCK-CYC transcription factor and the PER–TIM complex. J Neurosci 20: 1746–1753.

Berndt A, Kottke T, Breitkreuz H, Dvorsky R, Hennig S, Alexander M, and Wolf E. (2007). A novel photoreaction mechanism for the circadian blue light photoreceptor *Drosophila* cryptochrome. J Biol Chem 282: 13011–13021.

Bowen MF, Saunders DS, Bollenbacher WE, and Gilbert LI. (1984). In vitro reprogramming of the photoperiodic clock in an insect brain retrocerebral complex. Proc Natl Acad Sci USA 81: 5881–5884.

Bradshaw WE and Holzapfel CM. (2001a). Genetic shift in photoperiodic response correlated with global warming. Proc Natl Acad Sci USA 98: 14509–14511.

Bradshaw WE and Holzapfel CM. (2001b). Phenotypic evolution and the genetic architecture underlying photoperiodic time measurement. J Insect Physiol 47: 809–820.

Bradshaw WE and Holzapfel CM. (2007a). Evolution of animal photoperiodism. Annu Rev Ecol Evol Syst 38: 1–25.

Bradshaw W and Holzapfel C. (2007b). Tantalizing *timeless*. Science 316: 1851–1852.

Bradshaw WE, Holzapfel CM, and Mathias D. (2006). Circadian rhythmicity and photoperiodism in the pitcher-plant mosquito: Can the seasonal timer evolve independently of the circadian clock? Am Nat 167: 601–605.

Bradshaw WE, Quebodeaux MC, and Holzapfel CM. (2003). Circadian rhythmicity and photoperiodism in the pitcher-plant mosquito: Adaptive response to the photic environment or correlated response to the seasonal environment? Am Nat 161: 735–748.

Bünning E. (1936). Die endonome Tagesrhythmik als Grundlage der photoperiodischen Reaktion. Ber Dtsch Bot Ges 54: 590–607.

Carr AJ, Johnston JD, Semikhodskii AG, Nolan T, Cagampang FR, Stirland JA, and Loudon AS. (2003). Photoperiod differentially regulates circadian oscillators in central and peripheral tissues of the Syrian hamster. Curr Biol 13: 1543–1548.

Claret J. (1966). Mise en évidence du rôle photorécepteur du cerveau dans l'induction de la diapause, chez *Pieris brassicae* (Lepido.). Ann Endocrinol 27(suppl): 311–320.

Claret J. (1989). Vitamine A et induction photopériodique ou thermopériodique de la diapause chez *Pieris brassicae* (Lepidoptera). C R Acad Sci Ser III 308: 347–352.

Claret J and Volkoff N. (1992). Vitamin A is essential for two processes involved in the photoperiodic reaction in *Pieris brassicae*. J Insect Physiol 38: 569–574.

Cymborowski B, Lewis RD, Hong S-F, and Saunders DS. (1994). Circadian locomotor activity rhythms and their entrainment to light-dark cycles continue in flies (*Calliphora vicina*) surgically deprived of their optic lobes. J Insect Physiol 40: 501–510.

Danks HV. (1987). *Insect Dormancy: An Ecological Perspective*. Ottawa: Biological Survey of Canada.

Denlinger DL, Yocum GD, and Rinehart JP. (2005). Hormonal control of diapause. In *Comprehensive Molecular Insect Science*, vol 3, *Endocrinology* (LI Gilbert, K Iatrou, and SS Gill, eds). Amsterdam: Elsevier, pp. 615–650.

De Wilde J, Duintjer CS, and Mook L. (1959). Physiology of diapause in the adult Colorado beetle (*Leptinotarsa decemlineata* Say)—I. The photoperiod as a controlling factor. J Insect Physiol 3: 75–85.

Doležel D, Šauman I, Košt'ál V, and Hodková M. (2007). Photoperiodic and food signals control expression pattern of the clock gene, *Period*, in the linden bug, *Pyrrhocoris apterus*. J Biol Rhythms 22: 335–342.

Dunlap JC, Loros JJ, and Decoursey P. (2003). *Chronobiology: Biological Timekeeping*. Sundeland, MA: Sinauer Associates.

Egan ES, Franklin TM, Hiderbrand-Chae MJ, McNeil GP, Roberts MA, Schroeder AJ, Zhang X, and Jackson FR. (1999). An extraretinally expressed insect cryptochrome with similarity to the blue light photoreceptors of mammals and plants. J Neurosci 19: 3665–3673.

Emery P, So WV, Kaneko M, Hall JC, and Rosbash M. (1998). CRY, a *Drosophila* clock and light-regulated cryptochrome, is a major contributor to circadian rhythm resetting and photosensitivity. Cell 95: 669–679.

Ferenz H-J. (1975). Photoperiodic and hormonal control of reproduction in male beetles, *Pterostichus nigrita*. J Insect Physiol 21: 331–341.

Gao N, Foster RG, and Hardie J. (2000). Two opsin genes from the vetch aphid, *Megoura viciae*. Insect Mol Biol 9: 197–202.

Gao N, von Schantz M, Foster RG, and Hardie J. (1999). The putative brain photoperiodic photoreceptors in the vetch aphid, *Megoura viciae*. J Insect Physiol 45: 1011–1019.

Geldiay S. (1969). *Anacridium aegyptium* L. (Misir çekirgesi) de fotoperiodik resoptorlarin lokalizasyonu ve farkli fotoperiodlarda neurosekresyon hücrelerinin faaliyeti. Ege Üniv Fen Fak Ilmi Rap Ser 89: 1–37.

Gillanders SW and Saunders DS. (1992). A coupled pacemaker-slave model for the insect photoperiodic clock: Interpretation of ovarian diapause data in *Drosophila melanogaster*. Biol Cybern 67: 451–459.

Goto SG and Denlinger DL. (2002). Short-day and long-day expression patterns of genes involved in the flesh fly clock mechanism: *period*, *timeless*, *cycle* and *cryptochrome*. J Insect Physiol 48: 803–816.

Goto SG, Han B, and Denlinger DL. (2006). A nondiapausing variant of the flesh fly, *Sarcophaga bullata*, that shows arrhythmic adult eclosion and elevated expression of *period* and *timeless*. J Insect Physiol 52: 1213–1218.

Guo H, Yang H, Mockler TC, and Lin C. (1998). Regulation of flowering time by *Arabidopsis* photoreceptors. Science 279: 1360–1363.

Hamasaka Y, Watari Y, Arai T, Numata H, and Shiga S. (2001). Retinal and extraretinal pathways for entrainment of the circadian activity rhythm in the blow fly, *Protophormia terraenovae*. J Insect Physiol 47: 867–875.

Hasegawa K and Shimizu I. (1987). *In vivo* and *in vitro* photoperiodic induction of diapause using isolated brain suboesophageal ganglion complexes of the silkworm, *Bombyx mori*. J Insect Physiol 33: 959–966.

Hayes DK. (1971). Action spectra for breaking diapause and absorption spectra of insect brain tissue. In *Biochronometry* (M Menaker, ed). Washington, DC: National Academy of Science Press, pp. 392–402.

Helfrich-Förster C. (2001). The locomotor activity rhythm of *Drosophila melanogaster* is controlled by a dual oscillator system. J Insect Physiol 47: 877–887.

Helfrich-Förster C. (1995). The period clock gene is expressed in CNS neurons which also produce a neuropeptide that reveals the projections of circadian pacemaker cells within the brain of *Drosophila melanogaster*. Proc Natl Acad Sci USA 92: 612–616.

Helfrich-Förster C, Winter C, Hofbauer A, Hall JC, and Stanewsky R. (2001). The circadian clock of fruit flies is blind after elimination of all known photoreceptors. Neuron 30: 249–261.

Hodková M, Syrová Z, Doležel D, and Šauman I. (2003). *Period* gene expression in relation to seasonality and circadian rhythms in the linden bug, *Pyrrhocoris apterus* (Heteroptera). Eur J Entomol 100: 267–273.

Iwai S, Fukui Y, Fujiwara Y, and Takeda M. (2006). Structure and expression of two circadian clock genes, *period* and *timeless* in the commercial silk moth, *Bombyx mori*. J Insect Physiol 52: 625–637.

Kaneko M and Hall JC. (2000). Neuroanatomy of cells expressing clock genes in *Drosophila*: Transgenic manipulation of the *period* and *timeless* genes to mark the perikarya of circadian pacemaker neurons and their projections. J Comp Neurol 422: 66–94.

Kim W-Y, Fujiwara S, Suh S-S, Kim J, Kim Y, Han L, David K, Putterill J, Nam HG, and Somers DE. (2007). ZEITLUPE is a circadian photoreceptor stabilized by GIGANTEA in blue light. Nature 449: 356–360.

Kono Y. (1970). Photoperiodic induction of diapause in *Pieris rapae crucivora* Boisduval (Lepidoptera: Pieridae). Appl Entomol Zool 5: 213–224.

Kono Y. (1973). Photoperiodic sensitivity of the implanted brain of *Pieris rapae crucivora* and ultrastructural changes of its neurosecretory cells [in Japanese with English summary]. Jpn J Appl Entomol Zool 17: 203–209.

Konopka RJ and Benzer S. (1971). Clock mutants of *Drosophila melanogaster*. Proc Natl Acad Sci USA 68: 2112–2116.

Koštál V and Shimada K. (2001). Malfunction of circadian clock in the *non-photoperiodic-diapause* mutants of the drosophilid fly, *Chymomyza costata*. J Insect Physiol 11: 1269–1274.

Lankinen P and Forsman P. (2006). Independence of genetic geographical variation between photoperiodic diapause, circadian eclosion rhythm and Thr-Gly repeat region of the *period* gene in *Drosophila littoralis*. J Biol Rhythms 21: 3–12.

Lankinen P and Riihimaa AJ. (1992). Weak circadian eclosion rhythmicity in *Chymomyza costata* (Diptera: Drosophilidae), and its independence of diapause type. J Insect Physiol 38: 803–811.

Lees AD. (1964). The location of the photoperiodic receptors in the aphid *Megoura viciae* Buckton. J Exp Biol 41: 119–133.

Lees AD. (1981). Action spectra for the photoperiodic control of polymorphism in the aphid *Megoura viciae*. J Insect Physiol 27: 761 771.

Majercak J, Sidote D, Hardin PE, and Edery I. (1999). How a circadian clock adapts to seasonal decreases in temperature and day length. Neuron 24: 219–230.

Mathias D, Jacky L, Bradshaw WE, and Holzapfel CM. (2007). Quantitative trait loci associated with photoperiodic response and stage of diapause in the pitcher-plant mosquito, *Wyeomyia smithii*. Genetics 176: 391–402.

Mathias D, Reed LK, Bradshaw WE, and Holzapfel CM. (2006). Evolutionary divergence of circadian and photoperiodic phenotypes in the pitcher-plant mosquito, *Wyeomyia smithii*. J Biol Rhythms 21: 132–139.

Miyawaki R, Tanaka SI, and Numata H. (2003). Photoperiodic receptor in the nymph of *Poecilocoris lewisi* (Heteroptera: Scutelleridae). Eur J Entomol 100: 301–303.

Morita A and Numata H. (1997). Distribution of photoperiodic receptors in the compound eyes of the bean bug, *Riptortus clavatus*. J Comp Physiol A 180: 181–185.

Morita A and Numata H. (1999). Localization of the photoreceptor for photoperiodism in the stink bug, *Plautia crossota stali*. Physiol Entomol 24: 190–196.

Myers MP, Wagner-Smith K, Rothenfluh-Hilfikerm A, and Young MW. (1996). Light-induced degradation of TIMELESS and entrainment of the *Drosophila* circadian clock. Science 271: 1736–1740.

Nakamura K and Hodkova M. (1998). Photoreception in entrainment of rhythms and photoperiodic regulation of diapause in a hemipteran, *Graphosoma lineatum*. J Biol Rhythms 13: 159–166.

Numata H. (1985). Photoperiodic control of adult diapause in the bean bug, *Riptortus clavatus*. Mem Fac Sci Kyoto Univ Ser Biol 10: 29–48.

Numata H and Hidaka T. (1983). Compound eyes as the photoperiodic receptors in the bean bug. Experientia 39: 868–869.

Numata H, Shiga S, and Morita A. (1997). Photoperiodic receptors in arthropods. Zool Sci 14: 187–197.

Pavelka J, Shimada K, and Koštál V. (2003). TIMELESS: A link between fly's circadian and photoperiodic clocks? Eur J Entomol 100: 255–266.

Pittendrigh CS, Eichhorn JH, Minis DH, and Bruce VG. (1970). Circadian systems, VI. Photoperiodic time measurement in *Pectinophora gossypiella*. Proc Natl Acad Sci USA 66: 758–764.

Price JL. (2004). *Drosophila melanogaster*: A model system for molecular chronobiology. In *Molecular Biology of Circadian Rhythms* (A Sehgal, ed). Hoboken, NJ: Wiley-Liss, pp. 33–74.

Riihimaa AJ and Kimura MT. (1988). A mutant strain of *Chymomyza costata* (Diptera: Drosophilidae) insensitive to diapause-inducing action of photoperiod. Physiol Entomol 13: 441–445.

Rubin EB, Shemesh Y, Cohen M, Elgavish S, Robertson HM, and Bloch G. (2006). Molecular and phylogenetic analyses reveal mammalian-like clockwork in the honey bee (*Apis mellifera*) and shed new light on the molecular evolution of the circadian clock. Genome Res 16: 1352–1365.

Rubio V and Deng ZW. (2007). Standing on the shoulders of GIGANTEA. Science 318: 206–207.

Rutila JE, Suri V, Le M, So WV, Rosbash M, and Hall JC. (1998). CYCLE is a second bHLH-PAS clock protein essential for circadian rhythmicity and transcription of *Drosophila period* and *timeless*. Cell 93: 805–814.

Sakamoto K, Asai R, Okada A, and Shimizu I. (2003). Effect of carotenoid depletion on the hatching rhythm of the silkworm, *Bombyx mori*. Biol Rhythm Res 34: 61–71.

Sakamoto K and Shimizu I. (1994). Photosensitivity in the circadian hatching rhythm of the carotenoid-depleted silkworm, *Bombyx mori*. J Biol Rhythms 9: 61–70.

Sakamoto T and Tomioka K. (2007). Effects of unilateral compound-eye removal on the photoperiodic responses of nymphal development in the cricket *Modycogryllus siamensis*. Zool Sci 24: 604–610.

Sandrelli F, Tauber E, Pegoraro M, Mozzotta G, Cisotto P, Landskron J, Stanewsky R, Piccin A, Rosato E, Zordan M, Costa R, and Kyriacou CP. (2007). A molecular basis for natural selection at the *timeless* locus in *Drosophila melanogaster*. Science 316: 1898–1900.

Sauman I and Reppert SM. (1996a). Circadian clock neurons in the silkmoth *Antheraea pernyi*: Novel mechanisms of Period protein regulation. Neuron 17: 979–990.

Sauman I and Reppert SM. (1996b). Molecular characterization of prothoracicotropic hormone (PTTH) from the giant silkmoth *Antheraea pernyi*: Developmental appearance of PTTH-expressing cells and relationship to circadian clock cells in central brain. Dev Biol 178: 418–429.

Saunders DS. (1990). The circadian basis of ovarian diapause regulation in *Drosophila melanogaster*: Is the period gene causally involved in photoperiodic time measurement? J Biol Rhythms 5: 315–331.

Saunders DS. (2002). *Insect Clocks*, 3rd ed. Amsterdam: Elsevier.

Saunders DS and Cymborowski B. (1996). Removal of optic lobes of adult blow flies (*Calliphora vicina*) leaves photoperiodic induction of larval diapause intact. J Insect Physiol 42: 807–811.

Saunders DS, Henrich VC, and Gilbert LI. (1989). Induction of diapause in *Drosophila melanogaster*: Photoperiodic regulation and the impact of arrhythmic clock mutations on time measurement. Proc Natl Acad Sci USA 86: 3748–3752.

Sawa M, Nusinow DA, Kay SA, and Imaizumi T. (2007). FKF1 and GIGANTEA complex formation is required for day-length measurement in *Arabidopsis*. Science 318: 261–265.

Schmidt PS, Matzkin L, Ippolito M, and Eanes WF. (2005a). Geographic variation in diapause incidence, life-history traits, and climatic adaptation in *Drosophila melanogaster*. Evolution 59: 1721–1732.

Schmidt PS, Paaby AB, and Heschel MS. (2005b). Genetic variance for diapause expression and associated life histories in *Drosophila melanogaster*. Evolution 59: 2616–2625.

Seugé J and Veith K. (1976). Diapause de *Pieris brassicae*: Rôle des photorécepteurs céphaliques, étude des caroténoides cérébraux. J Insect Physiol 22: 1229–1235.

Shafer OT, Levine JD, Truman JW, and Hall JC. (2004). Flies by night: Effects of changing day length on *Drosophila's* circadian clock. Curr Biol 14: 424–432.

Shiga S and Numata H. (1996). Effects of compound eye removal on the photoperiodic response of the band legged ground cricket, *Pteronemobius nigrofasciatus*. J Comp Physiol A 179: 625–633.

Shiga S and Numata H. (1997). The adult blow fly (*Protophormia terraenovae*) perceives photoperiod through the compound eyes for the induction of reproductive diapause. J Comp Physiol A 181: 35–40.

Shiga S, Numata H, and Yoshioka E. (1999). Localization of the photoreceptor and pacemaker for the circadian activity rhythm in the band-legged ground cricket, *Dianemobius nigrofasciatus*. Zool Sci 16: 193–201.

Shimizu I. (1982). Photoperiodic induction in the silkworm, *Bombyx mori*, reared on artificial diet: Evidence for extraretinal photoreception. J Insect Physiol 28: 841–846.

Shimizu I and Hasegawa K. (1988). Photoperiodic induction of diapause in the silkworm, *Bombyx mori*: Location of the photoreceptor using a chemiluminescent paint. Physiol Entomol 13: 81–88.

Shimizu I and Kato M. (1984). Carotenoid functions in photoperiodic induction in the silkworm, *Bombyx mori*. Photobiochem Photobiophys 7: 47–52.

Shimizu I and Matsui K. (1983). Photoreceptions in the eclosion of the silkworm, *Bombyx mori*. Photochem Photobiol 37: 409–413.

Shimizu I, Yamanaka Y, Shimazaki Y, and Iwasa T. (2001). Molecular cloning of *Bombyx* cerebral opsin (boceropsin) and cellular localization of the expression in the silkworm brain. Biochem Biophys Res Commun 287: 37–34.

Song S-H, Öztürk N, Denaro TR, Özlem Arat N, Kao Y-T, Zhong D, Reppert SM, and Sancar A. (2007). Formation and function of flavin anion radical in cryptochrome 1 blue-light photoreceptor of monarch butterfly. J Biol Chem 282: 17608–17612.

Stavenga DG. (1992). Eye regionalization and spectral tuning of retinal pigments in insects. Trends Neurosci 15: 213–218.

Steel CGH and Lees AD. (1977). The role of neurosecretion in the photoperiodic control of polymorphism in the aphid *Megoura viciae*. J Exp Biol 67: 117–135.

Syrová Z, Doležel D, Šauman I, and Hodková M. (2003). Photoperiodic regulation of diapause in linden bugs: Are *period* and *Clock* gene involved? Cell Mol Life Sci 60: 2510–2515.

Takeda M. (1978). Photoperiodic time measurement and seasonal adaptation of the southwestern corn borer, *Diatraea grandiosella*. PhD dissertation, University of Missouri, Columbia.

Tanaka Y. (1950). Studies on hibernation with special reference to photoperiodicity and breeding of Chinese tussar silkworm (III) [in Japanese]. J Sericult Sci Jpn 19: 580–590.

Tatar MS, Chien S, and Priest NK. (2001). Negligible senescence during reproductive dormancy in *Drosophila melanogaster*. Am Nat 158: 248–258.

Tauber E and Kyriacou BP. (2001). Insect photoperiodism and circadian clocks: Models and mechanisms. J Biol Rhythms 16: 381–390.

Tauber E, Zordan M, Sandrelli F, Pegoraro M, Osterwalder N, Breda C, Daga A, Selmin A, Monger K, Benna C, Rosato E, Kyriacou CP, and Costa R. (2007). Natural selection favors a newly derived *timeless* allele in *Drosophila melanogaster*. Science 316: 1895–1898.

Tournier BB, Menet JS, Dardente H, Poirel VJ, Malan A, Masson-Pévet M, Pévet P, and Vuillez P. (2003). Photoperiod differentially regulates clock genes' expression in the suprachiasmatic nucleus of Syrian hamster. Neuroscience 118: 317–322.

Truman JW. (1972). Physiology of insect rhythms. II. The silkmoth brain as the location of the biological clock controlling eclosion. J Comp Physiol 81: 99–114.

Vaz Nunez M and Saunders D. (1999). Photoperiodic time measurement in insects: A review of clock models. J Biol Rhythms 14: 84–104.

Veerman A. (1980). Functional involvement of carotenoids in photoperiodic induction of diapause in the spider mite, *Tetranychus urticae*. Physiol Entomol 5: 291–300.

Veerman A. (2001). Photoperiodic time measurement in insects and mites: A critical evaluation of the oscillator-clock hypothesis. J Insect Physiol 47: 1097–1109.

Veerman A and Helle W. (1978). Evidence for the functional involvement of carotenoids in the photoperiodic reaction of spider mites. Nature 275: 234.

Veerman A and Veenendaal RL. (2003). Experimental evidence for a non-clock role of the circadian system in spider mite photoperiodism. J Insect Physiol 49: 727–732.

Weitzel G and Rensing L. (1981). Evidence for cellular circadian rhythms in isolated fluorescent dye-labelled salivary glands of wild type and an arrhythmic mutant of *Drosophila melanogaster*. J Comp Physiol B 143: 229–235.

Williams CM. (1969). Photoperiodism and the endocrine aspects of insect diapause. Symp Soc Exp Biol 23: 285–300.

Williams CM and Adkisson PL. (1964). Physiology of insect diapause. XIV. An endocrine mechanism for the photoperiodic control of pupal diapause in the oak silkworm, *Antheraea pernyi*. Biol Bull 127: 511–525.

Williams JA and Sehgal A. (2001). Molecular components of the circadian system in *Drosophila*. Annu Rev Physiol 63: 729–755.

Yuan Q, Metterville D, Briscoe AD, and Reppert SM. (2007). Insect cryptochromes: Gene duplication and loss define diverse ways to construct insect circadian clocks. Mol Biol Evol 24: 948–955.

Zeng H, Qian Z, Myers MP, and Rosbash M. (1996). A light-entrainment mechanism for the *Drosophila* circadian clock. Nature 380: 129–135.

Zhu H, Sauman I, Yuan Q, Casselman A, Emery-Le M, Emery P, and Reppert SM. (2008). Cryptochromes define a novel circadian clock mechanism in monarch butterflies that may underlie sun compass navigation. PLoS Biol 6: E4.

Zhu H, Yuan Q, Briscoe AD, Froy O, Casselman A, and Reppert SM. (2005). The two CRYs of the butterfly. Curr Biol 15: R953–954.

12

Photoperiodism in Insects: Molecular Basis and Consequences of Diapause

Karen D. Williams, Paul S. Schmidt, and Marla B. Sokolowski

Diapause is a state of dormancy that allows an insect to escape in time. In diapause, the insect postpones development or reproduction in response to specific cues that anticipate the onset of hazardous conditions. For example, changes in day length are often used as cues to anticipate the onset of winter conditions. A glimpse of the molecular mechanisms underlying diapause has begun to emerge in a number of insect species providing us with tantalizing directions for future research. From a genetic perspective, it appears that genes involved in clock function, insulin signaling, stress resistance and development have been coopted into insect diapause pathways. Diapause has consequences for growth, reproduction, survival, and longevity; this, too, provides exciting avenues for future research. Finally, there are similarities between insect diapause and other dormancies such as dauer formation in nematodes, hibernation in mammals, and mammalian embryonic diapause. Together, these new findings provide ample fodder for a review of what is known about the molecular mechanisms involved in the onset, maintenance, and emergence from diapause in insects.

Traditionally, the genetic analysis of diapause involved the analysis of the progeny of crosses between strains that differed in diapause propensity (reviewed in Lees, 1955; Saunders, 1976; Lumme and Lakovaara, 1983). Some studies suggested that natural variation in diapause involved many genes with small effects (Hoy, 1994), while others suggested that genes of large effect could explain natural variation in diapause (Williams and Sokolowski, 1993). Exciting advances in genetics, molecular biology, and genomics have shown that it is now possible to identify actual genes and pathways involved in diapause (Hoy, 1994). Our ability to identify and characterize genes involved in insect diapause in combination with the wealth of information from physiological studies of diapause is leading us toward a deeper understanding of this fascinating phenomenon.

ONSET OF DIAPAUSE: THE PERCEPTION OF TRIGGERS FOR DIAPAUSE

Insects enter diapause at a species-specific stage and exhibit particular arrests of development: insects in egg, larval, or pupal diapause exhibit growth arrests, whereas adult insects in diapause exhibit reproductive arrests (Tauber et al., 1986). Within species individual differences in diapause are found throughout the insect world. Insects with a photoperiodic diapause vary in their response to the cues that trigger diapause (Lumme and Lakovaara, 1983). Geographic variations in the incidence of diapause and species specific sensitive stages for the induction of diapause are also common (Tauber et al., 1986). Below, we expand upon two examples from the literature to address the question of how diapause is induced. In the first example, we examine the role of the circadian rhythm gene *timeless* (*tim*) in the induction of adult diapause in the fruit fly *Drosophila melanogaster*, while in the second example, we explore the role of hormones in the induction of embryonic diapause in the silk moth *Bombyx mori*.

Drosophila species predominantly exhibit an adult reproductive diapause where photoperiod and temperature function to control the induction, maintenance, and termination of diapause (Carson and Stalker, 1948; Lumme et al., 1974; Lumme and Oikarinen, 1977; Lumme, 1978; Lumme and Lakovaara, 1983; Kimura, 1984; Saunders et al., 1989). In *D. melanogaster* diapause, a young adult female is placed at low-temperature and short-day conditions, and she is scored as being in diapause some weeks later if her ovaries remain previtellogenic (Saunders et al., 1989). The induction of diapause in many insects is triggered in part by a change in photoperiod; these shorter day lengths predict the onset of winter conditions (Tauber and Kyriacou, 2001). As expected, in the northern hemisphere, the percentage of *D. melanogaster* in diapause increases with latitude (Williams and Sokolowski, 1993; Schmidt et al., 2005a; Tauber et al., 2007). This type of relationship between diapause propensity and latitude is found for many insect species (Geyspits and Simonenko, 1970; Bradshaw, 1976; Lumme and Oikarinen, 1977; Minami and Kimura, 1980; Kimura, 1984; Kimura et al., 1994; Mathias et al., 2005).

Photoperiod, the amount of light in a day, is perceived by both the visual and nonvisual systems (Saunders, 1976, chapter 11). Photoreceptors in the insect eyes and brain play important roles in the perception and integration of information about day length. Since time measurement is affected by photoperiod, it was hypothesized that genes involved in circadian rhythms and clock function might also play a role in seasonal cycles such as diapause (Saunders, 1976). Studies from the Kyriacou lab have shown a role for the clock gene *tim* in *D. melanogaster* reproductive diapause (Sandrelli et al., 2007; Tauber et al., 2007). Prior to discussing these studies, we review what is known about the molecular basis of circadian rhythms.

The genetic and molecular analysis of the circadian system was pioneered in *D. melanogaster* (Dunlap, 1999). Much is known about the *Drosophila* genes (e.g., *period* [*per*], *tim*, *clock* [*clk*], *cycle* [*cyc*], and *cryptochrome* [*cry*]) involved in circadian rhythms as well as the molecular cascade by which light cues are transmitted to the optic lobes in the *Drosophila* brain (reviewed in Dunlap, 1999; Hall, 2003). Light stimulates CRY, which binds to TIM and PER (Ceriani et al., 1999; Tauber and Kryiacou, 2001; Peschel et al., 2006). CRY proteins that are important in the perception of low light levels increase during the night and decline with approaching daylight (Hall, 2000; Tauber and Kyriacou, 2001). CRY is predominantly found in the *Drosophila* deep brain photoreceptors (Emery, 2000). The binding of CRY to the TIM-PER complex promotes its degradation and reduces the transcription of *per* and *tim* genes via the transcription factors CLK and CYC (Hall, 2000; Tauber and Kyriacou, 2001; Hardin, 2005). The nuclear localization of the PER protein is *tim* dependent (Vosshall et al., 1994). The TIM protein is degraded by light driving the cyclical nature of abundance and concomitant nuclear localization of the TIM-PER complex (Hunter-Ensor et al., 1996; Lee et al., 1996). Flies that possess the more stable form of TIM (L-TIM) have impaired light-dependent degradation of the PER-TIM-CRY complex; L-TIM flies are less responsive to short light pulses (Sandrelli et al., 2007; see below). Mutations in the *period* gene do not directly affect diapause in *D. melanogaster* (Saunders et al., 1989; Saunders, 1990), but as mentioned above, PER forms a complex with TIM, and *tim* is important for diapause (Tauber et al., 2007). The role of *tim* in diapause is discussed below.

Northern European populations of *D. melanogaster* enter diapause almost two months earlier than those from southern populations, even at the same low temperatures (Tauber et al., 2007; Kyriacou et al., 2008). Geographic variation in the incidence of diapause in the natural populations of *D. melanogaster* is mediated by differences in their light response as mediated by different forms of *tim* (Tauber et al., 2007). Rosato et al. (1997) described a natural polymorphism in *tim* in *D. melanogaster* populations. The *tim* gene encodes two protein products that differ in their amino acid length and are denoted as long (L-TIM) and short (S-TIM) (Peschel et al., 2006). Natural populations of flies with the *s-tim* allele have a single nucleotide deleted at position 294 and exclusively make the short form of the protein, whereas natural populations with the *ls-tim* allele lack this deletion and have two start codons so they make the long (L-TIM: 1,421 amino acids) and the short (S-TIM: 1,398 amino acids) forms of the protein (Tauber et al., 2007). In natural populations, flies with *ls-tim*, respond less to day length than flies with *s-tim* (Sandrelli et al., 2007). A positive association between the *s-tim* allele frequency and the proportion of flies that express diapause was found in natural populations from Europe (Tauber et al., 2007). Diapause is enhanced by the presence of L-TIM because the fly, bearing the *ls-tim* allele, is less light sensitive, is unable to respond to the effects of increased day length and continues to have immature ovaries (Sandrelli et al., 2007; Tauber et al., 2007). Finally, the interaction of *ls-tim* with a gene *Veela* that modifies clock function and mediates the action of *tim* in the light (Peschel et al., 2006), supports the idea that other clock genes may be involved in diapause.

During diapause induction in *D. melanogaster,* photoperiodic signals are transmitted to the tissues to block development and induce adult reproductive diapause (Sandrelli et al., 2007; Tauber et al., 2007; see figure 12.1). In *D. melanogaster,* young females undergo reproductive diapause under low temperatures and short day lengths and the development of their ovaries remains previtellogenic (Saunders et al., 1989). Early studies of insect diapause hypothesized the existence of a brain factor that transmitted an induction signal for the initiation of diapause (Lees, 1955). Where might this brain factor lie? In the drosophilid *Chymomyza costata,*

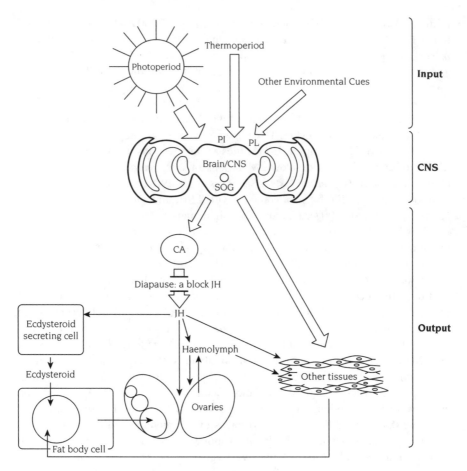

FIGURE 12.1.

Model of insect diapause based on models by Postlethwait and Shirk (1981), Allen (2007), and Zhu et al. (2008). During growth, development cues (predominantly photoperiod, but also thermoperiod and nutrition) trigger cells in the pars intercerebralis (PI), pars lateralis (PL), or subesophageal ganglion (SOG) to elicit the release of hormone(s). For example, in adults the release of juvenile hormone (JH) from the corpora allata (CA) promotes the development of ovaries. The growth promoting hormone may also increase the secretion of other compounds (e.g., ecdysteroids) or may promote lipogenesis in the fat body. By contrast, during diapause, photoperiodic cues perceived by TIM- and CRY-expressing cells (and other cues such as thermoperiod and nutrition) trigger cells in the brain to cause an inhibition of hormone release. For example, in adult diapause neurons in the PI or PL inhibit the secretion of insulin-like peptides and cause an inhibition of the CA release of JH. Thus diapause is maintained by the inhibition of factors that are important for growth and development and may be terminated by factors that promote development.

290

neurons that express TIM in the larval brain are important for the induction of larval diapause (Stehlik et al., 2008). *D. melanogaster* genes, such as *tim* and *cry*, are expressed in tissues that are part of the light perception mechanism important to the induction of diapause (Shiga and Numata, 2007). In *D. melanogaster*, CRY is expressed in the photoreceptor cells most associated with circadian photoreception: lateral neurons (LNvs) and dorsal neurons (DNs) (Emery, 2000). The LNvs and DNs connect to the cells of the pars intercerebralis (PI) region of the brain known to be important for circadian locomotor activity (Kaneko and Hall, 2000; Helfrich-Forster et al., 2001). In the Colorado potato beetle *Leptinotarsa decimlineata*, diapause is characterized by a reduction of juvenile hormone (JH) and the neurosecretory cells in the PI act on the corpora allata (CA), the structure that synthesizes JH, to affect the control of JH secretion (Vermunt et al., 1999; Shiga and Numata, 2007). The PI region of the brain is also important for adult diapause in other flies such as *Protophormia terranovae* (Shiga and Numata, 1997, 2000). Thus, the hypothesized brain factor important for the diapause-inducing photoperiodic signal may reside in the PI region of the brain.

We do not yet know what molecules are important for conveying the photoperiodic information from the brain to the effectors of diapause in *D. melanogaster*, but there are hints from other insects. The cell bodies of neurons that project to the PI and the pars lateralis (PL) are important for adult diapause in many insects (reviewed in Shiga and Numata, 2007). Severance of neurosecretory neurons from the PL in the brain of blow flies promotes ovary development during diapause, suggesting that the PL neurons inhibit the production of JH from the CA (Shiga and Numata, 2000). The PI neurons may electrically stimulate the CA and inhibit the synthesis of JH as has been demonstrated for diapause in *Locusta migratoria* (Horseman et al., 1994). In the male accessory glands of *Drosophila*, JH action is mediated by calcium and protein kinase C (Yamamoto et al., 1988). Thus, a calcium signaling pathway may be involved in the neural regulation of the CA and JH production. The CA has voltage-gated calcium channels, and it is thought that an increase in calcium levels stimulates JH production whereas the reduction in calcium levels inhibits JH production (Tobe, 1999). Thus, it appears that the perception of photoperiod is important for the induction of diapause and that the brain is involved in triggering diapause through hormonal mechanisms.

In our second example, we continue our discussion of the possible mechanisms involved in the transmission of the diapause induction signal by examining the induction of diapause in the silkworm. A factor from the head is essential for conveying the environmental signal triggering diapause in silkworms (Lees, 1955). In the silkworm *Bombyx mori* L., embryonic diapause is characterized by the delay between the perception of the environmental signals and the onset of diapause (Lees, 1955). Photoperiod and temperature are perceived by the mother while in the egg stage, and interestingly, it is the eggs of the subsequent generation that express diapause (Kogure, 1933, as cited in Andrewartha 1952; Lees, 1955). Thus, the sensitive stage for the induction of diapause is separated from the expression of diapause in the next generation. It is thought that hormones are important for conveying the photoperiodic information from the areas of perception to the ovaries. The *B. mori* diapause hormone (BomDH), a 24-amino-acid hormone that affects the maternal ovary of the silk worm, promotes embryonic diapause in the offspring (Hasegawa 1957; Yamashita, 1996; Imai et al., 1998). Thus, the mother's brain controls whether BomDH is secreted, and it is the release of a BomDH that determines whether diapausing or nondiapausing eggs are laid (Fukuda, as cited in Lees, 1955; Hasegawa, 1957). The gene that encodes BomDH is expressed in 12 cells in the subesophageal ganglion of the silkworm brain (Kawano et al., 1992; Sato et al., 1993, 1994). It is not yet known how the perception of the photoperiod in the brain signals the secretion of BomDH. However, BomDH is the hormonal factor from the head that is important for the induction of diapause in the silkworm. But BomDH's role in diapause may be unique to *B. mori* as application of BomDH in other insects does not induce diapause (Xu and Denlinger, 2003; Zhang et al., 2004).

Nevertheless, it is informative to consider how BomDH triggers diapause because the mechanisms involved in diapause induction share similarities with other insects. BomDH is produced from a pheromone biosynthesis activating neuropeptide (PBAN)-containing precursor protein, and it is the addition of an amide domain at the C-terminus that renders the hormone active (Sato et al., 1993; Imai, 1998; Yamashita et al., 1998). Only the amidated hormone is recognized by the receptor BmDHR, a seven-transmembrane G-protein–coupled receptor and thus only the amidated active hormone triggers the action of ovary trehalase and diapause in the developing embryo (Homma et al., 2006). Ovary trehalase converts sugars to glycogen, and when the mother's ovaries are exposed to BomDH, the ovary trehalase triggers embryonic diapause (Homma et al., 2006). The receptor for BomDH is expressed in the mother's ovaries, and the transcript level of diapause hormone receptor (DHR) is highest at the mid-pupal stage of development when the ovaries are most sensitive to the hormone (Ikeda et al., 1993; Homma et al., 2006). Interestingly, calcium and protein kinase C must be present for BomDH and DHR to stimulate ovary trehalase, which implies once again that calcium signaling might be involved in the induction of diapause (Homma et al., 2006).

Overall, the induction of egg diapause in *B. mori* and adult diapause in *D. melanogaster* share common characteristics. In both cases, the photoperiod information that cues the induction of diapause is transmitted to the brain and these brain regions (PI, PL, and subesophageal ganglion) then elicit the action of hormones that trigger diapause.

MAINTENANCE AND EMERGENCE FROM DIAPAUSE

When we compare across insect species, we find that certain molecular pathways are repeatedly involved in diapause. Diapause is not simply a general repression of all developmental pathways; it is appears to be an alternate pathway or set of pathways that enhances survival under stressful conditions (Tauber and Tauber, 1992; Denlinger, 2002). Below, we use the dynamics of diapause expression in pupal diapause of the flesh fly *Sarcophaga crassipalpis* to develop this idea. We also discuss molecules and pathways important for maintaining diapause and for coordinating ovary development and nutrition in the adult diapause of a variety of insects. The overall message is that adult diapause, like egg, larval and pupal diapause, is thought to be maintained by molecules that inhibit growth, and terminated by molecules important for the resumption of development.

A closer look at genes differentially expressed in diapause sheds some light on the particular pathways important during diapause. A subtractive hybridization screen for genes that showed differences in RNA abundance between brains of diapausing and nondiapausing *Sarcophaga crassipalpis* pupae identified four genes that were up-regulated and seven that were down-regulated during pupal diapause (Flannagan et al., 1998). Two of the four genes that were up-regulated during diapause have been identified. pScD14 *Heat shock protein* (*hsp23*) and *hsp70* (inducible) (Flannagan et al., 1998). Both are involved in stress resistance. Yocum et al. (1998) showed that transcription of *hsp23*, a small heat-shock protein, is up-regulated during pupal diapause of *S. crassipalpis*. In *D. melanogaster*, the expression of small heat-shock proteins is correlated with protection from thermal stress (Berger and Woodward, 1983) and cell cycle arrest (Ireland and Berger, 1982). In *S. crassipalpis,* the up-regulation of the small heat-shock proteins during diapause has been associated with cold hardiness, a process where stress resistant genes come into play (Joplin et al., 1990). (The consequences of diapause and diapause-associated traits are discussed in detail below.) Low expression of proliferating cell nuclear antigen (PCNA), a gene involved in cell proliferation, is associated with cell cycle arrest during the pupal diapause of *S. crassipalpis* (Tammariello and Denlinger, 1998). Interestingly, the levels of PCNA increase throughout the brain after the termination of pupal diapause by hexane treatment (Tammariello and Denlinger, 1998). Hexane is thought to break diapause in flesh flies by activating the mitogen-activated protein (MAP) kinase pathway (Fujiwara and Denlinger, 2007); thus, examination of specific genes in this pathway could further facilitate the mechanistic analysis of the maintenance and termination of diapause. Further, investigation of patterns of gene expression in the nervous system using microarray analyses will be particularly informative, as the insect brain has long been considered a center for diapause regulation (Denlinger, 2002).

The genes and proteins expressed during pupal diapause provide us with clues for elucidating molecular pathways that maintain and terminate diapause. Pupal diapause is affected by prothoracicotropic hormone (PTTH) released from the brain and is characterized by a reduction in ecdysteroid synthesis (Williams, 1946; Fraenkel and Hsaio, 1968; Denlinger, 1985). Like BomDH, *Helicoverpa amigera* diapause hormone (HarDH) is expressed in the subesophageal ganglion of the pharate adult brain and promotes ecdysteroid synthesis (Zhang et al., 2004). The DNA sequences of HarDH and the H-PBAN (the *H. amigera* pheromone biosynthesis activating neuropeptide) are homologues of PBAN, the *B. mori* diapause hormone precursor protein (Zhang et al., 2004). Similar PBAN peptides from *Spodoptera littoralis* activate MAP kinases and calcium signaling

(Zheng et al., 2007). In *Bombyx mori*, factors known to promote growth (e.g., ecdysteriods and certain kinases) are down-regulated in diapause and up-regulated when diapause is terminated (Fujiwara et al., 2006). In the goldenrod gall insect *Epiblema scudderiana*, a protein tyrosine phosphatase decreases adenosine 3',5'-cylic adenosine monophosphate (cAMP) and promotes diapause (Pfister and Storey, 2006). Thus, kinase pathways and calcium signaling are recognized as important players in reducing growth during diapause and in promoting growth at the end of diapause.

Natural variation in the incidence of *D. melanogaster* diapause maps to the *Drosophila Dp110* gene that encodes phosphoinositol 3-OH kinase (PI3K), a component of the insulin signaling pathway (Williams et al., 2006). PI3K overexpression in the nervous system inhibits diapause, implying that diapause termination and ovary development is promoted by the increase in PI3K mediated insulin signaling in the neurons (Williams et al., 2006). Reducing the gene dosage of PI3K promotes diapause, while increasing the expression of PI3K promotes ovary development. The ovarian diapause of *Drosophila* is an extension of the period of posteclosion immaturity in the adult fly (Saunders et al., 1989). As is typical of adult diapause in other species, the diapause in *D. melanogaster* may be broken by application of methoprene, a JH analog (Saunders et al., 1989, 1990; Saunders and Gilbert, 1990; Tatar et al., 2001b). The model of adult diapause as a block in JH secretion is relevant also for *D. melanogaster* since reduced insulin signaling affects JH synthesis (see figure 12.1; Tu et al., 2005). Future research should investigate if adult diapause in other insects is terminated by an increase in insulin signaling and if all members of the insulin pathway are involved in insect reproductive diapause as they appear to be in nematode dormancy (Tatar and Yin, 2001; Williams et al., 2006).

Insulin signaling coordinates nutrient availability with growth and is essential for development and growth in many organisms (Wu and Brown, 2006). In *D. melanogaster*, the regular progression of yolk deposition requires a functional insulin signaling pathway (Drummond-Barbosa and Spradling, 2001). Diapausing *D. melanogaster* females have ovaries that are previtellogenic, similar in many ways to the undeveloped ovaries characteristic of flies experiencing poor nutrition. Brain derived insulin-like peptides control the response of ovaries to changing nutrition (LaFever and Drummond-Barbosa, 2005). The insulin signaling pathway ties nutrition, feeding, and fat storage to reproduction in many organisms (reviewed by Hahn and Denlinger, 2007). Interestingly, flies in diapause have been reported to accumulate lipids (*D. triauraria*: Kimura et al., 1992; Ohtsu et al., 1992; *D. melanogaster*: Schmidt et al., 2005b). Furthermore females of *D. subauraria* and *D. triauraria* that have been in diapause for more than 12 days, show reduced feeding in comparison to nondiapausing females (Matsunaga et al., 1995). There is an association between immature ovaries and lipid accumulation in some insects, and intriguingly, in the cricket *Gryllus firmus* a reduction in ovarian growth is associated with lipid accumulation (Zera and Larsen, 2001; Zhao and Zera, 2002; Zera and Zhao, 2003). How lipid levels impact the induction, maintenance, and/or termination of diapause is a subject for future studies.

Feeding may play a role in the maintenance and termination of diapause by affecting the transcription of genes important for egg development in adult insects. Most striking is the effect of feeding on diapause in insects that mature their ovaries in response to the increased protein intake via a blood meal. Evidence for the role of food in the termination of diapause comes from studies on feeding and diapause in *Culex pipiens*, where protein intake is important for ovary maturation (Mitchell and Briegel, 1989; Robich and Denlinger, 2005). In some mosquitoes, a blood meal is necessary for the initiation of vitellogenesis (Riehle and Brown, 2002). Among these mosquitoes, proteins are important for the termination of reproductive arrest and the initiation of egg laying (Riehle and Brown, 2002). The target of rapamycin (TOR) pathway is important for the transduction of the protein signal to the fat body, where the vitellogenin genes are transcribed in mosquitoes (Hansen et al., 2004). The TOR and insulin signaling pathways in fruit flies are linked by key genes important for lipid metabolism such as the forkhead transcription factor FOXO (Luong et al., 2006). In the mosquito *Culex pipiens*, transcripts of genes involved in the metabolism of fatty acids persist in diapausing females longer than in the nondiapausing ones (Robich and Denlinger, 2005). The insulin signaling pathway has been shown to be important for diapause in *Culex pipiens*, through FOXO; this demonstrates a link between nutrition, fat storage, and diapause (Sim and Denlinger, 2008).

The connection between food response and photoperiodic response is intriguing because of the previously mentioned relationship between nutrition and the incidence of diapause (Danks, 1987; Denlinger and Tanaka, 1999). The connections between nutrition, resource accumulation, growth, and reproduction are clearest among species where diapause is related to food response, such as in lacewings (Tauber and Tauber, 1987, 1992). The mechanisms by which photoperiods control the response to food and reproduction in the *Crysopa carnea* lacewings is unknown. However, in some beetles, proteins important for diapause also affect feeding behavior and the transcription of genes promoting ovary development. Ohtsu et al. (2003) showed that ladybird beetles treated with chitinase had reduced oviposition relative to controls, whereas in other beetles, peptides with chitinase activity were important for diapause termination (Fujita et al., 2006). Chitinases are enzymes that hydrolyze the amino polysaccharide chitin (Merzendorfer and Zimoch, 2003). It is possible that chitinase proteins in the gut inhibit ovary development and maintain diapause by inducing a starvation-like response (Merzendorfer and Zimoch, 2003). Peptides with chitinase activity (active phase associated proteins [APAP I and II]) are critical for diapause termination in leaf beetles (*Gastrophysa atrocyanea*) (Fujita et al., 2006). Diapausing beetles cease feeding and have low levels of *apap1* transcripts in comparison with beetles that are prediapausing or reproductive (Fujita et al., 2006). Diapausing beetles injected with *apap1* interference RNA (RNAi) occasionally feed but do not undergo diapause termination in response to JH (Fujita et al., 2006). APAP I activity is essential for the JH-mediated termination of diapause (Fujita et al., 2006). Furthermore, chitinase proteins regulated by ecdysone promote growth (Merzendorfer and Zimoch, 2003), and as previously discussed, proteins that stimulate growth hormones and promote growth also terminate diapause.

Diapause may be maintained by compounds that inhibit molecules important for the continuation of growth. Adult diapause is terminated by JH. In *Leptinotarsa decimlineata* transcript levels of the *juvenile hormone esterase (jhe)* gene differed between fat bodies of beetles maintained in diapause inducing conditions and those kept under nondiapausing conditions (Vermunt et al., 1997a, 1997b, 1999). Similarly, in the fat bodies of diapausing *Riportus clavatus* bean bugs, the transcript and protein levels of the JH repressible *transferrin (RcTf)* were higher than in bugs treated with the JH analog methoprene (Hirai et al., 1998, 2000). Furthermore the hexamerin storage products that characterize the diapause of *Polistes metricus* may act in the maintenance of diapause by reducing the effects of JH (Hunt et al., 2007), as similar hexamerins in termites have been shown to repress the action of JH (Zhou et al., 2006). Thus, adult diapause, similar to egg, larval, and pupal diapause, is thought to be maintained by molecules that inhibit growth and terminated by molecules important for the resumption of development.

CONSEQUENCES OF DIAPAUSE

The effects of diapause expression on organismal behavior, physiology, life histories, and patterns of gene expression are many and varied; this reflects the complexity and variation in seasonal responses across taxa as well as the pleiotropic effects of underlying neuroendocrine signaling. Diapause induction is generally associated with reduced rates or lack of feeding, decreased mating and receptivity to mating, and reduced metabolic rates. The preparatory phase or onset of diapause may also be associated with migration and selection of dormancy sites (e.g., Urquhart and Urquhart, 1977). Thus, the onset of diapause is often preceded by increased energy acquisition and the accumulation of metabolic reserves (e.g., Ohtsu et al., 1992; Ding et al., 2003; Hahn and Denlinger, 2007; Khani et al., 2007); a subset of these compounds may also confer increased resistance to environmental stress (see below).

The consequences of diapause may also extend beyond its actual termination. Postdormancy effects on mating behavior are commonly observed (e.g., Hidaka, 1977), and diapause may also be associated with a subsequent decrease in individual fecundity (e.g., Denlinger, 1981; Bradshaw and Holzapfel, 1996; Kroon and Veenendaal, 1998). Such a decline in fecundity of postdiapause individuals has been called a "cost of diapause" (Leather et al., 1993) and may stem from the depletion of energy reserves and/or the consequences of hormonal changes that accompany the diapause syndrome. Furthermore, the duration and intensity of diapause are negatively correlated with postdiapause reproductive output (Matsuo, 2006). In *D. melanogaster*, for example, increasing the time spent in reproductive diapause resulted in a progressive decline in subsequent fecundity (Tatar et al., 2001a). Such a pattern has also been observed in the spider mite *Tetranychus kanzawai*. In this species, there exists substantial within-population genetic variance for diapause duration as well as postdiapause fecundity, and these traits exhibit significant and pronounced negative genetic correlations (Ito, 2007).

In contrast to its effects on reproduction, diapause expression commonly results in life span extension and a reduction in age-specific mortality rates (Tatar et al., 2001a). This has been extensively studied in *C. elegans*, in which mutations in genes of the dauer formation (*daf*) pathway result in delayed senescence and longer life span (see below). Similarly, mutations of homologous genes in the insulin–insulin-like signaling (IIS) pathway in *D. melanogaster* also prolong life and delay aging (Clancy et al., 2001; Tatar et al., 2001b; Hwangbo et al., 2004). The recent identification (Williams et al., 2006) of diapause regulation in *D. melanogaster* by *Dp110*, a component of the IIS pathway, suggests that the effects of diapause on life span and aging may also be mediated by insulin signaling (Tatar and Yin, 2001; Denlinger, 2002).

Mechanisms that facilitate an escape from adverse conditions also promote enhanced resistance to the variety of environmental stresses often experienced in seasonal environments. Accordingly, the impact of diapause on the stress response and underlying physiology has received much attention. Diapause generally results in a pronounced increase in cold hardiness, heat-stress tolerance, and desiccation resistance (Tauber et al., 1986). However, a variety of taxa are not characterized by such tolerance of environmental stress, such as those that avoid cold stress by migration (e.g., Herman, 1981) or selection of overwintering sites (e.g., Pitts and Wall, 2006). Cold hardiness can be generated by a variety of mechanisms (e.g., Lee and Denlinger, 1991; Danks, 2006). Taxa are generally classified as freezing tolerant or freezing resistant, depending on whether extracellular ice crystals are tolerated by the organism. While many arctic species are freezing tolerant (Danks, 2004), the majority of temperate taxa are freezing resistant (Tauber et al., 1986). The prevention of ice crystal formation may be conferred by the accumulation of a variety of antifreeze proteins (Duman, 2001; Walker et al., 2001) and/or cryoprotectants such as glycerol, sorbitol, other polyols, and trehalose (Lee, 1991; Koštál et al., 2007). Many of these compounds are regulated as a function of diapause or are at higher concentrations in diapausing versus nondiapausing individuals (e.g., Koštál et al., 2004; Bennett et al., 2005; Hunt et al., 2007; Khani et al., 2007). In addition to their role in supercooling, elevated concentrations of cryoprotectants can also affect response to hormones (Singtripop et al., 2002), enhance the effectiveness of antifreeze proteins (Duman and Serriani, 2002), and increase resistance to desiccation while in diapause (e.g., Danks, 2000; Bennett et al., 2001, 2005; Williams et al., 2004; Yoder et al., 2006). The elevated resistance to environmental stress as a function of diapause may also involve up-regulation of enzymes with antioxidant properties (e.g., Jovanovic-Galovic et al., 2007), remodeling of membrane phospholipids (Hodkova et al., 2002; Koštál et al., 2003; Michaud and Denlinger, 2006; Tomcala et al., 2006), other changes in metabolites or metabolic pools (Ding et al., 2003; Michaud and Denlinger, 2007), and the structure of the cuticle (e.g., Kelty and Lee, 2000).

Furthermore, patterns of gene and protein expression as a function of diapause have been elucidated in a variety of taxa. In *S. crassipalpis*, Joplin et al. (1990) examined the relative abundance of proteins expressed in the brains of diapausing and nondiapausing flesh flies. Assuming that the protein assays reflected global transcriptional patterns, diapause results in the down-regulation or silencing of an appreciable percentage of genes, whereas a relatively smaller number of genes are uniquely expressed in diapause.

Interestingly, some of the transcripts up-regulated or expressed as a function of diapause are directly related to a subset of the phenotypes that characterize diapause expression: life span extension, increased stress resistance, and reproductive quiescence. Ecdysteroids have been well characterized in the regulation of diapause (e.g., Denlinger, 1985), and the functional ecdysone receptor (a heterodimer of the ecdysone receptor, *EcR*, and ultraspiracle, *Usp*) changes in expression pattern as a function of diapause (Fujiwara et al., 1995; Rinehart et al., 2001). In addition to widespread effects on the diapause syndrome, such expression variation may influence life histories and resistance to stressors. In *D. melanogaster*, heterozygous *EcR* mutants are characterized by longer life span, delayed senescence, and elevated resistance to oxidative, heat, and starvation stress (Simon et al., 2003).

Similarly, a series of cellular chaperones also vary in expression pattern as a consequence of diapause. In *S. crassipalpis*, the inducible proteins Hsp23 and Hsp70 are up-regulated in diapause (Yocum et al., 1998; Rinehart et al., 2000; Li et al., 2007), Hsp90 is down-regulated (Rinehart and Denlinger, 2000), and the constitutive Hsc70 is expressed at relatively constant levels (Rinehart et al., 2000). The expression of these heat-shock proteins is not required for diapause expression but is required for resistance to cold while in diapause (Rinehart et al., 2007). Increased expression of Hsp70 during diapause has also been observed in *Leptinotarsa decemlineata* (Yocum, 2001), and Hsp70 levels increase following temperature stress exposure in diapausing *D. triauraria*. Alternatively, levels of Hsp90 are elevated in diapause in *Delia antiqua* (Chen et al., 2005b) and *Chilo supressalis* (Sonoda et al., 2006a). While heat-shock protein levels do not change as a function of diapause in some taxa (e.g., Goto and Kimura, 2004; Tachibana et al., 2005; Rinehart et al., 2006), these data suggest a widespread role of chaperones in diapause-associated stress resistance. Presumably, the positive association between expression level and stress resistance is directly due to the functional role of heat-shock protein (and other chaperones, e.g., the DaTCP-1 chaperonin in *D. antiqua*; Kayukawa et al., 2005) in protein binding and the maintenance of protein conformation during and subsequent to stress exposure. Elevated levels of heat-shock proteins may also directly influence life histories exhibited in diapause, as increasing Hsp70 copy number in *D. melanogaster* has been shown to extend life and delay senescence (Tatar et al., 1997).

In addition to cellular chaperones, enhanced resistance to stress during diapause may be associated with the up-regulation of other genes. The acquisition of elevated resistance to cold may be associated with up-regulation of genes involved in actin dynamics (Kayukawa et al., 2005; Kim et al., 2006); the reorganization of polymerized actin or actin monomers may contribute to stress resistance by cytoskeletal fortification. Changes in lipid membranes have been hypothesized as central for resistance to cold induced injury, and the up-regulation of genes encoding enzymes that increase the relative proportion of unsaturated fatty acids may contribute to cold resistance while in diapause (Kayukawa et al., 2007). Given the established links between cryoprotectants and diapause-associated phenotypes, the increased activity of enzymes involved in polyol metabolism (Koštál et al., 2004) or genes involved in polyol synthesis (Pfister and Storey, 2006) also occur as a consequence of diapause expression. Similarly, changes in the expression of antifreeze proteins (Qin et al., 2007) and hexamerin storage proteins (Sonoda et al., 2006b, 2007) are also associated with diapause expression. While these compounds may have direct effects on phenotypic expression while in diapause, they may also have indirect effects through interactions with hormone signaling (e.g., Zhou et al., 2006; Hunt et al., 2007).

In summary, the expression of any form of diapause has widespread and predictable effects on many aspects of biological function, ranging from patterns of gene expression to organismal persistence. While the diapause syndrome is highly variable among taxa, several unifying themes and mechanisms have emerged. Commonalities such as hormonal mediation, insulin signaling, and the reliance on metabolic pools suggest that future research on diapause will facilitate the understanding of dormancy in other organisms.

MOLECULAR MECHANISMS OF DORMANCY IN OTHER ORGANISMS

Mechanisms involved in dormancy in organisms other than insects can also provide us with candidate genes and pathways important to diapause in insects. For example, genes important for coordinating growth, nutrition, and life span have been shown to be important in dauer formation, a type of dormancy in nematodes. Molecular processes integral to reducing metabolism and increasing stress tolerance are important to hibernation in mammals. Genes and pathways involved in mammalian embryonic diapause may be important in the seasonal coordination of reproduction characteristic of diapause in some insects.

Many of the genes important for the dauer developmental arrest of *Caenorhabditis elegans* have been identified (see table 12.1). Dauer formation is induced by environmental factors such as pheromones, food, temperature, and population density. The enduring dauer state occurs at stage 3 of nematode larval development and is, like diapause, an alternate developmental pathway (Riddle and Albert, 1997). The life cycle of *C. elegans* proceeds throughout four larval stages to adulthood unless the factors that trigger dauer formation (lack of food, particular pheromones, or high temperature) are present (Riddle, 1988). Dauer larvae are more resistant to stress, do not feed or store lipids, and are morphologically recognizable as being thin, with cuticle striations and a constricted pharynx

TABLE 12.1. Genes known to play a role in various types of dormancies.

Molecular function or biological process	GO term	Diapause gene name	Diapause stage	Species	References
Circadian rhythm	GO:0007623	timeless	Adult	Drosophila melanogaster	Sandrelli et al. (2007), Tauber et al. (2007)
		Clock (Clk)	Adult	Pyrrhocoris apterus	Syrova et al. (2003)
		period	Adult	Pyrrhocoris apterus	Hodkova et al. (2003), Dolezel et al. (2007)
		Bombyx sorbitol dehydrogenase	Egg	Bombyx mori	Niimi et al. (1993)
		COX 1	Pupae	Agrius convolvuli	Uno et al. (2004)
		fatty acid synthase	Adult	Culex pipiens	Robich and Denlinger (2005)
		60S ribosomal protein AP endonuclease	Pupae	Sarcophaga crassipalpis	Craig and Denlinger (2000)
		Ommochrome binding protein-like (obp1 and obp2)	Larval	Ostrinia nubilalis	Coates et al. (2005)
		Leukemia inhibitory factor receptor	Blastocyst implantation	Spilogale putorius latifrons	Hirzel et al. (1999), Passavant et al. (2000)
		daf-1	Dauer	Caenorhabditis elegans	Georgi et al. (1990)
		daf-4	Dauer	Caenorhabditis elegans	Estevez et al. (1993)
		daf-11	Dauer	Caenorhabditis elegans	Birnby et al. (2000), Riddle and Albert (1997)
		Daf-16 forkhead transcription factor P21Cip1	Dauer / Blastocyst implantation	Caenorhabditis elegans / Mus musculus	Ogg et al. (1997) / Hamatani et al. (2004)
Enzyme-linked receptor protein signaling pathway	GO:0007167	daf-9 cytochrome P450	Dauer	Caenorhabditis elegans	Gerisch and Antebi (2004)
		daf-7	Dauer	Caenorhabditis elegans	Ren et al. (1996)

	Dp110, phosphatidylinositol 3-kinase	Adult	*Drosophila melanogaster*	Williams et al. (2006)
	age-1, phosphatidylinositol 3- kinase	Dauer	*Caenorhabditis elegans*	Morris et al. (1996)
	PDK1, phosphoinositide dependent kinase	Dauer	*Caenorhabditis elegans*	Paradis et al. (1999)
	Insulin receptor (Daf-2)	Dauer	*Caenorhabditis elegans*	Antebi et al. (2000), Kimura et al. (1997)
G-protein–coupled receptor signaling pathway GO:0007186	ERK/p38 MAPK	Egg	*Atrachya menetriesi*	Kidokoro et al. (2006)
	ERK/p38 MAPK	Egg	*Bombyx mori*	Fujiwara et al. (2006)
	Pheromone biosynthesis activating neuropeptide	Egg	*Bombyx mori*	Kawano et al. (1992)
	Pyrokinin/PBAN	Pupae	*Helicoverpa armigera*	Sun et al. (2005)
	PBAN	Pupae	*Manduca sexta*	Xu and Denlinger (2004)
	sNPF I and II	Adult	*Leptinotarsa decemlineata*	Huybrechts et al. (2004)
	gpa-2 and -3	Dauer	*Caenorhabditis elegans*	Zwaal et al. (1997)
Cellular lipid metabolic process GO:0044255	Acyl-CoA desaturase homolog	Egg	*Bombyx mori*	Yoshiga et al. (2000)
Receptor binding GO:0005102	Juvenile hormone esterase (JHE)	Adult	*Leptinotarsa decemlineata*	Vermunt et al. (1999)
	APAP I and II	Adult	*Gastrophysa atrocyanea*	Fujita et al. (2006)
	Prothoracicotrophic hormone	Pupae	*Helicoverpa armigera*	Wei et al. (2005)
Response to stress GO:0006950	pScD14 Heat shock protein 23	Pupae	*Sarcophaga crassipalpis*	Flannagan et al. (1998), Yocum et al. (1998)
	Heat shock protein 90	Pupae	*Sarcophaga crassipalpis*	Rinehart and Denlinger (2000)
	Hsp70 (inducible)	Pupae	*Sarcophaga crassipalpis*	Rinehart et al. (2000)

(Continued)

301

TABLE 12.1. (*Continued*)

Molecular function or biological process	GO term	Diapause gene name	Diapause stage	Species	References
		Proliferating-cell nuclear antigen	Pupae	*Sarcophaga crassipalpis*	Tammariello and Denlinger (1998)
Serine-type peptidase activity	GO:0008236	*DaTrypsin*	Pupae	*Delia antiqua*	Chen et al. (2005a)
		chymotrypsin-like serine protease	Adult	*Culex pipiens*	Robich and Denlinger (2005)
		trypsin	Adult	*Culex pipiens*	Robich and Denlinger (2005)
Cellular iron ion homeostasis	GO:0006879	*RcTf* (*Transferrin*)	Adult	*Riptortus clavatus*	Hirai et al. (2000)
Voltage-gated potassium channel activity	GO:0005249	*tax-2 and -4*	Dauer	*Caenorhabditis elegans*	Coburn and Bargmann (1996), Komatsu et al. (1996)

Genes taken from research on insect diapause, dauer formation in *Caenorhabditis elegans*, mammalian hibernation, and blastocyst implantation. The annotation was performed on homologs or orthologs in *Drosophila melanogaster*, and clustering was done using the Affymetrix or Flybase gene ID numbers using David Annotation Clustering (Dennis et al., 2003).

(Riddle and Albert, 1997). Genes involved in dauer formation fall into two main signaling pathways, the transforming growth factor β (TGF-β) pathway and the insulin signaling pathways. Both are important for the perception of environmental triggers, transduction of the inducing stimuli, and the control of transcription (Riddle et al., 1981; Riddle and Albert, 1997). Both the TGF-β and insulin signaling pathways act through the hormone receptor DAF-12 and affect growth, dauer formation, and life span (Antebi et al., 2000; Gerisch et al., 2001). Mutations in the insulin receptor homolog, *age-1* phosphoinositide-3 kinase, of *akt-1* and *akt-2*, akt/ PKB homologs, cause dauer arrest of development (Morris et al., 1996; Kimura et al., 1997; Paradis and Ruvkun, 1998). These components of the insulin signaling pathway transduce insulin signals from the membrane to DAF-16, a Forkhead transcription factor (Ogg et al., 1997; Paradis and Ruvkun, 1998). Thus, the mecha nisms underlying dauer formation point to the involvement of the insulin signaling pathway in coordinating growth, nutrition, and life span in the worm. This signaling pathway is also known to play a role in insect diapause as discussed above.

Mammalian hibernation is a characteristic response to winter conditions. It is a period of dormancy in which the core temperature of the animal drops and the metabolic rate is reduced with no adverse affects on the cells (Storey and Storey, 2007). As hibernators survive prolonged hypothermia and a lack of oxygen, the molecular mechanisms that protect the animal from cell damage can provide us with clues for what might be similarly protective during insect diapause. One common mechanism by which cell damage is prevented is by phosphorylation of the enzymes required for glycolysis that results in a depression of the metabolic rate (reviewed in Storey and Storey, 2004; Storey, 2005). In the 13-lined ground squirrel, phosphorylation of AKT (protein kinase B) is reduced during hibernation (Cai et al., 2004). Chemicals that reduce phosphorylation may also cause a reduction in metabolic rate and body temperature, as was found for the hydrogen sulfide induction of suspended animation in mice (Blackstone et al., 2005). Interestingly, in hibernating mammals, glycolytic enzymes play a role in the switch from using carbohydrates for metabolism to using lipids (Storey and Storey, 2004; Dark, 2005). Two of the genes down-regulated during hibernation (reviewed in Storey and Storey, 2007) are *insulin growth factor binding protein (IGFBP)* (Schmidt and Kelley, 2001) and *pyruvate dehydrogenase kinase (PDK)* (Andrews et al., 1998). Both are involved in the insulin signaling pathway, a pathway already known to be important for the metabolism of lipids and dormancy in other organisms (Saltiel and Kahn, 2001; Storey and Storey, 2004).

Components of the insulin signaling pathway have also been implicated in the control of cell cycle arrest important to mammalian embryonic diapause (Lopes et al., 2004). Embryonic diapause in mammals is an arrest of preimplantation development where the cells of the embryo enter a quiescent state where proliferation is virtually halted until arrest is terminated. Embryonic diapause is terminated when favorable reproductive conditions have returned (Renfree and Shaw, 2000). Temperature is known to be important for delayed implantation in bats and long photoperiods are important for implantation in the spotted skunk (Mead, 1993).

How do these environmental cues signal the change to embryonic diapause at the blastocyst stage? A photoperiod responsive factor such as melatonin is thought to be involved in the transmission of the environmental signals with important subsequent effects on prolactin and embryonic diapause (Mead, 1993; Lincoln et al., 2003). In the western spotted skunk *Spilogale putorius latifrons*, the leukemia inhibitory factor (LIF) receptor is up-regulated by prolactin, and LIF transcripts in mice are required for ending the mitotic arrest associated with delayed embryo implantation (Stewart et al., 1992; Hirzel et al., 1999; Paria et al., 2001). Cytokines (e.g., LIF) were shown to be important in preparing the uterus for the implantation of mouse embryos (Daikoku et al., 2004). However in a microarray study of dormant and activated mouse embryos the strongest differences between the two types of embryos occurred among the transcripts of genes associated with cell cycle arrest (Hamatani et al., 2004). For example, there were significantly more *p21* transcripts among embryos that were dormant by comparison to those that were activated (Hamatani et al., 2004). Lopes et al. (2004) proposed that, given that a forkhead transcription factor (FOXO) increases p21 in *C. elegans* and high expression of FOXO in mammalian cells triggers cell cycle arrest, there may be a role for FOXO and p21 in the cell cycle arrest associated with embryonic diapause in mammals, but this remains to be determined. Wang and Dey (2006) showed that although the molecular mechanisms that underlie delayed implantation in mammals have not yet been fully elucidated, specific pathways important for embryo diapause have been identified (e.g., cytokine growth factor, cell cycle arrest, MAPK, and phosphoinositide- 3-kinase/Ca^{2+} signaling pathways). In summary, studies of dormancies such as nematode dauer formation, mammalian hibernation, and delayed mammalian embryo implantation provide us with candidate genes and pathways for insect diapause.

CONCLUSIONS AND FUTURE DIRECTIONS

Diapause is a complex phenomenon in which circadian information is transmitted to output tissues triggering changes in physiology and behavior. Table 12.1 shows many of the genes important for diapause, hibernation, and preimplantation arrest. The circadian rhythm pathway was identified when the genes were clustered based on their molecular or biological function. The functional groups identified in table 12.1 suggest the possibility that the circadian pathway is linked to output pathways by genes important for metabolism, nutrient sensing, and energy utilization. Interestingly, nutrient acquisition has the effect of resetting circadian rhythms in mice (Froy and Miskin, 2007). Future studies should investigate links between circadian and metabolic pathways in insects. Furthermore, the genes in table 12.1 could serve to coordinate nutrition, energy, growth, and reproduction and affect cellular functions important for stress resistance perhaps by linking metabolic and circadian cycles as suggested by Tu and McKnight (2006). Circadian phenomena, metabolism, and nutrition are also linked in mammalian hibernation

(Andrews, 2004). There is also an intriguing link between nutrition, diapause, and stress resistance in paper wasps where differential nutrition determines whether a larva develops into a worker or a reproductive wasp (Hunt et al., 2007). In these wasps, a diapause pathway characterized by the accumulation of storage protein serves to bias development from the worker pathway to the pathway of the longer lived, more stress resistant, reproductive gyne (Hunt et al., 2007).

Genes important for stress resistance have been identified through proteomic studies on diapause in insects (Li et al., 2007) and in brine shrimp (Wang et al., 2007). These proteomic studies substantially support the picture of diapause as an alternate developmental pathway, as discussed above. The picture emerging is one in which particular genetic or signaling pathways are used across a variety of species. Future investigations of the genes and pathways integral to the induction, maintenance, and termination of diapause would be greatly benefited by microarrays, proteomic, and metabolomic analyses as have been employed in other research areas such as aging (e.g., Pletcher et al., 2002) and immunity (e.g., Levy et al., 2004).

This review suggests that diapause pathways are often shared between divergent species. Thus, results from studies of the molecular mechanisms of diapause in model genetic organisms such as *D. melanogaster* can inform us about the genes important for diapause in insects that are food crop pests or that carry human pathogens. For example, in *Cydia pomonella* moths that infest Bartlett pears, an increase in the day length that induces diapause was found in moths that have become resistant to insecticide (Boivin et al., 2004; van Steenwyk et al., 2004). Similarly, prolonged diapause among the root worm pest *Diabrotica barberi* affects its resistance to *Bacillus thuringiensis* (Bt) corn (Mitchell and Onstad, 2005). Understanding the mechanisms that moths use to measure day length and rootworms use to maintain diapause may help explain the increases in stress resistance resulting from diapause. This will ultimately aid in both the development of products and biological control measures that can be used to control these pests.

Some diapause pathways important for growth and development are known to be important in human growth, for example, the insulin signaling pathway. Human cells important for insulin signaling arrest development and can be considered to show signs of "diapause." They switch from carbohydrate to lipid metabolism and show a decreased response to agents that increase cAMP and a reduction in their insulin signaling (Buteau et al., 2007). Notably, increased FOXO in the pancreatic beta cells triggers this "metabolic arrest" by increasing transcription of genes involved in lipid metabolism (Buteau et al., 2007). It would be of interest to determine whether human "metabolic arrest" is affected by photoperiod as is insect diapause.

Overall, what can be gleaned from recent discoveries of genes and molecular pathways involved in insect diapause? The molecular basis of diapause in insects involves signal transduction pathways important for transmitting the induction signal, pathways integral to development, and pathways important for stress resistance. Genes involved in the perception of time are involved in the onset of diapause. Some of them are developmental genes whose molecules function in the perception and transmission of photoperiodic signals. However, diapause is not simply the repression of genes required for normal development. Some genes are more highly expressed in diapause, while others are down-regulated. Of the genes and proteins involved in the maintenance of diapause, some play a role in stress resistance pathways, while others are involved in developmental and cellular arrests. It is fascinating to speculate how genes involved in all of these processes might have been "recruited" during the evolution of insect diapause. To be sure, diapause is a physiological escape from hazardous conditions. Interestingly, molecules associated with dormancies in noninsect species are also associated with stress resistance and delayed development and reproduction and therefore provide us with candidate genes for future studies of insect diapause. Future research on the mechanisms of diapause will help us understand the function and evolution of this fascinating overwintering strategy in insects.

Acknowledgments   This chapter is dedicated to the memory of our dear friend Lionel Peypelut, whose research interests lay in the area of insect diapause. We also thank Bianco Marco for drawing the figure, Clement Kent for suggesting the use of the David clustering tool, and Hiwote Belay for comments on the manuscript. Research was funded by a Discovery Grant from the Natural Sciences and Engineering Research Council of Canada to M.B.S.

References

Allen MJ. (2007). What makes a fly enter diapause? Fly 1: 307–310.

Andrewartha HG. (1952). Diapause in relation to the ecology of insects. Biol Rev Cambr Philos Soc 27: 50–107.

Andrews MT. (2004). Genes controlling the metabolic switch in hibernating mammals. Biochem Soc Trans 32: 1021–1024.

Andrews MT, Squire TL, Bowen CM, and Rollins MB. (1998). Low-temperature carbon utilization is regulated by novel gene activity in the heart of a hibernating mammal. Proc Natl Acad Sci USA 95: 8392–8397.

Antebi A, Yeh WH, Tait D, Hedgecock EM, and Riddle DL. (2000). daf-12 encodes a nuclear receptor that regulates the dauer diapause and developmental age in *C. elegans*. Genes Dev 14: 1512–1527.

Bennett VA, Lee RE, and Kukal O. (2001). Abiotic and biotic factors affecting water loss rates in the polar desert caterpillar *Gynaephora groenlandica* (Lepidoptera : Lymantriidae) and the temperate caterpillar, *Pyrrharctia isabella* (Lepidoptera : Arctiidae). Am Zool 41: 1388–1389.

Bennett VA, Sformo T, Walters K, Toien O, Jeannet K, Hochstrasser R, Pan Q, Serianni AS, Barnes BM, and Duman JG. (2005). Comparative overwintering physiology of Alaska and Indiana populations of the beetle *Cucujus clavipes* (Fabricius): Roles of antifreeze proteins, polyols, dehydration and diapause. J Exp Biol 208: 4467–4477.

Berger EM and Woodward MP. (1983). Small heat-shock proteins in *Drosophila* may confer thermal tolerance. Exp Cell Res 147: 437–442.

Birnby DA, Link EM, Vowels JJ, Tian H, Colacurcio PL, and Thomas JH. (2000). A transmembrane guanylyl cyclase (DAF-11) and Hsp90 (DAF-21) regulate a common set of chemosensory behaviors in *Caenorhabditis elegans*. Genetics 155: 85–104.

Blackstone E, Morrison M, and Roth MB. (2005). H2S induces a suspended animation-like state in mice. Science 308: 518–518.

Boivin T, Bouvier JC, Beslay D, and Sauphanor B. (2004). Variability in diapause propensity within populations of a temperate insect species: Interactions between insecticide resistance genes and photoperiodism. Biol J Linnean Soc 83: 341–351.

Bradshaw WE. (1976). Geography of photoperiodic response in diapausing mosquito. Nature 262: 384–386.

Bradshaw WE and Holzapfel CM. (1996). Genetic constraints to life-history evolution in the pitcher-plant mosquito, *Wyeomyia smithii*. Evolution 50: 1176–1181.

Buteau J, Shlien A, Foisy S, and Accili D. (2007). Metabolic diapause in pancreatic beta-cells expressing a gain-of-function mutant of the forkhead protein Foxo1. J Biol Chem 282: 287–293.

Cai D, McCarron RM, Yu EZ, Li Y, and Hallenbeck J. (2004). Akt phosphorylation and kinase activity are down-regulated during hibernation in the 13-lined ground squirrel. Brain Res 1014: 14–21.

Carson HL and Stalker HD. (1948). Reproductive diapause in *Drosophila robusta*. Proc Natl Acad Sci USA 34: 124–129.

Ceriani MF, Darlington TK, Staknis D, Mas P, Petti AA, Weitz CJ, and Kay SA. (1999). Light-dependent sequestration of TIMELESS by CRYPTOCHROME. Science 285: 553–556.

Chen B, Kayukawa T, Jiang H, Monteiro A, Hoshizaki S, and Ishikawa Y. (2005a). DaTrypsin, a novel clip-domain serine proteinase gene up-regulated during winter and summer diapauses of the onion maggot, *Delia antiqua*. Gene 347: 115–123.

Chen B, Kayukawa T, Monteiro A, Ishikawa Y. (2005b). The expression of the HSP90 gene in response to winter and summer diapause and thermal-stress in the onion maggot, *Delia antiqua*. Insect Mol Biol 14: 697–702.

Clancy DJ, Gems D, Harshman LG, Oldham S, Stocker H, Hafen E, Leevers SJ, and Partridge L. (2001). Extension of life-span by loss of CHICO, a *Drosophila* insulin receptor substrate protein. Science 292: 104–106.

Coates BS, Hellmich RL, and Lewis LC. (2005). Two differentially expressed ommochrome-binding protein-like genes (obp1 and obp2) in larval fat body of the European corn borer, *Ostrinia nubilalis*. J Insect Sci 5: 19.

Coburn CM and Bargmann CI. (1996). A putative cyclic nucleotide-gated channel is required for sensory development and function in *C. elegans*. Neuron 17: 695–706.

Craig TL and Denlinger DL. (2000). Sequence and transcription patterns of 60S ribosomal protein P0, a diapause-regulated AP endonuclease in the flesh fly, *Sarcophaga crassipalpis*. Gene 255: 381–388.

Daikoku T, Song H, Guo Y, Riesewijk A, Mosselman S, Das SK, and Dey SK. (2004). Uterine Msx-1 and Wnt4 signaling becomes aberrant in mice with the loss of leukemia inhibitory factor or Hoxa-10: Evidence for a novel cytokine-homeobox-Wnt signaling in implantation. Mol Endocrinol 18: 1238–1250.

Danks HV. (1987). *Insect Dormancy: An Ecological Perspective*. Ottawa: Biological Survey of Canada.

Danks HV. (2000). Dehydration in dormant insects. J Insect Physiol 46: 837–852.

Danks HV. (2004). Seasonal adaptations in arctic insects. Integr Comp Biol 44: 85–94.

Danks HV. (2006). Insect adaptations to cold and changing environments. Can Entomol 138: 1–23.

Dark J. (2005). Annual lipid cycles in hibernators: Integration of physiology and behavior. Annu Rev Nutr 25: 469–497.

Denlinger DL. (1981). Basis for a skewed sex-ratio in diapause-destined flesh flies. Evolution 35: 1247–1248.

Denlinger DL. (1985). Hormonal control of diapause. In *Comprehensive Insect Physiology, Biochemistry and Pharmacology* (GA Kerkut and LI Gilbert, eds). Oxford: Pergamon Press, pp. 353–412.

Denlinger DL. (2002). Regulation of diapause. Annu Rev Entomol 47: 93–122.

Denlinger DL and Tanaka S. (1999). Diapause. In *Encyclopedia of Reproduction* (E Knobil and JD Neill, eds). San Diego: Academic Press, pp. 863–872.

Dennis G, Sherman BT, Hosack DA, Yang J, Gao W, Lane HC, and Lempicki RA. (2003). DAVID: Database for annotation, visualization, and integrated discovery. Genome Biol 4: P3.

Ding L, Li Y, and Goto M. (2003). Physiological and biochemical changes in summer and winter diapause and non-diapause pupae of the cabbage armyworm, *Mamestra brassicae* L. during long-term cold acclimation. J Insect Physiol 41153–1159.

Dolezel D, Sauman I, Koštál V, and Hodkova M. (2007). Photoperiodic and food signals control expression pattern of the clock gene, *period*, in the linden bug, *Pyrrhocoris apterus*. J Biol Rhythms 22: 335–342.

Drummond-Barbosa D and Spradling AC. (2001). Stem cells and their progeny respond to nutritional changes during *Drosophila* oogenesis. Dev Biol 231: 265–278.

Duman JG. (2001). Antifreeze and ice nucleator proteins in terrestrial arthropods. Annu Rev Physiol 63: 327–357.

Duman JG and Serianni AS. (2002). The role of endogenous antifreeze protein enhancers in the hemolymph thermal hysteresis activity of the beetle *Dendroides canadensis*. J Insect Physiol 48: 103–111.

Dunlap JC. (1999). Molecular bases for circadian clocks. Cell 96: 271–290.

Emery P, Stanewsky R, Helfrich-Forster C, Emery-Le M, Hall JC, and Rosbash M. (2000). *Drosophila* CRY is a deep brain circadian photoreceptor. Neuron 26: 493–504.

Estevez M, Attisano L, Wrana JL, Albert PS, Massague J, and Riddle DL. (1993). The daf-4 gene encodes a bone morphogenetic protein receptor controlling *C. elegans* dauer larva development. Nature 365: 644–649.

Flannagan RD, Tammariello SP, Joplin KH, Cikra-Ireland RA, Yocum GD, and Denlinger DL. (1998). Diapause-specific gene expression in pupae of the flesh fly *Sarcophaga crassipalpis*. Proc Natl Acad Sci USA 95: 5616–5620.

Fraenkel G and Hsiao C. (1968). Morphological and endocrinological aspects of pupal diapause in a fleshfly *Sarcophaga argyrostoma*. J Insect Physiol 14: 707–718.

Froy O and Miskin R. (2007). Interrelations among feeding, circadian rhythms and ageing. Prog Neurobiol 82: 142–150.

Fujita K, Shimomura K, Yamamoto K, Yamashita T, and Suzuki K. (2006). A chitinase structurally related to the glycoside hydrolase family 48 is indispensable for the hormonally induced diapause termination in a beetle. Biochem Biophys Res Commun 345: 502–507.

Fujiwara H, Jindra M, Newitt R, Palli SR, Hiruma K, and Riddiford LM. (1995). Cloning of an ecdysone receptor homolog from *Manduca sexta* and the developmental profile of its mRNA in wings. Insect Biochem Mol Biol 25: 881–897.

Fujiwara Y and Denlinger DL. (2007). High temperature and hexane break pupal diapause in the flesh fly, *Sarcophaga crassipalpis*, by activating ERK/MAPK. J Insect Physiol 53: 1276–1282.

Fujiwara Y, Tanaka Y, Iwata K, Rubio RO, Yaginuma T, Yamashita O, and Shiomi K. (2006). ERK/MAPK regulates ecdysteroid and sorbitol metabolism for embryonic diapause termination in the silkworm, *Bombyx mori*. J Insect Physiol 52: 569–575.

Georgi LL, Albert PS, and Riddle DL. (1990). daf-1, a *C. elegans* gene controlling dauer larva development, encodes a novel receptor protein kinase. Cell 61: 635–645.

Gerisch B and Antebi A. (2004). Hormonal signals produced by DAF-9/cytochrome P450 regulate *C. elegans* dauer diapause in response to environmental cues. Development 131: 1765–1776.

Gerisch B, Weitzel C, Kober-Eisermann C, Rottiers V, and Antebi A. (2001). A hormonal signaling pathway influencing *C. elegans* metabolism, reproductive development, and life span. Dev Cell 1: 841–851.

Geyspits KF and Simonenko NP. (1970). An experimental analysis of seasonal changes in the photoperiodic reaction of *Drosophila phalerata* Meig. (Diptera Drosphilidae). Entomol Rev 49: 46–54.

Goto SG and Kimura MT. (2004). Heat-shock-responsive genes are not involved in the adult diapause of *Drosophila triauraria*. Gene 326: 117–122.

Hahn DA and Denlinger DL. (2007). Meeting the energetic demands of insect diapause: Nutrient storage and utilization. J Insect Physiol 53: 760–773.

Hall JC. (2000). Cryptochromes: Sensory reception, transduction, and clock functions subserving circadian systems. Curr Opin Neurobiol 10: 456–466.

Hall JC. (2003). Genetics and molecular biology of rhythms in *Drosophila* and other insects. Adv Genet 48: 1–280.

Hamatani T, Daikoku T, Wang H, Matsumoto H, Carter MG, Ko MS, and Dey SK. (2004). Global gene expression analysis identifies molecular pathways distinguishing blastocyst dormancy and activation. Proc Natl Acad Sci USA 101: 10326–10331.

Hansen IA, Attardo GM, Park JH, Peng Q, and Raikhel AS. (2004). Target of rapamycin-mediated amino acid signaling in mosquito anautogeny. Proc Natl Acad Sci USA 101: 10626–10631.

Hardin PE. (2005). The circadian timekeeping system of *Drosophila*. Curr Biol 15: R714–R722.

Hasegawa K. (1957). Diapause hormone of the silkworm, *Bombyx mori*. Nature 179: 1300–1301.

Helfrich-Forster C, Winter C, Hofbauer A, Hall JC, and Stanewsky R. (2001). The circadian clock of fruit flies is blind after elimination of all known photoreceptors. Neuron 30: 249–261.

Herman WS. (1981). Studies on the adult reproductive diapause of the monarch butterfly, Biol Bull 160: 89–106.

Hidaka T (1977). Ucographical differentiation of rice gall midge, *Orseolia oryza* (Wood-Mason) (Diptera: Cecidomyiidae). Appl Entomol Zool 12: 4–8.

Hirai M, Watanabe D, and Chinzei Y. (2000). A juvenile hormone-repressible transferrin-like protein from the bean bug, *Riptortus clavatus*: CDNA sequence analysis and protein identification during diapause and vitellogenesis. Arch Insect Biochem Physiol 44: 17–26.

Hirai M, Yuda M, Shinoda T, and Chinzei Y. (1998). Identification and cDNA cloning of novel juvenile hormone responsive genes from fat body of the bean bug, *Riptortus clavatus* by mRNA differential display. Insect Biochem Mol Biol 28: 181–189.

Hirzel DJ, Wang J, Das SK, Dey SK, and Mead RA. (1999). Changes in uterine expression of leukemia inhibitory factor during pregnancy in the western spotted skunk. Biol Reprod 60: 484–492.

Hodkova M, Berkova P, and Zahradnickova H. (2002). Photoperiodic regulation of the phospholipid molecular species composition in thoracic muscles and fat body of *Pyrrhocoris apterus* (Heteroptera) via an endocrine gland, corpus allatum. J Insect Physiol 48: 1009–1019.

Hodkova M, Syrova Z, Dolezel D, and Sauman I. (2003). *Period* gene expression in relation to seasonality and circadian rhythms in the linden bug, *Pyrrhocoris apterus* (Heteroptera). Eur J Entomol 100: 267–273.

Homma T, Watanabe K, Tsurumaru S, Kataoka H, Imai K, Kamba M, Niimi T, Yamashita O, and Yaginuma T. (2006). G protein-coupled receptor for diapause hormone, an inducer of *Bombyx* embryonic diapause. Biochem Biophys Res Commun 344: 386–393.

Horseman G, Hartmann R, Virant-Doberlet M, Loher W, and Huber F. (1994). Nervous control of juvenile hormone biosynthesis in *Locusta migratoria*. Proc Natl Acad Sci USA 91: 2960–2964.

Hoy MA. (1994). *Insect Molecular Genetics*. San Diego: Academic Press.

Hunt JH, Kensinger BJ, Kossuth JA, Henshaw MT, Norberg K, Wolschin F, and Amdam GV. (2007). A diapause pathway underlies the gyne phenotype in *Polistes* wasps, revealing an evolutionary route to caste-containing insect societies. Proc Natl Acad Sci USA 104: 14020–14025.

Hunter-Ensor M, Ousley A, and Sehgal A. (1996). Regulation of the *Drosophila* protein Timeless suggests a mechanism for resetting the circadian clock by light. Cell 84: 677–685.

Huybrechts J, De Loof A, and Schoofs L. (2004). Diapausing Colorado potato beetles are devoid of short neuropeptide F I and II. Biochem Biophys Res Commun 317: 909–916.

Hwangbo DS, Gershman B, Tu MP, Palmer M, and Tatar M. (2004). *Drosophila* dFOXO controls lifespan and regulates insulin signalling in brain and fat body. Nature 429: 562–566.

Ikeda M, Su ZH, Saito H, Imai K, Sato Y, Isobe M, and Yamashita O. (1993). Induction of embryonic diapause and stimulation of ovary trehalase activity in the silkworm, *Bombyx mori*, by synthetic diapause hormone. J Insect Physiol 39: 889–895.

Imai K, Nomura T, Katsuzaki H, Komiya T, and Yamashita O. (1998). Minimum structure of diapause hormone required for biological activity. Biosci Biotechnol Biochem 62: 1875–1879.

Ireland RC and Berger EM. (1982). Synthesis of low molecular weight heat shock peptides stimulated by ecdysterone in a cultured *Drosophila* cell line. Proc Natl Acad Sci USA 79: 855–859.

Ito K. (2007). Negative genetic correlation between diapause duration and fecundity after diapause in a spider mite. Ecol Entomol 32: 643–650.

Joplin KH, Yocum GD, and Denlinger DL. (1990). Diapause specific proteins expressed by the brain during the pupal diapause of the flesh fly, *Sarcophaga crassipalpis*. J Insect Physiol 36: 825–834.

Jovanovic-Galovic A, Blagojevic DP, Grubor-Lajsic G, Worland MR, and Spasic MB. (2007). Antioxidant defense in mitochondria during diapause and postdiapause development of European corn borer (*Ostrinia nubilalis*, Hubn.). Arch Insect Biochem Physiol 64: 111–119.

Kaneko M and Hall JC. (2000). Neuroanatomy of cells expressing clock genes in *Drosophila*: Transgenic manipulation of the *period* and *timeless* genes to mark the perikarya of circadian pacemaker neurons and their projections. J Comp Neurol 422: 66–94.

Kawano T, Kataoka H, Nagasawa H, Isogai A, and Suzuki A. (1992). cDNA cloning and sequence determination of the pheromone biosynthesis activating neuropeptide of the silkworm, *Bombyx mori*. Biochem Biophys Res Commun 189: 221–226.

Kayukawa T, Chen B, Hoshizaki S, and Ishikawa Y. (2007). Upregulation of a desaturase is associated with the enhancement of cold hardiness in the onion maggot, *Delia antiqua*. Insect Biochem Mol Biol 37: 1160–1167.

Kayukawa T, Chen B, Miyazaki S, Itoyama K, Shinoda T, and Ishikawa Y. (2005). Expression of mRNA for the t-complex polypeptide-1, a subunit of chaperonin CCT, is upregulated in association with increased cold hardiness in *Delia antiqua*. Cell Stress Chaperones 10: 204–210.

Kelty JD and Lee RE. (2000). Diapausing pupae of the flesh fly *Sarcophaga crassipalpis* (Diptera: Sarcophagidae) are more resistant to inoculative freezing than non-diapausing pupae. Physiol Entomol 25: 120–126.

Khani A, Moharramipour S, and Barzegar M. (2007). Cold tolerance and trehalose accumulation in overwintering larvae of the codling moth, *Cydia pomonella* (Lepidoptera : Tortricidae). Eur J Entomol 104: 385–392.

Kidokoro K and Ando Y. (2006). Effect of anoxia on diapause termination in eggs of the false melon beetle, *Atrachya menetriesi*. J Insect Physiol 52: 87–93.

Kim M, Robich RR, Rinehart JP, and Denlinger DL. (2006). Upregulation of two actin genes and redistribution of actin during diapause and cold stress in the northern house mosquito, *Culex pipiens*. J Insect Physiol 52: 1226–1233.

Kimura KD, Tissenbaum HA, Liu Y, and Ruvkun G. (1997). daf-2, an insulin receptor-like gene that regulates longevity and diapause in *Caenorhabditis elegans*. Science 277: 942–946.

Kimura MT. (1984). Geographic variation of reproductive diapause in the *Drosophila auraria* complex (Diptera, Drosophilidae). Physiol Entomol 9: 425–431.

Kimura MT, Awasaki T, Ohtsu T, and Shimada K. (1992). Seasonal changes in glycogen and trehalose content in relation to winter survival of 4 temperate species of *Drosophila*. J Insect Physiol 38: 871–875.

Kimura MT, Ohtsu T, Yoshida T, Awasaki T, and Lin FJ. (1994). Climatic adaptations and distributions in the *Drosophila takahashii* species subgroup (Diptera, Drosophilidae). J Nat Hist 28: 401–409.

Komatsu H, Mori I, Rhee JS, Akaike N, and Ohshima Y. (1996). Mutations in a cyclic nucleotide-gated channel lead to abnormal thermosensation and chemosensation in *C. elegans*. Neuron 17: 707–718.

Koštál V, Berkova P, and Simek P. (2003). Remodelling of membrane phospholipids during transition to diapause and cold-acclimation in the larvae of *Chymomyza costata* (Drosophilidae). Comp Biochem Physiol B Biochem Mol Biol 135: 407–419.

Koštál V, Tollarova M, and Sula J. (2004). Adjustments of the enzymatic complement for polyol biosynthesis and accumulation in diapausing cold-acclimated adults of *Pyrrhocoris apterus*. J Insect Physiol 50: 303–313.

Koštál V, Zahradnickova H, Simek P, and Zeleny J. (2007). Multiple component system of sugars and polyols in the overwintering spruce bark beetle, *Ips typographus*. J Insect Physiol 53: 580–586.

Kroon A and Veenendaal RL. (1998). Trade-off between diapause and other life-history traits in the spider mite *Tetranychus urticae*. Ecol Entomol 23: 298—304.

Kyriacou CP, Peixoto AA, Sandrelli F, Costa R, and Tauber E. (2008). Clines in clock genes: Fine-tuning circadian rhythms to the environment. Trends Genet 24: 124–132.

LaFever L and Drummond-Barbosa D. (2005). Direct control of germline stem cell division and cyst growth by neural insulin in *Drosophila*. Science 309: 1071–1073.

Leather SR ed. (1993). *The Ecology of Insect Overwintering*. Cambridge: Cambridge University Press.

Lee C, Parikh V, Itsukaichi T, Bae K, and Edery I. (1996). Resetting the *Drosophila* clock by photic regulation of PER and a PER-TIM complex. Science 271: 1740–1744.

Lee RE. (1991). Principles of insect low temperature tolerance. In *Insects at Low Temperature* (Lee RE and Denlinger DI, eds), New York. Chapman and Hall, pp. 17–46.

Lee RE Jr and Denlinger DL, eds. (1991). *Insects at Low Temperature*. New York: Chapman and Hall, p. 513.

Lees AD. (1955). *The Physiology of Diapause in Arthropods*. London: Cambridge University Press.

Levy F, Bulet P, and Ehret-Sabatier L. (2004). Proteomic analysis of the systemic immune response of *Drosophila*. Mol Cell Proteom 3: 156–166.

Li AQ, Popova-Butler A, Dean DH, and Denlinger DL. (2007). Proteomics of the flesh fly brain reveals an abundance of upregulated heat shock proteins during pupal diapause. J Insect Physiol 53: 385–391.

Lincoln GA, Andersson H, and Hazlerigg D. (2003). Clock genes and the long-term regulation of prolactin secretion: Evidence for a photoperiod/circannual timer in the pars tuberalis. J Neuroendocrinol 15: 390–397.

Lopes FL, Desmarais JA, and Murphy BD. (2004). Embryonic diapause and its regulation. Reproduction 128: 669–678.

Lumme J. (1978). Phenology and photoperiodic diapause in northern populations of *Drosophila*. In *Evolution of Insect Migration and Diapause* (H Dingle, ed). New York: Springer-Verlag, pp. 145–170.

Lumme J and Lakovaara S. (1983). Seasonality and diapause in drosophilids. In *Genetics and Biology of Drosophila* (M Ashburner, HL Carson, and JNJ Thompson, eds). London: Academic Press, pp. 177–220.

Lumme J and Oikarinen A. (1977). Genetic basis of geographically variable photoperiodic diapause in *Drosophila littoralis*. Hereditas 86: 129–141.

Lumme J, Oikarinen A, Lakovaara S, and Alatalo R. (1974). The environmental regulation of adult diapause in *Drosophila littoralis*. J Insect Physiol 20: 2023–2033.

Luong N, Davies CR, Wessells RJ, Graham SM, King MT, Veech R, Bodmer R, and Oldham SM. (2006). Activated FOXO-mediated insulin resistance is blocked by reduction of TOR activity. Cell Metab 4: 133–142.

Mathias D, Jacky L, Bradshaw WE, and Holzapfel CM. (2005). Geographic and developmental variation in expression of the circadian rhythm gene, *timeless*, in the pitcher-plant mosquito, *Wyeomyia smithii*. J Insect Physiol 51: 661–667.

Matsunaga K, Takahashi H, Yoshida T, and Kimura MT. (1995). Feeding, reproductive and locomotor activities in diapausing and nondiapausing adults of *Drosophila*. Ecol Res 10: 87–93.

Matsuo Y. (2006). Cost of prolonged diapause and its relationship to body size in a seed predator. Funct Ecol 20: 300–306.

Mead RA. (1993). Embryonic diapause in vertebrates. J Exp Zool 266: 629–641.

Merzendorfer H and Zimoch L. (2003). Chitin metabolism in insects: Structure, function and regulation of chitin synthases and chitinases. J Exp Biol 206: 4393–4412.

Michaud MR and Denlinger DL. (2006). Oleic acid is elevated in cell membranes during rapid cold-hardening and pupal diapause in the flesh fly, *Sarcophaga crassipalpis*. J Insect Physiol 52: 1073–1082.

Michaud MR and Denlinger DL. (2007). Shifts in the carbohydrate, polyol, and amino acid pools during rapid cold-hardening and diapause-associated cold-hardening in flesh flies (*Sarcophaga crassipalpis*): A metabolomic comparison. J Comp Physiol B Biochem Syst Environ Physiol 177: 753–763.

Minami N and Kimura MT. (1980). Geographical variation of photoperiodic adult diapause in *Drosophila auraria*. Jpn J Genet 55: 319–324.

Mitchell CJ and Briegel H. (1989). Fate of the blood meal in force-fed, diapausing *Culex pipiens* (Diptera: Culicidae). J Med Entomol 26: 332–341.

Mitchell PD and Onstad DW. (2005). Effect of extended diapause on evolution of resistance to transgenic *Bacillus thuringiensis* corn by northern corn rootworm (Coleoptera: Chrysomelidae). J Econ Entomol 98: 2220–2234.

Morris JZ, Tissenbaum HA, and Ruvkun G. (1996). A phosphatidylinositol-3-OH kinase family member regulating longevity and diapause in *Caenorhabditis elegans*. Nature 382: 536–539.

Niimi T, Yamashita O, and Yaginuma T. (1993). A cold-inducible *Bombyx* gene encoding a protein similar to mammalian sorbitol dehydrogenase. Yolk nuclei-dependent gene expression in diapause eggs. Eur J Biochem 213: 1125–1131.

Ogg S, Paradis S, Gottlieb S, Patterson GI, Lee L, Tissenbaum HA, and Ruvkun G. (1997). The Fork head transcription factor DAF-16 transduces insulin-like metabolic and longevity signals in *C. elegans*. Nature 389: 994–999.

Ohtsu T, Kimura MT, and Hori SH. (1992). Energy storage during reproductive diapause in the *Drosophila melanogaster* species group. J Comp Physiol [B]Biochem Syst Environ Physiol 162: 203–208.

Ohtsu Y, Mori H, Komuta K, Shimizu H, Nogawa S, Matsuda Y, Nonomura T, Sakuratani Y, Tosa Y, Mayama S, and Toyoda H. (2003). Suppression of leaf feeding and oviposition of phytophagous ladybird beetles (Coleoptera : Coccinellidae) by chitinase gene-transformed phylloplane bacteria and their specific bacteriophages entrapped in alginate gel beads. J Econ Entomol 96: 555–563.

Paradis S, Ailion M, Toker A, Thomas JH, and Ruvkun G. (1999). A PDK1 homolog is necessary and sufficient to transduce AGE-1 PI3 kinase signals that regulate diapause in *Caenorhabditis elegans*. Genes Dev 13: 1438–1452.

Paradis S and Ruvkun G. (1998). *Caenorhabditis elegans* Akt/PKB transduces insulin receptor-like signals from AGE-1 PI3 kinase to the DAF-16 transcription factor. Genes Dev 12: 2488–2498.

Paria BC, Song H, and Dey SK. (2001). Implantation: Molecular basis of embryo-uterine dialogue. Int J Dev Biol 45: 597–605.

Passavant C, Zhao X, Das SK, Dey SK, and Mead RA. (2000). Changes in uterine expression of leukemia inhibitory factor receptor gene during pregnancy and its up-regulation by prolactin in the western spotted skunk. Biol Reprod 63: 301–307.

Peschel N, Veleri S, and Stanewsky R. (2006). Veela defines a molecular link between Cryptochrome and Timeless in the light-input pathway to *Drosophila*'s circadian clock. Proc Natl Acad Sci USA 103: 17313–17318.

Pfister TD and Storey KB. (2006). Responses of protein phosphatases and cAMP-dependent protein kinase in a freeze-avoiding insect, *Epiblema scudderiana*. Arch Insect Biochem Physiol 62: 43–54.

Pitts KM and Wall R. (2006). Cold shock and cold tolerance in larvae and pupae of the blow fly, *Lucilia sericata*. Physiol Entomol 31: 57–62.

Pletcher SD, Macdonald SJ, Marquerie R, Certa U, Stearns SC, Goldstein DB, and Partridge L. (2002). Genome-wide transcript profiles in aging and calorically restricted *Drosophila melanogaster*. Curr Biol 12: 712–723.

Postlethwait JH and Shirk PD. (1981). Genetic and endocrine regulation of vitellogenesis in *Drosophila*. Am Zool 21: 687–700.

Qin W, Doucet D Tyshenko MG, and Walker VK. (2007). Transcription of antifreeze protein genes in *Choristoneura fumiferana*. Insect Mol Biol 16: 423–434.

Ren P, Lim CS, Johnsen R, Albert PS, Pilgrim D, and Riddle DL. (1996). Control of *C. elegans* larval development by neuronal expression of a TGF-beta homolog. Science 274: 1389–1391.

Renfree MB and Shaw G. (2000). Diapause. Annu Rev Physiol 62: 353–375.

Riddle DL. (1988). The Dauer Larva. In *The Nematode Caenorhabditis elegans* (W Wood, ed). New York: Cold Spring Harbor Laboratory Press, pp. 393–412.

Riddle DL and Albert PS. (1997). Genetic and Environmental Regulation of Dauer Larva Development. In *C. elegans II* (DL Riddle, T Blumenthal, BJ Meyer, and JR Priess, eds). New York: Cold Spring Harbor Laboratory Press, pp. 739–768.

Riddle DL, Swanson MM, and Albert PS. (1981). Interacting genes in nematode dauer larva formation. Nature 290: 668–671.

Riehle MA and Brown MR. (2002). Insulin receptor expression during development and a reproductive cycle in the ovary of the mosquito *Aedes aegypti*. Cell Tissue Res 308: 409–420.

Rinehart JP and Denlinger DL. (2000). Heat-shock protein 90 is down-regulated during pupal diapause in the flesh fly, *Sarcophaga crassipalpis*, but remains responsive to thermal stress. Insect Mol Biol 9: 641–645.

Rinehart JP, Cikra-Ireland RA, Flannagan RD, and Denlinger DL. (2001). Expression of ecdysone receptor is unaffected by pupal diapause in the flesh fly, *Sarcophaga crassipalpis*, while its dimerization partner, USP, is downregulated. J Insect Physiol 47: 915–921.

Rinehart JP, Li A, Yocum GD, Robich RM, Hayward SA, and Denlinger DL. (2007). Up-regulation of heat shock proteins is essential for cold survival during insect diapause. Proc Natl Acad Sci USA 104: 11130–11137.

Rinehart JP, Robich RM, and Denlinger DL. (2006). Enhanced cold and desiccation tolerance in diapausing adults of *Culex pipiens*, and a role for Hsp70 in response to cold shock but not as a component of the diapause program. J Med Entomol 43: 713–722.

Rinehart JP, Yocum GD, and Denlinger DL. (2000). Developmental upregulation of inducible hsp70 transcripts, but not the cognate form, during pupal diapause in the flesh fly, *Sarcophaga crassipalpis*. Insect Biochem Mol Biol 30: 515–521.

Robich RM and Denlinger DL. (2005). Diapause in the mosquito *Culex pipiens* evokes a metabolic switch from blood feeding to sugar gluttony. Proc Natl Acad Sci USA 102: 15912–15917.

Rosato E, Trevisan A, Sandrelli F, Zordan M, Kyriacou CP, and Costa R. (1997). Conceptual translation of *timeless* reveals alternative initiating methionines in *Drosophila*. Nucleic Acids Res 25: 455–458.

Saltiel AR and Kahn CR. (2001). Insulin signalling and the regulation of glucose and lipid metabolism. Nature 414: 799–806.

Sandrelli F, Tauber E, Pegoraro M, Mazzotta G, Cisotto P, Landskron J, Stanewsky R, Piccin A, Rosato E, Zordan M, Costa R, and Kyriacou CP. (2007). A molecular basis for natural selection at the *timeless* locus in *Drosophila melanogaster*. Science 316: 1898–1900.

Sato Y, Ikeda M, and Yamashita O. (1994). Neurosecretory cells expressing the gene for common precursor for diapause hormone and pheromone biosynthesis-activating neuropeptide in the suboesophageal ganglion of the silkworm, *Bombyx mori*. Gen Comp Endocrinol 96: 27–36.

Sato Y, Oguchi M, Menjo N, Imai K, Saito H, Ikeda M, Isobe M, and Yamashita O. (1993). Precursor polyprotein for multiple neuropeptides secreted from the suboesophageal ganglion of the silkworm *Bombyx mori*: Characterization of the cDNA encoding the diapause hormone precursor and identification of additional peptides. Proc Natl Acad Sci USA 90: 3251–3255.

Saunders DS. (1976). *Insect Clocks*. Oxford: Pergamon Press.

Saunders DS. (1990). The circadian basis of ovarian diapause regulation in *Drosophila melanogaster*: Is the *period* gene causally involved in photoperiodic time measurement? J Biol Rhythms 5: 315–331.

Saunders DS and Gilbert LI. (1990). Regulation of ovarian diapause in *Drosophila melanogaster* by photoperiod and moderately low temperature. J Insect Physiol 36: 195–200.

Saunders DS, Henrich VC, and Gilbert LI. (1989). Induction of diapause in *Drosophila melanogaster*: Photoperiodic regulation and the impact of arrhythmic clock mutations on time measurement. Proc Natl Acad Sci USA 86: 3748–3752.

Saunders DS, Richard DS, Applebaum SW, Ma M, and Gilbert LI. (1990). Photoperiodic diapause in *Drosophila melanogaster* involves a block to the juvenile hormone regulation of ovarian maturation. Gen Comp Endocrinol 79: 174–184.

Schmidt KE and Kelley KM. (2001). Down-regulation in the insulin-like growth factor (IGP) axis during hibernation in the golden-mantled ground squirrel, *Spermophilus lateralis*: IG-F-I and the IGF-binding proteins (IGFBPs). J Exp Zool 289: 66–73.

Schmidt PS, Matzkin L, Ippolito M, and Eanes WF. (2005a). Geographic variation in diapause incidence, life-history traits, and climatic adaptation in *Drosophila melanogaster*. Evolution 59: 1721–1732.

Schmidt PS, Paaby AB, and Heschel MS. (2005b). Genetic variance for diapause expression and associated life histories in *Drosophila melanogaster*. Evolution 59: 2616–2625.

Shiga S and Numata H. (1997). Induction of reproductive diapause via perception of photoperiod through the compound eyes in the adult blow fly, *Protophormia terraenovae*. J Comp Physiol A Sens Neural Behav Physiol 181: 35–40.

Shiga S and Numata H. (2000). The role of neurosecretory neurons in the pars intercerebralis and pars lateralis in reproductive diapause of the blowfly, *Protophormia terraenovae*. Naturwissenschaften 87: 125–128.

Shiga S and Numata H. (2007). Neuroanatomical approaches to the study of insect photoperiodism. Photochem Photobiol 83: 76–86.

Sim C and Denlinger DL. (2008). Insulin signaling and FOXO regulate the overwintering diapause of the mosquito *Culex pipiens*. Proc Natl Acad Sci USA 105: 6777–6781.

Simon AF, Shih C, Mack A, and Benzer S. (2003). Steroid control of longevity in *Drosophila melanogaster*. Science 299: 1407–1410.

Singtripop T, Oda Y, Wanichachewa S, and Sakurai S. (2002). Sensitivities to juvenile hormone and ecdysteroid in the diapause larvae of *Omphisa fuscidentalis* based on the hemolymph trehalose dynamics index. J Insect Physiol 48: 817–824.

Sonoda S, Fukumoto K, Izumi Y, Ashfaq M, Yoshida H, and Tsumuki H. (2006b). Methionine-rich storage protein gene in the rice stem borer, *Chilo suppressalis*, is expressed during diapause in response to cold acclimation. Insect Mol Biol 15: 853–859.

Sonoda S, Fukumoto K, Izumi Y, Ashfaq M, Yoshida H, and Tsumuki H. (2007). Expression profile of arylphorin gene during diapause and cold acclimation in the rice stem borer, *Chilo suppressalis* Walker (Lepidoptera: Crambidae). Appl Entomol Zool 42: 35–40.

Sonoda S, Fukumoto K, Izumi Y, Yoshida H, and Tsumuki H. (2006a). Cloning of heat shock protein genes (hsp90 and hsc70) and their expression during larval diapause and cold tolerance acquisition in the rice stem borer, *Chilo suppressalis* Walker. Arch Insect Biochem Physiol 63: 36–47.

Stehlik J, Zavodska R, Shimada K, Sauman I, and Koštál V. (2008). Photoperiodic induction of diapause requires regulated transcription of *timeless* in the larval brain of *Chymomyza costata*. J Biol Rhythms 23: 129–139.

Stewart CL, Kaspar P, Brunet LJ, Bhatt H, Gadi I, Kontgen F, and Abbondanzo SJ. (1992). Blastocyst implantation depends on maternal expression of leukaemia inhibitory factor. Nature 359: 76–79.

Storey KB. (2005). Hibernating mammals: Can natural cryoprotective mechanisms help prolong lifetimes of transplantable organs? In *Extending the Lifespan: Biotechnical, Gerontological, and Social Problems* (K Sames, S Sethe, and A Stolzing, eds). Munster: LIT Verlag, pp. 219–228.

Storey KB and Storey JM. (2004). Metabolic rate depression in animals: Transcriptional and translational controls. Biol Rev Cambr Philos Soc 79: 207–233.

Storey KB and Storey JM. (2007). Tribute to P. L. Lutz: Putting life on "pause"—molecular regulation of hypometabolism. J Exp Biol 210: 1700–1714.

Sun JS, Zhang QR, Zhang TY, Zhu ZL, Zhang HM, Teng MK, Niu LW, and Xu WH. (2005). Developmental expression of FXPRLamide neuropeptides in peptidergic neurosecretory cells of diapause- and nondiapause-destined individuals of the cotton bollworm, *Helicoverpa armigera*. Gen Comp Endocrinol 141: 48–57.

Syrova Z, Dolezel D, Saumann I, and Hodkova M. (2003). Photoperiodic regulation of diapause in linden bugs: Are *period* and *clock* genes involved? Cell Mol Life Sci 60: 2510–2515.

Tachibana S, Numata H, and Goto SG. (2005). Gene expression of heat-shock proteins (Hsp23, Hsp70 and Hsp90) during and after larval diapause in the blow fly *Lucilia sericata*. J Insect Physiol 51: 641–647.

Tammariello SP and Denlinger DL. (1998). G0/G1 cell cycle arrest in the brain of *Sarcophaga crassipalpis* during pupal diapause and the expression pattern of the cell cycle regulator, proliferating cell nuclear antigen. Insect Biochem Mol Biol 28: 83–89.

Tatar M, Bartke A, and Antebi A. (2003). The endocrine regulation of aging by insulin-like signals. Science 299: 1346–1351.

Tatar M, Chien S, and Priest NK. (2001a). Negligible senescence during reproductive dormancy in *Drosophila melanogaster*. Am Nat 158: 248–258.

Tatar M, Khazaeli AA, and Curtsinger JW. (1997). Chaperoning extended life. Nature 390: 30–30.

Tatar M, Kopelman A, Epstein D, Tu MP, Yin CM, and Garofalo RS. (2001b). A mutant *Drosophila* insulin receptor homolog that extends life-span and impairs neuroendocrine function. Science 292: 107–110.

Tatar M and Yin C. (2001). Slow aging during insect reproductive diapause: Why butterflies, grasshoppers and flies are like worms. Exp Gerontol 36: 723–738.

Tauber CA and Tauber MJ. (1987). Inheritance of seasonal cycles in *Chrysoperla* (Insecta, Neuroptera). Genet Res 49: 215–223.

Tauber CA and Tauber MJ. (1992). Phenotypic plasticity in *Chrysoperla*—Genetic variation in the sensory mechanism and in correlated reproductive traits. Evolution 46: 1754–1773.

Tauber E and Kyriacou BP. (2001). Insect photoperiodism and circadian clocks: Models and mechanisms. J Biol Rhythms 16: 381–390.

Tauber E, Zordan M, Sandrelli F, Pegoraro M, Osterwalder N, Breda C, Daga A, Selmin A, Monger K, Benna C, Rosato E, Kyriacou CP, and Costa R. (2007). Natural selection favors a newly derived *timeless* allele in *Drosophila melanogaster*. Science 316: 1895–1898.

Tauber MJ, Tauber CA, and Masaki S. (1986). *Seasonal Adaptations of Insects*. New York: Oxford University Press.

Tobe SS. (1999). Allatostatins. In *Encyclopedia of Reproduction* (E Knobil and JD Niell, eds). San Diego: Academic Press, pp. 97–106.

Tomcala A, Tollarova M, Overgaard J, Simek P, and Koštál V. (2006). Seasonal acquisition of chill-tolerance and restructuring of membrane glycerophospholipids in an overwintering insect: Triggering by low temperature, desiccation and diapause progression. J Exp Biol 209: 4102–4114.

Tu BP and McKnight SL. (2006). Metabolic cycles as an underlying basis of biological oscillations. Nat Rev Mol Cell Biol 7: 696–701.

Tu MP, Yin CM, and Tatar M. (2005). Mutations in insulin signaling pathway alter juvenile hormone synthesis in *Drosophila melanogaster*. Gen Comp Endocrinol 142: 347–356.

Uno T, Nakasuji A, Shimoda M, and Aizono Y. (2004). Expression of cytochrome c oxidase subunit 1 gene in the brain at an early stage in the termination of pupal diapause in the sweet potato hornworm, *Agrius convolvuli*. J Insect Physiol 50: 35–42.

Urquhart FA and Urquhart NR. (1977). Overwintering areas and migratory routes of monarch butterfly (*Danaus plexippus*, Lepidoptera Danaidae) in North America, with special reference to western populations. Can Entomol 109: 1583–1589.

Van Steenwyk RA, Fouche CF, and Collier TR. (2004). Seasonal susceptibility of 'Bartlett' pears to codling moth (Lepidoptera: Tortricidae) infestation and notes on diapause induction. J Econ Entomol 97: 976–980.

Vermunt AM, Koopmanschap AB, Vlak JM, and de Kort CA. (1997a). Cloning and sequence analysis of cDNA encoding a putative juvenile hormone esterase from the Colorado potato beetle. Insect Biochem Mol Biol 27: 919–928.

Vermunt AM, Koopmanschap AB, Vlak JM, and de Kort CA. (1999). Expression of the juvenile hormone esterase gene in the Colorado potato beetle, *Leptinotarsa decemlineata*: Photoperiodic and juvenile hormone analog response. J Insect Physiol 45: 135–142.

Vermunt AM, Vermeesch AM, and de Kort CA. (1997b). Purification and characterization of juvenile hormone esterase from hemolymph of the Colorado potato beetle. Arch Insect Biochem Physiol 35: 261–277.

Vosshall LB, Price JL, Sehgal A, Saez L, and Young MW. (1994). Block in nuclear localization of Period protein by a 2nd clock mutation, *timeless*. Science 263: 1606–1609.

Walker VK, Kuiper MJ, Tyshenko MG, Doucet D, Graether SP, Liou YC, Sykes BD, Jia Z, Davies PL, and Graham LA. (2001). Surviving winter with antifreeze proteins: Studies on budworms and beetles. In *Insect Timing: Circadian Rhythmicity to Seasonality* (Denlinger DL, Giebultowiz J, and Saunders DS, eds). Amsterdam: Elsevier Science, pp. 199–212.

Wang H and Dey SK. (2006). Roadmap to embryo implantation: Clues from mouse models. Nature Rev Genet 7: 185–199.

Wang WW, Meng B, Chen WH, Ge XM, Liu S, and Yu J. (2007). A proteomic study on postdiapaused embryonic development of brine shrimp (*Artemia franciscana*). Proteomics 7: 3580–3591.

Wei ZJ, Zhang QR, Kang L, Xu WH, and Denlinger DL. (2005). Molecular characterization and expression of prothoracicotropic hormone during development and pupal diapause in the cotton bollworm, *Helicoverpa armigera*. J Insect Physiol 51: 691–700.

Williams CM. (1946). Physiology of insect diapause—the role of the brain in the production and termination of pupal dormancy in the giant silkworm, *Platysamia cecropia*. Biol Bull 90: 234–243.

Williams JB, Ruehl NC, and Lee RE Jr. (2004). Partial link between the seasonal acquisition of cold-tolerance and desiccation resistance in the goldenrod gall fly *Eurosta solidaginis* (Diptera: Tephritidae). J Exp Biol 207: 4407–4414.

Williams KD and Sokolowski MB. (1993). Diapause in *Drosophila melanogaster* females: A genetic analysis. Heredity 71 (Pt 3): 312–317.

Williams KD, Busto M, Suster ML, So AK, Ben-Shahar Y, Leevers SJ, and Sokolowski MB. (2006). Natural variation in *Drosophila melanogaster* diapause due to the insulin-regulated PI3-kinase. Proc Natl Acad Sci USA 103: 15911–15915.

Wu, Qi and Brown MR. (2006). Signaling and function of insulin-like peptides in insects. Annu Rev Entomol 51: 1–24.

Xu WH and Denlinger DL. (2003). Molecular characterization of prothoracicotropic hormone and diapause hormone in *Heliothis virescens* during diapause, and a new role for diapause hormone. Insect Mol Biol 12: 509–516.

Xu WH and Denlinger DL. (2004). Identification of a cDNA encoding DH, PBAN and other FXPRL neuropeptides from the tobacco hornworm, *Manduca sexta*, and expression associated with pupal diapause. Peptides 25: 1099–1106.

Yamamoto K, Chadarevian A, and Pellegrini M. (1988). Juvenile hormone action mediated in male accessory glands of *Drosophila* by calcium and kinase C. Science 239: 916–919.

Yamashita O. (1996). Diapause hormone of the silkworm, *Bombyx mori*: Structure, gene expression and function. J Insect Physiol 42: 669–679.

Yamashita O, Imai K, Saito H, Shiomi K, and Sato Y. (1998). Phe-X-Pro-Arg-Leu-NH(2) peptide producing cells in the central nervous system of the silkworm, *Bombyx mori*. J Insect Physiol 44: 333–342.

Yocum GD. (2001). Differential expression of two HSP70 transcripts in response to cold shock, thermoperiod, and adult diapause in the Colorado potato beetle. J Insect Physiol 47: 1139–1145.

Yocum GD, Joplin KH, and Denlinger DL. (1998). Upregulation of a 23 kDa small heat shock protein transcript during pupal diapause in the flesh fly, *Sarcophaga crassipalpis*. Insect Biochem Mol Biol 28: 677–682.

Yoder JA, Benoit JB Denlinger DL, and Rivers DB. (2006). Stress-induced accumulation of glycerol in the flesh fly, *Sarcophaga bullata*: Evidence indicating anti-desiccant and cryoprotectant functions of this polyol and a role for the brain in coordinating the response. J Insect Physiol 52: 202–214.

Yoshiga T, Okano K, Mita K, Shimada T, and Matsumoto S. (2000). cDNA cloning of acyl-CoA desaturase homologs in the silkworm, *Bombyx mori*. Gene 246: 339–345.

Zera AJ and Larsen A. (2001). The metabolic basis of life history variation: Genetic and phenotypic differences in lipid reserves among life history morphs of the wing-polymorphic cricket, *Gryllus firmus*. J Insect Physiol 47: 1147–1160.

Zera AJ and Zhao Z. (2003). Morph-dependent fatty acid oxidation in a wing-polymorphic cricket: Implications for the trade-off between dispersal and reproduction. J Insect Physiol 49: 933–943.

Zhang T-Y, Sun J-S, Zhang L-B, Shen J-L, and Xu W-H. (2004). Cloning and expression of the cDNA encoding FXPRL family of peptides and a functional analysis of their effect on breaking pupal diapause in *Helicoverpa armigera*. J Insect Physiol 50: 25–33.

Zhao Z and Zera AJ. (2002). Differential lipid biosynthesis underlies a tradeoff between reproduction and flight capability in a wing-polymorphic cricket. Proc Natl Acad Sci USA 99: 16829–16834.

Zheng L, Lytle C, Njauw CN, Altstein M, and Martins-Green M. (2007). Cloning and characterization of the pheromone biosynthesis activating neuropeptide receptor gene in *Spodoptera littoralis* larvae. Gene 393: 20–30.

Zhou X, Oi FM, and Scharf ME. (2006). Social exploitation of hexamerin: RNAi reveals a major caste-regulatory factor in termites. Proc Natl Acad Sci USA 103: 4499–4504.

Zhu H, Sauman I, Yuan Q, Casselman A, Emery-Le M, Emery P, and Reppert SM. (2008). Cryptochromes define a novel circadian clock mechanism in monarch butterflies that may underlie sun compass navigation. PLoS Biol 6: E4.

Zwaal RR, Mendel JE, Sternberg PW, and Plasterk RH. (1997). Two neuronal G proteins are involved in chemosensation of the *Caenorhabditis elegans* dauer-inducing pheromone. Genetics 145: 715–727.

13

Photoperiodism in Insects: Effects on Morphology

H. Frederik Nijhout

Photoperiod can have profound effects on insect morphology. Many insects have distinctive seasonal adult morphs that are controlled by the photoperiod experienced by the larvae. Sometimes these seasonal morphs are so different that they are easily mistaken for different species, as in the case of the butterfly *Araschnia levana*, many species of the butterfly genus *Precis* (figures 13.1 and 13.2), and many species of aphids (see chapter 14). Understanding the ability of species to develop such extreme alternative morphologies without genetic differentiation poses interesting challenges to both evolutionary and developmental biology.

Photoperiod is an accurate predictor of seasonal change, and both animals and plants have evolved mechanisms that take advantage of this predictive power to alter their development or physiology to produce form or function appropriate for the season. Plants flower or drop their leaves, mammals and birds develop seasonally appropriate pelage or plumage, cervids develop antlers, and birds migrate. Insects are able to alter their developmental trajectory in response to the length of day or night and develop alternative adult forms that are adaptations to the different environmental conditions and the challenges that characterize different seasons.

Normal summer form   Normal spring form

FIGURE 13.1.
Seasonal forms of the European map butterfly, *Araschnia levana*. Normal spring and summer forms are shown on right and left, respectively. Between them are two morphological intermediates that were produced in the laboratory.

318

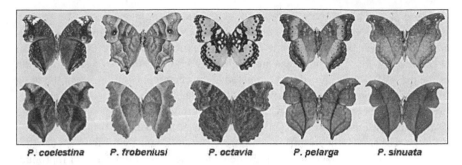

FIGURE 13.2.
Seasonal forms of selected *Precis* species. Top row, wet-season forms; bottom row, dry-season forms.

The sensitivity of insect morphology to photoperiod is an outstanding example of phenotypic plasticity (West-Eberhard, 2003). Photoperiod-induced phenotypic differences are a special case of polyphenism: the ability to develop discrete alternative morphologies in response to environmental variables (Nijhout, 1999, 2003). In addition to seasonal morphs, polyphenisms include caste differentiation in social insects, the solitary and gregarious forms of locusts, the sexual and parthenogenetic forms of aphids, and the horned and hornless forms of some scarab beetles. Portions of the mechanism by which insect development responds to environmental variables such as photoperiod are now well understood.

SEASONAL MORPHS

There are some general patterns of expression of seasonal morphs in insects. Seasonal morph differentiation in temperate-zone insects is most commonly associated with dispersal, with diapause, and with color adaptation to a seasonally varying environment. Many Orthoptera, Hemiptera, Homoptera, and thrips have seasonal alate/apterous or long-wing/short-wing morphs (Harrison, 1980; Roff and Fairbairn, 1991). The winged and long-wing morphs are specialized for dispersal and migration and are typically produced in the autumn, or when host plants senesce and food becomes restrictive. Multivoltine insects that have a diapausing stage often have distinct adult morphs, one associated with diapause and the other not. As in most biological systems, these generalizations are not universal, and there are a lot of exceptions and a lot of diversity in their expression. This is probably because many seasonal polyphenisms are independently evolved adaptations and thus have emerged within different ecological contexts and genetic backgrounds and use different developmental and physiological mechanisms to control the expression of the alternative phenotypes.

The adaptive significance of long-winged and short-winged alternative phenotypes seems obvious. Short-winged morphs often do not develop flight muscles and tend also to have larger abdomens and a higher reproductive capacity. Thus, it is tempting to conclude that the resources that would have gone into large wings are instead used for reproduction. This simple explanation, however, fails to take account of the fact that the wings and flight muscles make up only a small fraction of the mass of eggs produced by a female, that the chemical components of the wing (mostly carbohydrates and proteins) are dissimilar from those used for reproduction (mostly lipoproteins), that there is a long period of time over which a female can acquire resources for reproduction (i.e., she does not have to rely exclusively on resources accumulated during wing development), and that a female can alter her nutritive input to match the requirements for energy and material for reproduction (Zera and Harshman, 2001). Mole and Zera (1993) have suggested that in the wing polymorphic cricket *Gryllus rubens* (where short- vs. long-winged morphs are genetically determined), there is a trade-off within the adult between energy requirements for the maintenance of flight muscle in the long-winged morph, and ovarian growth in the short-winged morph.

Many butterflies have a pupal green/brown polyphenism. Green color is typically found in the summer generations and brown is typically found in the autumn and winter and is often associated with diapause (Shapiro, 1976). The adaptive significance of such a polyphenism is that green pupae tend to occur while foliage is green, and brown pupae occur during the winter when foliage is either brown or absent. Hazel and West (1983) studied the effects of photoperiod and substrate on the development of pupal coloration in four species of North American swallowtail butterflies: *Eurytides marcellus*, *Battus philenor*, *Papilio polyxenes*, and *Papilio troilus*. When reared on long-day photoperiod all four species produce dimorphic (green or brown) pupae, depending on the brightness, color, or texture on the substrate on which they pupate. In *Papilio polyxenes* and *P. troilus*, short-day photoperiods induce diapause and brown pupal coloration. In *Battus philenor* and *Euritydes marcellus*, short-day photoperiods do not affect the pupal color response but induce pupal diapause. Thus pupal coloration and diapause are linked in some species but not in others. Other cases, and the control of pupal color polymorphism, are discussed further below.

Many butterflies have seasonal adult forms. In the temperate zone, summer and autumn forms often differ in pigmentation and in the distinctness of the color pattern. With a few notable exceptions, the overall pattern does not change, but the spots and lines differ in size or width, and the color contrast between pattern elements is greater or less, in different seasons. Typically, summer forms are more brightly colored and have a more distinct pattern than autumn forms, which are generally darker and duller in coloration. In the tropics, the dry/cool-season forms are often darker and duller in pattern than the paler and more brightly colored wet/warm-season forms. In both tropical and temperate-zone butterflies, the polyphenic color pattern is typically only expressed on the ventral wing surfaces. All species in the *Precis/Junonia* group, a large supergenus of tropical and warm-temperate zone butterflies, have a polyphenic color pattern. The dry/cool-season forms have darker and duller ventral wings surfaces, and the shape of the wings of these forms is also much more angular and indented than that of the wet/warm-season forms. Many dry-season forms have a dead-leaf–mimicking ventral pattern and wing shape (figure 13.2). More angular wings and a more cryptic ventral pattern are characteristic of the autumn/winter form of some other butterflies, such as the angle-wings and commas (Wiklund and Tullberg, 2004). Among the Pieridae, short-day cool-weather forms are more darkly colored than the long-day summer forms, particularly along the wing veins, and this darker wing pigmentation helps raise body temperature by basking in the sun (Shapiro, 1976; Kingsolver and Wiernasz, 1991).

Some seasonal effects are sex specific. The Japanese vapourer moth *Orgyia thyellina* has a sex limited seasonal polyphenism. Males are always long-winged. Females reared under long-day photoperiod as larvae form pale colored pupae and develop as long-winged adults that lay direct-developing eggs. When larvae are reared under short-day conditions, females produce darker pupae and emerge as short-winged adults that lay diapausing eggs (Kimura and Masaki, 1977).

In the hemipteran bug *Pyrrocoris apterus*, an interesting maternal effect mediates the photoperiodic signal. In Central Europe, *Pyrrhocoris* is polyphenic and develops a long-winged morph under long-day photoperiods and a short-winged morph that enters reproductive diapause when larvae are reared under short-day photoperiod. Through artificial selection, Honek (1980) obtained a strain that was predominantly long winged under both long-day and short-day conditions, demonstrating that the phenotype could be uncoupled from the environmental signal. Interestingly, if adult females of this strain are kept under short-day conditions, then their offspring will also develop short wings if reared under short-day conditions (and, as expected, long wings if reared under long-day conditions). Thus, there is a maternal effect in this otherwise predominantly long-winged strain that sensitizes the next generation to photoperiod. The mediator of this maternal effect is not known, but in other insects hormones of the mother can affect the pattern of hormone secretion of her offspring and affect their development (Cassier and Papillion, 1968; Willis, 1969; Riddiford, 1970; Nijhout, 1994).

SEASONAL CUES: THE INTERACTION OF PHOTOPERIOD AND TEMPERATURE

The environment that induces development of the alternative seasonal forms is usually not the environment to which the polyphenism is an adaptation. Because of the time lag between the inductive signal and the development of the adapted phenotype, insects typically use token stimuli that are good predictors of future seasonal change. In temperate zones, photoperiod is the most commonly used cue. But at low latitudes, the seasonal differences in day length are small, and the "seasons" are wet/dry or hot/cool. These seasons do not correspond temporally to summer/winter at temperate latitudes, but are similar in the sense that one of the seasons (wet/cool) is generally favorable for growth and the other is not. Thus, at low latitudes, insects often use temperature as a seasonal cue. In temperate zones, the water and nitrogen content of leaves declines as the season progresses, as do the concentrations of many phytochemicals (Danilevskii, 1965; Scriber and Slansky, 1981; Tauber et al., 1986). So, the nutrient quality of leaves, as well as specific chemical signals, can provide seasonal cues. Food plant quality may also be an important seasonal cue in tropical regions with distinct wet and dry seasons.

In some temperate-zone species temperature and photoperiod can substitute for each other (Rountree and Nijhout, 1995a; Smith, 1993), so either short day length or low temperature can induce diapause and/or seasonal morphology. But it is more common for the two to synergize each other. For instance, low temperatures tend to enhance the inductive effects of short days (Smith, 1993; Sasaki et al., 2002). It is well established that temperature affects the critical day length of diapause induction: a 5°C change in rearing temperature can change the critical photoperiod by as much as 1.5 h, depending on the species (Bünning, 1973; Beck, 1980). Some of this effect of temperature may occur because the summation of a certain number of photoperiod cycles is required to induce a seasonal response, and low temperatures slow development and thus increase the number of inductive cycles experienced; likewise, at high temperatures, development is fast and the duration of the photosensitive period too brief to experience a sufficient number of inductive cycles (Mousseau and Dingle, 1991). Alternatively, it is possible that insects can use photoperiod and temperature cues independently, as the observations of Shapiro (1976, 1978, 1982), Rountree and Nijhout (1995a, 1995b), and Smith (1993) suggest, so the effects of these two cues are essentially additive.

But why can temperature and photoperiod substitute for each other? What is the common mechanism? In the tropics seasonal polyphenisms are primarily induced by temperature alone in species whose congeners in temperate zones use photoperiod or a combination of photoperiod and temperature. Temperature slows down most biochemical reactions and developmental processes, but the rhythm of circadian clocks is either temperature compensated or entrained by the photoperiod and is therefore unaffected by temperature (Bünning, 1973; Beck, 1980; Saunders, 2002). Experimental manipulation of the absolute and relative amounts of darkness and light during a photoperiodic cycle have shown that in most cases it is the length of the dark phase (the scotophase) that gives the cue whether days are "short" or "long" (Beck, 1980; Saunders, 2002). A number of authors have suggested hypothetical mechanisms by which a circadian clock could be used to "measure" day length and how this information could be summed and stored to serve as a seasonal indicator for, say, the control of diapause (Pittendrigh, 1972; Gibbs, 1975; Zaslavskii, 1988; Pittendrigh et al., 1991). Most of these models offer variations on the basic assumption that animals measure day length by the accumulation of some molecule during the dark phase. If this molecule accumulates from day to day, then more will accumulate during long-night (short day) photoperiods. This leads to a simple hypothesis about the reason for the apparent additivity of photoperiod and temperature. More of this hypothetical molecule will accumulate at low temperatures because many more day–night cycles will occur as development is slowed. Even under long day lengths, low temperature may slow development down sufficiently to allow the accumulation of this molecule above the threshold required to evoke a developmental switch. It is necessary, of course, that the environment-sensitive period during which the photoperiodic effects accumulate be sufficiently long to prevent above-threshold accumulation during brief and irrelevant environmental cold periods. Such a mechanism could explain why in temperate zones short day lengths and cool temperatures can substitute for each other and why in tropical regions, where there is little seasonal variation in the scotophase, polyphenism are typically controlled by seasonal differences in temperature.

In some cases photoperiod and temperature are not simply additive (Saunders, 2002). Nice examples of the complex interaction of temperature and photoperiod come from the work of Arthur Shapiro. Shapiro (1968, 1973, 1978, 1980, 1982, 1984) has studied the seasonal polyphenism of pierid butterflies in both high and low latitude locales in the Americas. In many temperate zone pierids, short-day photoperiods induce pupal diapause. Diapausing pupae give rise to a pale-colored summer morph, and nondiapausing pupae give rise to a much darker autumn morph. Interestingly, when long-day pupae are chilled during the first day after pupation, they develop into adults of the summer phenotype. In *Pieris napi* and *Pieris protodice*, short days and high temperatures inhibit diapause but induce the summer form (Shapiro, 1968, 1973, 1978). In *Pieris occidentalis, Pieris protodice*, and *Colias eurytheme*, chilling a pupa that was reared on a long-day photoperiod during the first 24 h after pupation induces the dark autumn phenotype (Shapiro, 1978, 1982). Thus, in these species, a brief period of low temperature during the first day after pupation can completely abolish the effect of photoperiod on seasonal morphology. Moreover, these results show that the development of seasonal morphology can be readily uncoupled from diapause.

HOW PHOTOPERIOD AFFECTS DEVELOPMENT AND MORPHOLOGY

Photoperiod does not have a direct effect on developmental processes. Its effect is mediated by central nervous system (CNS) and, as far as we know, exclusively via the neural control of hormone secretion. A variety of developmental hormones have been discovered that affect the outcomes of development. Most common among these are ecdysone (20E) and juvenile hormone (JH), and these are joined by an ever growing number of neurosecretory hormones that affect diverse aspects of development. The secretion of JH and 20E is controlled by the CNS via tropic or inhibitory neurosecretory hormones. Ecdysone secretion is stimulated by the prothoracicotropic hormone (PTTH), and JH secretion can be under positive and negative control by neurosecretions (allatotropins and allatostatins, respectively). The CNS thus controls the secretion of JH and 20E and, obviously, all the neurosecretory hormones. The CNS thus acts as a central regulator of development and provides the means by which development can adjust to environmental contingencies.

Because the insect cuticle is nonliving, changes in form are necessarily associated with molting and metamorphosis. In larval polyphenisms, successive larval instars can have different pigmentation and surface morphology associated with seasonal change. Most seasonal polyphenisms, however, are only expressed in the adult and thus arise as alternative outcomes of metamorphosis. Many, but by no means all, seasonal polyphenisms are associated with diapause or estivation of the pupal or adult stage.

Studies on induction of diapause have revealed that insects are only susceptible to diapause induction during more or less well-defined environment-sensitive periods (Danilevskii, 1965; Beck, 1980; Tauber et al., 1986). Environment-sensitive periods tend to be quite long, many days or weeks, during which environmental signals are accumulated and integrated. The environment-sensitive period can precede the actual onset of diapause by weeks or months, the exact time interval being an adaptation to a particular life cycle.

Studies on the control of metamorphosis and the control of diapause have proven indispensable in developing an understanding of the control of seasonal polyphenism (Nijhout, 1994, 2003). These studies have shown that in metamorphosis, JH and 20E act during well-defined hormone-sensitive periods. Different tissues undergo different changes in commitment and prospective fate depending on whether one or both hormones are above or below a tissue-specific threshold during a hormone-sensitive period. Hormone sensitive periods are often about a day in duration (Nijhout, 1983, 1994, 2003), but their limits have only been studied in very few cases, so we do not know how much diversity there is. The causes of the hormone-sensitive periods to 20E are beginning to be understood. Expression of the various isoforms of the ecdysteroid receptor (figure 13.3) varies in complex tissue-specific temporal patterns (Fujiwara et al., 1995; Jindra et al., 1996; Nijhout, 1999, 2003). Insofar as tissues can only respond to a hormone when the receptor is present, periods of receptor expression correspond to hormone-sensitive periods. Periods when insects are sensitive to JH often coincide with, or follow shortly after, pulses of 20E secretion. Although a specific receptor for JH has not been found (Wheeler and Nijhout, 2003), JH is known to bind to the ecdysteroid receptor (Jones and Sharp, 1997; Jones et al., 2001), so some of its effects may be functionally tied to the action of 20E (Nijhout, 1994). The sensitive periods for neurosecretory hormones have been little studied.

Seasonal polyphenisms, like diapause, are programmed during environment-sensitive periods. Depending on the environment that is sensed, a developmental switch is activated that alters either the pattern of hormone secretion or of hormone-sensitive periods at the time of metamorphosis so that development takes one of two alternative pathways, resulting in the alternative forms (Nijhout, 2003). Figure 13.4 illustrates in a diagrammatic form the current model for the combined environmental and endocrine control of insect polyphenism. This model applies to all polyphenisms whose developmental control has been elucidated, including seasonal polyphenisms. The environmental signal can have four different effects on the endocrine switching mechanism: (1) it can raise or lower the hormonal signal

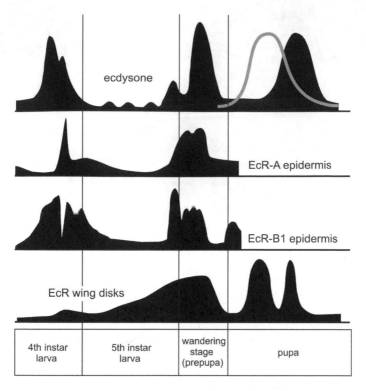

FIGURE 13.3.

Temporal patterns of 20E (ecdysone) titers and 20E receptor (EcR) expression in *Manduca sexta*. (Top) Titers of 20-hydroxyecdysone in black, and of 20E in gray. (Middle) These two panels show the expression of two different isoforms of the 20E receptor in the abdominal epidermis. Different isoforms of the 20E receptor (EcR-A, EcR-B1) activate different genes (Cherbas and Cherbas, 1996; Truman, 1996). (Bottom) Expression of joint isoforms of the 20E receptor in the wing imaginal disks. Evidently the pattern of receptor expression varies greatly over time and is tissue specific. The work of Koch et al. (2003) has shown that in the wing imaginal disk of *Precis*, there is also a spatially varying pattern of EcR expression. After Fujiwara et al. (1995) and Jindra et al. (1996).

relative to a threshold; (2) it can alter the value of the threshold; (3) it can move a hormone secretion peak away from a particular hormone-sensitive period; (4) it can move the hormone sensitive period. In each of these alternatives, hormone secretion occurs in two different contexts, and this leads to two alternative patterns of gene expression and different subsequent developmental trajectories.

In the development of (seasonal) polyphenisms the CNS is in effect in control of development, and this provides two alternatives for ways the environment affects developmental processes. Environmental variables such as temperature and nutrition can effect development directly, by differential alteration in the rate or timing of biochemical reactions and metabolic and developmental processes.

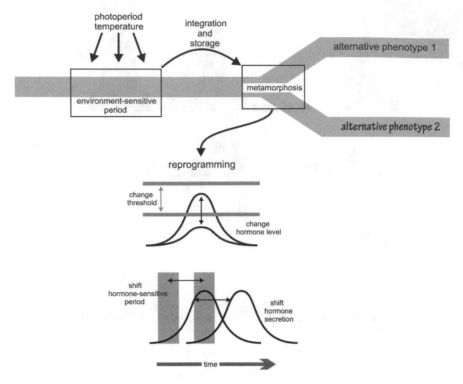

FIGURE 13.4.
Schematic diagram of the environmental and endocrine control of polyphenism. The gray forked bar represents a developmental trajectory that can result on two alternative phenotypes. At some time, usually during larval life, there is a more or less prolonged environment-sensitive period, during which information about photoperiod and temperature is accumulated. This information is integrated and stored and used later, during metamorphosis, to redirect development along one of two alternative pathways. The alternative switch is controlled by hormones as shown in the diagrams below, as follows. A hormone can rise above or fall below a threshold of sensitivity, depending on the environment perceived, and this results in a different pattern of gene expression. Alternatively, the threshold can rise or fall; or the timing of hormone secretion can shift into or away from a hormone sensitive period, or the hormones sensitive period can shift relative to the timing of hormone secretion. Examples of each of these modes have been documented (Nijhout, 2003).

The environment can also affect development indirectly. Specific environmental signals such as photoperiod can be sensed and integrated by the CNS, and this information can be used at a later time to switch the secretion of developmental hormones between two alternative temporal patterns. It must also be possible for environmental signals to alter the spatial and temporal patterns of expression of receptors signaling pathways for the developmental hormones, but exactly how this could be accomplished is as yet unclear.

DEVELOPMENTAL HORMONES IN PHOTOPERIODIC RESPONSES

A number of different hormones control the switch between alternative developmental pathways in seasonal polyphenisms. These are conveniently divided into two categories: glandular hormones and neurosecretory hormones. The former are two lipid hormones, 20E and JH, that also control molting and metamorphosis; the latter are all small polypeptides.

Glandular Hormones

Ecdysone Control of Seasonal Polyphenism

Ecdysone controls seasonal polyphenism in several species of butterflies. The European map butterfly *Araschnia levana* exhibits one of the most dramatic seasonal polyphenisms known. Larvae reared on long-day photoperiod develop into the black-and-white summer form, while larvae reared under short-day photoperiods enter diapause upon pupation and after overwintering emerge as the orange-and-brown spring form. The morphological differences are so dramatic that the two forms are easily mistaken for different species. When 20E is injected into short-day diapausing pupae, they commence adult development immediately and develop into the summer form, an intermediate form, or the spring form, depending entirely on the timing of the injection. If 20E is injected within three days after pupation they develop into summer forms, and if injected more than 10 days after pupation they develop into spring forms (Koch and Bückmann, 1987). Injections between 3 and 10 days result in varying degrees of intermediacy between the summer and spring phenotypes (figure 13.1). Thus, the hormone-sensitive period is about a week long, starting about three days after pupation. Long-day direct-developing pupae begin to secrete 20E to start adult development shortly after pupation, whereas short-day diapausing pupae do not secrete 20E until many months after pupation. The Asian sibling species, *Araschnia burejana*, has an identical seasonal polyphenism that is also controlled by 20E, as demonstrated by Keino and Endo (1973), who showed that 20E injections into diapausing pupae on the first day after pupation resulted in the development of summer-form adults, whereas injections five or more days after pupation produced normal spring-form adults.

The North American buckeye butterfly *Precis coenia* also has a seasonal polyphenism. In the summer, under long days the ventral wing surface of the adults is a pale straw-yellow color. In the autumn, under short-day conditions, the ventral wing surface is a dark reddish brown. Ecdysone injections into short-day pupae can result in development of the summer phenotype, and the 20E-sensitive period in this case is relatively brief, only 20 h long, from 28 to 48 h after pupation (Rountree and Nijhout, 1995a). The natural secretion of 20E that initiates adult development begins about 24 h after pupation under long-day conditions, but not until 48 h after pupation under short day conditions (Rountree and Nijhout, 1995a). The pupal stage is, correspondingly, about 24 h longer in short-day than in long-day pupae. This means that the rate and duration of adult development are not changed, and that the difference between the forms lies entirely in the timing of 20E secretion relative to the hormone-sensitive period.

Seasonal form in *Bicyclus anynana* is also controlled by the timing of 20E secretion, though in this species the environmental cue is temperature. At low temperatures, the ventral hind wing is dark with small eyespots, whereas at warm temperatures the eyespots on the ventral hind wing are large and the wing has a distinctive pale banding pattern (Brakefield and Reitsma, 1991). Injection of 20E early in the pupal stage of cool-temperature pupae induces development of the warm-season phenotype (Brakefield et al., 1998). When *Bicyclus* were selected for different development times, the short-development (fast) line developed much larger ventral hind wing eyespots than those selected for slow development. Ecdysone injection early in the pupal stage caused the slow-selected line to develop the larger eyespots characteristic of the fast line (Zijlstra et al., 2004).

Regulation of Polyphenism by JH
Juvenile hormone (JH) is involved in the regulation of many polyphenisms, including the caste polyphenisms in social insects and the various polyphenisms of aphids (Nijhout, 1994, 1999, 2003; see also chapter 14), but there are rather few cases in which photoperiodic effects are mediated by JH. Among the seasonal polyphenisms, many wing-length diphenisms are controlled by JH. In the grasshopper *Zonocerus variegatus*, allatectomy of nymphs results in the development of long-winged adults, and injection of JH causes the development of short-winged adults (McCaffery and Page, 1978). JH has also been implicated in the control of wing length dimorphism in crickets (Fairbairn and Roff, 1990).

Neurosecretory Hormones

Diapause Hormone
The diapause hormone (DH) is secreted by females of *Bombyx mori* and controls whether the eggs she lays will enter embryonic diapause. DH secretion is under photoperiodic control, but with an extraordinarily long interval between the environment-sensitive and hormone-sensitive periods. Embryos are sensitive to day length, and under long-day conditions develop into females that secrete DH and whose eggs will enter diapause (Fukuda, 1952; Hasegawa, 1957; Denlinger, 1985). Short-days experienced during embryonic life results in females that lay direct developing eggs. Diapausing eggs are darker than direct developing eggs due to the incorporation of ommochromes in the chorion. DH also controls the adult color pattern in the bivoltine Daizo race of *Bombyx mori* (Tsurumaki et al., 1999; Yamanaka et al., 2000). Adults that develop from larvae reared under short-day photoperiod and cool temperature develop into a pale-colored adult summer form and lay non-diapausing eggs. Larvae reared under long-day and warm conditions developed into

adults of the autumn form with a much darker coloration and a distinctive banding pattern on their wings. Injection of DH into short-day pupae causes them to develop into autumn forms or intermediates (Yamanaka et al., 2000).

Summer Morph-Producing Hormone

The Asian comma butterfly *Polygonia c-aureum* has distinctive summer and autumn forms that are induced by photoperiod and temperature. Under long-day conditions and warm temperatures, the adults that develop have a yellowish tan ground color (the summer form), and under short days and low temperatures, the adult wings are reddish brown to brown (the autumn form). The autumn form also has more sharply indented and angular wings than the summer form. When the brain was excised, or the medial neurosecretory cell were removed from the brain of long-day animals on the day of pupation, they developed into dark colored adults of the autumn phenotype. A small polypeptide can be extracted from the brains of *Polygonia* that, when injected into short-day pupae causes them to develop into summer-form adults (Endo et al., 1988; Masaki et al., 1989). The active peptide, called summer morph-producing hormone (SMPH), has been purified but not yet sequenced. It has an approximate molecular weight of 4.5 kDa, which is close to that of bombyxin, an insulin-like hormone of Lepidoptera. Tanaka et al. (1997) have shown that SMPH and bombyxin activity can be separated by ion exchange chromatography, so they are unlikely to be the same molecule. SMPH activity can be extracted from the brains of *Bombyx mori* (Inoue et al., 2005), a species that does not show a color response to this hormone, indicating that SMPH may have different functions in other species.

The Japanese swallowtail butterfly *Papilio xuthus* also has a photoperiodically determined adult polyphenism, but in this species it is associated with diapause. Long-day photoperiods during the larval stage leads to direct development and the summer adult form, which has a much larger body size and broader dark bands on a yellow background than does the spring form, which emerges after pupal diapause when larvae are reared on short days. When long-day pupae are decerebrated on day 0, they develop into adults of intermediate morphology. When short-day pupae are decerebrated on day 0, they enter permanent diapause but can be caused to develop with an injection of 20E. All such animals develop into the spring form (Endo et al., 1985). But when such pupae are injected with an extract from brains and subesophageal ganglia, they develop into the summer form (Ito et al., 2001). These experiments indicate that a brain factor that is not PTTH (which stimulates 20E secretion) that causes the development of the summer morph. This brain factor has also been called SMPH (Ito et al., 2001), by analogy to the factor that induces wing polyphenism in *Polygonia*.

The seasonal polyphenism of the Japanese small copper butterfly *Lycaena phlaeas daimio* is also controlled by photoperiod and mediated by a hormone that has also been called SMPH (Endo and Funatsu, 1985; Usui et al., 2004). It is not known whether the three SMPHs from *Polygonia*, *Papilio*, and *Lycaena* are the same molecule. In *Lycaena phlaeas*, injections of 20E together with a preparation that has SMPH activity have a synergistic effect in stimulating development of the summer form (Endo and Funatsu, 1985), so it is possible that the putative SMPH also has prothoracicotropic activity. *Lycaena phlaeas* also has a pupa color polyphenism that is controlled by photoperiod: long days induce beige pupae that emerge as brown-winged adults, and short days induce black diapausing pupae that emerge as red-winged adults (Usui et al., 2004).

Pupal Melanization Reducing Factor, Orange Pupa Inducing Factor, and Pupal Cuticle Melanizing Hormone

The pupae of many Lepidoptera exhibit a color polyphenism that adapts them to the background. The environmental cue is typically the color and/or texture of the substrate on which the larva pupates, and there is sometimes an effect of photoperiod because short days can induce pupal diapause and diapausing pupae occasionally have different pigmentation than nondiapausing pupae. The alternative colors are often controlled by neurosecretory hormones, and three endocrine factors that affect pupal coloration have been discovered in different species of butterflies: PMRF, OPIF, and PCMH (Bückmann and Maisch, 1987; Yamanaka et al., 1999, 2004, 2006; Jones et al., 2007).

In addition to the adult size and color polyphenism, *Papilio xuthus* also has a pupal color polyphenism. Larvae reared on long-day develop into green or brown pupae, depending on color and texture of the environment in which they pupate. Short days induce pupal diapause, and most (60–95%) larvae reared on short days develop into orange pupae, with the remainder split between green and brown coloration. If short-day pharate pupae are ligated, the parts posterior to the ligature become green upon pupation, and the parts anterior to the ligature become orange (Yamanaka et al., 2004). Interestingly, injection of extracts from brain and subesophageal ganglia into the posterior parts of ligated larvae shortly before pupation caused them to develop the brown phenotype, but extracts from the thoracic and abdominal ganglia caused development of the normal orange phenotype. These findings suggest that an orange-inducing factor (termed OPIF) is produced by the postcephalic ganglia, while the cephalic nervous system produced a brown-inducing factor, called PCMH. PCMH has also been described as the neurohormone that controls pupal coloration in *Inachis io* and *Papilio polyxenes* (Starnecker and Hazel, 1999). As in the case of SMPH discussed above, it is not known whether the activity that has been called PCMH in different species actually represents the same molecule.

Jones et al. (2007) have shown that in *Inachis io* PMRF, acting via cAMP, inhibits pupal melanization, but that in *Papilio polyxenes* PMRF stimulates brown (ommochrome) coloration. This suggests that the control of pupal color by PMRF has evolved independently in the Nymphalidae and Papilionidae. Interestingly, in *Papilio glaucus*, a species that only produces brown pupae, midbody ligation causes the posterior compartment to become green, and the brown color can be rescued with an injection of PMRF. The observation indicates that environmentally cued pupal color could evolve by facultative inhibition of PMRF release (Jones et al., 2007).

The Taxon-Specific Action of Neurohormones

Experimental studies have shown that a given neurosecretory hormone often produces quite different effects in different species. Conversely, the same physiological effect can be evoked by different hormones in different species. This suggests that there has been much adaptive evolution in the use of endocrine seasonal signal mediators (Koch et al., 1990; Starnecker and Hazel, 1999). The named neurohormones do not have a defined function; their function is taxon specific, and it is not safe to assume that when one observes the same effect in different species that one is dealing with the same molecule. A sobering example has been the misidentification of bombyxin, an insulin-like growth factor, as PTTH, the prothoracicotropic hormone, because bombyxin from *Bombyx mori* has PTTH activity when assayed in *Samia cynthia*. Bombyxin does not stimulate ecdysteroidogenesis in *Bombyx* itself (Kiriishi et al., 1992).

Among other instances of taxon-specific effects of neurohormones are the effects of SMPH and PMRF outlined above. Another, more elaborate, example are the melanization and reddish coloration hormone (MRCH), which controls larval coloration in response to crowding in the armyworm *Leucania separata* (Ogura, 1975), and the pheromone biosynthesis activating neuropeptides (PBAN). These two hormones are members of a large family of closely related neurohormones with diverse functions. Interestingly, the MRCH-I of *Leucania* (= *Pseudaletia*) *separata* is identical to PBAN-I of *Bombyx mori*, whereas the PBAN of *Leucania* is a much smaller member of the family, of abut half the molecular weight. The diapause hormone (DH) of *Bombyx mori*, which is a maternal hormone that induces diapause in the eggs, is also a member of the MRCH/PBAN family. It is worth noting that the DH is particularly enriched in the brains of male moths, which indicates it may have an additional role besides the control of egg diapause.

Another groups of structurally related neuropeptides is the AKH/RPCH family. Carbohydrate and lipid metabolism and storage are regulated by adipokinetic hormones (AKH) and hyperthrehalosemic hormones (HTH), which are structurally closely related. The are also very similar to the crustacean red pigment concentrating hormone (RPCH) and the corazonins, which control pigmentation in orthopteroids. The corazonins themselves are a family of small neurosecretory peptides that were, as their name indicates, initially identified as cardioactive peptides in *Periplaneta americana* (Veenstra, 1989). Corazonins were subsequently shown to be involved in the control of red-brown-black chromatic adaptation and color polyphenism in *Locusta migratoria* (Tanaka, 2000, 2006). In locusts, corazonin interacts with JH (which is required to produce a green color) to generate an extraordinarily broad range of color phenotypes ranging from pale tan to gray, brown, black, and red. Peptides that have corazonin activity in a *Locusta* bioassay can be extracted from the nervous system of many orders of insects that do not exhibit color differences.

In many studies, the physiological and developmental effects of neurohormones have been studied using semipurified extracts of the CNS, without clear knowledge of the actual molecules such an extract might contain. Since many neuropeptides are fairly similar in size and in physicochemical properties, such preparations may well contain multiple active factors, and observing the same physiological effect on different species does not guarantee they are responding to the same molecule.

HOW HORMONES CONTROL MORPHOLOGICAL DIFFERENCES

The most common seasonal effects on morphology are on body size, wing length, wing shape, wing pattern, wing pigmentation, and pupal pigmentation, either singly but more commonly in some combination. Insofar as hormones control postembryonic development, and differences in hormonal signaling account for phenotypic differences in seasonal polyphenisms, it is worth asking by what mechanism hormones can affect and alter morphology.

Ecdysone and bombyxin (or insulin-like hormones) are both required for cell division and growth (Nijhout and Grunert, 2002; Oldham et al., 2002; Nijhout et al., 2007). Thus, differences in the timing and amount of these hormones, or in the expression of their receptor and signaling pathway peptides can account for differences in body size. Differences in the proportion of body parts in alternative morphs can come about by tissue-specific differences in the receptor and response pathways. For instance, reduced expression of the insulin or 20E receptor in an imaginal disk will lead to reduced growth and a smaller final size of the corresponding adult appendage. Programmed cell death, or apoptosis, plays an important role in insect morphogenesis. The shape of lepidopteran wings is determined largely by the pattern of apoptosis of the outer margin of the wing imaginal disk during the pupal stage (Nijhout, 1991). Changes in the spatial pattern of apoptosis presumably account for the different wing shapes in some seasonal forms. Brachypterous and apterous seasonal morphs are common, and although the developmental mechanism has not been studied, it is likely that selective apoptosis plays a role, by analogy to the regulation of genetic wing polymorphisms. Winglessness in dimorphic alate/wingless moths is due to apoptosis of wing imaginal disks during metamorphosis. The wing imaginal disks grow normally during the larval sage, but after pupation they undergo selective apoptosis (Niitsu, 2001; Lobbia et al., 2003).

The *Precis* polyphenism involves a change in the kinds of pigments synthesized on one surface of the wing. The dark brown color of the short-day form is due to a mixture of the ommochrome pigments dihydroxanthommatin and ommatin-D, whereas the yellowish pigmentation of the long-day form is due to xanthommatin (Nijhout, 1997). So the color polyphenism involves the differential expression of a single enzyme to take xanthommatin to ommatin-D, and a mechanism, presumably a protein, to stabilize the alternative redox forms of xanthommatin. Pupal color polyphenisms appear to involve primarily differences in ommochrome and melanin synthesis.

Lepidopteran color pattern polyphenisms involve changes in the locations where pigments are synthesized, and in the dimensions of pattern elements like eyespots. The dry-season form of *Bicyclus*, and the short-day form of *Precis* have smaller eyespots than their respective wet-season and long-day forms. Short-day and dry-season forms have more cryptic patterns than their counterparts, and some species have a distinctive seasonal dry-leaf–mimicking pattern. In *Arashnia*, the changes involve dramatic differences in the color pattern and distinctive though somewhat more subtle differences in body size and various morphometric features (Fric and Konvicka, 2002).

Seasonal color pattern differences appear to be controlled by the timing of 20E. Early exposure to 20E in the pupal stage tends to cause development of the long-day and wet-season form in *Araschnia*, *Precis*, and *Bicyclus*, respectively. Injection of 20E early in the pupal stage of cool-temperature pupae induces development of the warm-season phenotypes whereas a delayed exposure to 20E causes develop ment of the short-day/dry-season forms (Koch and Bückmann, 1987; Rountree and Nijhout, 1995a; Brakefield et al., 1998; Zijlstra et al., 2004). An indication of the underlying mechanisms by which a timing shift in 20E section can cause a change in spatially patterned gene expression is given by the work of Koch et al. (2003), who showed that there is a changing spatial pattern of expression of the 20E receptor in the developing wings of *Precis coenia*. If this pattern of variation is independent of the timing of 20E secretion, then a shift in the time of 20E secretion would impinge on a different spatial pattern of the 20E receptor, and thus would result in a spatially altered pattern of 20E stimulated gene expression.

EVOLUTION OF SEASONALLY PLASTIC TRAITS

Genetic and developmental processes must obey the laws of chemistry and physics and are thus likely to be sensitive to environmental variables like temperature, pH, hydration, and the availability of substrates for energy, growth, and maintenance. This means that most traits are environment sensitive, or plastic, unless some mechanism has evolved to eliminate the environmental variable or to buffer developmental sensitivity (Nijhout, 2003). Physiological homeostatic mechanisms are examples of the former; genetic and developmental robustness mechanisms are examples of the latter (Nijhout, 2002; Wagner, 2005).

In some cases plasticity is advantageous because the different phenotypes are more suited to different environments. In such cases it is possible for selection to gradually improve the "fit" of a range of plastic phenotypes to their respective environments (Via and Lande, 1985). If the temporal variation in the environment is coarse-grained relative to the life cycle, so that successive generations experience different environments, then a seasonal polyphenism will be favored over monophenism (Roff, 1992). Quantitative theories for the evolution and maintenance of alternative phenotypes have been developed by Hazel et al. (1990) and Roff (1992). These theories all assume that there is initially a reaction norm, or a continuously variable phenotype that is a function of an environmental variable, that through selection becomes a threshold with only two discrete alternative phenotypes over a range of environments (Roff, 1996; Schlichting and Pigliucci, 1998). Further selection can shift the position of his threshold relative to the environmental gradient (Hazel et al., 1990; Schlichting and Pigliucci, 1998; Moczek and Nijhout, 2003; Tomkins et al., 2005).

Other hypotheses about the evolutionary path that could give rise to a polyphenism are those of Bradshaw (1973), who suggested that polyphenisms might arise from preexisting genetic polymorphisms, and Shapiro (1976), who suggested they may arise by genetic assimilation from phenotypically plastic traits. The idea that has the most traction today is that polyphenisms arise via the Baldwin effect, by genetic accommodation (West-Eberhard, 2003). Genetic accommodation is a process by which a novel phenotype becomes established in a population through changes in the genetic background, rather than changes in the genes that specifically lead to a trait. The effect is polygenic and is driven by selection on phenotypic variants that are produced by either mutation or environmental change. The key difference between genetic accommodation and traditional theories of the evolution of traits is that the origin of a trait does not require initially favorable mutations whose frequency in a population is increased through selection. Rather, an event is required that reveals preexisting but "hidden" genetic variation, that in the wild type is buffered by some kind of homeostatic mechanism that stabilizes the phenotype. This event can be a mutation or an environmental change that disturbs the homeostatic mechanism and that allows the cryptic genetic variation to be revealed as phenotypic variation. Selection on extreme variants will cause a polygenic change in the genetic background that can shift the phenotypic mean and eventually stabilize a novel phenotype. Genetic assimilation (Waddington, 1953) is a special case of genetic accommodation (West-Eberhard, 2003; Braendle and Flatt, 2006). Evolution of a polyphenism from an initially monophenic species by means of genetic accommodation was demonstrated by Suzuki and Nijhout (2006, 2008).

Many biologists have wondered whether the alternative forms in a seasonal polyphenism can provide the basis for speciation through the genetic fixation and subsequent genetic isolation of the two alternative forms (reviewed in West-Eberhard, 2003). Although there do not appear to be any cases of species divergence, there are a number of instances in which one of the alternative forms of a polyphenism has become fixed in a population. In the Pieridae, there is evidence that one of the morphs in a photoperiodically induced seasonal polyphenisms can become fixed in a population. In western North America, *Pieris occidentalis* has a photoperiodically controlled polyphenism, with a pale-colored morph in warm seasons and a darker colored morph in cool seasons. In Alaska, *Pieris occidentalis* only occurs as the cool-season form, but when brought into the laboratory and reared under warm long-day conditions, the Alaska population was able to develop the long-day warm-season form never found in Alaska (Shapiro, 1976).

More important, Shapiro (1971) showed that *Pieris virginiensis*, a species that is entirely monophenic in nature, can be made to develop the summer phenotype characteristic of its polyphenic sister species *Pieris napi*, by rearing larvae under continuous light and high temperatures in the laboratory. It is therefore possible that

P. virginiensis evolved from a polyphenic ancestor by fixation of the cool-season form, but evidently maintained a latent ability to develop the alternate form, which could be evoked under extreme conditions.

Another example in which a seasonal polyphenism has acquired a genetic basis is found in *Precis coenia*, where the seasonal wing color polyphenism that is controlled by day length also occurs as a genetic polymorphism that is controlled by a single gene, with the recessive allele giving the dark (short-day) phenotype (Rountree and Nijhout, 1995a, 1995b). The phenotype of this dark genetic morph is not sensitive to photoperiod or temperature, so if this gene would go to fixation, the population would become genetically monophenic and monomorphic. Zijlstra et al. (2004) have shown that in *Bicyclus anynana* alternative phenotypes in a seasonal polyphenism can become fixed by selection on development time.

Thus, laboratory studies have shown that it is possible for polyphenic alternative phenotypes to become genetically fixed via three mechanisms: by selection on expressed genetic variation, as in *Bicyclus*, by genetic accommodation though polygenic selection on cryptic genetic variation, as in *Manduca* (Suzuki and Nijhout, 2006, 2008), and by mutations at a single gene, as in *Precis* (Rountree and Nijhout, 1995a, 1995b). Whether these processes actually play a role in the fixation of polygenic alternatives in nature, and whether this can be the basis of speciation are open questions.

FUTURE DIRECTIONS

The general mechanism by which photoperiodic cues induce morphological change is now well understood. In all cases that have been studied so far, the brain is required and acts via the secretion of neurohormones. In some cases, these neurohormones appear to act directly on developmental processes, and in others they act indirectly by controlling the secretion of other hormones like 20E, and presumably, JH. But the brain-neurohormone axis resides somewhere in the middle of a long causal chain that links photoperiod to morphology, and many of the steps upstream and downstream of the endocrine axis remain unknown. What is missing upstream is an understanding of how photoperiodic signals are stored in the brain, and how this information controls the release of neurohormones. Downstream, we are still largely in the dark about the mechanisms by which hormones induce specific alternative developmental pathways, other than to say that they induce changes in gene expression. Resolving these questions will require a massive research effort, and the answer to some may well be beyond the reach of current experimental technology. Intermediate goals are achievable, however, and I outline a few that I see as most pressing and accessible.

1. It is necessary to resolve interaction between photoperiod and temperature. Most previous work on the interaction between temperature and photoperiod has been done in the context of diapause induction, but the same principles

should apply to the induction of alternative phenotypes, since both involve changes in hormone secretion. Although many studies have shown that temperature and photoperiod can substitute for each other, or that the two act additively, or synergistically, the quantitative data are rather spotty and do not clearly differentiate between these three possible modes of interaction. A multilevel factorial design (at least 4×4) across a reasonably broad range of photoperiods and temperatures would be desirable. In species that have a broad latitudinal range, the critical photoperiod varies with the latitude to which a given population is adapted. It would be nice to know whether southern populations are also more sensitive to temperature induction than northern ones. In the extreme one would expect that northern populations are not sensitive to temperature and southern populations are not sensitive to photoperiod. If such crossing gradients of temperature and photoperiod responsiveness could be demonstrated in a single species, it would open the door to a genetic dissection of their interaction.

2. It would be useful to have a good model organism for photoperiodically cued morphological change. Like other models systems it should be easy to rear in large numbers and have a short generation time. Obviously, it should have simple but unambiguous alternative forms, and preferably be associated with diapause to give it the broadest possible appeal for research.

3. To understand what exactly happens during a critical period, we need tissue-specific gene expression data of high temporal resolution during and after a hormone-induced developmental switch. To date, gene expression and proteomic studies have been done mostly on whole-body preparations. These studies have revealed that there are major changes in gene expression that accompany these developmental switches, but the data are overwhelming in their detail and conflate the responses of different tissues, so it is impossible to infer a causal chain of events. Improved methods that can use ever smaller tissue samples are now becoming available, and these will allow us dissect developmental switches with the requisite spatial and temporal resolution.

4. A number of neurohormones appear to play a role in the control of seasonal polyphenisms. Most of these have been characterized as polypeptides, but none have so far been sequenced. Insect neurosecretory hormones can have different effects in different species, so it will be important to tie the names and activities of the known putative hormones to specific peptide sequences. Receptors need to be identified, and the kinetics of a broad panel of ligand–receptor interactions needs to be done to establish specificity. Surveys of the taxonomic distribution of ligands and receptors need to be done to elucidate evolution of function. Function does not appear to be associated with molecular structure, which suggests a recent evolution of current function. These neurohormones often act at the same time that 20E is secreted at the time of the metamorphic molt. It will therefore be important to determine whether they are stimulated by the same factors that stimulate PTTH

secretion, and to elucidate how these neurohormones interact with 20E to induce the development of alternative phenotypes.

Acknowledgments I thank Bernd Koch for the *Araschnia* specimens, and Philip Ackery of the British Museum for taking the photographs of the seasonal forms of *Precis*. Supported by grants IBN-9728727 and IBN-0315897 from the National Science Foundation.

References

Beck SD. (1980). *Insect Photoperiodism*. New York: Academic Press.

Bradshaw WE. (1973). Homeostasis and polymorphism in vernal development of *Chaoborus americanus*. Ecology 54: 1247–1259.

Braendle C and Flatt T. (2006). A role for genetic accommodation in evolution? BioEssays 28: 868–873.

Brakefield PM, Kesbeke F, and Koch PB. (1998). The regulation of phenotypic plasticity of eyespots in the butterfly *Bicyclus anynana*. Am Nat 152: 853–860.

Brakefield PM and Reitsma N. (1991). Phenotypic plasticity, seasonal climate and the population biology of *Bicyclus* butterflies (Satyridae) in Malawi. Ecol Entomol 16 : 291–303.

Bünning E. (1973). *The Physiological Clock*. New York: Springer-Verlag.

Bückmann D and Maisch A. (1987). Extraction and partial purification of the pupal melanization reducing factor (PMRF) from *Inachis io* (Lepidoptera). Insect Biochem 17: 841–844.

Cassier P and Papillion M. (1968). Effects des implantations des corps alates sur la reproduction de femelles groupées de *Schistocerca gregaria* (Forsk.) et sur le polymorphisme de leur descendance. C R Acad Sci 266D: 1048–1051.

Cherbas P and Cherbas L. (1996). Molecular aspects of ecdysteroids hormone action. In *Metamorphosis* (LI Gilbert, JR Tata, and BG Atkinson, eds). New York: Academic Press, pp. 175–221.

Danilevskii AS. (1965). *Photoperiodism and Seasonal Development of Insects*. Edinburgh: Oliver and Boyd.

Denlinger DL. (1985). Hormonal control of diapause. In *Comprehensive Insect Physiology Biochemistry and Pharmacology*, vol 8 (GA Kerkut and LI Gilbert, eds). Oxford: Pergamon Press, pp. 353–412.

Endo K and Funatsu S. (1985). Hormonal control of seasonal morph determination in the swallowtail butterfly, *Papilio xuthus* L. (Lepidoptera: Papilionidae). J Insect Physiol 31: 669–674.

Endo K, Masaki T, and Kumagai K. (1988). Neuroendocrine regulation of the development of seasonal morphs in the Asian comma butterfly, *Polygonia c-aureum* L.: Difference in activity of summer morph-producing hormone from brain-extracts of long-day and short-day pupae. Zool Sci 5: 145–152.

Endo K, Yamashita I, and Chiba Y. (1985). Effect of photoperiodic transfer and brain surgery on the photoperiodic control of pupal diapause and seasonal morphs in the swallowtail *Papilio xuthus*. Appl Entomol Zool 20: 470–478.Fairbairn DJ, and Roff, DA. (1990). Genetic correlations among traits determining migratory tendency in the sand cricket, *Gryllus firmus*. Evolution 44: 1787–1795.

Fric Z and Konvicka M. (2002). Generations of the polyphenic butterfly *Araschnia levana* differ in body design. Evol Ecol Res 4: 1017–1032.

Fujiwara H, Jindra M, Newitt R, Palli SR, Kiyoshi Hiruma, and Riddiford LM. (1995). Cloning of an ecdysone receptor homolog from *Manduca sexta* and the developmental profile of its mRNA in wings. Insect Biochem Mol Biol 25: 845–856.

Fukuda S. (1952). Function of the pupal bran and subesophageal ganglion in the production of diapause and non-diapause eggs in the silkworm. Annot Zool Jpn 25: 149–155.

Gibbs D. (1975). Reversal of pupal diapause in *Sarcophaga argyrostoma* by temperature shifts after puparium formation. J Insect Physiol 21: 1179–1186.

Harrison RG. (1980). Dispersal polymorphisms in insects. Annu Rev Ecol Syst 11: 95–118.

Hasegawa K. (1957). The diapause hormone of the silkworm, *Bombyx mori.* Nature 179: 1300–1301.

Hazel WN, Smock R, and Johnson MD. (1990). A polygenic model for the evolution and maintenance of conditional strategies. Proc R Soc Lond B 242: 181–187.

Hazel WN and West DA. (1983). The effect of larval photoperiod on pupal coloration and diapause in swallowtail butterflies. Ecol Entomol 8: 37–42.

Honek A. (1980). Maternal regulation of wing polymorphism in *Pyrrhocoris apterus*: Effect of cold activation. Experientia 36: 418–419.

Inoue M, Sakamioto S, Okuhira R, Yamanaka A, Islam ATMF, and Endo K. (2005). Purification of *Bombyx* neuropeptide showing summer-morph-producing-hormone (SMPH) activity in the Asian comma butterfly, *Polygonia c-aureum* [abstract]. J Insect Sci 5: 22.

Ito T, Yamanaka A, Tanaka H, Watanabe M, and Endo K. (2001). Evidence for the presence of summer-morph-producing hormone in the swallowtail butterfly, *Papilio xuthus* L. (Lepidoptera, Papilionidae). Zool Sci 18: 1117–1122.

Jindra M, Malone F, Hiruma K, and Riddiford LM. (1996). Developmental profiles and ecdysteroid regulation of the mRNAs for two ecdysone receptor isoforms in the epidermis and wings of the tobacco hornworm, *Manduca sexta.* Dev Biol 180: 258–272.

Jones G and Sharp PA. (1997). Ultraspiracle: An invertebrate nuclear receptor for juvenile hormones. Proc Natl Acad Sci USA 94: 13499–13503.

Jones G, Wozniak M, Chu YX, Dhar S, and Jones D. (2001). Juvenile hormone III dependent conformational changes of the nuclear receptor ultraspiracle. Insect Biochem Mol Biol 32: 33–49.

Jones M, Rakes L, Yochum M, Dunn G, Wurster S, Kinney K, and Hazel W. (2007). The proximate control of pupal color in swallowtail butterflies: Implications for the evolution of environmentally cued pupal color in butterflies (Lepidoptera: Papilionidae). J Insect Physiol 53: 40–46.

Keino H and Endo K. (1973). Studies on the determination of seasonal forms in the butterfly, *Araschnia burejana* Bermer. Zool Mag 82: 48–62.

Kimura T and Masaki S. 1977 Brachypterism and seasonal adaptation in Orgyia thyellina Butler (Lepidoptera, Lymantriidae). Jpn J Entomol 45: 97–106.

Kingsolver JG and Wiernasz DC. (1991). Seasonal polyphenism in wing-melanin pattern and thermoregulatory adaptation in *Pieris* butterflies. Am Nat 137: 816–830.

Kiriishi S, Nagasawa H, Kataoka H, Suzuki A, and Sakurai S. (1992). Comparison of the in vivo and in vitro effects of bombyxin and prothoracicotropic hormone on prothoracic glands of the silkworm, *Bombyx mori.* Zool Sci 9: 149–155.

Koch PB and Bückmann D. (1987). Hormonal control of seasonal morphs by the timing of ecdysteroid release in *Araschnia levana* (Nymphalidae: Lepidoptera). J Insect Physiol 33: 823–829.

Koch PB, Merk R, Reinhardt R, and Weber P. (2003). Localization of ecdysone receptor protein during colour pattern formation in wings of the butterfly *Precis coenia* (Lepidoptera: Nymphalidae) and co-expression with Distal-less protein. Dev Genes Evol 212: 571–584.

Koch PB, Starnecker G, and Bückmann D. (1990). Interspecific effects of the pupal melanization reducing factor on pupal colouration in different lepidopteran families. J Insect Physiol 36: 159–161.

Lobbia S, Niitsu S, and Fujiwara H. (2003). Female-specific wing degeneration caused by ecdysteroid in the tussock moth, *Orgyia recens*: Hormonal and developmental regulation of sexual dimorphism. J Insect Sci 3: 11.

Masaki T, Endo K, and Kumagai K. (1989). Neuroendocrine regulation of the development of seasonal morphs in the Asian comma butterfly, *Polygonia c-aureum* L.: Stage-dependent changes in the activity of summer-morph-producing hormone of the brain-extracts. Zool Sci 6: 113–119.

McCaffery AR and Page WW. (1978). Factors influencing the production of long-winged *Zonoceros variegatus.* J Insect Physiol 24: 465–472.

Moczek AP and Nijhout HF. (2003). Rapid evolution of a polyphenic threshold. Evol Dev 5: 259–268.

Mole S and Zera AJ. (1993). Differential allocation of resources underlies the dispersal-reproduction trade-off in the wing-dimorphic cricket, *Gryllus rubens*. Oecologia 93: 121–127.

Mousseau TA and Dingle H. (1991). Maternal effects in insect life histories. Annu Rev Entomol 36: 511–534.

Niitsu S. (2001). Wing degeneration due to apoptosis in the female of the winter moth *Nyssiodes lefuarius* (Lepidoptera, Geometridae). Entomol Sci 4: 1–7.

Nijhout HF. (1983). Definition of a juvenile hormone sensitive period in *Rhodnius prolixus*. J Insect Physiol 29: 669–677.

Nijhout HF. (1994). *Insect Hormones*. Princeton, NJ: Princeton University Press.

Nijhout HF. (1997). Ommochrome pigmentation of the linea and rosa seasonal forms of *Precis coenia* (Lepidoptera: Nymphalidae). Arch Insect Biochem Physiol 36: 215–222.

Nijhout HF. (1999). Control mechanisms of polyphenic development in insects. BioScience 49: 181–192.

Nijhout HF. (2002). The nature of robustness in development. BioEssays 24: 553–563.

Nijhout HF. (2003). Development and evolution of adaptive polyphenisms. Evol Dev 5: 9–18.

Nijhout HF and Grunert LW. (2002). Bombyxin is a growth factor for wing imaginal disks in Lepidoptera. Proc Natl Acad Sci USA 99: 15446–15450.

Nijhout HF, Smith WA Schachar, I, Subramanian S, Tobler A, and Grunert LW. (2007). The control of growth and differentiation of the wing imaginal disks of *Manduca sexta*. Dev Biol 302: 569–576.

Ogura N. (1975). Hormonal control of larval coloration in the armyworm, *Leucania separata*. J Insect Physiol 21: 559–576.

Oldham S, Stocker H, Laffargue M, Wittwer F, Wymann M, and Hafen E. (2002). The *Drosophila* insulin/IGF receptor controls growth and size by modulating PtdInsP3 levels. Development 129: 4103–4109.

Pittendrigh CS. (1972). Circadian surfaces and the diversity of possible roles of circadian organization in photoperiodic induction. Proc Natl Acad Sci USA 69: 2734–2737.

Pittendrigh CS, Kyner WT, and Takamura T. (1991). The amplitude of circadian oscillations: Temperature dependence, latitudinal clines, and photoperiodic time measurement. J Biol Rhythms 6: 299–313.

Riddiford LM. (1970). Prevention of metamorphosis by exposure of insect eggs to juvenile hormone analogs. Science 167: 287–288.

Roff DA. (1992). *The Evolution of Life Histories*. New York: Chapman and Hall.

Roff DA. (1996). The evolution of threshold traits in animals. Q Rev Biol 71: 3–35.

Roff DA and Fairbairn DJ. (1991). Wing dimorphisms and the evolution of migratory polymorphisms among the Insecta. Am Zool 31: 243–251.

Rountree DB and Nijhout HF. (1995a). Hormonal control of a seasonal morph in *Precis coenia* (Lepidoptera: Nymphalidae). J Insect Physiol 41: 987–992.

Rountree DB and Nijhout HF. (1995b). Genetic control of a seasonal morph in *Precis coenia* (Lepidoptera: Nymphalidae). J Insect Physiol 41: 1141–1145.

Sasaki R, Nakasuji F, and Fujisaki K. (2002). Environmental factors determining wing form in the lygaeid bug, *Dimorphopterus japonicus* (Heteroptera: Lygaeidae). Appl Entomol Zool 37: 329–333.

Saunders DS. (2002). *Insect clocks*. Amsterdam: Elsevier.

Schlichting CD and M. Pigliucci. (1998). *Phenotypic Evolution: A Reaction Norm Perspective*. Sunderland, MA: Sinauer Associates.

Scriber JM and Slansky F. (1981). The nutritional ecology of immature insects. Annu Rev Entomol 26: 183–211.

Shapiro AM. (1968). Photoperiodic induction of vernal phenotype in *Pieris protodice* Boisduval and Le Conte (Lepidoptera: Pieridae). Wasmann J Biol 26: 137–149.

Shapiro AM. (1971). Occurrence of a latent polyphenism in *Pieris virginiensis* (Lepidoptera: Pieridae). Entomol News 82: 13–16.

Shapiro AM. (1973). Generational "carryover" and the suppression of submarginal pattern elements in vernal phenotypes of *Pieris protodice* (Lepidoptera: Pieridae). Entomol News 84: 294–298.

Shapiro AM. (1976). Seasonal polyphenisms. Evol Biol 9: 259–333.

Shapiro AM. (1978). The evolutionary significance of redundancy and variability in phenotypic-induction mechanisms of pierid butterflies (Lepidoptera). Psyche 85: 275–283.

Shapiro AM. (1980). Convergence in pierine polyphenisms (Lepidoptera). J Nat Hist 14: 781–802.

Shapiro AM. (1982). Redundancy in pierid polyphenisms: Pupal chilling induces vernal phenotype in *Pieris occidentalis* (Pieridae). J Lepidopt Soc 36: 174–177.

Shapiro AM. (1984). The genetics of seasonal polyphenism and the evolution of "general purpose genotypes" in butterflies. In *Population Biology and Evolution* (K Woermann and V Loeschcke, eds). Berlin: Springer-Verlag, pp. 16–30.

Smith KC. (1993). The effects of temperature and daylight on the rosa polyphenism in the buckeye butterfly, *Precis coenia* (Lepidoptera: Nymphalidae). J Res Lepidopt 30: 225–236.

Starnecker G and Hazel W. (1999). Convergent evolution of neuroendocrine control of phenotypic plasticity in pupal colour in butterflies. Proc R Soc Lond B 266: 2409–2412.

Suzuki Y and Nijhout HF. (2006). Evolution of a polyphenism by genetic accommodation. Science 311: 650–652.

Suzuki Y and Nijhout HF. (2008). Genetic basis of adaptive evolution of a polyphenism by genetic accommodation. J Evol Biol 21: 57–66.

Tanaka D, Sakurama T, Mitsumasu K, Yamanaka A, and Endo K. (1997). Separation of bombyxin from a neuropeptide of *Bombyx mori* showing summer-morph-producing hormone (SMPH) activity in the Asian comma butterfly, *Polygonia c-aureum* L. J Insect Physiol 43: 197–201.

Tanaka S. (2000). Hormonal control of body-color polymorphism in *Locusta migratoria*: Interaction between [His^7]-corazonin and juvenile hormone. J Insect Physiol 46: 1535–1544.

Tanaka S. (2006). Corazonin and locust phase polyphenism. Appl Entomol Zool 41: 179–193.

Tauber MJ, Tauber CA, and Masaki S. (1986). *Seasonal Adaptations of Insects*. Oxford: Oxford University Press.

Tomkins JL, Kotiaho JS, and LeBas NR. (2005). Matters of scale: Positive allometry and the evolution of male dimorphisms. Am Nat 165: 389–402.

Truman JW. (1996). Metamorphosis of the insect nervous system. In *Metamorphosis* (LI Gilbert, JR Tata, and BG Atkinson, eds). New York: Academic Press, pp. 283–320.

Tsurumaki J, Ishiguro J, Yamanaka A, and Endo K. (1999). Effects of photoperiod and temperature on seasonal morph development and diapause egg oviposition in a bivoltine race (Daizo) of the silkmoth, *Bombyx mori* L. J Insect Physiol 45: 101–106.

Usui Y, Yamanaka A, Islam ATMF, Shahjahan R, and Endo K. (2004). Photoperiod-and temperature-dependent regulation of pupal beige/black polymorphism in the small copper butterfly, *Lycaena phlaeas daimio* Seitz. Zool Sci 21: 835–839.

Veenstra J. (1989). Isolation and structure of corazonin, a cardioactive peptide from the American cockroach. FEBS Lett 250: 231–234.

Via S and Lande R. (1985). Genotype-environment interaction and the evolution of phenotypic plasticity. Evolution 39: 505–522.

Waddington CH. (1953). Genetic assimilation of an acquired character. Evolution 7: 118–126.

Wagner A. (2005). *Robustness and Evolvability in Living Systems*. Princeton, NJ: Princeton University Press.

West-Eberhard MJ. (2003). *Developmental Plasticity and Evolution*. Oxford: Oxford University Press.

Wheeler DE and Nijhout HF. (2003). A perspective for understanding the modes of juvenile hormone action as a lipid signaling system. BioEssays 25: 994–1001.

Wiklund C and Tullberg BS. (2004). Seasonal polyphenism and leaf mimicry in the comma butterfly. Anim Behav 68: 621–627.

Willis JH. (1969). The programming of differentiation and its control by juvenile hormone in saturniids. J Embryol Exp Morphol 22: 27–44.

Yamanaka A, Adachi M, Imai H, Uchiyama T, Inoue M, Islam ATMF, Kitazawa C, and Endo K. (2006). Properties of orange-pupa-inducing factor (OPIF) in the swallowtail butterfly, *Papilio xuthus* L. Peptides 27: 534–538.

Yamanaka A, Endo K, Nishida H, Kawamura N, Hatase Y, Kong W, Kataoka H, and Suzuki A. (1999). Extraction and partial characterization of pupal-cuticle-melanizing hormone (PCMH) in the swallowtail butterfly, *Papilio xuthus* L. (Lepidoptera, Papilionidae). Zool Sci 16: 261–268.

Yamanaka A, Imai H, Adachi M, Komatsu M, Islam ATMF, Kodama I, Kitazawa C, and Endo K. (2004). Hormonal control of the orange coloration of diapause pupae in the swallowtail butterfly, *Papilio xuthus* L. (Lepidoptera: Papilionidae). Zool Sci 21: 1049–1055.

Yamanaka A, Tsurumaki J, and Endo K. (2000). Neuroendocrine regulation of seasonal morph development in a bivoltine race (Daizo) of the silkmoth, *Bombyx mori* L. J Insect Physiol 46: 803–808.

Zaslavskii VA. (1988). *Insect Development, Photoperiodic and Temperature Control*. Berlin: Springer Verlag.

Zera AJ and Harshman LG. (2001). The physiology of life history trade-offs in animals. Annu Rev Ecol Syst 32: 95–126.

Zijlstra WG, Steigenga MJ, Koch PB, Zwaan BJ, and Brakefield PM. (2004). Butterfly selected lines explore the hormonal basis of interactions between life histories and morphology. Am Nat 163: E76–E87.

14

Photoperiodism in Insects: Aphid Polyphenism

Jim Hardie

Although the intricacies of the aphid life cycle had been reported in the mid-1700s, the role of the environment in controlling phenotypic expression was not realized until much later (Lees, 1966). Almost 90 years ago, Garner and Allard (1920) first showed that plants responded to day length, and working in the same institute was S. Marcovitch, an agricultural entomologist pondering the life history of the strawberry root aphid *Aphis forbesi*. Just a few years later, Marcovitch published the first papers on the photoperiodic response in animals (1923, 1924). Those studies showed that, despite this being a root-dwelling aphid, artificially shortening summer day lengths led to the early appearance of sexual morphs, while experimentally extending the short days of autumn prevented the formation of sexual morphs, and asexual morphs continued to be produced. As a root feeder, it seemed possible that the photoperiodic effect was indirect and operating via the plant but in 1928 Shull showed that photoperiod directly affected the potato aphid *Macrosiphum euphorbiae*. These events pioneered the study of animal photoperiodism.

Aphids are hemipteran bugs and belong to the series Sternhorryncha, superfamily Aphidoidea, which in turn comprises three main groupings. Adelgidae and Phylloxeridae are different from other aphids and are solely oviparous, and perhaps the best known aphids reside in the family Aphididae, where parthenogenetic forms are viviparous but the sexual females (oviparae) are oviparous (Blackman and Eastop, 2000). Many species within Aphididae show cyclical parthenogenesis where asexual and sexual reproduction alternate at different times of the year (Moran, 1992; Hales et al., 1997). The ancestral aphid appears to have been an egg-laying, sexually reproducing winged insect originating in the Carboniferous/Permian period (359–351 million years ago), with the oldest fossil from the Triassic period (200–250 million years ago; Dixon, 1998). Parthenogenesis probably appeared prior to this fossil but was initially via oviposition with viviparity developing sometime later. Viviparity is associated with parthenogenesis in extant Aphididae species, none of which reproduce solely by sexual means (Dixon 1998). Under continuous long-day conditions, many aphid species will continue to reproduce parthenogenetically indefinitely. Thus, there are clones in our laboratory that have been reared parthenogenetically for 50–60 years—*Megoura viciae* (from at least 1953 when obtained from Rothamsted Research in the United

Kingdom; Lees, 1959) and *Aphis fabae* (collected from Cambridge in 1946; Kennedy and Booth, 1951).

Like many other insect photoperiodic responses, the final short-day effect in aphids is the production of a diapause condition that allows the insect to survive adverse winter conditions. However, due to the complexities of the life cycle, it is not usual to look for diapause per se. Diapause in the egg stage is most common in aphids, with the egg being the result of sexual reproduction in autumn. Diapause appears similar to other insects (Shingleton et al. 2003) but has reduced the effectiveness of aphids as a genetic tool because diapause can be broken only by a period of six to eight weeks of chilling, and then an interval timer prevents the induction of further sexual stages by short days for another few months (e.g., Lees 1960, Campbell and Treigidga, 2006; see "Endocrine/ Neuroendocrine Effector System," below). Thus, rearing aphids through the sexual cycle is very slow. On the other hand, parthenogenetic reproduction is fast with telescoping of generations, where the most mature embryos in the ovaries of a fourth-stadium (last nymphal stage) aphid already contain developing embryos of the next generation (3 in 1). This phenomenon, together with viviparity and rapid, albeit temperature dependent, development time does allow for particularly manageable experimental studies; an advantage here is that the offspring of a single mother are genetically identical (Blackman, 1987), and polyphenic studies can be based on a single genotype. Random mutation will, of course, affect the aphid genotype, but this will progress quite slowly. Thus, the genomes of our laboratory clones may not be precisely identical to those of the original females (see above), but when placed into short-day conditions, they still readily respond and produce sexual forms.

The life cycles of aphids are extremely variable and often highly polyphenic. Although some species/clones reproduce solely parthenogenetically, both winged and wingless adult forms can develop. Where cyclical parthenogenesis occurs, the sexual phenotypes are also represented, and some 10% of aphid species show an alternation between a winter or primary host plant, usually a tree or shrub, and a summer or secondary herbaceous host (Eastop, 1977; Moran, 1992). Such host alternation (heteroecy) offers the insect sequential access to nutrients as well as an appropriate situation for oviposition of the diapause eggs. With such life cycle arrangements, up to seven different phenotypes can occur. An example is the life cycle of the damson-hop aphid, *Phorodon humuli*, showing cyclical parthenogenesis and host alternation (figure 14.1). *Phorodon humuli* is unusual in that it can overwinter on a number of *Prunus* species but has a single summer host plant, *Humulus lupulus*. Other host-alternating species tend to be substantially monophagous for the winter host but oligophagous/polyphagous during the summer. *Phorodon humuli* is also unusual in that the summer generations on the hop are all wingless with no wing dimorphism (Campbell, 1985). Winged females colonize the summer host in spring, and the winged males and gynoparae (winged parthenogenetic females that migrate from summer host to the winter host, where they give birth to sexual females, oviparae; see figure 14.2) leave in

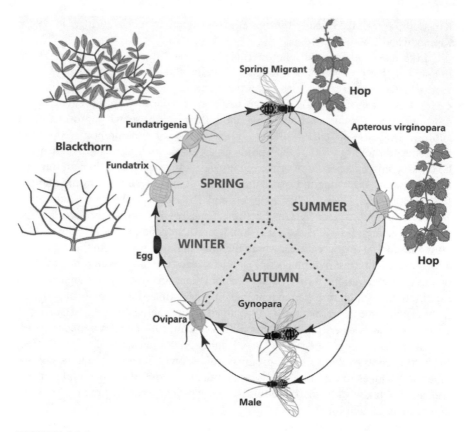

FIGURE 14.1.
The life cycle of the damson-hop aphid *Phordon humuli* showing the overwintering egg laid in buds of *Prunus* trees/shrubs, for example, *P. domestica* (plum), *P. insititia* (damson), *P. mahaleb* (St. Lucie cherry), or *P. spinosa* (blackthorn). The eggs hatch in spring to give rise to the first parthenogenetic, viviparous generation (fundatrix or stem mother, and a distinct morphological phenotype). These in turn give birth to the second generation on the winter host, termed the fundatrigeniae (generation born to the fundatrices). Toward the end of spring, winged adult females appear and fly to the summer host, hop (*Humulus lupulus*), and give birth to wingless forms. A number of generations then occur on hops depending upon temperature and nutritional conditions before the short days of autumn induce the formation of the winged male and the winged gynopara generations. These fly to the winter host where the oviparae give birth to the sexual females, which mate with males and lay fertilized, diapausing eggs close to next year's buds before leaf fall. Drawing courtesy of Tom Pope.

autumn. On the winter host, males locate the sexual females using sex phero-mones (Campbell et al., 1990), and after copulation (figure 14.3) fertilized eggs are laid.

In developmental terms, sex is determined initially in parthenogenetic reproduction at the first cell division, diapausing eggs are always female. Sex

FIGURE 14.2.
Damson-hop aphid, *Phorodon
humuli*: winged female
gynopara that, in autumn,
gives birth to sexual females
on a winter host. Photo
courtesy of Tom Pope.

FIGURE 14.3.
Damson-hop aphid,
Phorodon humuli: winged male
copulating with wingless
sexual female. Photo courtesy
of Tom Pope.

determination is of the XX:X0 type, whereby females have two X chromosomes but males only one (Blackman, 1987). If male, there are no further environmental influences on phenotype and, depending on species/clone, males develop as wingless or winged adults. If the female genetic composition is determined then the environment can influence whether the female will become a sexual (oviparous) or an asexual (viviparous) form. The earliest visible detection of this is when the ovaries differentiate. In both sexual and asexual females, the ovaries are telotrophic, meroistic with trophic chords linking nurse cells in the germaria to the first one or two developing eggs/embryos in the ovarioles. Ovulation and parthenogenetic embryo development begin when their mothers are in the fourth stadium (e.g., Hardie, 1987a). The sexual female germaria are eventually larger, due to hypertrophy of the nurse cells, but ovulation does not occur until later in postnatal development. If the females are developing along the viviparous/parthenogenetic pathway, then environmental influences such as day length, nutrition, and crowding determine a winged, or wingless adult form either prenatally, postnatally or both pre- and postnatally (Müller et al., 2001). Sexual females are invariably wingless.

In aphids, the photoperiodic effect is maternally transmitted to the embryos with regard to induction of the parthenogenetic or sexual forms. In host-alternating aphids where winged females move from summer to winter hosts, the winged presexual form may be the result of photoperiodic effects on both maternal and individual systems. For example, exposure of parthenogenetic black bean aphid *Aphis fabae* females to short days influences the progeny to develop as winged gynoparae. However, if the presumptive gynoparae are transferred to long days at birth, they invariably develop to wingless adults and give birth to asexual, not sexual, females (Lees, 1977; Hardie 1980). Precise control of this phenotype is thus due to both maternal and individual effects such that at least five prenatal (maternal) short days are required, in addition to the postnatal short days, for wing induction. However, presumptive gynoparae are also directly responsive to long days as mature embryos, as revealed by localized extension of day length to the abdomen of their mothers (Hardie and Lees, 1983).

As for all organisms responding to day length, aphids require a photoreceptor system that can distinguish light from dark, a clock mechanism that can measure the duration of the light, or in most cases the dark period, a counter or photoperiodic memory that accumulates the number of long or short days, and an endocrine/neuroendocrine effector system that modifies the developmental processes associated with either long- or short-day development. Each of these systems has been studied in aphids.

LIGHT SENSITIVITY

Aphids, like other insects, are remarkably sensitive to light, with respect to the photoperiodic response. Thus Lees (1981) showed that just 0.2 or 3–10 mW/m² of monochromatic blue to red light (450–500 nm) induced 50% or 100%, respectively, of the vetch aphid *Megoura viciae* to produce asexual females after early or late night interruption with 1 h illumination. Both *M. viciae* and *A. fabae* were photoperiodically sensitive to 7.7 mW/m² integrated across approximately 450–694 nm wavelengths measured from quartz-halogen lamps at the start and end of simulated civil twilight (Vaz Nunes et al. 1996). Taking account of the luminosity function, these figures are closely matched. In field experiments at Ascot (51°5'N), the pea aphid *Acyrthosiphon pisum* (Lees, 1989), *M. viciae*, and *A. fabae* (Vaz Nunes et al., 1996) appear to be sensitive to light levels extending well into natural civil twilight. Civil twilight dusk has an illuminance of approximately 0.1–1.0 lux across the human visible spectrum (Johnsen et al., 2006), equivalent to approximately 0.4 mW/m² at 450 nm and 2 mW/m² at 500 nm, indicating that all of civil twilight may be photoperiodically active.

PHOTOPERIODIC PHOTORECEPTORS

Spectral Sensitivity

The most comprehensive study of aphid photoperiodic photoreceptors was under-taken by Lees (1981) in *M. viciae*. Both the early and late light-sensitive segments of the scotophase were investigated using monochromatic light at various energy levels. Thus, for an early night interruption, which was interpreted as resetting the dark measurement response, 1 h of light was placed 1.5 h after dark in an overt 13.5/10.5-h light/dark (LD 13.5:10.5) cycle and a late night interruption, which was interpreted to terminate dark measurement, comprised 0.5 h light at 7.5 or 8 h after the onset of darkness. Both showed maximum sensitivities in the blue region of the visible spectrum (450–470 nm), but the later interruptions showed a broader peak with sensitivity extending into green/orange/red regions (550–650 nm). Using spectral light for a main 8-h photophase revealed an action spectrum close to that of the late night interruption. The action spectra were compatible with the idea that the photoreceptor was a caretenoprotein. Subsequent studies on the filtering action of the head capsule cuticle (photoperiodic photoreceptor region is located in the brain; see below) had minimal effect on the shape of the action spectra but showed that the receptors were more sensitive to shorter wavelengths than indicated by the whole-insect response (Hardie et al., 1981).

Location

In the 1960s, Lees used eye occlusion and localized illumination to elicit different day lengths on different parts of the body of *M. viciae* (Lees, 1964). The study showed that the photoperiodic photoreceptor mechanism was located in the brain, not the eyes, and light was transmitted through the semitransparent head cuticle and brain tissue. This result was later supported by Steel and Lees (1977) using microcautery to ablate various brain regions. However, microspectrophotometry of brain tissue did not reveal any possible regions of absorption at relevant wave-lengths (Hardie et al., 1981).

Gao et al. (1999) showed that a neuropil region in the anterior ventral protocer-ebrum of *M. viciae* was labeled by a number of antibodies raised to visual proteins. These included some raised against *Drosophila* rhodopsin and a number of verte-brate opsins, and supported the idea that the protein moiety of photoperiodic photo-receptor could be an opsin. No cell bodies were labeled, but Steel (1976) speculated that neurons lateral to the group I (medial) neurosecretory cells (see "Endocrine/ Neuroendocrine Effector System," below) in the dorsal anterior protocerebrum of *M. viciae* might provide the input to the group I cells and act as clock and photo-period receptors. Efforts to isolate and sequence the brain opsin failed, and only two visual opsins were located (Gao et al., 2000). However, a brain-located opsin, termed boceropsin and possibly involved in the photoperiodic response, has since been identified in *Bombyx mori* (Shimizu et al., 2001).

In *A. fabae*, the photoperiodic photoreceptors involved in the induction of winged gynoparae are located in the head of the mother and in the head of individual newly born presumptive gynoparae (Hardie and Lees, 1983). Nothing further is known of the identity of the photoreceptor mechanism of *A. fabae* or other aphids.

PHOTOPERIODIC CLOCK MECHANISM

Many attempts have been made to elucidate the mechanism that measures photoperiodic time in insects and aphids have played a role, in particular the work of Tony Lees on *M. viciae* has provided an important input. At the outset, it should be mentioned that experimental strategies invariably involve whole-insect protocols and thus the data collected are subject to the clock mechanism, photoperiodic counter, and endocrine effector. Dissecting the contribution of each, particularly when little is known of these mechanisms, is difficult. In many ways the photoperiodic response of *M. viciae* is similar to those of other insects, and photoperiodic protocols produce similar data. The photoperiodic response curve matches those of other insects with a 9.5 h critical night length which is relatively temperature insensitive between 6°C and 20°C (Lees, 1973, 1986), although not all aphid photoperiodic responses are (e.g., Dixon and Glen, 1971; Vaz Nunes et al., 1996; Vaz Nunes and Hardie, 2000). Scanning a scotophase longer than critical with light pulses reveals early and late periods that are sensitive, while the central portion is light insensitive (see "Spectral Sensitivity," above). What is different is the fact that experimental protocols that would normally reveal a circadian influence over the photoperiodic response, namely, the Nanda-Hamner protocol (a short photophase associated with constant but varying scotophases to give a variety of cycle lengths: $T = 72$ h or more) and Bünsow protocol (extended dark periods are interrupted at different times by light pulses of 1–2 h) do not do so in the vetch aphid (Lees, 1973). Positive responses in the form of peaks and troughs of induction at regular approximately 24-h intervals failed to appear. After discussions with David Saunders on fly responses, Lees repeated the Nanda-Hamner protocols at higher temperatures, but again, no evidence of an oscillatory input was revealed (Lees, 1986). Up to that time the photoperiodic mechanism of the vetch aphid appeared to differ from other studied insects and the mechanism for night-length measurement was held to be an hourglass, rather than oscillator-based mechanism that was reset by light. This oddly out-of-step insect finally came into the fold when fully dampened oscillator mechanisms proposed for scotophase measurement and the *M. viciae* hourglass could be considered as such (Lewis and Saunders, 1987; Saunders, 2005). Indeed, Saunders (2005) argues eloquently that all photoperiodic clocks should be considered to be based upon circadian oscillatory mechanisms which differ only in the degree of persistence in constant conditions (damped or free running), and that this parameter will vary with external conditions such as temperature. However, there is still a school of thought that proposes an hourglass mechanism for photoperiodic time measurement (e.g., Bradshaw et al., 2003, 2006; Veerman and Veenendaal, 2003).

Only after Lees' death in 1992 was a seemingly circadian influence revealed in the photoperiodic response of *M. viciae* (Vaz Nunes and Hardie, 1993). The Veerman and Vaz Nunes (1987) protocol was applied to a 10-day sensitive period, and the results of exposing insects to variable photophases coupled to 12-h scotophases were compared to the results of varying numbers of 24-h, LD 12:12 cycles. This protocol was used to discriminate between single (hourglass) and repetitive (circadian oscillatory) measurement of long nights in extended scotophases. For example, an 84-h scotophase would be measured as one long night by a photoperiodic clock with an hourglass mechanism but as four long nights by a clock with a circadian mechanism. Where the light-dark cycles did not cover the 10-day experimental exposure period, continuous light (which has a weak long-day effect) made up the initial portion. Predictions based upon a circadian oscillatory clock mechanism, where night length was measured repeatedly in extended scotophases beyond the critical night length, closely matched the data, while those predictions based upon an hourglass mechanism where long scotophases were measured only once whatever the length, did not. The data were interpreted in terms of the photoperiodic clock, but again, the results could be influenced by other aspects of the overall photoperiodic response.

One other aphid has been investigated in some detail with different results to *M. viciae*. Thus, Nanda-Hamner studies of wing induction in presumptive gynoparae of *A. fabae* did reveal a positive response with peaks 20 and 24 h apart at 20°C but not at 15°C (Hardie, 1987b).

Critical night length is a property of the photoperiodic clock but it can be affected by the clone of aphid chosen (e.g., Lushai et al., 1996), and inheritance is complicated as clonal crosses showed considerable overall variation. Nevertheless, it was concluded that the ability to survive winter was unlikely to be affected even if individuals moved long latitudinal distances (Lushai et al., 1996). Critical night length can also be affected by other factors such as rearing conditions. Lees (1986) showed that starvation for 4 h during the first or last part of the scotophase shifted the critical night length from 9.5 to 9 h in *M. viciae*. The same treatment during the first part of the photophase had no effect (Lees, 1986). It is also known that the critical night length can vary between winged and wingless parthenogenetic forms. In a pink English clone of the pea aphid, *A. pisum*, winged forms had a critical night length of 9.5–10 h for sexual female production compared to 10.5 h for wingless females (Vaz Nunes and Hardie, 1996). Winged females also showed a critical night length of 9–9.5 h for male production compared to 10.5 h the wingless form. Thus, even females with the same genotype but different morphological phenotypes can vary in the detail of their photoperiodic response.

The location of the clock mechanism remains unknown, but from the microcautery experiments of the brain of *M. viciae* (see "Endocrine/Neuroendocrine Effector System," below), Steel has proposed that neurons with cell bodies positioned laterally to the group I, medial neurosecretory cells possibly house the clock mechanism (Steel, 1976).

PHOTOPERIODIC COUNTER

In general, a number of light/dark cycles are needed to induce a photoperiodic response, and the accumulation of this information is referred to as the photoperiodic memory or counter (Saunders, 1981). Noncyclical conditions such as continuous light or dark are usually relatively neutral photoperiodic stimuli but often produce variable responses and can be used as background conditions. It was recognized early on in Lees's studies of *M. viciae* that night length was crucial, day length, above a minimal length of approximately 4 h, was less influential on the outcome (Lees, 1973). It was also apparent that short nights were a more powerful influence that long nights, and Lees (1973) showed that the promotion of asexual forms was affected more by the number of short nights, regardless of intervening long nights that appeared to be neutral. Fewer short nights (LD 16:8 h) were required to effect 50% asexual reproduction (required day/cycle number) when placed into a background of long nights (LD 12:12 h) than vice versa (3 vs. 7–8; Hardie, 1990). However, when examining data from non-24-h light/dark cycles Hardie (1990) showed that long, 12-h scotophases were accumulated in a straightforward fashion, irrespective of the accompanying photophase length. Despite the apparently stronger effects of 8-h short scotophases, these were not accumulated in a straightforward fashion except when accompanied by a 16-h photophase ($T = 24$). With photophases longer or shorter than this, there was a reduction in photoperiodic strength (figure 14.4). It seems that although short (8-h) nights are photoperiodically stronger, they merely modify the counter's accumulation of long (12-h) scotophases.

It is recognized, however, that long- (> critical) and short- (< critical) nights can have different quantitative effects on the counter; that is, not all long nights have the same photoperiodic strength. This was observed in a Russian clone of *M. viciae* (Zaslavski and Fomenko, 1983) as well as the English clone used by Lees (Hardie, 1990; Vaz Nunes and Hardie, 1993) and is also true for *A. fabae*, and is related to the photophase length (Vaz Nunes and Hardie, 1994). Thus, using 24-h LD cycles with *M. viciae*, an 8-h scotophase is the most potent, reducing rapidly in strength as it extends toward the 9.5-h critical night length and is most potent over 6–8 h (Hardie, 1990).

It was also shown that the accumulation of short nights was temperature sensitive while that for long nights was temperature stable in vetch aphid and black bean aphid (Hardie 1990; Vaz Nunes and Hardie 1994). Thus, for example, in *M. viciae* long 12-h nights show a required day number of 7–8 at 15°C and 9–10 at 12°C, while the required day number for a short 8-h night was 3 at 15°C but 8 at 12°C. These observations led to the idea that long- and short-night measurement differed in some way and to a double-oscillator model of photoperiodic responses (Vaz Nunes, 1998).

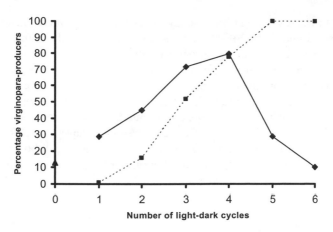

FIGURE 14.4.
Data from a series of photoperiodic transfer experiments at 15°C with the vetch aphid *Megoura viciae*. Short-day (LD 12:12)-born aphids were transferred shortly after birth during the photophase for a 96-h period and returned to short days afterward. During this photoperiodically sensitive period, the insects were exposed to a single LD 88:8 cycle, two 40:8 cycles, three LD 24:8 cycles, four LD 16:8 cycles, five LD 11.2:8 cycles, or six LD 8:8 cycles (solid line). The y-axis shows the numbers of insects responding to the short, 8-h "night" by producing virginoparae (asexual females; a long-day response). The dashed line shows the results of transfer experiments to 24-h cycles (LD 16:8). The accumulation of information derived from 8-h nights affects the photoperiodic mechanism differently when applied in 24-h and non-24-h LD cycles (redrawn from Hardie, 1990).

As stated above, inherent in any photoperiodic response is the problem of teasing apart the various mechanisms from an overall, whole-organism experiment— the "black-box" of Saunders (1984).

ENDOCRINE/NEUROENDOCRINE EFFECTOR SYSTEM

Neurosecretion

Following from the studies of where the photoperiodic photoreceptors were located, Lees' laboratory began to explore how destruction of various regions of the brain would affect polyphenism associated with day-length changes. These studies revealed that the medial neurosecretory cell group (called the group I cells in aphids) were necessary for the production of parthenogenetic, long-day aphids (Steel and Lees, 1977). The two groups of approximately 10 cell bodies in total were located in the anterior protocerebrum, and their destruction in long-day conditions led to a switch in the type of progeny produced from parthenogenetic to sexual females (a short-day effect). Similarly, a successful cauterization would prevent the switch to the production of parthenogenetic forms when the insects were reared in short days and then transferred to long days.

Initial ideas using light microscopy and selective staining of neurosecretory material suggested that while the cell bodies contained little stainable material, their axons projected ventrally through the brain toward the posterior and into the thoracic ganglionic mass. It was proposed that they might continue along the ventral nerve cord and there could be a direct release of a neurosecretory product onto/into the vicinity of the ovaries. This would act directly upon the developing embryos and induce parthenogenetic forms. Later, this proposed product was termed "virginoparin" by Steel (1976), but these ideas have not stood the test of time. While electron microscopy confirmed that the group I cell bodies contained little neurosecretory material, the ventral nerve cord was also surprisingly devoid of neurosecretory material (Hardie, 1987c, 1987d).

Juvenile Hormones

Subsequently the spotlight fell on the possibility that juvenile hormone was involved in the determination of parthenogenetic forms. A paper by Mittler et al. (1976) showed that treatment of adult gynopare of *A. fabae* and *Myzus persicae* with a juvenile hormone analogue induced the formation of winged progeny with ovaries containing parthenogenetic embryos. The progeny should have been almost exclusively wingless oviparae with ovaries containing haploid yolky eggs. Further investigations showed that similar effects were possible with insect juvenile hormones in other aphid species and in ovipara producers of nonhost alternating species (Hardie, 1981a; Corbitt and Hardie, 1985; Hardie and Lees, 1985). Indeed, judicious application enabled short-day induction of sexual females to be countered by juvenile hormone to produce "normal" parthenogenetic insects that would go on to reproduce normally. Juvenile hormone would also induce changes in host-plant preference associated with the long- and short-day forms of *A. fabae* in such a way that they mimicked long-day influences; that is, summer-host preferences were promoted in winter-host–preferring gynoparae (Hardie, 1981b; Hardie and Glascodine 1990). The flight behavior of gynoparae could also be changed by juvenile hormone applications prior to, but not after, the final molt such that migratory flight could be switched to foraging flight (Hardie et al., 1989). That is, during the initial stages of maiden flight gynoparae ignore a laterally presented green target, representing a plant-like visual stimulus, and continue to fly toward an over head "sky" light. Winged long-day forms do not show this behavior and readily move toward the green targets soon after takeoff. Thus again, juvenile hormone application mimicked the effect of long days, in these instances behavioral rather than morphological/reproductive phenotypic traits.

Further evidence that juvenile hormone could be involved in aphid photoperiodism comes from a study of corpus allatum size and its relationship with day length, and the group I cells (Hardie, 1987e). The corpora allata of parthenogenetic female *M. viciae* reared in short days and producing sexual females are significantly larger than those reared in long days and giving birth to parthenogenetic females, at least from 10 days after the final molt. If the group I cells are destroyed in long-day–reared aphids, and the success of that procedure judged by the switch of progeny to sexual females, then there is a corresponding change in the size of the corpora allata (see "Neurosecretion," above).

Attempts to measure aphid juvenile hormone titers have proved difficult, but the presence of JH III has been confirmed albeit in low levels in *A. fabae*, *A. pisum*, and *M. viciae* (Hardie et al., 1985; Westerlund and Hoffmann, 2004; Chen et al., 2007). Although titers were low, 0.02–0.28 ng/g in *M. viciae*, higher titers were found in long-day rather than short-day reared aphids (Hardie et al. 1985) and were higher at 25°C than 10°C in *A. pisum* (~95 vs. 45 ng/g), but day length was not stated (Chen et al., 2007). The biological significance of these data is still uncertain, and the possibility of an as yet unidentified juvenile hormone in aphids remains a possibility.

Melatonin

Another possible player in the endocrine control of photoperiodism in aphids is melatonin (*N*-acetyl-5-methoxytryptamine), which is a ubiquitous biogenic amine known to be involved in transducing day-length effects in vertebrates (Arendt, 1995). It has been found in insects, and in some cases there are higher levels during the scotophase than the photophase, for example, the face fly *Musca autumnali* (Wetterberg et al., 1987), the damselfly *Ischnura verticalis* (Tilden et al., 1994), the silk moth *Bombyx mori* (Itoh et al., 1995), the cabbage looper *Trichoplusia ni* (Linn et al., 1995), and the blood-sucking bug *Rhodnius prolixus* (Gorbet and Steel, 2003). It has also been shown to effect release of neurohormones in insects, for example, prothoracicotropic hormone in the cockroach *Periplaneta americana* (Richter et al., 2000) and peptides from the corpora cardiaca of locusts (Huybrechts et al., 2005).

Melatonin has been found in *A. pisum*, but there were no significant differences between whole-body titers during photophase and scotophase (Gao and Hardie, 1997). This did not bode well for melatonin as a mediator of the photoperiodic response in aphids, but insects fed on artificial diets containing melatonin produced females that were intermediate between sexual and parthenogenetic females, that is, with ovaries containing haploid yolky eggs and embryos, and males in a dose-dependent fashion (Gao and Hardie, 1997). The latter observations indicated that melatonin was mimicking, to an extent, short days (long nights) as it does in vertebrates.

MALE PRODUCTION

So far, the discussion has mainly concentrated on the photoperiodic control of females morphs. A number of authors have examined sex determination and the induction of male aphids that, like the induction of the sexual females or their precursors—gynoparae or sexuparae—is dependent upon day length. As mentioned above, males are chromosomally different from females lacking one X chromosome (XX:X0 system). This takes place at the first maturation division of the oocyte within the germarium and just prior to the oocyte entering the ovariole, with the end-end joining of one X chromosome and its subsequent loss (Blackman, 1980).

The most detailed investigation of male induction has been done on an English clone of *A. pisum*, where the photoperiodic response curves differ drastically between ovipara and male induction (Lees, 1989, 1990; Vaz Nunes and Hardie, 1996). Thus, ovipara induction is zero for night lengths from 0 to 10.5 h, with a critical night length of 11 h and ovipara production remaining at 100% with scotophases up to 20 h and then falling to 12% in continuous darkness. Male induction, however, shows a peak at scotophases of 10.5–11 h; no males appear in night lengths below 9 h, and at night lengths greater than 12 h, few or no males are produced (figure 14.5). These data suggest that different photoperiodic clock mechanisms are involved in sex determination (male vs. female) and parthenogenetic/sexual female determination. Indeed, further experiments using Nanda-Hamner resonance protocols and night interruption also show differences between ovipara and male induction, but neither reveals a positive response for circadian involvement. The counter mechanisms differ also as required day number for 50% male induction is 4 but is 8 for ovipara production. This difference most likely reflects the fact that male determination is genetic, and once determined, any change in environmental conditions

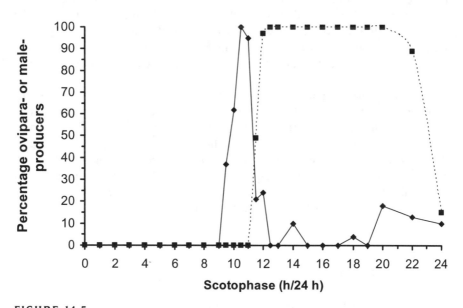

FIGURE 14.5.

The photoperiodic response curve for male (solid line) and ovipara (sexual female; dashed line) for 24-h cycles at 15°C (redrawn from Lees, 1990).

cannot effect a switch to female, while female form determination is more fluid and reversible by a change in conditions. Not all species show such a peak of male production between restricted night lengths; for example, in *R. padi* male induction remains high well beyond the critical night length (Lushai et al., 1996).

Endocrine involvement in male induction is not understood, although Hales and Mittler (1983) showed that male production by *M. persicae* could be suppressed or reduced by topical application of the juvenile hormone analogue, kinoprene, and induced precocene III, which selectively destroys corpus allatum tissue and so reduces juvenile hormone synthesis. Thus, it is possible that day length induces both male and sexual female production by a similar mechanism, perhaps utilizing differential sensitivities and temporal organization to effect different developmental pathways. However, it is difficult to envisage a simple effective system as there would be times when both sex and female morph determination would be taking place simultaneously.

As a working hypothesis, with the information currently available, a system could be envisaged where darkness is detected by photoreceptors in the anterior protocerebrum that are themselves part of clock neurons or signal to clock neurons. In turn these clock neurons interface with the group I neurosecretory cells, which in short nights release an allatotropic factor(s) (melatonin?) that stimulate the synthesis of juvenile hormone from the corpus allatum and direct a parthenogenetic reproductive strategy. In long nights, allatotropic factors are not produced and low juvenile hormone titers effect the production of morphs associated with sexual reproduction.

OTHER OVERWINTERING STRATEGIES AND SUMMER DIAPAUSE

It is possible for parthenogenetic females to overwinter under certain conditions. Thus, tropical aphids tend to remain asexual and not undergo diapause, even though scotophase lengths constantly exceed critical night lengths for temperate species. In temperate regions, some species may overwinter as parthenogenetic forms in mild winters and/or in protected environments such as inside glass houses. One unusual strategy occurs in the leaf-curling plum aphid *Brachycaudus helichrysi*, where sexual reproduction, most likely induced by short days, gives rise to an egg stage but the egg hatches early on, November/December in the United Kingdom, and first-instar fundatrices overwinter (Bennett, 1955).

A different strategy is used by the host-alternating lettuce root aphid *Pemphigus bursarius*, which utilizes a number of summer hosts, mainly members of Compositae, and normally overwinters as diapausing eggs in bark crevices on poplar trees (*Populus* spp.) (Blackman and Eastop, 2000). Here, the winged forms moving from summer to winter host are termed the sexuparae, which give birth to both males and sexual females. However, *P. bursarius* also develops a wingless asexual form, the hiemalis, which can overwinter in soil. The summer hosts are annuals and germination of the following generation of summer hosts in spring allows for root recolonization. The hiemalis forms are more cold and starvation tolerant than are the summer wingless forms on lettuce roots, and have elevated triglyceride levels (Phillips et al., 2000). Development of the sexupara and hiemalis forms is influenced by day length and temperature (Phillips et al., 1999).

In some tree-dwelling aphid species there may be a summer diapause. The sycamore resident *Periphyllus testudinaceus* shows a developmental arrest as a distinct first-instar morph lasting from May to August in the United Kingdom. Initiation of the diapause state appears to be endogenous but may be linked to nutrition and can be artificially shortened by exposing these insects to short days (Lees, 1966; Lees, personal communication). The diapausing larvae eventually develop into adult sexuparae.

Adult winged females of the sycamore aphid *Drepanosiphum platanoidis* also undergo a summer diapause for four to six weeks, but this takes the form of an adult reproductive hiatus (Dixon, 1963). Induction may again be nutritional and perhaps a crowding response, but long days are necessary for diapause maintenance, with short days initiating reproduction. It is possible that the symbiotic *Buchnera* bacteria are involved (Douglas, 2000).

Spring Interval Timers

Diapause eggs hatch in spring when day lengths are still short, but sexual forms are not produced until much later in the year. Indeed, Marcovitch in his original observations of day-length effects on polyphenism in *A. forbesi* noted that although eggs normally hatched in February, sexual females could not be artificially elicited until May (Marcovitch, 1924). The same phenomenon has been noted in a number of other aphid species and was termed the "time factor" (Wilson, 1938) or the "facteur fundatrice" (Bonnemaison, 1949). Lees (1960) used the term "interval timer" and, by taking early and late-born progeny from clonal lines, showed that in *M. viciae*, the period during which short days were noninductive depended upon time elapsed since egg hatch and not upon the number of generations. Other aphid species show a similar phenomenon, for example, *Drepanosiphum platanoidis* (Dixon, 1971), *Eucalliptus tilliae* (Dixon, 1972), *Rhopalosiphum padi* (Dixon and Glen, 1971; Lushai et al., 1996), *Aphis rubicola* (Brodel and Schaefers, 1979), *Myzus persicae* (Margaritopoulos et al., 2002), and *Phorodon humuli* (Campbell and Tregidga, 2006). The mechanism of the interval timer is unknown.

DAY LENGTH EFFECT ON WING DEVELOPMENT IN THE FEMALE PEA APHID *ACRYRTHOSIPHON PISUM*

While investigating the photoperiodic responses of a pink English clone of *A. pisum*, it was noted that day length affected the production of crowd-induced winged forms

(Vaz Nunes and Hardie, 1996). *Acyrthosiphon pisum* is a complex of biotypes and subspecies that Müller (1962, 1980) divided into two groups: those that host alternate between pea in the summer and perennial herbaceous hosts in the winter, and those that live year-round on perennial host and occasionally on pea. The former tend to be green in color and have winged males with winged viviparous females in autumn, while the latter can be green, yellow, or pink and produce wingless males and winged females in spring and summer (Müller, 1962, 1980). Wing induction in long days can be initiated by artificially crowded conditions in the laboratory (Sutherland, 1966), and this is how Vaz Nunes and Hardie (1996) induced the different forms to examine the photoperiodic effects on winged and wingless forms (see "Photoperiodic Clock Mechanism," above). However, it was noticed that if adult, reproducing aphids were crowded in short days or were transferred to short days prior to crowding, they produced few or no winged offspring compared to 60–70% produced in long days. This effect was shown to be maternal with a critical night length of approximately 9.5 h. As a non-host-alternating biotype, it may be that short-day inhibition of wing induction prevents the aphids moving away from the natal plant in autumn when seeking fresh herbaceous hosts will carry a higher risk than earlier in the year (Vaz Nunes and Hardie, 1996).

FUTURE DIRECTIONS

With groundwork in place, as with seasonal timers in other insects and, indeed, circadian timing mechanisms, there should now be a push to understand the molecular mechanisms of aphid photoperiodism. The difficulty will, of course, be dissecting the mechanism from the downstream gene expression. Work has already begun in aphids with recent papers, examining differences in gene expression in long- and short-day–reared *A. pisum*. Thus, using differential display reverse transcriptase polymerase chain reaction Ramos et al. (2003) identified a gene product (ApSDI-1: showing sequence similarities to a protein involved in amino acid transport in GABAergic neurons) that was overexpressed in short days compared to long days. This differential expression appears to occur in aphids exposed to as little as four short days but continues with increasing amounts over two generations (Ramos et al., 2003). As indicated by the authors, using whole aphids as starting material complicates the system, because developing embryos are contained within the mother at birth, and developing embryos are present in the larger embryos of adult females. Each of these three generations could finally represent different aphid morphs (e.g., long-day parthenogenetic vivipara, short-day sexual female producer, and sexual female, respectively). The whole insect, and the more mature embryos, also contain the bacterial symbionts. One possible way around this complexity would be dissection of the ovaries prior to RNA extraction of the corpse, but the ovaries of fourth-instar and adult aphids could still contain the embryos of two generations, thus complicating any ovary analysis. Another possibility would be to decapitate the aphids and use the head as starting material. As mentioned above, the photoperiodic photoreceptors appear to be in the head and interact with the cerebral neurosecretory system. Depending upon the precise point of severance, the retrocerebral endocrine system, including corpora cardiaca and corpus allatum, could be intact. Using the same clone of *A. pisum* as Ramos et al. (2003), Le Trionnaire et al. (2007) used the decapitation approach to extract RNA from long- and short-day insects. Here, a 3,321 cDNA microarray was used to compare the extracted RNAs and revealed 59 transcripts that were differentially regulated. Of these, 19 were identified as cuticular proteins with 9 up-regulated and 10 down-regulated in short days. The idea that this finding might be related to head capsule cuticle, through which LD cycles are initially detected by the brain-located photoperiodic photoreceptors, changing to be more or less transparent seems unlikely as day-length measurement occurs in long and short days. It will be interesting to know if there is a real phenotypic difference in cuticular proteins between the parthenogenetic and sexual females. There are substantial differences in the adults, in the shape of the cauda, and the hind tibiae of sexual females have raised scent glands from which sex pheromones are released, but whether such differences reflect different protein components, and whether these would be recognizable in the embryonic insect remains to be established. Other transcripts identified and assigned a function included proteins associated with metabolism and cell signaling (Le Trionnaire et al., 2007). The discovery of marker transcripts/proteins that would provide an early indication of the developmental pathway would be very useful and reduce experimental time from weeks, for whole animal experiments, to days or perhaps less if the crucial long-day/short-day signaling process takes place over a short period of time. If the techniques are sensitive enough, I would encourage a microarray approach to differences in gene expression just prior to and just after the critical night length to catch the very initial stages of a photoperiodic switch. The photoperiodic timer is accurate and a 15-min difference can be detected by *M. viciae* (Lees, 1966), allowing for a precise experimental protocol. However, it is possible that this very initial stage involves neuro/neuroedocrine transmitters and is independent of changes in gene expression.

For the dissection of aphid photoperiodic responses, the present time offers great excitement. Techniques such as RNA interference are successfully being applied to aphids such that specific genes can be selectively down-regulated (e.g., Jaubert-Possamai et al., 2007). However, the major excitement is the near completion of the pea aphid genome (Brisson and Stern, 2006; see the Pea Aphid Genome Project at http://www.hgsc.bcm.tmc.edu/projects/aphid/), which should allow rapid progress in understanding the molecular mechanisms of the eccentricities of these major pests that exhibit such an interesting biology.

APPENDIX: GLOSSARY OF TERMS

Fundatrigeniae: specifically offspring of the fundatrix but often used to include all spring generations on the winter host of host-alternating species

Fundatrix: parthenogenetic female hatching from a fertilized egg, the first parthenogenetic generation of the annual cycle; also known as the stem mother

Gynoparae: parthenogenetic females that give birth to sexual females and in host-alternating species are winged and fly from summer to winter hosts

Heteroecious: host alternating, annual movement between winter and summer host plants

Oviparae: sexual females that lay fertilized eggs after mating

Sexuparae: parthenogenetic aphids that produce both males and sexual females

Virginoparae: parthenogenetic females

Viviparae: parthenogenetic aphids that give birth to live young

References

Arendt J. (1995). *Melatonin and the Vertebrate Pineal Gland.* London: Chapman and Hall.

Bennett SH. (1955). The biology, life history and methods of control *Brachycaudus helichrysi.* J Hortic Sci 30: 252–259.

Blackman RL. (1980). Chromosomes and parthenogenesis in aphids. In *Insect Cytogenetics,* (RL Blackman, GM Hewitt, and M. Ashburner, eds), 10th Symposium of the Royal Entomological Society. London: Blackwell, pp. 133–148.

Blackman RL. (1987). Reproduction, cytogenetics and development. In *Aphids, Their Biology, Natural Enemies and Control,* vol A (AK Minks and P Harrewijn, eds). Amsterdam: Elsevier, pp. 163–195.

Blackman RL and Eastop VF. (2000). *Aphids on the World's Crops: An Identification and Information Guide,* 2nd ed. Chichester, UK: John Wiley and Sons.

Bonnemaison L. (1949). Sur l'existance d'un facteur inhibitant l'apparition des formes sexuées chez les Aphidinae. C R Hebd Seances Acad Sci 229: 386–388.

Bradshaw WE, Holzapfel CM, and Mathias D. (2006). Circadian rhythmicity and photoperiodism in the pitcher-plant mosquito: Can the seasonal timer evolve independently of the circadian clock? Am Nat 167: 601–605.

Bradshaw WE, Quebodeaux MC, and Holzapfel CM. (2003). The contribution of an hourglass timer to the evolution of photoperiodic response in the pitcher-plant mosquito, *Wyeomyia smithii.* Evolution 57: 2342–2349.

Brisson JA and Stern DL. (2006). The pea aphid, *Acyrthosiphon pisum*: an emerging genomic model system for ecological, developmental and evolutionary studies. Bioessays 28: 747–755.

Brodel CF and Schaefers GA. (1979). An 'interval timer' for the production of oviparae in *Aphis rubicola* (Homoptera: Aphididae). Entomol Exp Appl 25: 1–8.

Campbell CAM. (1985). Has the damson-hop aphid an alate alienicolous morph? Agric Ecosyst Environ 12: 171–180.

Campbell CAM, Dawson GW, Griffiths DC, Pettersson J, Pickett JA, Wadhams LJ, and Woodcock C. (1990). Sex attractant pheromone of damson-hop aphid *Phorodon humuli* (Homoptera, Aphididae). J Chem Ecol 16: 3455–3464.

Campbell CAM and Tregidga EL. (2006). A transgenerational interval timer inhibits unseasonal sexual morph production in damson-hop aphid, *Phorodon humuli.* Physiol Entomol 31: 394–397.

Chen Z, Linse KD, Taub-Montemayor TE, and Rankin MA. (2007). Comparison of radioimmunoassay and liquid chromatography tandem mass spectrometry for determination of juvenile hormone titres. Insect Biochem Mol Biol 37: 799–807.

Corbitt TS and Hardie J. (1985). Juvenile hormone effects on polymorphism in the pea aphid, *Acyrthosiphon pisum.* Entomol Exp Appl 38: 131–135.

Dixon AFG. (1963). Reproductive activity of the sycamore aphid, *Drepanosiphum platanoidis* (Schr.). J Anim Ecol 32: 33–38.

Dixon AFG. (1971). The interval timer and photoperiod in the determination of parthenogenetic and sexual morphs in the aphid, *Drepanosiphum platanoidis.* J Insect Physiol 17: 251–260.

Dixon AFG. (1972). The interval timer, photoperiod and temperature in seasonal development of parthenogenetic and sexual forms in the lime aphid, *Eucallipterus tiliae.* Oecologia 9: 301–310.

Dixon AFG. (1998). *Aphid Ecology,* 2nd ed. London: Chapman and Hall.

Dixon AFG and Glen DM. (1971). Morph determination in the bird cherry-oat aphid, *Rhopalosiphum padi* L. Ann Appl Biol 68: 11–21.

Douglas AE. (2000). Reproductive diapause and the bacterial symbiosis in the sycamore aphid *Drepanosiphum platanoidis* . Ecol Entomol 25: 256–261.

Eastop VF. (1977). Worldwide importance of aphids as virus vectors. In *Aphids as Virus Vectors* (KF Harris and K Maramorosch, eds). London: Academic Press, pp. 3–62.

Gao N, Foster RG, and Hardie J. (2000). Two opsin genes from the vetch aphid, *Megoura viciae.* Insect Mol Biol 9: 197–202.

Gao N and Hardie J. (1997). Melatonin and the pea aphid, *Acythosiphon pisum.* J Insect Physiol 43: 615–620.

Gao N, Von Schantz M, Foster RG, and Hardie J. (1999). The putative brain photoperiodic photoreceptors in the vetch aphid, *Megoura viciae.* J Insect Physiol 45: 1011–1019.

Garner WW and Allard HA. (1920). Effect of the relative length of day and night and other factors of the environment on growth and reproduction in plants. J Agric Res 18: 553–606.

Gorbet DJ and Steel CGH. (2003). A miniature radioimmunoassay for melatonin for use with small samples from invertebrates. Gen Comp Endocrinol 134: 193–197.

Hales DF and Mittler TE. (1983). Precocene causes male determination in the aphid *Myzus persicae.* J Insect Physiol 28: 819–823.

Hales DF, Tomiuk J, Wohmann K, and Sumnnocks P. (1997). Evolutionary and genetic aspects of aphid biology: A review. Eur J Entomol 94: 1–55.

Hardie J. (1980). Reproductive, morphological and behavioural affinities between the alate gynopara and virginopara of the aphid, *Aphis fabae.* Physiol Entomol 5: 385–396.

Hardie J. (1981a). Juvenile hormone and photoperiodically controlled polymorphism in *Aphis fabae*: Prenatal effects on presumptive oviparae. J Insect Physiol 27: 257–265.

Hardie J. (1981b). The effect of juvenile hormone on host-plant preference in the black bean aphid, *Aphis fabae.* Physiol Entomol 6: 369–374.

Hardie J. (1987a). Juvenile hormone stimulation of oocyte development and embryogenesis in the parthenogenetic ovaries of an aphid, *Aphis fabae.* Int J Invertebr Reprod Dev 11: 189–202.

Hardie J. (1987b). The photoperiodic control of wing development in the black bean aphid, *Aphis fabae.* J Insect Physiol 33: 543–549.

Hardie J. (1987c). Nervous system. In *Aphids, Their Biology, Natural Enemies and Control,* vol A (AK Minks and P Harrewijn, eds). Amsterdam: Elsevier, pp. 131–138.

Hardie J. (1987d). Neurosecretory and endocrine systems. In *Aphids, Their Biology, Natural Enemies and Control,* vol A (AK Minks and P Harrewijn, eds). Amsterdam: Elsevier, pp. 139–152.

Hardie J. (1987e). The corpus allatum, neurosecretion and photoperiodically controlled polymorphism in an aphid. J Insect Physiol 33: 201–205.

Hardie J. (1990). The photoperiodic counter, quantitative day-length effects and scotophase timing in the vetch aphid *Megoura viciae.* J Insect Physiol 36: 939–949.

Hardie J, Baker FC, Jamieson GC, Lees AD, and Schooley DA. (1985). The identification of an aphid juvenile hormone, and its titre in relation to photoperiod. Physiol Entomol 10: 297–302.

Hardie J and Glascodine J. (1990). Polyphenism and host-plant preference in the black bean aphid, *Aphis fabae.* Acta Phytopathol Entomol Hung 25: 323–320.

Hardie J and Lees AD. (1983). Photoperiodic regulation of the development of winged gynoparae in the aphid, *Aphis fabae.* Physiol Entomol 8: 385–391.

Hardie J and Lees AD. (1985). The induction of normal and teratoid viviparae by a juvenile hormone and kinoprene in two species of aphid. Physiol Entomol 10: 65–74.

Hardie J, Lees AD, and Young S. (1981). Light transmission through the head capsule of an aphid, *Megoura viciae*. J Insect Physiol 27: 773–777.

Hardie J, Poppy GM, and David CT. (1989). Visual responses of flying aphids and their chemical modification. Physiol Entomol 14: 41–51.

Huybrechts J, Loof A de and Schoofs L. (2005). Melatonin-induced neuropeptide release from isolated locust corpora cardiaca. Peptides 26: 73–80.

Itoh MT, Hattori A, Nomura T, Sumi Y, and Suzuki T. (1995). Melatonin and arylalkylamine *N*-acetyltransferase activity in the silkworm, *Bombyx mori*. Mol Cell Endocrinol 115: 59–64.

Jaubert-Possamai S, Le Trionnaire G, Bonhomme J, Christophides GK, Rispe C, and Tagu D. (2007). Gene knockdown by RNAi in the pea aphid *Acyrthosiphon pisum*. BMC Biotechnol 7: Article 63.

Johnsen S, Kelber A, Warrant E, Sweeney AM, Widder EA, Lee RL Jr, and Hernández-Andrés J. (2006). Crepuscular and nocturnal illumination and its effects on color perception by the nocturnal hawkmoth *Deilephila elpenor*. J Exp Biol 209: 789–800.

Kennedy JS and Booth CO. (1951). Host alternation in *Aphis fabae* Scop. I. Feeding preferences and fecundity in relation to the age and kind of leaves. Ann Appl Biol 38: 25–64.

Le Trionnaire G, Jaubert S, Sabater-Muñoz B, Benedetto A, Bonhomme J, Prunier-Leterme N, Martinez-Torres D, Simon J-C, and Tagu D. (2007). Seasonal photoperiodism regulates the expression of cuticular and signalling protein genes in the pea aphid. Insect Biochem Mol Biol 37: 1094–1102.

Lees AD. (1959). The role of photoperiod and temperature in the determination of parthenogenetic and sexual forms in the aphid *Megoura viciae* Buckton—I. The influence of these factors on apterous virginoparae and their progeny. J Insect Physiol 3: 92–117.

Lees AD. (1960). The role of photoperiod and temperature in the determination of parthenogenetic and sexual forms in the aphid *Megoura viciae* Buckton—II. The operation of an interval timer in young clones. J Insect Physiol 4: 154–175.

Lees AD. (1964). Location of the photoperiodic receptors in the aphid *Megoura viciae* Buckton. J Exp Biol 41: 119–133.

Lees AD. (1966). The control of polymorphism in aphids. Adv Insect Physiol 3: 207–277.

Lees AD. (1973). Photoperiodic time measurement in the aphid, *Megoura viciae*. J Insect Physiol 19: 2279–2316.

Lees AD. (1977). Action of juvenile hormone mimics on the regulation of larval-adult and alary polymorphism in aphids. Nature 267: 46–48.

Lees AD. (1981). Action spectra for the photoperiodic control of polymorphism in the aphid, *Megoura viciae*. J Insect Physiol 27: 761–771.

Lees AD. (1986). Some effects of temperature on the hour glass photoperiod timer in the aphid *Megoura viciae*. J Insect Physiol 32: 79–89.

Lees AD. (1989). The photoperiodic responses and phenology of an English strain of the pea aphid *Acyrthosiphon pisum*. Ecol Entomol 14: 69–78.

Lees AD. (1990). Dual photoperiodic timers controlling sex and female morph determination in the pea aphid *Acyrthosiphon pisum*. J Insect Physiol 36: 585–591.

Lewis RD and Saunders DS. (1987). A damped circadian oscillator model of an insect photoperiodic clock. I. Description of the model based on a feedback control system. J Theor Biol 128: 47–59.

Linn CE, Poole KR, Roelofs WL, and Wu W-Q. (1995). Circadian changes in melatonin in the nervous system and hemolymph of the cabbage looper moth, *Trichoplusia ni*. J Comp Physiol A 176: 761–771.

Lushai G, Hardie J, and Harrington R. (1996). Inhibition of sexual morph production in the bird cherry aphid, *Rhopalosiphum padi*. Entomol Exp Appl 81: 117–119.

Marcovitch S. (1923). Plant lice and light exposure. Science 58: 537–538.

Marcovitch S. (1924). The migration of the Aphididae and the appearance of the sexual forms is affected by the relative length of the daily light exposure. J Agric Res 27: 513–522.

Margaritopoulos JT, Tsitsipis JA, and Prophetou-Athanasiadou DA. (2002). An interval timer controls the morphs in *Myzus persicae* (Homoptera: Aphididae). Physiol Entomol 27: 251–255.

Mittler TM, Nassar SG, and Staal GB. (1976). Wing development and parthenogenesis induced in progenies of kinoprene-treated gynoparae of *Aphis fabae* and *Myzus persicae*. J Insect Physiol 22: 1717–1725.

Moran NA. (1992). The evolution of aphid life cycles. Annu Rev Entomol 37: 321–348.

Müller CB, Williams IS, and Hardie J. (2001). The role of nutrition, crowding and interspecific interactions in the development of winged aphids. Ecol Entomol 26: 330–340.

Müller FP. (1962). Biotypen und Unterarten der 'Erbsenlaus' *Acyrthosiphon piusm* (Harris). Z Pflanzenkr Pflanzenschutz 69: 129–136.

Müller FP. (1980). Wirtzpflanzen Generationenfolge und reproductive Isolation infraspezifischer Formen von *Acyrthosiphon pisum*. Entomol Exp Appl 28: 145–157.

Phillips SW, Bale JS, and Tatchell GM. (1999). Escaping an ecological dead end: Asexual overwintering and morph determination in the lettuce root aphid *Pemphigus bursarius* L. Ecol Entomol 24: 336–344.

Phillips SW, Bale JS, and Tatchell GM. (2000). Overwintering adaptations in the lettuce root aphid *Pemphigus bursarius* (L.). J Insect Physiol 46: 353–363.

Ramos S, Moya A, and Martinez-Torres D. (2003). Identification of a gene overexpresssed in aphids reared under short photoperiod. Insect Biochem Mol Biol 33: 289–298.

Richter K, Peschke E, and Peschke D. (2000). A neuroendocrine releasing effect of melatonin in the brain of an insect, *Periplaneta americana*. J Pineal Res 28: 129–135.

Saunders DS. (1981). Insect photoperiodism—the clock and the counter. Physiol Entomol 6: 99–116.

Saunders DS. (1984). Introduction: The links between "wet" and "dry" physiology. CIBA Found Symp 104: 2–6.

Saunders DS. (2005). Erwin Bünning and Tony Lees, two giants of chronobiology, and the problem of time measurement in insect photoperiodism. J Insect Physiol 51: 599–608.

Shimizu I, Yamakawa Y, Shimazaki Y, and Iwasa T. (2001). Molecular cloning of *Bombyx* cerebral opsin (boceropsin) and cellular localization of its expression in the silkworm brain. Biochem Biophys Res Commun 287: 27–34.

Shingleton AW, Sisk GC, and Stern DL. (2003). Diapause in the pea aphid *(Acyrthosiphon pisum)* is a slowing but not a cessation of development. Dev Biol 3: 7–15.

Shull AF. (1928). Duration of light and the wings of *Macrosipum solanifolii*. Wilhelm Roux' Arch Entwickl Mech Org 113: 210–239.

Steel CGH. (1976). Neurosecretory control of polymorphism in aphids. In *Phase and Caste Determination in Insects: Endocrine Aspects* (M Lüscher, ed). Oxford: Pergamon Press, pp. 117–130.

Steel CGH and Lees AD. (1977). The role of neurosecretion in the photoperiodic control of polymorphism in the aphid *Megoura viciae*. J Exp Biol 67: 117–135.

Sutherland OW. (1969). The role of crowding in the production of winged forms by two strains of the pea aphid, *Acyrthosiphon pisum*. J Insect Physiol 15: 1385–1410.

Tilden AR, Anderson WJ, and Hutchison VH. (1994). Melatonin in two species of damselfly *Ischnura verticalis* and *Enallagma civile*. J Insect Physiol 40: 775–780.

Vaz Nunes M. (1998). A double circadian oscillator model for quantitative photoperiodic time measurement in insects and mites. J Theor Biol 194: 299–311.

Vaz Nunes M and Hardie J. (1993). Circadian rhythmicity is involved in photoperiodic time measurement in the aphid *Megoura viciae*. Experientia 49: 711–713.

Vaz Nunes M and Hardie J. (1994). The photoperiodic counter in the black bean aphid, *Aphis fabae*. J Insect Physiol 40: 827–834.

Vaz Nunes M and Hardie J. (1996). Differential photoperiodic responses in genetically identical winged and wingless pea aphids, *Acyrthosiphon pisum*, and the effect of day length on wing development. Physiol Entomol 21: 339–343.

Vaz Nunes M and Hardie J. (2000). The effect of temperature on the photoperiodic "clock" and "counter" of a Scottish clone of the vetch aphid, *Megoura viciae*. J Insect Physiol 46: 727–733.

Vaz Nunes M, Young S, and Hardie J. (1996). Laboratory-simulated naturally-decreasing day lengths, twilight and aphid photoperiodism. Physiol Entomol 21: 231–241.

Veerman A and Vaz Nunes M. (1987). Analysis of the operation of the photoperiodic counter provides evidence for hourglass time measurement in the spider mite *Tetranychus urticae*. J Comp Physiol 160A: 421–430.

Veerman A and Veenendaal RL. (2003). Experimental evidence for a non-clock role of the circadian system in spider mite photoperiodism. J Insect Physiol 49: 727–732.

Westerlund SA and Hoffmann KH. (2004). Rapid quantification of juvenile hormones and their metabolites in insect haemolymph by liquid chromatography-mass spectrometry (LC-MS). Anal Bioanal Chem 379: 540–543.

Wetterberg L, Hayes DK, and Halberg F. (1987). Circadian rhythms of melatonin in the brain of the face fly, *Musca autumnalis*, De Greer. Chronobiologia 14: 377–381

Wilson I. (1938). Some experiments on the influence of environment upon the forms of *Aphis chloris* Koch (Aphididae). Trans R Entomol Soc Lond 87: 165–180.

Zaslavski VA and Fomenko RB. (1983). Quantitative photoperiod perception in the aphid *Megoura viciae* Buckt. (Homoptera, Aphidiae). Entomol Rev 62: 1–10.

Part III

Photoperiodism in Vertebrates

OVERVIEW

Randy J. Nelson

> In Winter I get up at night,
> And dress by yellow candle-light,
> In Summer, quite the other way,
> I have to go to bed by day.
>
> —*Robert Louis Stevenson*

Photoperiodism has evolved in virtually all taxa of plants and animals that experience seasonal changes in their habitats. Changes in day length, though probably of little direct importance to most animals, provide the most error-free indication of time of year and thus enable individuals to anticipate seasonal conditions. Among vertebrates photoperiodism is linked to a number of seasonal adaptations, including reproductive, metabolic, immunological, and morphological adaptations to cope with seasonal changes in ambient conditions. The first demonstration of photoperiodic time measurement regulating vertebrate biology was established in birds (Rowan, 1925). During the winter, juncos (*Junco hyemalis*) were maintained in outdoor aviaries in Edmonton, Alberta, and exposed to several minutes of artificial light after the onset of dark each day (lights were illuminated at sunset). Under these artificial lighting conditions, these birds came into reproductive condition despite the harsh Canadian winter temperatures. In comparison, juncos living in the wild in the relatively mild "Riviera-like" climate of Berkeley, California, but exposed to normal winter day lengths, remained in nonreproductive condition (Rowan, 1925). Thus, it was concluded that the number of hours of light per day, not ambient temperature or food availability, regulated the annual breeding cycle of juncos. The initial demonstration of photoperiod regulating mammalian reproduction was reported for the European field vole *Microtus agrestis* (Baker and Ranson, 1932). Currently, the role of photoperiod in mediating seasonal adaptations has been documented for hundreds, if not thousands, of vertebrate species. Because the same photoperiod occurs twice a year (e.g., 21 March and 21 September), it is advantageous for individuals to be able to discriminate between these two dates. Many photoperiodic vertebrates have solved this problem by developing an annual alteration between two physiological states termed photoresponsiveness and photorefractoriness.

HOURGLASS MODEL OF PHOTOPERIODISM

The critical problem for any photoperiodic individual is to discriminate between long and short day lengths. Two broad hypotheses have been advanced to explain how animals respond to photoperiodic changes: (1) the response depends on the total number of hours of light per day or (2) the response depends upon the phase of the light exposure relative to some internal rhythm of photoresponsiveness. According to the first model, animals monitor the accumulation (or depletion) of some physiological agent during one part of the light–dark (LD) cycle; this process is reversed during another portion of the cycle. This time measurement hypothesis is referred to as the hourglass model. The hourglass model appears to be used by several invertebrate species, but only by a very few vertebrate species.

The hourglass model presumes that the absolute duration of either the light or dark period is monitored, and when some threshold value is attained, a photoperiodic "adjustment" is made. Alternatively, the absolute ratio between the amount of light and dark per day could also be monitored. To determine whether vertebrates use an hourglass timing mechanism to measure photoperiod, the first step is to determine the critical day length necessary to mediate the photoperiodic response. Male anole lizards (*Anolis carolinensis*) undergo gonadal regression during the autumn and then display gonadal recrudescence during early spring. Housing male anole lizards in a variety of photoperiods revealed that day lengths less than 13.5 h of light per day evoked gonadal regression (Underwood and Goldman, 1987). This critical day length corresponds to the day lengths of mid-August in the natural habitat of these lizards. To determine whether anole lizards monitor and respond to the length of the daily dark period (i.e., night lengths >11.5 h/day should induce gonadal regression), male anole lizards were maintained in cycles of LD 8:16, LD 16:8, and LD 16:20. If anole lizards turn off reproductive function in response to night lengths longer than 11.5 h, then males housed in both LD 8:16 and LD 16:20 should display regressed reproductive systems. However, only male lizards maintained in LD 8:16 photoperiods underwent gonadal regression. Males housed in LD 16:8 or LD 16:20 photoperiods maintained reproductive function, suggesting that these lizards were monitoring absolute day length (Underwood and Goldman, 1987).

In common with plants and invertebrates, alternative models exist to explain how vertebrates respond to day length. These so-called "coincidence models" rely on phase relations between internal circadian oscillators or phase relations between external factors and endogenous circadian cycles of receptive states. There are two types of coincidence models of photoperiodic time measurement: (1) the external coincidence model and (2) the internal coincidence model.

COINCIDENCE MODELS OF PHOTOPERIODISM

External Coincidence Model

As noted previously in this volume, the second hypothesis of photoperiodic time measurement was originally formulated to explain flowering in plants. According to Bünning, "the physiological basis of photoperiodism.... lies with the endogenous daily rhythms." (1973, p. 69). The crux of Bünning's hypothesis is the assumption of an endogenous circadian rhythm of subjective day (photoinsensitivity/noninducibility, with a duration of about 12 h) and subjective night (photosensitivity/inducibility, also with a duration of about 12 h). Light was postulated to serve two functions: (1) light entrains (i.e., synchronizes) the endogenous circadian rhythm of subjective day and subjective night, and (2) light stimulates photoperiodic responses if it is coincident with the subjective night (i.e., the photoinducible phase). Thus, when first exposed to light, the 12 h subjective day is set; light intruding during the next 12 h will not maintain the reproductive system. For example, an LD 6:18 cycle will not maintain the reproductive system of a hamster because light will only coincide with the subjective night of the cycle. The inverse photoperiod (i.e., LD 18:6) would maintain the reproductive system because 6 h of light would coincide with the photosensitive/inducible part of the cycle. Many studies across different taxa have produced results consistent with Bünning's hypothesis. In principle, a few seconds of light per day, appropriately timed to coincide with an individual's photosensitive/inducible phase, would evoke a "long-day" response.

There have been several tests of Bünning's hypothesis in vertebrates. One type of test of circadian involvement in photoperiodic time measurement makes use of the resonance paradigm (Nanda and Hamner, 1958; Elliott, 1976). In resonance studies, groups of animals are maintained on photoperiod cycles that couple a fixed light phase (e.g., 6 or 8 h) with various durations of darkness resulting in LD cycles of varying lengths. For example, hamsters can be maintained in LD 6:18, 6:24, 6:30 6:42, and 6:54 photoperiods. If hamsters use an hourglass timer for photoperiodic time measurement, then animals maintained in each of these photoperiods should have regressed reproductive systems; the duration of light in each instance is less than 12 h/day. Alternatively, if hamsters employ a circadian timer for photoperiodic time measurement, then only animals housed in the LD 6:18 and 6:42 photoperiods should have regressed reproductive systems because only during these photoregimens is light restricted to the putative photoinsensitive/noninducible phase. Animals exposed to LD 6:30 and 6:54 photoperiods ought to have comparably sized reproductive systems to long-day males because light would coincide with the subjective night every other day or every third day, respectively. The results of these types of studies are consistent with Bünning's hypothesis of photoperiodic time measurement.

Skeleton photoperiods have also been used to test Bünning's hypothesis. In skeleton photoperiods, the full light cycle is usually replaced by appropriately timed light pulses. For example, the reproductive system of male hamsters maintained in

LD 16:8 photoperiods could be sustained in LD 2:22 photoperiods if one hour of light occurred at 0800–0900 h and another hour of light occurred at 2300–24:00 h. This skeleton photoperiod would provide a pulse of light in the morning to entrain the cycle of photo-noninducible/photoinducible phases and a second pulse of light near the middle of the photoinducible phase. Thus, an appropriately timed, very short photoperiod of LD 2:22 could "trick" the reproductive system into a long-day response. Of course, it is possible for the individual to entrain in such a way that the 1-h pulse of light between 2300 and 2400 h would entrain the cycle of photo-noninducible/photoinducible phases and the second pulse 8 h later would coincide with the photo-noninducible phase resulting in short-day responses. Thus, predictions about specific skeleton photoperiods depend on entrainment patterns; monitoring locomotor activity or other circadian cycles is important in explaining photoperiodic responsiveness to skeleton photoperiods.

Internal Coincidence Model

This model of photoperiodic time measurement assumes the existence of two or more internal oscillators that change their phase relationship during the annual change in day length (Pittendrigh and Minis, 1964). For example, if the peak blood concentration of glucocorticoids was coupled to a "dawn" oscillator and the peak blood concentration of prolactin was coupled to a "dusk" oscillator, the difference in the timing of the peak values of these two hormones might range from 15 to 9 h throughout the year among temperate zone animals. According to the internal coincidence model, a photoperiodic response should be observed when certain internal phase relations are attained. Although the existence of internal coincidence processes in vertebrate photoperiodic time measurement systems has not been firmly established, it continues to be an attractive hypothesis because this model is based on entrainment theory and does not rely on a special photoinducible oscillator.

Two themes emerge from the third section of this volume on photoperiodism in vertebrates. First, vertebrate mechanisms and processes underlying the measurement of photoperiod conserve (and in some cases improve upon) the basic mechanisms and processes underlying photoperiodism that have evolved in plants and invertebrates. Second, despite the precision with which photoperiod responses have permitted vertebrate animals to anticipate the onset and departure of food availability to fuel reproduction and other energetically expensive activities, climatic change and light pollution at night are changing the timing of production of plants and invertebrates (Navara and Nelson, 2007; Charmantier et al., 2008). This divergence in timing can easily cause photoperiodic vertebrates to misjudge the temporal peaks in food availability because their seasonal adaptations are based on historical food production of plants and invertebrate animals.

For example, breeding records maintained for nearly 50 years in the United Kingdom indicate that great tits (*Parus major*) have phase-advanced the onset of their breeding season by two weeks in Wytham Woods (Charmantier et al., 2008).

Within this population, there is remarkably little variation among individuals in the new onset time of breeding. This contrasts with a study of tits in the Netherlands in which significant variation among the population exists—some individuals have advanced the onset of breeding to match the earlier availability of caterpillars, whereas other individuals have not shifted the timing of reproductive (Visser et al., 1998, 2006; Nussey et al., 2005). The net effect is reduced population fitness among the Dutch birds. What accounts for the populationwide phenotypic plasticity among the U.K. birds and the limited phenotypic plasticity among the Dutch birds remains unspecified.

Certainly, selection for responsiveness to photoperiod occurs very rapidly. In a study of *Peromyscus*, mice were selected to ignore photoperiod in just two generations (Bronson et al., 1986). In a study of vernal reproductive activation in Siberian hamsters (*Phodopus sungorus*), the interval timer was approximately five weeks longer and twice as variable in females relative to males (Prendergast et al., 2004). Heritability estimates from parent offspring regression analyses were significant for female but not for male offspring, suggesting strong additive genetic components for this trait, primarily in females. In nature, individual differences, both within and between sexes, in the timekeeping properties of seasonal interval timers, and a strong heritable basis thereof, would provide ample substrate for selection to rapidly influence seasonal clocks. Balancing selection in environments where the onset of spring conditions varies from year to year could maintain genetic variance in interval timers and yield interval timers tuned to the local environment (Prendergast et al., 2004). Thus, the potential exists for populations to adjust to changing temporal habitats, although the speed with which these changes can occur requires further study.

In chapter 15, Bertil Borg reviews how photoperiod controls seasonal fluctuations in fishes. Consistent with the literature, this chapter focuses on reproduction, but the effects of photoperiod on migration are also described where available. In chapter 16, Zachary Weil and David Crews discuss the role of photoperiod on seasonality among amphibians and reptiles. Again, the main focus of this chapter is seasonal breeding, which represents the studies conducted on photoperiodism in these taxa. George Bentley examines the role of photoperiod and avian reproductive biology in chapter 17. The genetic and molecular mechanisms underlying avian photoperiodism are reviewed in chapter 18 by Takashi Yoshimura and Peter Sharp. Nonreproductive mammalian adaptations driven by photoperiod are reviewed in chapter 19 by Gregory Demas, Zachary Weil, and Randy Nelson. In chapter 20, Lance Kriegsfeld and Eric Bittman document the role of photoperiod in mammalian reproduction. In chapter 21, the genetic and molecular mechanisms of mammalian photoperiodism are described by David Hazlerigg.

References

Baker JR and Ranson RM. (1932). Factors affecting the breeding of the field mouse (*Microtus agrestis*). Part I. Light. Proc R Soc Lond 110: 313–323.

Bünning E. (1973). *The Physiological Clock, Circadian Rhythms and Biological Chronometry.* New York: Springer-Verlag.

Charmantier A, McCleery RH, Cole LR, Perrins C, Kruuk LEB, and Sheldon BC. (2008). Adaptive phenotypic plasticity in response to climate change in a wild bird population. Science 320: 800–803.

Desjardins C, Bronson FH, and Blank JL. (1986). Genetic selection for reproductive photoresponsiveness in deer mice. Nature: 172–173.

Garner WW and Allard HA. (1922). Photoperiodism, the response of the plant to relative length of day and night. Science 60: 582–583.

Hyde LL and Underwood H. (1993). Effects of nightbreak, T-cycle, and resonance lighting schedules on the pineal melatonin rhythm of the lizard *Anolis carolinensis*: Correlations with the reproductive response. J Pineal Res 15: 70–80.

Nanda KK and Hamner KC. (1958). Studies on the nature of the endogenous rhythm affecting photoperiodic response of Biloxi soybean. Bot Gaz 120: 14–28.

Navara KJ and Nelson RJ. (2007). The dark side of light at night: Physiological, epidemiological, and ecological consequences. J Pineal Res 43: 215–315.

Nussey DH, Postma E, Gienapp P, and Visser ME. (2006). Selection on heritable phenotypic plasticity in a wild bird population. Science 310: 304–306.

Pittendrigh CS and Minis DH. (1964). The entrainment of circadian oscillations by light and their role as photoperiodic clocks. Am Nat 98: 261–299.

Prendergast BJ, Renstrom RA, and Nelson RJ. (2004). Genetic analysis of a seasonal interval timer. J Biol Rhythms 19: 298–311.

Rowan W. (1925). Relation of light to bird migration and developmental changes. Nature 115: 494–496.

Underwood H and Goldman BD. (1987). Vertebrate circadian and photoperiodic systems. Role of the pineal gland and melatonin J Biol Rhythms 2: 279–315.

Visser ME, van Noordwijk AJ, Tinbergen JM, and Lessells CM. (1998). Warmer springs lead to mistimed reproduction in great tits (*Parus major*). Proc R Soc Lond B Biol Sci 265: 1867–1870.

Visser ME, Holleman LJ, and Gienapp P. (2006). Shifts in caterpillar biomass phenology due to climate change and its impact on the breeding biology of an insectivorous bird. Oecologia 147: 164–172.

15

Photoperiodism in Fishes

Bertil Borg

The aim of this chapter is to provide an overview of how the photoperiod controls seasonal cycles in fishes. Most attention is given to reproduction, but migration is also included. The chapter mainly deals with experimental studies; investigations where biological events are correlated with environmental factors (e.g., sampling series from the field) are given less attention. Even so, the literature on photoperiodic effects on fish is huge, and my choice in what to include is necessarily subjective.

Fishes comprise cyclostomes, cartilagenous fishes, and bony fishes. The modern bony fishes or teleosts have been estimated to have close to 27,000 described living species or about 96% of all extant fishes (Nelson, 2006). There are very few studies relevant to photoperiodic effects on nonteleost fishes; consequently, this chapter deals with teleosts alone unless otherwise specifically stated.

FISH REPRODUCTION

Sex Hormones

Testosterone (T), the main male sex steroid hormone in mammals, is also present in fish. However, the most important androgens in male fish are 11-oxygenated androgens, especially, 11-ketotestosterone (11KT), that are much more effective than T in stimulating male secondary sexual characters and male reproductive behavior and in controlling spermatogenesis than T (Borg, 1994). Levels of 11-androgens are considerably higher in male than in female fish. On the other hand, T concentrations in females may be equivalent or even higher than in males. It is not known whether T has a biological role in female fish, or whether it is only important as a precursor to estrogens. Estrogens are considered mainly as female hormones, the most important of which is estradiol (E2). The main function of E2 in fish is to induce production of vitellogenin (VG) in the liver. Unlike in mammals, E2 levels do not peak close to ovulation. Progestins are produced both by ovaries and testes. Different fishes often produce different progestin hormones, the most studied of which is 17α,20β-dihydroxy-4-pregnen-3-one (17,20P), which is found in salmonids. In females a major role of progestins is to stimulate final oocyte maturation, whereas in males progestins stimulate spermiation, the production of running milt.

371

Gonadotropins

Gonadal production of sex hormones and gametes is stimulated by gonadotropic hormones (GTHs) from the pituitary, which in turn is controlled from hypothalamus. This comprises the brain-pituitary-gonadal axis. Similarly as in tetrapods, fish have two GTHs: follicle-stimulating hormone (FSH, GTH I) and luteinizing hormone (LH, GTH II) (e.g., Suzuki et al., 1988). Both GTHs are glycoproteins and consist of two chains: an α-chain that is identical for FSH and LH and a β-chain that is separate but related.

The biological effects of the two hormones have been particularly well studied in salmonids (e.g., Swanson, 1991). LH is much more effective than FSH in stimulating the production of 17,20P in males and females, whereas both GTHs can stimulate the production of androgens and estrogens. Usually FSH levels in the pituitary and in the circulation rise earlier than the LH levels (e.g., Swanson, 1991) in both sexes. In salmonids, the FSH peak stimulates the production of androgens and estrogens (LH levels are very low in this phase) and thus indirectly gonadal growth, spermatogenesis, secondary sexual characters, and vitellogenesis. The LH peak stimulates the production of progestins and thus indirectly ovulation and spermiation.

Gametogenesis

Fish gonads undergo large changes in size according to the reproductive state. The ovarian weights often reach 20–30% of the total body weight at full maturity. When maturing, oocytes accumulate large quantities of yolk, which increase their sizes and the ovarian weights drastically. The yolk precursor protein, VG, is produced in the liver under stimulation by E2 and reaches the oocytes via the circulation. Plasma levels of VG can be studied by using radioimmunoassay, but also the measurement of calcium levels can be useful because calcium increases markedly at vitellogenesis as VG contains protein-bound calcium. In the ovaries, VG is taken up in the oocytes.

At final maturation, high LH levels induce synthesis of progestin, which stimulates final oocyte maturation (germinal vesicle breakdown). Via induction of prostaglandin synthesis, the progestin also induces ovulation proper; that is, the follicles break up and release the oocytes. The ovaries of teleosts are usually hollow and the ovulated eggs are stored in ovarian fluid in the lumen. In some fishes, for example, salmonids, the ovaries are not closed and the ovulated eggs are stored in the abdominal cavity. Ovulated eggs can easily be expelled by a gentle pressure on the abdomen; the fish is then running ripe. This offers a convenient and clear-cut test for attainment of full maturity in females. In most fishes, except salmonids, ovulated eggs remain viable for only a short period, a few hours to a few days. Following this, they may be laid spontaneously, absorbed, or hardened. Fish eggs develop in batches, each consisting of a cohort of eggs at a similar stage of development.

Some fishes (e.g., salmonids) develop only one batch of eggs per spawning season, whereas others (e.g., sticklebacks and medaka, *Oryzias latipes*) can develop several and spawn repeatedly.

Sperm are produced via spermatogonial multiplication followed by meiosis and spermiogenesis; the transformation of spermatids into spermatozoa. Spermatogenesis, which is a rather slow process, is usually completed when spawning starts. At that time the only germ cells present are numerous spermatozoa and a few spermatogonia. Seasonally spawning fishes usually display either of two spermatogenetic patterns. In prespawning spermatogenesis, the rapid spermatogonial multiplication starts some months before spawning and the process is finished as spawning approaches. This pattern can be found in, for example, salmonids and many cyprinids. The other pattern is postspawning spermatogenesis during which spermatogonia start to multiply as the breeding season ends and where the process can be finished several months before spawning. This pattern can be found in sticklebacks, for example. Spermatogenesis is usually stimulated by androgens, though there are exceptions.

At spawning, spermiation takes place; that is, spermatozoa become detached from the Sertoli cells and the walls of the seminiferous lobules, and running milt is produced. In many fishes, the presence of running milt can be assessed by lightly stroking the belly and is a convenient indication of maturity. As mentioned, spermiation is stimulated by progestins.

The choice of parameters from which maturation can be estimated is often limiting in studies on seasonal reproduction. Gonadal weights can only be determined once in each fish, though it may be possible to take repeated gonadal biopsies to study the development histologically. For measurement of circulating hormone concentrations, repeated blood samples can be taken from large fishes, but not from small ones. The possibility of measuring steroid hormones present in urine or excreted into water opens a possibility to nondestructive, repeated sampling also from small fishes. It is also possible to follow the development of maturation by repeatedly testing fish for running roe or milt by a light stroking of the belly, but not all fishes produce running milt even when fully mature.

Semelpari

Animals can reproduce iteroparously or semelparously. Iteroparous animals can reproduce repeatedly. Semelparous animals, on the other hand, always die after breeding. Semelpari is common in fish; the most well-known example is the Pacific salmon. In semelpari, all available resources are mobilized for reproducing a single time, allowing a maximal reproductive output but also leading to the death of the individual. It is advantageous when the individuals' chances for surviving to reproduce a second time are small and when long spawning migrations impose a heavy cost just to arrive in the right place.

GENERAL BREEDING PATTERNS AND EFFECTS
OF PHOTOPERIODS IN SOME FISHES

Depending on the environment, offspring are more likely to survive in some seasons than at others and the selective advantage in choosing the best time to breed is high. In order to time reproduction and other biological events, animals both use endogenous cycles and environmental cues. Photoperiod is often an important cue in fishes, although there are also others. Temperature effects have often been often found to be important (e.g., Bullough, 1939; Baggerman, 1957; Borg, 1982b; Borg and van Veen, 1982) but are outside the scope of this chapter.

Photoperiodic experiments have mainly been conducted with small fishes that are convenient to work with in the laboratory or with fishes that are of aquacultural interest. Aquacultural goals for photoperiodic treatments include attainment of off-seasonal breeding to get a more continuous production of fry, but also suppression of maturation because this diminishes growth and often lowers flesh quality. Most studies has been done on fishes living in temperate regions, but photoperiod is important for many fishes (e.g., stinging catfish *Heteropneustes fossilis*) living as far south as India. Equatorial fishes are not likely to influenced by photoperiod, even though there may be seasonal reproduction related to wet and dry seasons. Neither fishes living in caves nor in the deep sea are likely directly affected by photoperiod.

Salmonids

Salmoninae contains several species that are aquaculturally important, both for production of consumption fish and for rearing of young fish to be released as compensation for destruction of natural spawning habitats by hydropower dams. Extensively studied species include rainbow trout (*Oncorhynchus mykiss* [in older literature *Salmo gairdneri* or *S. irideus*]), masu salmon (*O. masou*), and Atlantic salmon (*Salmo salar*). All Salmoninae spawn in freshwater, and many also live there their entire lives. However, there are also many anadromous species that return from the sea to the rivers where they were born in order to breed and migrate out to the sea where a richer food supply allows faster growth. Some species contain both freshwater and andromous forms, and there can be mixed strategies even in the same populations. Salmoninae generally spawn in autumn–winter, and attainment of complete maturation is often stimulated by short/decreasing photoperiods (e.g., Amano et al. 1994). Effects of photoperiod on fish reproduction was first reported in postspawning brook trout (*Salvelinus fontinalis*), which were exposed from February to either an accelerated photoperiod, gradually increasing to a maximum light–dark (LD) cycle of LD 21.5:2.5 in May followed by a rapid decline, or to a natural photoperiod (maximum LD 15.5:8.5 in late June; Hoover, 1937). The experimental fish reached full maturity (running milt, ovulated eggs) in August, about three months earlier than the controls. Most salmonids spawn in running water, and the eggs are deposited in "nests" dug by the female. They are covered with gravel, but there is no further parental care. After hatching, the young fish spend stationary lives in the river beds, first as fry-alevins and then as parr. The parr stage can last for a number of years for Atlantic salmon in cold northern areas, but is usually shorter. In anadromous fish, the young fish then undergo a parr-smolt transformation (see below), after which they migrate to the sea or sometimes to large lakes. After having grown large, the fish return to their river of birth to spawn. Migration into freshwater often takes place in late spring–summer several months before spawning. Many males mature already as parr, also known as precocious males. These small males achieve fertilizations by sneaking up on large spawning fish and releasing their sperm. Pacific salmons are semelparous, whereas others as the Atlantic salmon can breed more than once.

Three-Spined Stickleback

Seasonal reproductive cycles have been extensively studied in the three-spined stickleback (*Gasterosteus aculeatus*) whereas other sticklebacks are almost unstudied. In this chapter, "stickleback" always refers to the three-spined stickleback. Levels of 11KT are at their highest in the breeding season (Mayer et al. 1990), when the male develops androgen-dependent secondary sexual characters such as breeding colors and a hypertrophied kidney. The male constructs a nest of plant material, which he glues together with threads of the kidney-produced protein spiggin (after *spigg*, "stickleback" in Swedish; Jakobsson et al., 1999). Kidney hypertrophy can been used as a measure for maturation in sticklebacks (e.g., Borg, 1982a, 1982b; Mayer et al., 1997b). Usually the kidney epithelium height (KEH) (i.e., the height of the secondary proximal tubules—where the glue is been produced) has been studied on histological sections. The females can produce several batches of eggs over the breeding season. After one or more females have laid eggs in his nest, the male takes care of the eggs, which are ventilated by fanning, and of the young fry. The male may then nest again.

The stickleback displays postspawning spermatogenesis. Spermatogenesis is completed in late autumn–early winter and is quiescent during the breeding season when androgen levels are high and starts afterwards when secondary sexual characters and androgen levels decline (Borg, 1982b; Mayer et al., 1990). The onset of spermatogenesis can be inhibited by administration of androgens (Andersson et al., 1988). In some areas, sticklebacks are semelparous.

Effects of photoperiod have been extensively studied in the stickleback (e.g., van den Eeckhoudt, 1946; Baggerman, 1957, 1972, 1980, 1985; Borg, 1982a, 1982b; Borg and van Veen, 1982). Long photoperiods, especially in combination with high temperature, stimulate maturation. Short photoperiods are less effective even when combined with high temperature. Early in the season, only long photoperiods are able to stimulate maturation, whereas later in winter–spring when the natural breeding season approached, maturation can take place also under short photoperiods.

This indicates a cycle in photosensitivity with a decreasing threshold for maturation, studying as the attainment of running roe in females and nest building in males, from autumn to spring (Baggerman, 1972, 1980, 1985). The ecological importance of the photoperiodic response in the stickleback is probably more to prevent unseasonal breeding in autumn, than to control maturation in spring, when also short photoperiods are increasingly effective. The proportion of sticklebacks that mature at high temperature changes gradually with season and photoperiod (Baggerman, 1980, 1985), but each individual either builds nests/oviposits rather quickly or not at all. For those fish that do mature, the time this takes is similar under different photoperiods, but gets shorter as the season progresses (Baggerman, 1985). Fish that do not mature completely under a nonstimulatory photoperiod in combination with high temperature are strongly suppressed (Borg, 1982b; Borg and van Veen, 1982; Borg et al., 1987a), suggesting that the photoperiodic response at high temperature is of an all-or-nothing type. Indeed, KEHs are lower and atretic oocytes are more common in the ovaries under high temperature combined with short photoperiod than under short photoperiod and low temperature or under natural winter conditions. Spermatogenesis, on the other hand, is activated under short photoperiod combined with high temperature (Borg, 1982b, Borg et al., 1987a), probably due to the absence of otherwise suppressing androgens, but despite low expressions of LH-β and FSH-β (Hellqvist et al. 2004). Following breeding, the sticklebacks display a nonbreeding, refractory period.

Cyprinids

Some cyprinids such as the goldfish (*Carassius auratus*) and valuable aquacultural species such as carp (*Cyprinus carpio*) have been extensively studied. They usually spawn in spring/early summer. Long photoperiods in combination with high temperature have in many cases been found to promote maturation. An effect by light conditions on cyprinid reproduction was first observed by Bullough (1939); he reported that LD 17:7 could induce gonadal maturation in winter in the spring-spawning Eurasian minnows (*Phoxinus phoxinus*) if combined with high temperature (17°C), whereas high temperature or long photoperiod alone were not effective.

Closely Related Fish Spawning at Different Seasons

Some species such as sticklebacks and cyprinids spawn in spring in temperate regions, whereas many salmonids spawn in autumn. However, there are also several fishes were stocks spawning at markedly different seasons occur in the same species. This has been extensively studied in the Atlantic herring (*Clupea harengus*) where there are several autumn- and spring-spawning stocks (e.g., McQuinn, 1997). Both types of stocks can occur in same general area (e.g., off the coast of Newfoundland and in the Baltic) and may even mix outside the spawning season.

In the Baltic area, spring and winter spawners in the Lillebaelt area were found to be genetically distinct from each other as studied using microsatellite DNA analysis, whereas this was not the case for spring and winter spawners at Rügen (Bekkevold et al., 2007). Further, age analysis based on otoliths indicated that winter spawners at Rügen had spring-spawning parents (Bekkevold et al., 2007). Thus, it appears that large shifts in spawning seasons and presumably in the mechanisms controlling reproduction in fish may even be phenotypic.

Vendace (*Coregonus albula*) usually spawns in autumn and the eggs hatch first the next year. However, in Northern Europe spring-spawning vendaces (*C. fontanae, luciensis, trybomi*) have independently evolved in different lakes, where they may occur sympatrically with genetically distinct autumn-spawning vendace (e.g., Schulz et al., 2006).

Cod (*Gadus morhua*) in the Baltic spawn much later (March to August) than the cod in Skagerrak on the Swedish west coast (January to March) (Vallin and Nissling, 1994). Stocks spawning at widely different time of the year may offer interesting models for the study the mechanisms in photoperiodic control. Little experimental work has been done, though maturation in winter-spawning Norwegian cod has been found to be strongly influenced by light. Karlsen et al. (1990) exposed previously immature two-year-old cod to either natural photoperiod or constant light from June to July the next year. All males and females in the former treatment matured, with gonadosomatic indices (GSIs) peaking in January, whereas no fish matured under constant light.

ENDOGENOUS REPRODUCTIVE CYCLES

Many biological phenomena display endogenous cyclicity. Under constant conditions, such a cycle is free-running, whereas a cycle is entrained when it is controlled by an environmental cue or *Zeitgeber*, such as photoperiod. Rhythms are circannual if the endogenous cycle is close to one year and are self-sustaining under constant environmental conditions. Endogenous, circannual rhythms in reproduction have been reported in many animals kept under constant photoperiod and temperature (for review, see Turek and van Cauter, 1994).

Refractory Periods

Breeding only lasts for a rather limited time in many species, also among iteroparous repeat spawners. Then comes a refractory period when reproductive activity ceases even under otherwise stimulatory environmental conditions. Refractory periods have been extensively studied in birds, but they also occur in fishes, including sticklebacks. A territorial male stickleback builds a nest, and if that is destroyed he continues to build new ones as long as he is in reproductive condition. Under a constant long photoperiod of LD 16:8 and constant temperature, the breeding season ceases after one or a few months, followed by nonbreeding period of several months (Baggerman, 1957; Bornestaf and Borg, 2000a). However, the cessation of breeding is not entirely independent of environmental conditions. Under constant light, the nesting period lasted three times as long as compared to LD 16:8 (Bornestaf and Borg, 2000a). Constant light was ecologically valid, considering that Swedish sticklebacks were used and that the midsummer nights in Scandinavia never get dark.

Circannual Cycles

In stickleback males kept under constant LD 16:8 and 20°C, the mean duration of the breeding (nest-building) period was 91 days, followed by a mean nonbreeding period of 108 days, that is, a free-running period of 199 days (Baggerman, 1957). The second breeding period lasted only 47 days. However, because the few animals that survived to mature a third time displayed a 149-day interval between the second and third nest-building period, the second cycle had a length of 47 + 149 = 196 days, or similar to the first cycle. Females displayed a pattern similar as that of the males. In males exposed to LD 8:16, the breeding period lasted shorter than under long photoperiod and no fish matured again (Baggerman, 1957). From the hypothesis of a seasonal cycle in threshold of photosensitivity (Baggerman, 1980; see below) it can be predicted that the first breeding period should be shorter, the total breeding cycles longer, and the proportion of time spent in breeding condition lower under a shorter than under a longer photoperiod. However, no differences were detected in length of breeding cycle between males kept under constant LD 12:12 and LD 16:8 (both being shorter than a year), suggesting that this is not the entire explanation for the endogenous control of breeding in sticklebacks (Bornestaf and Borg, 2000a). Following the first cycle(s), increasing numbers of males, particularly under LD 12:12, displayed irregular cycles with shortened nonbreeding periods. This pattern suggests a desynchronization of internal oscillators (Bornestaf and Borg, 2000a). Surprisingly, a few males maintained under LD 12:12 continued their first nesting period for very extensive periods of time (294–511 days; Bornestaf and Borg, 2000a), whereas no first nesting period under LD 16:8 exceeded 91 days. It is tempting to compare this with the situation in some birds, where a longer photoperiod is needed to induce the onset of postreproductive refractoriness than to stimulate the onset of the reproductive period. These observations contrast with the observation that the first breeding period was very long in sticklebacks kept under constant LD 24:0 (Bornestaf and Borg, 2000a).

Duston and Bromage (1986) kept female rainbow trout under constant photoperiods of LL, LD 18:6, and LD 6:18 at 9°C for two to four years beginning in February. Reproductive cycles were evaluated using stripping of roe (trout only produce one batch of roe per season), as well as by measuring serum T, E2, and calcium (indicating vitellogenin). In the first cycle LL and LD 18:6, fish spawned in October–November, up to two months ahead of the natural time, whereas the

LD 6:18 fish spawned approximately two months after the natural time. Following this, the LD 18:6 and LL fish matured at intervals of approximately 160 days, whereas the LD 6:18 group matured twice more with intervals of one year.

Female stinging catfish were maintained under LL or DD for more than 2.5 years; ovarian weights and oocyte development were examined in samples obtained a few months apart (Sundararaj et al., 1982). Marked cycles with one year between peaks were found under both regimes and in fish exposed to LD 12:12 for 1.5 years. Thus, unlike in the stickleback and rainbow trout, the circannual cycle was largely independent of the length of the constant photoperiod.

Baggerman (1980) tested the attainment of maturity (nest building/oviposition) in sticklebacks under different skeleton photoperiods (see below) and 20°C starting at different points of time from autumn to spring. As the season progressed, increasing number of fish became stimulated also by photoperiodic regimes that were nonstimulatory earlier. This change took place not only in animals from natural conditions, but also in animals stored under constant LD 8:16 and 15°C. Thus, there was also an endogenous shift in photosensitivity.

LUNAR CYCLES IN REPRODUCTION

Reproduction in fishes and other animals in coastal areas is often strongly influenced by the lunar cycle (for review, see Taylor, 1984). Mummichogs (*Fundulus heteroclitus*) spawn at high tides and leave the eggs to develop aerially, which may save them from oxygen-poor water. Mummichogs kept in groups under a constant photoperiod of LD 14:0, showed endogenous cycles in spawning (number of laid eggs present in aquaria) with periods between 14.4 and 16.0 days in different groups (Hsiao et al., 1994), compared to 14.8 days in natural semilunar cycles. Most studies on lunar cycles in fishes are, however, based on natural correlations rather than experimental treatments. It is usually impossible to say whether moonlight or tidal factors, such as water levels, entrain the natural cycle. A laboratory study on mummichogs suggested the former since dim (0.1 lux) artificial lunar light could entrain spawning cycles (Taylor, 1991).

MIGRATION AND OSMOREGULATION

Many fishes undertake migrations between freshwater and saltwater. Anadromous fishes (e.g., many salmonids) spawn in freshwater and migrate out to the sea to grow, whereas catadromous fishes (e.g., freshwater eels) spawn in the sea and migrate to freshwater to grow. The shift between life in freshwater and in the sea is a complicated process; there is a need not only for changing osmoregulation but also for changes in feeding habits, coloration, sense organs, and so forth. Migratory events usually take place at specific times of year and can be controlled by the photoperiod.

In anadromous salmonids, juvenile fish undergo a transformation from the freshwater parr stage to the marine smolt stage in spring. The stationary, spotted, largely insect-feeding, freshwater-adapted parr changes into a pelagic, silvery, largely fish-eating, saltwater-adapted fish. At smoltification, the chloride cells in the gills enlarge and the Na^+-K^+-ATPase activity in the gills increases (Scholz et al., 1983). The increased ability to secrete salts improves hypoosmoregulation, which can be tested by seawater challenge. Fish are transferred directly from freshwater into full or dilute seawater, and the survival or, in the less drastic tests, osmolality and/or ion levels in the blood are assessed. A smolt is much better able than a parr to tolerate seawater. Osmolality and ion levels increase little or not at all in seawater challenged smolt, but drastically in parr. Several hormones, of which thyroxine-triiodothyronine, cortisol, and growth hormone have been most extensively studied, are involved in the control of smoltification (Scholz et al., 1983). If smolts are not allowed to migrate downstream, then they desmoltify and the hypo-osmoregulatory ability is soon lost (Scholz et al., 1983).

A large number of studies have shown that smoltification is under photoperiodic control. Increasing day lengths accelerate smoltification and a delay of the normal increase retards the process (for review, see Scholz et al., 1983). Accelerated photoperiod can advance smoltification several months (Thrush et al., 1994). This practice, which permits young salmon to be transferred early to pens in the sea, is used commercially to a large extent. There is also a circannual cycle in smoltification. Individually tagged young Atlantic salmon were kept under constant photoperiod (LD 12:12) and temperature (11°C) for more than a year (Eriksson and Lundqvist, 1982). Indications of smoltification (i.e., low condition factors—smolts are more elongated than parr, and silvery color with dark fins) appeared, disappeared, and reappeared with a period of about 10 months.

Apart from the seasonal changes in day length, there is often also a strong influence of lunar phases on smoltification and downstream migration. The development of plasma thyroxine in young coho salmon (*Oncorhynchus kisutch*), and anadromous trout was examined over a number of years and at several localities, and it was discovered that levels peaked at new moon (Grau et al., 1981). In California, this occurred at the new moon closest to vernal equinox, whereas further north this took place at the following new moon.

Several environmental factors, such as lunar phases, temperature, water flux, and turbidity, have been implicated in the control of downstream migration, although different studies often do not agree. The descent of Atlantic salmon smolts in a Norwegian river in different years could be correlated with temperature, but not with other factors such as lunar phase (Jonsson and Ruud-Hansen, 1985). On the other hand, the entry of smolts of chinook (*Oncorhynchus tshawytscha*), coho, and sockeye salmon (*O. nerka*) into an estuary in Washington was not related to temperature but to lunar phases, most closely to lunar apogee (deVries et al., 2004).

There are many other anadromous fishes than salmonids. In many areas, the stickleback is anadromous, though there are also many landlocked populations and sticklebacks that live in the sea all year. Baggerman (1957) worked with sticklebacks that reproduce in spring in freshwater in the Netherlands and spend the winter in the North Sea. Salinity preference was tested using troughs with compartments filled with water with different salinities (fresh-fresh, ~0.8–1.2% salinity) and a layer of freshwater above them. Fish collected from the field preferred freshwater in March to May and brackish water from the end of June (postbreeding) to November. Fish hatched in the laboratory in April-May and reared in freshwater under natural photoperiod, preferred brackish water when tested in June to August. Exposure to LD 16:8 in early October induced a clear freshwater preference from late October. In young fish exposed to LD 8:16 from June, brackish water preference persisted a whole year. Adult fish treated with LD 16:8 in the postbreeding period shifted the initial brackish water preference to freshwater preference after three months, whereas under LD 8:16 brackish water preference remained for more than half a year.

European eels (*Anguilla anguilla*) spawn in oceanic areas and grow large as yellow eels in freshwater or coastal areas, followed by metamorphosis into migratory silver eels. The main migratory period in European waters is in the autumn and the eels migrate (are caught) mainly during the night. Migration peaks in the later part of the lunar cycle, when nights get darker (for review, see Tesch, 1973). Eels kept under constant conditions in the laboratory also show a peak in migratory behavior at this time (Tesch, 1973).

ROLE OF DIFFERENT PHOTORECEPTORS

In order for light to influence biological cycles, it must first be sensed. Apart from the retina nonmammalian vertebrates have extraretinal photoreceptors in the brain. Fish also have photoreceptors in the skin (for review, see Oshima, 2001). These photoreceptors influence chromatophore pigment migration locally, but no other function is known.

Retinal versus Extraretinal Photoreceptors

The role of different photoreceptors in controlling seasonal events can only be critically tested by removing or blocking them, but also these experiments can have pitfalls. Although fish often behave remarkably normal also after major surgery, blinding (for which there is no appropriate sham operation) may have general traumatic effects. Therefore, suppression of reproduction after blinding is not alone a reliable indication for the importance of eyes in mediating photoperiodic effects. Serum estrogen concentrations decrease in female goldfish after optic tract section under simulated natural photoperiods both in fall and in spring, whereas there was no effect on GSI (Delahunty et al., 1979). Pinealectomy had no effect in these experiments. Photoperiodic effects on reproduction can persist in the absence of the eyes. Borg (1982a) exposed intact stickleback males and males where the eyes had been removed to LD 8:16 or 16:8 in winter. Both intact and blinded males matured (i.e., developed kidney hypertrophy) under LD 16:8, but not under LD 8:16. Thus, in this case extraretinal photoreception was able to mediate the photoperiodic effect.

At least in some fishes (see also below), extraretinal photoreception is able to mediate photoperiodic effects on reproduction. However, does the extraretinal photoreception normally do this, or is this function usually carried out by the retinas? The lateral eyes are not involved in photoperiodism in house sparrows (*Passer domesticus*) (Menaker et al., 1970). Even though the eyes could see as usual, testes weights remained low in males where India ink had been injected under the skull skin before exposure to low light intensities at a photoperiod of LD 16:8, whereas males with the head feathers plucked matured under these conditions. In another study, covered sparrows did not mature at low light intensity even though the eyes could see, but matured at high intensity (McMillan et al., 1975). This may be explained by that sufficient light could leak into the skull to stimulate the receptors.

Bornestaf and Borg (2000b) stitched pieces on black (shielded) and transparent (control) plastic foil on top of the heads of stickleback males which were then exposed to LD 16:8. Shielding suppressed kidney hypertrophy at low light intensities, whereas there were no significant differences at high light intensities in which both groups matured, or in darkness, in which both groups remained immature. Similarly as in the bird studies above, this indicates that the eyes are less important than the extraretinal photoreceptors in stimulating reproduction. However, it is not possible to decide whether the maturation of shielded fish under higher light intensities is due to light leaking in amounts sufficient to stimulate the extraretinal receptors or if the eyes are able to mediate the response at the higher intensity.

Pineal Organ

The best-known extraretinal photoreceptor in fish is the pineal organ (for review, see Ekström and Meissl, 1997), which is formed at the diencephalic roof. The organ has been studied at the electron microscopic level in many fishes, and contains well developed photoreceptor cells (e.g., Oksche and Kirschstein, 1976; van Veen et al., 1980). The photosensitivity has also been demonstrated using electrophysiology (e.g., Dodt, 1963). The pineal organ innervates many brain areas in fish (e.g., Hafeez and Zerihun, 1974), unlike in mammals, where it instead receives sympathetic innervation.

Pinealectomy influences reproduction in many fishes. It often suppresses reproduction under long photoperiod in long-day–breeding fishes. In goldfish (de Vlaming and Vodicnik, 1978; Vodicnik et al., 1978; Hontela and Peter, 1980) pinealectomy under long photoperiod suppressed maturation, as studied by GSI and histology in males and females and by plasma GTH levels in females (Vodicnik et al., 1978). Similarly, pinealectomy inhibited reproduction under long photoperiod in medaka (e.g., Urasaki, 1972a, 1973), cyprinid golden shiners (*Notemigonus chrysoleucas*;

de Vlaming, 1975; de Vlaming and Vodicnik, 1977), and walking catfish (*Clarias batrachus*; Nayak and Singh, 1988). Under short photoperiods, pinealectomy instead generally stimulates the otherwise suppressed gonads in long-day breeders, for example, goldfish (Fenwick, 1970b; de Vlaming and Vodicnik, 1978), golden shiner (de Vlaming, 1975; de Vlaming and Vodicnik, 1977), and walking catfish (Nayak and Singh, 1988). Pinealectomy did not influence goldfish maturation under short photoperiods (Vodicnik et al., 1978) and had no effect on goldfish reproduction (Peter, 1968; Delahunty et al., 1979). Pinealectomized or sham-operated goldfish were tested for the ovulation response to an increase in temperature one week after the operation (Kezuka et al., 1989b). A lowered proportion of the pinealectomized females ovulated and their ovulations were not synchronized.

The effects of pinealectomy have also been studied in short-day-breeding fishes. Long-term pinealectomy delays spawning in rainbow trout (Popek et al., 1992; Bromage et al., 1995; Randall et al., 1995b). In gray mullets (*Liza ramada*), which are also short-day breeders, pinealectomy resulted in accelerated oocyte development under short photoperiod (Sagi et al., 1983).

Deep Encephalic Photoreceptors

Apart from eyes and pineal complex, fishes also have deep encephalic photoreceptors. This was first discovered by von Frisch (1911) in the Eurasian minnow. By changes in the skin melanophores both intact and blinded fish became dark under light conditions and pale under darkness. The reaction also persisted after pinealectomy. Lighting of small spots showed that the reaction depended on light reaching a small area of the brain close to the pineal organ (von Frisch, 1911). However, the precise location of the deep encephalic receptor long remained unknown. Using antibodies against a vertebrate ancient opsin, Kojima et al. (2000) localized immunoreactive cells close to the third ventricle in diencephalic central posterior thalamic nucleus in zebrafish (*Danio rerio*). This localization is consistent with the experiments by von Frisch (1911).

Day and Taylor (1983) exposed mummichogs to LD 9:15 or LD 15:9 at 20°C for six weeks after having carried out pinealectomy, blinding or the combined treatment. In December–January neither photoperiod nor operations had any effect on GSI, whereas in experiments carried out in July–August and in March–April, fish kept under LD 15:9 had markedly higher ovarian weights than did fish under LD 9:15, irrespective of previous surgery. This indicates that extraretinal and extrapineal photoreception could mediate the photoperiodic response. The lack of differences between the surgical treatments and the controls suggests, but does not prove, that the eyes and pineal organ are not involved in this species.

Both eyes and the pineal complex were removed from male and female ayu (*Plecoglossus altivelis*), which were then exposed to LD 4:20 or 20:4 at 19°C for three weeks starting in November (Masuda et al., 2005). The absence of the pineal complex, including the pineal stalk, was carefully controlled using histology and immunohistochemistry with an antiserum against opsin. Maturation was evaluated using gonadal weights, steroid hormone levels, and gonadal histology. In all studied parameters, both intact and operated fish exposed to LD 4:20 were more mature than those exposed to LD 20:4, demonstrating that extraretinal and extrapineal photoreceptors could mediate the photoperiodic response. However, in operated females kept under LD 4:20, gonadal weights and steroid levels were lower than in intact fish. Thus, it is possible that retinal and/or pineal photoreception can also have a role.

MELATONIN

Secretion Patterns

Higher circulating levels of melatonin during the night /darkness than during the day has been observed in all studied fishes (e.g., Gern et al., 1978; Kezuka et al., 1988, 1992; Falcón et al., 1989; Randall et al., 1991a, b, 1995a; Alvariño et al., 1993, 2001; Mayer et al., 1997b; Shi, 2005; Maitra and Chattoraj, 2007). Melatonin levels are elevated for a longer time under short than under long photoperiod (Kezuka et al., 1988, 1992; Randall et al., 1991a, 1995). A diurnal melatonin cycle with higher levels during the night than during the day has also been found in a cyclostome, the river lamprey (*Lampetra fluviatilis*; Mayer et al. 1998).

In most studies on fishes (e.g., Kezuka et al. 1992; Bromage et al., 1995; Porter et al., 1995), pinealectomy resulted in drastically reduced melatonin levels and a disappearance of daily melatonin cycles. In some studies, also the eyes have been found to contribute to circulating melatonin levels. Levels decreased 30–40% after enucleation in the ricefield eel (*Monopterus albus*) (Shi, 2005). In other studies, this has not been case (e.g., goldfish, Kezuka et al. 1992). The effects of light on melatonin secretion are exerted on the pineal organ itself, which has been demonstrated by light effects *in vitro* in different fishes (Gern and Greenhouse, 1988; Falcón et al., 1989; Kezuka et al., 1989a; Cahill, 1996).

Melatonin secretion often displays an endogenous circadian cycle under constant darkness, with high levels under the expected night and low during the "day." This rhythmicity is often also displayed by pineal organs *in vitro* (zebrafish, Cahill, 1996; northern pike [*Esox lucius*], Falcón et al., 1989; goldfish, Kezuka et al., 1989a). Under constant light, melatonin levels are always low (e.g., Kezuka et al., 1989b). However, no support for endogenous melatonin cycles has been found either *in vivo* or *in vitro* in salmonids (Gern and Greenhouse, 1988; Randall et al., 1991b, 1995a; Alvariño et al., 1993). Similarly as for the pineal organ, the perfused zebrafish retinas produced more melatonin in the night than during the day, a pattern that followed the reversal of the photoperiod (Cahill, 1996). However, the retinal melatonin rhythm rapidly damped out under constant darkness.

The major factor controlling melatonin levels is light. There are, however, also other factors that can have an influence: melatonin levels have been found to be higher under high than under low temperatures in goldfish (Iigo and Aida, 1995) and in ricefield eels (Shi, 2005). The amplitude of changes in melatonin levels was higher in the summer months than at other seasons in Atlantic salmon (Randall et al., 1995a). Melatonin levels increased after transfer in coho salmon to seawater (Gern et al., 1984). Starvation increased melatonin levels in the ricefield eel (Shi, 2005).

Effects of Melatonin Injections

The marked diurnal cycles of melatonin makes it very difficult to administer in a physiologically relevant manner. In many studies, antigonadal effects of melatonin injections have been found in fishes where reproduction is stimulated by long photoperiods (Fenwick, 1970a; Urasaki, 1972b; de Vlaming et al., 1974; Sundararaj and Keshavanath, 1976; Saxena and Anand, 1977; Iwamatsu, 1978; Borg and Ekström, 1981; Maitra and Chattoraj, 2007). This is consistent with melatonin mediating photoperiodic effects on reproduction. However, the injections probably led to high, but transient, melatonin levels. For this reason, these effects cannot be regarded as physiological.

Effects of Continuous Melatonin Treatments

Another way to administer melatonin is to use permanent implants from which it continuously diffuses out. This has in many cases been very effective in stimulating reproduction in short-day–breeding mammals (e.g., in sheep, Staples et al., 1992). Similar experiments have, however, generally failed to influence reproduction in fishes. Melatonin administrated via Silastic implants to stinging catfish under LD 14:10 or 9:15, did not influence ovary weights, vitellogenin levels, or oocyte development (Garg, 1989). However, there was also little effect of photoperiod in this study. Melatonin implants administered to Atlantic salmon (Bromage et al., 1995) or rainbow trout (Randall et al., 1995b) had no significant effect on timing of spawning, nor did they affect gonadotropin secretion in female Arctic charr (*Salvelinus alpinus*; Gillet et al., 1998). Slow-release microspheres raised daytime plasma melatonin to levels similar to natural nighttime levels, but did not affect time of spawning, gonad size, or plasma vitellogenin in female rainbow trout kept under natural photoperiod (Nash et al., 1995). The lack of effects in these studies docs not support a physiological role of melatonin in the photoperiodic control of reproduction. However, constantly high melatonin levels, as opposed to diurnal cycles, are unnatural and may fail to be effective due to effects such as receptor down regulation. Alvariño et al. (2001) implanted melatonin into male and female turbot (*Scophthalmus maximus*) approximately one month before winter solstice, trying to mimic an advance short-day signal. This treatment accelerated gonadal maturation in both sexes. However, because the plasma melatonin levels achieved by the implants were dramatically higher (first weeks about 100-fold) than the control nighttime levels, this cannot be regarded as a physiological effect. To summarize, no implant studies support the concept that melatonin should mediate the effects of long nights on reproduction in fish.

Porter et al. (1998) studied the effects of pinealectomy and melatonin implants starting at winter solstice on smoltification in Atlantic salmon. Pinealectomy delayed smoltification as shown by poorer survival in seawater tests and by a higher condition index (smolts otherwise become more elongated than parr) than in controls. Melatonin treatment (resulting in levels about three times higher than normal nighttime levels) respectively reversed and diminished these effects. However, because smoltification is stimulated by increasing day lengths a physiological role of melatonin should be expected to be inhibitory.

Effects of Melatonin Given in Daily Cycles

In order to study possible physiological effects, melatonin should ideally be administered in a pattern similar to the natural diurnal cycle. This has been achieved in some elegant studies on mammals. Djungarian hamsters, *Phodopus sungorus* (Carter and Goldman, 1983) and sheep (e.g., Bittman et al. 1983) have been cannulated and infused with melatonin following precise schedules. Melatonin thus administered suppressed reproduction in the long-day–breeding hamsters and stimulated reproduction in the short-day–breeding sheep. The number of hours melatonin levels are elevated was found to be the most important parameter for the effectiveness of melatonin.

Cannulation experiments are, of course, not feasible to carry out on fishes. However, there are other ways in which melatonin can be given in a simulated daily cycle. Mayer et al. (1997b) gave stickleback males melatonin via the water at doses (20 and 80 µg/L) resulting in plasma levels respectively similar and a few times higher than those found naturally at night. By adding melatonin and changing water at set times, fish were exposed to 16 h of melatonin treatment but kept under photoperiods of LD 16:8 or 24:0. In the LD 24:0 experiment, there was also a LD 8:16 control group. There was no effect of melatonin on maturation (KEH). Whereas LD 8:16 suppressed maturation, both solvent-treated controls and melatonin-treated fish matured under long photoperiods. Bornestaf et al. (2001) treated female sticklebacks similarly for two months under LD 16:8 and LD 24:0. Maturation pace was studied by recording the appearance of running ripe roe (ovulation), and ovarian weights and oocyte maturation was studied at the end of the experiment. Under LD 24:0, females treated with the higher melatonin dose displayed a slower maturation than controls, though the proportion of females that matured over the experiment was not lower. Apart from this, there was no significant effect of melatonin in any studied parameter, whereas LD 8:16 strongly suppressed them all.

A prolonged daily period of elevated plasma melatonin levels was achieved by feeding precocious male masu salmon melatonin-sprayed pellets, calculated to give a dose of 0.5 mg/kg body weight/day (Amano et al., 2000). The fish were kept under LD 16:8 and treated from June to October. Fish were sampled at several times and GSI, spermiation, plasma T, 11KT, pituitary FSH, and LH levels were studied. These maturity parameters displayed similar, increasing, patterns over time in both treated males and controls in this autumn-spawning fish. There were no consistent significant differences in spermiation or in levels of T, 11KT, or LH between the groups. Melatonin-treated fish displayed generally higher gonadal weights and pituitary FSH levels than did the controls, though there were no differences in the temporal patterns. The absence of an effect on the latter are in contrast to the marked differences in temporal patterns for both parameters (and several others) seen between young masu salmon exposed to LD 8:16 and 16:8 from June to October (Amano et al., 1994, 1995), with accelerated maturation under LD 8:16. In a later experiment, the melatonin levels were increased by feeding pellets (calculated to give 7 mg/kg/day) from June to October to precocious masu salmon males kept under natural photoperiod (Amano et al., 2004). This resulted in high circulating levels of melatonin at all times of the day and in lower body weight and GSI and T levels than in controls. Neither the timing of spermiation nor testes histology was influenced. It was concluded that a daily melatonin profile is important for mediating photoperiodic effects on reproduction.

Does Melatonin Mediate Photoperiodic Effects?

To summarize, there are few reliable data consistent with a major physiological role for melatonin mediating photoperiodic effects on reproduction in fishes. This is in marked contrast to situation in mammals, but agrees with other nonmammalian vertebrates (for review, see Mayer et al., 1997a). It should be emphasized that effects of pinealectomy on fish reproduction do not necessarily suggest a role of melatonin. If melatonin mediated photoperiodic effects in fishes, then pinealectomy under short photoperiods should stimulate maturation in long-day–breeding fishes and retard it in short-day–breeding fishes. This has often (e.g., de Vlaming and Vodicnik, 1977, 1978; Popek et al., 1994 1992; Bromage et al., 1995; Randall et al., 1995b), but not always (e.g., Vodicnik et al., 1978; Sagi et al., 1983), been found.

However, it should also be expected that pinealectomy should have little effect on reproduction under long photoperiod, because pinealectomy should (from the perspective of melatonin) correspond roughly to constant light. The inhibitory effects of pinealectomy on reproduction under long-day conditions (e.g., Urasaki, 1973; de Vlaming and Vodicnik, 1977, 1978; Hontela and Peter, 1980) or under constant light (Urasaki, 1972a) in long-day–spawning fish do not support an important role of melatonin in this respect. The pineal organ may also produce other bioactive substances and there is also a pineal innervation of extensive brain areas in fishes (e.g., Hafeez and Zerihun, 1974; Ekström, 1984), which opens other possibilities for transmission of photic information.

Short-Term Melatonin Effects

Melatonin may influence reproduction in other ways than by mediating effects of day length. In many fishes (e.g., in medaka, Iwamatsu, 2004), ovulation and spawning occur at specific times on the day. Some well-documented short-term effects of melatonin in fish may be related to the control of these daily rhythms. Single intraperitoneal injections of melatonin at doses down to 5 ng/g body weight increased plasma LH levels in mature female Atlantic croaker (*Micropogonias undulatus*; Khan and Thomas, 1996). Using 50 ng/g, plasma LH levels were raised after 30 min and 1 h but declined after 2 h. Melatonin injections of 500 ng/g were found to be effective from the late-light to mid-dark period, but not at other times. It was suggested that melatonin could be responsible for stimulating an early nighttime rise in plasma LH.

PROPERTIES OF THE PHOTOPERIOD

Is Change in or Absolute Length of Photoperiod Important?

A change in photoperiod often has a strong impact on reproduction and other seasonal events. However, is it the absolute photoperiod or the presence of a change that is critical? In autumn-spawning Atlantic salmon, mature parr males attained final maturation (spermiation) more rapidly under gradually declining photoperiods than after an abrupt transition from a long to a short photoperiod (Eriksson and Lundqvist, 1980). Female rainbow trout were maintained under LD 18:6 from January until May, when photoperiods were changed to LD 6:18; LD 10:14, or LD 14:10 and then kept constant (Duston and Bromage, 1987). In all these treatments, full maturity (running roe) was attained at a similar time in September–October, earlier than in females that continued to be kept under LD 18:6 (October–January) or had been under LD 6:18 from the start (March–April). The results were confirmed by serum levels of estradiol, testosterone, and calcium. Because the change of the light period advanced maturation similarly whether the shortening was 4, 8, or 12 h, it appears that the shortening per se, rather than the absolute length of the photoperiod reached, is important. In contrast, Baggerman (1957) found a slower maturation of sticklebacks under gradually increasing photoperiod than under a sudden change from LD 8:16 to LD 16:8.

Circadian Rhythms in Photosensitivity

Two main models by which the photoperiod can affect organisms are possible: hourglass mechanism(s) can measure the length of the light and/or dark period, or endogenous clocks/oscillators can provide a rhythm of sensitivity where light at certain phases, but not at others, is effective (Duston and Bromage, 1986). In order to distinguish between these models, experiments using skeleton and/or resonance photoperiods are needed.

Baggerman (1972) exposed sticklebacks to skeleton photoperiods at 20°C in autumn and studied the proportion of fish that matured, that is, nest building in males and oviposition in females. All photoperiodic regimes comprised 8 h of light and 16 h of darkness: 8L-16D (A), 6L-4D-2L-12D (B), 6L-6D-2L-10D (C), 6L-8D-2L-8D (D), and 6L-11D-2L-5D (E). The highest maturity rate was observed in treatment D (78%), when the additional light was given 14–16 h after the onset of the main light period; about half of the fish matured under C and E, less than 10% in B, and none in A. The results were interpreted as indicating a daily cycle in sensitivity to light. However, they could also be interpreted as the fish responding primarily to the length of uninterrupted darkness. In a follow-up study, Baggerman (1985) used a slightly different set of treatments: 8L-16D (*a*), 6L-2D-2L-14D (*b*), 6L-4D-2L-12D (*c*), 6L-6D-2L-12D (*d*), and 6L-8D-2L-8L (*e*), at five different times of the year. The relative effectiveness in inducing maturation was always $a \leq b \leq c \leq d \leq e$, though there was a large seasonal change. In experiments starting in October, only *e* led to a large proportion of the fish maturing. As the season progressed, however, additional fish matured also under the other regimes until in March only a few fish in *a* did not mature. This was interpreted as an extension of the photoinducible phase but is also consistent with a progressively longer period of uninterrupted darkness being needed to inhibit maturation.

Duston and Bromage (1986) exposed female rainbow trout to LD 6:18 and to the skeleton photoperiods 6L-4D-2L-12D, 6L-6D-2L-10D, and 6L-8D-2L-8D and studied the effect on the attainment of full maturation (eggs stripped). The spawning was advanced in the 6L-6D-2L-10D group compared to the other groups, which suggested a sensitive phase 12–14 h after the onset of the main light period. However, although the results were significant, spawning dates overlapped widely between groups, and the maximum differences between groups was only about one month after a treatment period of about two years.

Resonance photoperiods are "diurnal" light cycles with a total length different from 24 h, for example, LD 6:48 for a 54 h cycle, and are considered more reliable than skeleton photoperiods in demonstrating a circadian sensitive phase. Female rainbow trout exposed to LD 6:42, 6:48, or 6:54 advanced their spawning period compared to controls under LD 6:18 (Duston and Bromage, 1986). The sensitive period should have a maximum sensitivity 12–14 h after lights on, as suggested by the skeleton-photoperiod experiment above.

A sensitive period should be expected to occur in light under LD 6:54 and 6:48, but in darkness under LD 6:42. Thus, it is questionable whether the similar effectiveness of LD 6:42, 6:48, and 6:54 in advancing maturation supports the concept of a circadian sensitive period for photostimulation. Baggerman (1972) mentions, but gives no data for, an experiment where exposure of sticklebacks to LD 16:32 was almost as stimulatory as LD 16:8 and much more stimulatory than LD 8:16, which is more consistent with a circannual sensitive time than with an hourglass model for night hours or with a response to the proportion of hours of light and darkness. However, if a single long day is effective, that could also explain the effect. To summarize, the evidence for a circadian sensitive period in fish are rather tentative.

Wavelength

Sticklebacks of both sexes were exposed to a simulated natural increase in photoperiod from January to June, but with light of different spectral composition obtained with filters: 388–466 nm (purple), 455–518 nm (blue), 513–583 nm (green), 586–653 nm (red), or light from fluorescent daylight tubes with neutral filter alone (McInerney and Evans, 1970). The intensities were adjusted to 370 erg/cm^2/s (corresponding to red, 87 lux; green, 230 lux; blue, 41 lux; purple, 5 lux). Maturation in both sexes was studied by means of gonad weights and histology, as well as the appearance of breeding colors in the male. The fish matured under all light regimes, with no or at most marginal differences between them.

PHOTOPERIOD AND THE BRAIN-PITUITARY-GONADAL AXIS

As in other vertebrates, the gonads in fish are primarily controlled by GTHs produced in the pituitary. Secretion of LH and FSH is in turn controlled from the brain, in particular from the hypothalamus. GTH secretion is stimulated by gonadotropin-releasing hormone (GnRH) and in many cases inhibited by dopamine, but also many other substances can be effective.

Exposure of precocious male masu salmon to LD 8:16 in summer–autumn not only accelerated the rise in GSI, attainment of spermiation, and pituitary levels of FSH and LH compared to fish treated with LD 16:8 (Amano et al., 1994), but also resulted in increases in pituitary salmon GnRH (sGnRH) content (Amano et al., 1994) and in sGnRH expression in the preoptic area and in the ventral telencephalon, as studied using in situ hybridization (Amano et al., 1995).

Feedback Mechanisms

The brain-pituitary-gonadal axis is feedback regulated: gonadal hormones act on the hypothalamus and/or the pituitary and control the secretion of GTHs. In fishes, there are both inhibitory (negative) and stimulatory (positive) feedback effects by gonadal hormones on the brain-pituitary-gonadal axis (e.g., Crim et al., 1981; de Leeuw et al., 1986; Borg et al. 1998; Antonopoulou et al., 1999; Hellqvist et al., 2008). In mammals and birds there are many studies (e.g., Turek, 1977; Urbanski and Follett, 1982; Rosa and Bryant, 2003) indicating that changes in the feedback systems are important components in the mediation of photoperiodic effects on reproduction. Negative feedback effects can be stronger under nonstimulatory than under stimulatory photoperiod. This effect has been observed when either long photoperiod (Turek, 1977) or short photoperiod (Rosa and Bryant, 2003) stimulates breeding, thus suppressing maturation under nonappropriate light conditions.

Only a few studies on the effects of photoperiod on feedback mechanisms have been conducted in fish. Pituitary FSH content was increased after castration in precociously mature masu salmon males kept under stimulatory short photoperiods, but not under long photoperiods (Amano et al., 2001), whereas LH content was not affected. However, it is difficult to interpret what relevance this could have for the control of natural maturation, since there was no effect of photoperiod on pituitary FSH levels in control fish.

The number of sGnRH-immunoreactive cells in sham-operated and castrated precocious masu salmon was studied in autumn (Amano et al., 1999). Thirty days after surgery, there were more sGnRH-immunoreactive cells in the preoptic area in males kept under stimulatory short photoperiods than under long photoperiods, whereas the number in the ventral telencephalon was unaffected by light. There was no effect of castration on the number of sGnRH-immunoreactive cells or on the expression of sGnRH, as studied using in situ hybridization, in either photoperiod.

Castrated and androgen-treated stickleback males were maintained to either LD 8:16 or 16:8 during winter and the expression of pituitary LH-β and FSH-β was evaluated (Hellqvist et al., 2008). Sham-operated controls matured (i.e., displayed kidney hypertrophy) under long, but not under short photoperiod. The expression of both GTHs in the control groups was also higher under LD 16:8 than under LD 8:16. The feedback effects on the expression of LH-β were similar under both photoperiods. Expression was low in castrated control males under both photoperiods, but increased with androgen treatments, indicating positive feedback. The control of FSH-β expression was, however, markedly different between the two photoperiods. Under LD 8:16, expression was strongly increased after castration and suppressed when castration was combined by androgen treatment, indicating negative feedback. Under LD 16:8, on the other hand, castration diminished expression, which has instead increased by androgen treatment, indicating a positive feedback. As in many other fishes, FSH increases earlier in the season than LH in the stickleback (Hellqvist et al., 2006) and is thus more likely than LH to control the onset of maturation. A negative feedback under the nonstimulatory short photoperiod may suppress maturation, whereas a positive feedback under the stimulatory long photoperiod may accelerate maturation.

Aromatase

Some androgens, for example, T, can be aromatized into estrogens. This conversion can take place in the brain and is much higher in bony fishes than in other vertebrates (e.g., Callard et al., 1981). The aromatase activity is generally high in the pituitary and in the hypothalamus (e.g., stickleback, Borg et al., 1987b). In some cases, aromatization is important both for positive (Crim et al., 1981) and negative (de Leeuw et al., 1986) feedback effects on the brain-pituitary-gonadal axis in fishes.

Implantation of Silastic capsules containing the aromatase inhibitors 1,4,6-androstatriene-3,17-dione or fadrozole (CGS 16949) did not influence maturation (kidney hypertrophy) in male sticklebacks kept under long photoperiod (LD 16:8) in winter but increased maturation under short photoperiod (LD 8:16) when controls did not mature (Bornestaf et al., 1997). These results indicate that aromatization has a role in the mechanisms by which short photoperiod inhibits reproduction in this species. It is not known at what level(s) these effects of aromatase inhibitors are exerted. Though it is tempting to suggest that they interfere with the feedback mechanisms on the brain-pituitary-gonadal axis, this is not necessarily so. The feedback effects by androgens on LH-β and FSH-β (Hellqvist et al., 2008) reviewed above do not appear to be aromatase dependent since they could be exerted both by the aromatizable androgen T and the nonaromatizable 11-ketoandrostenedione.

SUMMARY AND FUTURE PERSPECTIVES

Photoperiods often have a strong influence on reproduction in fishes, in species spawning both under long-day conditions in spring–summer and under short-day conditions in autumn–winter. A role of extraretinal photoreception has been shown in a number of cases, whereas a physiological role of melatonin in seasonal control of reproduction is doubtful.

From an applied point of view, developments in photoperiodic treatments are likely to further improve aquacultural practices of an increasing number of species. From a scientific point of view, intriguing questions remain in the mechanisms by which photoperiods control reproduction and other seasonal events in fishes. The roles of pinealofugal innervation, deep encephalic receptors, organization of feedback on the brain-pituitary-gonadal axis, and mechanisms in the control of lunar cycles are topics that need to be addressed. The recent characterizations of the entire genomes of stickleback and medaka (Hubbard et al., 2007) are bound to yield addition tools for such studies.

The role of photoperiods in the lives of cartilagenous fishes and cyclostomes is almost entirely unknown. Especially from a comparative point of view, an increased knowledge would be valuable. However, since none of these fishes show the full life cycle under normal laboratory conditions or are of aquacultural importance, progress is unfortunately likely to be slow.

References

Alvariño JMR, Randall CF, and Bromage NR. (1993). Effects of skeleton photoperiods in melatonin secretion in the rainbow trout, *Oncorhynchus mykiss* (Walbaum). Aquacult Fish Manag 24: 157–162.

Alvariño JMR, Rebollar PG, Olmedo M, Alvarez-Blázquez B, Ubilla E, and Peleteiro JB. (2001). Effects of melatonin implants on reproduction and growth of turbot broodstock. Aquacult Int 9: 477–487.

Amano M, Hyodo S, Kitamura SN, Ikuta K, Suzuki Y, Urano A, and Aida K. (1995). Short photoperiod accelerates preoptic and ventral telencephalic salmon GnRH synthesis and precocious maturation in underyearling male masu salmon. Gen Comp Endocrinol 99: 22–27.

Amano M, Iigo M, Ikuta K, Kitamura S, Okusawa K, Yamada H, and Yamamori K. (2004). Disturbance of plasma melatonin profile by high dose melatonin administration inhibits testicular maturation of precocious male masu salmon. Zool Sci 21: 79–85.

Amano M, Iigo M, Ikuta K, Kitamura S, Yamada H, and Yamamori K. (2000). Roles of melatonin in gonadal maturation of underyearling precocious male masu salmon. Gen Comp Endocrinol 120: 190–197.

Amano M, Ikuta K, Kitamura S, and Aida K. (1999). Effects of photoperiod on salmon GnRH mRNA levels in brain of castrated underyearling precocious male masu salmon. Gen Comp Endocrinol 115: 70–75.

Amano M, Ikuta K, Kitamura SN, and Aida K. (2001). Effects of photoperiod on pituitary gonadotropin levels in masu salmon. J Exp Zool 289: 449–455.

Amano M, Okumoto, Kitamura SN, Ikuta K, Suzuki Y, and Aida K. (1994). Salmon gonadotropin-releasing hormone and gonadotropin are involved in precocious maturation induced by photoperiod manipulation in underyearling male masu salmon, *Oncorhynchus masou*. Gen Comp Endocrinol 95: 368–373.

Andersson E, Mayer I, and Borg B. (1988). Inhibitory effect of 11-ketoandrostenedione and androstenedione on spermatogenesis in three-spined stickleback, *Gasterosteus aculeatus* L. J Fish Biol 33: 835–840.

Antonopoulou E, Swanson P, Mayer I, and Borg B. (1999). Feedback control of gonadotropins in Atlantic salmon, *Salmo salar*, male parr. II. Aromatase inhibitor and androgen effects. Gen Comp Endocrinol 114: 142–150.

Baggerman B. (1957). An experimental study on the timing of breeding and migration in the three-spined stickleback (*Gasterosteus aculeatus* L.). Arch Neerl Zool 12: 105–318.

Baggerman B. (1972). Photoperiodic responses in the stickleback and their control by a daily rhythm of photosensitivity. Gen Comp Endocrinol 3(suppl): 466–476.

Baggerman B. (1980). Photoperiodic and endogenous control of the annual reproductive cycle in teleost fishes. In *Environmental Physiology of Fishes* (MA Ali, ed). New York: Plenum, pp. 533–567.

Baggerman B. (1985). The role of biological rhythms in the photoperiodic regulation of seasonal breeding in the stickleback *Gasterosteus aculeatus* L. Neth J Zool 35: 14–31.

Bekkevold D, Clausen LAW, Mariani S, André C, Christensen TB, and Moegaard H. (2007). Divergent origins of sympatric herring population components determined using genetic mixture analysis. Mar Ecol Prog Ser 337: 187–196.

Bittman EL, Demsey RJ, and Karsch FJ. (1983). Pineal melatonin secretion drives the reproductive response to daylength in the ewe. Endocrinology 113: 2276–2283.

Borg B. (1982a). Extraretinal photoreception involved in photoperiodic effects on reproduction in male three-spined sticklebacks, *Gasterosteus aculeatus*. Gen Comp Endocrinol 47: 84–87.

Borg B. (1982b). Seasonal effects of photoperiod and temperature on spermatogenesis and male secondary sexual characters in the three-spined stickleback, *Gasterosteus aculeatus* L. Can J Zool 60: 3377–3386.

Borg B. (1994). Androgens in teleost fishes. Comp Biochem Physiol 109C: 219–245.

Borg B, Antonopoulou E, Mayer I, Andersson E, Berglund I, and Swanson P. (1998). Effects of gonadectomy and androgen treatments on pituitary and plasma levels of gonadotropins in mature male Atlantic salmon, *Salmo salar*, parr—positive feedback control of both gonadotropins. Biol Reprod 58: 814–820.

Borg B and Ekström P. (1981). Gonadal effects of melatonin in the three-spined stickleback, *Gasterosteus aculeatus* L, during different seasons and photoperiods. Reprod Nutr Dev 21: 919–927.

Borg B, Peute J, Reschke M, and van den Hurk R. (1987a). Effects of photoperiod and temperature on testes, renal epithelium and pituitary gonadotropic cells of the threespine stickleback, *Gasterosteus aculeatus* L. Can J Zool 65: 14–19.

Borg B, Timmers RJM, and Lambert JGD. (1987b). Aromatase activity in the brain of the three-spined stickleback, *Gasterosteus aculeatus*. I. Distribution and effects of season and photoperiod. Exp Biol 47: 63–68.

Borg B and van Veen T. (1982). Seasonal effects of photoperiod and temperature on the ovary of the three-spined stickleback, *Gasterosteus aculeatus* L. Can J Zool 60: 3387–3393.

Bornestaf C, Antonopoulou E, Mayer I, and Borg B. (1997). Effects of aromatase inhibitors of reproduction in male three-spined sticklebacks, *Gasterosteus aculeatus,* exposed to long and short photoperiods. Fish Physiol Biochem 16: 419–423.

Bornestaf C and Borg B. (2000a). Endogenous breeding cycles in male threespine sticklebacks, *Gasterosteus aculeatus*. Behaviour 137: 921–932.

Bornestaf C and Borg B. (2000b). Extraretinal photoreception is more important than retinal photoreception for sexual maturation in the three-spined stickleback. In: Bornestaf, C. Mechanisms in the Photoperiodic Control of Reproduction in the Three-Spined Stickleback, *Gasterosteus aculeatus*. PhD thesis, Stockholm University.

Bornestaf C, Mayer I, and Borg B. (2001). Melatonin and maturation pace in female three-spined stickleback, *Gasterosteus aculeatus*. Gen Comp Endocrinol 122: 341–348.

Bromage NR, Randall CF, Porter MJR, and Davies B. (1995). How do photoperiod, the pineal gland, melatonin, and circannual rhythms interact to co-ordinate seasonal reproduction in salmonid fish? In *Proceedings of the Fifth International Symposium on Reproductive Physiology of Fish* (F Goetz and P Thomas, eds), Austin, TX, pp. 164–166.

Bullough WS. (1939). A study of the reproductive cycle of the minnow in relation to the environment. Proc Zool Soc Lond A 109: 79–102.

Cahill GM. (1996). Circadian regulation of melatonin production in cultured zebrafish pineal organ. Brain Res 708: 177–181.

Callard GV, Petro Z, and Ryan KJ. (1981). Estrogen synthesis in vitro and in vivo in the brain of the marine teleost (*Myoxocephalus*). Gen Comp Endocrinol 29: 14–20.

Carter DS and Goldman BD. (1983). Antigonadal effects of timed melatonin infusion in pinealectomized male Djungarian hamsters (*Phodopus sungorus sungorus*): Duration is the critical parameter. Endocrinology 113: 1261–1267.

Crim LW, Peter RE, and Billard R. (1981). Onset of gonadotropic hormone accumulation in the immature trout pituitary gland in response to estrogen or aromatizable androgen steroid hormones. Gen Comp Endocrinol 44: 374–381.

Day JR and Taylor MH. (1983). Environmental control of the annual gonadal cycle of *Fundulus heteroclitus* L.: The pineal organ and eyes. J Exp Zool 227: 453–458.

Delahunty G, Schreck C, Specker J, Olcese J, Vodicnik MJ, and de Vlaming VL. (1979). The effects of light reception on circulating estrogen levels in female goldfish, *Carassisus auratus*: Importance of retinal pathways versus the pineal. Gen Comp Endocrinol 38: 148–152.

de Leeuw R, Wurth YA, Zandbergen MA, Peute J, and Goos HJT. (1986). The effects of aromatizable androgens, non-aromatizable androgens, and estrogens on gonadotropin release in castrated African catfish, *Clarias gariepinus* (Burchell). Zool Sci 243: 587–594.

de Vlaming V. (1975). Effects of pinealectomy on gonadal activity in the cyprinid teleost, *Notemigonus crysoleucas*. Gen Comp Endocrinol 26: 36–49.

de Vlaming VL, Sage M, and Charlton CB. (1974). The effects of melatonin treatment on gonadosomatic index in the teleost, *Fundulus similis,* and the frog, *Hyla cinerea*. Gen Comp Endocrinol 22: 433–438.

de Vlaming V and Vodicnik MJ. (1977). Effects of pinealectomy on pituitary gonadotrophs, pituitary gonadotropin potency and hypothalamic gonadotropin releasing activity in *Notemigonus crysoleucas*. J Fish Biol 10: 73–86.

de Vlaming V and Vodicnik MJ. (1978). Seasonal effects of pinealectomy on gonadal activity in the goldfish, *Carassius auratus*. Biol Reprod 19: 57–63.

DeVries P, Goetz F, Fresh K, and Seiler D. (2004). Evidence of a lunar gravitational cue on timing of estuarine entry by Pacific salmon smolts. Trans Am Fish Soc 133: 1379–1395.

Dodt E. (1963). Photosensitivity of the pineal organ in the teleost, *Salmo irideus* (Gibbons). Experientia 19: 642–643.

Duston J and Bromage N. (1986). Photoperiodic mechanisms and rhythms of reproduction in the female rainbow trout. Fish Physiol Biochem 2: 35–51.

Duston J and Bromage N. (1987). Constant photoperiod regimes and the entrainment of the annual cycle of reproduction in the female rainbow trout (*Salmo gairdneri*). Gen Comp Endocrinol 65: 373–384.

Ekström P. (1984). Central nervous connections of the pineal organ and retina in the teleost *Gasterosteus aculeatus* L. J Comp Neurol 262: 321–335.

Ekström P and Meissl H (1997). The pineal organ of teleost fishes. Rev Fish Biol Fish 7: 199–284.

Eriksson L-O and Lundqvist H. (1980). Photoperiod entrains ripening by its differential effect in salmon. Naturwissenschaften 67: 202–203.

Eriksson L-O and Lundqvist H. (1982). Circannual rhythms and photoperiod regulation of growth and smolting in Baltic salmon (*Salmo salar*). Aquaculture 28: 113–121.

Falcón J, Marmillon JB, Claustrat B, and Collin J-P. (1989). Regulation of melatonin secretion in a photoreceptive pineal organ: An in vitro study in the pike. J Neurosci 9: 1943–1950.

Fenwick JC. (1970a). Demonstration and effect of melatonin in fish. Gen Comp Endocrinol 14: 86–97.

Fenwick JC. (1970b). The pineal organ: Photoperiod and reproductive cycles in the goldfish, *Carassius auratus* L. J Endocrinol 46: 101–111.

Garg SK. (1989). Effects of pinealectomy, eye enucleation, and melatonin treatment on ovarian activity and vitellogenin levels in the catfish exposed to short or long photoperiods. J Pineal Res 7: 91–104.

Gern WA, Dickhoff WW, and Folmar LC. (1984). Increases in plasma melatonin titers accompanying seawater adaptation of coho salmon (*Oncorhynchus kisutch*). Gen Comp Endocrinol 55: 458–462.

Gern WA and Greenhouse SS. (1988). Examination of in vitro melatonin secretion from superfused trout (*Salmo gairdneri*) pineal organs maintained under diel illumination or continous darkness. Gen Comp Endocrinol 71: 163–174.

Gern A, Owens DW, and Ralph CL. (1978). Plasma melatonin in the trout: Day night change demonstrated by radioimmunoassay. Gen Comp Endocrinol 34: 453–458.

Gillet C, Rideau I, and Breton B. (1998). Effets du conditionnement en jours longs a la fin du cycle reproducteur sur la periode d'ovulation et les secretions gonadotropes chez l'omble chevalier (*Salvelinus alpinus*). Bull Franc Peche Piscicult Paris 350–351: 241–253.

Grau EG, Dickhoff WW, Nishioka RS, Bern HA, and Folmar LC. (1981). Lunar phasing of the thyroxine surge preparatory to seaward migation of salmonid fish. Science 211: 607–609.

Hafeez MA and Zerihun L. (1974). Studies on central projections of the pineal nerve in rainbow trout, *Salmo gairdneri* Richardson, using cobalt iontophoresis. *Cell Tissue Res* 154: 485–510.

Hellqvist A, Bornestaf C, Borg B, and Schmitz M. (2004). Cloning and sequencing of the FSH-β and LH β-subunit in the three-spined stickleback, *Gasterosteus aculeatus*, and effects of photoperiod and temperature on LH-β and FSH-β mRNA expression. Gen Comp Endocrinol 135: 167–174.

Hellqvist A, Schmitz M, and Borg B. (2008). Effects of castration and androgen-treatment on the expression of FSH-β and LH-β in the threespine stickleback, *Gasterosteus aculeatus*— Feedback differences mediating the photoperiodic response? Gen Comp Endocrinol. 158: 178–182.

Hellqvist A, Schmitz M, Mayer I, and Borg B. (2006). Seasonal changes in expression of LH-β and FSH-β in male and female three-spined stickleback, *Gasterosteus aculeatus*. Gen Comp Endocrinol 145: 263–269.

Hontela AL and Peter RE. (1980). Effects of pinealectomy, blinding and sexual condition on serum gonadotropin levels in goldfish. Gen Comp Endocrinol 40: 168–179.

Hoover EE. (1937). Experimental modification of the sexual cycle in trout by control of light. Science 86: 425–426.

Hsiao S-M, Greely MS Jr, and Wallace RA. (1994). Reproductive cycling in female *Fundulus heteroclitus*. Biol Bull 186: 271–284.

Hubbard TJP, Aken BL, Beal K, Ballester B, Caccamo M, Chen Y, Clarke L, Coates G, Cunningham F, Cutts T, Down T, Dyer SC, Fitzgerald S, Fernandez-Banet J, Graf S, et al. (2007). Ensembl 2007. Nucleic Acids Res 35: D610–D617.

Iigo M and Aida K. (1995). Effects of season, temperature, and photoperiod on plasma melatonin rhythms in the goldfish, *Carassius auratus*. J Pineal Res 18: 62–68.

Iwamatsu T. (1978). Studies on oocyte maturation of the medaka, *Oryzias latipes*. VII. Effects of pinealectomy and melatonin on oocyte maturation. Annot Zool Jpn 51: 198–203.

Iwamatsu T. (2004). Stages of normal development in the medaka *Oryzias latipes*. Mech Dev 121: 605–618.

Jakobsson S, Borg B, Haux C, and Hyllner J. (1999). An 11-ketotestosterone induced kidney-secreted protein: The nest building glue from male three-spined stickleback, *Gasterosteus aculeatus*. Fish Physiol Biochem 20: 79–85.

Jonsson B and Ruud-Hansen J. (1985). Water temperature as the primary influence on timing of seaward migrations of Atlantic salmon (*Salmo salar*) smolts. Can J Fish Aquat Sci 42: 593–595.

Karlsen Ø, Taranger GL, Dahle R, and Norberg B. (2000). Effects of exercise and continuous light on early sexual maturation in farmed Atlantic cod (*Gadus morhua* L.). In *Reproductive Physiology of Fish (6th International Symposium on the Reproductive Physiology of Fish, Bergen (Norway), 4–9 Jul 1999* (Norberg B, Kjesbu OS, Taranger GL, Andersson E, and Stefansson SO, eds). Bergen, Norway: University of Bergen, Department of Fisheries and Marine Biology, pp. 328–330.

Kezuka H, Furukawa K, Aida K, and Hanyu I. (1988). Daily cycle in plasma melatonin levels under long and short photoperiod in the common carp, *Cyprinus carpio*. Gen Comp Endocrinol 72: 296–302.

Kezuka H, Aida K, and Hanyu I. (1989a). Melatonin secretion from goldfish pineal gland in organ culture. Gen Comp Endocrinol 75: 217–221.

Kezuka H, Kobayashi M, Aida K, and Hanyu I. (1989b). Effects of photoperiod and pinealectomy on the gonadotropin surge and ovulation in goldfish *Carassius auratus*. Nippon Suisan Gakkai Shi 55: 2099–2103.

Khan IA and Thomas P. (1996). Melatonin influences gonadotropin II secretion in the Atlantic croaker (*Micropogonias undulatus*). Gen Comp Endocrinol 104: 231–242.

Kojima D, Mano H, and Fukada Y. (2000). Vertebrate ancient-long opsin: A green-sensitive photoreceptive molecule present in zebrafish deep brain and retinal horizontal cells. J Neurosci 20: 2843–2851.

Maitra SK and Chattoraj A. (2007). Role of photoperiod and melatonin in the regulation of ovarian functions in Indian carp *Catla catla*: Basic information for future application. Fish Physiol Biochem 33: 367–382.

Masuda T, Iigo M, and Aida K. (2005). Existence of an extra-retinal and extra-pineal photoreceptive organ that regulates photoperiodism in gonadal development of an osmerid teleost, ayu (*Plecoglossus altivelis*). Comp Biochem Physiol 140A: 414–422.

Mayer I, Borg B, and Schulz R. (1990). Seasonal changes in and effect of castration/androgen replacement on the plasma levels of five androgens in the male three-spined stickleback, *Gasterosteus aculeatus* L. Gen Comp Endocrinol 79: 23–30.

Mayer I, Bornestaf C, and Borg B. (1997a). Melatonin in non-mammalian vertebrates: Physiological role in reproduction? Comp Biochem Physiol A 118: 525–531.

Mayer I, Bornestaf C, Wetterberg L, and Borg B. (1997b). Melatonin does not prevent long photoperiod stimulation of secondary sexual characters in the male three-spined stickleback *Gasterosteus aculetus*. Gen Comp Endocrinol 108: 368–394.

Mayer I, Elofsson H, Bergman U, and Bengtson J. (1998). Diel melatonin profile in a cyclostome, the river lamprey. J Fish Biol 53: 906–909.

McInerney JE and Evans DO. (1970). Action spectrum of the photoperiod mechanism controlling sexual maturation in the threespine stickleback, *Gasterosteus aculeatus*. J Fish Res Bd Can 27: 749–763.

McMillan JP, Underwood HA, Elliott JA, Stetson MH, and Menaker M. (1975). Extraretinal light perception in the sparrow, IV: Further evidence that the eyes to not participate in photoperiodic photoreception. J Comp Physiol 97: 205–213.

McQuinn IH. (1997). Metapopulations and the Atlantic herring. Rev Fish Biol Fish 7: 297–329.

Menaker M, Roberts R, Elliott J, and Underwood H. (1970). Extraretinal light perception in the sparrow, III: The eyes do not participate in photoperiodic photoreception. Proc Natl Acad Sci USA 67: 320–325.

Nash J, Kime DE, Holtz W, and Steinberg H. (1995). Has melatonin a role in reproductive seasonality in the female rainbow trout, *Oncorhynchus mykiss*? In *Proceedings of the Fifth International Symposium on Reproductive Physiology of Fish* (F Goetz and P Thomas, eds), Austin, TX, p. 193.

Nayak PK and Singh TP. (1988). Effect of pinealectomy on testosterone, estradiol-17β, estrone, and 17α-hydroxyprogesterone levels during the annual reproductive cycle in the freshwater catfish, *Clarias batrachus*. J Pineal Res 5: 419–426.

Nelson JS. (2006). *Fishes of the world*, 4th ed. Hoboken, NJ: John Wiley and Sons.

Oksche A and Kirschstein H. (1967). Die Ultrastruktur der Sinneszellen im Pinealorgan von *Phoxinus laevis* L. Z Zellforsch Mikrosk Anat 78: 151–166.

Oshima N. (2001). Direct reception of light by chromatophores of lower vertebrates. Pigment Cell Res 14: 312–319.

Peter RE. (1968). Failure to detect an effect of pinealectomy in goldfish. Gen Comp Endocrinol 10: 443–449.

Popek W, Bieniarz K, and Epler P. (1994 1992). The role of the pineal gland in sexual maturation of female rainbow trout (*Oncorhynchus mykiss* Walbaum). J Pineal Res 13: 97–100.

Porter M, Randall C, and Bromage NR. (1995). The effect of pineal removal and enucleation on circulating melatonin levels in Atlantic salmon parr. In *Proceedings of the Fifth International Symposium on Reproductive Physiology of Fish* (F Goetz and P Thomas, eds). Austin, TX, p. 75.

Porter MJR, Randall CF, Bromage NR, and Thorpe JE. (1998). The role of melatonin and the pineal gland on development and smoltification of Atlantic salmon (*Salmo salar*) parr. Aquaculture 168: 139–155.

Randall CF, Bromage NR, Thorpe J E, Miles MS, and Muir JS. (1995a). Melatonin rhythms in Atlantic salmon (*Salmo salar*) maintained under natural and out-of-phase photoperiods. Gen Comp Endocrinol 98: 73–86.

Randall CF, Bromage NR, Thrush MA, and Davies B. (1991a). Photoperiodism and melatonin rhythms in salmonid fish. In *Proceedings of the Fourth International Symposium on Reproductive Physiology of Fish* (AP Scott, JP Sumpter, DE Kime, and MS Rolfe, eds). Sheffield, UK, pp. 136–138.

Randall C, Porter, M, Bromage NR, and Davies B. (1995b). Preliminary observations on the effects of melatonin implants and pinealectomy in the timing of reproduction in rainbow trout. In *Proceedings of the Fifth International Symposium on Reproductive Physiology of Fish* (F Goetz and P Thomas, eds), Austin, TX, p. 196.

Randall C, Thrush M, and Bromage N. (1991b). Absence of an endogenous component regulating melatonin secretion in the rainbow trout. Adv Pineal Res 5: 279–281. Rosa HJD, and Bryant MJ. (2003). Seasonality of reproduction in sheep. Rev Small Rumin Res 48: 155–171.

Sagi G, Abraham M, and Hilge V. (1983). Pinealectomy and ovarian development in the grey mullet, *Liza ramada*. J Fish Biol 23: 339–345.

Saxena PK and Anand K. (1977). A comparison of ovarian recrudescence in the catfish, *Mystus tengara* (Ham.), exposed to long photoperiods and to melatonin. Gen Comp Endocrinol 33: 506–511.

Scholz AT, Goy RW, and Hasler AD. (1983). Hormonal regulation of smolt transformation and olfactory imprinting in salmon. In *Olfactory Imprinting and Homing in Salmon* (AD Hasler and AT Scholz, eds). Berlin: Springer, pp. 43–110.

Schulz M, Freyhof J, Saint-Laurent R, Østbye K, Mehner T, and Bernatchez L. (2006). Evidence for independent origin of two spring-spawning ciscoes (Salmoniformes: Coregonidae) in Germany. J Fish Biol 68: 119–135.

Shi Q. (2005). Melatonin is involved in sex change of the ricefield eel, *Monopterus albus* Zuiew. Rev Fish Biol Fish 15: 23–36.

Staples LD, McPhee S, Kennaway DJ, and Williams AH. (1992). The influence of exogenous melatonin on the seasonal patterns of ovulation and oestrus in sheep. Anim Reprod Sci 30: 185–223.

Sundararaj BI, Vasal S, and Halberg F. (1982). Circannual rhythmic ovarian recrudescence in the catfish *Heteropneustes fossilis* (Bloch). Adv Biosci 41: 319–337.

Sundararaj BI and Keshavanath P. (1976). Effects of melatonin and prolactin treatment on the hypophysial-ovarian system in the catfish, *Heteropneustes fossilis* (Bloch). Gen Comp Endocrinol 29: 84–96.

Suzuki K, Kawauchi H, and Nagahama Y. (1988). Isolation and characterization of two distinct gonadotropins from chum salmon pituitary glands. Gen Comp Endocrinol 71: 292–301.

Swanson P. (1991). Salmon gonadotropins: Reconciling old and new ideas. In *Proceedings of the Fourth International Symposium on Reproductive Physiology of Fish* (AP Scott, JP Sumpter, DE Kime, and MS Rolfe, eds). Sheffield, UK, pp. 2–7.

Taylor MH. (1984). Lunar synchronization of fish reproduction. Trans Am Fish Soc 113: 484–493.

Taylor MH. (1991). Entrainment of the semilunar reproductive cycle of *Fundulus heteroclitus*. In *Proceedings of the Fourth International Symposium on the Reproductive Physiology of Fish* (AP Scott, JP Sumpter, DE Kime, and MS Rolfe, eds). Sheffield, UK, pp. 157–159.

Tesch F-W. (1973). *Der Aal—Biologie und Fischerei*. Berlin: Paul Parey.

Turek F. (1977). The interaction of photoperiod and testosterone in regulating serum gonadotropin levels in castrated male hamsters. Endocrinology 101: 1210–1215.

Turek F and van Cauter E (1994). Rhythms in reproduction.In *The Physiology of Reproduction*, 2nd ed., vol. 2 (E Knobil, and J-D Neill, eds). New York: Raven Press, pp. 487–540.

Urasaki H. (1972a). Effect of pinealectomy on gonadal development in the Japanese killifish (medaka), *Oryzias latipes*. Annot Zool Jpn 45: 10–15.

Urasaki H. (1972b). Effects of restricted photoperiods and melatonin administration on gonadal weights in Japanese killifish. J Endocrinol 55: 619–620.

Urasaki H. (1973). Effect of pinealectomy and photoperiod on oviposition and gonadal development in the fish, *Oryzias latipes*. J Exp Zool 185: 241–246.

Urbanski HF and Follett BK. (1982). Androgen feedback-dependent and—independent control of photoinduced LH secretion in male tree sparrows (*Spizella arborea*). J Endocrinol 105: 141–152.

Vallin L and Nissling A. (1994). Estimation of egg quality at early blastula stages in eggs from Skagerrak cod and Baltic cod (*Gadus morhua*). ICES Council Meeting Publications 1994/J:23 Copenhagen: International Council for Exploration of the Sea.

van den Eeckhoudt JP. (1946). Recherches sur l'influence de la lumière sur le cycle sexuel de l'Epinoche (*Gasterosteus aculeatus* L.). Ann Soc Roy Zool Belg 77: 83–89.

van Veen T, Ekström P, Borg B, and Møller M. (1980). The pineal complex of the three-spined stickleback, *Gasterosteus aculeatus* L. Cell Tissue Res 209: 11–28.

Vodicnik MJ, Kral RE, and de Vlaming VL. (1978). The effects of pinealectomy on pituitary and plasma gonadotropin levels in *Carassius auratus* exposed to various photoperiod-temperature regimes. J Fish Biol 12: 187–196.

von Frisch K. (1911). Beiträge zur Physiologie der Pigmentzellen in der Fischhaut. Pflugers Archiv 138: 319–387.

16

Photoperiodism in Amphibians and Reptiles

Zachary M. Weil and David Crews

As animals moved from the aquatic to terrestrial habitats, novel adaptations developed to cope with the more pronounced seasonal variations in environmental conditions. This is especially true for tetrapod vertebrates inhabiting temperate and boreal zones that evolved mechanisms to time reproduction to coincide with the time of year during which environmental conditions are mild and conducive to offspring (and parental) survival. Specifically, most small terrestrial animals breed during the spring and summer when temperatures are mild and food is relatively abundant. Most temperate-zone amphibians and reptiles display marked annual cycles in breeding. As noted throughout this volume, organisms across taxa have evolved mechanisms to attend to photoperiod to monitor seasonal time, presumably because it is a consistent and relatively noise-free environmental signal.

Photoperiodic signals can be decoded by organisms in essentially two broad ways. The first relatively straightforward system is an hourglass or interval timer. These systems function by measuring the total period of light (or dark), and the buildup (or breakdown) of photoproducts regulates the downstream neuroendocrine systems. This type of photoperiodic system exists in some insect species (Lees, 1966; Truman, 1971). The other general class of photoperiodic systems are those based on the circadian clock (Bünning, 1936). Circadian theories of photoperiodism propose that organisms have endogenous daily rhythms of responsiveness and nonresponsiveness to the effects of light. Said another way, light must fall during the circadian phase of photoresponsiveness in order to interact with the photoperiod detection circuitry. This model has been called "external coincidence" because the circadian rhythm of photosensitivity has to coincide with external stimuli (light). An alternative model suggests that light entrains two separate oscillators (e.g., a dawn and dusk oscillator) and the phase relationship between the two rhythms or "internal coincidence" determines the responses to the observed day length (Pittendrigh and Minis, 1964; Pittendrigh, 1972).

In contrast to mammals and most birds, however, ambient temperature is an important variable for poikilothermic organisms such as amphibians and reptiles in addition to photoperiod in regulating seasonal physiology. The principle aim of this chapter is to provide a broad overview of photoperiodic regulation of reproductive and nonreproductive responses in amphibians and reptiles. This chapter focuses mostly on laboratory investigations of photoperiodic and seasonal phenomena where environmental factors can be precisely controlled and the respective contributions of day length and environmental temperature can be elucidated.

AMPHIBIANS

The class Amphibia consists of three orders, anurans (frogs and toads), urodeles (salamanders and newts), and caecilians (legless snakelike or wormlike amphibians), together comprising thousands of species (Duellman and Trueb, 1986). Little is known about the caecilians, so we focus on the anuran and urodele amphibians. Evolutionarily, amphibians represent an intermediate between fish and fully terrestrial vertebrates and thus can provide important insights into the evolution of photoperiodism in higher vertebrates. Additionally, amphibian species utilize a substantial variety of reproductive strategies, including the dramatic indirect development; indeed, the term "amphibian" is derived from the Greek *αμφις*, "double," and *βιος*, "life," as many species are born in larval states and undergo a metamorphosis into an adult phenotype.

As a general rule, day length appears to be less of an important player in the regulation of reproduction in amphibians than it is in other vertebrate taxa. Few data are available in controlled laboratory studies on the role of photoperiod in biological timing in amphibians. Further, many of the studies have been conducted used "unnatural" extreme photoperiods (e.g., constant light, or a 1/23-h light/dark cycle [LD 1:23]) that do not recapitulate the day lengths experienced in nature. However, in some temperate-zone species, day length can play an important modulatory role in the regulation of reproductive responses to other stimuli (Rastogi et al., 1976).

Amphibians use a variety of environmental conditions (temperature, rainfall, and photoperiod) to regulate the timing of reproduction and metamorphosis (Lofts, 1984). Reproduction in tropical amphibians tends to be continuous or restricted to the rainy season with most species minimally responsive to day length. It is important to note, however, that this characteristic of "continuous reproduction" refers to the population and not to the individual. That is, reproduction at some level is always apparent in the population, but this is due to the lack of reproductive synchrony among the individuals in reproductive readiness. Temperate amphibians, on the other hand, tend to fall into one of two categories in terms of the timing of breeding, which we will designate here as determinate and indeterminate breeding strategies. An example of the first category is the European common frog (*Rana temporaria*) that breeds after emerging from hibernation in the spring and then undergoes gonadal regression and ceases spermatogenesis and steroidogenesis for several months, followed by spontaneous regression and induction of gonadal activity in preparation for the next breeding season. During the quiescent period, the gonads are refractory to stimulation with environmental or endocrine variables (Van Oordt, 1956). However, most species of amphibians are capable of breeding at any time of the year if conditions permit. Such indeterminate breeding species may exhibit seasonal patterns of reproduction in the field but are physiologically capable of breeding indefinitely as long as the environmental conditions remain favorable. This dichotomy thus suggests the lack of a circannual rhythm of sensitivity to environmental cues.

Anurans

Most instances of photoperiodism have been reported in temperate-zone anurans that exhibit an indeterminate breeding strategy and species with a determinate strategy in which gonadal recrudescence and regression are regulated by a circannual clock that renders the gonadal responses sensitive and refractory, respectively, to day-length and temperature signals.

Determinate breeders utilize day length and other environmental cues to time breeding behavior. Perhaps the best studied in this regard are green frog (*Rana esculenta*). The breeding season begins as temperatures increase in late March and early April. The cessation of breeding is followed by a decrease in circulating androgens and a concomitant increase in all aspects of spermatogenesis. Over the following winter, androgen concentrations rise in preparation for breeding in the spring, and the spermatogenic tissues degenerate (Rastogi et al., 1976). Importantly, in this species, androgens appear to be important for the regulation of breeding behavior, but not necessary for the early stages of spermatogenesis (Rastogi et al., 1972). However, androgens appear to inhibit mitotic division of spermatogonia and also be necessary for spermatidogenesis (meiotic division of secondary spermatocytes into spermatids). In the laboratory exposure to intermediate day lengths (12:12 LD) and mild temperatures (20°C) can maintain summer-active gonadal activity indefinitely (at least up to 60 days) (Rastogi et al., 1978). During winter, LD 12:12 cycles maintain gonadal activity at 28°C, but if temperatures are lowered to 4°C for as few as seven days, spermatogenesis is markedly inhibited. Short day lengths (LD 3:21) inhibit gonadal activity in late summer (Rastogi et al., 1978). Conversely, winter gonadal activity can persist in warm temperatures even in short day lengths. Taken together, it appears that gonadal regression in winter is mediated by low temperatures, and these frogs exploit a warm fall–early winter with a second breeding event, as the short day lengths cannot solely mediate spermatogenic arrest (Rastogi et al., 1978). Conversely, spring recrudescence of the gonads requires both warm temperatures and permissive day lengths (>12L) suggesting that day length prevents inappropriate early breeding in response to early spring increases in temperature (summarized in Figure 16.1).

It is important to emphasize that the reproductive axis of green frogs can be transformed from a determinate to an indeterminate breeding strategy. Such flexibility is inherent in biological systems and usually revealed under specific laboratory conditions. Here the direction of the flexibility can be instructive. For example, a female rat normally exhibits a four- to five-day estrous cycle and spontaneously ovulates at a predictable time, but when placed under constant light she will go

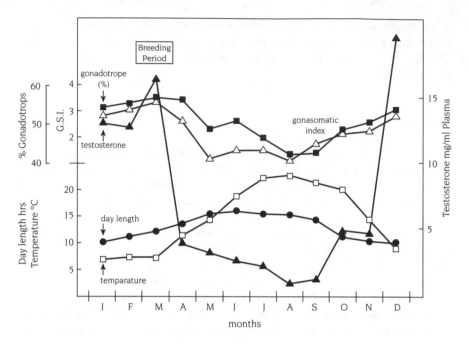

FIGURE 16.1.
Seasonal patterns of photoperiod, ambient temperatures, testes size, circulating androgens and gonadotropins, and spermatogenic activity in green frogs. Reprinted with permission of Wiley-Liss from Rastogi et al. (1976).

into constant estrus and ovulate in response to copulation. However, it is not possible to transform an induced ovulating species into one that ovulates spontaneously. In the case of green frogs, breeding will become continuous breeding in mild conditions if individuals are removed from the inhibitory effects of short day lengths. For instance, exposure to LD 21:3 in midsummer induced spermatogenic arrest after 60 days, but LD 12:12 maintained gonadal activity throughout. It is likely that these extremely long photoperiods have various nonspecific effects on the neuroendocrine and circadian systems of these animals. In their natural habitats these animals never experience day lengths greater than LD 16:8 or LD 8:16, and so it is difficult to determine what would happen with more natural photoperiods. Whereas such studies are common in the amphibian literature, they provide evidence that the neuroendocrine-reproductive axis can process and respond to changes in day length, albeit in a permissive rather than direct causal fashion (Rastogi et al., 1976, 1978).

Iberian water frogs (*Rana perezi*) exhibit a reproductive pattern that combines an indeterminate breeding strategy with a temporal dissociation between spermatogenesis and steroidogenesis (Delgado et al., 1989). Reproduction in this species appears to be regulated by both exogenous environmental conditions and an endogenous clock. Frogs captured in the winter and exposed to either warm temperatures (25°C) or long photoperiods (LD 18:6) increase gonadal size and the number of primary spermatocytes. However, in the early spring, testicular recrudescence is sensitive only to high temperatures; frogs exposed to long days in low temperatures are unresponsive (Delgado et al., 1992). Spermatogenesis is reduced by warm temperatures in the winter and enhanced by short days in the late spring but is refractory to low temperatures following breeding. Unlike green frogs, Iberian water frogs appear to have a reproductive cycle that is the product of an interaction between environmental light and temperature. In the winter, low temperatures and short photoperiods stimulate androgen production and inhibit spermatogenesis. Rising photoperiods stimulate further release of androgens in low temperatures in the late winter and through breeding, indicating that rising photoperiods in the spring are necessary for mating behavior. Following breeding, warm temperatures and long photoperiods inhibit androgen production and stimulate spermatogenesis. Once started, spermatogenesis proceeds to completion independent of photoperiod signaling (Delgado et al., 1992). Taken together, these data indicate that day length interacts with temperature and an endogenous oscillator to time breeding and spermatogenesis.

Urodeles

Much as with anuran amphibians, early experiments with urodeles consistently failed to detect photoperiodism in the regulation of reproduction (e.g., Ifft, 1942). Since then, there have been few studies that have examined the question of photoperiodism in urodeles. Red-backed salamanders (*Plethodon cinereus*) breed twice a year, once in early spring and again in late September and early October. Between the two breeding periods, the gonads grow and produce sperm. In spring, when spermatogenesis is just beginning, warm temperatures greatly increase gonadal growth, and this effect is slightly potentiated by exposure to long days. Similarly, warm temperatures stimulate spermatogenesis with maximal effect in long days (Werner, 1969). Early in the winter the gonads are refractory to both warm temperatures and long photoperiods, but later in the winter quiescent period, long days and warm temperatures stimulate gonadal growth and slightly increase spermatogenesis. Neither warm temperatures nor photoperiod alone stimulate either gonadal growth or spermatogenesis (Werner, 1969). Similar results have been reported for marbled newts (*Triturus marmoratus*), which also exhibit gonadal growth and enhanced spermatogenesis following exposure to long days at warm temperatures (Fraile et al., 1988).

Regeneration

A striking characteristic of urodeles is their ability to regenerate limbs. In vermillion spotted newts (*T. viridiscens*), day length alters the rate of forelimb regeneration after bilateral amputation. Exposure to constant light increases, while constant darkness inhibits, limb regeneration, with LD 5:9 cycles falling intermediate (Maier and Singer, 1977). The effects of light appears to interact with circulating prolactin concentrations, as treating newts in constant light with prolactin has no effect on forelimb regeneration. However, exogenous prolactin does markedly speed regeneration in newts housed in constant darkness (Maier and Singer, 1981).

RECEPTION OF PHOTOPERIOD INFORMATION

The anatomical site of photoperiod transduction in amphibians remains somewhat unspecified. The lateral eyes, pineal gland, and associated frontal organ are all competent photoreceptors (Oksche and Hartwig, 1979; Oksche, 1984). Removal of either the eyes or the pineal gland eliminated the stimulatory effects of long day lengths in green frogs (Rastogi et al., 1976) and enhanced ovarian function in Iberian water frogs (Alonso-Gomez et al., 1990). In skipping frogs (*R. cyanophlictis*), blinding or parietal shielding (blocking pineal photoreceptors directly) each increased ovarian size alone and in combination (Udaykumar and Joshi, 1996). On the other hand, marbled newts lacking eyes and intact animals both respond identically to photoperiod stimulation (Fraile et al., 1988).

The role of the pineal melatonin system in the interpretation of day-length information in amphibians has yet to be fully explicated even though melatonin was originally described as the hormone responsible for blanching of leopard frog (*Rana pipiens*) skin in the darkness (McCord and Allen, 1917). The pineal gland does not seem to be the principle source of circulating melatonin in amphibians. Most amphibians exhibit strong daily rhythms of melatonin in the blood with peaks during the scotophase (Gern and Norris, 1979; Delgado and Vivien-Roels, 1989; Rawding and Hutchison, 1992; d'Istria et al., 1994) and both temperature and day length contribute to the pattern of melatonin production (Delgado and Vivien-Roels, 1989). However, in anurans the concentration of melatonin in the pineal itself is markedly lower than in the retina. Further blood dynamics of circulating melatonin most closely mirror the concentrations of melatonin in the retina (Delgado and Vivien-Roels, 1989). In green frogs daily rhythms in the expression of the rate limiting enzyme *N*-acetyl transferase do not relate to circadian rhythms of melatonin content in the pineal (d'Istria et al., 1994). However, pinealectomy abolished nighttime peaks in melatonin rhythms in melatonin, but not diurnal concentrations in tiger salamanders (*Ambystoma tigrinum;* Gern and Norris, 1979).

Melatonin administration to amphibians is generally associated with gonadal inhibition. The gonadosomatic index (GSI) decreases with daily injections of melatonin in tree frogs (*Hyla cinera;* De Vlaming et al., 1974) and marsh frogs (*Rana ridibunda;* Delgado et al., 1983). Exogenous melatonin counteracts the stimulatory effects of blinding on the GSI in skipper frogs (*Rana cyanophlyctis;* Joshi and Udaykumar, 2000). Exogenous melatonin induces gonadal regression in black spined toads (*B. melanosticus*) when administered either in the morning or the evening, but not at both phases of the day (Chanda and Biswas, 1982). Similarly, in Indian green frogs (*R. hexadactlya*) melatonin inhibited spermatogenesis under some dosing schedules but not others (Kasinathan and Gregalatchoumi, 1988). Melatonin also inhibits *in vitro* ovulation of Argentine common toads (*Bufo arenarum*) (de Atenor et al., 1994). However, other in vivo and in vitro studies have reported no effects or stimulatory actions of exogenous melatonin, including in seasonally breeding frogs such as Iberian water frog and European common frogs (Alonso-Bedate et al., 1988, 1990).

REPTILES

Reptiles are the most diverse of the vertebrate taxa, including the snakes, lizards, turtles, Tuatara, and crocodilians. Ancestral reptiles gave rise to both birds and mammals 250–300 million years ago (Padian and Chiappe, 1998; Godwin and Crews, 2002). The study of reptilian neuroendocrine systems is guided by the implicit assumption that modern reptiles exhibit phenotypic properties similar to the putative systems of ancestral amniotes. Extant traits including poilkiothermy, temperature-dependent sex determination, oviparity, and underdeveloped cerebral cortices likely also occurred in common ancestral amniotes (Godwin and Crews, 2002). Additionally, many reptilian species restrict breeding to a specific time of the year, but substantial diversity exists in the mechanisms that underlie these rhythms (Crews, 1999). Studying reptilian brains is therefore advantageous for researchers interested in photoperiodism and seasonality in general, and also important to the understanding of the evolutionary processes that shaped seasonality across vertebrate taxa.

Photoperiodic control of reproduction in reptiles was initially reported in the mid-1930s. Long day lengths produced out-of-season gonadal development in both green anoles (*Anolis carolinensis*) (Clausen and Poris, 1937) and red-eared slider turtles (*Trachemys scripta elegans*) (Burger, 1937). The overall goal of this section, therefore, is to review what has been learned in the 80 years since those initial discoveries and to also address remaining unanswered questions facing the field.

The control of seasonal breeding in reptiles is accomplished by a complex interplay between photoperiod and a variety of nonphotic cues including temperature, food, and water availability. Additionally, some species of reptiles appear to exhibit a circannual clock that interacts with other proximate cues. In general, the reproductive patterns of reptiles are characterized by those species that exhibit an associated reproductive strategy in which gonadal growth, steroidogenesis, and gametogenesis precedes and are temporally associated with breeding or a dissociated reproductive strategy in which gonadal growth, steroidogenesis, and gametogenesis follow breeding; in the latter instance, the gametes produced are then used in the next breeding season (Crews, 1984; Whittier and Crews, 1987). We will consider the control of seasonal breeding in both of these general types of breeders.

PHOTOPERIODISM AND BREEDING
CYCLES IN ASSOCIATED BREEDERS

There have been reports of photoperiodic modulation of reproductive and nonreproductive traits in all reptile groups (Burger, 1937; McPherson, 1981; Mendonca and Licht, 1986; Haldar and Pandey, 1989). Both the annual breeding cycle and the control of these rhythms by photoperiodic and nonphotoperiodic cues in reptiles have been best studied in the green anole lizard, so we will consider this species in detail. The green anole is a small iguanid lizard that inhabits the temperate zone throughout the southeastern United States (Conant and Collins, 1998). Although the timing of reproduction will vary with latitude, in general male green anoles are reproductively quiescent from late September until late in January. At that point the males emerge from hibernation and begin establishing and defending territories. Females become active approximately one month later (see below), and breeding commences shortly thereafter. The breeding season lasts until approximately the middle of August, at which point testicular collapse occurs.

Early studies based on the experimental methodology that had established a role for the circadian clock in plant and insect photoperiodism reported no evidence of a similar mechanism in reptiles. For example, night interruption (presentation of relatively brief light pulses during various parts of the daily cycle) can distinguish between hourglass and circadian models because the timing but not the absolute amount of light is the important parameter; resonance (light:dark cycles with periods of multiples of 12 such as 6:18, 6:30, 6:42) can detect the presence of a circadian rhythm in photoresponsiveness because without changing the amount of light increasing dark phases can render these rhythms inductive; and T-cycle experiments (similar to resonance experiments but with light:dark cycles within the range of entrainment; Nanda and Hamner, 1958; Pittendrigh and Minis, 1964; see below) initially failed to detect circadian involvement in the photoperiodic response of the green anole lizard (i.e., the total amount of light, rather than the timing in which it was administered, appeared to be the key determinant of reproductive responses to day length). An hour of light placed at various phases of the circadian clock did not alter gonadal responses to short (LD 6:18) photoperiods in either the regressive or recrudescent phases. T-cycles of various lengths (24–60 h) failed to induce gonadal development with 8-h photophases. Additionally, photocycles with 6 h of light and increasing dark periods were all nonstimulatory, while photoperiods with 16 h of light and increasing dark cycles were all inductive. These data were interpreted as support for the hypothesis that the anoles utilized an hourglass-type system (Underwood, 1978; Underwood and Hall, 1982). However, further studies on the anole circadian system indicated that locomotor rhythms became disorganized or exhibit splitting under very short photoperiods or the resonance and T-cycle experiments in the earlier experiments (Underwood, 1983a). This disorganization apparently rendered detection of a circadian regulation of photoperiod transduction impossible (Underwood and Hyde, 1990).

When later studies used longer photoperiods and night break manipulations, it became apparent that the circadian system controlled the reception of photoperiod information as long as the perceived length of the photoperiod was 10–11 h or greater in night break, resonance, or T-cycle experiments. Again, as in previous experiments, shorter photoperiods were not permissive to a circadian involvement in day-length measurement (Underwood and Hyde, 1990). Light pulses late in the subjective night either from night break, resonance, or T-cycle experiments were associated with induction of reproductive recrudescence and fat deposition (Ferrell and Meier, 1981; Underwood and Hyde, 1990).

Temperature and Photoperiod Acting at Different Times of the Cycle

The exogenous and endogenous control of the green anole reproductive cycle is summarized in Figure 16.2. Temperature is the critical regulator of testicular recrudescence, with temperatures a few degrees below the preferred body temperature impairing spermatogenesis, whereas a few degrees over the body temperature induces rapid testicular degeneration (Licht and Basu, 1967). Stimulation of reproductively inactive adult male green anoles with a LD 14:10 photoperiod enhances

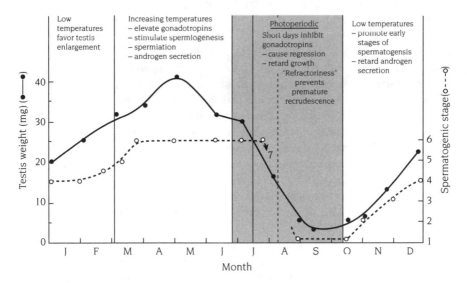

FIGURE 16.2.
Summary of the relationships among photoperiod, temperature, and photoperiod in the regulation of reproduction in green anole lizards. Shaded bars indicate the part of the year when the lizards are most responsive to photoperiod in the wild. Reproduced by permission of the Ecological Society of America.

testicular recrudescence when animals are housed at 32°C but not when maintained at 20°C; shorter photoperiods do not accelerate testicular recrudescence even at warmer temperatures (Licht, 1967a; see also Fox and Dessauer, 1958). Interestingly, a 12-h cycle alternating between 32°C and 20°C was stimulatory when the warmer temperatures coincided with the light cycle, but not when the elevated temperatures occurred at night (Licht, 1966, 1967a).

Toward the end of the breeding season as fall approaches photoperiod becomes the critical factor regulating testicular activity, with shortening day lengths at the end of the breeding season responsible for testicular collapse. Studies indicate that a photoperiod below LD 13.5:10.5 inhibits testicular activity, whereas photoperiods above LD 13.5:10.5 maintain testicular activity (Licht, 1970). Several lines of evidence suggest that induction of testicular recrudescence during the winter when animals normally are in hibernation is independent of day length. First, during the early part of the fall when day lengths are still long (but decreasing), green anoles exhibit a relative refractoriness to the stimulatory effects of photoperiod (Licht, 1967b; Crews and Licht, 1974). Additionally, the low winter temperatures are permissive to photoperiod responsiveness (Licht, 1966). Finally, once gonadal regeneration begins it is independent of further photoperiod stimulation (Licht, 1967b).

In addition to photoperiodic and temperature effects on gonadal activity, there is a circannual rhythm in the responsivity of green anoles to both environmental factors. Early in the fall, long day lengths maintain gonadal recrudescence, but this effect is greater when the photoperiod treatment started in October then if it is introduced several weeks earlier, suggesting a relative photorefractoriness (see Figure 16.3). Additionally, high temperatures accelerate testicular recrudescence independent of day length after mid-October. Together, these data indicate that soon after the initial regression of the reproductive tract, there is a relative refractory period to both temperature and day length (Licht, 1967b; Crews and Licht, 1974). Finally, photoperiod and temperature regimens that do not accelerate gonadal recrudescence do eventually result in gonadal development the following spring. Full spermatogenesis, however, never occurs in green anoles without exposure to warm temperatures for at least part of the day (Licht, 1967a, 1967b). Gonadal quiescence appears to be mediated by reduced hypothalamic gonadotropin secretion rather than at the level of the pituitary or gonads themselves (Crews and Licht, 1974). A similar finding has been reported in mammals (Kriegsfeld et al., 1999; Shanbhag et al., 2000).

Aside from the pronounced dependence on environmental temperature, this type of rhythm is similar to those seen in warm-blooded vertebrates. For instance, hamsters can breed indefinitely under stimulatory day lengths; gonads regress in response to short days and either will "spontaneously" recrudesce the following spring or can be stimulated to do so sooner with exposure to long days (Zucker and Morin, 1977; Goldman, 2001). Additionally, reptiles display the familiar vertebrate pattern of an endogenous oscillator that interacts with environmental information to regulate breeding cycles. However, the relative contribution of day-length signaling and circannual rhythms are difficult to tease apart based on the extant

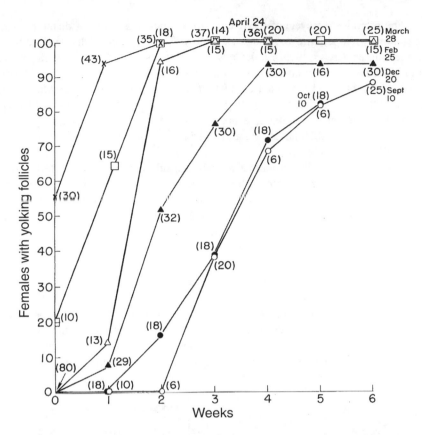

FIGURE 16.3.
An underlying circannual rhythm in environmental sensitivity is indicated by the more rapid rate of ovarian recrudescence in winter dormant female green anole lizard to a stimulatory environmental regimen as the period of normal breeding nears. Dates indicate time of year. In all studies, reproductively inactive females were exposed to an unseasonal environmental conditions (LD 14:10 photic cycle with a corresponding 32:23°C daily thermal cycle and constant 60–70% relative humidity. Reprinted from Crews and Garrick (1980). Courtesy Society for the Study of Amphibians and Reptiles.

literature. Fence lizards (*Sceloropus undulatus*) and parthenogenetic whiptail lizards (*Cnemidophorous uniparens*) appear to regulate reproductive cycles independently of photoperiod and via endogenous fluctuations in the sensitivity to temperature (Marion, 1970; Cuellar and Cuellar, 1977; Moore et al., 1984), whereas the response of checkered water snakes (*Natrix piscator*) to photoperiod is dependent on ambient humidity (Haldar and Pandey, 1989). The green anole studies presented here were conducted on lizards that had been captured in the weeks after the conclusion of the breeding season (e.g., Fox and Dessauer, 1958; Licht, 1966). Therefore, it is difficult to tell whether endogenous oscillators were entrained by photoperiod exposure prior to the previous breeding season or a true circannual clock exists.

The reproductive cycle of the female green anole is somewhat different in its regulation by temperature and photoperiod cues. The cycle can be divided into three distinct phases (Crews, 1975, 1980; Crews and Garrick, 1980): (1) Previtellogenesis from November to February is marked by inactive ovaries with small translucent unyolked follicles and regressed oviducts. (2) Vitellogenesis starts in March and continues until the end of the breeding season. This stage is characterized by follicular development and ovulation every 10–14 days. (3) The third stage regression is associated with follicular atresia and ovarian collapse (Licht, 1973; Crews and Licht, 1974; Crews, 1975).

Long day lengths can stimulate ovarian recrudescence late in the fall or during the winter, but immediately after breeding there is a period of photo- and thermorefractoriness (Licht, 1973; Crews and Licht, 1974). This is very similar to the pattern of reproductive cyclicity in side-blotched lizards (Uta stansburiana) (Tinkle and Irwin, 1965) and to mammalian patterns. The period of refractoriness may be related to the presence of large highly vascularized atretic follicles during the postbreeding period. The atretic follicles may directly inhibit the ovarian sensitivity to gonadotropin-induced by exogenous cues (Crews and Licht, 1974; Figure 16.4). In addition to photoperiod, temperature, humidity, and social stimuli are significant (Crews et al., 1974). For example, long days and warm temperatures during the photophase induce ovarian recrudescence during the winter, and this effect is potentiated by exposure to intact males (Crews et al., 1974; Crews, 1975). High relative humidity is essential important for this photothermal regimen to

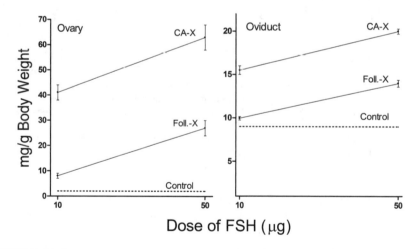

Dose of FSH (µg)

FIGURE 16.4.
Average ovarian and oviducal responses (± SEM) to either 10 or 50 µg of exogenous follicle-stimulating hormone following removal of an atretic or nonatretic previtellogenic follicle in the green anole lizard. Dashed lines indicate responses in saline-injected animals. Atretic follicles may mediate the photo- and thermorefractoriness exhibited by female green anoles after the mating season. Reprinted with permission from Crews and Licht (1974). Copyright 1974, The Endocrine Society.

stimulate gonadal activity in anole lizards and the checkered water snake (Haldar and Pandey, 1989).

DETECTION AND TRANSDUCTION
OF PHOTOPERIOD SIGNALS

The circuitry through which photoperiod information is communicated to reptile tissues is not well defined. The pineal gland seems critical to the expression of photoperiodic responsiveness, and this is likely secondary to the role of the pineal as both photoreceptor and endogenous oscillator (see below). The lizard pineal gland differs from mammalian pineal glands in several important ways. First, the principal pineal cells (analogous to the pinealocytes of birds and mammals) are both secretory in nature (as in mammals), but also are rudimentary photoreceptors. Additionally, there are populations of nerve cells that project axons to tectal and tegmental structures and receives synaptic input from various brain regions (Quay, 1979).

Pineal glands in lizards appear to be the critical site for the detection of photoperiod information. Retinal photoreceptors are not necessary for the regulation of circadian rhythms or day-length determinations in green anole lizards. Extraretinal photoreceptors maintained gonadal responses to day length, and the phase response curve for circadian responses to environmental light was actually enhanced by blinding (Underwood, 1985b; Underwood and Calaban, 1987b). Removal of the parietal eye also does not alter the reception of photoperiodic information (Underwood, 1985c). Conversely, circadian rhythms of locomotor activity can be entrained in pinealectomized and/or blinded lizards by 24-h cycles of either light or temperature (Underwood and Groos, 1982; Underwood, 1986a), indicating that extraretinal and extrapineal photoreceptors exist in lizards. However, shielding the heads of blinded-pinealectomized Texas spiny lizards (*Sceloporous olivaceous*) causes them to free run (Underwood and Menaker, 1976). The anatomical sites of these receptors have not been determined, but it is known that the parietal eye cannot be the location because pinealectomy includes removal of the parietal eye.

The lizard pineal is a rhythmic oscillator that is directly responsive to light and maintains daily rhythms of melatonin production in culture under constant lighting conditions (Menaker and Wisner, 1983; Underwood, 1983b), although the pineal gland of desert iguanas (*Diposaurus dorsalis*) becomes arrhythmic under constant conditions (Janik and Menaker, 1990). Additionally, removal of the pineal gland leads to splitting of locomotor rhythms of anoles, suggesting that the pineal is a master oscillator that when removed allows other subordinate oscillators to be expressed (Underwood, 1983a, 1983b). Melatonin may be the key output signal for the circadian system particularly because exogenous melatonin can shift the circadian rhythm of locomotor activity (Underwood, 1986a; Hyde and Underwood, 1995) possibly by binding to receptors in the suprachiasmatic nucleus (Bertolucci et al., 2000).

As in mammals, the pineal is necessary for the proper transduction of photo-period signals in reptiles. However, melatonin replacement cannot recapitulate the physiological effects of day length as it can in mammals. In green anole lizards males will undergo testicular recrudescence regardless of day length at all times of the year, including during the photorefractory period (Underwood, 1985a). A similar pattern is observed in Indian garden lizards (*Calotes versicolor*). Injections of melatonin during the photo- or scotophase (or both) will not reverse the effect of pinealectomy (Haldar and Thapliyal, 1977; Thapliyal and Haldar, 1979) Administration of melatonin to intact animals also did not alter gonadal responses (Underwood, 1985a). Chronic-release melatonin capsules, however, will block the progonadal effects of pinealectomy in short inhibitory day lengths. This is consis-tent with data from ruin lizards that exhibit bimodal activity patterns with peaks in the early morning and late afternoon; in the autumn and spring, however, activity is unimodal (Foa et al., 1992, 1994). Pinealectomy or constant-release melatonin in constant conditions induces the unimodal spring/fall phenotype (Bertolucci et al., 2000). Additionally, phase response curves to melatonin are not apparent in seasons other than summer (Bertolucci and Foa, 1998). Considered together, these data sug-gest that in addition to being a key regulator of photoperiodic responses, the pineal gland in lizards also is involved in the expression of circadian rhythms, although this function varies during different times of the year.

It is apparent that the pineal gland is the key receptor for photoperiod informa-tion, but that melatonin is not the sole neuroendocrine signal of day-length infor-mation. Although melatonin replacement does not rescue the apparent photoperiod insensitivity in pinealectomized green anoles under all conditions, a potential role for pineal melatonin in the regulation of intact reptile photoperiodism cannot be ruled out. At a minimum, the pineal or circulating melatonin content can provide information about the phase of the pineal oscillator. Indeed, the pineal melatonin rhythm is a powerful transducer of environmental information (particularly light and temperature) into a physiological one. The phase, amplitude, and duration of the pineal melatonin rhythm in iguanid lizards can be affected by photoperiod and thermoperiod duration, light intensity, and amplitude of temperature rhythms (Underwood, 1985d; Underwood and Calaban, 1987a; Vivien-Roels et al., 1988). Under appropriate photoperiod and temperature cycles (e.g., LD 12:12 and cooler temperatures during the lights on period) strong daily peaks of melatonin can occur during lights on (Underwood and Calaban, 1987a).

Light at night does not acutely suppress melatonin production in green anole lizards (Underwood, 1986b) or box turtles (*Terrapene carolina triunguis*) (Vivien-Roels et al., 1988). When daily melatonin rhythms are assessed under various T-cycle, resonance, or night-break conditions, neither the amplitude nor the dura-tion of the melatonin signal predicts whether the treatment would stimulate gonadal regrowth (Hyde and Underwood, 1993). Instead, the only aspect of the melatonin rhythm that covaries significantly with reproductive responses is the phase at which the melatonin peak occurred. Photoperiod (and temperature) regimens that produce melatonin peaks during the early or late phase of the lights-off period stimulated gonadal growth, whereas those regimens that produced a peak toward the middle of the dark period are inhibitory (Hyde and Underwood, 1993). Finally, the same types of manipulations (e.g., temperature amplitude, light intensity) that alter melatonin rhythms also alter the expression of circadian clock genes (Magnone et al., 2005; Malatesta et al., 2007; Vallone et al., 2007). Together, these data suggest that the phase relationship between melatonin and some endogenous oscillator (e.g., an internal coincidence system) determines reproductive responses to day length and temperature. However, an external coincidence system cannot be ruled out, nor can a direct role of pineal synaptic connections on other central nervous system targets.

DISSOCIATED BREEDERS

In most seasonal vertebrates increases in gonadal steroids and spermatogenesis occur shortly before the onset of the breeding season and are necessary for the full expression of reproductive behaviors. However, there are examples of vertebrates with dissociated reproductive patterns in which the gonads are quiescent during winter hibernation and the breeding season and become activated only after the conclusion of the mating season (Crews, 1984; Whittier and Crews, 1987). Red-sided garter snakes (*Thamnophis sirtalis*) display such a dissociated pattern as the gonads do not become active until after the brief summer breeding season; sex behavior is not dependent on circulating androgens. Instead, the increase in sex steroid hormones the previous summer appear to program the brain and regulate mating the following spring (Crews, 1991; Crews et al., 1993). These animals are not photoperiodic per se as the major cue for the expression of spring mating behavior is prolonged exposure to low temperatures (Whittier et al., 1987; Lutterschmidt et al., 2006).

However, a role for pineal melatonin has been suggested in the regulation of spring breeding. Depending upon the time of year the pineal is removed, spring courtship behavior is either disrupted or enhanced. Pinealectomy prior to hibernation prevents males from courting females on spring emergence, but if conducted on males that are actively courting, there is no effect on behavior (Nelson et al., 1987; Crews et al., 1988; Mendonca et al., 1996a, 1996b). Interestingly, in those few males that fail to court on emergence, pinealectomy induces robust courtship behavior. Measurements of circulating concentrations of melatonin reveal significantly higher levels in males that are actively courting but in males that fail to court, the circadian rhythm is absent. Shielding the eyes disrupts the retinal melatonin rhythm and abolishes the effect of pinealectomy (Nelson et al., 1987; Crews et al., 1988; Mendonca et al., 1996a, 1996b). Exogenous melatonin fails to reinstate mating behavior (Mendonca et al., 1996b). It is likely that melatonin communicates temperature rather than photoperiod information, as these animals are nonresponsive to day length. Circulating melatonin also varies with temperature in the diamondback water snake (*Nerodia rhombifera*) (Tilden and Hutchinson, 1993).

CONCLUSIONS AND FUTURE DIRECTIONS

The study of amphibian and reptile photoperiodism has revealed evolutionary processes that underlie seasonality in all vertebrate taxa. Changes in the photoperiodic machinery over evolutionary time, including a reduced number of pineal efferents and a greater reliance on the pineal itself for the production of melatonin, could not be appreciated by studying mammalian or avian systems alone. Of particular interest is the manner by which day-length signals are communicated to all target tissues. In mammals, melatonin productions is under the direct control of a multisynaptic pathway from the retina, and it serves as the principle, if not the only, cue that transduces photoperiodic information from an environmental signal into a physiological one. However, in all other vertebrate taxa, the source, targets, and control of production and release of melatonin are all more equivocal and subject to regulation by nonphotic cues. In any case, to fully appreciate the way in which photoperiodic measurement of seasonal time evolved, it is critical to studies taxa that retain traits similar to those of ancestral vertebrates.

There is certainly much more work to do as the physiological substrates that underlie photoperiodism in reptiles and particularly amphibians have not been fully described. Additionally, research on photoperiodism in these taxa has dropped off in recent years, so many of the modern neuroendocrine and molecular techniques have not been used in these species, although there are some notable exceptions (e.g., Inai et al., 2003; Magnone et al., 2005; Izzo et al., 2006; Malatesta et al., 2007; Neal and Wade, 2007). Finally, the demonstrated diversity in the reproductive strategies present in these two vertebrate taxa combined with studies of additional species with differing life history characteristics will deepen our understanding of the causes and constraints of reproduction in vertebrate in general.

References

Alonso-Bedate M, Carballada R, and Delgado MJ. (1990). Effects of melatonin on gonadal steroids and glucose plasma levels in frogs (*Rana perezi* and *Rana temporaria*). J Pineal Res 8: 79–89.

Alonso-Bedate M, Delgado MJ, and Carballada R. (1988). In vivo effect of melatonin and gonadotropin-releasing hormone on testicular function in *Rana temporaria*. J Pineal Res 5: 323–332.

Alonso-Gomez AL, Tejera M, Alonso-Bedate M, and Delgado MJ. (1990). Response to pinealectomy and blinding in vitellogenic female frogs (*Rana perezi*) subjected to high temperature in autumn. Can J Physiol Pharmacol 68: 94–98.

Bertolucci C and Foa A. (1998). Seasonality and role of SCN in entrainment of lizard circadian rhythms to daily melatonin injections. Am J Physiol Reg Integr Comp Physiol 43: R1004–R1014.

Bertolucci C, Sovrano VA, Magnone MC, and Foa A. (2000). Role of suprachiasmatic nuclei in circadian and light-entrained behavioral rhythms of lizards. Am J Physiol Reg Integr Comp Physiol 279: R2121–R2131.

Bünning E. (1936). Die endogene tagesrhythmik als Grundlage der photoperiodischen Reaktion. Ber Dtsch Bot Ges: 590–607.

Burger JW. (1937). Experimental sexual photoperiodicity in the male turtle, *Pseudemys elegans* (Wied). Am Nat 71: 481–487.

Chanda S and Biswas NM. (1982). Effect of morning and evening injections of melatonin on the testis of toad (*Bufo melanostictus*). Endocrinol Jpn 29: 483–485.

Clausen HJ and Poris EG. (1937). The effect of light upon sexual activity in the lizard *Anolis carolinensis*, with special reference to the pineal body. Anat Rec 69: 39–50.

Conant R and Collins JT. (1998). *A Field Guide to Reptiles and Amphibians: Eastern and Central North America*. New York: Houghton-Mifflin.

Crews D. (1975). Psychobiology of reptilian reproduction. Science 189: 1059–1065.

Crews D. (1980). Interrelationships among ecological, behavioral and neuroendocrine processes in the reproductive cycle of *Anolis carolinensis* and other reptiles. Adv Stud Behav 11: 1–74.

Crews D. (1984). Gamete production, sex hormone secretion, and mating behavior uncoupled. Horm Behav 18: 22–28.

Crews D. (1991). Trans-seasonal action of androgen in the control of spring courtship behavior in male red-sided garter snakes. Proc Natl Acad Sci USA 88: 3545–3548.

Crews D. (1999). Reptilian reproduction overview. In *Encylopedia of Reproduction* (E Knobil and JD Neill, eds). New York: Academic Press, pp. 254–259.

Crews D and Garrick LD. (1980). Methods for inducing reproduction in captive reptiles. In *Reproductive Biology and Diseases of Captive Reptiles* (JB Murphy and JT Collins, eds). St. Louis, Mo. Society for the Study of Amphibians and Reptiles, pp. 49–70.

Crews D, Hingorani V, and Nelson RJ. (1988). Role of the pineal-gland in the control of annual reproductive behavioral and physiological cycles in the red-sided garter snake (*Thamnophis-sirtalis-parietalis*). J Biol Rhythms 3: 293–302.

Crews D and Licht P. (1974). Inhibition by corpora atretica of ovarian sensitivity to environmental and hormonal stimulation in the lizard, *Anolis carolinensis*. Endocrinology 95: 102–106.

Crews D, Robker R, and Mendonca M. (1993). Seasonal fluctuations in brain nuclei in the red-sided garter snake and their hormonal control. J Neurosci 13: 5356–5364.

Crews D, Rosenblatt JS, and Lehrman DS. (1974). Effects of unseasonal environmental regime, group presence, group composition and males' physiological state on ovarian recrudescence in the lizard, *Anolis carolinensis*. Endocrinology 94: 541–547.

Cuellar H and Cuellar O. (1977). Evidence for endogenous rhythmicity in the reproductive cycle of the parthenogenetic lizard *Cnemidophorus uniparens* (Reptilia: Teiidae). Copeia 1977: 554–557.

d'Istria M, Monteleone P, Serino I, and Chieffi G. (1994). Seasonal variations in the daily rhythm of melatonin and NAT activity in the Harderian gland, retina, pineal gland, and serum of the green frog, *Rana esculenta*. Gen Comp Endocrinol 96: 6–11.

de Atenor MSB, de Romero IR, Brauckmann E, Pisanó A, and Legname AH. (1994). Effects of the pineal gland and melatonin on the metabolism of oocytes in vitro and on ovulation in *Bufo arenarum*. J Exp Zool 268: 436–441.

De Vlaming VL, Sage M, and Charlton CB. (1974). The effects of melatonin treatment on gono-somatic index in the teleost, *Fundulus similis,* and the tree frog, *Hyla cinerea*. Gen Comp Endocrinol 22: 433–438.

Delgado MJ, Alonso-Gomez AL, and Alonso-Bedate M. (1992). Role of environmental temperature and photoperiod in regulation of seasonal testicular activity in the frog, *Rana perezi*. Can J Physiol Pharmacol 70: 1348–1352.

Delgado MJ, Gutierrez P, and Alonso-Bedate M. (1989). Seasonal cycles in testicular activity in the frog, *Rana perezi*. Gen Comp Endocrinol 73: 1–11.

Delgado MJ, Gutierrez P, and Alonso-Bedate M. (1983). Effects of daily melatonin injections on the photoperiodic gonadal response of the female frog *Rana ridibunda*. Comp Biochem Physiol AComp Biol 76: 389–392.

Delgado MJ and Vivien-Roels B. (1989). Effect of environmental temperature and photoperiod on the melatonin levels in the pineal, lateral eye, and plasma of the frog, *Rana perezi:* Importance of ocular melatonin. Gen Comp Endocrinol 75: 46–53.

Duellman WE and Trueb L. (1986). *Biology of Amphibians*. Baltimore, MD: Johns Hopkins University Press.

Ferrell BR and Meier AH. (1981). Photo-thermoperiodic effects on fat stores in the green anole, *Anolis-carolinensis.* J Exp Zool 217: 353–359.

Foa A, Monteforti G, Minutini L, Innocenti A, Quaglieri C, and Flamini M. (1994). Seasonal-changes of locomotor-activity patterns in ruin lizards *Podarcis-sicula.* 1. Endogenous control by the circadian system. Behav Ecol Sociobiol 34: 267–274.

Foa A, Tosini G, and Avery R. (1992). Seasonal and diel cycles of activity in the ruin lizard, *Podarcis-sicula.* Herpetol J 2: 86–89.

Fox W and Dessauer HC. (1958). Response of the male reproductive system of lizards (*Anolis carolinensis*) to unnatural day-lengths in different seasons. Biol Bull 115: 421–439.

Fraile B, Paniagua R, and Rodriguez MC. (1988). Long day photoperiods and temperature of 20 degrees C induce spermatogenesis in blinded and non-blinded marbled newts during the period of testicular quiescence. Biol Reprod 39: 649–655.

Gern WA and Norris DO. (1979). Plasma melatonin in the neotenic tiger salamander (*Ambystoma tigrinum*): Effects of photoperiod and pinealectomy. Gen Comp Endocrinol 38: 393–398.

Godwin J and Crews D. (2002). Hormones, brain, and behavior in reptiles. In *Hormones, Brain and Behavior,* vol 2 (DW Pfaff, AP Arnold, AM Etgen, SE Fahrbach, and RT Rubin, eds). New York: Academic Press, pp. 545–585.

Goldman BD. (2001). Mammalian photoperiodic system: Formal properties and neuroendocrine mechanisms of photoperiodic time measurement. J Biol Rhythms 16: 283–301.

Haldar C and Pandey R. (1989). Effect of pinealectomy on the testicular response of the fresh-water snake natrix-piscator to different environmental-factors. Can J Zool 67: 2352–2357.

Haldar C and Thapliyal JP. (1977). effect of pinealectomy on annual testicular cycle of *Calotes-versicolor.* Gen Comp Endocrinol 32: 395–399.

Hyde LL and Underwood H. (1993). Effects of nightbreak, T-cycle, and resonance lighting schedules on the pineal melatonin rhythm of the lizard *Anolis carolinensis:* Correlations with the reproductive response. J Pineal Res 15: 70–80.

Hyde LL and Underwood H. (1995). Daily melatonin infusions entrain the locomotor activity of pinealectomized lizards. Physiol Behav 58: 943–951.

Ifft JD. (1942). The effect of environmental factors on the sperm cycle of *Triturus viridescens.* Biol Bull 83: 111–128.

Inai Y, Nagai K, Ukena K, Oishi T, and Tsutsui K. (2003). Seasonal changes in neurosteroid concentrations in the amphibian brain and environmental factors regulating their changes. Brain Res 959: 214–225.

Izzo G, d'Istria M, Ferrara D, Serino I, Aniello F, and Minucci S. (2006). Connexin 43 expression in the testis of the frog *Rana esculenta.* Zygote 14: 349–357.

Janik DS and Menaker M. (1990). Circadian locomotor rhythms in the desert iguana. I. The role of the eyes and the pineal. J Comp Physiol A Neuroethol Sens Neural Behav 166: 803–810.

Joshi BN and Udaykumar K. (2000). Melatonin counteracts the stimulatory effects of blinding or exposure to red light on reproduction in the skipper frog *Rana cyanophlyctis.* Gen Comp Endocrinol 118: 90–95.

Kasinathan S and Gregalatchoumi S. (1988). Effect of melatonin on the spermatogenesis of *Rana hexadactyla* (Lesson). Endocrinol Jpn 35: 255.

Kriegsfeld LJ, Drazen DL, and Nelson RJ. (1999). Effects of photoperiod and reproductive responsiveness on pituitary sensitivity to GnRH in male prairie voles (*Microtus ochrogaster*). Gen Comp Endocrinol 116: 221–228.

Lees AD. (1966). Photoperiodic timing mechanisms in insects. Nature 210: 986–989.

Licht P. (1966). Reproduction in lizards: Influence of temperature on photoperiodism in testicular recrudescence. Science 154: 1668–1670.

Licht P. (1967a). Environmental control of annual testicular cycles in the lizard *Anolis carolinensis.* I. Interaction of light and temperature in the initiation of testicular recrudescence. J Exp Zool 165: 505–516.

Licht P. (1967b). Environmental control of annual testicular cycles in the lizard *Anolis carolinensis*. II. Seasonal variations in the effects of photoperiod and temperature on testicular recrudescence. J Exp Zool 166: 243–253.

Licht P. (1970). Regulation of the annual testis cycle by photoperiod and temperature in the lizard *Anolis carolinensis*. Ecology 52: 240–252.

Licht P. (1973). Influence of temperature and photoperiod on the annual ovarian cycle in the lizard *Anolis carolinensis*. Copeia 1973: 465.

Licht P and Basu SL. (1967). Influence of temperature on lizard testes. Nature 213: 672–674.

Lofts B. (1984). Amphibians. In *Marshall's Physiology of Reproduction* (GE Lamming, ed). Edinburgh: Churchill Livingstone, pp. 127–205.

Lutterschmidt DI, LeMaster MP, and Mason RT. (2006). Minimal overwintering temperatures of red-sided garter snakes (*Thamnophis sirtalis parietalis*): A possible cue for emergence? Can J Zool 84: 771–777.

Magnone MC, Incohmrier D, Bertolucci C, Foa A, and Albrecht U. (2005). Circadian expression of the clock gene Per2 is altered in the ruin lizard (*Podarcis sicula*) when temperature changes. Brain Res Mol Brain Res 133: 281–285.

Maier EC and Singer M. (1977). The effect of light on forelimb regeneration in the newt. J Exp Zool 202: 241–244.

Maier EC and Singer M. (1981). The effect of prolactin on the rate of forelimb regeneration in newts exposed to photoperiod extremes. J Exp Zool 216: 395–397.

Malatesta M, Frigato E, Baldelli B, Battistelli S, Foa A, and Bertolucci C. (2007). Influence of temperature on the liver circadian clock in the ruin lizard *Podarcis sicula*. Microsc Res Tech 70: 578–584.

Marion K. (1970). Temperature as the reproductive cue for the female fence lizard *Sceloporus undulatus*. Copeia 1970: 562–564.

McCord CP and Allen FP. (1917). Evidences associating pineal gland function with alterations in pigmentation. J Exp Zool 23: 207–224.

McPherson RJ. (1981). Seasonal testicular cycle of the stinkpot turtle (*Sternotherus odoratus*) in central Alabama. Herpetologica 37: 33–40.

Menaker M and Wisner S. (1983). Temperature-compensated circadian clock in the pineal of *Anolis*. Proc Natl Acad Sci USA 80: 6119–6121.

Mendonca MT and Licht P. (1986). Photothermal effects on the testicular cycle in the musk turtle, *Sternotherus odoratus*. J Exp Zool 239: 117–130.

Mendonca MT, Tousignant AJ, and Crews D. (1996a). Courting and noncourting male red-sided garter snakes, *Thamnophis sirtalis parietalis:* Plasma melatonin levels and the effects of pinealectomy. Horm Behav 30: 176–185.

Mendonca MT, Tousignant AJ, and Crews D. (1996b). Pinealectomy, melatonin, and courtship behavior in male red-sided garter snakes (*Thamnophis sirtalis parietalis*). J Exp Zool 274: 63–74.

Moore MC, Whittier JM, and Crews D. (1984). Environmental-control of seasonal reproduction in a parthenogenetic lizard *Cnemidophorus uniparens*. Physiol Zool 57: 544–549.

Nanda KK and Hamner KC. (1958). Studies on the nature of the endogenous rhythm affecting photoperiodic response of Biloxi soybean. Bot Gaz 120: 14–25.

Neal JK and Wade J. (2007). Effects of season, testosterone and female exposure on c-fos expression in the preoptic area and amygdala of male green anoles. Brain Res 1166: 124–131.

Nelson RJ, Mason RT, Krohmer RW, and Crews D. (1987). Pinealectomy blocks vernal courtship behavior in red-sided garter snakes. Physiol Behav 39: 231–233.

Oksche A. (1984). Evolution of the pineal complex: Correlation of structure and function. Ophthalmic Res 16: 88.

Oksche A and Hartwig HG. (1979). Pineal sense organs—components of photoneuroendocrine systems. Progr Brain Res 52: 113.

Padian K and Chiappe LM. (1998). The origin of birds and their flight. Sci Am 278: 38–47.

Pittendrigh CS. (1972). Circadian surfaces and diversity of possible roles of circadian organization in photoperiodic induction. Proc Natl Acad Sci USA 69: 2734–2737.

Pittendrigh CS and Minis DH. (1964). The entrainment of circadian oscillations by light and their role as photoperiodic clocks. Am Nat 98: 261–294.

Quay WB. (1979). The parietal eye-pineal complex. In *Biology of the Reptilia* (C Gans, ed). New York: Academic Press, pp. 245–406.

Rastogi RK, Chieffi G, and Iela L. (1972). Effects of antiestrogen and antiandrogen in amphibia. Gynecol Invest 2: 271–275.

Rastogi RK, Iela L, Delrio G, Dimeglio M, Russo A, and Chieffi G. (1978). Environmental influence on testicular activity in green frog, *Rana esculenta.* J Exp Zool 206: 49–63.

Rastogi RK, Iela L, Saxena PK, and Chieffi G. (1976). Control of spermatogenesis in green frog, *Rana-esculenta.* J Exp Zool 196: 151–165.

Rawding RS and Hutchison VH. (1992). Influence of temperature and photoperiod on plasma melatonin in the mudpuppy, *Necturus maculosus.* Gen Comp Endocrinol 88: 364–374.

Shanbhag BA, Radder RS, and Saidapur SK. (2000). GnRH but not warm temperature induces recrudescence of quiescent testes in the tropical lizard *Calotes versicolor* (Daud.) during post-breeding phase Gen Comp Endocrinol 119: 232–238.

Thapliyal JP and Haldar CMN. (1979). Effect of pinealectomy on the photoperiodic gonadal response of the Indian garden lizard, *Calotes versicolor.* Gen Comp Endocrinol 39: 79–86.

Tilden AR and Hutchinson VH. (1993). Influence of photoperiod and temperature on serum melatonin in the diamondback water snake, *Nerodia rhombifera.* Gen Comp Endocrinol 92: 347–354.

Tinkle DW and Irwin LN. (1965). Lizard reproduction: Refractory period and response to warmth in *Uta stansburiana* females. Science 148: 1613–1614.

Truman JW. (1971). Hour-glass behavior of the circadian clock controlling eclosion of the silkmoth *Antheraea pernyi.* Proc Natl Acad Sci USA 68: 595–599.

Udaykumar K and Joshi BN. (1996). Ovarian follicular kinetics in response to enucleation and parietal shielding in *Rana cyanophlyctis.* Gen Comp Endocrinol 103: 244–248.

Underwood H. (1978). Photoperiodic time measurement in male lizard *Anolis carolinensis.* J Comp Physiol 125: 143–150.

Underwood H. (1983a). Circadian organization in the lizard *Anolis-carolinensis*—a multioscillator system. J Comp Physiol 152: 265–274.

Underwood H. (1983b). Circadian pacemakers in lizards: Phase-response curves and effects of pinealectomy. Am J Physiol 244: R857–R864.

Underwood H. (1985a). Annual testicular cycle of the lizard *Anolis carolinensis:* Effects of pinealectomy and melatonin. J Exp Zool 233: 235–242.

Underwood H. (1985b). Extraretinal photoreception in the lizard *Sceloporus occidentalis:* Phase response curve. Am J Physiol 248: R407–R414.

Underwood H. (1985c). Parietalectomy does not affect testicular growth in photoregulating lizards. Comp Biochem Physiol A Comp Physiol 80: 411–413.

Underwood H. (1985d). Pineal melatonin rhythms in the lizard *Anolis carolinensis:* Effects of light and temperature cycles. J Comp Physiol A Sens Neural Behav Physiol 157: 57–65.

Underwood H. (1986a). Circadian rhythms in lizards: Phase response curve for melatonin. J Pineal Res 3: 187–196.

Underwood H. (1986b). Light at night cannot suppress pineal melatonin levels in the lizard *Anolis carolinensis.* Comp Biochem Physiol A Comp Physiol 84: 661–663.

Underwood H and Calaban M. (1987a). Pineal melatonin rhythms in the lizard *Anolis carolinensis:* I. Response to light and temperature cycles J Biol Rhythms 2: 179–193.

Underwood H and Calaban M. (1987b). Pineal melatonin rhythms in the lizard *Anolis carolinensis:* II. Photoreceptive inputs. J Biol Rhythms 2: 195–206.

Underwood H and Groos G. (1982). Vertebrate circadian rhythms: Retinal and extraretinal photoreception. Experientia 38: 1013–1021.

Underwood H and Hall D. (1982). Photoperiodic control of reproduction in the male lizard *Anolis carolinensis*. J Comp Physiol 146: 485–492.

Underwood H and Hyde LL. (1990). A circadian clock measures photoperiodic time in the male lizard *Anolis-carolinensis*. J Comp Physiol A Sens Neural Behav Physiol 167: 231–243.

Underwood H and Menaker M. (1976). Extraretinal photoreception in lizards. Photophysiology 23: 227–243.

Vallone D, Frigato E, Vernesi C, Foa A, Foulkes NS, and Bertolucci C. (2007). Hypothermia modulates circadian clock gene expression in lizard peripheral tissues. Am J Physiol Reg Integr Comp Physiol 292: R160–R166.

Van Oordt, PGWJ. (1956). The role of temperature in regulating the spermatogenetic cycle in the common frog (*Rana temporaria*). Acta Endocrinol 23: 251.

Vivien-Roels B, Pevet P, and Claustrat B. (1988). Pineal and circulating melatonin rhythms in the box turtle, *Terrapene carolina triunguis:* Effect of photoperiod, light pulse, and environmental temperature. Gen Comp Endocrinol 69: 163 173.

Werner JK. (1969). Temperature-photoperiod effects on spermatogenesis in the salamander *Plethodon cinereus*. Copeia 1969: 592–602.

Whittier JM and Crews D. (1987). Seasonal reproduction: Patterns and controls. In *Hormones and Reproduction in Fishes, Amphibians and Reptiles* (DO Norris and RE Jones, eds). New York: Plenum, 385–409.

Whittier JM, Mason RT, Crews D, and Licht P. (1987). Role of light and temperature in the regulation of reproduction in the red-sided garter snake, *Thamnophis-Sirtalis-parietalis*. Can J Zool 65: 2090–2096.

Zucker I and Morin LP. (1977). Photoperiodic influences on testicular regression, recrudescence and the induction of scotorefractoriness in male golden hamsters. Biol Reprod 17: 493–498.

17

Photoperiodism and Reproduction in Birds

George E. Bentley

In northern temperate zones, most bird species have evolved mechanisms to coincide their breeding with periods when environmental conditions are optimal for the raising of young. Often this depends on seasonal food supply and/or climatic conditions. In general, the higher the latitude, the more limited the time window in which conditions are conducive to breeding. Thus, it is imperative to time reproductive activity precisely. Equally important is the ability to avoid breeding during less suitable periods. Although such factors as rainfall, food availability, song, and temperature may advance or delay the breeding period slightly (Gibb, 1950; Kluijver, 1951; Immelman, 1971; Hinde and Steel, 1978; Wingfield et al., 1983; Wingfield and Farner, 1993), the most reliable proximate factor is the changing length of photoperiod throughout the annual cycle. Thus, most nontropical birds have a very well-defined annual breeding season that is regulated by changing photoperiod. This was first demonstrated in juncos (*Junco hyemalis*) by Rowan in 1925.

This chapter describes mechanisms thought to be involved in the regulation of photoperiodism in birds and major discoveries since Rowan's initial findings. The key hormones involved in avian photoperiodism are discussed explicitly. Note that gonadotropins are not discussed per se, as seasonal changes in gonadotropins are regulated as a result of changes in the neuroendocrine photoperiodic machinery and are not thought to participate themselves in the timing of the photoperiodic response. Thus, in a sense, changes in gonadotropins are more a symptom of photoperiodism rather than a cause.

AVIAN PHOTOPERIODIC RESPONSE

Often it is perceived that avian photoperiodism results from seasonal changes in responsiveness to day length, but this cannot be true. Birds are able to measure day length year-round. Thus, photoperiodism in birds involves a change in the *quality* of the response of the bird's physiology to changing day length. At one time of the year, the neuroendocrine system is able to respond to long days with a physiological cascade that includes a dramatic increase in gonadotropin secretion, gonadal growth, and a range of hormone-dependent processes, including behavior. A long day is defined as one for which the duration of the light period is over and above a "critical" day length, that being the length of day below which photorefractoriness cannot be induced. The state of responsiveness to long day lengths in birds is known as photosensitivity, and the actual response to long days is known as photostimulation. However, temperate-zone birds develop a change in response to these stimulatory day lengths over time, so that at other times of the year the same long day length maintains a state of reproductive inactivity (for reviews, see Nicholls et al., 1988a; Wilson and Donham, 1988; Wingfield and Farner, 1993; Dawson et al., 2001). In birds, this state is known as photorefractoriness. Gonadotropin-releasing hormone (GnRH) cell bodies in the brain shrink and fibers emanating from these toward the median eminence (ME) show a marked decrease in immunocytochemical labeling (Foster et al., 1987; Goldsmith et al., 1989; Boulakoud and Goldsmith, 1991). Very recently, it was shown that GnRHmRNA production is halted after the onset of photo refractoriness (Ubuka et al., 2009). Thus, pituitary release of the gonadotropins luteinizing hormone (LH) and follicle-stimulating hormone (FSH) is reduced to nondetectable levels (Dawson and Goldsmith, 1982, 1983), and the gonads undergo marked regression. Increased synaptic input to the GnRH neurons is coincident with the long-term photorefractory state (Parry and Goldsmith, 1993). Other physiological effects associated with photorefractoriness are a peak in plasma prolactin and a postnuptial molt (Goldsmith and Nicholls, 1984a, 1984b). When birds are housed in laboratory conditions, long day lengths maintain photorefractoriness indefinitely. Without a long-day stimulus, photorefractory birds develop what is termed photosensitivity, so that they are once again able to grow their gonads in response to a long-day stimulus. Seen this way, it seems that the terms "photosensitive," "photostimulated," and "photorefractory" are somewhat misleading, given that the reproductive system of photoperiodic birds is sensitive to, and exhibits a response to, the ambient photoperiod no matter what it is; it is just the nature of the response that changes from one time of year to another. In the wild, photosensitivity is acquired in the autumn, when day length falls below approximately 11.5 h—resulting in some species what has been termed "autumnal sexuality" (Murton and Westwood, 1977; Lincoln et al., 1980; Dawson, 1983).

One might posit that a decrease in food intake is responsible for the dramatic changes in reproductive state from one time of year to another, but this is not the case for most species. For example, in European starlings (*Sturnus vulgaris*), photorefractoriness ensues as day length is still increasing (Dawson and Goldsmith, 1982), and therefore, the length of time per day available to feed is increasing as photorefractoriness occurs. Also, experimental manipulation with ad libitum food supply contradicts this hypothesis. An example of this is a study carried out by Dawson (1986), in which starlings housed on long days but with restricted food supply still exhibited robust gonadal responses to long days. In addition, if photosensitive starlings are exposed to only seven long days and then returned to short days (and thus have a reduced time to feed), then this is often sufficient to provide the photoperiodic drive for photorefractoriness to ensue some weeks later (Dawson et al., 1985). Despite these observations, food is an important cue for gonadal growth in zebra finches and is possibly a more potent cue than long day lengths (Perfito et al., 2008).

The rate of onset of photorefractoriness is proportional to the length of the photoperiod (Hamner, 1971; Dawson and Goldsmith, 1983). Even though photorefractoriness does not become apparent for several weeks after exposure to long days, it is initiated rapidly and the reproductive system continues to proceed toward a photorefractory state regardless of subsequent photoperiod once initiation occurs (Dawson et al., 1985). More recently, Dawson provided evidence that a single long day can initiate progression toward photorefractoriness (Dawson, 2001). In the same way, a single long day causes activation of the reproductive system in quail (Nicholls et al., 1983; Meddle and Follett, 1995, 1997). Just as the onset of photorefractoriness is proportional to the length of the long-day stimulus, the regaining of photosensitivity is a gradual process, and the rate of recovery is proportional to the "shortness" of short days and to the length of time held on short days (Vaugien, 1955; Farner and Follett, 1966; Steel et al., 1975; Turek, 1975; Nicholls and Storey, 1977; Gwinner et al., 1988; Dawson, 1991; Boulakoud and Goldsmith, 1994, 1995).

PHOTOPERIODISM IN TROPICAL SPECIES AND OPPORTUNISTIC BREEDERS

Birds that breed within the tropics (where 80% of birds live) have traditionally been considered to be nonphotoperiodic, since the annual change in photoperiod is slight (Dittami and Gwinner, 1985). Nevertheless, tropical birds have been shown to be photoperiodic, for example, African stonechats (Gwinner and Scheuerlein, 1999), zebra finches (Bentley et al., 2000), and rufous-collared sparrows (Miller, 1965), although the experimental amplitude in photoperiod used in these studies exceeded that of the tropics. The latter study on zebra finches might have been confounded by increased food availability; it is now known that food cues are at least equally potent as photoperiod cues in terms of reproductive activation in this species (Perfito et al., 2008). Two studies suggest that another tropical species, spotted antbirds, can respond to tropical changes in photoperiod (Hau et al., 1998; Beebe et al., 2005). The strategy adopted by tropical birds appears to be to remain in a physiological state of "readiness to breed" for a large proportion of the year and then to use nonphotoperiodic cues, such as rainfall or food abundance, to determine the exact time of breeding. In zebra finches, for example, breeding can occur at any time of the year, but breeding bouts are more intense during periods of rainfall (Zann et al., 1995). Stonechats may respond to low light intensity as a predictive cue for rainfall (Gwinner and Scheuerlein, 1998), and this response might be mediated via the melatonin signal (Kumar et al., 2007). Rufous-collared sparrows appear to respond to rainfall and possibly food availability (Moore et al., 2006). There is evidence that the hypothalamo-pituitary-gonadal (HPG) axis of such birds remains for much of the year in a state somewhere between that which is characteristic of photosensitivity and photostimulation in nontropical seasonally breeding birds, and that full functionality is triggered by the relevant proximate cues (Perfito et al., 2006). Photorefractoriness does not seem to occur in spotted antbirds (Beebe et al., 2005).

Gonadal regression and molt frequently follow breeding, but the extent to which the GnRH system is inactivated during this period is mostly unknown. One study on Ecuadorean rufous-collared sparrows (*Zonotrichia capensis*) that experience photoperiod fluctuations of only 3 min over the year suggests that there are changes in the GnRH system of a magnitude similar to that seen in obligately photoperiodic birds (Moore et al., 2006).

Nontropical northern hemisphere opportunistic breeders such as crossbills can breed at most times of the year (mainly between January and August). Crossbills feed on conifer seeds, the availability of which appears to synchronize reproductive effort. However, these species do have a fixed nonbreeding period during fall, when the HPG axis appears to be photorefractory and when molt occurs (Hahn et al., 1995; Hahn, 1998). Even if opportunistic breeders are photoperiodic to a degree (Bentley et al., 2000), then the relative importance of nonphotoperiodic cues in "fine-tuning" the timing of breeding varies and appears to be greater than that in obligately photoperiodic species (Wingfield, 1980; Perfito et al., 2008).

ABSOLUTE AND RELATIVE PHOTOREFRACTORINESS

In birds that exhibit absolute photorefractoriness, for example, European starling (Nicholls et al., 1988a) and European rook, *Corvus frugilegus* (Lincoln et al., 1980), breeding attempts are halted at a specific time of year and at a very specific time after initial photoperiodic stimulation. At this point, the gonads spontaneously regress, molt is initiated and hormonal changes occur concurrently. There is a rise in plasma prolactin coincident with the onset of molt (Dawson and Goldsmith, 1982, 1983) and plasma LH and FSH decrease to a minimum as photorefractoriness occurs (Goldsmith and Nicholls, 1984a). Such changes occur even though day lengths are longer than those that were originally stimulatory and often still increasing. In other species where absolute photorefractoriness occurs, for example, American crowned sparrows (*Zonotrichia sp.*), mallard (*Anas platyrhynchos*), canary (*Serinus canarius*), and American tree sparrows (*Spizella arborea*), changes associated with photorefractoriness happen when day length is decreasing but is still longer than those at the time of reproductive stimulation. Such a condition is termed *absolute* photorefractoriness because no increase in day length, even up to 24 h of light per day (in those species tested), will cause a change in the reproductive state once a photorefractory condition has been induced.

A condition similar to absolute photorefractoriness is *relative* photorefractoriness. The main difference is that once relative photorefractoriness has been induced and the gonads have regressed, a subsequent substantial increase in day length will once more initiate reproductive maturation, without the need for a sensitization, or photosensitive, stage (Robinson and Follett, 1982). For example, if Japanese quail (*Coturnix coturnix japonica*) experience day lengths of longer than 11.5 h, rapid gonadal development occurs. After about three months, and when (in the wild) day length decreases below 14.5 h, complete gonadal regression occurs, in a manner similar to absolute photorefractoriness (Nicholls et al., 1988). However, if the day length is subsequently artificially increased further, then a full return to reproductive maturity occurs. Further, if quail are maintained on any constant long day length, no form of photorefractoriness will be elicited unless they experience a decrease in day length, for example, from 23 h to 16 h (Nicholls et al., 1988). This suggests a shift in the duration of the critical day length in birds that are relatively photorefractory—a shift that appears to depend on the photoperiodic history of such birds (Robinson and Follett, 1982).

Evidence for a Shift in Critical Day Length in Response to a Changing Environment

Experimental evidence from laboratory studies implies that there is no change in critical day length in birds that exhibit absolute photorefractoriness, regardless of their photoperiodic history (Follett and Nicholls, 1984, 1985; Dawson, 1987). However, there is recent evidence of different populations of blue tits (*Parus caeruleus*) at the same latitude shifting their photoperiodic response to coincide with flushes of food (caterpillars) with which they feed their young (Caro et al., 2005). The two study populations occupy different oak habitats and are separated by only 25 km. Despite their close proximity, they show a one-month difference in onset of egg laying, even after controlling for altitude. This difference in timing of breeding appears to be mediated solely through the female reproductive system (Caro et al., 2006). In addition, there is latitudinal variation in the threshold, or critical day length for photoperiodic induction (Silverin et al., 1993).

Thoughts on the Evolution of the Photoperiodic Response

The fact that there is such an apparent difference between bird species that exhibit absolute photorefractoriness and those that exhibit relative photorefractoriness, and that breeding seasons vary so widely among bird species to suit their ecological needs has led to two schools of thought as to how mechanisms controlling the termination of the breeding season have evolved. The first school argues for the separate evolution of similar mechanisms in different bird groups (Farner, 1964; Farner et al., 1977). The idea behind such views is that although circadian systems became established very early in evolution, the photoperiodic systems that are regulated in a circadian manner evolved later and from multiple origins. The alternative view suggests that photorefractoriness has a single origin, but simple modifications have made the phenomenon appear to be very different in separate bird species (Follett and Nicholls, 1984, 1985). Whatever their evolutionary history, both types of photorefractoriness involve long-day–induced changes in physiology that will not occur without a functional thyroid gland (Follett and Nicholls, 1984, 1985). Because of their dependence upon the presence of thyroid hormones, it is thought that there are similarities between photorefractoriness in juvenile and adult starlings (Dawson et al., 1987; Williams et al., 1987). For the same reason, homology has also been suggested for the neural mechanisms controlling seasonal breeding in Welsh mountain ewes and starlings (Nicholls et al., 1988b).

THE DETECTION AND MEASUREMENT OF DAY LENGTH

Unlike other vertebrates, seasonally breeding mammals generally use a combination of their eyes and pineal gland for photoreception and transduction of the light signal (for review, see Foster et al., 1989). For example, if sexually mature hamsters are blinded while held under long days, then their gonads regress—even though they are still receiving the long-day light stimulus (Reiter, 1978). This indicates that ocular photoreceptors are responsible for light detection and transduction. Although it has been unclear as to the nature of the photoreceptors that fulfill this function (Foster, et al., 1991), it appears that melanopsin is most likely involved (Hattar et al., 2003).

Nonmammalian vertebrates transduce seasonal changes in photoperiod via photoreceptors that are extraretinal and extrapineal (Nelson and Zucker, 1981 and for review see Groos, 1982). The first evidence for avian extraretinal photoreceptors came from Benoit (1935a, 1935b), who demonstrated that simultaneous photostimulation of blinded and sighted ducks resulted in equal testicular growth rates. Furthermore, the testicular response could be abolished by covering the ducks' heads with black caps. Since then, further experiments involving ducks (Benoit and Ott, 1944), house sparrows (Menaker and Keatts, 1968), and tree sparrows (Wilson, 1991) have demonstrated convincingly that an extraretinal photoreceptor participates in the reproductive responses of birds, and that there is little or no retinal involvement in reproductive responses (Underwood and Menaker, 1970; see also Foster and Follett, 1985). Thus in birds, the detection of light for reproductive purposes occurs via an extraretinal, hypothalamic pathway (Menaker, 1971).

Despite numerous studies, it is still unclear precisely as to where the photoreceptors lie in the avian brain. There is some evidence that extraretinal photoreceptors may reside in the infundibular region of the hypothalamus (Oliver and Baylé, 1982) and in the parolfactory lobe of quail (Sicard et al., 1983). Additionally, Foster et al. (1985) demonstrated the involvement of a rhodopsin-like photopigment, which has a maximum spectral photosensitivity at 492 nm (Foster et al., 1985). More recent research involving immunostaining of opsin (a protein involved in signal transduction by light activation) indicates that the deep brain photoreceptors are located in the tuberal hypothalamus of the ring dove, duck, and quail (Saldanha et al., 2001; Silver et al., 1988). Thus, it appears that in birds, light must pass through the skull and into the hypothalamus for transduction of its signal to occur.

For a long-day signal to result in photostimulation, day length must be measured accurately. Birds, unlike mammals, do not seem to require the melatonin time signal (from the pineal gland) to control reproductive changes in response to day length (but see "Melatonin," below). The physiological mechanism(s) underlying measurement of day length are still unknown, but there are two main models for which there are some supporting experimental data. The first is the "hourglass" model, which hypothesizes an hourglass-like timer set in motion by the onset of dusk or dawn, and a photochemical that accumulates during either the light or the dark phase. If a sufficient amount of the photochemical accumulates, then a photoperiodic response is initiated. The fundamental property of the hourglass model is that it requires constant resetting by the light/dark cycle. Thus, it would not operate under conditions of constant light or constant dark. There is evidence that such a system operates in some insects (for examples, see Beck, 1968; Lees, 1973; Vaz Nunes and Veerman, 1984; 1986; Veerman et al., 1988). The second theory, which is certainly true for birds, was first proposed by Bünning in 1936 (see also Pittendrigh and Minis, 1964, 1971; Follett, 1973; Elliott, 1976). It assumes that there is a circadian rhythm of responsiveness to light; that is, for the first part of the cycle (subjective day) the organism is insensitive to light, whereas during the second part (subjective night) it becomes "photoinducible." Should light fall during the subjective night, then a long-day photoperiodic response is initiated. Evidence that this "external coincidence model" applies to avian photoperiodicity was first supplied by Hamner's (1963, 1964) night-interruption experiments on the house finch *Carpodacus mexicanus*, which has been reinforced by experiments on other bird species (for examples, see Follett and Sharp, 1969; Follett et al., 1974, 1992; Turek, 1974). Another model of how photoperiodic responses are initiated by day length is termed the "internal coincidence model" (Pittendrigh and Minis, 1964). This model assumes that photoperiod entrains two circadian oscillators, one entrained by dusk and the other by dawn, with photoperiodic time measurement a function of the phase relationship between the two clocks.

Thus, when the phase of the two oscillators coincides, photoinduction occurs. However, there is little experimental evidence to help distinguish between the latter two hypotheses, so it is impossible to conclude which is correct. Evidence against that external coincidence model comes from Bentley et al. (1998a), in an experiment where light intensity, rather than length of photoperiod, was manipulated. Few investigations have incorporated manipulation of light intensity rather than day length to affect reproductive responses since Bissonnette's study in 1931. It is difficult to draw definitive conclusions from most early experiments on this subject, as mixtures of natural and artificial light were used in the same days of treatment (Bissonnette, 1931); others omitted to control for different wavelengths of light from sources of differing intensities (e.g., Bissonnette and Wadlund, 1933). (The importance of wavelength in the avian photoperiodic response has been established for some time [Ringoen, 1942; Benoit and Ott, 1944].) There are even reports of both low and high light intensities causing convulsions and death in the African weaver finch (Rollo and Domm, 1943). So far, the most comprehensive experiment of this type to have been carried out was by Bartholomew (1949), in which sparrows (*Passer domesticus*) were subjected to different light intensities under the same photoperiod. His study concluded that light intensity can certainly modify the reproductive response of sparrows to day length, but that there is a minimum intensity below which no photoperiodic response can be evoked. In addition, there is an upper intensity threshold above which there is no increase in rate of response, and intensity cannot be substituted for day length per se.

Work carried out on mammals indicates that there are also light intensity effects on the mammalian endocrine system, such as in ferrets (Marshall and Bowden, 1934) and pigs (Griffith and Minton, 1992). In the experiment by Bentley et al. (1998a), photosensitive starlings transferred from short days to long days of different light intensities underwent graded reproductive responses according to the light intensity they experienced. Testes size in the group in the lowest intensity (3 lux) increased faster than that in controls on short days of normal intensity, but they did not become photorefractory. Testes size increased in the groups on 13, 45, and 108 lux and subsequently became photorefractory. However, the 13- and 45-lux groups required more time to become photorefractory than did the 108-lux group. The responses observed were similar to those seen in starlings exposed to different photoperiods (e.g., 11 h light:13 h dark [11L:13D], 13L:11D, 16L:8D, 18L:6D), even though all were on the same 18L:6D photoperiod. Initially, the results appear to challenge the external coincidence model for photoperiodic time measurement, but consideration of the phase response curve of the circadian rhythm of photoinducibility in starlings and the way in which it might be affected by low light intensities refute this challenge and reinforce the external coincidence model (for discussion, see Bentley et al., 1998a). See chapter 18 for further details on recent advances in our understanding of the molecular and cellular intricacies of the avian circadian system.

ENDOCRINOLOGICAL ASPECTS OF THE AVIAN PHOTOPERIODIC RESPONSE

Gonadotropin-Releasing Hormone

In the 1980s, it was discovered that the hypothalamus of the domestic fowl had two distinct GnRH molecules, designated chicken GnRH-I and GnRH-II (cGnRH-I and -II) that differed from mammalian GnRH by one and three amino acid substitutions, respectively (King and Millar, 1982a, 1982b; Miyamoto et al., 1984). The two chicken GnRHs are also found in the hypothalamus of wild species, European starling *Sturnus vulgaris* and song sparrow *Melospiza melodia*, with c-GnRH-I predominating (Sherwood et al., 1988). The cDNA encoding for the pre-pro-GnRH molecule has been sequenced for the domestic fowl (Dunn et al., 1993); cDNA encoding passerine GnRH has also been sequenced (Ubuka and Bentley, 2009; Ubuka et al., 2009). In chickens, cGnRH-I and -II had different potencies in releasing LH, with GnRH-II being more potent (Sharp et al., 1987). Note that as yet there is no good evidence for the existence of kisspeptin, a neuropeptide considered to increase activation of the GnRH system in mammals, or its receptor in any avian species. Nor is there good evidence for the action of kisspeptin on the avian GnRH system.

In general, cGnRH-I is located in the preoptic region of the brain, with projections to the ME allowing for regulation of pituitary gonadotropin release. The oculomotor complex of the midbrain is the location of the cGnRH-II neurons, which are typically less polar in shape than the cGnRH-I neurons and, as a result, have few processes extending from the cell bodies (for distribution, see Millam et al., 1993, Millam et al., 1998; Teruyama and Beck, 2000). Despite a long-held acceptance of the notion that cGnRH-II neurons are not hypophysiotropic, there have been some reports in quail that cGnRH-II–immunoreactive (-ir) fibers are present in the preoptic area (POA), the lateral septum (Millam et al., 1993), and the ME (van Gils et al., 1993; Millam et al., 1998; D'Hondt et al., 2001; Clerens et al., 2003). However, there has only been one published report of cGnRH-II in the ME of songbirds (Stevenson and MacDougall-Shackleton, 2005), while others find no evidence of GnRH-II in the same area (Meddle et al., 2006), and experiments in chickens suggest that the ME is a site of release of cGnRH-I, but not cGnRH-II, into the hypophysial portal vasculature (Sharp et al., 1990). Thus, the distribution and function(s) of cGnRH-II have remained somewhat enigmatic, although central administration of cGnRH-II, but not cGnRH-I, enhances copulation solicitation in female white-crowned sparrows (Maney et al., 1997). Thus, cGnRH-II might act in a neurotransmitter role to influence reproductive behaviors, as has also been postulated for some mammalian species (Temple et al., 2003; Kauffman and Rissman, 2004).

Gonadotropin-Inhibitory Hormone

Gonadotropin-inhibitory hormone (GnIH) was initially located in the quail hypothalamus with projections to the ME. Thus, the correct neuroanatomical infrastructure was present for this dodecapeptide to regulate pituitary hormone release. In vitro, it decreased gonadotropin release from cultured quail anterior pituitary in a dose-dependent manner, with no effect upon prolactin (Tsutsui et al., 2000). This novel RFamide peptide was therefore named gonadotropin-inhibitory hormone (Tsutsui et al., 2000).

Using immunocytochemistry, clusters of distinct GnIH-ir neurons were found in the paraventricular nucleus (PVN) in the hypothalamus. In addition to the PVN, some scattered small cells were immunoreactive in the septal area (Tsutsui et al., 2000; Ubuka et al., 2003; Ukena et al., 2003). In contrast to the highly localized clusters of cell bodies, GnIH-ir nerve fibers were widely distributed in the diencephalic and mesencephalic regions. Dense networks of immunoreactive fibers were found in the ventral paleostriatum, septal area, POA, hypothalamus, and optic tectum. The most prominent fibers were seen in the ME of the hypothalamus, and in the dorsal motor nucleus of the vagus in the medulla oblongata.

Addition of a physiological dose (10–7 M) of GnIH to short-term (120 min) cultures of diced pituitary glands from adult cockerels suppressed common α and

FSH β subunit mRNAs, with no effect on LH β subunit mRNA. The suppressive effect of GnIH on gonadotropin mRNA was associated with an inhibition of both LH and FSH release in the adult chicken (Ciccone et al., 2004). When administered intraperitoneally to quail in vivo via osmotic pumps, GnIH significantly reduced gonadotropin common α and LH β subunit mRNAs, as well as reducing plasma LH (Ubuka et al., 2006).

Following its discovery, GnIH-ir peptide was localized in the brains of seasonally breeding songbird species (Bentley et al., 2003; Osugi et al., 2004). Dense populations of GnIH-ir neurons were found in the PVN of song sparrows (*Melospiza melodia*), house sparrows (*Passer domesticus*), white-winged crossbills (*Loxia leucoptera*), pine siskins (*Carduelis pinus*), redpolls (*Carduelis flammea*), rufous-collared sparrows (*Zonotrichla capensis*), and Gambel's white-crowned sparrows (*Zonotrichia leucophrys gambelii*). The PVN was the only location where immunoreactive neurons were located, regardless of sex or species (data published only for three species in this list: Bentley et al., 2003; Osugi et al., 2004). Thus, the presence of GnIH in the PVN appears to be a conserved property among several families and at least two orders of birds (Galliformes and Passeriformes). In addition to the dense population of GnIH-ir neurons within the hypothalamus of all the avian species studied so far, there were extensive networks of branching beaded fibers emanating from those cells, presumably transporting GnIH. Some of the fibers extended to terminals in the ME, consistent with a role for GnIH in pituitary gonadotropin regulation, as in quail. In house sparrows, song sparrows, and Gambel's white-crowned sparrows, other fibers extended through the brain caudally at least as far as the brainstem and possibly into the spinal cord, consistent with multiple regulatory roles for GnIH. In situ hybridization further revealed the cellular localization of GnIH mRNA solely in the PVN of quail and sparrow hypothalami (Ukena et al., 2003; Osugi et al., 2004). Thus, only the PVN expresses GnIH, and in birds, the immunoreactive peptide found in fibers in multiple brain areas appears to originate solely from the PVN (Ukena et al., 2003; Osugi et al., 2004). So far, no colchicine studies have been performed in birds to reaffirm the finding that the PVN is the sole source of GnIH, but the in situ hybridization findings are quite convincing in this regard.

As already discussed, the cell bodies for each population of neurons containing cGnRH-I, GnIH, or cGnRH-II are in discrete locations. The POA contains cGnRH-I–ir neurons, the PVN contains GnIH-ir neurons, and the midbrain contains cGnRH-II neurons (Millam et al., 1993). There is putative contact of GnIH fibers to the cGnRH-I neurons and fibers in the POA in songbirds. The GnIH-ir fibers are also in putative contact with cGnRH-II neurons in the midbrain (Bentley et al., 2003). Further, confocal microscopy indicates that the cGnRH-I and GnIH proteins are located within the same 0.2-μm optical plane and possibly in contact with one another, although electron microscopy will be necessary to determine the presence of a functional synapse. Thus, contact appears to occur between GnIH fibers and GnRH-I and -II neurons, and between GnIH and GnRH-I fibers in the ME (Bentley et al., 2003). These data imply a functional interaction between GnIH and the GnRH system. Similar data have been published for mammals (Bentley et al., 2006a; Kriegsfeld et al., 2006). Intracerebroventricular (ICV) infusion of GnIH inhibits LH release and copulation solicitation in estradiol-primed, photostimulated female Gambel's white-crowned sparrows, and this effect upon behavior is thought to be mediated by GnIH binding to GnRH-II neurons (Bentley et al., 2006b). Administration of ICV GnIH also inhibits LH release in rodents (Kriegsfeld et al, 2006; Johnson et al., 2007). More recently, Ubuka et al. (2008) demonstrated that GnRH-I and -II neurons in European starlings express mRNA for GnIH receptor.

Taken together, there is potential for GnIH to influence the GnRH system at the neuron and fiber terminal levels. Furthermore, when song sparrows were subjected to a simulated annual cycle of changing photoperiod, GnIH-ir neuron area was significantly greater at the onset of photorefractoriness when compared to photo-sensitive or photostimulated birds. Thus, there is potential for dynamic interactions of GnIH and GnRH peptides in different reproductive states and neuroanatomical locations. Seasonal modulation of the GnIH system was also seen in a study on the regulation of luteinizing hormone (LH) in rufous-winged sparrows (*Aimophila carpalis*; Small et al., 2008). This species is a resident of the Sonoran desert that breeds after irregular summer rains. Although the testes develop in March due to increasing photoperiod and regress in September due to decreasing photoperiod, LH does not consistently increase in the spring as in other photoperiodic birds but rather is correlated with rainfall in the summer monsoon season. Compared to pre-monsoon birds, birds caught during the monsoon season had larger GnRH-I cell bodies as well as fewer, less densely labeled GnIH cell bodies. Further, there were fewer GnIH fibers in the POA of birds caught during the monsoon season, when LH was high. Thus, GnIH could directly inhibit GnRH neuronal activity prior to the monsoon season.

Avian gonads express GnIH and its receptor (Bentley et al., 2008), and it is thought that GnIH might modulate gonadal steroid release. It is also possible that gonadal GnIH is involved in seasonal changes in gonadal activity and in the regulation of follicular hierarchy during the breeding season. Overall, GnIH appears to be a modulator of gonadotropin release in vivo as well as in vitro, but its precise role in avian seasonality is as yet unclear. There are certainly temporal changes in GnIH content of the brain (Bentley et al., 2003). In addition, there is a seasonal component to the response of the GnIH system to stress (Calisi et al., 2008), with baseline GnIH in house sparrows being lower at the start of the breeding season and the elevation above baseline greater at this time of year than at the end of the breeding season. It is clear that further study is needed to elucidate the seasonal dynamics of GnIH synthesis and release and its relation to seasonal changes in GnRH and gonadotropins in songbirds.

Thyroid Hormones

The state of photorefractoriness can also be dissipated by the removal of circulating thyroid hormones, that is, thyroidectomy. The initiation and maintenance of photorefractoriness are dependent upon the presence of thyroxine (T_4) (Wieselthier and van Tienhoven, 1972; Goldsmith and Nicholls, 1984b). Termination of the breeding condition by thyroidectomy has also been observed in other vertebrates, for example, tree sparrows (Wilson and Reinert, 1993, 1995a, 1995b), sheep (Moenter et al., 1991; Dahl et al., 1994; Parkinson and Follett, 1994; Parkinson et al., 1995), and red deer (Shi and Barrell, 1992).

Although the presence of T_4 is required, and plasma T_4 levels rise markedly upon transfer from short to long days in starlings (Dawson, 1984) and in quail (Sharp and Klandorf, 1981), it is likely that thyroid hormones are not the cause of photorefractoriness. It appears that T_4 is a permissive factor allowing the mechanisms that cause the reproductive system to "switch off" under long days to function, and may be involved in the perception of day length (Dawson, 1989a, 1989b; Bentley et al., 1997a, 1997b). Similar suggestions have been made for the termination of breeding in sheep (Dahl et al., 1994). It has not been clear whether the long-day–induced rise in circulating T_4 is a necessary precursor of the photorefractory response, or if this rise is associated with other long-day responses, such as increased metabolic rate. Importantly, thyroidectomy does not affect the daily or circadian pattern of circulating melatonin concentrations (Dawson and King, 1994).

Some potentially exciting findings on the role of thyroid hormones in the avian photoperiodic response have been published in recent years. The first of these findings was that long day lengths induce the gene for type 2 iodothyronine deiodinase (*DIO2*), an enzyme that activates thyroid hormone (Yoshimura et al., 2003). Thus, long days increase conversion of T_4 into its bioactive form, triiodothyronine (T_3). Under long-day conditions, the hypothalamic content of T_3 was about 10-fold higher than under short-day conditions. In addition, the ICV infusion of T_3 induced testicular growth in quail held under nonstimulatory short days. The second finding was that there is high expression of *DIO3* (type 3 deiodinase, an enzyme that inactivates thyroid hormone via conversion of T_4 and T_3 to reverse T_3 and T_2, respectively) and low expression of *DIO2* under short-day conditions. Conversely, there was low expression of *DIO3* and high expression of *DIO2* under long-day conditions, indicating increased activation and decreased inactivation of thyroid hormone in response to long days. The implication of these findings is that thyroid hormone gene switching is one of the earliest events detected in the photoperiodic cascade and that it must occur at or before 16 h of a long day, as a single 16-h day can induce LH secretion (Meddle and Follett, 1995, 1997). These opposite effects on gene activation are thought to amplify the localized action of thyroid hormones and lead to neuroendocrine changes that cause GnRH secretion a few hours later. Exactly how signals from this local regulation of thyroid hormones are processed and transferred to the GnRH system is not yet known, but there is possible involvement of mechanical actions of glia in the ME (Yamamura et al., 2004). More recently, a wave of gene expression was identified at 14 h of a single long day, and included increased thyrotropin (TSH) β-subunit expression in the pars tuberalis (Nakao et al., 2008). Central administration of TSH to short-day quail stimulated gonadal growth and expression of *Dio2*; thus, there appears to be a role for TSH in the regulation of the avian photoperiodic response (chapter 18). These recent findings are quite stunning and have the potential to revolutionize our understanding of verte-brate seasonal breeding. It is not yet known how the photoperiod signal is relayed to the hypothalamus to regulate *Dio2/Dio3*, but there is some evidence that the avian pars tuberalis expresses melatonin receptors (Bentley, unpublished data). In theory, the daily cycle of melatonin in the circulation could provide day-length informa-tion to the hypothalamus via effects upon TSH release from the pars tuberalis. However, care needs to be taken when interpreting the effects of TSH on gonadal growth because of the possibility of cross-reactivity of TSH with LH receptors. In addition, TSH is chronically elevated in thyroidectomized animals, but if birds are thyroidectomized soon after the transfer to long days, they subsequently become photorefractory and exhibit gonadal regression in the normal time course (Wilson and Reinert, 1995a, 1995b). Photorefractory starlings that are held on long days and thyroidectomized do not exhibit gonadal growth until four to eight weeks after thy-roidectomy (and thus high TSH). As few as seven long days are sufficient to induce the onset of photorefractoriness some weeks later in thyroid-intact starlings, even when those birds are transferred back to short days after the week-long photostimu-lation and thus presumably experience short-day expression of *Dio2* (Dawson et al., 1985). Finally, short-day doses of T_4 to thyroidectomized starlings held on long days permit the onset of photorefractoriness in the same time course as thyroid-intact birds (Bentley et al., 1997a; see figure 17.1). Thus, the early photoperiodic events described by Yoshimura and his team may well be the initial steps in the pathway that regulates seasonality in birds, but several thyroid-dependent processes that take several weeks to occur once initiated likely remain to be identified.

Melatonin

In diurnal bird species melatonin is secreted at night (i.e., light inhibits release) and injections of melatonin can reverse activity rhythms. Transplants of pineal glands from individuals entrained to one rhythm can induce a similar rhythm in the recipi-ent, suggesting some autonomy. Pineals cultured in vitro can retain rhythmicity and even entrain to changing photoperiod (Cassone and Menaker, 1984; Brandstätter et al., 2000, 2003). Note, however, that the retina can also produce melatonin (up to 30% of circulating melatonin in quail); consequently, the eyes may also be important for melatonin profiles in birds (Underwood et al., 1984). Even in blinded and pinea-lectomized quail, there is a residual melatonin rhythm from unidentified sources (Underwood et al., 1984)—perhaps the gut? As for other vertebrates studied, the avian gut can produce and bind melatonin (Lee and Pang, 1993; Bubenik, 2002). Thus, the gut might be a rich source of extrapineal melatonin, but it is unclear what role, if any, gut melatonin plays in circadian and reproductive rhythmicity. It is pos-sible that gut melatonin simply plays an autocrine or paracrine role.

Unlike mammals, there is little evidence of a significant direct role of the pineal in the regulation of gonadal function in birds. However, it is possible that melatonin is, to some degree, involved in regulation of reproduction. Pinealectomy of male

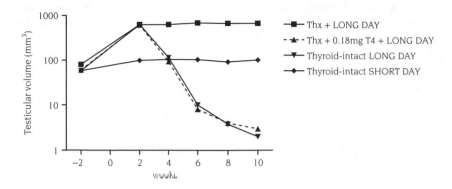

FIGURE 17.1.

Testicular responses of European starlings to different photoperiod and thyroid hormone treatments. Thx + LONG DAY = thyroidectomized birds provided with vehicle solution and exposed to 18L:6D; Thx + 0.18mg T_4 + LONG DAY = thyroidectomized birds exposed to 18L:6D and provided with a dose of T4 in their drinking water that resulted in a plasma T_4 concentration that was equivalent to that seen on short days (~7.5 nmol/liter); Thyroid-intact LONG DAY = thyroid-intact birds provided with vehicle and exposed to 18L:6D (these birds had ~14 nmol/liter circulating T_4 as a result of long-day exposure); Thyroid-intact SHORT DAY = thyroid-intact birds provided with vehicle and exposed to 8L:16D (these birds had identical plasma T_4 concentrations to the Thx + 0.18mg T4 + LONG DAY group). Note that the Thx + 0.18mg T4 + LONG DAY group underwent gonadal regression in the same time course as the Thyroid-intact LONG DAY group, despite having much lower (short-day) plasma T_4 concentrations. Thus, the long-day–induced rise in plasma T_4 does not induce photorefractoriness in this species. Redrawn from Bentley et al. (1997).

Baya weavers resulted in increased rate of photoperiodically induced testicular growth (Balasubramanian and Saxena, 1973), and when mature males were transferred to short days, the pinealectomized group showed partial testicular regression in comparison with complete regression in the sham-operated group. The authors interpreted these results as effects of removal of an inhibitory effect of melatonin on the hypothalamo-hypophysial-gonadal axis. However, because the pineal body is a component of the avian circadian system (Cassone and Menaker, 1984), it is possible that a shift in phase angle between a daily cycle in photosensitivity and the daily photoperiod could explain the results (e.g. Gwinner et al., 1981).

More recently, there has been a renaissance of interest in the possibility that melatonin is involved in seasonal processes in birds, including regulation of the reproductive axis. In a study involving the use of melatonin antiserum, Ohta et al. (1989) concluded that melatonin is involved in at least the initial stages of photoperiodism in quail (i.e., early in the dark phase), and that the timing of suppression of plasma melatonin is critical to gonadal development. In a contradictory study, injecting melatonin into quail exposed to long day lengths (and thereby simulating a short-day melatonin signal) did not inhibit the long-day–induced gonadal growth (Juss et al., 1993). The observation that one study took the approach of shortening the melatonin signal during short days (Ohta et al., 1989) and another lengthened the melatonin signal during long days (Juss et al., 1993) might prove fundamental to the differences in results, especially if it is true that the timing of the melatonin signal is critical to the ensuing effect. It is possible that studies on melatonin and the reproductive axis in birds have been hampered by the modes and timing administration of melatonin, making interpretation of the findings difficult at best. At any rate, it seems obvious that the role of melatonin in reproduction in birds is somewhat more complex than its role in mammalian reproduction.

Whatever its actions are, it is likely that melatonin affects a suite of physiological systems in addition to the reproductive system, such as the song control and immune systems (Bentley et al., 1998b, 1999; Cassone et al., 2008). Guyomarc'h et al. (2001) suggest that melatonin supplementation might affect food intake and fat deposition and thereby be responsible for the slight inhibition of sexual development that they observed in European quail (*Coturnix coturnix*).

Perhaps the most convincing, and straightforward, demonstration of effects of melatonin upon the avian reproductive system stems from two recent studies. One study indicates that in chickens the effects of melatonin on plasma LH can be described very simply: Rozenboim et al. (2002) demonstrated very large reductions in plasma LH (up to 70%) after injections of relatively high doses of melatonin into castrated white leghorn roosters. Additionally, melatonin administration inhibited plasma LH in a time- and dose-dependent manner, and the effects persisted for as long as melatonin was administered. The other study addressed dose-dependent regulation of gonadotropin inhibitory hormone by melatonin: Ubuka et al. (2005) investigated the action of melatonin on GnIH expression in the quail brain. Pinealectomy combined with orbital enucleation (Px+Ex) decreased the expression of GnIH precursor mRNA and the mature peptide GnIH in the diencephalon, including the PVN and ME. Melatonin administration to Px+Ex birds caused a dose-dependent increase in expression of GnIH precursor mRNA and production of mature peptide. The expression of GnIH was photoperiodically controlled and increased under short photoperiods (Ubuka et al., 2005), when the duration of melatonin secretion increases (Cockrem and Follett, 1985; Kumar and Follett, 1993). Finally, Mel$_{1c}$, a melatonin receptor subtype, was expressed in GnIH-ir neurons in the PVN (Ubuka et al., 2005). Thus, melatonin appears to act directly on GnIH neurons via its receptor to induce GnIH expression. The potential therefore exists for photoperiodic regulation of the reproductive axis via action of the changing melatonin signal on the GnIH system.

It is possible that the avian hypothalamus can synthesize melatonin *de novo* (Kang et al., 2007). Melatonin was detected in turkey hypothalamus along with tryptophan hydroxylase 1 and 5-HT *N*-acetyltransferase, key enzymes in melatonin biosynthesis. If this finding is applicable to birds in general, then it could explain the lack of effect of pinealectomy on the avian reproductive system. Clearly, further work is needed to identify this intriguing possibility.

In sum, although many bird species are photoperiodic, a dogma has existed that birds do not use seasonal changes in melatonin secretion to time their reproductive effort, and a role for melatonin in birds has remained enigmatic (Wilson, 1991; Juss et al., 1993). Despite the accepted dogma, there is strong evidence that melatonin is involved in regulation of several seasonal processes, including gonadal activity and gonadotropin secretion (Ohta et al., 1989; Bentley et al., 1999; Bentley and Ball, 2000; Bentley, 2001; Guyomarc'h et al., 2001; Rozenboim et al., 2002).

Prolactin

For most species, there is a peak in the level of plasma prolactin coinciding with the onset of photorefractoriness (Dawson and Goldsmith, 1982, 1983; Ebling et al., 1982; Goldsmith, 1983). Thus, prolactin was long thought to be a strong candidate for the inhibition of reproduction at the end of the breeding season. However, Goldsmith (1985) found that the administration of exogenous prolactin did not on its own cause the onset of photorefractoriness, although this particular experiment was carried out using heterologous prolactin. Prolactin itself is not responsible for gonadal regression. If photosensitive starlings are kept under an 11L:13D photoperiod, slow but complete reproductive maturity ensues (Hamner, 1971; Goldsmith and Nicholls, 1984c). With 11 h of light (11L), photorefractoriness does not occur because this is below the critical day length for this species. When the birds are subsequently transferred back to short days (6–8 h of light), the gonads regress in size, but there is no increase in plasma prolactin levels (Goldsmith and Nicholls, 1984c). If such 11L birds are put in to long days (18L) instead of being returned to short days, photorefractoriness occurs, along with the associated rise in plasma prolactin. Thus, prolactin is not responsible for gonadal regression in short days but is associated somehow (perhaps only temporally) with regression during photorefractoriness. The timing of high levels of circulating prolactin is also closely linked to plumage molt, and, where it occurs, premigratory fattening and migration (Meier, 1972; Meier and MacGregor, 1972; Dawson and Goldsmith, 1983).

Prolactin is not responsible for photorefractoriness (Dawson and Sharp, 1998). European starlings were actively immunized against vasoactive intestinal polypeptide (VIP), the prolactin-releasing hormone in birds, or against prolactin, during a photoinduced breeding cycle. VIP-immunized birds became photorefractory but the rate of gonadal regression was markedly slowed, and the photoinduced increase in prolactin was completely suppressed in 50% of these birds. Molt was prevented in those birds in which prolactin release was completely suppressed. In those VIP-immunized birds in which the photoinduced increase in prolactin was inhibited but not completely suppressed, gonadal regression was delayed, but molt occurred normally. The same was true for prolactin-immunized birds. Thus, in European starlings the associated increase in prolactin accelerates gonadal regression during the onset of photorefractoriness but does not itself cause photorefractoriness. In addition, the increase in prolactin associated with photorefractoriness is required for the induction of the postnuptial molt (Dawson and Sharp, 1998).

Gonadal Steroid Hormones

It was initially proposed that the termination of the avian breeding season involved a seasonal change in hypothalamic sensitivity to gonadal steroid feedback (Cusick and Wilson, 1972; Sharp and Moss, 1977; Stokkan and Sharp, 1980a, 1980b). The idea was that as gonadal steroid feedback increases under long days, the hypothalamus becomes increasingly sensitive to this feedback and "shuts down" completely, resulting in photorefractoriness. This was shown not to be the case by Dawson and Goldsmith (1984). Gonadectomized starlings became photorefractory in the same way, and within the same time course as intact birds. Furthermore, Wilson (1985) has demonstrated that in tree sparrows, hypothalamic sensitivity to inhibition by gonadal steroids decreases, rather than increases, following exposure to long days. Thus, although gonadal steroids might play a role in fine-tuning gonadotropin release, especially early in the breeding season, there is no good evidence that they are involved in the termination of the breeding season.

CONCLUSIONS

Since Rowan's initial findings on photoperiodism in juncos, we have developed an in-depth understanding of photoperiodism in birds. However, our understanding has been limited to knowledge of the environmental factors and the hormones involved in the timing of avian seasonal breeding. We still do not know precisely which brain cells possess photoreceptors for the photoperiodic response, nor do we know the molecular mechanisms involved. Despite its precise circadian cycle, melatonin has been ruled out for many years as a participant in timing of the annual reproductive cycle, and our knowledge of thyroid hormone control of seasonal reproduction has been limited to knowing simply that it needs to be present for the correct gonadal response to changing day length to occur. Fortunately, some major advances have occurred in very recent years, with Yoshimura's work on thyroid hormone involvement in early photoperiodic responses most likely being a major component of avian photoperiodism (see chapter 18). There is also the potential that melatonin can still emerge as another key component of the photoperiodic timing mechanism, especially if it becomes clear that the avian hypothalamus can synthesize melatonin de novo in a rhythmic fashion.

References

Balasubramanian KS and Saxena RN. (1973). Effect of pinealectomy and photoperiodism in the reproduction of Indian weaver birds, *Ploceus phillipinus*. J Exp Zool 185: 333–340.

Bartholomew GA Jr. (1949). The effect of light intensity and day length on reproduction in the English sparrow. Bull Mus Comp Zool 101: 433–476.

Beck SD. (1968). *Insect Photoperiodism*. New York: Academic Press.

Beebe K, Bentley GE, and Hau M. (2005). A tropical rainforest bird lacks photorefractoriness in the wild, despite high photosensitivity. Funct Ecol 19: 505–512.

Benoit J. (1935a). Le rôle des yeux dans l'action stimulante de la lumière sur le développement testiculaire chez le canard. CR Soc Biol (Paris) 118: 669–671.

Benoit J. (1935b). Stimulation par la lumière artificielle du développement testiculaire chez les canards aveuglés par section du nerf optique. CR Soc Biol (Paris) 120: 133–136.

Benoit J and Ott L. (1944). External and internal factors in sexual activity. Effect of irradiation with different wavelengths on the mechanisms of photostimulation of the hypophysis and on testicular growth in the immature duck. Yale J Biol Med 17: 27–46.

Bentley GE. (2001). Unraveling the enigma: The role of melatonin in seasonal processes in birds. Micros Res Tech 53: 63–71.

Bentley GE and Ball GF. (2000). Photoperiod-dependent and -independent regulation of melatonin receptors in the forebrain of songbirds. J Neuroendocrinol 12: 745–752.

Bentley GE, Demas GE, Nelson RJ, and Ball GF. (1998b). Melatonin, immunity and cost of reproductive state in male European starlings. Proc R Soc Biol Sci 265: 1191–1195.

Bentley GE, Goldsmith AR, Dawson A, Briggs C, and Pemberton M. (1998a). Decreased light intensity alters the perception of day length by male European starlings (*Sturnus vulgaris*). J Biol Rhythms 13: 148–158.

Bentley GE, Goldsmith AR, Dawson A, Glennie L, Talbot RT, and Sharp PJ. (1997a). Photorefractoriness in European starlings (*Sturnus vulgaris*) is not dependent upon the long-day-induced rise in plasma thyroxine. Gen Comp Endocrinol 107: 428–438.

Bentley GE, Goldsmith AR, Juss TS, and Dawson A. (1997b). The effects of nerve growth factor and anti-nerve growth factor antibody on the neuroendocrine reproductive system in the European starling *Sturnus vulgaris*. J Comp Physiol A 181: 133–141.

Bentley GE, Jensen JP, Kaur GJ, Wacker DW, Tsutsui K, and Wingfield JC. (2006b). Rapid inhibition of female sexual behavior by gonadotropin-inhibitory hormone (GnIH). Horm Behav 49: 550–555.

Bentley GE, Kriegsfeld LJ, Osugi T, Ukena K, O'Brien S, Perfito N, Moore IT, Wingfield JC, and Tsutsui K. (2006a). Interactions of gonadotropin-releasing hormone (GnRH) and gonadotropin-inhibitory hormone (GnIH) and in birds and mammals. J Exp Zool 305A: 807–814.

Bentley GE, Perfito N, Ukena K, Tsutsui K, and Wingfield JC. (2003). Gonadotropin-inhibitory peptide in song sparrows (*Melospiza melodia*) in different reproductive conditions, and in house sparrows (*Passer domesticus*) relative to chicken-gonadotropin-releasing hormone. J Neuroendocrinol 15: 794–802.

Bentley GE, Spar BD, MacDougall-Shackleton SA, Hahn TP, and Ball GF. (2000). Photoperiodic regulation of the reproductive axis in male zebra finches, *Taeniopygia guttata*. Gen Comp Endocrinol 117: 449–455.

Bentley GE, Ubuka T, McGuire NL, Chowdhury VS, Morita Y, Yano T, Hasunuma I, Binns M, Wingfield JC, and Tsutsui K. (2008). Gonadotropin-inhibitory hormone and its receptor in the avian reproductive system. Gen Comp Endocrinol 156: 34–43.

Bentley GE, Van't Hof TJ, and Ball GF. (1999). Seasonal neuroplasticity in the songbird telencephalon: A role for melatonin. Proc Natl Acad Sci USA 13: 4674–4679.

Bissonnette TH. (1931). Studies on the sexual cycle in birds. V. Effects of light of different intensities upon the testis activity of the European starling (*Sturnus vulgaris*). Physiol Zool 4: 542–574.

Bissonnette TH and Wadlund APR. (1933). Testis activity in *Sturnus vulgaris* in relation to artificial sunlight and to electric lights of equal heat and luminous intensities. Bird Banding 4: 8–18.

Boulakoud MS and Goldsmith AR. (1991). Thyroxine treatment induces changes in hypothalamic gonadotrophin-releasing hormone characteristic of photorefractoriness in starlings (*Sturnus vulgaris*). Gen Comp Endocrinol 82: 78–85.

Boulakoud MS and Goldsmith AR. (1994). Acquisition of photosensitivity in castrated male starlings (*Sturnus vulgaris*) under short daily photoperiods. J Reprod Fert 100: 77–79.

Boulakoud MS and Goldsmith AR. (1995). The effect of duration of exposure to short days on the gonadal response to long days in male starlings (*Sturnus vulgaris*). J Reprod Fert 104: 215–217.

Brandstätter R. (2003). Encoding time of day and time of year by the avian circadian system. J Neuroendocrinol 15: 398–404.

Brandstätter R, Kumar V, Abraham U, and Gwinner E. (2000). Photoperiodic information acquired and stored in vivo is retained in vitro by a circadian oscillator, the avian pineal gland. Proc Natl Acad Sci USA 97: 12324–12328.

Bubenik GA. (2002). Gastrointestinal melatonin: Localization, function, and clinical relevance. Dig Dis Sci 47: 2336–2348.

Bünning E. (1936). Die endogene Tagesrhythmik als Grundlage der photoperiodischen Reaktion. Ber Dt Bot Ges 54: 590–607.

Calisi RM, Rizzo NO, and Bentley GE. (2008). Seasonal differences in hypothalamic EGR-1 and GnIH expression following capture-handling stress in House sparrows (*Passer domesticus*). Gen Comp Endocrinol.157: 283–287.

Caro SP, Balthazart J, Thomas DW, Lacroix A, Chastel O, and Lambrechts MM. (2005). Endocrine correlates of the breeding asynchrony between two Corsican populations of blue tits (*Parus caeruleus*). Gen Comp Endocrinol 140: 52–60.

Caro SP, Lambrechts MM, Chastel O, Sharp PJ, Thomas DW, and Balthazart J. (2006). Simultaneous pituitary-gonadal recrudescence in two Corsican populations of male blue tits with asynchronous breeding dates. Horm Behav 50: 347–360.

Cassone VM, Bartell PA, Earnest BJ, and Kumar V. (2008). Duration of melatonin regulates seasonal changes in song control nuclei of the house sparrow, *Passer domesticus*: Independence from gonads and circadian entrainment. J Biol Rhythms 23: 49–58.

Cassone VM and Menaker M. (1984). Is the avian circadian system a neuroendocrine loop? J Exp Zool 232: 539–549.

Ciccone NA, Dunn IC, Boswell T, Tsutsui K, Ubuka T, Ukena K, and Sharp PJ. (2004). Gonadotrophin inhibitory hormone depresses gonadotrophin α and follicle-stimulating hormone β subunit expression in the pituitary of the domestic chicken. J Neuroendocrinol 16: 999–1006.

Clerens S, D'Hondt E, Berghman LR, Vandesande F, and Arckens L. (2003). Identification of cGnRH-II in the median eminence of Japanese quail (*Coturnix coturnix japonica*). Gen Comp Endocrinol 131: 48–56.

Cockrem JF and Follett BK. (1985). Circadian rhythm of melatonin in the pineal gland of the Japanese quail (*Coturnix coturnix japonica*). J Endocrinol 107: 317–324.

Cusick EK and Wilson FE. (1972). On the control of spontaneous testicular regression in tree sparrows (*Spizella arborea*). Gen Comp Endocrinol 19: 441–456.

Dahl GE, Evans NP, Moenter SM, and Karsch FJ. (1994). The thyroid gland is required for reproductive responses to photoperiod in the ewe. Endocrinology 135: 10–15.

Dawson A. (1983). Plasma gonadal steroid levels in wild starlings (*Sturnus vulgaris*) during the annual cycle and in relation to the stages of breeding. Gen Comp Endocrinol 49: 286–294.

Dawson A. (1984). Changes in plasma thyroxine concentration in male and female starlings (*Sturnus vulgaris*) during a photo-induced gonadal cycle. Gen Comp Endocrinol 56: 193–197.

Dawson A. (1986). The effect of restricting the daily period of food availability on testicular growth of starlings *Sturnus vulgaris*. Ibis 128: 572–575.

Dawson A. (1987). Photorefractoriness in European starlings: Critical daylength is not affected by photoperiodic history. Physiol Zool 60: 722–729.

Dawson A. (1989a). Pharmacological doses of thyroxine simulate the effects of increased daylength, and thyroidectomy, decreased daylength on the reproductive system of European starlings. J Exp Zool 249: 62–67.

Dawson A. (1989b). The involvement of thyroxine and daylength in the development of photorefractoriness in European starlings. J Exp Zool 249: 68–75.

Dawson A. (1991). Effect of daylength on the rate of recovery of photosensitivity in male starlings (*Sturnus vulgaris*). J Reprod Fert 93: 521–524.

Dawson A. (2001). The effect of a single long photoperiod on induction and dissipation of reproductive photorefractoriness in European starlings. Gen Comp Endocrinol 121: 316–324.

Dawson A and Goldsmith AR. (1982). Prolactin and gonadotrophin secretion in wild starlings (*Sturnus vulgaris*) during the annual cycle and in relation to nesting, incubation and rearing young. Gen Comp Endocrinol 48: 213–221.

Dawson A and Goldsmith AR. (1983). Plasma prolactin and gonadotrophins during gonadal development and the onset of photorefractoriness in male and female starlings (*Sturnus vulgaris*) on artificial photoperiods. J Endocrinol 97: 253–260.

Dawson A and Goldsmith AR. (1984). Effects of gonadectomy on seasonal changes in plasma LH and prolactin concentrations in male and female starlings (*Sturnus vulgaris*). J Endocrinol 100: 213–218.

Dawson A, Goldsmith AR, and Nicholls TJ. (1985). Development of photorefractoriness in intact and castrated male starlings (*Sturnus vulgaris*) exposed to different periods of long daylengths. Physiol Zool 58: 253–261.

Dawson A and King V. (1994). Thyroidectomy does not affect the daily or free-running rhythms of plasma melatonin in European starlings. J Biol Rhythms 9: 137–144.

Dawson A, King VM, Bentley GE, and Ball GF. (2001). Photoperiodic control of seasonality in birds. J Biol Rhythms 16: 365–380.

Dawson A and Sharp PJ. (1998). The role of prolactin in the development of reproductive photorefractoriness and postnuptial molt in the European starling (*Sturnus vulgaris*). Endocrinology 139: 485–490.

Dawson A, Williams TD, and Nicholls TJ. (1987). Thyroidectomy of nestling starlings appears to cause neotenous sexual maturation. J Endocrinol 112: R5–R6.

D'Hondt E, Billen J, Berghman L, Vandesande F, and Arckens L (2001). Chicken luteinizing hormone-releasing hormone-I and -II are located in distinct fiber terminals in the median eminence of the quail: a light and electron microscopic study. Belg J Zool 131: 137–144.

Dittami JP and Gwinner E. (1985). Annual cycles in the African stonechat *Saxicola torquata* axillaris and their relationship to environmental factors. J Zool Lond 207: 357–370.

Dunn IC, Chen C, Hook C, Sharp PJ, and Sang HM. (1993). Characterization of the chicken preprogonadotrophin-releasing hormone-I gene. J Mol Endocrinol 11: 19–29.

Ebling FJP, Goldsmith AR, and Follett BK. (1982). Plasma prolactin and luteinizing hormone during photoperiodically induced testicular growth and regression in starlings (*Sturnus vulgaris*). Gen Comp Endocrinol 48: 485–490.

Elliott JA. (1976). Circadian rhythms and photoperiodic time measurement in mammals. Fed Proc 35: 2339–2346.

Farner DS. (1964). The photoperiodic control of reproductive cycles in birds. Am Sci 52: 137–156.

Farner DS, Donham RS, Lewis RA, Mattocks PW, Darden TR, and Smith JP. (1977). The circadian component in the photoperiodic mechanism of the house sparrow, *Passer domesticus*. Physiol Zool 50: 247–268.

Farner DS and Follett BK. (1966). Light and other environmental factors affecting avian reproduction. J Animal Sci 25(suppl): 90–118.

Follett BK. (1973). Circadian rhythms and photoperiodic time measurement in birds. J Reprod Fert 19(suppl): 5–18.

Follett BK, Kumar V, and Juss TS. (1992). Circadian nature of the photoperiodic clock in Japanese quail. J Comp Physiol A 171: 533–540.

Follett BK, Mattocks PW Jr, and Farner DS. (1974). Circadian function in the photoperiodic induction of gonadotropin secretion in the white-crowned sparrow, *Zonotrichia leucophrys gambelii*. Proc Natl Acad Sci USA 71: 1666–1669.

Follett BK and Nicholls TJ. (1984). Photorefractoriness in Japanese quail: Possible involvement of the thyroid gland. J Exp Zool 232: 573–580.

Follett BK and Nicholls TJ. (1985). Influences of thyroidectomy and thyroxine replacement on photoperiodically controlled reproduction in quail. J Endocrinol 107: 211–221.

Follett BK and Sharp PJ. (1969). Circadian rhythmicity in photoperiodically induced gonadotrophin secretion and gonadal growth in quail. Nature 223: 968–971.

Foster RG and Follett BK. (1985). The involvement of a rhodopsin-like photopigment in the photoperiodic response of the Japanese quail. J Comp Physiol A 157: 519–528.

Foster RG, Follett BK, and Lythgoe JN. (1985). Rhodopsin-like sensitivity of extra-retinal photoreceptors mediating the photoperiodic response in quail. Nature 313: 50–52.

Foster RG, Plowman G, Goldsmith AR, and Follett BK. (1987). Immunohistochemical demonstration of marked changes in the LHRH system of photosensitive and photorefractory European starlings (*Sturnus vulgaris*). J Endocrinol 115: 211–220.

Foster RG, Provencio I, Hudson D, Fiske S, DeGrip W, and Menaker M. (1991). Circadian photoreception in the retinally degenerate mouse (rd/rd). J Comp Physiol A 169: 39–50.

Foster RG, Timmers AM, Schalken JJ, and DeGrip WJ. (1989). A comparison of some photoreceptor characteristics in the pineal and retina: II. The Djungarian hamster (*Phodopus sungarus*). J Comp Physiol A 165: 565–572.

Gibb J. (1950). The breeding biology of the great and blue titmice. Ibis 92: 507–539.

Goldsmith AR. (1983). Prolactin in avian reproductive cycles. In *Hormones and Behaviour in Higher Vertebrates* (J Balthazart, E Pröve, and R Gilles, eds). Toronto: Springer-Verlag, pp. 375–387.

Goldsmith AR. (1985). Prolactin in avian reproduction: Incubation and the control of seasonal breeding. In *Fidia Research Series*, vol 1, *Prolactin* (RM Macleod, U Scapagnini, and MO Thorner, eds). Padova: Liviana Press/Berlin: Springer-Verlag, pp. 411–426.

Goldsmith AR, Ivings WE, Pearce-Kelly AS, Parry DM, Plowman G, Nicholls TJ, and Follett BK. (1989). Photoperiodic control of the development of the LHRH neurosecretory system of European starlings (*Sturnus vulgaris*) during puberty and the onset of photorefractoriness. J Endocrinol 122: 255–268.

Goldsmith AR and Nicholls TJ. (1984a). Thyroidectomy prevents the development of photorefractoriness and the associated rise in plasma prolactin in starlings. Gen Comp Endocrinol 54: 256–263.

Goldsmith AR and Nicholls TJ. (1984b). Changes in plasma prolactin in male starlings during testicular regression under short days compared with those during photorefractoriness. J Endocrinol 102: 353–356.

Goldsmith AR and Nicholls TJ. (1984c). Thyroxine induces photorefractoriness and stimulates prolactin secretion in European starlings (*Sturnus vulgaris*). J Endocrinol 101: R1–R3.

Griffith MK and Minton JE. (1992). Effect of light intensity on circadian profiles of melatonin, prolactin, ACTH, and cortisol in pigs. J Anim Sci 70: 492–498.

Groos G. (1982). The comparative physiology of extraocular photoreception. Exp Gen 38: 989–1128.

Guyomarc'h C, Lumineau S, Vivien-Roels B, Richard J, and Deregnaucourt S. (2001). Effect of melatonin supplementation on the sexual development in European quail (*Coturnix coturnix*). Behav Processes 53: 121–130.

Gwinner E, Dittami JP, and Beldhuis HJA. (1988). The seasonal development of photoperiodic responsiveness in an equatorial migrant, the garden warbler *Sylvia borin*. J Comp Physiol A 162: 389–396.

Gwinner E and Scheuerlein A. (1998). Seasonal changes in day-light intensity as a potential zeitgeber of circannual rhythms in equatorial stonechats. J Ornithol 139: 407–412.

Gwinner E and Scheuerlein A. (1999). Photoperiodic responsiveness of equatorial and temperate-zone stonechats. Condor 101: 347–359.

Gwinner E, Wozniak J, and Dittami J. (1981). The role of the pineal organ in the control of annual rhythms in birds. In *The Pineal Organ: Photobiology—Biochronometry—Endocrinology* (A Oksche and P Pévet, eds). Amsterdam: Elsevier/North-Holland, pp. 99–121.

Hahn TP. (1998). Reproductive seasonality in an opportunistic breeder, the red crossbill, *Loxia curvirostra*. Ecology 79: 2365–2375.

Hahn TP, Wingfield JC, Mullen R, and Deviche PJ. (1995). Endocrine bases of spatial and temporal opportunism in Arctic-breeding birds. Am Zool 35: 259–273.

Hamner WM. (1963). Diurnal rhythm and photoperiodism in testicular recrudescence of the house finch. Science 142: 1294–1295.

Hamner WM. (1964). Circadian control of photoperiodism in the house finch demonstrated by interrupted-night experiments. Nature 203: 1400–1401.

Hamner WM. (1971). On seeking an alternative to the endogenous reproductive rhythm hypothesis in birds. In *Biochronometry* (M Menaker, ed). Washington, DC: National Academy of Sciences, pp. 448–461.

Hattar S, Lucas RJ, Mrosovsky N, Thompson S, Douglas RH, Hankins MW, Lem J, Biel M, Hofmann F, Foster RG, and Yau KW. (2003). Melanopsin and rod-cone photoreceptive systems account for all major accessory visual functions in mice. Nature 424: 76–81.

Hau M, Wikelski M, and Wingfield JC. (1998). A neotropical forest bird can measure the slight changes in tropical photoperiod. Proc R Soc Lond Biol Sci 1391: 89–95.

Hinde RA and Steel E. (1978). The influence of day length and male vocalizations on the estrogen-dependent behavior of female canaries and budgerigars, with discussion of data from other species. Adv Stud Behav 8: 39–73.

Immelman K. (1971). Ecological aspects of periodic reproduction. In *Avian Biology*, vol 1 (DS Farner and JR King, eds). New York: Academic Press. pp 341–389.

Johnson MA, Tsutsui K, and Fraley GS. (2007). Rat RFamide-related peptide-3 stimulates GH secretion, inhibits LH secretion, and has variable effects on sex behavior in the adult male rat. Horm Behav 51: 171–180.

Juss TS, Meddle SL, Servant RS, and King VM. (1993). Melatonin and photoperiodic time measurement in Japanese quail (*Coturnix coturnix japonica*). Proc R Soc Lond B Biol Sci 254: 21–28.

Kang SW, Thayananuphat A, Bakken T, and El Halawani ME. (2007). Dopamine-melatonin neurons in the avian hypothalamus controlling seasonal reproduction. Neuroscience 150: 223–233.

Kauffman AS and Rissman EF. (2004). A critical role for the evolutionarily conserved gonadotropin-releasing hormone II: Mediation of energy status and female sexual behavior. Endocrinology 145: 3639–3646.

King JA and Millar RP. (1982a). Structure of chicken hypothalamic luteinizing hormone-releasing hormone. I. Structural determination on partially purified material. J Biol Chem 257: 10722–10732.

King JA and Millar RP. (1982b). Structure of chicken hypothalamic luteinizing hormone-releasing hormone. II. Isolation and characterization. J Biol Chem 257: 10729–10732.

Kluijver HN. (1951). The population ecology of the Great tit *Parus m.* major. Ardea 39: 1–135.

Kriegsfeld LJ, Feng-Mei D, Bentley GE, Ukena K, Tsutsui K, and Silver R. (2006). Identification and characterization of a gonadotropin-inhibitory system in the brains of mammals. Proc Natl Acad Sci USA 103: 2410–2415.

Kumar V and Follett BK. (1993). The circadian nature of melatonin secretion in Japanese quail (*Coturnix coturnix japonica*). J Pineal Res 14: 192–200.

Kumar V, Rani S, Malik S, Trivedi AK, Schwabl I, Helm B, and Gwinner E. (2007). Daytime light intensity affects seasonal timing via changes in the nocturnal melatonin levels. Naturwissenschaften 94: 693–696.

Lee PP and Pang SF. (1993). Melatonin and its receptors in the gastrointestinal tract. Biol Signals 2: 181–193.

Lees AD. (1973). Photoperiodic time measurement in the aphid *Megoura viciae*. J Insect Physiol 19: 2279–2316.

Lincoln GA, Racey PA, Sharp PJ, and Klandorf H. (1980). Endocrine changes associated with spring and autumn sexuality of the rook (*Corvus frugilegus*). J Zool 190: 137–153.

Maney DL, Richardson RD, and Wingfield JC. (1997). Central administration of chicken gonadotropin-releasing hormone-II enhances courtship behavior in a female sparrow. Horm Behav 32: 11–18.

Marshall FHA and Bowden FP. (1934). The effect of irradiation with different wave-lengths on the oestrus cycle of the ferret, with remarks on the factors controlling sexual periodicity. J Exp Biol 11: 409–422.

Meddle SL, Bush S, Sharp PJ, Millar RP, and Wingfield JC. (2006). Hypothalamic pro-GnRH-GAP, GnRH-I and GnRH-II during the onset of photorefractoriness in the white-crowned sparrow (*Zonotrichia leucophrys gambelii*). J Neuroendocrinol 18: 217–226.

Meddle SL and Follett BK. (1995). Photoperiodic activation of fos-like immunoreactive protein in neurones within the tuberal hypothalamus of Japanese quail. J Comp Physiol A 176: 79–89.

Meddle SL and Follett BK. (1997). Photoperiodically driven changes in Fos expression within the basal tuberal hypothalamus and median eminence of Japanese quail. J Neurosci 17: 8909–8918.

Meier AH. (1972). Temporal synergism of prolactin and adrenal steroids in the regulation of fat storage. Gen Comp Endocrinol 3(suppl): 499–508.

Meier AH and MacGregor R. (1972). Temporal organization in avian reproduction. Am Zool 12: 257–271.

Menaker M. (1971). In *Biochronometry: Proceedings of a Symposium: Synchronization with the Photic Environment via Extraretinal Receptors in the Avian Brain.* (M Menaker, ed). Washington, DC: National Academy of Sciences, pp. 315–332.

Menaker M and Keatts H. (1968). Extraretinal light perception in the sparrow, II. Photoperiodic stimulation of testis growth. Proc Nat Acad Sci USA 60: 146–151.

Millam JR, Faris PL, Youngren OM, El Halawani ME, and Hartman BK. (1993). Immunohistochemical localization of chicken gonadotropin-releasing hormones I and II (cGnRH-I and -II) in turkey hen brain. J Comp Neurol 333: 68–82.

Millam JR, Ottinger MA, Craig-Veit CB, Fan Y, Chaiseha Y, and El Halawani ME. (1998). Multiple forms of GnRH are released from perifused medial basal hypothalamic/preoptic area (MBH/POA) explants in birds. Gen Comp Endocrinol 111: 95–101.

Miller AH. (1965). Capacity for photoperiodic response and endogenous factors in the reproductive cycles of an equitorial sparrow. Proc Natl Acad Sci USA 54: 97–101.

Miyamoto K, Hasegawa Y, Nomura M, Igarashi M, Kangawa K, and Matsuo H. (1984). Identification of the second gonadotropin-releasing hormone in chicken hypothalamus: Evidence that gonadotropin secretion is probably controlled by two distinct gonadotropin-releasing hormones in avian species. Proc Natl Acad Sci USA 81: 3874–3878.

Moenter SM, Woodfill CJI, and Karsch FJ. (1991). Role of the thyroid gland in seasonal reproduction: Thyroidectomy blocks seasonal suppression of reproductive neuroendocrine activity in ewes. Endocrinology 128: 1337–1344.

Moore IT, Bentley GE, Wotus C, and Wingfield JC. (2006). Photoperiod-independent changes in immunoreactive brain gonadotropin-releasing hormone (GnRH) in a free-living, tropical bird. Brain Behav Evol 68:37–44.

Murton RK and Westwood NJ. (1977). *Avian Breeding Cycles.* Oxford: Clarendon.

Nakao N, Ono H, Yamamura T, Anraku T, Takagi T, Higashi K, Yasuo S, Katou Y, Kageyama S, Uno Y, Kasukawa T, Iigo M, Sharp PJ, Iwasawa A, et al. (2008). Thyrotrophin in the pars tuberalis triggers photoperiodic response. Nature 452: 317–322.

Nelson RJ and Zucker I. (1981). Absence of extraocular photoreception in diurnal and nocturnal rodents exposed to direct sunlight. Comp Biochem Physiol 69A:145–148.

Nicholls TJ, Follett BK, Goldsmith AR, and Pearson H. (1988b). Possible homologies between photorefractoriness in sheep and birds: The effect of thyroidectomy on the length of the ewe's breeding season. Reprod Nutr Dev 28: 375–385.

Nicholls TJ, Follett BK, and Robinson JE. (1983). A photoperiodic response in gonadectomized Japanese quail exposed to a single long day. J Endocrinol 97: 121–126.

Nicholls TJ, Goldsmith AR, and Dawson A. (1988a). Photorefractoriness in birds and comparison with mammals. Physiol Rev 68: 133–176.

Nicholls TJ and Storey CR. (1977). The effect of duration of the daily photoperiod on recovery of photosensitivity in photorefractory canaries (*Serinus canarius*). Gen Comp Endocrinol 31: 72–74.

Ohta M, Kadota C, and Konishi H. (1989). A role of melatonin in the initial stage of photoperiodism in the Japanese quail. Biol Reprod 40: 935–941.

Oliver J and Baylé JD. (1982). Brain photoreceptors for the photo-induced testicular response in birds. Experientia 38: 1021–1029.

Osugi T, Ukena K, Bentley GE, O'Brien S, Moore IT, Wingfield JC, and Tsutsui K. (2004). Gonadotropin-inhibitory hormone in Gambel's white-crowned sparrows: CDNA identification, transcript localization and functional effects in laboratory and field experiments. J Endocrinol 182: 33–42.

Parkinson TJ, Douthwaite JA, and Follett BK. (1995). Responses of prepubertal and mature rams to thyroidectomy. J Reprod Fert 104: 51–56.

Parkinson TJ and Follett BK. (1994). Effect of thyroidectomy upon seasonality in rams. J Reprod Fert 101: 51–58.

Parry DM and Goldsmith AR. (1993). Ultrastructural evidence for changes in synaptic input to the hypothalamic luteinizing hormone-releasing hormone neurons in photosensitive and photorefractory starlings. J Neuroendocrinol 5: 387–395.

Perfito N, Bentley GE, and Hau M. (2006). Tonic activation of brain GnRH immunoreactivity despite reduction of peripheral reproductive parameters in opportunistically breeding zebra finches. Brain Behav Evol 67: 123–134.

Perfito N, Kwong JM, Bentley GE, and Hau M. (2008). Cue hierarchies and testicular development: Is food a more potent stimulus than day length in an opportunistic breeder (*Taeniopygia g. guttata*)? Horm Behav 53: 567–572.

Pittendrigh CS and Minis DH. (1964). The entrainment of circadian oscillators by light and their role as photoperiodic clocks. Am Nat 98: 261–294.

Pittendrigh CS and Minis DH. (1971). The photoperiodic time measurement in *Pectinophora gossypiella* and its relation to the circadian system in that species. In *Biochronometry* (M Menaker, ed). Washington, DC: National Academy of Science, pp. 212–250.

Reiter RJ. (1978). Interaction of photoperiod, pineal and seasonal reproduction as exemplified by findings in the hamster. Prog Reprod Biol 4: 169–190.

Ringoen AR. (1942). Effects of continuous green and red light illumination on gonadal response in the English sparrow, *Passer domesticus* (Linnaeus). Am J Anat 71: 99–116.

Robinson JE and Follett BK. (1982). Photoperiodism in Japanese quail: The termination of seasonal breeding by photorefractoriness. Proc R Soc Lond B Biol Sci 215: 95–116.

Rollo M and Domm LV. (1943). Light requirements of the weaver finch. 1. Light period and intensity. Auk 60: 357–367.

Rowan W. (1925). Relation of light to bird migration and developmental changes. Nature 115: 494–495.

Rozenboim I, Aharony T, and Yahav S. (2002). The effect of melatonin administration on circulating plasma luteinizing hormone concentration in castrated White Leghorn roosters. Poult Sci 81: 1354–1359.

Saldanha CJ, Silverman AJ, and Silver R. (2001). Direct innervation of GnRH neurons by encephalic photoreceptors in birds. J Biol Rhythms 16: 39–49.

Sharp PJ, Dunn IC, and Talbot RT. (1987). Sex differences in the LH responses to chicken LHRH-I and -II in the domestic fowl. J Endocrinol 115: 323–331.

Sharp PJ and Klandorf H. (1981). The interaction between day length and the gonads in the regulation of levels of plasma thyroxine and triiodothyronine in the Japanese quail. Gen Comp Endocrinol 45: 504–512.

Sharp PJ and Moss R. (1977). The effect of castration on plasma LH levels in photorefractory red grouse (*Lagopus lagopus scoticus*). Gen Comp Endocrinol 32: 289–293.

Sharp PJ, Talbot RT, Main GM, Dunn IC, Fraser HM, and Huskisson NS. (1990). Physiological roles of chicken LHRH-I and -II in the control of gonadotrophin release in the domestic chicken. J Endocrinol 124: 291–299.

Sherwood NM, Wingfield JC, Ball GF, and Dufty AM. (1988). Identity of gonadotropin-releasing hormone in passerine birds: Comparison of GnRH in song sparrow (*Melospiza melodia*) and starling (*Sturnus vulgaris*) with five vertebrate GnRHs. Gen Comp Endocrinol 69: 341–351.

Shi ZD and Barrell GK. (1992). Requirement of thyroid function for the expression of seasonal reproductive and related changes in red deer (*Cervus elaphus*) stags. J Reprod Fert 94: 251–259.

Sicard V, Oliver J, and Baylé JD. (1983). Gonadotrophic and photosensitive abilities of the lobus parolfactorius: Electrophysiological study in quail. Neuroendocrinology 36: 81–87.

Silver R, Witkovsky P, Horvath P, Alones V, Barnstable CJ, and Lehman MN. (1988). Coexpression of opsin-like and VIP-like immunoreactivity in CSF-contacting neurons of the avian brain. Cell Tiss Res 253: 189–198.

Silverin B, Massa R, and Stokkan KA. (1993). Photoperiodic adaptation to breeding at different latitudes in great tits. Gen Comp Endocrinol 90: 14–22.

Small TW, Sharp PJ, Bentley GE, Millar RP, Tsutsui K, Mura E, and Deviche P. (2008). Photoperiod-independent hypothalamic regulation of luteinizing hormone secretion in a free-living Sonoran desert bird, the rufous-winged Sparrow (*Aimophila carpalis*). Brain Behav Evol 71: 127–142.

Steel E, Follett BK, and Hinde RA. (1975). The role of short days in the termination of photorefractoriness in female canaries (*Serinus canarius*). J Endocrinol 64: 451–464.

Stevenson TJ and Macdougall-Shackleton SA. (2005). Season- and age-related variation in neural cGnRH-I and cGnRH-II immunoreactivity in house sparrows (*Passer domesticus*). Gen Comp Endocrinol 143: 33–39.

Stokkan K-A and Sharp PJ. (1980a). The roles of daylength and the testes in the regulation of plasma LH levels in photosensitive and photorefractory willow ptarmigan (*Lagopus lagopus lagopus*). Gen Comp Endocrinol 41: 520–526.

Stokkan K-A and Sharp PJ. (1980b). The development of photorefractoriness in willow ptarmigan (*Lagopus lagopus lagopus*) after the suppression of photoinduced LH release with implants of testosterone. Gen Comp Endocrinol 41: 527–530.

Temple JL, Millar RP, and Rissman EF. (2003). An evolutionarily conserved form of gonadotropin-releasing hormone coordinates energy and reproductive behavior. Endocrinology 144: 13–19.

Teruyama R and Beck MM. (2000). Changes in immunoreactivity to anti-cGnRH-I and -II are associated with photostimulated sexual status in male quail. Cell Tiss Res 300: 413–426.

Tsutsui K, Saigoh E, Ukena K, Teranishi H, Fujisawa Y, Kikuchi M, Ishii S, and Sharp PJ. (2000). A novel avian hypothalamic peptide inhibiting gonadotropin release. Biochem Biophys Res Commun 275: 661–667.

Turek FW. (1974). Circadian rhythmicity and the initiation of gonadal growth in sparrows. J Comp Physiol 92: 59–64.

Turek FW. (1975). The termination of the avian photorefractory period and the subsequent gonadal response. Gen Comp Endocrinol 26: 562–564.

Ubuka T and Bentley GE. (2009). Identification, localization, and regulation of passerine GnRH-1 messenger RNA. J Endocrinol. 201: 81–87.

Ubuka T, Cadigan PA, Wang A, Liu J, and Bentley GE. (2009). Identification of European starling GnRH-1 precursor mRNA and its seasonal regulation. Gen Comp Endocrinol. 162: 301–305.

Ubuka T, Bentley GE, Ukena K, Wingfield JC, and Tsutsui K. (2005). Melatonin induces the expression of gonadotropin-inhibitory hormone in the avian brain. Proc Natl Acad Sci USA 102: 3052–3057.

Ubuka T, Ueno M, Ukena K, and Tsutsui K. (2003). Developmental changes in gonadotropin-inhibitory hormone in the Japanese quail (*Coturnix japonica*) hypothalamo-hypophysial system. J Endocrinol 178: 311–318.

Ubuka T, Ukena K, Sharp PJ, Bentley GE, and Tsutsui K. (2006). Gonadotropin-inhibitory hormone inhibits gonadal development and maintenance by decreasing gonadotropin synthesis and release in male quail. Endocrinology 147: 1187–1194.

Ukena K, Ubuka T, and Tsutsui K. (2003). Distribution of a novel avian gonadotropin-inhibitory hormone in the quail brain. Cell Tissue Res 312: 73–79.

Underwood H, Binkley S, Siopes T, and Mosher K. (1984). Melatonin rhythms in the eyes, pineal bodies, and blood of Japanese quail (*Coturnix coturnix japonica*). Gen Comp Endocrinol 56: 70–81.

Underwood H and Menaker M. (1970). Photoperiodically significant photoreception in sparrows: Is the retina involved? Science 167: 299–301.

van Gils J, Absil P, Grauwels L, Moons L, Vandesande F, and Balthazart J. (1993). Distribution of luteinizing hormone-releasing hormones I and II (LHRH-I and -II) in the quail and chicken brain as demonstrated with antibodies directed against synthetic peptides. J Comp Neurol 334: 304–323.

Vaz Nunes M and Veerman A. (1984). Light-break experiments and photoperiodic time measurement in the spider mite *Tetranychus urticae*. J Insect Physiol 30: 891–897.

Vaz Nunes M and Veerman A. (1986). A "dusk" oscillator affects photoperiodic induction of diapause in the spider mite *Tetranychus urticae*. J Insect Physiol 32: 605–614.

Vaugien L. (1955). Sur les réactions testiculaires du jeune moineau domestique illuminé à diverses époques de la mauvais saison. Bull Biol Fran Belg 89: 218–244.

Veerman A, Beckman M, and Veenedaal RL. (1988). Photoperiodic induction of diapause in the large white butterfly, *Pieris brassicae*—evidence for hour glass time measurement. J Insect Physiol 34: 1063–1069.

Wieselthier AS and van Tienhoven A. (1972). The effect of thyroidectomy on testicular size and on the photorefractory period in the starling (*Sturnus vulgaris* L.) J Exp Zool 179: 331–338.

Williams TD, Dawson A, Nicholls TJ, and Goldsmith AR. (1987). Short days induce premature reproductive maturation in juvenile starlings, *Sturnus vulgaris*. J Reprod Fert 80: 327–333.

Wilson FE. (1985). Androgen feedback-dependent and independent control of photoinduced LH secretion in male tree sparrows (*Spizella arborea*). J Endocrinol 105: 141–152.

Wilson FE. (1991). Neither retinal nor pineal photoreceptors mediate photoperiodic control of seasonal reproduction in American tree sparrows (*Spizella arborea*). J Exp Zool 259: 117–127.

Wilson FE and Donham RS. (1988). Daylength and control of seasonal reproduction in male birds. In *Processing of Environmental Information in Vertebrates* (MH Stetson, ed). Berlin: Springer-Verlag, pp. 101–120.

Wilson FE and Reinert BD. (1993). The thyroid and photoperiodic control of seasonal reproduction in American tree sparrows (*Spizella arborea*). J Comp Physiol B 163: 563–573.

Wilson FE and Reinert BD. (1995a). The photoperiodic control circuit in euthyroid American tree sparrows (*Spizella arborea*) is already programmed for photorefractoriness by week 4 under long days. J Reprod Fert 103: 279–284.

Wilson FE and Reinert BD. (1995b). A one-time injection of thyroxine programmed seasonal reproduction and postnuptial moult in chronically thyroidectomised male American tree sparrows *Spizella arborea* exposed to long days. J Avian Biol 26: 225–233.

Wingfield JC. (1980). Fine temporal adjustment of reproductive functions. In *Avian Endocrinology* (A Epple and MH Stetson, eds). New York: Academic Press, pp. 367–389.

Wingfield JC and Farner DS. (1993). The endocrinology of wild species. In *Avian Biology*, vol 9 (DS Farner, JR King, and KC Parkes, eds). New York: Academic, pp. 163–327.

Wingfield JC, Moore MC, and Farner DS. (1983). Endocrine responses to inclement weather in naturally breeding populations of white-crowned sparrows (*Zonotrichia leucophrys pugetensis*). Auk 100: 56–62.

Yamamura T, Hirunagi K, Ebihara S, and Yoshimura T. (2004). Seasonal morphological changes in the neuro-glial interaction between gonadotropin-releasing hormone nerve terminals and glial endfeet in Japanese quail. Endocrinology 145: 4264–4267.

Yoshimura T, Yasuo S, Watanabe M, Iigo M, Yamamura T, Hirunagi K, and Ebihara S. (2003). Light-induced hormone conversion of T_4 to T_3 regulates photoperiodic response of gonads in birds. Nature 426: 178–181.

Zann RA, Morton SR, Jones KR, and Burley NT. (1995). The timing of breeding by zebra finches in relation to rainfall in central Australia. Emu 95: 208–222.

18

Genetic and Molecular Mechanisms of Avian Photoperiodism

Takashi Yoshimura and Peter J. Sharp

Many birds have highly sophisticated photoperiodic mechanisms and show robust responses to changing photoperiod. William Rowan is generally credited with the first demonstration of the avian photoperiodic response when he photostimulated dark eyed juncos (*Junco hyemalis*) during the Canadian winter and observed stimulation of testicular growth (Rowan, 1925). The application of molecular biological techniques to understanding the mechanisms controlling the avian photoperiodic response continue Rowan's pioneering tradition of using birds to study vertebrate photoperiodism. This chapter focuses on current understanding of genetic and molecular mechanisms underlying the avian photoperiodic response.

CLONING OF AVIAN CIRCADIAN CLOCK GENES

It is well established that a biological clock is involved in photoperiodic time measurement and that this clock is a function of circadian rhythmicity (Pittendrigh, 1972). In the past decade it has been shown that circadian rhythms are generated by a transcription-translation-based oscillatory loop involving "circadian clock genes" (Young and Kay, 2001; Reppert and Weaver, 2002) that are conserved across the animal kingdom from fruit flies to humans. Avian orthologs of circadian clock genes, including *Clock*, *Per2* and *-3*, *Bmal1* and *-2*, *Cry1* and *-2*, and *E4bp4*, have been cloned in Japanese quail and chicken (Yoshimura et al., 2000; Doi et al., 2001; Okano et al., 2001; Yamamoto et al., 2001) and are expressed in the quail pineal gland, hypothalamic suprachiasmatic nucleus (SCN) (Yoshimura et al., 2001; Yasuo et al., 2003), medial basal hypothalamus (MBH) (Yasuo et al., 2003), and pars tuberalis of the pituitary gland (Yasuo et al., 2004). The MBH is considered to be the center for photoperiodic time measurement in birds (Sharp and Follett, 1969; Juss, 1993; Saldanha et al., 1994; Meddle and Follett, 1997), whereas in mammals, the pars tuberalis is thought to respond to changes in the duration and amplitude of daily rhythms of circulating melatonin through abundant melatonin receptor to mediate the photoperiodic control of prolactin secretion (Morgan and Williams, 1996; Wittkowski et al., 1999). The expression profiles of circadian clock genes in the avian pars tuberalis, the pineal gland, and the SCN reflect the duration of the daily photoperiod. These observations suggested that avian pars tuberalis may relay photoperiodic signal information to the anterior pituitary, as in mammals. But because the avian pars tuberalis appears not to contain melatonin receptors, the mechanism through which photoperiodic information is relayed to the avian pars tuberalis is unknown. In contrast to the pars tuberalis, diurnal changes in expression profiles of clock genes in the quail MBH do not reflect changing day lengths and therefore maintain a stable time relationship with dawn (Yasuo et al., 2003). The circadian clock localized in the MBH therefore appears to be the long-sought biological clock responsible for photoperiodic time measurement (Ball and Balthazart, 2003; Yasuo et al., 2003).

IDENTIFICATION OF PHOTOPERIOD-REGULATED GENES: EVENTS DOWNSTREAM OF THE CIRCADIAN CLOCK

When short-day quail are photostimulated, an increase in plasma gonadotropin (luteinizing hormone [LH]) occurs before the end of the first long day (Nicholls et al., 1983; Follett et al., 1998). This photoperiodic response is the core feature of the so-called "first day release model" of photoperiodic induction and has been exploited to unravel the underlying neural and molecular mechanisms. Using this model, Fos-like immunoreactivity increases in the MBH by 18 h after dawn (Meddle and Follett, 1995, 1997). Fos is encoded by the immediate-early gene c-*fos*, which is transiently expressed in neural tissues in response to a wide range of stimuli. Photoinduced Fos-like immunoreactivity is localized in glial cells in the median eminence and neuronal cells in the adjacent infundibular nucleus. Importantly, this Fos-like immunoreactivity appears in the MBH before the initiation of photoinduced LH release. This observation suggests that gonadotropin-releasing hormone (GnRH) release may be controlled by altering glial cell function through a photoresponse mechanism involving increased c-*fos* expression.

The observation that light pulses given to short-day quail between 12 and 16 h after dusk, during the "photoinducible phase" of the circadian rhythm of photoinducibility, stimulate LH secretion and testicular growth (Follett and Sharp, 1969) has been exploited to discover the key molecular responses occurring in the MBH when quail are first photostimulated. Differential subtractive hybridization analysis of gene expression in the quail MBH demonstrates a dramatic increase in expression of the gene encoding type 2 iodothyronine deiodinase (*DIO2*) in response to a light pulse given during the "photoinducible phase" (Yoshimura et al., 2003). Importantly, induction of *DIO2* is not observed when the light pulse is given outside the photoinducible phase. This acute induction of *DIO2* gene by a light pulse is observed in the dorsal part of the MBH, whereas up-regulation of *DIO2* gene by a continuous long-day stimulus is observed in the basal part of the MBH. Electrolytic lesions in either the dorsal or basal quail MBH block the photoperiodic response without destroying the GnRH terminals in the median eminence (Sharp and Follett, 1969; Juss, 1993); this is consistent with a critical role for *DIO2* expression at these two loci in photoperiodic signal transduction. In addition, differential subtraction analysis of MBH gene expression in quail demonstrates that expression of the gene encoding type 3 iodothyronine deiodinase (*DIO3*) is up-regulated under short days and down-regulated under long days (Yasuo et al., 2005). When the temporal gene expression profiles of *DIO2* and *DIO3* in the MBH are examined in quail after photostimulation, reciprocal switching of *DIO2* and *DIO3* expression is observed at around 16 h of the first long day (Yasuo et al., 2005).

INVOLVEMENT OF THYROID HORMONES

It is well known that thyroid hormones are involved in the regulation of the photoperiodic response in birds and mammals (Nicholls et al., 1988; Dawson et al., 2001). Thyroxine (T_4) is generally considered to be a prohormone, and triiodothyronine (T_3) to be the biologically active thyroid hormone (figure 18.1). DIO2 converts T_4 to T_3, whereas DIO3 converts T_4 and T_3 to the inactive metabolites reverse T_3 and T_2, respectively. The amounts of T_3 and T_4 in the quail MBH increase markedly under long-day conditions (Yoshimura et al., 2003). However, parallel changes are not observed in the plasma or in the other brain regions such as cerebellum and optic tectum. These results support the view that photoperiodically regulated expression of deiodinases in the MBH result in the local activation of thyroid hormones. The possible functional significance of increased thyroid hormones in the MBH in photoperiodic signaling has been investigated using intracerebroventricular (icv) infusion of thyroid hormones or DIO2 inhibitor. Administration of T_3 under short-day conditions induces testicular growth, whereas DIO2 inhibitor treatment decreases testicular weight under long-day conditions (Yoshimura et al., 2003). These findings suggest that a long-day–induced local increase in the metabolism of T_4 to T_3 in the MBH plays a critical role in the regulation of seasonal reproduction in birds. The discovery that local activation of thyroid hormone metabolism in the MBH is the central to the regulation of photoperiodic response has been confirmed in various mammalian species including long-day breeders: Djungarian hamsters (Watanabe et al., 2004, 2007; Barrett et al., 2007; Freeman et al., 2007), Syrian hamsters (Revel et al., 2006; Yasuo et al., 2007a), rats (Yasuo et al., 2007b), and short-day–breeding goats (Yasuo et al., 2006).

Events Downstream of Thyroid Hormone Action

A target site of action for the photoinduced increase in the quail MBH T_3 is suggested by the presence of gene expression for thyroid hormone receptors (*TRα*, *TRβ*, and *RXRα*) in the median eminence (Yoshimura et al., 2003). In order to understand the functional significance of these receptors, the ultrastructure of quail median eminence was examined under short- and long-day conditions (Yamamura et al.,

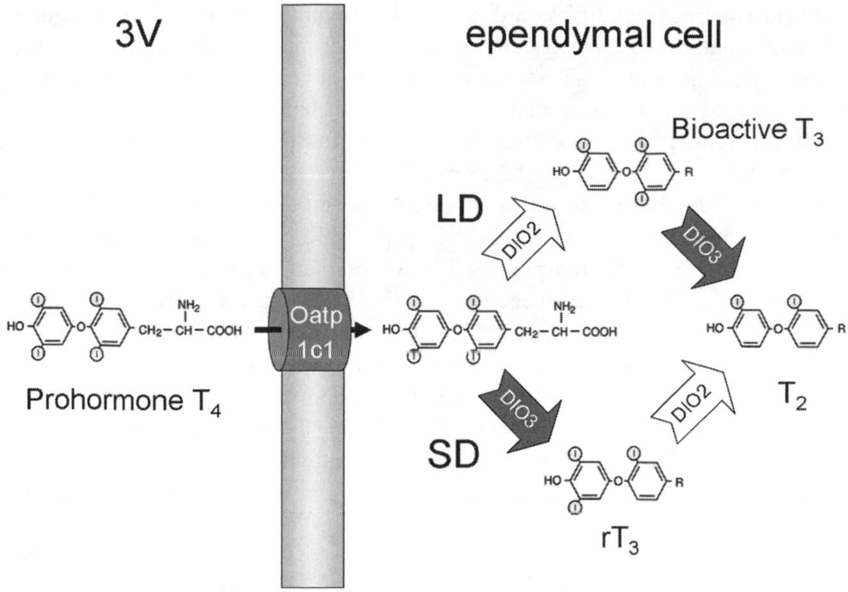

FIGURE 18.1.
A model to show how the prohormone thyroxine (T_4) is proposed to be transported from the third ventricle (3V) to the quail medial basal hypothalamus via type 1c1 organic anion transporting polypeptide (Oatp1c1) in the ependymal cells lining third ventricle, and subsequently metabolized. Under long-day conditions (LD), T_4 is metabolized in these cells through increased type 2 iodothyronine deiodinase (DIO2) and decreased type 3 iodothyronine deiodinase (DIO3) to biologically active triiodothyronine (T_3). Under short-day conditions (SD) DIO2 is low and DIO3 is high, resulting in T_4 being principally metabolized to biologically inactive reverse T_3 (rT_3). Both T_3 and rT_3 are further metabolized to biologically inactive diiodothyronine (T_2).

2004). This study demonstrated that GnRH nerve terminals are in close proximity to the basal lamina under long days, that is, with immediate access to the hypophyseal portal vasculature, but not under short-day conditions when the GnRH terminals are more extensively encased in the end feet of glial cell processes. Because retraction of glial cell end feet around nerve terminals in the median eminence is seen in short-day quail treated with T_3 to stimulate testicular growth (Yamamura et al., 2006), it appears that this morphological change in neuron–glial interaction is involved in the regulation of photoinduced GnRH secretion.

Thyroid Hormone Transporters

If a local increase in metabolism of T_4 to T_3 in the MBH plays a key role in the avian photoperiodic response, then there must be a mechanism to transport T_4 into the MBH. Although thyroid hormones were long thought to traverse plasma membranes by passive diffusion due to their lipophilic nature, it is now believed that a membrane transport system for thyroid hormone exists. Although expression of thyroid hormone transporters such as transthyretin, T_4-binding globulin, and albumin have been demonstrated to be altered by changing day length in Siberian hamsters (Prendergast et al., 2002), expression of these genes is not observed in the MBH of quail. Recently, some of the organic anion transporting polypeptide (Oatp) family have been shown to transport thyroid hormones in mammals (Abe et al., 2002; Hagenbuch and Meier, 2004) and the possibility that a member of this family might be involved in transporting T_4 into the quail brain has been investigated (Nakao et al., 2006). The gene encoding of *Oatp1c1* is highly expressed in the ventrolateral walls of the quail basal MBH, while four *Oatp* genes are expressed in the choroid plexus. Functional expression of chicken *Oatp1c1* in Chinese hamster ovary cells reveals that Oatp1c1 is a highly specific transporter for T_4. These observations suggest that Oatps could be involved in transporting T_4 from the general circulation to cerebrospinal fluid and from there to the ependymal cells lining ventrolateral walls of third ventricle in the MBH, where *DIO2* and *DIO3* genes are expressed (figure 18.1).

TGFα Mediates a Thyroid-Hormone–Independent Photoinduction Pathway

Although changes in MBH T_3 are likely to be key element in the regulation of the vertebrate photoperiodic response, the existence of a thyroid-hormone–independent regulatory mechanism cannot be discounted (Dawson et al., 2001). For example, although European starlings, house sparrows, and sheep become photoperiodically blind after thyroidectomy, thyroidectomized quail still can respond to changing day length (Follett and Nicholls, 1984, 1985; Nicholls et al., 1988; Dawson, 1993, 1998; Dawson and Thapliyal, 2001). The possible existence of a thyroid-hormone–independent regulatory pathway transducing photoperiodic information in the quail has been investigated using differential subtractive hybridization analysis of gene expression in the MBH (Takagi et al., 2007). An increase in the expression of *TGFα* (transforming growth factor α) in the basal MBH is observed after exposure to one long day, reminiscent of the photoinduced increase in *DIO2* expression. A role for TGFα in the regulation of GnRH release is proposed in mammals since TGFα appears to be involved in controlling the onset of puberty (Ma et al., 1994). A site of action for quail MBH TGFα is suggested by the expression of receptors for TGFα (epidermal growth factor receptors: *erbB1, -2, -4*) in the quail median eminence (Takagi et al., 2007). In order to establish whether MBH TGFα is involved in transducing photoperiodic information in quail, short-day birds received icv TGFα. This treatment increased LH secretion and testicular growth but did not increase MBH *DIO2* (Takagi et al., 2007).

Thus, icv TGFα does not stimulate reproductive neuroendocrine function by activating the conversion of T_4 to T_3. Conversely, photoinduced quail MBH TGFα is not a consequence of a photoinduced increase in MBH T$_3$ because T$_3$ implantation into the MBH of short-day quail, which stimulates testicular growth, does not affect the expression of *TGFα* (Takagi et al., 2007). Long-day–induced activation of the TGFα signaling pathway in the quail MBH appears to mediate a thyroid-hormone–independent pathway for the photoperiodic regulation of reproduction. TGFα is known in mammals to activate EGF receptors in median eminence glial cells with a subsequent release of prostaglandin E$_2$, which then acts on GnRH neurons to induce GnRH secretion (Ojeda et al., 1990; Ma et al., 1997; Rage et al., 1997). A similar mechanism may occur in quail; preliminary research observations suggest that testicular growth induced by icv administration of TGFα in short-day quail does not induce morphological changes in neuroglial interactions in the median eminence (Yamamura and Yoshimura, unpublished data).

INSULIN RECEPTOR MAY ENHANCE PHOTOPERIODIC RESPONSE

Differential subtractive hybridization analysis of gene expression in the MBH in long- and short-day quail has identified a long-day–induced expression of insulin receptor (*IR*) (Anraku et al., 2007). However, in contrast to the rapid photoinduction of *DIO2* and *TGFα* expression, photoinduction of *IR* gene expression is slow, being first statistically significant after 10 days of photostimulation. Because photoinduced testicular growth in quail is maximal after 10 days photostimulation, it appears that long-day induction of *IR* expression in the MBH may depend on increased concentrations of circulating gonadal steroids. Consistent with this hypothesis, MBH *IR* gene expression in castrated quail MBH is not induced after photostimulation, whereas it is induced in short-day quail treated with testosterone (Anraku et al., 2007). This observation suggests that, in birds, insulin may play a role in enhancing the photoperiodic responsiveness. Some support for this view comes from the finding in mammals that the central administration of insulin increases LH secretion (Miller et al., 1995; Kovacs et al., 2002), and that neuron-specific disruption of *IR* gene expression causes impaired gonadal function due to the dysregulation of LH secretion (Bruning et al., 2000). Further, photoinduced changes in *IR* expression in the arcuate nucleus (which is anatomically homologous to avian infundibular nucleus) have been recently reported in the hamster (Tups et al., 2006).

MICROARRAY ANALYSIS OF PHOTOINDUCTION PATHWAY IN THE MEDIAL BASAL HYPOTHALAMUS

Differential subtraction analysis of gene expression in the quail MBH allows only a limited analysis of the genes involved in transducing photoperiodic information. A more complete analysis has been made possible through the recent availability of chicken genomic sequence information and a high-density chicken oligonucleotide microarray (Affymetrix Chicken Genome Array) suitable for use in the quail. This microarray has been used to identify changes in gene expression in the quail MBH during the first long day in relation to the timing of the initiation of photoinduced LH secretion (Nakao et al., 2008). Two waves of gene expression are induced during the first day of photostimulation: the first centers around 14 h, and the second centers around 18 h after dawn (figure 18.2). Both waves of gene expression precede the initiation of increased LH secretion at 22 h after dawn. The first wave comprises two genes encoding TSHβ (thyroid-stimulating hormone beta subunit) and EYA3 (eyes absent 3), and the second wave comprises 11 genes, including *DIO2* and *DIO3*, which show inversely related changes in expression. Analysis of the spatiotemporal expression profiles of these genes by *in situ* hybridization shows that the two first-wave genes (*TSHB* and *EYA3*) are expressed in the pars tuberalis of the pituitary gland, whereas second-wave genes including *DIO2* and *DIO3* are expressed in the ependymal cells lining the ventrolateral walls of third ventricle in the basal MBH (figure 18.3). The gene encoding the common pituitary glycoprotein

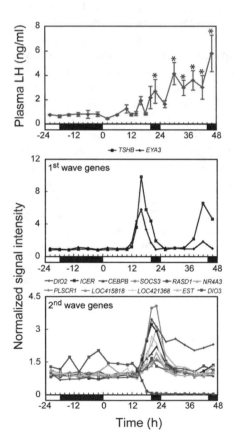

FIGURE 18.2.

Changes in concentrations of plasma LH (top) in male Japanese quail during the first day of photostimulation in relation to changes in gene expression in the medial basal hypothalamus (middle and bottom). Time 0 h is dawn of the first long day. Two waves of gene expression occur: the first peaking at around 14 h after dawn, comprising genes encoding TSHβ and EYA3 (middle), and the second peaking at 4 h later, comprising 11 genes, including *DIO2*. The photo increase in *DIO2* is associated with a decrease in the expression of *DIO3*. Modified from Nakao et al. (2008).

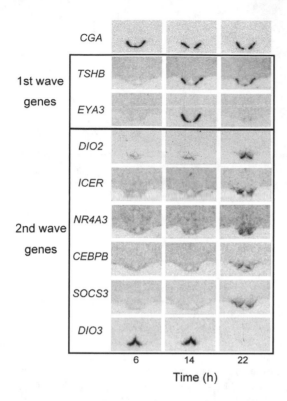

FIGURE 18.3.
An in situ hybridization analysis of the spatiotemporal expression of first-wave genes (*TSHB* and *EYA3*) and second-wave genes (*DIO2, ICER, NR4A3, CEBPB, SOCS3,* and *DIO3*) expressed in the medial basal hypothalamus of Japanese quail during the first day of photostimulation. Note that the first wave genes *TSHB* and *EYA3* are visible in the pars tuberalis after 14 h but not after 6 or 22 h of photostimulation, whereas the second-wave genes, including *DIO2*, are visible in the ependyma of the basal medial hypothalamus after 22 h but not after 6 or 14 h photostimulation. Note that *DIO3* expression occurs in the ependymal layer of the medial basal hypothalamus at 6 and 14 h, but not after 22 h of photostimulation. Modified from Nakao et al. (2008).

alpha subunit (CGA or α-GSU) is expressed in the pars tuberalis and its expression is rhythmic. The expression of this gene together with that encoding the TSHβ suggests that the encoded proteins associate in the pars tuberalis to form thyrotropin (TSH). This deduction is supported by the observation that TSHβ immunoreactivity is present in the pars tuberalis in long-day quail (Nakao et al., 2008). The other first-wave gene, *EYA3*, is a transcriptional coactivator and forms a nuclear complex with the SIX (sine oculis) DNA-binding homeodomain factor and DACH (dachshund) nuclear cofactors (Rebay et al., 2005). Because the sites of expression of *EYA3* and second-wave genes are different, *EYA3* was unlikely to regulate expression of second-wave genes.

TSH REGULATION OF *DIO2* EXPRESSION

The median eminence is outside the blood–brain barrier (Ganong, 2000), and consequently a long-day–induced increase TSH in the pars tuberalis may result in TSH entering the MBH to regulate the expression of second-wave genes, including *DIO2*, and induce gonadal growth. This view is supported by the observation that TSH receptor gene (*TSHR*) expression occurs in quail MBH ependymal cells with weaker

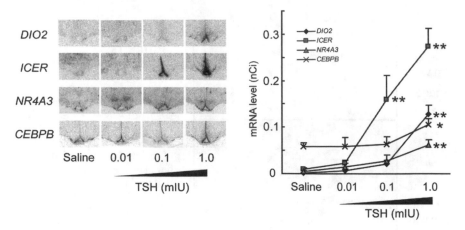

FIGURE 18.4.
The effect of icv injection of TSH in short-day quail on the expression of four second-wave genes in the medial basal hypothalamus (see figure 18.2), including *DIO2*, as demonstrated by in situ hybridization (left) and densitometric quantification of autoradiograms (right). Note that the expression of the four second-wave genes is induced by TSH treatment in a dose-dependent manner, and that gene induction is not observed in control birds given icv injection of saline vehicle. Modified from Nakao et al. (2008).

expression in the adjacent infundibular nucleus (Nakao et al., 2008). The presence of *TSHR* at these loci has been confirmed by showing specific binding of ^{125}I-TSH. Further, icv injection of TSH in short-day quail induces *DIO2* expression in the MBH in a dose-dependent manner (figure 18.4) and testicular growth, whereas icv injection of chicken TSHβ antibody, in long-day quail suppresses *DIO2* expression (Nakao et al., 2008). The expression of *DIO2* in the human thyroid gland is regulated through a TSHR-Gsα-cAMP regulatory cascade (Murakami et al., 2001a, 2001b), and several putative cAMP-responsive elements (CREs) occur in the 1.5 kb 5′ upstream region of quail *DIO2*. To determine whether these CRE sequences are involved in the regulation quail *DIO2* by TSH, the promoter activity of the quail *DIO2* gene has been analyzed using a luciferase reporter construct transfected into the 293 cell line (figure 18.5) (Nakao et al., 2008). Treatment of the transfected cells with TSH induces *DIO2* reporter activity in a dose-dependent manner, provided that the cells are cotransfected with *TSHR*. If CRE sites are mutated in the *DIO2* reporter constructs, DIO2 reporter activity is not induced by TSH. It is therefore deduced that *DIO2* expression in the quail MBH is likely to be induced after photostimulation by increased pars tuberalis TSH signaling through a TSHR-Gsα-cAMP cascade targeted to CREs in the promoter region of *DIO2*. In addition to *DIO2*, the expression of some other second-wave genes are also regulated by TSH (figure 18.4), and it is possible that all second-wave genes are similarly regulated. The functional significance of these genes in the photoperiodic signaling pathway remains to be explored.

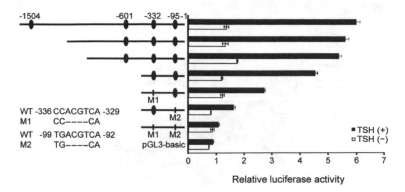

FIGURE 18.5.
A demonstration of a cAMP signaling pathway in TSH induction of quail *DIO2* expression. The sequence of the 5′ region of quail *DIO2* was fused to a luciferase gene reporter and transfected into a cell line with a TSH receptor gene. Various deletions and mutations at CRE sites were made in the reporter constructs, as shown. Treatment of the reporter-construct–infected cells with TSH induced luciferase activity when the wild-type 5′ sequence was part of the reporter construct, but this response was greatly reduced when the CRE sites (ovals) were mutated (vertical bars). Modified from Nakao et al. (2008).

PHOTOPERIODICALLY REGULATED OUTPUT GENES

A cDNA microarray analysis of the chronic effects of photostimulation on gene expression in the quail MBH has identified 124 up-regulated and 59 down-regulated genes (Nakao et al., 2008). The up-regulated genes include *DIO2*, and the down-regulated genes include *DIO3*. In addition, a high expression of *TSHB* and *CGA* is observed, indicating that the photoinduced increase in pars tuberalis TSH not only play a role in initiating photoinduced LH secretion, but is necessary to maintain it through the expression of other genes required to support a full reproductive response.

CONCLUSION AND PERSPECTIVES

Although photoperiodic regulation of TSH in the pars tuberalis and expression of *TSHR* in the mammalian brain has been known for some time (Hojvat et al., 1982; Wittkowski et al., 1988; Bockmann et al., 1997a, 1997b; Crisanti et al., 2001), the physiological function of TSH in photoperiodic signaling has been uncertain. This uncertainty has now been resolved by the demonstration in the quail that pars tuberalis TSH triggers and probably maintains, the reproductive photoperiodic response (figure 18.6). However, this discovery still leaves many unanswered questions, including the identity of the molecular basis underlying the photoinducible phase of the circadian rhythm photoinducibly and the development of photorefractoriness.

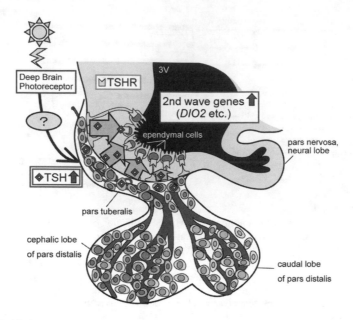

FIGURE 18.6.
A diagram to illustrate a photoperiodic signal transduction pathway through which a photoinduced increase in pars tuberalis TSH acts through TSHR in the ependymal cells lining the walls of the third ventricle in the medial basal hypothalamus to induce a second wave of gene expression, including *DIO2*. The resulting local increase in T_3 is thought to facilitate GnRH release by causing the retraction of tanycyte end feet encasement of GnRH terminals in the median eminence. The functional significance of second-wave genes in the photoperiodic signal transduction pathway, other than those regulating thyroid hormone metabolism, remains to be determined. The pathway through which photoperiodic information is relayed to the avian pars tuberalis also remains to be discovered. Modified from Nakao et al. (2008).

But perhaps the most urgent question requiring an answer is the identity of the photoperiodic signaling pathway that activates the avian pars tuberalis TSH cell (Figure 18.6).

References

Abe T, Suzuki T, Unno M, Tokui T, and Ito S. (2002). Thyroid hormone transporters: Recent advances. Trends Endocrinol Metab 13: 215–220.

Anraku T, Takagi T, Nakao N, Watanabe M, Yasuo S, Katou Y, Ueda Y, Murai A, Iigo M, Ebihara S, and Yoshimura T. (2007). Photoperiodic changes in hypothalamic insulin receptor gene expression are regulated by gonadal testosterone. Brain Res 1163: 86–90.

Ball GF and Balthazart J. (2003). Birds return every spring like clockwork, but where is the clock? Endocrinology 144: 3739–3741.

Barrett P, Ebling FJ, Schuhler S, Wilson D, Ross AW, Warner A, Jethwa P, Boelen A, Visser TJ, Ozanne DM, Archer ZA, Mercer JG, and Morgan PJ. (2007). Hypothalamic thyroid hormone catabolism acts as a gatekeeper for the seasonal control of body weight and reproduction. Endocrinology 148: 3608–3617.

Bockmann J, Bockers TM, Winter C, Wittkowski W, Winterhoff H, Deufel T, and Kreutz MR. (1997b). Thyrotropin expression in hypophyseal pars tuberalis-specific cells is 3,5,3′-triiodothyronine, thyrotropin-releasing hormone, and Pit-1 independent. Endocrinology 138: 1019–1028.

Bockmann J, Winter C, Wittkowski W, Kreutz MR, and Bockers TM. (1997a). Cloning and expression of a brain-derived TSH receptor. Biochem Biophys Res Commun 238: 173–178.

Bruning JC, Gautam D, Burks DJ, Gillette J, Schubert M, Orban PC, Klein R, Krone W, Muller-Wieland D, and Kahn CR. (2000). Role of brain insulin receptor in control of body weight and reproduction. Science 289: 2122–2125.

Crisanti P, Omri B, Hughes E, Meduri G, Hery C, Clauser R, Jacquemin C, and Saunier B. (2001). The expression of thyrotropin receptor in the brain. Endocrinology 142: 812 822.

Dawson A. (1993). Thyroidectomy progressively renders the reproductive system of starlings (*Sturnus vulgaris*) unresponsive to changes in daylength. J Endocrinol 139: 51–55.

Dawson A. (1998). Thyroidectomy of house sparrow (*Passer domesticus*) prevents photo-induced testicular growth but not the increased hypothalamic gonadotrophin-releasing hormone. Gen Comp Endocrinol 110: 196–200.

Dawson A King VM, Bentley GE, and Ball GF. (2001). Photoperiodic control of seasonality in birds. J Biol Rhythms 16: 365–380.

Dawson A and Thapliyal JP. (2001). The thyroid and photoperiodism. In *Avian Endocrinology* (A Dawson and CM Chatruvedi, eds). New Delhi: Narosa, pp. 141–151.

Doi M, Nakajima Y, Okano T, and Fukada Y. (2001). Light-induced phase-delay of the chicken pineal circadian clock is associated with the induction of cE4bp4, a potential transcriptional repressor of cPer2 gene. Proc Natl Acad Sci USA 98: 8089–8094.

Follett BK and Nicholls TJ. (1984). Photorefractoriness in Japanese quail: Possible involvement of the thyroid gland. J Exp Zool 232: 573–580.

Follett BK and Nicholls TJ. (1985). Influences of thyroidectomy and thyroxine replacement on photoperiodically controlled reproduction in quail. J Endocrinol 107: 211–221.

Follett BK, King VM, and Meddle SL. (1998). Rhythms and photoperiodism in birds. In *Biological Rhythms and Photoperiodism in Plants* (PJ Lumsden and AJ Miller, eds). Oxford: BIOS Scientific, pp. 231–242.

Follett BK and Sharp PJ. (1969). Circadian rhythmicity in photoperiodically induced gonadotrophin release and gonadal growth in the quail. Nature 223: 968–971.

Freeman DA, Teubner BJ, Smith CD, and Prendergast BJ. (2007). Exogenous T_3 mimics long day lengths in Siberian hamsters. Am J Physiol Regul Integr Comp Physiol 292: R2368–R2372.

Ganong WF. (2000). Circumventricular organs: Definition and role in the regulation of endocrine and autonomic function. Clin Exp Pharmacol Physiol 27: 422–427.

Hagenbuch B and Meier PJ. (2004). Organic anion transporting polypeptides of the OATP/SLC21 family: Phylogenetic classification as OATP/SLCO superfamily, new nomenclature and molecular/functional properties. Eur J Physiol 447: 653–665.

Hojvat S, Baker G, Kirsteins L, and Lawrence AM. (1982). TSH in the rat and monkey brain: Distribution, characterization and effect of hypophysectomy. Neuroendocrinology 34: 327 332.

Juss T. (1993). Neuroendocrine and neural changes associated with the photoperiodic control of reproduction. In *Avian Endocrinology* (PJ Sharp, ed). Bristol, UK: Society for Endocrinology, pp. 47–60.

Kovacs P, Parlow AF, and Karkanias GB. (2002). Effect of centrally administered insulin on gonadotropin-releasing hormone neuron activity and luteinising hormone surge in the diabetic female rat. Neuroendocrinology 76: 357–365.

Ma YJ, Berg-von der Emde K, Rage F, Wetsel WC, and Ojeda SR. (1997). Hypothalamic astrocytes respond to transforming growth factor-α with the secretion of neuroactive substances that stimulate the release of luteinising hormone-releasing hormone. Endocrinology 138: 19–25.

Ma YJ, Dissen GA, Merlino G, Coquelin A, and Ojeda SR. (1994). Overexpression of a human transforming growth factor-α (TGFα) transgene reveals a dual antagonistic role TGFα in female sexual development. Endocrinology 135: 1392–1400.

Meddle SL and Follett BK. (1995). Photoperiodic activation of Fos-like immunoreactive protein in neurones within the tuberal hypothalamus of Japanese quail. J Comp Physiol A 176: 79–89.

Meddle SL and Follett BK. (1997). Photoperiodically driven changes in Fos expression within the basal tuberal hypothalamus and median eminence of Japanese quail. J Neurosci 17: 8909–8918.

Miller DW, Blache D, and Martin GB. (1995). The role of intracerebral insulin in the effect of nutrition on gonadotropin secretion in mature male sheep. J Endocrinol 147: 321–329.

Morgan PJ and Williams LM. (1996). The pars tuberalis of the pituitary: A gateway for neuroendocrine output. Rev Reprod 1: 153–161.

Murakami M, Araki O, Hosoi Y, Kamiya Y, Morimura T, Ogiwara T, Mizuma H, and Mori M. (2001b). Expression and regulation of type II iodothyronine deiodinase in human thyroid gland. Endocrinology 142: 2961–2967.

Murakami M, Kamiya Y, Morimura T, Araki O, Imamura M, Ogiwara T, Mizuma H, and Mori M. (2001a). Thyrotropin receptors in brown adipose tissue: Thyrotropin stimulates type II iodothyroine deiodinase and uncoupling protein-1 in brown adipocytes. Endocrinology 142: 1195–1201.

Nakao N, Ono H, Yamamura, T, Anraku T, Takagi T, Higashi K, Yasuo S, Katou Y, Kageyama S, Uno Y, Kasukawa T, Iigi M, Sharp PJ, Iwasawa A, Suzuki Y, et al. (2008). Thyrotropin in the pars tuberalis triggers photoperiodic response. Nature 452: 317–322.

Nakao N, Takagi T, Iigo M, Tsukamoto T, Yasuo S, Masuda T, Yanagisawa T, Ebihara S, and Yoshimura T. (2006). Possible involvement of organic anion transporting polypeptide 1c1 in the photoperiodic response of gonads in birds. Endocrinology 147: 1067–1073.

Nicholls TJ, Follett BK, and Robinson JE. (1983). A photoperiodic response in gonadectomised Japanese quail exposed to a single long day. J Endocrinol 97: 121–126.

Nicholls TJ, Follett BK, Goldsmith AR, and Pearson H. (1988). Possible homologies between photorefractoriness in sheep and birds: The effect of thyroidectomy on the length of the ewe's breeding season. Reprod Nutr Dev 28: 375–385.

Ojeda SR, Urbanski HF, Costa ME, Hill DF, and Moholt-Siebert M. (1990). Involvement of transforming growth factor α in the release of luteinising hormone-releasing hormone from the developing female hypothalamus. Proc Natl Acad Sci USA 87: 9698–9702.

Okano T, Yamamoto K, Okano K, Hirota T, Kasahara T, Sasaki M, Takanaka Y, and Fukada Y. (2001). Chicken pineal clock genes: Implication of BMAL2 as a bidirectional regulator in circadian clock oscillation. Genes Cells 6: 825–836.

Pittendrigh CS. (1972). Circadian surfaces and the diversity of possible roles of circadian organization in photoperiodic induction. Proc Natl Acad Sci USA 69: 2734–2737.

Prendergast BJ, Mosinger B Jr, Kolattukudy PE, and Nelson RJ. (2002). Hypothalamic gene expression in reproductively photoresponsive and photorefractory Siberian hamsters. Proc Natl Acad Sci USA 99: 16291–16296.

Rage Y, Lee BJ, Ma YJ, and Ojeda SR. (1997). Estradiol enhances prostaglandin E2 receptor gene expression in luteinising hormone-releasing hormone (LHRH) neurons and facilitates the LHRH response to PGE2 by activating a glia-to-neuron signalling pathway. J Neurosci 17: 9145–9156.

Rebay I, Silver SJ, and Tootle TL. (2005). New vision from eyes absent: Transcription factors as enzymes. Trends Genet 21: 163–171.

Reppert SM and Weaver DR. (2002). Coordination of circadian timing in mammals. Nature 418: 935–941.

Revel FG, Saboureau M, Pevet P, Mikkelsen JD, and Simonneaux V. (2006). Melatonin regulates type 2 deiodinase gene expression in the Syrian hamster. Endocrinology 147: 4680–4687.

Rowan W. (1925). Relation of light to bird migration and developmental changes. Nature 115: 494–495.

Saldanha CJ, Leak RK, and Silver R. (1994). Detection and transduction of daylength in birds. Psychoneuroendocrinology 19: 641–656.

Sharp PJ and Follett BK. (1969). The effect of hypothalamic lesions on gonadotrophin release in Japanese quail (*Coturnix coturnix japonica*). Neuroendocrinology 5: 205–218.

Takagi T, Yamamura T, Anraku T, Yasuo S, Nakao N, Watanabe M, Iigo M, Ebihara S, and Yoshimura T. (2007). Involvement of transforming growth factor α in the photoperiodic regulation of reproduction in birds. Endocrinology 148: 2788–2792.

Tups A, Helwig M, Stohr S, Barrett P, Mercer JG, and Klingenspor M. (2006). Photoperiodic regulation of insulin receptor mRNA and intracellular insulin signalling in the arcuate nucleus of the Siberian hamster, *Phodopus sungorus*. Am J Physiol Regul Integr Comp Physiol 291: R643–R650.

Watanabe M, Yasuo S, Watanabe T, Yamamura T, Nakao N, Ebihara S, and Yoshimura T. (2004). Photoperiodic regulation of type 2 deiodinase gene in Djungarian hamster: Possible homologies between avian and mammalian photoperiodic regulation of reproduction. Endocrinology 145: 1546–1549.

Watanabe T, Yamamura T, Watanabe M, Yasuo S, Nakao N, Dawson A, Ebihara S, and Yoshimura T. (2007). Hypothalamic expression of thyroid hormone-activating and –inactivating enzyme genes in relation to photorefractoriness in birds and mammals. Am J Physiol Regul Integr Comp Physiol 292: R568–R572.

Wittkowski W, Bergmann M, Hoffmann K, and Pera F. (1988). Photoperiod-dependent changes in TSH-like immunoreactivity of cells in the hypophysial pars tuberalis of the Djungarian hamster, *Phodopus sungorus*. Cell Tissue Res 251: 183–187.

Wittkowski W, Bockmann J, Kreutz MR, and Bockers TM. (1999). Cell and molecular biology of the pars tuberalis of the pituitary. Int Rev Cytol 185: 157–194.

Yamamoto K, Okano T, and Fukada Y. (2001). Chicken pineal *Cry* genes: Light-dependent up-regulation of *cCry1* and *cCry2* transcripts. Neurosci Lett 313: 13–16.

Yamamura T, Hirunagi K, Ebihara S, and Yoshimura T. (2004). Seasonal morphological changes in the neuro-glial interaction between gonadotropin- releasing hormone nerve terminals and glial endfeet in Japanese quail. Endocrinology 145: 4264–4267.

Yamamura T, Yasuo S, Hirunagi K, Ebihara S, and Yoshimura T. (2006). T_3 implantation mimics photoperiodically reduced encasement of nerve terminals by glial processes in the median eminence of Japanese quail. Cell Tissue Res 324: 175–179.

Yasuo S, Nakao N, Ohkura S, Iigo M, Hagiwara S, Goto A, Ando H, Yamamura T, Watanabe M, Watanabe T, Oda S-I, Maeda K-I, Lincoln GA, Okamura H, Ebihara S, and Yoshimura T. (2006). Long-day suppressed expression of type 2 deiodinase gene in the mediobasal hypothalamus of the Saanen goat, a short-day breeder: Implication for seasonal window of thyroid hormone action on reproductive neuroendocrine axis. Endocrinology 147: 432–440.

Yasuo S, Watanabe M, Iigo M, Nakamura TJ, Watanabe T, Takagi T, Ono H, Ebihara S, and Yoshimura T. (2007b). Differential response of type 2 deiodinase gene expression to photoperiod between photoperiodic Fischer 344 and nonphotoperiodic Wistar rats. Am J Physiol Regul Integr Comp Physiol 292: R1315–R1319.

Yasuo S, Watanabe M, Nakao N, Takagi T, Follett BK, Ebihara S, and Yoshimura T. (2005). The reciprocal switching of two thyroid hormone-activating and -inactivating enzyme genes is involved in the photoperiodic gonadal response of Japanese quail. Endocrinology 146: 2551–2554.

Yasuo S, Watanabe M, Okabayashi N, Ebihara S, and Yoshimura T. (2003). Circadian clock genes and photoperiodism: Comprehensive analysis of clock genes expression in the mediobasal hypothalamus, the suprachiasmatic nucleus and the pineal gland of Japanese quail under various light schedules. Endocrinology 144: 3742–3748.

Yasuo S, Watanabe M, Tsukada A, Takagi T, Iigo M, Shimada K, Ebihara S, and Yoshimura T. (2004). Photoinducible phase-specific light induction of *Cry1* gene in the pars tuberalis of Japanese quail. Endocrinology 145: 1612–1616.

Yasuo S, Yoshimura T, Ebihara S, and Korf HW. (2007a). Temporal dynamics of type 2 deiodinase expression after melatonin injections in Syrian hamsters. Endocrinology 148: 4385–4392.

Young MW and Kay SA. (2001). Time zones: A comparative genetics of circadian clocks. Nature Rev Genet 2: 702–715.

Yoshimura T, Suzuki Y, Makino E, Suzuki T, Kuroiwa A, Matsuda Y, Namikawa T, and Ebihara S. (2000). Molecular analysis of avian circadian clock genes. Mol Brain Res 78: 207–215.

Yoshimura T, Yasuo S, Suzuki Y, Makino E, Yokota Y, and Ebihara S. (2001). Identification of the suprachiasmatic nucleus in birds. Am J Physiol Regul Integr Comp Physiol 280: R1185–R1189.

Yoshimura T, Yasuo S, Watanabe M, Iigo M, Yamamura T, Hirunagi K, and Ebihara S. (2003). Light-induced hormone conversion of T_4 to T_3 regulates photoperiodic response of gonads in birds. Nature 426: 178–181.

19

Photoperiodism in Mammals: Regulation of Nonreproductive Traits

Gregory E. Demas, Zachary M. Weil, and Randy J. Nelson

As noted in previous chapters, many plants and animals are exposed to annual fluctuations in the deterioration and renewal of their environments. Organisms tend to restrict energetically expensive processes and activities to a specific time of the year. Animals migrate or limit activity when food availability is low; for example, reproduction, preparation for migration, and other energetically demanding activities have evolved to coincide with abundant local food resources or other environmental conditions that promote survival. Thus, precise timing of physiology and behavior is critical for individual reproductive success and subsequent fitness. Importantly, vertebrate animals have evolved to fill both temporal and spatial niches.

Some physiological and behavioral adjustments occur in direct response to environmental fluctuations that have an obvious and immediate adaptive function. For example, reduced food or water availability can inhibit breeding (Nelson, 1987; Bronson, 1988). Such environmental factors have been termed the "ultimate factors" underlying seasonality (Baker, 1938). Many animals need to forecast the optimal time to breed so that spermatogenesis, territorial defense, or any other time-consuming adaptations can be developed prior to the onset of the breeding season. Thus, seasonally breeding vertebrate animals often must detect and respond to environmental cues that accurately signal, well in advance, the arrival or departure of seasons favoring reproductive success. The environmental cues used to anticipate environmental change may or may not have direct survival value. Such cues are called "proximate factors" (Baker, 1938). Photoperiod (day length) is the most notable example of a proximate factor. The annual changes in day length serve as a precise reference for the time of year. Under some circumstances, proximate and ultimate factors are identical (Nelson, 1987). For example, some individuals may not begin breeding until food cues are detected (Bronson, 1988).

In many cases, there are trade-offs (i.e., direct or indirect antagonistic interactions between two physiological processes that may have fitness consequences for organisms) among expensive physiological activities. Often individuals trade off energy between reproductive functions during mild seasons and survival functions during winter. For example, investment in reproduction and growth is curtailed, whereas investment in immune function is bolstered during winter.

This chapter addresses the physiological and cellular mechanisms underlying the detection of and response to environmental factors in regulating nonreproductive seasonal adaptations. Although the majority of the research within the area of mammalian seasonality has focused on seasonal changes in reproduction (see chapter 20), pronounced fluctuations in other nonreproductive responses, including changes in energy balance, immune function, and behavior, occur as well. Most research has focused on the role of photoperiod; presumably, with only two bits of data, length of day and direction of change in the photoperiod, individuals can precisely determine time of year and might then use this information to anticipate subsequent seasonal environmental changes.

PHOTOPERIODIC REGULATION OF ENERGY BALANCE

Introduction to Seasonal Variation in Energy Requirements

In many photoperiodic species, substantial changes in both metabolism and food intake occur when animals are transferred from long "summerlike" to short "winterlike" days, leading to appreciable changes in body weight and total body fat. Although a wide variety of mammalian species undergo seasonal cycles of body fat, the vast majority of research on these seasonal responses has focused on Siberian hamsters (*Phodopus sungorus*) and Syrian hamsters (*Mesocricetus auratus*) (Wade and Bartness, 1984b; Bartness et al., 2002; Morgan et al., 2003). For example, adult male Siberian hamsters housed in long days (16/8 h light–dark cycle [LD 16:8]) display relatively constant body masses; transfer to short days (LD 8:16), however, results in gradual and progressive loss in body weight (Wade and Bartness, 1984b; Bartness et al., 2002; Morgan et al., 2003). Although some of this weight loss is driven by decreased testis and muscle mass, the majority of the weight loss occurs in the form of decreased adiposity (Mercer et al., 2001).

Approximately 30–40% of the initial long-day body fat is lost by approximately 12–16 weeks in short days. Indeed, if long-day Siberian hamsters are transferred to short days and subsequently food restricted during their progressive decline in body mass, then these animals lose a greater amount of body mass than do short-day animals fed ad libitum. When short-day food-restricted hamsters are allowed to refeed, they substantially increase their food intake to compensate for food restriction. Interestingly, however, these animals do not return to their pre-food restriction levels. Rather, they regain a stable body mass at the reduced level that is consistent with the progressive short-day–induced decrease in body mass (Steinlechner et al., 1983). This and subsequent studies have confirmed the notion that body mass (and body fat) is a highly regulated, photoperiod-dependent, physiological response.

Neuroendocrine Mechanisms

Steroid Hormones

In a large number of rodent species, including hamsters (Hoffman et al., 1965), voles (Dark et al., 1983), deer mice (*Peromyscus maniculatus*) (Whitsett et al., 1983), and lemmings (Hasler et al., 1976), the gonads regress in response to exposure to short days. Rodents also display marked differences in body mass adjustments in response to photoperiod. For example, prairie voles (*Microtus ochrogaster*) (Kriegsfeld and Nelson, 1996), Syrian hamsters (Bartness and Wade, 1984; Campbell et al., 1983), and collared lemmings (*Dicrostonyx groenlandicus*) (Gower et al., 1994) increase body mass in short days, whereas Siberian hamsters (Hoffmann, 1973; Steinlechner and Heldmaier, 1982; Wade and Bartness, 1984a), meadow voles (*Microtus pennsylvanicus*) (Dark et al., 1983; Dark and Zucker, 1984), and deer mice (Blank and Freeman, 1991; Nelson et al., 1992) display short-day decreases in body mass. Furthermore, in many cases, the direction of the short-day change in body mass can be mimicked by gonadectomies, suggesting that photoperiodic changes in body mass are driven by changes in gonadal steroids. For example, gonadectomized long-day Syrian hamsters, prairie voles, and collared lemmings display increased body (and fat) masses comparable to those seen when intact animals are housed in short days (Morin and Fleming, 1978; Gower et al., 1994; Bartness, 1996). In contrast, gonadectomized Siberian hamsters, deer mice and meadow voles display short-day-like decreases in body mass and adiposity (Dark and Zucker, 1984, 1986; Wade and Bartness, 1984b; Blank and Freeman, 1991; Bartness, 1996). Short-day changes in gonadal steroid concentrations, however, cannot account for the total change in total body fat following short-day exposure. For example, exposure of gonadectomized Siberian hamsters to short days results in a further reduction in body mass (Wade and Bartness, 1984b). Furthermore, animals that experience prolonged exposure to short days experience "spontaneous recrudescence" during which they become refractory to the short-day signal and return to long-day gonadal and body masses (Dark and Zucker, 1984; Wade and Bartness, 1984b). Body masses return to long-day levels even in animals that were gonadectomized before short-day exposure (Hoffman et al., 1982). These findings suggest that, although gonadal steroids play an important role in regulating photoperiod changes in energy balance, other factors contribute to these changes.

Pancreatic Peptides

In most vertebrate species, peptides produced by the alpha and beta cells of endocrine pancreas act in a yin-yang fashion to coordinate storage (insulin) or liberation (glucagon) of glucose to be used as energy. Despite extensive research on these hormones in nonphotoperiodic "model systems" (e.g., rats, mice), much less is known about the role of these peptide in the photoperiodic regulation of body fat. In fact, among photoperiodic animals, seasonal fluctuations have only been investigated in two species of hamsters, Syrian and Siberian hamsters. As expected due to their differential responses to short days, Siberian hamsters display decreased serum insulin concentrations (Bartness et al., 1991), whereas Syrian hamsters increase serum insulin in response to short day lengths (Cincotta et al., 1991, 1993; Cincotta and Meier, 1995). In both species, insulin levels correlate with the level of adiposity. Siberian hamsters experiencing experimentally induced diabetes mellitus (via streptozotocin) and transferred to short days display normal decreases in body mass, body fat and food intake, suggesting that insulin does not play a critical role in photoperiod changes in energy, at least in this species. Insulin levels have been directly manipulated in Syrian hamsters via subdiaphragmatic vagotomies, which block insulin secretion indirectly via disruption of parasympathetic control of insulin release. Short-day hamsters that received subdiaphragmatic vagotomies continued to display gonadal regression, but the typical short-day decrease in body fat was blocked in these animals (Miceli et al., 1989). Although these data suggest that insulin may play a role in short-day increases in body fat, they are difficult to interpret because vagotomies generate many nonspecific physiological effects. Thus, the failure of short-day vagotomized animals to display photoperiodic changes in total body fat does not necessarily support the idea that insulin contributes to short-day–induced changes in body fat in this or other species.

In contrast to insulin, the pancreatic peptide glucagon plays an important role in stimulating lipolysis of white adipose tissue and thermogenesis in brown adipose tissue. This latter effect plays an important role in the thermogenic response to exposure to low ambient temperatures; thus, it seems plausible that increases in glucagon may play a role in regulating photoperiodic changes in energy balance, at least in species that undergo short-day decreases in body fat. Despite this intriguing possibility, a role for glucagon in regulating seasonal adiposity has not been examined in any photoperiodic species studied to date. Future studies will be needed to answer this question.

Leptin (OB Protein)

All mammals that undergo seasonal changes in reproduction and body mass display seasonal/photoperiodic changes in leptin, a peptide hormone produced almost exclusively by adipose tissue (Woods and Seeley, 2000; Drazen et al., 2001). Importantly, in both photoperiodic and nonphotoperiodic species, circulating levels of leptin are highly correlated with total body fat, suggesting that leptin serves an important role as a peripheral signal of adiposity. For example, serum leptin concentration correlates positively with body fat in Siberian hamsters over their yearly cycle (Drazen et al., 2000b; Horton et al., 2000). In addition, white adipose tissue leptin gene expression, circulating leptin concentrations, and leptin receptor gene expression are all reduced in short days compared with long days, consistent with short-day–induced decreases in body fat (Klingenspor et al., 1996a; Drazen et al., 2000b; Mercer et al., 2000b; Demas et al., 2002). Given that reduced leptin receptor gene expression contributes to a decrease in sensitivity to leptin, reduced gene expression in short days may reduce leptin sensitivity in short-day hamsters, and this indeed seems to be the case in Siberian hamsters (Mercer et al., 2000b). Note, however, that the dogma associated with the regulation of body fat by leptin states that when body fat levels decrease, the decrease in leptin triggers increases in food intake. The data above support the first portion of this dogma (i.e., short-day–induced decreases in body fat are associated with decreases in leptin gene expression by white fat; Klingenspor et al., 1996b) and circulating leptin concentrations (Klingenspor et al., 1996b; Drazen et al., 2000b; Horton et al., 2000); however, food intake is *decreased* in short photoperiods *not* increased, especially when body fat is at its seasonal nadir (Wade and Bartness, 1984b).

At least one of the energy-related short-day–induced changes by Siberian hamsters is reversed by peripheral chronic administration of leptin; the increase in food intake occurring when these animals are switched from short to long days is blocked by exogenous leptin (Drazen et al., 2001). Unlike other species, however, such as standard strains of laboratory rats and mice, leptin administration did not affect food intake when Siberian hamsters were at the body and lipid mass peaks in long days (Drazen et al., 2001). This result contrasts with the findings of two earlier studies of Siberian hamsters where peripheral leptin injections decreased food intake to the same extent in both long and short days, but reduced body and fat pad mass to a greater extent in short days (Atcha et al., 2000; Klingenspor et al., 2000). The precise reasons for these discrepancies are unknown, but in part may be due to differences in leptin administration, as well as other methodological considerations (for discussion, see Drazen et al., 2001). The most robust effect of leptin on food intake in rats and mice is when it is given intracerebroventricularly, and to our knowledge, this has not been done in Siberian hamsters. Although Siberian hamsters do not increase food intake after a fast, release from a less than complete food restriction can stimulate food intake (Fine and Bartness, 1996; Rousseau et al., 2003), and chronic peripheral leptin administration does not block this increase, nor does it have any effect on body or lipid mass in these animals (Rousseau et al., 2002). Despite the varied leptin-induced responses across these experiments, there is the tendency for leptin to act differentially between the photoperiods to affect energy balance and food intake. Therefore, seasonal changes in circulating leptin concentrations, coupled with changes in leptin sensitivity, may serve as part of an adaptive mechanism for increasing the odds of winter survival when food availability is decreased and adipose tissue stores are at their nadir (for review, see Rousseau et al., 2003).

Lastly, receptors for leptin (Ob-Rb) are located in several regions of the hypothalamus, and leptin binding to these receptors can contribute to photoperiodic changes in body mass and adiposity. As demonstrated for nonphotoperiodic rodents, photoperiodic rodents such as the Siberian hamster displays Ob-Rb in the hypothalamus, particularly in the areas of the arcuate nucleus (Arc), the ventromedial nucleus (VMN), the dorsomedial nucleus (DMN), the paraventricular nucleus (PVN), and the lateral hypothalamic area (LH) (Mercer et al., 2000a). In addition, Ob-Rb also respond to photoperiod; short-day hamsters display decreases in hypothalamic Ob-Rb compared with long-day animals (Mercer et al., 2000a, 2001). Importantly for the regulation of body mass, the neurons containing these receptors colocalize with other hypothalamic peptides that are involved in the regulation of energy balance; these peptides are described in the following section.

Hypothalamic Neuropeptides

A large number of neuropeptides are localized within specific regions of the hypothalamus in mammals which appear to play a critical role in the regulation food
intake (reviewed in Adam and Mercer, 2004). Among these peptides, a core group
of peptides have been identified and these peptides have been categorized into those
that stimulate food intake (orexigenic) and those that inhibit food intake (anorexigenic). Those peptides receiving most experimental attention are the orexigenic
peptides neuropeptide Y (NPY) and agouti gene–related peptide (AGRP), and the
anorexigenic peptides pro-opiomelanocortin (POMC) and cocaine- and amphet-
amine-related transcript (CART). Considerable evidence exists in support of the
idea that changes in one or more of these peptides are critical to the regulation of
food intake and thus energy balance. Although considerably less is known about the
potential role of these peptides in the seasonal regulation of body mass and energy
balance, recent evidence suggests that these peptides play an import role in photoperiodic species (figure 19.1). Recently, hypothalamic peptide gene expression has
been examined in long- and short-day–housed Siberian hamsters (Mercer et al.,
2000a, 2001). In general, these studies consistently report decreases in POMC and
increased expression in CART in short days, whereas both AGRP and NPY are
unaffected by photoperiod.

As with other rodent species, the hypothalamic peptide NPY acts as an orexigenic molecule by triggering robust increases in food intake (Boss-Williams and
Bartness, 1996). In addition, fasting-induced decreases in body fat in nonphotoperiodic species suggest increases in NPY gene expression (Adam and Mercer,
2004). In contrast, photoperiod has no effect on NPY gene expression (Adam
and Mercer, 2004). Additionally, based on food restriction studies, one would
expect increased CART expression in short days, consistent with that seen in
fasted animals. As with NPY, however, CART expression does not fit this simple
prediction. In fact, CART gene expression actually changes in the opposite direction as predicted, with decreased expression seen in short-day–housed animals.
Similar findings have been reported in the seasonally breeding sheep. Unlike
rodents, sheep are short-day breeders, breeding during autumn and winter
and inhibiting breeding during long days. As with rodents, however, there are
pronounced decreases in leptin in short days. Food restriction or food deprivation
in sheep results in increases in NPY and AGRP gene expression, and decreases
in CART and POMC gene expression, a finding consistent with rodents and with
the decrease in leptin levels. However, when leptin is decreased via exposure to
short days, there is no change in NPY or Ob-Rb, and up-regulated CART and
POMC expression (Marie et al., 2001). A subsequent study in sheep confirmed
the up-regulation in POMC gene expression, but reported increases in Ob-Rb
and NPY expression in short-day–housed sheep compared with long-day animals (Clarke et al., 2003). Thus, it is clear that some discrepancies exist in the
literature. Despite these discrepancies, however, taken together, the data from
food restriction studies combined with data from photoperiodic analyses, in both

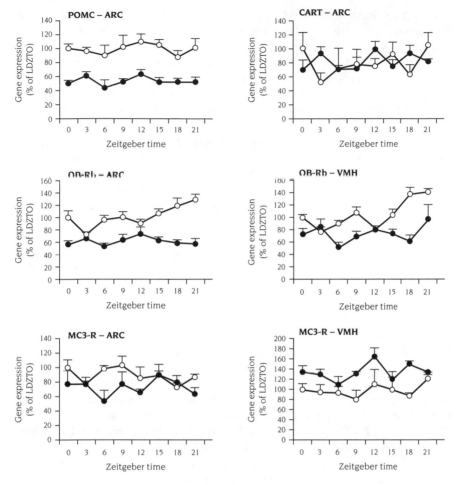

FIGURE 19.1.
Twenty-four–hour profiles of hypothalamic gene expression from adult male Siberian hamsters housed in long days (open circles) or short days (solid circles) for 12 weeks and killed at 3-h intervals throughout the respective LD cycles. Modified from Ellis et al. (2008).

sheep and hamsters, suggest that short-day animals, unlike food-restricted animals, maintain a state of energy balance despite decrease food intake and loss of body fat (Adam and Mercer, 2004).

Central Nervous System Regulation

Both white adipose tissue (WAT) and brown adipose tissue (BAT) receive innervation from the autonomic nervous system (ANS). Although sympathetic nervous system (SNS) innervation of BAT has been well established, SNS control of WAT has been more recently elucidated. Unlike SNS, no convincing evidence in support of parasympathetic innervation of WAT has been documented to date (reviewed in Bartness et al., 2002). Studies employing a variety of monosynaptic and trans-synaptic tract tracers, as well as functional studies involved physical or pharmacological denervations of SNS nerves innervating WAT, have demonstrated that SNS input plays an important role in the regulation of lipid mobilization. More important for this chapter, seasonal or photoperiodic changes in SNS outflow may play a critical role in photoperiodic regulation of body fat. For example, exposure to short days increases adipocyte sensitivity to noradrenergic stimulation in Siberian hamsters (Bowers et al., 2005). Furthermore, SNS denervations of WAT, coupled with adrenal demedullations (which eliminate noradrenergic outflow), can block short-day–induced decreases in body fat in Siberian hamsters (Demas and Bartness, 2001). These short-day–induced changes in lipid mobilization, like changes in reproduction, are likely driven my pineal melatonin; central nervous system (CNS) neurons that are part of the sympathetic outflow to WAT express melatonin (MEL1a) receptors. Thus, seasonally coded melatonin signals can act directly on SNS targets to regulated photoperiodic changes in today body fat, at least in hamsters, and likely other seasonally breeding rodents.

PHOTOPERIODIC REGULATION OF IMMUNE FUNCTION

Seasonal variation is observed in most major classes of diseases among vertebrates (Nelson et al., 2002; Nelson, 2004). Although much of the seasonality, for instance, in infectious diseases is related to specific environmental conditions and life-history traits of pathogens and vectors, there are also prominent fluctuations in host immune function that can affect disease parameters. In common with other nonreproductive adaptations to day length, changes in the immune system appear to be the result of "eavesdropping" on the neuroendocrine signals that tie photoperiod information to the reproductive system. Over evolutionary time adjusting immunological function in concert with changes in the reproductive system must have provided a competitive advantage. As the two systems are largely under the control of the pineal melatonin rhythm (Nelson and Demas, 1997), marked adjustments in immunological function can be induced in the laboratory by altering day length.

Why should immune function vary seasonally? Logically, it would seem that organisms would favor strategies in which they maintained the maximal immune defenses possible without inducing autoimmunity (Nelson et al., 2002). However, immune defenses are energetically costly to maintain and utilize (see below; see also Demas et al., 1997a; Martin et al., 2003), and so competing physiological processes reduce the energy available for immune defense. In a seasonal context, this is usually conceptualized via energetic trade-offs between the reproductive and immune systems (Deerenberg et al., 1997; Prendergast et al., 2004a; Martin et al.,

2008). Said another way, maximal reproductive output is not fully compatible with high levels of immune defenses. Indeed, there are many examples of short winterlike photoperiods that suppress reproduction also enhancing various immunological processes across vertebrate taxa (reviewed in Nelson and Demas, 1996; Nelson et al., 2002). From a life history perspective, individuals invest in survival mechanisms prior to puberty, whereas they invest in reproduction after puberty. Seasonally breeding animals fluctuate between reproductive condition and retrogression to a prepubertal state. We have conceptualized enhanced immune function outside of the breeding season in short days among small mammals and birds as representing individuals investment in overwinter survival (Nelson and Demas, 1996).

Previously, we outlined a scheme in which harsh winter conditions could both directly kill small vertebrates via starvation or hypothermia and indirectly kill by rendering them more susceptible to infectious diseases (Nelson and Demas, 1996), particularly via chronic-stress–induced glucocorticoid secretion (McEwen et al., 1997) and other immunosuppressive mediators. The primary hypothesis that has driven research in our laboratories has centered on the assumption that animals boost immune function during the winter in order to maintain immune function at viable levels despite the effects of harsh winter conditions and the associated stress responses (Nelson and Demas, 1996; Sinclair and Lochmiller, 2000; Nelson et al., 2002). As changes in day length reliably predict a set of environmental conditions, including low temperatures and reduced food availability, a boosting of immune responses can prophylactically block compromised immune function. This hypothesis suggests that the short-day enhancement of immune function seen in the laboratory would be limited in the wild during particularly harsh winters during conditions of reduced food availability, pronounced stressors, and higher thermoregulatory demands. Our research has largely borne out these predictions as low temperatures or food restriction can reduce short-day enhancement of immune activity among rodents (Demas and Nelson, 1996; Bilbo and Nelson, 2002). An alternative explanation is that maintenance and use of the reproductive system shunts energy away from the immune system and short-day enhancement of immune function actually represents a disinhibition mediated by regression of the reproductive tract. A thorough review of these two competing hypotheses is beyond the scope of this chapter (but see section below on sex steroids; for a review, see Martin et al., 2008), but most data in mammals support the former rather than the latter hypothesis.

The argument requires significantly more nuance as the vertebrate immune system is complex and consists of multiple partially redundant subsystems that do not necessarily covary (Demas and Nelson, 1998a; Martin et al., 2007). Thus, it would be an oversimplification and inaccurate to state that all aspects of immune defenses are enhanced in the nonbreeding season. Some (perhaps most) immunological processes are enhanced by short day lengths, but others are inhibited (Nelson and Demas, 1996; Yellon et al., 1999a; Drazen et al., 2001; Bilbo et al., 2002a).

The Vertebrate Immune System

The vertebrate immune system is a complex set of interacting tissues, cells, and soluble proteins diffusely distributed thoughout the body. Collectively, these systems serve to prevent infection and also control and expel pathogens if infections do take place (Janeway et al., 1999). Broadly, the immune system can be broken down into two basic categories that differ in their evolutionary and developmental origins. The innate immune system consists of a variety of cells (e.g., macrophages, granulocytes, and natural killer cells) that are principally responsible for rapid and nonspecific antimicrobial actions and for the removal of extracellular pathogens. Innate immune cells contain pattern recognition receptors for common microbial components and are capable of initiating an immune response rapidly and also can directly destroy many types of invaders. However, activation of the innate immune system can be particularly damaging to host tissue as the responses tend to be extremely nonspecific. Importantly, there are strong interconnections between the innate immune system and the nervous and neuroendocrine systems, as we discuss below.

In contrast, the acquired or adaptive immune system consists of cell-mediated (T-cell) and humoral (antibody-mediated; B-cell) arms. Cell-mediated immune function is primarily responsible for controlling intracellular infection (cytotoxic T-cells) and for coordinating immune responses in B-cells and other immunological cells (T-helper cells). B-cells produce antibodies to foreign cells or proteins and thus target invaders for destruction by other arms of the immune system. The adaptive immune system is relatively recent evolutionarily and is characterized by membrane bound receptors on B- and T-cells, which can detect foreign antigens. These receptors are generated in development by a process of gene recombination. Adaptive immunity is distinct from innate mechanisms because (1) they are slower acting than innate immunity and (2) adaptive immune cells retain immunological "memory" and thus can respond vigorously to a foreign antigen that it has previously encountered. Immunologists traditionally assail the distinctions between the two systems and new interconnections are continually discovered, but for our purposes it is advantageous to discuss the systems separately (e.g., Chan et al., 2006). Various measures of cell-mediated immune function are nearly always enhanced by short day lengths in a variety of reproductively photoperiodic rodent species (Demas and Nelson, 1998a; Bilbo et al., 2002a; Weil et al., 2006b).

Photoperiodic Modulation of Immune Function

Historically, many ecological immunology investigators have measured the size of various immunological tissues in early studies of the seasonality of immune function. This approach, though relatively uninformative about specific immune defenses, supports the notion that there is reception of photoperiod information by immunological tissues. Major organs of the immune system include the spleen, thymus, Bursa of Fabricius (in birds), bone marrow, and lymph nodes. These tissues perform various immunological processes including supporting leukocyte (white blood cell) development, filtering blood-borne antigens, and providing anatomical loci for interaction between innate and adaptive immune cells. Implicit in the measurement of these tissues is the argument that increased size of these tissues represents greater immune activity. This perspective is taken from seasonal reproductive function in which larger gonads are consistent with increased function. Short days increase the mass of these tissues in a variety of free-living species. For instance, splenic masses are larger in short days in Norway rats (*Rattus norvegicus*) and Syrian hamsters (Wurtman and Weisel, 1969; Vriend and Lauber, 1973; Vaughan et al., 1985). Similar patterns are evident in birds as spleen and thymus mass are at their nadir during vernal recrudescence (Lange and Silverin, 1985; John, 1994; Silverin et al., 1999). In other rodents, however, such as the short-tailed vole (*Microtus agrestis*) and European ground squirrel (*Spermophilus citellus*), splenic mass is reduced during winter. These data indicate the importance of comparing laboratory and fieldwork and also emphasize the need for more sensitive immunological measures. However, it is highly suggestive that immunological tissues (and thus putatively some aspects of immune function) are responsive to the changing seasons.

In addition to increasing immune activity, short day lengths also appear to buffer the immune system from suppression by metabolic stressors. 2-Deoxy-D-glucose (2-DG) interferes with the cellular uptake and utilization of glucose a key energy source for the immune system. Leukocyte proliferation is inhibited by 2-DG in female deer mice housed in long but not short days. 2-DG also elevated corticosterone concentrations in long-day but not short-day mice (Demas et al., 1997b). In contrast, Siberian hamsters enhanced delayed-type hypersensitivity (DTH) responses in short days; this effect is potentiated by access to a running wheel. However, the enhancing effects of both short days and exercise availability are blocked by food restriction. Together, these data indicate that short-day animals increase and defend enhanced immune responses against suppression by metabolic stressors (Bilbo and Nelson, 2004).

Adaptive Immune Function

The adaptive immune system, especially the cell-mediated arm of the system, is typically responsive to changes in day length. Cell-mediated immune function is generally measured by assessing leukocyte counts, lymphatic tissue masses, leukocyte proliferation to a mitogen in vitro (a putative index of the responsivity of the cells to stimulation in vivo), and also DTH responses (in vivo assay of cell-mediated immune function that integrates multiple immune parameters including antigen processing and presentation, leukocyte targeting, and extravasation, as well as T-cell–mediated inflammation). DTH represents an ecologically valid response to antigenic challenge. As a general rule, functional assays of immune activity (e.g., DTH or proliferative responses) are more meaningful than morphological measures such as spleen size (Martin et al., 2006).

Cotton rats (*Sigmodon hispidus*) enhance splenocyte (spleen lymphocytes) proliferation in response to the mitogen, concanavalin A during February relative to other months of year. In the laboratory, enhanced cell-mediated immune function in short day lengths has been reported in a variety of rodent species including several *Peromyscus* species, Siberian hamsters, meadow voles, and collared lemmings (Demas and Nelson, 1998b; Bilbo et al., 2002a; Pyter et al., 2005c; Weil et al., 2006b). DTH responses in Siberian hamsters are particularly responsive to changes in day length (figure 19.2). Short days increase total circulating leukocytes, lymphocytes, and T-cells and also increase cutaneous immune function, and these effects are boosted by acute stress. Similarly, short days enhance wound healing in deer mice (Nelson and Blom, 1994); stress enhances cutaneous wound healing in short- but not long-day hamsters (Kinsey et al., 2003), suggesting that neuroendocrine modulation of cell-mediated immunity is photoperiod responsive.

Humoral immune function tends to be more variable than cell-mediated responses. Photoperiod differences depend on the species and specific end points measured, but antigen-specific immune antibody responses are blunted in Siberian hamsters (e.g., Yellon et al., 1999b; Drazen et al., 2001; Yellon, 2007). Further, exogenous melatonin does not enhance antibody production in deer mice or Syrian

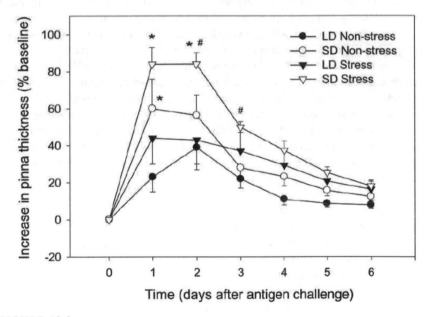

FIGURE 19.2.
Short-day (SD) hamsters exhibited an enhanced DTH response percentage during both stress and nonstress conditions compared to long-day (LD) hamsters. Restraint stress significantly increased the inflammatory response during the two to three days after challenge with the allergen dinitrofluorobenzene in short-day hamsters only. Reprinted with permission from Bilbo et al. (2002a).

hamsters (Demas and Nelson, 1998a; Drazen et al., 2002). Antigen-specific antibody production, however, was enhanced in Syrian hamsters by exposure to short days; total circulating antibodies (natural antibodies) showed a similar response in deer mice (Demas and Nelson, 1996; Drazen et al., 2002), although antigen-specific antibody production did not vary by photoperiod (Nelson and Blom, 1994). In mice, melatonin can enhance antibody responses in vivo and this effect is mediated by the MT2 (Drazen and Nelson, 2001), the melatonin receptor missing in Siberian hamsters (Weaver et al., 1996). Photoperiod adjustments in adaptive immune responses are more common and distinct in the cell-mediated rather than humoral arm of the immune system. Presumably, this reflects investments in the immune defenses best suited for host defense at different times of the year (Martin et al., 2008).

In addition to regulating immune function acutely, several studies have reported that past photoperiod exposure can modulate or organize immune responses later. Prenatal photoperiod information is transferred to developing fetuses in utero, and this information is processed by the immune system; immune cell counts were altered even when mice were cross-fostered to dams in the opposite photoperiod (Horton, 1984, 1985; Blom et al., 1994). Siberian hamsters housed in short days both perinatally and postweaning exhibit strong DTH responses. However, exposure to long days during either of those developmental periods prevented the enhanced DTH responses (Weil et al., 2006c), indicating that early-life photoperiod organizes and directs how photoperiod exposure later in life alters the immune system. Additionally, adaptive immune responses produce immunological memory that persists far beyond the initial response. To determine whether photoperiod altered the induction, retention, or expression of immunological memories, Siberian hamsters were acclimated to either long or short days and then exposed to the T-cell antigen dinitrofluorobenzene. Some hamsters were maintained in their current photoperiod or transferred to the opposite one. Day length modulated both the acquisition and retrieval of immunological memory. Short day lengths during both the initial exposure to an antigen or during a subsequent challenge boosted cell-mediated immune responses. This effect was mitigated by exposure to long days after the initial exposure. Similarly, short days during the secondary exposure enhanced responses, but to a lesser degree if the hamster had been in long days during the initial exposure (Prendergast et al., 2004b).

Innate Immune Function
The innate immune system is a broadly effective but nonspecific defense system that can readily protect hosts against invading microbes. The innate arm of the immune system is relatively understudied compared to the adaptive side. Our research groups have generally observed a suppression of most aspects of innate immune function in short day lengths. It seems possible that the full investment in energetically expensive innate immune activity such as fever is prohibited by the tight energy budget associated with winter (Lochmiller and Deerenberg, 2000). In Siberian hamsters, short days appear to suppress most aspects of the innate immune system. For instance, phagocytosis (engulfing bacteria) and oxidative burst activity

(index of cytotoxic potential) by granulocytes and monocytes were inhibited by short day lengths (Yellon et al., 1999a). We sought to expand on these findings using an integrative measure of innate immune activity. We challenged Siberian hamsters with lipopolysaccharide (LPS), a component of gram-negative bacterial cell walls that activates the immune system and recorded body temperatures and food intake over the subsequent days (Bilbo et al., 2002b). These measures provide an index of a large number of immunological processes, including LPS recognition, cytokine production, and neuroimmune communication. Additionally, peripheral inflammation also induces a suite of behavioral responses, including anorexia, lethargy, and reduced social interactions, that have been termed "sickness behaviors" and are mediated by cytokine signaling in the brain (Hart, 1988; Dantzer, 2001). Short day lengths suppressed both the amplitude and duration of febrile responses, weight loss, and reduction in food and sweetened milk intake associated with simulated infection (figure 19.3). This effect was associated with attenuation of both

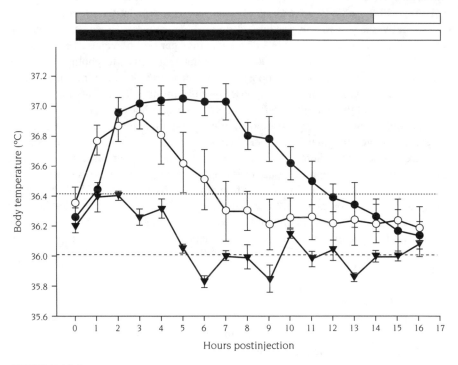

FIGURE 19.3.
Body temperature from 0 to 16 h post-LPS injections in long-day (black circles) and short-day (open circles) hamsters, and after saline injections in control hamsters (triangles). Black and gray bars above the graph indicate the active (dark) phase of the light–dark cycle in long- versus short-day hamsters, respectively. Horizontal dashed and dotted lines represent mean baseline body temperatures during the inactive (36°C) versus active (36.4°C) phases, respectively. Reprinted with permission from Bilbo et al. (2002b).

in vivo and in vitro proinflammatory cytokine production at both the protein and mRNA levels (Bilbo et al., 2002b; Pyter et al., 2005b). Additionally, expression of cyclooxygenase-2, the rate-limiting enzyme in the production of the inflammatory mediator prostaglandin E_2, was also reduced (Bilbo et al., 2003). High doses of LPS produce a condition termed endotoxemia that is similar to the clinical condition of sepsis. Short day lengths protected hamsters from lethal endotoxemia and reduced the dose of LPS that was lethal in 50% of animals by more than 90%. This was associated with reduced expression of the proinflammatory cytokine tumor necrosis factor α (Prendergast et al., 2003a).

Several possible explanations exist for these data. First, reduced cytokine production in short day lengths could be the key to a reduction in all downstream events. Treatment with interleukin-1β produces a sickness response similar to that created with LPS (Wen and Prendergast, 2007). However, short days attenuated the febrile response to an equivalent amount of interleukin-1β, indicating that differential cytokine production cannot completely explain photoperiod differences in inflammatory responses. Additionally, there are no photoperiod differences in gene expression of the principal LPS detector, toll-like receptor 4 (TLR4), in the spleen or peritoneal macrophages of Siberian hamsters (Navara et al., 2007). Taken together, these data suggest that the intracellular signaling pathways connecting TLR and cytokine receptor to gene expression changes are blunted by exposure to short days. Future studies should examine the nuclear factorκB signaling pathway as it is involved in both TLR and cytokine receptor signaling (Carmody and Chen, 2007). Finally, the possibility has to be considered that melatonin acting in the CNS mediates photoperiod modulation of sickness responses. Melatonin administration in the suprachiasmatic nucleus reduced the behavioral, but not febrile or cytokine, responses to acute LPS (Freeman et al., 2007).

Proximate Mediators of Photoperiod Changes in the Immune System

Researchers have consistently reported large effects of day length on various aspects of immune activity. In trying to identify proximate mediators of photoperiod changes in the immune system, our research quickly centered on the pineal indole melatonin. Photoperiod changes in immune function, as with most day-length–dependent traits, can be reproduced with the administration of appropriately timed exogenous melatonin (Bilbo and Nelson, 2002; Hotchkiss and Nelson, 2002; Prendergast et al., 2003b); similarly, immune adjustments to chronic short-day exposure are not evident in pinealectomized animals (Yellon, 2007). However, it was not immediately evident in studies of photoperiodic modulation of immune function whether melatonin was acting directly on the immune system or alternatively whether melatonin-induced changes in other neuroendocrine systems (e.g., neuropeptides, sex steroid hormones, or glucocorticoids) could be the key mediator of photoperiod adjustments in the immune system.

Melatonin

The principal physiological mediator of day length is the neurohormone melatonin. Pineal melatonin is secreted at night, and its production is under the control of a multisynaptic pathway originating in photosensitive cells in the retina. Melatonin interacts with high-affinity G-protein–coupled membrane receptors on a variety of tissues. Most mammals have two melatonin receptor subtypes: melatonin receptor 1a (Mel 1a or mt1) and melatonin receptor 1b (Mel 1b or mt2) that are widely distributed on a large variety of cell types in the CNS and in the periphery. In birds, a third melatonin receptor subtype (1c) has also been identified (Reppert et al., 1995). Importantly, melatonin receptors are present on a variety of immune cell types and tissues (Garcia-Perganeda et al., 1997; Pozo et al., 1997; Barjavel et al., 1998; Konakchieva et al., 1999). Melatonin is a potent immune modulator in a variety of species across taxa, including laboratory rodents, ungulates, and passerines (Bentley et al., 1998; Bilbo and Nelson, 2002; Dahl et al., 2002; Guerrero and Reiter, 2002).

In domesticated laboratory animals, melatonin consistently enhances nearly all aspects of immune function (Guerrero and Reiter, 2002). It should be noted that many laboratory mouse strains have a defect in the *N*-acetyltransferase gene (the rate-limiting enzyme in melatonin synthesis) (Goto et al., 1989; Stehle et al., 2002), and as such, nighttime melatonin rhythms are blunted or absent, suggesting that exogenous melatonin effects on the immune system must be interpreted carefully. However, in general, surgical or functional (i.e., housing in constant light which suppresses pineal melatonin production) pinealectomy reduces thymic and splenic masses in laboratory rodents (Vaughan and Reiter, 1971; McKinney et al., 1975), an effect that is mediated largely by a reduction in lymphocytes (Maestroni et al., 1986). Additionally, both cell-mediated and humoral immune function are impaired by pinealectomy, an effect that is blocked by exogenous melatonin (Vermeulen et al., 1993; Yellon et al., 1999b). Exogenous melatonin enhances such diverse immune responses as lymphocyte proliferation, antibody-dependent cellular cytotoxicity, antigen presentation, and cytokine production; evidence suggests that these effects are mediated at least in part by interaction with endogenous opioids (Maestroni et al., 1987; Vermeulen et al., 1993; Guerrero and Reiter, 2002; Hotchkiss and Nelson, 2002). Further, melatonin is particularly effective in enhancing immune function in immunosuppressed organisms and appears to antagonize some of the immunosuppressive effects of glucocorticoids (Maestroni et al., 1988).

In photoperiodic rodents, however, exogenous melatonin tends to recapitulate photoperiodic differences in immune function rather than being universally immunoenhancing. As discussed above, short day lengths and the lengthened melatonin rhythm are associated with suppression of some immunological parameters such as humoral immunity and cytokine-mediated sickness behaviors in Siberian hamsters (Bilbo et al., 2002b; Drazen et al., 2001). Melatonin can mimic these suppressive effects (Bilbo and Nelson, 2002).

Photoperiodic adjustments in the mammalian immune system require alterations in melatonin secretion as pinealectomy blocks photoperiod-induced enhancement of humoral immunity and exogenous melatonin can recapitulate short day lengths

(Demas and Nelson, 1998a; Yellon et al., 1999a, 2005; Bilbo and Nelson, 2002; Yellon, 2007). Further, melatonin and photoperiod share similar temporal characteristics. For instance, photoperiodic rodents typically become refractory to the suppressive effects of short day lengths on the reproductive system and will spontaneously regrow their testes after 20–24 weeks in photoperiod treatment (Prendergast et al., 2002a). Immunological adjustments associated with short day lengths also revert to the long-day pattern once animals become reproductively refractory to short days (Prendergast and Nelson, 2001). Melatonin does not alter immune function either in vivo or in vitro in refractory animals (Prendergast et al., 2002b).

Latitude of origin is an important determinant of reproductive responses to day length (animals from lower latitudes tend to be less responsive to day length than those from higher latitudes) (Bronson, 1985). Similarly, immunological adjustments to both day length and exogenous melatonin were linked to reproductive responses to photoperiod (Demas et al., 1996). Said another way, individuals that did not regress their gonads in short days also did not adjust cell-mediated or humoral immune function in response to changes in day length or exogenous melatonin.

Although lengthened melatonin rhythms are necessary for photoperiod changes in the immune system, it is difficult to parse direct effects of melatonin on immune cells and tissues from indirect immunomodulatory effects of melatonin that occur via changes in other neuroendocrine axes. To test the hypothesis that melatonin directly altered proliferation responses, we treated cultured prairie vole lymphocytes with melatonin and observed an enhanced proliferative response (Drazen et al., 2000a; Kriegsfeld et al., 2001). Indirect immune effects of melatonin on the reproductive or hypothalamic-pituitary-adrenal (HPA) axes are not possible as those tissues were no longer present. Siberian hamsters reduce lymphocyte proliferation following exposure to short day lengths; addition of melatonin to lymphocyte cultures suppresses proliferative responses in long but not short days (presumably because endogenous melatonin has already suppressed proliferative responses) (Prendergast et al., 2001b). Importantly, in house mice, the melatonin 2 (mt2), but not mt1, receptor is required for enhancement of lymphocyte proliferation (Drazen and Nelson, 2001). However, Siberian hamsters lack a functional mt2 receptor, and this may underlie the divergent effects of photoperiod on the hamster immune system as compared to other photoperiodic rodents (Weaver et al., 1996). These data indicate that at least some aspects of immune photoperiodic modulation of the immune system are mediated directly by acute changes in circulating melatonin. In most cases, however, chronic long-duration melatonin treatment is necessary for the expression of short-day-like patterns of immune activity (Bilbo and Nelson, 2002; Drazen et al., 2002). Taken together, these data indicate that melatonin is necessary for the expression of photoperiod changes in immune activity, and to a large extent, it is also sufficient even in the absence of other neuroendocrine cues.

Sex Steroid Hormones

One of the key downstream targets of long-duration melatonin is the neuroendocrine reproductive axis. As sex steroids are also potent immunomodulators, it was reasonable to hypothesize that melatonin acted on the immune system at least in part by modulation of gonadal steroids. In general, females have stronger immune systems and are less prone to infection than males (Klein, 2000). This sex difference is mediated largely by sex steroids with androgens being generally immunosuppressive (Folstad and Karter, 1992), whereas estrogens tend to have the opposite effect (Klein, 2004); although many counterexamples are available (Roberts et al., 2004). Yet, it has been known since the nineteenth century that prepubertal castration of rabbits increased thymic mass (Calzolari, 1898), and important ecological theories have been based on the idea that the support of secondary sex characteristics by androgens obligately suppresses immune activity in males (e.g., Folstad and Karter, 1992).

Short day lengths decrease circulating androgens and enhance many aspects of immune function in *Peromyscus* (Blom et al., 1994). This set of findings led to an early hypothesis that short day lengths enhanced immune activity by removing (disinhibiting) the suppressive effects of androgens (Nelson and Demas, 1996). A further prediction of that hypothesis is that if estrogens are immunoenhancing and short day lengths suppress estrogen release, then short days should suppress immune activity in females but enhance immune activity in males. However, this explanation has fallen out of favor because photoperiod differences in immune activity are relatively similar in both sexes (Demas and Nelson, 1998b) and gonadectomy (or androgen replacement) does not significantly alter photoperiod differences in immune activity in either sex (Demas and Nelson, 1998b; Prendergast et al., 2005). Nonetheless, sex steroids may have some lesser modulatory role that is typically masked by the larger effects of melatonin (Bilbo and Nelson, 2001; Prendergast et al., 2002b, 2008).

Further evidence against sex steroid mediation of the immune system comes from a series of studies where reproductive and immunological responses to day length can be dissociated. First, castrated animals exhibit the expected photoperiod differences in immune function (Demas and Nelson, 1998b; Prendergast et al., 2005), with only relatively small contributions of sex steroids to photoperiod differences in the immune system (Prendergast et al., 2008). Second, intermediate and perinatal photoperiods can dissociate reproductive and immunological responses to day length (Prendergast et al., 2004a; Weil et al., 2006c) such that regression of the reproductive tract can occur in the absence of enhanced immune function. Housing Siberian hamster males with ovariectomized females in short days suppresses both immune activity and the reproductive tract (Weil et al., 2007b). Finally, photoperiodic nonresponders (animals that fail to respond to day length with regression of the reproductive tract; Prendergast et al., 2001a) enhance immune function (Drazen et al., 2000b). Taken together, these data suggest that the immune and reproductive systems are both affected by photoperiod, but there is minimal direct interaction between the two systems, in a seasonal context, at least when food is available ad libitum.

Glucocorticoids

The HPA axis is another immunomodulatory system that is regulated by day length (Nelson and Demas, 1996; Ronchi et al., 1998). Glucocorticoids are the end product,

primary effectors, and principal negative regulators of the HPA axis and are intimately and bidirectionally connected with the immune system. Glucocorticoids
receptors are present on most types of immune cells (Smith et al., 1977; Armanini
et al., 1988; Miller et al., 1998). Cytokines produced by cells of the innate immune
system are a potent driver of HPA activity (Turnbull and Rivier, 1995); the resulting
glucocorticoids then feedback to inhibit cytokine production and modulate other
aspects of leukocyte physiology. Glucocorticoids have thus been conceptualized as
brakes on the neuroendocrine-immune circuit that have evolved to prevent runaway
inflammation (McEwen et al., 1997; Sapolsky et al., 2000; Sapolsky, 2002). The
downside of this putative adaptation is that chronic elevation or high doses of glucocorticoids can suppress the immune system generally, and inflammatory responses
specifically (McEwen et al., 1997; Padgett and Glaser, 2003).

HPA axis physiology (basal glucocorticoid concentrations, stress-evoked concentrations, negative feedback dynamics, and receptor distribution, etc.) is altered
by day length (Ronchi et al., 1998; Pyter et al., 2007); the effects of photoperiod
on HPA physiology are not nearly as large as those on the reproductive axis (Pyter
et al., 2007), and the direction of these adjustments varies across experimental procedures and species (Ronchi et al., 1998; Bilbo et al., 2002a; Pyter et al., 2005a).
Thus, the stressors associated with harsh winter conditions and concomitant adjustments in HPA axis dynamics could suppress immune function and therefore may
have led to the evolution of photoperiod-mediated mechanisms for prophylactic
enhancement of immune activity (Demas and Nelson, 1996; Nelson and Demas,
1996).

Basal circulating corticosteroids do not appear to mediate photoperiod differences in immune activity. For example, photoperiod differences in immune activity have been reported in rodents that do not differ in circulating corticosteroids
(e.g., Demas and Nelson, 1998b; Weil et al., 2006a). Furthermore, many studies
have reported enhanced immune activity in short days in animals with *higher* basal
glucocorticoids (Bilbo et al., 2002a; Pyter et al., 2005c; Weil et al., 2006c). Such
studies, however, do not fully consider the importance of 24-h dynamics of glucocorticoids, potential differences in receptor distribution, or downstream signaling events. Studies on animals with clamped HPA axes will be necessary to fully
uncover the role of basal glucocorticoids in photoperiod differences in immune
activity.

In recent years there has been a partial reconceptualization of glucocorticoid-
immune interactions. Although chronic exposure to or high doses of glucocorticoids
can certainly be immunosuppressive, there is mounting evidence that glucocorticoids can also induce redistribution rather than death of immune cells (Dhabhar
and McEwen, 1997). Acutely stressing laboratory rats leads to a marked reduction
in circulating leukocytes; the cells are not dying, but rather are trafficking to the
front lines of immune defense in the skin, gut, and lymphatics to mediate enhanced
immune activity in these tissues (Dhabhar et al., 1995; Dhabhar and McEwen,
1999). This effect is mediated by acute elevation of glucocorticoids (Dhabhar et al.,
1996) and is sensible from an evolutionary perspective as acute stressors would

often have been associated with wounding and potential infection; thus, it would not be adaptive for acute increases in glucocorticoids to suppress the immune system. Short day lengths enhanced both basal and stress-induced trafficking of immune cells and skin immune function; the increase in immune cell trafficking was associated with higher glucocorticoid concentrations both prior to and after restraint stress (Bilbo et al., 2002a). Future studies will address whether short-day patterns of exogenous glucocorticoids can produce the short-day magnitude of leukocyte trafficking in long-day hamsters. We predict that this experiment would yield leukocyte-trafficking responses intermediate between the typical patterns of stress-induced trafficking of long- and short-day hamsters. In sum, these results suggest that although glucocorticoids may modulate immune activity differentially at different times of the year, glucocorticoids are probably not the key proximate mediator of photoperiod differences in immune activity.

Prolactin

Prolactin is a protein hormone produced in the anterior pituitary with many and varied roles in the regulation of growth, reproduction, development, and water and electrolyte balance, as well as immune function (Goffin et al., 1999; Yu-Lee, 2002). Prolactin receptors are expressed in the cells and tissues of the immune system (Leite De Moraes et al., 1995), and adenohypophysectomy (removal of the anterior pituitary) leads to thymic involution and deficits in cell-mediated and humoral immunity that can be blocked with exogenous prolactin (Smith, 1930; Reber, 1993). Consistent with these effects on immune tissue, prolactin generally enhances immune responses in laboratory animals. For instance, prolactin increases lymphocyte proliferation (Matera et al., 1992), although these studies typically use doses outside of the physiological range. High concentrations of the hormone can increase immune activity to the point of autoimmunity, and prolactin release inhibitors are being used clinically in cases of organ rejection and autoimmunity (Vera-Lastra et al., 2002). Prolactin antagonizes glucocorticoid-induced lymphocyte apoptosis and may be part of a network that maintains immunological homeostasis (Fletcher-Chiappini et al., 1993; Dorshkind and Horseman, 2001).

A reduced concentration of prolactin in short day lengths is one of the most consistent findings in mammalian physiology (Goldman and Nelson, 1993). Despite the generally immunoenhancing effects of prolactin in laboratory model species, prolactin concentrations are generally elevated in long days and are associated with blunted immune responses. This is likely related to variations in experimental procedures (e.g., doses and the use of highly domesticated laboratory animals). In more recently captive outbred animals such as deer mice, treatment with the chemical carcinogen dimethylbenzathracene causes squamous cell carcinomas to develop only if the animals are housed in long day lengths; no short-day mice developed the tumor. However, when long-day mice were treated with the prolactin release inhibitor bromocriptine, the incidence of tumors declined nearly 50% (Nelson and Blom, 1994), although the role of immunological processes in the development of these tumors is relatively small. In steers, short photoperiods increase lymphocyte proliferation and neutrophil chemotaxis. However, exogenous prolactin, administered to increase short-day animals to long-day levels, blocked the short-day enhancement of these responses (Auchtung and Dahl, 2004). More data are required, but it seems possible that reduced circulating prolactin in short days may be necessary for photoperiodic plasticity in the immune system.

Leptin

As described, leptin is an adipocyte-derived peptide hormone that was originally described as the product of the *OB* gene (Zhang et al., 1994). OB mice are obese due to both overeating and decreased energy expenditure (Coleman, 1978). Leptin serves at least in part as a signal to various tissues of adiposity and energy availability. Immune cells express leptin receptors, so leptin can interact directly with them, and the leptin receptor shares many signaling properties with the cytokine receptor interleukin-6R (Baumann et al., 1996). However, immunomodulatory effects of leptin may be, at least in part, mediated by changing the availability of metabolic fuels. Leptin-deficient mice have impaired wound healing, fewer T-cells, and impaired macrophage responses (Lord et al., 1998). These effects can be recapitulated in wild-type mice by severe food restriction and reversed in both starved and leptin-deficient mice with exogenous leptin (Lord et al., 1998). In wild-type animals, leptin increases phagocytosis and T-cell proliferation (Baumann et al., 1996).

No consistent relationship exists between day length and leptin concentrations across species. This probably reflects the various strategies that different species utilize in terms of winter energy storage (i.e., weight loss or weight gain). Photoperiodic differences have been reported in Siberian hamsters and woodchucks (*Marmota monax*) (Klingenspor et al., 1996b; Concannon et al., 2001). Siberian hamsters reduce antibody production in short day lengths. Treatment with exogenous leptin increased antibody production in short-day animals but had no effect on long-day animals (Drazen et al., 2001). This effect was mediated by increased food intake in leptin treated hamsters (Drazen et al., 2001). Surgical removal of body fat decreases humoral immune function (Demas et al., 2003), and this effect can be antagonized by exogenous leptin (Demas and Sakaria, 2005). Leptin appears to provide information to the immune system about energy stores and thus may regulate the expression of energetically costly immune responses (Demas, 2004; Demas and Sakaria, 2005).

PHOTOPERIODIC REGULATION OF BEHAVIOR

Affective Behaviors

Higher mental functions in human and nonhuman animals can be divided into cognition, behavior, and affect/mood (Rubin et al., 2002). Affect is defined as the "observable emotional state of individuals," and affective disorders as "abnormal states of feeling, primarily excessive sadness or elation" (Rubin et al., 2002).

Although traditionally considered to be maladaptive in humans, behaviors similar to symptoms of depression and anxiety disorders persist in other species and may be adaptive under specific environmental circumstances, including those associated with the changing seasons. For example, symptoms of affective disorders, such as lethargy, altered food intake, loss of sexual motivation, and fearfulness may actually conserve energy during the winter, a time when many organisms experience marked energetic bottlenecks (Nesse and Williams, 1996; Nesse, 2000; Wehr et al., 2001). Affective behaviors, and the neuroanatomical and neurochemical regulation of these behaviors, may be modified by seasonal information. Studies examining the effects of photoperiod on affective behaviors in nonhumans, however, are limited. One study conducted in laboratory rats (*Rattus rattus*) demonstrated that exposure of animals to long days reduced depressive-like behavior (Molina-Hernandez and Tellez-Alcantara, 2000). In addition, short day lengths decreased neophobia in two strains of *Mus musculus* (Kopp et al., 1999). Lastly, a recent study in rats demonstrated that rats housed in short days displayed higher levels of anxiety-like behavior on both open-field and elevated-plus tests (Benabid et al., 2008). Short-day–exposed rats that received a light pulse in the middle of the dark phase, however, did not display comparable increases in anxiety (Benabid et al., 2008). Interestingly, both house mice and laboratory rats have traditionally been considered nonresponsive to photoperiod, at least with respect to reproduction.

A recent study in a photoperiodic rodent reported elevated depressive- and anxiety-like peripubertal Siberian hamsters (Prendergast and Nelson, 2005). Short-day–exposed male hamsters spent less time in exposed areas of an elevated-plus maze relative to sheltered areas and exhibited behavioral despair more frequently in the Porsolt forced-swim test relative to long-day males (figure 19.4). These behaviors were seen after only two weeks of short-day exposure, well before changes in gonadal steroids and reproductive function. Subsequently, the effects of both perinatal and postweaning photoperiod on affective behavior were examined in Siberian hamsters (Pyter and Nelson, 2006). Hamsters exposed to short days (LD 8:16) perinatally displayed more anxiety-like behavior as adults in an elevated-plus maze test but displayed less anxiety-like behavior in the open-field and marble burying tests compared with hamsters born in long days (LD 16:8) (Pyter and Nelson, 2006). Hamsters exposed to short days postweaning displayed more anxiety-like behavior as adults in the elevated-plus maze and open-field tests and more depressive-like behavior in the Porsolt forced-swim test compared with those exposed to long days. A similar study was conducted in collared lemmings housed in long (LD 22:2), intermediate (LD 16:8) or short day lengths (LD 8:16) for nine weeks (Weil et al., 2007a). Specifically, lemmings housed in long days reduced anxiety-like responses in the elevated-plus maze. Depressive-like behaviors were decreases in animals housed in the intermediate photoperiod relative to both long- and short-day–housed lemmings (Weil et al., 2007a). Collectively, these results support the hypothesis that affective behaviors are organized early in life and can be maintained throughout adulthood. In addition, both anxiety- and depressive-like behavioral responses can be modulated by the postnatal environment, suggesting that photoperiodic

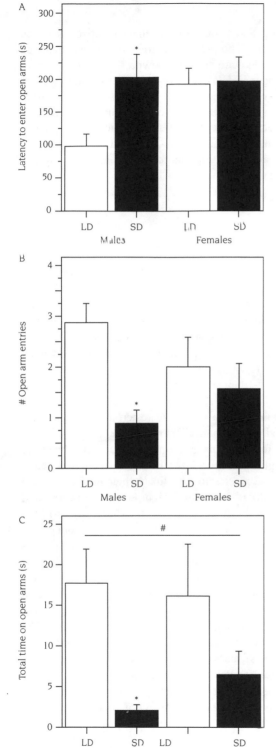

FIGURE 19.4.
Results of an elevated-plus maze test of male and female Siberian hamsters exposed to either long days (LD) or short days (SD) beginning at weaning (day 18). (A) Latency to first enter an exposed arm. (B) Number of entries onto exposed arms. (C) Total amount of time exploring exposed arms. Modified from Prendergast and Nelson (2005).

changes in human affective behaviors and disorders may reflect adaptive responses to a seasonally changing environment. The precise neuroendocrine mechanisms explaining these changes in behavioral state are not known at this time but may reflect seasonal/photoperiodic changes in HPA activity. In support of this idea, it has recently been demonstrated that exposure to short days increases corticosterone responses to environmental stressor and enhances glucocorticoid negative feedback of the HPA axis in white-footed mice (Pyter et al., 2007). Regardless of the precise mechanisms, the adaptive need for behavioral and affective adjustments to the seasonal environment suggests that resulting photoperiodic modifications of physiology and behavior are critical for survival.

Aggressive Behaviors

Aggression is a highly complex behavior displayed by virtually all living organisms that serves a wide range of adaptive functions. The possibility for aggressive behavior exists whenever the interests of two or more individuals are in conflict, typically involving limited resources (e.g., food, territories, and mates). Despite its importance, aggression is a notoriously nebulous concept that has been defined and categorized in a multitude of ways over the years. Aggression has traditionally been defined as overt behavior with the intention of inflicting physical damage upon another individual or "goal entity" (Moyer, 1971).

Photoperiod

Several studies have indirectly examined the role of testosterone (T) in aggression by manipulation of photoperiod. Many nontropical rodent species are seasonal breeders, maintaining reproductive function during summer and curtailing breeding during the winter. Ambient day length (photoperiod) is the proximal environmental cue used by individuals within these species to coordinate their reproduction to the appropriate season (Goldman, 2001). For example, reproductive function (and high levels of circulating T) is maintained during long "summerlike" days (>12.5 h of light per day), whereas reproductive regression, including virtual collapse of the gonads and marked decreases in T, occurs during the short "winterlike" days (<12.5 h/day) (Goldman, 2001). Interestingly, maintaining male Syrian hamsters in short days *increases* resident-intruder aggression compared with long-day hamsters (Garrett and Campbell, 1980). Specifically, adult male Syrian hamsters housed in short days for nine weeks display approximately twice the amount of aggression in a resident-intruder test compared with long-day controls when tested 4 h before dark, despite gonadal regression (Garrett and Campbell, 1980). After prolonged maintenance in short days (>15 weeks), hamsters typically undergo spontaneous gonadal recrudescence (i.e., increased testicular mass and circulating T), despite continued maintenance in short days. The short-day increase in aggressive behavior largely disappear in animals undergoing spontaneous recrudescence, returning to long-day levels of aggression by 21 weeks (Garrett and Campbell, 1980). More recently, short-day increases in aggression in male Syrian hamsters have been confirmed (Jasnow et al., 2002; Caldwell and Albers, 2004). For example, Syrian hamsters housed in short days (LD 10:14) for 10 weeks displayed a significantly greater number of attacks and a longer duration of attacks than did long-day hamsters when tested using a resident-intruder test (Jasnow et al., 2002). Furthermore, timed daily melatonin injections mimicking short-day patterns of the hormone in long-day, pineal-intact animals will produce short-day-like increases in aggression. Because these injections occurred for only 10 days, gonadal mass and circulating levels of T are unaffected, supporting the idea that photoperiodic changes in aggression are not mediated by changes in gonadal steroids in this species (Jasnow et al., 2002). In contrast, these results suggest that levels of aggressive behavior are mediated by changes in the pattern of melatonin secretion.

Photoperiodic changes in aggression have been demonstrated in females of at least one species, Syrian hamsters (Fleming et al., 1988; Badura and Nunez, 1989). Female hamsters were housed in long (LD 14:10) or short days (LD 6:18) for 12 weeks, and then both offensive and defensive aggression were tested (Fleming et al., 1988). Female hamsters maintained in short days displayed significantly less defensive aggression compared with long-day animals and thus had a higher ratio of offensive to defensive aggression than did long-day animals.

Melatonin

In virtually all mammals, photoperiodic responses are mediated by changes in the pineal indoleamine melatonin. Melatonin is secreted in abundance during darkness, whereas daylight inhibits pineal melatonin secretion (Goldman, 2001). Thus, changes in ambient day length result in changes in the pattern of secretion of melatonin. In this manner, it is the precise pattern of melatonin secretion, and not the amount of hormone per se, that provides the biochemical "code" for day length (Goldman, 2001).

Pinealectomy, which eliminates melatonin secretion and renders animals physiologically "blind" to day length, prevents the short-day increase in aggression in female Syrian hamsters, whereas treatment of long-day hamsters with exogenous short-day-like melatonin increases aggression in female Syrian hamster (Fleming et al., 1988). Ovariectomy, in contrast, has no effect of aggression. This finding suggests that photoperiodic changes in aggression are independent of changes in gonadal steroids in female Syrian hamsters (Fleming et al., 1988). A subsequent study in female Syrian hamsters confirmed these findings and provided further support for a role of pineal melatonin in mediating photoperiod changes in aggression. Specifically, a higher percentage of female hamsters housed in short days (LD 6:18) showed aggressive behavior compared with long-day–housed (LD 16:8) hamsters (Badura and Nunez, 1989). Consistent with previous findings, short-day aggression was attenuated by pinealectomy, but treatment with exogenous estradiol (alone or in combination with progesterone) had no effect on aggression. These results support the hypothesis that photoperiodic changes in aggression are mediated by pineal melatonin, but independent of gonadal steroids, at least in female Syrian hamsters.

In Syrian hamsters, unlike most rodent species, females are more aggressive than males (Marques and Valenstein, 1977; Ciaccio et al., 1979). Few studies have examined the role of photoperiod on male aggression in rodents displaying typical male-dominant aggression. Unlike Syrian hamsters, male Siberian hamsters display significantly more aggression than do females. It has been demonstrated that short-day male Siberian hamsters are significantly more aggressive than were long-day animals (Jasnow et al., 2000, 2002), consistent with previous studies in Syrian hamsters. Specifically, male Siberian hamsters housed in short days (LD 8:16) for 10 weeks display a greater number of attacks during a resident-intruder test and have a lower latency to initial attack, relative to long-day (LD 16:8) animals. As previously reported for many rodent species, prolonged maintenance on short days (i.e., 20 weeks) resulted in spontaneous reproductive recrudescence in which the gonads, and thus T, returned to normal long-day levels (Jasnow et al., 2000). Gonadally recrudesced hamsters displayed less aggression than gonadally regressed animals even though both groups experienced the same photoperiod and melatonin signal; levels of aggression in recrudesced hamsters were generally indistinguishable from long-day hamsters (Jasnow et al., 2000). These results support previous findings in male Syrian hamsters (Garrett and Campbell, 1980). When short-day Siberian hamsters were implanted with Silastic capsules containing T (to achieve long-day–like levels), aggression actually *decreased* compared with short-day control animals (Jasnow et al., 2000), suggesting that short-day increases in aggression may be inversely related to serum T concentrations.

Despite growing evidence that short-day increases in aggression are independent of (or inversely related to) circulating levels of T, much less is known about the precise neuroendocrine mechanisms underlying seasonal aggression in rodents. As previously described, several studies have implicated changes in the pineal hormone melatonin in mediating short-day aggression. More recent research in male Siberian hamsters (Demas et al., 2004) confirms previous findings that treatment of long-day animals with short-day-like levels of melatonin mimics photoperiodic changes in aggression; long-day hamsters given daily timed injections of melatonin 2 h before lights out to mimic short-day levels of the hormone displayed elevated aggression in a resident-intruder test compared with control animals. As with previous studies, these results were not likely due to changes in gonadal steroids, as serum T was unaffected by this injection protocol.

The effects of melatonin on aggression in rodents may be due to direct actions of this hormone on neural substrates mediating aggression (e.g., hypothalamus, limbic system). Alternatively, melatonin-induced aggression may be indirectly due to changes in HPA activity, as adrenal hormones have been implicated in aggressive behavior (Haller and Kruk, 2003). In support of the latter hypothesis, changes in both the size and function of the adrenal gland are associated with changes in aggression (Paterson and Vickers, 1981). In addition, male house mice housed in a LD 12:12 photoperiod and treated with melatonin display increased territorial aggression, but decreased adrenal masses compared to saline-treated animals (Paterson and Vickers, 1981). The increases in aggression displayed by melatonin-treated animals, however, can be blocked by adrenalectomy (Paterson and Vickers, 1981). Experimental reductions of both adrenomedullary catecholamines, as well as adrenocortical glucocorticoids, are associated with decreased aggression in rodents (Paterson and Vickers, 1981; Haller and Kruk, 2003) and reductions of glucocorticoids via pharmacological blockade of adrenocorticotropic hormone release can attenuate melatonin-induced increases in aggression in mice (Paterson and Vickers, 1981). Thus, exogenous melatonin, despite reducing adrenal mass, appears to increase aggression by stimulating adrenocortical steroid release. These results are particularly intriguing given that house mice have traditionally been assumed to be photoperiodically nonresponsive (Nelson, 1990).

Adrenal Steroids

More recently, research has implicated changes in adrenocortical hormones in mediating melatonin-induced and possibly short-day–induced aggression in Siberian hamsters. As described previously, long-day hamsters treated with short-day-like levels of melatonin displayed increased aggression, comparable to levels seen in short-day animals (Demas et al., 2004; figure 19.5). Melatonin-induced aggression could be blocked by bilateral adrenalectomy, consistent with previous results in house mice (Paterson and Vickers, 1981). Adrenal demedullation, which eliminates adrenal catecholamines (i.e., epinephrine) but leaves adrenocortical steroid release (i.e., cortisol, dehydroepiandrosterone [DHEA]) intact, had no effect on melatonin-induced aggression (Demas et al., 2004). Collectively, these results support the hypothesis that the effects of exogenous melatonin on aggression are mediated by the effects of this hormone on adrenocortical steroids. However, it is

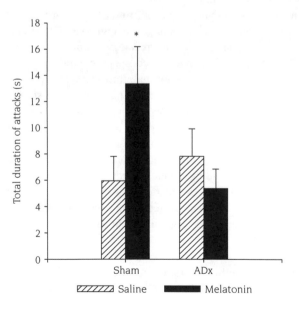

FIGURE 19.5.
Mean (+ standard error) total duration of attacks in adult male Siberian hamsters that received bilateral adrenalectomies (ADx) or sham operations and subsequently treated with either melatonin or control (saline) injections *p < 0.05. Modified from Demas et al. (2004).

currently not known which class of steroid hormones may mediate this effect, as adrenal androgens (e.g., DHEA) and glucocorticoids (e.g., cortisol) have both been implicated in aggression in rodents (Schlegel et al., 1985; Haller and Kruk, 2003). In laboratory rats and mice, corticosterone is the predominant adrenal glucocorticoid, and these species secrete little to no adrenal DHEA. In contrast, in hamsters, as in humans, cortisol is the primary adrenal glucocorticoid, and both hamsters and humans secrete measurable amounts of DHEA and its sulfated form, DHEA-S (Pieper and Lobocki, 2000; Mellon and Vaudry, 2001).

Evidence in avian species suggests that aggression in the nonbreeding season (i.e., winter) may be mediated by changes in DHEA (Soma et al., 2000). Although similar evidence suggesting a role for DHEA in mediating photoperiodic changes in aggression in rodents is lacking, studies in mice suggest that exogenous melatonin can stimulate DHEA production from cultured adrenal glands (Haus et al., 1996). Behavior was not examined in this study; however, these results are consistent with the hypothesis that short-day increases in melatonin may increase adrenal production of DHEA and thus affect aggression.

Several field studies published to date support the laboratory data discussed above suggesting differential dependence on gonadal steroids for animals during breeding season versus when they are not breeding. Specifically, male ratlike hamsters (*Cricetus triton*) in the field display elevated aggression during the winter nonbreeding season, despite low levels of plasma T (Zhang et al., 2001). Seasonal changes in aggression appear independent of seasonal changes in circulating T in wild wood rats (*Neotoma fuscipes*) (Caldwell et al., 1984). Pronounced seasonal changes in aggression are seen in male wood rats, with high levels during mid-breeding season and low levels during the nonbreeding season. Despite differences in circulating T levels at these two time points, castration has no affect on aggression, suggesting an independence of seasonal aggression from circulating levels of T (Caldwell et al., 1984). More recently, seasonal changes in aggressive encounters have been examined in free-living arctic ground squirrels (*Spermophilus parryii*) (Buck and Barnes, 2003). The effects of challenges by conspecific males on circulating T levels varied seasonally, with challenges by male intruders eliciting significant increases in circulating T during the spring breeding season, but similar challenges failed to trigger increase in androgen at the end of the summer after the breeding season. These results suggest that androgens play a more important role during the breeding season than during the nonbreeding season (i.e., late summer). Collectively, these studies fail to support the simple notion that all forms of aggression are mediated by circulating T by providing salient examples of T-*independent* aggression, at least with respect to circulating levels of the hormone. Unlike other forms of aggression, however, very little is known regarding the neuroendocrine mechanisms underlying seasonal changes in aggression in mammals.

Gonadal Steroids
In contrast to inbred laboratory mice, outbred *Peromyscus* mice show different patterns of aggression. In male California mice (*Peromyscus californicus*), aromatase

activity in the bed nucleus of the stria terminalis is negatively correlated with aggression (Trainor et al., 2004). Moreover, an aromatase inhibitor increases aggressive behavior (Trainor et al., 2004). Similarly, in male beach mice (*Peromyscus poliono-tus*) in long days, estradiol reduces aggression, but under short-day photoperiods, estradiol increases aggression. These effects are mediated by estrogen receptor-α as specific agonists for that receptor recapitulate the effect of estradiol. Photoperiod alters the expression of estrogen receptors in the beach mouse limbic system but the opposite effects of estrogen receptor activation on aggressive behavior appears to be independent of differential receptor expression (Trainor et al., 2007b). Rather, a photoperiod-mediated shift between genomic and nongenomic mechanisms of estrogen receptor signaling appears to mediate photoperiod differences in responses to estrogenic compounds. Microarray analysis indicated that a greater number of estrogen-responsive genes were up-regulated in the bed nucleus of stria termina-lis in long-day mice relative to short-day animals. Importantly, estrogen rapidly increased (within 15 min) aggressive behaviors in short- but not long-day mice. Such rapid effects of steroid hormones are unlikely to occur via genomic mechanisms given the very short time scale (Trainor et al., 2007a).

Nitric Oxide

Nitric oxide (NO) is an endogenous gas that has the biochemical properties of a free radical and was initially identified as regulator of blood vessel tone (Moncada and Higgs, 1993). Since its initial characterization, NO has also been identified as an important neuronal messenger in the CNS and peripheral nervous system (Dawson and Snyder, 1994). NO is labile, with a half-life of approximately 5 s; consequently, many studies have manipulated NO indirectly by affecting its synthetic enzyme, NO synthase (NOS), involved in the transformation of arginine into citrulline and NO. Three distinct isoforms of NOS have been discovered in rodents: eNOS is located in endothelial tissue of blood vessels, nNOS is localized in neurons, and an inducible form (iNOS) is found in macrophages. Male nNOS$^{-/-}$ mice display a marked decrease in behavioral inhibition and display persistent fighting despite submissive displays by other mice (Nelson et al., 1995). Female nNOS$^{-/-}$ mice, in contrast, do no display elevated aggressiveness. Similar increases in aggression are seen in normal wild-type mice following pharmacological blockade of nNOS (Demas et al., 1997c). Castration of nNOS$^{-/-}$ mice results in a marked reduction in aggression and testosterone replacement restores aggression to precastration levels, suggesting that testosterone is necessary, but not sufficient, to elevate aggression in nNOS$^{-/-}$ mice (Kriegsfeld et al., 1997). More recently, the role of NO in mediating seasonal aggression has been explored (Wen et al., 2004). Specifically, Siberian hamsters housed in short days displayed increased aggression and significantly fewer nNOS-immunoreactive cells in several amygdala regions compared to long-day animals. Interestingly, these short-day changes in aggression and NO staining were also seen in short-day "nonresponders" (i.e., the subset of animals that fail to inhibit their reproductive systems in short days), suggesting that these changes are independent of changes in T (Wen et al., 2004).

Much of research on the neuroendocrine mechanisms of aggression focused on the role of gonadal steroid hormones, and predominantly testosterone, as the primary factor regulating aggression. Recent studies examining seasonal changes in aggression described in this chapter, however, have demonstrated that aggression is more complex than a single hormonal mechanism. Furthermore, studies of seasonal aggression point out that the same behavior (i.e., aggression) can have markedly different underlying mechanisms depending on differences in environmental conditions.

PHOTOPERIODISM AS A "MODEL SYSTEM" FOR THE STUDY OF NONREPRODUCTIVE TRAITS

The primary goal of this chapter is to consider some of the exciting areas of research in which pronounced photoperiodic changes in physiology and behavior have been recently reported, including changes in immune function, energetics, and aggression/affective behavior. As discussed at the beginning of this chapter, the majority of research with respect to photoperiodism has traditionally focused on reproductive cycles. This is understandable, given the robust and reliable effects of changing day length on reproductive physiology and behavior, and its importance for reproductive success, in a large majority of mammalian species.

The considerable attention given to seasonal cycles in reproduction, however, has likely limited investigations of other interesting areas in which photoperiod may exert important effects on physiological and behavior. For example, considerable progress has been made in the last decade in understanding the neuroendocrine mechanisms regulating energy balance. Consistent with these findings, a large number of peptides have been identified within the last decade that appear to play a major role in the regulation of energy balance. Not surprisingly, these peptides, several of which were discussed above, are involved in the photoperiodic regulation of body weight in seasonally breeding animals. Thus, photoperiodic animals that display marked cycles in body fat on an annual basis serve as excellent animal models to uncover fundamental physiological mechanisms that regulate changes in adiposity. Further, an appreciation of these basic physiological mechanisms should contribute to the understanding and treatment of human metabolic disorders, including obesity.

Marked photoperiodic changes in immune function and sickness have also been demonstrated. The precise nature of these changes and whether they correlate with changes in other physiological systems (e.g., reproduction) or with changes in specific hormones (e.g., gonadal steroids, leptin, and melatonin) have provided important insights into the interactions among the endocrine, nervous, and immune systems. These studies have also changed the way we view epidemiological patterns of disease. By recognizing that seasonal changes in underlying immune responses, in addition to seasonal changes in pathogen prevalence, contribute to seasonal changes in disease prevalence, more effective approaches to the treatment of a variety illnesses can be developed.

Lastly, more recent investigations have demonstrated photoperiodic changes in aggression and affective behaviors. For example, some species of animals demonstrate marked aggression during short days despite low levels of testosterone, suggesting that the display of aggression is highly context specific and likely mediated by several neuroendocrine systems in addition to gonadal steroids. Similarly, photoperiodic changes in anxiety and depression have also been reported. As discussed above, these changes likely evolved as adaptive mechanisms to limit specific behaviors to the appropriate times of the year. Importantly, understanding the nature of these behavioral changes in nonhuman models will help us understand the physiological bases of clinical conditions such as seasonal affective disorder (SAD).

Despite the considerable progress that has been made in our understanding of the photoperiodic regulation of physiology and behavior, there are still important research questions that need to be addressed. For example, studies of photoperiodic changes in physiology and behavior have traditionally focused on specific characteristics (e.g., reproduction, immunity, energetic, aggression). Rather than focusing on specific systems in isolation, studies that employ an integrated approach to the study of photoperiodic changes in physiology and behavior will provide a more functional approach. For example, it is becoming increasingly clear that there are important links between seasonal changes and reproduction, energy balance, and immune function. In fact, recent evidence suggests the existence of important energetic trade-offs between reproduction and immune function such that one physiological system is curtailed while the other is up-regulated during times when energy is limited (i.e., winter). Furthermore, animals display increased depression-like changes in behavior during this same energetic bottleneck. These findings suggest that animals undergo a suite of coordinated adaptations, both behavioral and physiological, that maximize the chances for survival. Our understanding of these integrative responses would be considerably limited if each response was studied in isolation. In contrast, an integrative approach to understanding the coordinated photoperiodic responses of animals provides a deeper, more meaningful analysis of both the adaptive functions and the neuroendocrine mechanisms driving these responses. Lastly, photoperiodic species that display marked changes in physiology and behavior serve as ideal model systems to study naturally occurring changes in physiology and behavior in an environmentally relevant manner. In this era of translational research, the findings from these studies will contribute to our understanding and treatment of a wide range of clinical disorders, including obesity, autoimmune disorders, and SAD.

References

Adam CL and Mercer JG. (2004). Appetite regulation and seasonality: Implications for obesity. Proc Nutr Soc 63: 413–419.

Armanini D, Endres S, Kuhnle U, and Weber PC. (1988). Parallel determination of mineralocorticoid and glucocorticoid receptors in T- and B-lymphocytes of human spleen. Acta Endocrinol (Copenh) 118: 479–482.

Atcha Z, Cagampang FR, Stirland JA, Morris ID, Brooks AN, Ebling FJ, et al. (2000). Leptin acts on metabolism in a photoperiod-dependent manner, but has no effect on reproductive function in the seasonally breeding Siberian hamster (*Phodopus sungorus*). Endocrinology 141: 4128–4135.

Auchtung TL and Dahl GE. (2004). Prolactin mediates photoperiodic immune enhancement: Effects of administration of exogenous prolactin on circulating concentrations, receptor expression, and immune function in steers. Biol Reprod 71: 1913–1918.

Badura LL and Nunez AA. (1989). Photoperiodic modulation of sexual and aggressive behavior in female golden hamsters (*Mesocricetus auratus*): Role of the pineal gland. Horm Behav 23: 27–42.

Baker JR. (1938). The evolution of breeding seasonality. In *Evolution* (GR DeBeer, ed). Oxford: Clarendon Press, pp. 161–178.

Barjavel MJ, Mamdouh Z, Raghbate N, and Bakouche O. (1998). Differential expression of the melatonin receptor in human monocytes. J Immunol 160: 1191–1197.

Bartness TJ. (1996). Photoperiod, sex, gonadal steroids, and housing density affect body fat in hamsters. Physiol Behav 60: 517–529.

Bartness TJ, Demas GE, and Song CK. (2002). Seasonal changes in adiposity: The roles of the photoperiod, melatonin and other hormones, and sympathetic nervous system. Exp Biol Med 227: 363–376.

Bartness TJ, McGriff WR, and Maharaj MP. (1991). Effects of diabetes and insulin on photoperiodic responses in Siberian hamsters. Physiol Behav 49: 613–620.

Bartness TJ and Wade GN. (1984). Photoperiodic control of body weight and energy metabolism in Syrian hamsters (*Mesocricetus auratus*): Role of pineal gland, melatonin, gonads, and diet. Endocrinology 114: 492–498.

Baumann H, Morella KK, White DW, Dembski M, Bailon PS, Kim H, et al. (1996). The full-length leptin receptor has signaling capabilities of interleukin 6-type cytokine receptors. Proc Natl Acad Sci USA 93: 8374–8378.

Benabid N, Mesfioui A, and Ouichou A. (2008). Effects of photoperiod regimen on emotional behaviour in two tests for anxiolytic activity in Wistar rat. Brain Res Bull 75: 53–59.

Bentley GE, Demas GE, Nelson RJ, and Ball GF. (1998). Melatonin, immunity and cost of reproductive state in male European starlings. Proc R Soc Lond B Biol Sci 265: 1191–1195.

Bilbo SD, Dhabhar FS, Viswanathan K, Saul A, Yellon SM, and Nelson RJ. (2002a). Short day lengths augment stress-induced leukocyte trafficking and stress-induced enhancement of skin immune function. Proc Natl Acad Sci USA 99: 4067–4072.

Bilbo SD, Drazen DL, Quan N, He L, and Nelson RJ. (2002b). Short day lengths attenuate the symptoms of infection in Siberian hamsters. Proc R Soc Lond B Biol Sci 269: 447–454.

Bilbo SD and Nelson RJ. (2001). Sex steroid hormones enhance immune function in male and female Siberian hamsters. Am J Physiol Regul Integr Comp Physiol 280: R207–R213.

Bilbo SD and Nelson RJ. (2002). Melatonin regulates energy balance and attenuates fever in Siberian hamsters. Endocrinology 143: 2527.

Bilbo SD and Nelson RJ. (2004). Photoperiod influences the effects of exercise and food restriction on an antigen-specific immune response in Siberian hamsters. Endocrinology 145: 556–564.

Bilbo SD, Quan N, Prendergast BJ, Bowers SL, and Nelson RJ. (2003). Photoperiod alters the time course of brain cyclooxygenase-2 expression in Siberian hamsters. J Neuroendocrinol 15: 958–964.

Blank JL and Freeman DA. (1991). Differential reproductive response to short photoperiod in deer mice: Role of melatonin. J Comp Physiol A Sens Neural Behav Physiol 169: 501–506.

Blom JM, Gerber JM, and Nelson RJ. (1994). Day length affects immune cell numbers in deer mice: Interactions with age, sex, and prenatal photoperiod. Am J Physiol 267: R596–R601.

Boss-Williams KA and Bartness TJ. (1996). NPY stimulation of food intake in Siberian hamsters is not photoperiod dependent. Physiol Behav 59: 157–164.

Bowers RR, Gettys TW, Prpic V, Harris RB, and Bartness TJ. (2005). Short photoperiod exposure increases adipocyte sensitivity to noradrenergic stimulation in Siberian hamsters. Am J Physiol Regul Integr Comp Physiol 288: R1354–R1360.

Bronson FH. (1985). Mammalian reproduction: An ecological perspective. Biol Reprod 32: 1–26.

Bronson FH. (1988). Effect of food manipulation on the GnRH-LH-estradiol axis of young female rats. Am J Physiol Regul Integr Comp Physiol 254: 616–621.

Buck CL and Barnes BM. (2003). Androgen in free-living arctic ground squirrels: Seasonal changes and influence of staged male-male aggressive encounters. Horm Behav 43: 318–326.

Caldwell HK and Albers HE. (2004). Effect of photoperiod on vasopressin-induced aggression in Syrian hamsters. Horm Behav 46: 444–449.

Caldwell GS, Glickman SE, and Smith ER. (1984). Seasonal aggression independent of seasonal testosterone in wood rats. Proc Natl Acad Sci USA 81: 5255–5257.

Calzolari A. (1898). Recherches experimentales sur un rapport probable entre la function du thymus et celle des testicules. Arch Ital Biol 30: 71–77.

Campbell CS, Tabor J, and Davis JD. (1983). Small effect of brown adipose tissue and major effect of photoperiod on body weight in hamsters (*Mesocricetus auratus*). Physiol Behav 30: 349–352.

Carmody RJ and Chen YH. (2007). Nuclear factor-kappaB: Activation and regulation during toll-like receptor signaling. Cell Mol Immunol 4: 31–41.

Chan CW, Crafton E, Fan HN, Flook J, Yoshimura K, Skarica M, et al. (2006). Interferon-producing killer dendritic cells provide a link between innate and adaptive immunity. Nat Med 12: 207–213.

Ciaccio LA, Lisk RD, and Reuter LA. (1979). Prelordotic behavior in the hamster: A hormonally modulated transition from aggression to sexual receptivity. J Comp Physiol Psychol 93: 771–780.

Cincotta AH, MacEachern TA, and Meier AH. (1993). Bromocriptine redirects metabolism and prevents seasonal onset of obese hyperinsulinemic state in Syrian hamsters. Am J Physiol 264: E285–E293.

Cincotta AH and Meier AH. (1995). Bromocriptine inhibits in vivo free fatty acid oxidation and hepatic glucose output in seasonally obese hamsters (*Mesocricetus auratus*). Metabolism 44: 1349–1355.

Cincotta AH, Schiller BC, and Meier AH. (1991). Bromocriptine inhibits the seasonally occurring obesity, hyperinsulinemia, insulin resistance, and impaired glucose tolerance in the Syrian hamster, *Mesocricetus auratus*. Metabolism 40: 639–644.

Clarke IJ, Rao A, Chilliard Y, Delavaud C, and Lincoln GA. (2003). Photoperiod effects on gene expression for hypothalamic appetite-regulating peptides and food intake in the ram. Am J Physiol Regul Integr Comp Physiol 284: R101–R115.

Coleman D. (1978). Obese and diabetes: Two mutant genes causing diabetes-obesity syndromes in mice. Diabetologia 14: 141–148.

Concannon P, Levac K, Rawson R, Tennant B, and Bensadoun A. (2001). Seasonal changes in serum leptin, food intake, and body weight in photoentrained woodchucks. Am J Physiol Regul Integr Comp Physiol 281: R951–R959.

Dahl GE, Auchtung TL, and Kendall PE. (2002). Photoperiodic effects on endocrine and immune function in cattle. Reproduction 59(suppl): 191–201.

Dantzer R. (2001). Cytokine-induced sickness behavior: Where do we stand? Brain Behav Immun 15: 7–24.

Dark J and Zucker I. (1984). Gonadal and photoperiodic control of seasonal body weight changes in male voles. Am J Physiol Regul Integr Comp Physiol 247: 84–88.

Dark J and Zucker I. (1986). Photoperiodic regulation of body mass and fat reserves in the meadow vole. Physiol Behav 38: 851–854.

Dark J, Zucker I, and Wade GN. (1983). Photoperiodic regulation of body mass, food intake, and reproduction in meadow voles. Am J Physiol Regul Integr Comp Physiol 245: 334–338.

Dawson TM and Snyder SH. (1994). Gases as biological messengers: Nitric oxide and carbon monoxide in the brain. J Neurosci 14: 5147–5159.

Deerenberg C, Arpanius V, Daan S, and Bos N. (1997). Reproductive effort decreases antibody responsiveness. Proc R Soc Lond B Biol Sci 264: 1021–1029.

Demas GE. (2004). The energetics of immunity: A neuroendocrine link between energy balance and immune function. Horm Behav 45: 173–180.

Demas GE and Bartness TJ. (2001). Direct innervation of white fat and adrenal medullary catecholamines mediate photoperiodic changes in body fat. Am J Physiol Regul Integr Comp Physiol 281: R1499–R1505.

Demas GE, Bowers RR, Bartness TJ, and Gettys TW. (2002). Photoperiodic regulation of gene expression in brown and white adipose tissue of Siberian hamsters (*Phodopus sungorus*). Am J Physiol Regul Integr Comp Physiol 282: R114–R121.

Demas GE, Chefer V, Talan MI, and Nelson RJ. (1997a). Metabolic costs of mounting an antigen-stimulated immune response in adult and aged C57BL/6J mice. Am J Physiol 273: R1631–R1637.

Demas GE, DeVries AC, and Nelson RJ. (1997b). Effects of photoperiod and 2-deoxy-D-glucose-induced metabolic stress on immune function in female deer mice. Am J Physiol 272: R1762–R1767.

Demas GE, Drazen DL, and Nelson RJ. (2003). Reductions in total body fat decrease humoral immunity. Proc Biol Sci 270: 905–911.

Demas GE, Eliasson MJ, Dawson TM, Dawson VL, Kriegsfeld LJ, Nelson RJ, et al. (1997c). Inhibition of neuronal nitric oxide synthase increases aggressive behavior in mice. Mol Med 3: 610–616.

Demas GE, Klein SL, and Nelson RJ. (1996). Reproductive and immune responses to photoperiod and melatonin are linked in *Peromyscus* subspecies. J Comp Physiol A 179: 819–825.

Demas GE and Nelson RJ. (1996). Photoperiod and temperature interact to affect immune parameters in adult male deer mice (*Peromyscus maniculatus*). J Biol Rhythms 11: 94–102.

Demas GE and Nelson RJ. (1998a). Exogenous melatonin enhances cell-mediated, but not humoral, immune function in adult male deer mice (*Peromyscus maniculatus*). J Biol Rhythms 13: 245–252.

Demas GE and Nelson RJ. (1998b). Short-day enhancement of immune function is independent of steroid hormones in deer mice (*Peromyscus maniculatus*). J Comp Physiol B 168: 419–426.

Demas GE, Polacek KM, Durazzo A, and Jasnow AM. (2004). Adrenal hormones mediate melatonin-induced increases in aggression in male Siberian hamsters (*Phodopus sungorus*). Horm Behav 46: 582–591.Demas GE, and Sakaria S. (2005). Leptin regulates energetic tradeoffs between body fat and humoural immunity. Proc Biol Sci 272: 1845–1850.

Dhabhar FS and McEwen BS. (1997). Acute stress enhances while chronic stress suppresses cell-mediated immunity in vivo: A potential role for leukocyte trafficking. Brain Behav Immun 11: 286–306.

Dhabhar FS and McEwen BS. (1999). Enhancing versus suppressive effects of stress hormones on skin immune function. Proc Natl Acad Sci USA 96: 1059–1064.

Dhabhar FS, Miller AH, McEwen BS, and Spencer RL. (1995). Effects of stress on immune cell distribution. Dynamics and hormonal mechanisms. J Immunol 154: 5511–5527.

Dhabhar FS, Miller AH, McEwen BS, and Spencer RL. (1996). Stress-induced changes in blood leukocyte distribution. Role of adrenal steroid hormones. J Immunol 157: 1638–1644.

Dorshkind K and Horseman ND. (2001). Anterior pituitary hormones, stress, and immune system homeostasis. Bioessays 23: 288–294.

Drazen DL, Demas GE, and Nelson RJ. (2001). Leptin effects on immune function and energy balance are photoperiod dependent in Siberian hamsters (*Phodopus sungorus*). Endocrinology 142: 2768–2775.

Drazen DL, Jasnow AM, Nelson RJ, and Demas GE. (2002). Exposure to short days, but not short-term melatonin, enhances humoral immunity of male Syrian hamsters (*Mesocricetus auratus*). J Pineal Res 33: 118–124.

Drazen DL, Klein SL, Yellon SM, and Nelson RJ. (2000a). In vitro melatonin treatment enhances splenocyte proliferation in prairie voles. J Pineal Res 28: 34–40.

Drazen DL, Kriegsfeld LJ, Schneider JE, and Nelson RJ. (2000b). Leptin, but not immune function, is linked to reproductive responsiveness to photoperiod. Am J Physiol Regul Integr Comp Physiol 278: R1401–R1407.

Drazen DL and Nelson RJ. (2001). Melatonin receptor subtype MT2 (Mel 1b) and not mt1 (Mel 1a) is associated with melatonin-induced enhancement of cell-mediated and humoral immunity. Neuroendocrinology 74: 178–184.

Ellis C, Moar KM, Logie TJ, Ross AW, Morgan PJ, and Mercer JG. (2008). Diurnal profiles of hypothalamic energy balance gene expression with photoperiod manipulation in the Siberian hamster, *Phodopus sungorus*. Am J Physiol Regul Integr Comp Physiol 294: R1148–R1153.

Fange R and Silverin B. (1985). Variation of lymphoid activity in the spleen of a migratory bird, the pied flycatcher (*Ficedula-hypoleuca*—Aves, Passeriformes). J Morphol 184: 33–40.

Fine JB and Bartness TJ. (1996). Daylength and body mass affect diet self-selection by Siberian hamsters. Physiol Behav 59: 1039–1050.

Fleming AS, Phillips A, Rydall A, and Levesque L. (1988). Effects of photoperiod, the pineal gland and the gonads on agonistic behavior in female golden hamsters (*Mesocricetus auratus*). Physiol Behav 44: 227–234.

Fletcher-Chiappini S, Compton M, La Voie H, Day E, and Witorsch R. (1993). Glucocorticoid-prolactin interactions in Nb2 lymphoma cells: Antiproliferative versus anticytolytic effects. Proc Soc Exp Biol Med 202: 345–352.

Folstad I and Karter AJ. (1992). Parasites, bright males and the immunocompetence handicap hypothesis. Am Nat 139: 603–622.

Freeman DA, Kampf-Lassin A, Galang J, Wen JC, and Prendergast BJ. (2007). Melatonin acts at the suprachiasmatic nucleus to attenuate behavioral symptoms of infection. Behav Neurosci 121: 689–697.

Garcia-Perganeda A, Pozo D, Guerrero JM, and Calvo JR. (1997). Signal transduction for melatonin in human lymphocytes: Involvement of a pertussis toxin-sensitive G protein. J Immunol 159: 3774–3781.

Garrett JW and Campbell CS. (1980). Changes in social behavior of the male golden hamster accompanying photoperiodic changes in reproduction. Horm Behav 14: 303–318.

Goffin V, Binart N, Clement-Lacroix P, Bouchard B, Bole-Feysot C, Edery M, et al. (1999). From the molecular biology of prolactin and its receptor to the lessons learned from knockout mice models. Genet Anal 15: 189–201.

Goldman BD. (2001). Mammalian photoperiodic system: Formal properties and neuroendocrine mechanisms of photoperiodic time measurement. J Biol Rhythms 16: 283–301.

Goldman BD and Nelson RJ. (1993). Melatonin and seasonality in mammals. In *Melatonin: Biosynthesis, Physiological Effects, and Clinical Applications* (HS Yu and RJ Reiter, eds). Boca Raton, FL: CRC, pp. 225–252.

Goto M, Oshima I, Tomita T, and Ebihara S. (1989). Melatonin content of the pineal gland in different mouse strains. J Pineal Res 7: 195–204.

Gower BA, Hadjipanayis C, Nagy TR, and Stetson MH. (1994). Response of collared lemmings to melatonin: II. Infusions and photoperiod. J Pineal Res 17: 185–194.

Guerrero JM and Reiter RJ. (2002). Melatonin-immune system relationships. Curr Top Med Chem 2: 167–179.

Haller J and Kruk MR. (2003). Neuroendocrine stress responses and aggression. In *Neurobiology of Aggression* (MP Mattson, ed). Totawa, NJ: Humana Press.

Hamilton WD and Zuk M. (1982). Heritable true fitness and bright birds: A role for parasites? Science 218: 384–387.

Hart BL. (1988). Biological basis of the behavior of sick animals. Neurosci Biobehav Rev 12: 123.

Hasler JF, Buhl AE, and Banks EM. (1976). The influence of photoperiod on growth and sexual function in male and female collared lemmings (*Dicrostonyx groenlandicus*). J Reprod Fertil 46: 323–329.

Haus E, Nicolau GY, Ghinea E, Dumitriu L, Petrescu E, and Sackett-Lundeen L. (1996). Stimulation of the secretion of dehydroepiandrosterone by melatonin in mouse adrenals in vitro. Life Sci 58: PL263–PL267.

Hoffmann K. (1973). The influence of photoperiod and melatonin on testis size, body weight, and pelage colour in the Djungarian hamster (*Phodopus sungorus*). J Comp Physiol A Neuroethol Sens Neural Behav Physiol 85: 267–282.

Hoffman RA, Davidson K, and Steinberg K. (1982). Influence of photoperiod and temperature on weight gain, food consumption, fat pads and thyroxine in male golden hamsters. Growth 46: 150–162.

Hoffman RA, Hester RJ, and Towns C. (1965). Effect of light and temperature on the endocrine system of the golden hamster (*Mesocricetus auratus* Waterhouse). Comp Biochem Physiol 15: 525–533.

Horton TH. (1984). Growth and reproductive development of male *Microtus montanus* is affected by the prenatal photoperiod. Biol Reprod 31: 499–504.

Horton TH. (1985). Cross-fostering of voles demonstrates in utero effect of photoperiod. Biol Reprod 33: 934–939.

Horton TH, Buxton OM, Losee-Olson S, and Turek FW. (2000). Twenty-four-hour profiles of serum leptin in siberian and golden hamsters: Photoperiodic and diurnal variations. Horm Behav 37: 388–398.

Hotchkiss AK and Nelson RJ. (2002). Melatonin and immune function: Hype or hypothesis? Crit Rev Immunol 22: 351–371.

Janeway C, Travers P, Walport M, and Capra J. (1999). *Immunobiology: The Immune System in Health and Disease.* Garland, NY: Elsevier.

Jasnow AM, Huhman KL, Bartness TJ, and Demas GE. (2000). Short-day increases in aggression are inversely related to circulating testosterone concentrations in male Siberian hamsters (*Phodopus sungorus*). Horm Behav 38: 102–110.

Jasnow AM, Huhman KL, Bartness TJ, and Demas GE. (2002). Short days and exogenous melatonin increase aggression of male Syrian hamsters (*Mesocricetus auratus*). Horm Behav 42: 13–20.

John JL. (1994). The avian spleen: A neglected organ. Q Rev Biol 69: 327–351.

Kinsey SG, Prendergast BJ, and Nelson RJ. (2003). Photoperiod and stress affect wound healing in Siberian hamsters. Physiol Behav 78: 205–211.

Klein SL. (2000). The effects of hormones on sex differences in infection: From genes to behavior. Neurosci Biobehav Rev 24: 627–638.

Klein SL. (2004). Hormonal and immunological mechanisms mediating sex differences in parasite infection. Parasite Immunol 26: 247–264.

Klingenspor M, Dickopp A, Heldmaier G, and Klaus S. (1996a). Short photoperiod reduces leptin gene expression in white and brown adipose tissue of Djungarian hamsters. FEBS Lett 399: 290–294.

Klingenspor M, Dickopp A, Heldmaier G, and Klaus S. (1996b). Short photoperiod reduces leptin gene expression in white and brown adipose tissue of Djungarian hamsters. FEBS Lett 399: 290–294.

Klingenspor M, Niggemann H, and Heldmaier G. (2000). Modulation of leptin sensitivity by short photoperiod acclimation in the Djungarian hamster, *Phodopus sungorus*. J Comp Physiol B 170: 37–43.

Konakchieva R, Manchev S, Pevet P, and Masson-Pevet M. (1999). Autoradiographic detection of 2-(125I)-iodomelatonin binding sites in immune tissue of rats. Adv Exp Med Biol 460: 411–415.

Kopp C, Vogel E, Rettori MC, Delagrange P, Guardiola-Lemaitre B, and Misslin R. (1999). Effects of melatonin on neophobic responses in different strains of mice. Pharmacol Biochem Behav 63: 521–526.

Kriegsfeld LJ, Dawson TM, Dawson VL, Nelson RJ, and Snyder SH. (1997). Aggressive behavior in male mice lacking the gene for neuronal nitric oxide synthase requires testosterone. Brain Res 769: 66–70.

Kriegsfeld LJ, Drazen DL, and Nelson RJ. (2001). In vitro melatonin treatment enhances cell-mediated immune function in male prairie voles (*Microtus ochrogaster*). J Pineal Res 30: 193–198.

Kriegsfeld LJ and Nelson RJ. (1996). Gonadal and photoperiodic influences on body mass regulation in adult male and female prairie voles. Am J Physiol Regul Integr Comp Physiol 270: 1013–1018.

Leite De Moraes MC, Touraine P, Gagnerault MC, Savino W, Kelly PA, and Dardenne M. (1995). Prolactin receptors and the immune system. Ann Endocrinol (Paris) 56: 567–570.

Lochmiller RL and Deerenberg C. (2000). Trade-offs in evolutionary immunology: Just what is the cost of immunity? Oikos 88: 87–98.

Lord GM, Matarese G, Howard JK, Baker RJ, Bloom SR, and Lechler RI. (1998). Leptin modulates the T-cell immune response and reverses starvation-induced immunosuppression. Nature 394: 897–901.

Maestroni GJ, Conti A, and Pierpaoli W. (1986). Role of the pineal gland in immunity. Circadian synthesis and release of melatonin modulates the antibody response and antagonizes the immunosuppressive effect of corticosterone. J Neuroimmunol 13: 19–30.

Maestroni GJ, Conti A, and Pierpaoli W. (1987). The pineal gland and the circadian, opiatergic, immunoregulatory role of melatonin. Ann NY Acad Sci 496: 67–77.

Maestroni GJ, Conti A, and Pierpaoli W. (1988). Role of the pineal gland in immunity. III. Melatonin antagonizes the immunosuppressive effect of acute stress via an opiatergic mechanism. Immunology 63: 465–469.

Marie M, Findlay PA, Thomas L, and Adam CL. (2001). Daily patterns of plasma leptin in sheep: Effects of photoperiod and food intake. J Endocrinol 170: 277–286.

Marques DM and Valenstein ES. (1977). Individual differences in aggressiveness of female hamsters: Response to intact and castrated males and to females. Anim Behav 25: 131–139.

Martin LB 2nd, Scheuerlein A, and Wikelski M. (2003). Immune activity elevates energy expenditure of house sparrows: A link between direct and indirect costs? Proc Biol Sci 270: 153–158.

Martin LB, Weil ZM, and Nelson RJ. (2006). Refining approaches and diversifying directions in ecoimmunology. Integr Comp Biol 46: 1030–1039.

Martin LB 2nd, Weil ZM, and Nelson RJ. (2007). Immune defense and reproductive pace of life in *Peromyscus* mice. Ecology 88: 2516–2528.

Martin LB, Weil ZM, and Nelson RJ. (2008). Seasonal changes in vertebrate immune activity: Mediation by physiological trade-offs. Philos Trans R Soc Lond B Biol Sci 363: 321–339.

Matera L, Cesano A, Bellone G, and Oberholtzer E. (1992). Modulatory effect of prolactin on the resting and mitogen-induced activity of T, B, and NK lymphocytes. Brain Behav Immun 6: 409–417.

McEwen BS, Biron CA, Brunson KW, Bulloch K, Chambers WH, Dhabhar FS, et al. (1997). The role of adrenocorticoids as modulators of immune function in health and disease: Neural, endocrine and immune interactions. Brain Res Brain Res Rev 23: 79–133.

McKinney TD, Vaughan MK, and Reiter RJ. (1975). Pineal influence on intermale aggression in adult house mice. Physiol Behav 15: 213–216.

Mellon SH and Vaudry H. (2001). Biosynthesis of neurosteroids and regulation of their synthesis. Int Rev Neurobiol 46: 33–78.

Mercer JG, Moar KM, Logie TJ, Findlay PA, Adam CL, and Morgan PJ. (2001). Seasonally inappropriate body weight induced by food restriction: Effect on hypothalamic gene expression in male Siberian hamsters. Endocrinology 142: 4173–4181.

Mercer JG, Moar KM, Ross AW, Hoggard N, and Morgan PJ. (2000a). Photoperiod regulates arcuate nucleus POMC, AGRP, and leptin receptor mRNA in Siberian hamster hypothalamus. Am J Physiol Regul Integr Comp Physiol 278: R271–R281.

Mercer JG, Moar KM, Ross AW, and Morgan PJ. (2000b). Regulation of leptin receptor, POMC and AGRP gene expression by photoperiod and food deprivation in the hypothalamic arcuate nucleus of the male Siberian hamster (*Phodopus sungorus*). Appetite 34: 109–111.

Miceli MO, Post CA, and Woshowska Z. (1989). Abdominal vagotomy blocks hyperphagia and body weight gain in Syrian hamsters exposed to short photoperiod. Soc Neurosci Abstr 522:4.

Miller AH, Spencer RL, Pearce BD, Pisell TL, Azrieli Y, Tanapat P, et al. (1998). Glucocorticoid receptors are differentially expressed in the cells and tissues of the immune system. Cell Immunol 186: 45–54.

Molina-Hernandez M and Tellez-Alcantara P. (2000). Long photoperiod regimen may produce anti-depressant actions in the male rat. Prog Neuropsychopharmacol Biol Psychiatry 24: 105–116.

Moncada S and Higgs A. (1993). The L-arginine-nitric oxide pathway. N Engl J Med 329: 2002–2012.

Morgan PJ, Ross AW, Mercer JG, and Barrett P. (2003). Photoperiodic programming of body weight through the neuroendocrine hypothalamus. J Endocrinol 177: 27–34.

Morin LP and Fleming AS. (1978). Variation of food intake and body weight with estrous cycle, ovariectomy, and estradiol benzoate treatment in hamsters (*Mesocricetus auratus*). J Comp Physiol Psychol 92: 1–6.

Moyer KE. (1971). *The Physiology of Hostility*. Chicago: Markham.

Navara KJ, Trainor BC, and Nelson RJ. (2007). Photoperiod alters macrophage responsiveness, but not expression of Toll-like receptors in Siberian hamsters. Comp Biochem Physiol A Mol Integr Physiol 148: 354–359.

Nelson RJ. (1987). Gonadal regression induced by caloric restriction is not mediated by the pineal gland in deer mice (*Peromyscus maniculatus*). J Pineal Res 4: 339–345.

Nelson RJ. (1990). Photoperiodic responsiveness in house mice. Physiol Behav 48: 403.

Nelson RJ. (2004). Seasonal immune function and sickness responses. Trends Immunol 25: 187–192.

Nelson RJ and Blom JM. (1994). Photoperiodic effects on tumor development and immune function. J Biol Rhythms 9: 233–249.

Nelson RJ and Demas GE. (1996). Seasonal changes in immune function. Q Rev Biol 71: 511.

Nelson R and Demas G. (1997). Role of melatonin in mediating seasonal energetic and immunologic adaptations. Brain Res Bull 44: 423.

Nelson RJ, Demas GE, Huang PL, Fishman MC, Dawson VL, Dawson TM, et al. (1995). Behavioural abnormalities in male mice lacking neuronal nitric oxide synthase. Nature 378: 383–386.

Nelson RJ, Demas GE, Klein SL, and Kriegsfeld L. (2002). *Seasonal Patterns of Stress, Immune Function and Disease*. New York: Cambridge University Press.

Nelson RJ, Kita M, Blom JM, and Rhyne-Grey J. (1992). Photoperiod influences the critical caloric intake necessary to maintain reproduction among male deer mice (*Peromyscus maniculatus*). Biol Reprod 46: 226–232.

Nesse RM. (2000). Is depression an adaptation? Arch Gen Psychiatry 57: 14–20.

Nesse RM and Williams GC. (1996). *Why We Get Sick: The New Science of Darwininan Medicine*. New York: Vintage Books.

Padgett DA and Glaser R. (2003). How stress influences the immune response. Trends Immunol 24: 444–448.

Paterson AT and Vickers C. (1981). Melatonin and the adrenal cortex: Relationship to territorial aggression in mice. Physiol Behav 27: 983–987.

Pieper DR and Lobocki CA. (2000). Characterization of serum dehydroepiandrosterone secretion in golden hamsters. Proc Soc Exp Biol Med 224: 278–284.

Pozo D, Delgado M, Fernandez-Santos JM, Calvo JR, Gomariz RP, Martin-Lacave I, et al. (1997). Expression of the Mel1a-melatonin receptor mRNA in T and B subsets of lymphocytes from rat thymus and spleen. FASEB J 11: 466–473.

Prendergast BJ, Baillie SR, and Dhabhar FS. (2008). Gonadal hormone-dependent and -independent regulation of immune function by photoperiod in Siberian hamsters. Am J Physiol Regul Integr Comp Physiol 294: R384–R392.

Prendergast BJ, Bilbo SD, Dhabhar FS, and Nelson RJ. (2004a). Effects of photoperiod history on immune responses to intermediate day lengths in Siberian hamsters (*Phodopus sungorus*). J Neuroimmunol 149: 31–39.

Prendergast BJ, Bilbo SD, and Nelson RJ. (2004b). Photoperiod controls the induction, retention, and retrieval of antigen-specific immunological memory. Am J Physiol Regul Integr Comp Physiol 286: R54–R60.

Prendergast BJ, Bilbo SD, and Nelson RJ. (2005). Short day lengths enhance skin immune responses in gonadectomised Siberian hamsters. J Neuroendocrinol 17: 18–21.

Prendergast BJ, Hotchkiss AK, Bilbo SD, Kinsey SG, and Nelson RJ. (2003a). Photoperiodic adjustments in immune function protect Siberian hamsters from lethal endotoxemia. J Biol Rhythms 18: 51.

Prendergast BJ, Hotchkiss AK, Bilbo SD, and Nelson RJ. (2004c). Peripubertal immune challenges attenuate reproductive development in male Siberian hamsters (*Phodopus sungorus*). Biol Reprod 70: 813–820.

Prendergast BJ, Hotchkiss AK, and Nelson RJ. (2003b). Photoperiodic regulation of circulating leukocytes in juvenile Siberian hamsters: Mediation by melatonin and testosterone. J Biol Rhythms 18: 473–480.

Prendergast BJ, Kriegsfeld LJ, and Nelson RJ. (2001a). Photoperiodic polyphenisms in rodents: Neuroendocrine mechanisms, costs, and functions. Q Rev Biol 76: 293–325.

Prendergast BJ, Mosinger B Jr, Kolattukudy PE, and Nelson RJ. (2002a). Hypothalamic gene expression in reproductively photoresponsive and photorefractory Siberian hamsters. Proc Natl Acad Sci USA 99: 16291–16296.

Prendergast BJ and Nelson RJ. (2001). Spontaneous "regression" of enhanced immune function in a photoperiodic rodent *Peromyscus maniculatus*. Proc Biol Sci 268: 2221–2228.

Prendergast BJ and Nelson RJ. (2005). Affective responses to changes in day length in Siberian hamsters (*Phodopus sungorus*). Psychoneuroendocrinology 30: 438–452.

Prendergast BJ, Yellon SM, Tran LT, and Nelson RJ. (2001b). Photoperiod modulates the inhibitory effect of in vitro melatonin on lymphocyte proliferation in female Siberian hamsters. J Biol Rhythms 16: 224–233.

Prendergast BJ, Wynne-Edwards KE, Yellon SM, and Nelson RJ. (2002b). Photorefractoriness of immune function in male Siberian hamsters (*Phodopus sungorus*). J Neuroendocrinol 14: 318–329.

Pyter LM, Adelson JD, and Nelson RJ. (2007). Short days increase hypothalamic-pituitary-adrenal axis responsiveness. Endocrinology 148: 3402–3409.

Pyter LM, Neigh GN, and Nelson RJ. (2005a). Social environment modulates photoperiodic immune and reproductive responses in adult male white-footed mice (*Peromyscus leucopus*). Am J Physiol Regul Integr Comp Physiol 288: R891–R896.

Pyter LM and Nelson RJ. (2006). Enduring effects of photoperiod on affective behaviors in Siberian hamsters (*Phodopus sungorus*). Behav Neurosci 120: 125–134.

Pyter LM, Samuelsson AR, Quan N, and Nelson RJ. (2005b). Photoperiod alters hypothalamic cytokine gene expression and sickness responses following immune challenge in female Siberian hamsters (*Phodopus sungorus*). Neuroscience 131: 779–784.

Pyter LM, Weil ZM, and Nelson RJ. (2005c). Latitude affects photoperiod-induced changes in immune response in meadow voles (*Microtus pennsylvanicus*). Can J Zool 83: 1271–1278.

Reber PM. (1993). Prolactin and immunomodulation. Am J Med 95: 637–644.

Reppert SM, Weaver DR, Cassone VM, Godson C, and Kolakowski LF Jr. (1995). Melatonin receptors are for the birds: Molecular analysis of two receptor subtypes differentially expressed in chick brain. Neuron 15: 1003–1015.

Roberts ML, Buchanan KL, and Evans MR. (2004). Testing the immunocompetence handicap hypothesis: A review of the evidence. Anim Behav 68: 227–239.

Ronchi E, Spencer RL, Krey LC, and McEwen BS. (1998). Effects of photoperiod on brain corticosteroid receptors and the stress response in the golden hamster (*Mesocricetus auratus*). Brain Res 780: 348–351.

Rousseau K, Atcha Z, Cagampang FR, Le Rouzic P, Stirland JA, Ivanov TR, et al. (2002). Photoperiodic regulation of leptin resistance in the seasonally breeding Siberian hamster (*Phodopus sungorus*). Endocrinology 143: 3083–3095.

Rousseau K, Atcha Z, and Loudon AS. (2003). Leptin and seasonal mammals. J Neuroendocrinol 15: 409–414.

Rubin RT, Dinan TG, and Scott LV. (2002). The neuroendocrinology of affective disorders. In *Hormones, Brain and Behavior*, vol 5 (DW Pfaff, AP Arnold, AM Etgen, SE Fahrbach, and RT Rubin, eds). New York: Academic Press, pp. 467–514.

Sapolsky RM. (2002). Endocrinology of the Stress-response. In *Behavioral Endocrinology* (JB Becker, SM Breedlove, D Crews, and MM McCarthy, eds). Cambridge, MA: MIT Press.

Sapolsky RM, Romero LM, and Munck AU. (2000). How do glucocorticoids influence stress responses? Integrating permissive, suppressive, stimulatory, and preparative actions. Endocr Rev 21: 55–89.

Schlegel ML, Spetz JF, Robel P, and Haug M. (1985). Studies on the effects of dehydroepiandrosterone and its metabolites on attack by castrated mice on lactating intruders. Physiol Behav 34: 867–870.

Silverin B, Fange R, Viebke PA, and Westin J. (1999). Seasonal changes in mass and histology of the spleen in willow tits *Parus montanus*. J Avian Biol 30: 255–262.

Sinclair JA and Lochmiller RL. (2000). The winter immunoenhancement hypothesis: Associations among immunity, density, and survival in prairie vole (*Microtus ochrogaster*) populations. Can J Zool 78: 254–264.

Smith KA, Crabtree GR, Kennedy SJ, and Munck AU. (1977). Glucocorticoid receptors and glucocorticoid sensitivity of mitogen stimulated and unstimulated human lymphocytes. Nature 267: 523–526.

Smith PE. (1930). The effect of hypophysectomy upon the involution of the thymus in the rat. Anat Rec 47: 119–129.

Soma KK, Sullivan KA, Tramontin AD, Saldanha CJ, Schlinger BA, and Wingfield JC. (2000). Acute and chronic effects of an aromatase inhibitor on territorial aggression in breeding and nonbreeding male song sparrows. J Comp Physiol A 186: 759–769.

Stehle JH, von Gall C, and Korf HW. (2002). Organisation of the circadian system in melatonin-proficient C3H and melatonin-deficient C57BL mice: A comparative investigation. Cell Tissue Res 309: 173.

Steinlechner S and Heldmaier G. (1982). Role of photoperiod and melatonin in seasonal acclimatization of the Djungarian hamster, *Phodopus sungorus*. Int J Biometeorol 26: 329–337.

Steinlechner S, Heldmaier G, and Becker H. (1983). The seasonal cycle of body weight in the Djungarian hamster: Photoperiodic control and the influence of starvation and melatonin. Oecologia 60: 401–405.

Trainor BC, Bird IM, and Marler CA. (2004). Opposing hormonal mechanisms of aggression revealed through short-lived testosterone manipulations and multiple winning experiences. Horm Behav 45: 115–121.

Trainor BC, Lin S, Finy MS, Rowland MR, and Nelson RJ. (2007a). Photoperiod reverses the effects of estrogens on male aggression via genomic and nongenomic pathways. Proc Natl Acad Sci USA 104: 9840–9845.

Trainor BC, Rowland MR, and Nelson RJ. (2007b). Photoperiod affects estrogen receptor α, estrogen receptor β and aggressive behavior. Eur J Neurosci 26: 207–218.

Turnbull AV and Rivier C. (1995). Regulation of the HPA axis by cytokines. Brain Behav Immun 9: 253–275.

Vaughan MK, Brainard GC, and Reiter RJ. (1985). Photoperiodic and light spectral conditions which inhibit circulating concentrations of thyroxine in the male hamster. Life Sci 36: 2183–2188.

Vaughan MK and Reiter RJ. (1971). Transient hypertrophy of the ventral prostate and coagulating glands and accelerated thymic involution following pinealectomy in the mouse. Tex Rep Biol Med 29: 579–586.

Vera-Lastra O, Jara L, and Espinoza LR. (2002). Prolactin and autoimmunity. Autoimmun Rev 1: 360–364.

Vermeulen M, Palermo M, and Giordano M. (1993). Neonatal pinealectomy impairs murine antibody-dependent cellular cytotoxicity. J Neuroimmunol 43: 97–101.

Vriend J and Lauber JK. (1973). Effects of light intensity, wavelength and quanta on gonads and spleen of the deer mouse [letter]. Nature 244: 37–38.

Wade GN and Bartness TJ. (1984a). Effects of photoperiod and gonadectomy on food intake, body weight, and body composition in Siberian hamsters. Am J Physiol Regul Integr Comp Physiol 246: 26–30.

Wade GN and Bartness TJ. (1984b). Seasonal obesity in Syrian hamsters: Effects of age, diet, photoperiod, and melatonin. Am J Physiol 247: R328–R334.

Weaver DR, Liu C, and Reppert SM. (1996). Nature's knockout: The Mel1b receptor is not necessary for reproductive and circadian responses to melatonin in Siberian hamsters. Mol Endocrinol 10: 1478–1487.

Wehr TA, Duncan WC, Jr, Sher L, Aeschbach D, Schwartz PJ, Turner EH, et al. (2001). A circadian signal of change of season in patients with seasonal affective disorder. Arch Gen Psychiatr 58: 1108–1114.

Weil ZM, Bowers SL, and Nelson RJ. (2007a). Photoperiod alters affective responses in collared lemmings. Behav Brain Res 179: 305–309.

Weil ZM, Bowers SL, Pyter LM, and Nelson RJ. (2006a). Social interactions alter proinflammatory cytokine gene expression and behavior following endotoxin administration. Brain Behav Immun 20: 72–79.

Weil ZM, Martin LB 2nd, and Nelson RJ. (2006b). Photoperiod differentially affects immune function and reproduction in collared lemmings (*Dicrostonyx groenlandicus*). J Biol Rhythms 21: 384–393.

Weil ZM, Pyter LM, Martin LB 2nd, and Nelson RJ. (2006c). Perinatal photoperiod organizes adult immune responses in Siberian hamsters (*Phodopus sungorus*). Am J Physiol Regul Integr Comp Physiol 290: R1714–R1719.

Weil ZM, Workman JL, and Nelson RJ. (2007b). Housing condition alters immunological and reproductive responses to day length in Siberian hamsters (*Phodopus sungorus*). Horm Behav 52: 261–266.

Wen JC, Hotchkiss AK, Demas GE, and Nelson RJ. (2004). Photoperiod affects neuronal nitric oxide synthase and aggressive behaviour in male Siberian hamsters (*Phodopus sungorus*). J Neuroendocrinol 16: 916–921.

Wen JC and Prendergast BJ. (2007). Photoperiodic regulation of behavioral responsiveness to proinflammatory cytokines. Physiol Behav 90: 717–725.

Whitsett JM, Underwood H, and Cherry J. (1983). Photoperiodic stimulation of pubertal development in male deer mice: Involvement of the circadian system. Biol Reprod 28: 652–656.

Woods SC and Seeley RJ. (2000). Adiposity signals and the control of energy homeostasis. Nutrition 16: 894–902.

Wurtman RJ and Weisel J. (1969). Environmental lighting and neuroendocrine function: Relationship between spectrum of light source and gonadal growth. Endocrinology 85: 1218–1221.

Yellon SM. (2007). Melatonin mediates photoperiod control of endocrine adaptations and humoral immunity in male Siberian hamsters. J Pineal Res 43: 109–114.

Yellon SM, Fagoaga OR, and Nehlsen-Cannarella SL. (1999a). Influence of photoperiod on immune cell functions in the male Siberian hamster. Am J Physiol 276: R97–R102.

Yellon SM, Kim K, Hadley AR, and Tran LT. (2005). Time course and role of the pineal gland in photoperiod control of innate immune cell functions in male Siberian hamsters. J Neuroimmunol 161: 137–144.

Yellon SM, Teasley LA, Fagoaga OR, Nguyen HC, Truong HN, and Nehlsen-Cannarella L. (1999b). Role of photoperiod and the pineal gland in T cell-dependent humoral immune reactivity in the Siberian hamster. J Pineal Res 27: 243–248.

Yu-Lee LY. (2002). Prolactin modulation of immune and inflammatory responses. Rec Prog Horm Res 57: 435–455.

Zhang JX, Zhang ZB, and Wang ZW. (2001). Seasonal changes in and effects of familiarity on agonistic behaviors of rat-like hamsters (*Cricetulus triton*). Ecol Res 16: 309–317.

Zhang Y, Proenca R, Maffei M, Barone M, Leopold L, and Friedman JM. (1994). Positional cloning of the mouse obese gene and its human homologue. Nature 372: 425–432.

20

Photoperiodism and Reproduction in Mammals

Lance J. Kriegsfeld and Eric L. Bittman

The environment varies considerably throughout the seasons, with minimal food availability coinciding with maximal energetic demands during winter. Consequently, natural selection has acted to ensure that animals inhabiting temperate and boreal latitudes restrict breeding and other energetically costly activities to times of year when environmental conditions are optimal. Because day length (photoperiod) is the most reliable predictor of changing seasons, organisms use this signal to forecast local conditions and initiate adaptations well in advance of the appropriate time of year. This mechanism allows termination of reproduction and other nonessential processes prior to the onset of winter in order to promote energy conservation and survival. In some cases, this is accomplished through use of day length as a signal to entrain an endogenous timing mechanism to the annual cycle. This strategy is complementary to responses to more proximate factors (e.g., temperature, calories, and more specific nutritional signals), as it allows time for development of seasonal adaptations. Examination of seasonal breeders has the potential to provide valuable insight into the mechanisms controlling reproduction, fertility, and other aspects of neuroendocrine function. For example, understanding the mechanisms responsible for regrowth of the reproductive system each year may help to uncover the mechanisms driving puberty. Furthermore, because seasonal alterations in reproduction result, in part, from gross modifications in the function of the hypothalamo-pituitary-gonadal axis, seasonal breeders can be exploited as a tool to understand the negative feedback effects of gonadal steroid hormones. In this chapter, we provide a broad overview of seasonal breeding in mammals, emphasizing the role of photoperiod in changes in neuroendocrine function and the means by which day length affects the reproductive axis.

SEASONAL CYCLES AND TIMEKEEPING MECHANISMS

Two distinct kinds of mechanism allow mammals to synchronize breeding and other aspects of their physiology, metabolism, and behavior with the annual geophysical cycle. The first such mechanism is a circannual rhythm: an oscillation that persists with a period of approximately 12 months in organisms deprived of annual

503

time cues. The second type of pattern incorporates a refractory stage, such that an annual cycle cannot be completed unless the appropriate environmental changes occur. Each of these patterns is described in turn.

Circannual rhythms adopt the period of the sidereal year as a result of exposure to environmental cues (Zucker et al., 1991). Typically, photoperiod acts to entrain the annual cycle of reproduction (e.g., Karsch et al., 1984; Lee and Zucker, 1991; Woodfill et al., 1994). Although these rhythms persist in the absence of cues to the 12-month annual cycle, their expression may occur only within a restricted range of constant photoperiods and other environmental conditions. The timing of reproductive activity or inactivity is expected to depend both upon the period of the endogenous oscillator and that of the environmental cue. Thus entrainment is not equivalent to synchronization. The effectiveness of photoperiod-based cues in shifting the phase of the circannual rhythm may vary with the phase of the cycle, although strong resetting curves have been reported (Woodfill et al., 1994; Lincoln et al., 2006). Circannual rhythms occur in large, long-lived mammals such as sciurids and ungulates, although a few examples have been demonstrated in crayfish, cnidarians, and insects (Brock, 1974). These oscillations have been particularly well studied in birds (Gwinner, 2003).

Although seasonal breeding is the rule rather than the exception among mammals that occupy temperate habitats, experiments to establish photoperiodic control have not been completed in all orders. It is particularly difficult and time consuming to establish the persistence of seasonal rhythms in the absence of environmental cues, as is necessary to satisfy the criterion of a circannual oscillator. Thus it is not surprising that the capacity for such endogenous annual rhythmicity has been demonstrated only in a few eutherian orders. Among mammals, circannual rhythms were initially described in woodchucks (Pengelley and Asmundson, 1971) and deer (Goss, 1984). Striking progress in understanding physiological mechanisms that mediate circannual cycles in mammals has most recently been gained in sheep. These fall and winter breeders give birth and lactate during spring and summer months. Short photoperiod exposure stimulates the reproductive axis, whereas long days arrest the 17-day estrous cycle and precipitate testicular regression within several weeks (Legan and Karsch, 1980; Ortavant et al., 1988). As with other circannual breeders, sheep alternate between reproductive and nonbreeding states throughout the year as they become refractory to photoperiodic signals that initially turn the gonads off and on. As an example, ewes housed continuously under long, summer-like days beginning at the summer solstice enter the breeding season at the same time as animals exposed to naturally decreasing day lengths (Robinson and Karsch, 1984). This finding indicates that reproductive competence normally results from refractoriness to inhibitory long day lengths, rather than the stimulatory actions of short days. The converse is also true: reproductive arrest results from refractoriness to short days (Robinson et al., 1985). The mechanisms mediating these changes at the neuroendocrine level are discussed further below and in chapter 21.

The order Rodentia shows a remarkable diversity of breeding patterns. Among sciurids, including ground squirrels, marmots, and chipmunks, circannual rhythms are well documented. Among castorimorphs, seasonality is evident in beavers, pocket gophers, and kangaroo rats, but photoperiodic responses and circannual rhythmicity have apparently not been studied. Among hystricomorphs, seasonal breeding is common in some families (e.g., chinchillids, mole rats) but rare or absent in others (e.g., porcupines, guinea pigs, and nutria) and there is little evidence for circannual rhythms (Trillmich, 2000). There is also a wide diversity of patterns among myomorphs. In the domesticated rats and mice whose reproduction is most thoroughly studied in the laboratory, responsiveness to photoperiod is limited at best, and may be confined to very restricted points during development. In contrast, nondomesticated myomorphs are typically seasonal breeders. Although circannual rhythms occur in some muroid rodents including the European hamster (*Cricetus cricetus*; Canguilhem and Marx, 1973; Masson-Pevet et al., 1994), others, including Syrian and Siberian hamsters (*Mesocricetus auratus* and *Phodopus sungorus*), lemmings (*Dicrostonyx groenlandicus*), voles (e.g., *Microtus ochrogaster*), and field mice (e.g., *Peromyscus maniculatus* and *Peromyscus leucopus*), that typically breed during the long days of summer show a second type of seasonal reproductive pattern (reviewed in Goldman, 2001; Paul et al., 2008). In contrast to sheep, which become refractory to both long and short days, such rodents exhibit refractoriness only to short days. As autumn approaches and day lengths fall below a species-specific threshold, the gonads regress. After about 20 weeks, the reproductive axis becomes unresponsive to short days, and animals recrudesce their reproductive apparatus despite continued exposure to short day lengths (Reiter, 1973). Unlike sheep, which do not require alterations in photoperiod to regain sensitivity to day length (Yeates, 1949; Ducker et al., 1973), hamsters require exposure to long days to once again be capable of responding to short days (Stetson et al., 1977). Consequently, hamsters that are continuously exposed to such inhibitory photoperiods undergo only one seasonal arrest of reproductive function, whereas sheep and ground squirrels continue to show a circannual rhythm (Gaston and Menaker, 1967; Reiter, 1973; Zucker and Licht, 1983; Almeida and Lincoln, 1984).

PHOTOPERIODIC HISTORY AND NONRESPONSIVENESS

Most studies to date have used abrupt changes in photoperiod to alter reproductive physiology. It is clear, however, that photoperiodic history determines which day lengths precipitate gonadal regression (e.g., Gorman and Zucker, 1995b). This may be an adaptive response to the ambiguity of all photoperiods except for those that occur at the solstices: at intermediate temperate latitudes, a day length of 13 h of light and 11 h of dark (13L:11D) occurs both in early April, when favorable conditions for the survival of the young are likely to ensue, and in early September, when harsh conditions are soon to follow. Thus, individuals that interpret the current day length with reference to preceding conditions are more likely to experience reproductive success. Moreover, in at least some rodent species, a subset of individuals fails to undergo gonadal regression when exposed to day lengths that inhibit reproduction in the majority of the population (Prendergast and Freeman, 1999; Prendergast et al., 2001). The proportion of so-called nonresponders is affected by photoperiodic history (Gorman and Zucker, 1995a). Such nonresponsive animals have been found in the wild, suggesting an indirect fitness advantage of producing offspring with differing reproductive phenotypes (Nelson, 1987). Early work on photoperiodic nonresponders investigated whether this phenotype is genetically determined, or whether this differential response might be the result of unidentified experiential factors. Support for each of these possibilities has been uncovered in several species of rodents. Selective breeding experiments indicate a high degree of heritability for nonresponsiveness, whereas studies of the effects of photoperiodic history indicate that prior exposure to different environmental conditions can markedly influence the proportion of individuals that respond reproductively to short day lengths. Whether a common neuroendocrine mechanism responsible for nonresponsiveness is triggered by both genetic and experiential factors remains to be determined.

Genetic and Neuroendocrine Control of Nonresponsiveness

Mice in the genus *Peromyscus* have proven useful in investigations of nonresponsiveness, as selection has favored different reproductive phenotypes in the same species based on local environmental conditions. In deer mice (*P. maniculatus*) and white-footed mice (*P. leucopus*), the degree of nonresponsiveness varies by the latitude of origin and responds to selection (Gram et al., 1982; Dark et al., 1983; Carlson et al., 1989). For example, the majority of white-footed mice trapped in Connecticut respond to short day lengths with reproductive regression whereas animals collected from a Georgia population do not (Carlson et al., 1989). Similar effects of latitude and selection were reported in deer mice (Blank and Desjardins, 1986; Desjardins et al., 1986; Ruf et al., 1997; Heideman et al., 1999). In Siberian hamsters, bidirectional selection over four generations resulted in a population of animals in which more than 90% were unresponsive to short day lengths (Kliman and Lynch, 1992).

Environmental Factors Contributing to Nonresponsiveness

In male Siberian hamsters, 13L:11D is typically sufficient to stimulate the development of gonadal function in juveniles. However, photoperiods as long as 14L:10D can inhibit or stimulate gonadal development depending on prior day-length exposure during early (even prenatal) life (Stetson et al., 1989; Niklowitz et al., 1994). For animals gestated on day lengths longer than 14 h, a photoperiod of 14L:10D fails to stimulate gonadal development. Analogous responses have been observed in other species, including montane voles (Horton, 1984, 1985). Meadow voles housed in short days during pregnancy give birth to male pups that are only partially responsive to short day lengths, with reduced testis weight but maintenance of spermatogenesis (Lee and Zucker, 1988). Presumably, such influences of gestational photoperiod allow animals born very early in the spring to develop their reproductive apparatus shortly after the equinox.

Gorman and Zucker (1995a) used simulated natural photoperiods (SNPs), rather than a square-wave pattern in which day length is abruptly shifted, to examine photoperiodic responses in Siberian hamsters. They noted that almost half of the animals born around a 60°N simulated vernal equinox failed to undergo gonadal regression even when day lengths fell to 5h during simulated autumn. In contrast, virtually all animals in a 40°N SNP underwent gonadal regression during 9L:15D in the simulated fall. These findings raised the question of whether photoperiodic history can explain latitudinal differences in the degree of nonresponsiveness. Indeed, animals transferred to day lengths greater than 16L on the day of birth fail to undergo gonadal regression when transferred to 10L at 16 weeks of age (Gorman and Zucker, 1997).

PHOTOPERIODIC TIME MEASUREMENT IN MAMMALS AND THE ROLE OF MELATONIN

Measurement of Photoperiod: Comparative Perspective

Animals can use a variety of environmental stimuli to restrict reproductive effort to the appropriate season. Fluctuations in temperature and rainfall may have a direct impact on reproductive physiology, or can act indirectly by regulating the quantity and quality of food. Seasonal changes in the presence of specific compounds may trigger reproduction at particular times of year. The remarkable potency and common influence of day length in controlling reproduction presumably reflects its status as the most reliable and predictive of environmental cues. It is thus logical that photoperiodism is common among nonequatorial species. Even so, the widespread use of analogous circadian-based mechanisms for measuring day length is striking.

Photoperiod can regulate development and metabolism even in unicellular organisms. The absolute amount of light could conceivably limit availability of critical metabolites or nutrients. This would constitute a sensitivity to photoperiod, but it would not indicate that day length has informational value or is being measured as a cue in order to restrict cell division or other processes to the appropriate time of year. Evidence against this idea is provided by comparison of the response to equivalent numbers of photons delivered over different intervals to the unicellular alga *Chlamydomonas reinhardtii* (Mittag et al., 2005). When the fluence is held constant, development proceeds in a summer but not a winter day length. This finding indicates that a process of photoperiodic time measurement is executed, although the mechanism is unknown.

There are several ways in which organisms might discriminate short from long day lengths. For example, an hourglass-like mechanism might be employed: a light-sensitive product would accumulate steadily and reach a threshold for induction of an appropriate seasonal response only when the length of the day or the night exceeds a critical duration. Although there are some instances in which an hourglass mechanism may be used (Hyde and Underwood, 1993; Saunders, 2005), it is clear that a

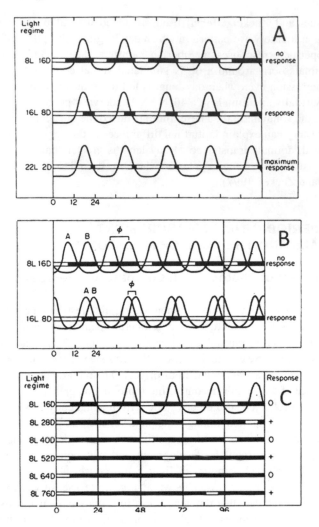

FIGURE 20.1.

Circadian basis of photoperiodic time measurement may be tested through use of resonance designs. (A) A putative rhythm of photosensitivity is illustrated as a waveform as postulated by Bünning (1936). In short days, the photosensitive phase does not coincide with light, so long-day responses are not induced by exposure to 8L:16D. In contrast, part of the photosensitive phase is illuminated upon exposure to 16L:8D, and more upon exposure to 22L:2D. Thus progressive increases in induction are seen as day length increases. (B) An internal coincidence interpretation of circadian-based photoperiodic time measurement. Two circadian oscillators are illustrated, one phase-locked to dusk (labeled A) and the other to dawn (labeled B). The phase relationship between the two entrained circadian oscillators (φ) depends upon day length; long-day effects are observed only when φ is small. (C) Design and results of a resonance experiment in which 8 h of light are alternated with various intervals of darkness ranging from 16 to 76 h. Organisms are typically exposed repeatedly to such "T cycles" for weeks or months before season-typical responses (reproduction, wide variety of organisms discriminate day length using one or more self-sustained endogenous daily (circadian) oscillator(s). Although their phase and period are normally set by environmental cues, principal among which are light–dark cycles, such circadian oscillations persist in constant conditions. Experiments in which summerlike seasonal responses are provoked by the interruption of prolonged nights by brief pulses of light rule out the possibility that the ratio of the durations of the light and dark phases is a critical cue. Even more powerful are "resonance" experiments, in which the period of the light–dark cycle (T) differs from 24 h (figure 20.1). In some of these designs, light is presented as infrequently as once every 84 h, yet longday effects are achieved (Nanda and Hamner, 1958; Elliott et al., 1972; Pittendrigh, 1981). Maxima and minima of reproductive competence result from cycles whose durations differ by 12 h. These peaks and nadirs recur at values of T that are multiples of 24 h, indicating resonance to the circadian period. This outcome cannot be explained by models in which either the absolute duration of darkness or the ratio of light to darkness is the critical variable. But the capacity of the circadian system to initiate successive cycles (free run) in prolonged darkness can explain such results. For example, an endogenous rhythm in which a photosensitive phase recurs at nearly 24-h intervals is consistent with these data. As originally suggested by Bünning (1936), the coincidence of an *external* event (light) with a particular phase of the endogenous circadian oscillation (the subjective night) is sufficient to trigger long-day responses even if the duration of light is short. In other words, it is not the duration of light but the circadian phase at which light falls that is critical in determining whether the summer or the winter state is achieved. An alternative formulation would stipulate that light present at dusk and dawn exerts entraining effects on different *internal* metabolic events, such that long- or short-day responses result from the phase relationship between oscillations locked to sunset and sunrise that is characteristic of a particular season. Regardless of whether the external or internal coincidence explanation is favored, resonance experiments argue forcefully for circadian-based photoperiodic time measurement in organisms as diverse as plants, insects, and vertebrates. These findings indicate either that the use of circadian oscillators to measure photoperiod is an ancestral feature that may have favored their evolution, or that remarkable convergence has led to the common strategy of utilizing circadian mechanisms that evolved for other purposes to measure day length.

Studies in diverse organisms raise the possibility that photoperiod discrimination depends upon the temporal relationship between circadian oscillations in the

---

diapause, flowering, etc.) are assessed; the response is illustrated at the right as "0" or "+." The theoretical circadian oscillation of photosensitivity is superimposed on the 8L:16D cycle at the top, but pertains to the other T-cycles, as well. Note that the photosensitive phase recurs (free runs) at circadian intervals when organisms are exposed to protracted dark phases, that is, when T exceeds 24 h. Thus, the photosensitive phase may be struck only every second or third cycle, but this is sufficient to induce long-day effects. Nadirs (0) and peaks (+) of longday responses recur every 24 h, so photoperiodic induction shows resonance to the circadian (24 h) period. Reprinted from Farner (1975) with permission.

expression of two or more genes. These examples provide a mechanistic explanation of the coincidence models initially proposed by Bünning and others. In *Arabidopsis thaliana*, a long-day plant, flowering depends upon the coincidence of a rise in concentrations of the zinc finger protein CONSTANS (CO) with the protein product of *Flowering locus T* (*ft*), which acts as a transcription factor (see chapter 1). Expression of *CO* is directly regulated by transcriptional-translational feedback loops that regulate circadian rhythms throughout the plant, and the stability of its protein product is determined by light through the action of the pigment, phytochrome. The FT protein directly induces flowering, and its transcription is induced by CO. Transcription of both *co* and *ft* show circadian oscillations, but as a result of its photic sensitivity the CO protein is present in sufficient concentrations to induce *ft* expression only in long days. Thus, flowering is restricted to long days as a result of the circadian control of gene expression, coupled with the regulation of protein stability by light. In rice, a short-day plant, a parallel interaction between orthologs of *co* (*Hd1*) and *ft* (*Hd3a*), determines whether flowering (heading) occurs (for reviews, see Imaizumi and Kay, 2006; Kobayashi and Weigel, 2007, and chapter 2).

In animals, day-length measurement may depend upon the phase relationship between evening and morning populations of oscillators that may be localized to different pacemaker cells. The tracing of neural pathways by which light cues entrain circadian rhythms has led to identification of central pacemakers in both invertebrates and vertebrates. Over the past decade, however, it has become apparent that circadian rhythmicity is a property not only of central pacemaker structures, but also of cells throughout the organism. In insects and fishes, the phase of circadian rhythms may be regulated directly by light striking peripheral structures (Plautz et al., 1997; Whitmore et al., 2000). In such organisms it is possible that day length discrimination can be achieved at the peripheral level without the necessity of participation of a central circadian pacemaker.

Within central pacemakers, separate groups of cells may be distinguished not only by location and phenotype, but also by responsiveness to dawn and dusk cues. Among insects, larval and ovarian diapause is controlled by photoperiod in *Drosophila melanogaster* and in blow flies (*Calliphora vicina* and *Protophormia terraenovae*). Measurement of day length is believed to depend upon brain-based mechanisms, although the master clock gene *Period* appears to be dispensable (Saunders, 1997; see chapter 9). Day length also regulates patterns of locomotor activity in *Drosophila melanogaster*, and it is believed that separate groups of pacemaker cells located between the central brain and the optic lobes are responsible for morning and evening bouts of activity (Stoleru et al., 2007). A small group of ventrolateral neurons, which are distinguished by their expression of pigment dispersing factor, represent the "M" cells. These cells act as master pacemakers in constant darkness, and they assume the dominant role in setting the phase of activity onset when flies are exposed to short day lengths. A group of dorsal neurons (LONds, DN1, and DN3) constitute the "E" cells, which dominate in constant light and control the phasing of activity in summer (long) photoperiods. The interaction of the E and M cells is believed to determine the seasonally appropriate pattern of locomotor activity, with changes in photoperiod determining which set of pacemaker cells has the predominant role. The participation of this system in regulation of diapause is uncertain.

In mammals, the master circadian pacemaker resides in the suprachiasmatic nucleus (SCN) of the hypothalamus. This structure is critical not only for the persistence of circadian rhythms of behavior and physiology, but also for photoperiodic regulation of reproduction (Stetson and Watson-Whitmyre, 1976; Morin et al., 1977). As in *Drosophila*, different populations of pacemaker cells may play predominant roles in determining morning and evening events, and the phase relationship between their activation may be used to discriminate photoperiod. The dominant and direct source of information about light that regulates SCN function is the retinohypothalamic tract (RHT), which is composed of the axons of a specialized subset of retinal ganglion cells distinguished by their expression of melanopsin. Photoperiod not only entrains the circadian pacemaker but also determines the pattern of electrical discharge of SCN cells. Such photoperiodic regulation of circadian function characterizes not only seasonal species but also animals whose reproduction and metabolism vary little with day length or time of year. Day length determines the duration of the subjective day as indicated by the interval of elevated action potential frequency, glucose utilization, expression of core clock genes, and induction of c-*fos* expression by light pulses (Travnickova et al., 1996; Nuesslein-Hildesheim et al., 2000; Schaap et al., 2003; de la Iglesia et al., 2004; Inagaki et al., 2007; VanderLeest et al., 2007, Yan and Silver, 2008).

Through its outputs, the SCN pacemaker regulates all known aspects of circadian function. Classic experiments in which SCN transplants restore locomotor rhythms when placed in polymer capsules that do not allow neurite outgrowth indicate that behavioral oscillations depend upon humoral outputs of the SCN (Silver et al., 1996). In contrast, axonal projections seem to be required in order for the SCN to regulate physiological rhythms, including hormonal secretion and reproductive responses to day length (Meyer-Bernstein et al., 1999). Projections of the SCN to the paraventricular region of the hypothalamus are of primary importance in seasonal responses (Pickard and Turek, 1983; Lehman et al., 1984; Perreau-Lenz et al., 2003). It is through this pathway that the SCN controls rhythms of the synthesis of melatonin (*N*-acetyl-5-methoxytryptamine [MEL]) in the pineal gland, and it is through the MEL secretion pattern that the pineal transmits information about photoperiod to the brain and the pars tuberalis of the pituitary, a highly vascularized modulatory region adherent to the infundibular stalk that contains abundant MEL receptors in all mammalian species studied to date (Bartness et al., 1993).

Role of Melatonin and the Pineal Gland in Mammalian Photoperiodic Responses

Interruption of the neural pathways by which the SCN drives the pineal gland, or pinealectomy, renders seasonal mammals incapable of responding appropriately to photoperiod. Administration of MEL to pinealectomized hamsters or sheep in

patterns that mimic day length–determined waveforms drives corresponding seasonal patterns of reproductive function. Thus a complete understanding of photoperiodic regulation of reproduction must include investigation of the links between
the circadian pacemaker and the pineal gland.

Although little work has been done in seasonal species, evidence gathered in
the rat indicates that GABAergic and glutamatergic efferents of the SCN to the
paraventricular nucleus (PVN) of the hypothalamus play the major role in shap-
ing the MEL secretory pattern (Kalsbeek et al., 2000; Perreau-Lenz et al., 2004).
Vasopressinergic and vasoactive intestinal polypeptidergic input during the night
may also have modulatory roles (Barassin et al., 2000). Descending projections of
the PVN regulate sympathetic preganglionic neurons that control noradrenergic
cells of the superior cervical ganglion. These cells, in turn, control the activity of
N-acetyltransferase (AANAT), the rate-limiting enzyme in MEL synthesis in pinealocytes. Melatonin production may also be modulated by peptides that are coreleased
along with norepinephrine by ganglionic fibers, and by inputs to the pineal from
the habenular complex. The level at which AANAT expression is regulated by such
input may differ between photoperiodic species: in rodents, transcription is the principally regulated step, but in sheep and primates important control of AANAT (and
hence melatonin synthesis) appears to occur at the posttranslational level (Klein,
2004). It has recently been argued that hydroxyindole-O-methyl transferase activity
is also light-regulated and limits MEL synthesis in rats (Liu and Borjigin, 2005).

MEL is not stored after its synthesis in the pineal gland but is released in a
pattern that corresponds to night length. Although melatonin concentrations are
highest in the ventricular compartment (Skinner and Malpaux, 1999), systemic infusion of MEL into pinealectomized animals at concentrations sufficient to replicate
intact serum levels is sufficient to restore appropriate seasonal reproductive patterns
(Bittman et al., 1983). The endogenous pattern of melatonin secretion changes with
the day length to which animals are entrained. Specifically, the nightly duration of
the rise of serum MEL appears to be the critical parameter by which photoperiod is
encoded in both long- and short-day breeders (Carter and Goldman, 1983; Bittman
and Karsch, 1984; Yellon et al., 1985).

Melatonin, Photoperiod History, and Responsiveness

A simple model would suggest that day lengths below a specific threshold lead
to MEL durations sufficiently long to signal winter onset and induce appropriate seasonal responses. Evidence that photoperiodic history affects threshold day
length (reviewed above) suggests that the actual mechanism is more complicated.
Furthermore, evidence that animals become refractory to day lengths that initially
stimulate or inhibit reproduction, and that some individuals are nonresponsive to
photoperiodic cues, may provide insight into the basis of MEL effects.

It has been proposed that variations in responsiveness arise at either prepineal
(changes in MEL duration driven by the circadian system) or postpineal (differential

response of MEL-sensitive targets) levels. Experimental support has been gathered to support each of these models. In the former case, nonresponsiveness occurs when the circadian clock drives a MEL signal that is of similar duration in long and short days. Conveniently, the duration of nightly locomotor activity (α) can be used as a behavioral assay reflecting the MEL signal: α is directly proportional to MEL duration, presumably because both are generated by the same circadian pacemaker (Elliott and Tamarkin, 1994). One can predict which individual Siberian hamsters will be nonresponsive upon transfer to short days based on the period of locomotor activity rhythms in constant darkness. Animals with free running periods greater than 24 h entrain to short day lengths with compression in α, presumably reflecting a short duration MEL signal (Puchalski and Lynch, 1986; Puchalski et al., 1988; Freeman and Goldman, 1997; Prendergast and Freeman, 1999). Measurements of neuronal activity in the SCN from responsive and nonresponsive Siberian hamsters support the hypothesis that nonresponsiveness occurs at the level of the master pacemaker (Margraf et al., 1991). Importantly, nonresponsive Siberian hamsters given long-duration MEL infusions regress their reproductive organs, indicating that systems downstream from the pineal respond appropriately to this signal (Puchalski et al., 1988). A compression of α occurs not only in nonresponsive Siberian hamsters transferred from long to short day lengths (Gorman et al., 1997; Prendergast et al., 2000b), but also in nonresponsive deer mice from southern latitudes following exposure to short day lengths. Apparently, the effects of prior photoperiod exposure take at least 2 weeks of long-day exposure to acquire, fade significantly 13 weeks after pinealectomy, and are completely abolished after 20 weeks (Prendergast et al., 2000b).

Other studies point to a postpineal mechanism for nonresponsiveness. For example, in deer mice (*P. maniculatus*) and white-footed mice (*P. leucopus*), α and MEL duration do not differ between responders and nonresponders when held in short days (Dark et al., 1983; Blank et al., 1988; Carlson et al., 1989; Ruf et al., 1997). MEL implants fail to inhibit reproduction in deer mice that are nonresponsive to short photoperiod (Blank and Freeman, 1991). Together, these findings suggest differences in MEL receptor affinity or an alteration in cellular events downstream of MEL receptors in relevant target loci. In one study of white-footed mice, however, this hypothesis received little support: neither ^{125}I-melatonin (IMEL) binding nor effects of MEL to antagonize forskolin-stimulated cAMP generation differed between individuals that varied in their response to short photoperiod (Weaver et al., 1990). Subsequent work, however, revealed greater IMEL binding in the preoptic area (POA) and bed nucleus of the stria terminalis (BNST) of nonresponders than in responders (Heideman et al., 1999). Responsive and nonresponsive morphs may also differ in the number or characteristics of gonadotropin-releasing hormone (GnRH) cells (Avigdor et al., 2005).

Although specific binding of IMEL has been reported in a variety of peripheral organs, receptors in the brain and pituitary complex are responsible for the effects of photoperiod on reproduction. Thus, it is important to localize and characterize MEL receptors in the central nervous system and pars tuberalis.

Melatonin Receptors and Their Localized Expression

In vertebrates, three G_i-coupled MEL receptor subtypes have been cloned. Evidence indicates that the MT1 (Mel_{1a}) receptor is most relevant to photoperiodic responses in mammals. The distribution of MT1 mRNA corresponds most closely to the pattern of IMEL binding (Duncan et al., 1989; Weaver et al., 1989; Morgan et al., 1994). Importantly, neither of the other two MEL receptors—the MT2 (Mel_{1b}) nor the MT3 (Mel_{1c})—are expressed in hamsters or sheep (Weaver et al., 1996; Pelletier et al., 2000). The MT2 receptor is expressed in mouse brain and may participate in other functions, including determination of circadian phase and hippocampal plasticity relevant to rhythms of learning and memory (Dubocovich et al., 1998; Wang et al., 2005). However, it is clearly dispensable for photoperiodic regulation of reproduction and thus will not be a focus of this review. The MT3 receptor is not expressed in mammals, but may be orthologous to a MEL-related receptor (G-protein–coupled receptor 50 [GPR50]; Dufourny et al., 2008). This receptor has little affinity for MEL, but may heterodimerize with MT1 receptors and modulate their affinity for MEL (Levoye et al., 2006). Expression of GPR50 in tanycytes of the median eminence is dramatically reduced by short photoperiod in Siberian hamsters (Barrett et al., 2006a). This orphan receptor may influence transport of substances into the ventricular system and ultimately participate in seasonal changes in metabolic function and body weight regulation (Morgan and Mercer, 2001).

Where does MEL act to regulate reproduction? Not only does there appear to be a diversity of sites between photoperiodic species, but within species MEL may act at disparate loci to influence secretion of different hormones and perhaps other seasonally changing responses. Considerable evidence favors the hypothesis that MEL acts primarily (but perhaps not exclusively) in the pars tuberalis of the pituitary complex to regulate secretion of a paracrine signal (tuberalin) that is relayed to the pars distalis, where it controls prolactin secretion (Morgan et al., 1996a). It is also speculated that MEL-dependent signals arising in the pars tuberalis gain access to the brain in order to convey photoperiodic information. In quail, photoperiodically elicited changes in function of the pars tuberalis may be rapidly communicated to the hypothalamus, perhaps using thyroid-stimulating hormone (thyrotropin) as a message (Nakao et al., 2008). A similar mechanism may exist in mammals (Ono et al, 2008). Seasonal reproduction of mustelids (including mink and ferrets) may rely on such a pathway. These animals are exquisitely sensitive to photoperiod, and pineal melatonin secretion appears indispensable to this response (Carter et al., 1982; Martinet et al., 1992). Although specific MEL binding sites have been found in the pars tuberalis of these species, there appear to be none in the brain (Weaver and Reppert, 1990). Thus the possibility that a photoperiod-dependent message originating in the mammalian pars tuberalis regulates GnRH secretion deserves exploration, at least in these carnivores.

Nevertheless, experiments in both sheep and hamsters make it clear that MEL acts directly in the brain of at least some species in order to regulate gonadotropin release (figure 20.2). Within the brain of rodents studied to date, MEL binding sites are highly concentrated in the SCN. This finding suggests that the SCN may

Sheep   Siberian Hamster

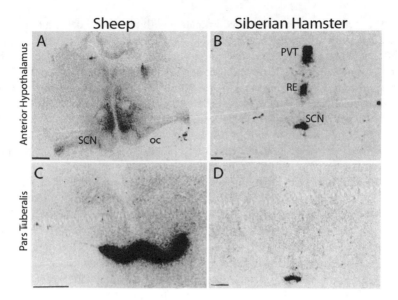

FIGURE 20.2.
Autoradiographic images illustrating total binding of 2-[^{125}I]iodomelatonin in the anterior
hypothalamus (A, B) and pars tuberalis (C, D) of a Suffolk ewe (A, C) and Siberian hamster
(B, D). Nonspecific binding is negligible in all cases (data not shown), so these images
are representative of specific binding. Notice that intense binding is evident in the
suprachiasmatic nuclei of the hamster, but absent in that of the ewe, despite the fact that
both are photoperiodic seasonal breeders. PVT, paraventricular nucleus; RE, nucleus
reunions; SCN, suprachiasmatic nucleus; oc, optic chiasma. Scale bars: 2 µm for A and C;
200 µm for B and D.

participate in photoperiodic responses not only by determining the pattern of MEL
secretion but also by serving as a critical target of MEL (Bittman et al., 1991).
However, this hypothesis is much better supported in some species than others.

The importance of the SCN as a MEL target is best documented in Siberian
hamsters. Destruction of the SCN eliminates gonadal responses to infusions of MEL
that occur in brain-intact pinealectomized animals (Bartness et al., 1991). Within
the SCN of Siberian hamsters, MT1 receptors are expressed in vasopressinergic
cells (Song et al., 2000). Other cell types within the SCN may also be responsive to
MEL but vasopressinergic efferents are most likely important for communicating
photoperiodic information to other brain regions. Other brain targets of MEL may
also participate, but they are less sensitive than the SCN and they fail to sustain
responses to MEL after the SCN has been destroyed. In particular, high concentrations of MEL receptors are found in the paraventricular and reuniens nuclei of the
thalamus, and infusion of MEL to these sites can block gonadal maturation (Badura
and Goldman, 1992). Interestingly, systemic administration of MEL can induce a
second collapse after long-term implants in the SCN have ceased to maintain reproductive inhibition (Freeman and Zucker, 2001). MEL receptor sites outside the SCN

may thus function independently or coordinately with the pacemaker, and may play different roles in various photoperiodic responses (Teubner and Freeman, 2007).

In contrast to Siberian hamsters, Syrian hamsters remain responsive to MEL after destruction of the SCN (Bittman et al., 1979; Maywood et al., 1990). Other forebrain structures, including the dorsomedial hypothalamus (DMH) and preoptic regions such as the BNST, may be more important in the response to MEL in this species (Maywood and Hastings, 1995; Maywood et al., 1996; Raitiere et al., 1997; Bae et al., 1999; Lewis et al., 2002). Only lesion studies and IMEL autoradiography of limited resolution have thus far been used to identify these sites. Thus, the phenotype of MEL responsive cells in these areas is unknown.

In sheep, it is clear that the premammillary hypothalamus functions as a particularly important target of MEL. This perifornical region has a high concentration of IMEL binding sites and expresses MT1 receptor (Malpaux et al., 1998). In contrast, the ovine SCN contains a lower concentration of MEL binding sites than other areas of the hypothalamus and POA (Bittman and Weaver, 1990; figure 20.2). Furthermore, electrolytic lesions that disrupt the caudal arcuate and premammillary region of Soay rams affect photoperiodically driven patterns of gonadotropin secretion. Finally, implants of MEL in this area stimulate the secretion of luteinizing hormone (LH). The premammillary hypothalamus is far more sensitive to MEL implants than the anterior hypothalamic-preoptic region in this regard (Lincoln, 1992; Lincoln and Maeda, 1992; Malpaux et al., 1993). The phenotype of MEL-responsive cells within this complex region is unknown, as are the projections through which they might regulate reproduction; candidates include estrogen receptor-α (ERα)-immunopositive, orexin-immunopositive, and thyroid-hormone receptor–immunopositive neurons (Beccavin et al., 1998; Sliwowska et al., 2004). Despite evidence for MEL binding sites in POA, brainstem and hippocampus of a variety of species, there is no evidence that the pineal hormone must act outside the hypothalamus and pars tuberalis in order to trigger seasonal reproductive changes.

Melatonin and Clock Gene Expression: A New Version of the Coincidence Model

Recent work has focused on the hypothesis that photoperiodic regulation arises from the control of temporal patterns of core clock gene expression by MEL. Circadian rhythms in mammals depend upon transcriptional-translational feedback loops that have been extensively reviewed (Ko and Takahashi, 2006). Briefly, the positive arm of the loop is comprised of the protein products of *Bmal1* and *Clock*. The BMAL1:CLOCK heterodimer acts at E-box motifs within the promoter regions of many genes to drive circadian rhythms of their transcription. Expression of at least four of the genes that lie at the core of the clock—*Per1*, *Per2*, *Cry1*, and *Cry2*—is regulated in just this way. The protein products of the *Per* and *Cry* genes dimerize and form a complex with BMAL1:CLOCK in order to terminate transcriptional activation. Thus, *Per* and *Cry* comprise the negative arm of the loop. While the maintenance of the oscillation in constant conditions is attributable to interactions of clock proteins at the E-box, light can shift the phase of circadian oscillations. In the SCN, this effect is mediated by inputs that ultimately trigger the phosphory-lation of the transcription factor CREB (cAMP-response element-binding protein), which induces expression of *Per1* upon binding to cAMP response elements (CREs) in its promoter. The basic transcriptional-translational loop is found not only in pacemaker structures, but also in tissues throughout the organism. BMAL1 concentrations typically oscillate in antiphase to those of PER1/2 and CRY1/2, although the specific time of day at which their peak is reached differs between the pacemaker and the periphery. The pathways by which the pacemaker regulates molecular rhythmicity in other structures are a subject of intense interest and investigation. It is clear that endocrine signals can have a major influence, and it is suggested that the action of MEL to regulate the timing of reproduction results from control of clock gene expression, at least in the pars tuberalis. Photoperiod also regulates clock gene expression in peripheral organs (Carr et al., 2003; Andersson et al., 2005), but this may be a consequence of seasonal changes in gonadal or other hormones rather than the result of a direct action of MEL.

Coincidence models that bear similarity to the Bünning hypothesis and current understanding of photoperiodism in *Arabidopsis* and *Drosophila* have been advanced to explain mammalian photoperiodic responses (see above and chapter 19). In hamsters and sheep, the amplitude and phase of clock gene expression in both the SCN and the pars tuberalis depend upon day length, although the downstream consequences of these changes have not been established (Messager et al., 2000; Tournier et al., 2003; de la Iglesia et al., 2004; von Gall et al., 2005; Johnston et al., 2006; Korf and von Gall, 2006a). Although MEL may regulate photic input and electrical activity in rodent SCN, it is not required for clock gene regulation in the circadian pacemaker (Korf and von Gallo, 2006b). Similarly, MEL has no apparent effect on clock gene expression in sheep SCN, as might be expected from the lack of IMEL binding in this structure (figure 20.2). In contrast, MEL may have important regulatory effects on clock gene expression in the pars tuberalis of hamsters and sheep that mediate at least some photoperiodic responses. These effects are presumably exerted through a regulation of CREB phosphorylation subsequent to occupation of the MT1 receptor. Although rhythms of clock gene expression in the ram pars tuberalis do not require pineal input, MT1 expression of *Cry1* in this tissue is induced by MEL. According to a current model (Lincoln, 2006), the phase relationship between peaks of *Per1* and *Cry1* expression in the pars tuberalis is set by MEL duration. This pattern may provide a circadian-based coincidence mechanism to drive seasonal rhythms of prolactin secretion. A critical test of this hypothesis awaits demonstration that direct manipulation of clock gene expression affects the ability of the pars tuberalis to regulate lactotrope function. It is unknown whether similar coincidence models may pertain to MEL action on targets in the brain. MEL regulation of type 2 iodothyronine deiodinase (*DIO2*) expression in hamster brain is not accompanied by changes in clock gene expression (Yasuo et al., 2008).

STEROID-DEPENDENT AND -INDEPENDENT EFFECTS OF PHOTOPERIOD

The striking seasonal changes in gonadal activity evident in mammals of the temperate zone can be traced to fluctuations in pituitary function. Serum concentrations of gonadal steroid hormones plunge during the nonbreeding season as the secretion of gonadotropins declines. The reduction in circulating androgens, estrogens, and/ or progestins might be expected to relax feedback inhibition on the hypothalamo-hypophyseal axis. However, such a homeostatic mechanism would not allow the maintenance of reproductive quiescence. From this point of view it is understandable that the capacity of gonadal steroid hormones to act as negative feedback regulators of LH and follicle-stimulating hormone (FSH) secretion changes seasonally. This mechanism ensures that gonadotropins remain available to stimulate gametogenesis during the breeding season, even in the face of high levels of gonadal steroid hormone secretion. Conversely, during the nonbreeding season even low concentrations of gonadal steroids are sufficient to suppress gonadotropin secretion, ensuring that reproductive capacity is held in check (Tamarkin et al., 1976; Legan et al., 1977; Turek, 1977). The nature of the gonadal signal that controls gonadotropin secretion may even change: in the ewe, estradiol acts as the principal negative feedback regulator of LH secretion during anestrus, whereas progesterone plays this role during the breeding season (Goodman and Karsch, 1980). Among gonad-intact animals, these "steroid-dependent" seasonal changes in gonadotropin secretion probably represent the rate-limiting and physiologically important mechanism by which reproductive capacity is controlled. Other changes in the hypothalamo-pituitary-gonadal axis appear to be secondary and less critical. Although photoperiod may affect pituitary responsiveness to exogenous GnRH, these changes are relatively minor and may reflect the synthesis or storage of LH and FSH as the pattern of GnRH secretion changes with photoperiod (Karsch et al., 1980; Goodman and Karsch, 1981; Bittman et al., 1985, 1990; Meredith et al., 1998). Sensitivity to gonadotropins changes with season (Legan et al., 1985; Bartlewski et al., 1999a, 1999b), but the gonads retain the ability to respond to pituitary hormones even during the nonbreeding season (Goodman et al., 1981; Legan et al., 1985; Wallace et al., 1986). Furthermore, the positive feedback action of estradiol to elicit a preovulatory gonadotropin surge seems largely unaffected by season in the ewe (although this may not be the case in goats; Mori et al., 1987).

Seasonal changes in the sensitivity of the negative feedback axis can be demonstrated by clamping estradiol or testosterone levels in gonadectomized animals. During the breeding season, gonadotropin secretion "escapes" from the negative feedback influence of steroid hormones. Transfer to inhibitory photoperiods reestablishes the potent negative feedback effects of testosterone or estradiol. Importantly, these fluctuations in gonadotropin secretion are photoperiodically driven when steroid hormone levels are held constant. These changes occur in both long- and short-day breeders. The effects of photoperiod are eliminated by pinealectomy and controlled by the duration of nightly MEL secretion (Karsch et al., 1984; Bartness et al., 1993).

In addition to these steroid-dependent changes in gonadotropin secretion, seasonal changes in LH and FSH release occur in castrated animals that are *not* given exogenous steroid hormones (Simpson et al., 1982; Urbanski et al., 1983). The rate at which LH and FSH levels rise after gonadectomy varies with season, and gonadotropin concentrations depend on photoperiod even when extragonadal (adrenal, dietary) sources of steroid hormones are removed in castrated animals (Bittman and Zucker, 1977). Another steroid-independent effect occurs in female Syrian hamsters upon exposure to short photoperiods. These animals experience a marked rise in gonadotropin secretion each afternoon (Bridges and Goldman, 1975; Seegal and Goldman, 1975; Jorgenson and Schwartz, 1984; Bittman et al., 1992; figure 20.3).

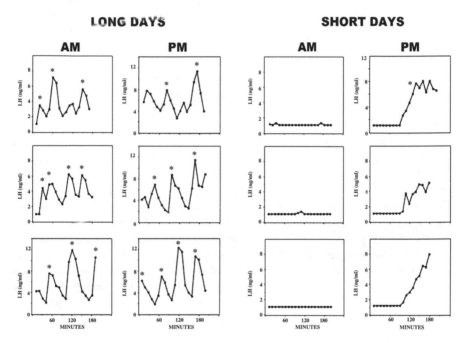

FIGURE 20.3.
Steroid-independent regulation of LH secretion by day length in female Syrian hamsters. Animals were ovariectomized, and half were transferred to short days (10L:14D; right) nine weeks prior to femoral vein cannulation and frequent (10 min) blood sampling. Control hamsters remained in 14L:10D (left). Three individuals housed in each photoperiod are shown. Statistically identified pulses are marked by asterisks. Note that high-amplitude LH pulses occur once per hour in long-day animals. Exposure to short days elicited two steroid-independent effects: LH pulses were absent or of extremely low amplitude in the morning (0930–1200 h) despite the long-term absence of the ovaries. In the afternoon, a sustained rise of serum LH occurred at approximately 1500 h. Effects of short photoperiod on LH pulse frequency result from a melatonin-dependent slowing of the GnRH pulse generator, whereas the timing of the afternoon rise of LH concentrations suggests that short days may allow steroid-independent activation of the surge mechanism in hamsters. Reprinted from Bittman et al. (1992), with permission.

The timing and apparent circadian basis of this secretion pattern are similar to those of the preovulatory gonadotropin surges that occur on the afternoon of proestrus in long-day animals. Unlike the preovulatory surge, however, the short-day surges are estrogen independent: they persist in ovariectomized hamsters. It is unknown whether this phenomenon represents an activation of the surge mechanism normally triggered by estradiol in long-day hamsters, but it seems that this pattern of gonadotropin secretion effectively blocks follicular maturation and estrous cycles in winter photoperiods. In this respect, it may complement or substitute for the enhancement of the negative feedback effects of estradiol. These "steroid-independent" effects of photoperiod may represent a change in the neural "drive" upon gonadotropin secretion. Insight into the mechanisms that underlie both the steroid-independent and steroid-dependent effects of photoperiod has been gained by frequent blood sampling that allows characterization of the hypothalamic-preoptic pulse generator that underlies the tonic mode of gonadotropin secretion. Measurement of LH in the systemic blood indicates that the operation of this neural pulse generator, which releases GnRH into the hypothalamo-hypophyseal portal vasculature, varies seasonally: upon exposure to photoperiods that stimulate reproduction, the frequency of LH pulses increases (Karsch et al., 1984; Bittman et al., 1985, 1992; Krey et al., 1989).

It is unclear whether the mechanisms that underlie the steroid-dependent and -independent effects overlap. For example, neurosteroid production in castrated animals may play a role in seasonally changing patterns of gonadotropin secretion (Baulieu and Robel, 1990; Sim et al., 2001; Inai et al., 2003; Schmidt et al., 2008). In addition, it has become apparent that steroid hormone receptors may be activated by neurotransmitter action and regulate transcription in the absence of cognate steroid hormones (Mani et al., 1996). Thus, it is possible that steroid-independent effects of photoperiod reflect changing impact of such unliganded steroid receptors. Alternatively, photoperiodic influences that differ in steroid dependence may differ in their physiological basis. Pharmacological studies in ewes indicate that steroid-independent seasonal changes in LH secretion depend on serotonergic mechanisms that differ from the dopamine-based steroid-dependent effects (Meyer and Goodman, 1986; Whisnant and Goodman, 1990).

Gonadal steroid hormones act on the brain to regulate not only GnRH and hence gonadotropin secretion, but also sexual and aggressive behaviors. Remarkably, the behavioral effects of gonadal steroid hormones also change with season. The mechanisms by which such changes are achieved are at least as mysterious as the feedback effects, and in a sense the behavioral fluctuations in steroid hormone action occur in the opposite direction. Specifically, the ability of androgens to activate male sexual behavior (including mounting, intromission, and ejaculation) wanes during the nonbreeding season just as the feedback effects increase. Similarly, the action of ovarian hormones to prime the brain and then trigger behavioral estrus diminishes during the nonbreeding season at the same time that negative feedback effects of the same hormones are magnified (Goodman et al., 1981; Badura et al., 1987; Bittman et al., 1990; Miernicki et al., 1990; Pospichal et al., 1991). The degree to which the opposite effects of season on behavioral and feedback actions differ in neuroanatomical locus or molecular mechanisms is unknown.

Various techniques have been utilized in attempts to pinpoint changes in the brain that might be responsible for seasonal changes in the action of gonadal steroid hormones. Local implants of estradiol in ovariectomized anestrous ewes implicate the ventromedial POA and retrochiasmatic area in the steroid-dependent suppression of LH pulse frequency (Gallegos-Sanchez et al., 1997). The negative feedback effects of estradiol during anestrus are apparently mediated by ERα, as the ERβ agonist genistein is ineffective (Hardy et al., 2003). Evidence for seasonal changes in the negative feedback action of gonadal steroid hormones led to hypotheses that the synthesis or stability of their cognate receptors changes with time of year. Over several decades, the experimental approaches to this question have gained in sophistication and specificity. Initial attempts to identify changes in the number or affinity of gonadal steroid hormone receptors in homogenates of various brain regions identified few major changes that could not be ascribed to occupancy as endogenous steroid hormone levels rise and fall with season, photoperiod, or MEL treatment (Bittman and Krey, 1988; Bittman and Blaustein, 1990). With the advent of techniques that allow better localization and, to some extent, quantification of steroid receptor expression it became possible to identify statistically significant effects of season. It is not clear whether any of the changes reported in steroid receptor expression can explain the dramatic effects of season on the efficacy of gonadal hormones as negative feedback signals. For example, Skinner and Herbison (1997) reported only a modest (20%) increase in the number of ERα-immunoreactive (-ir) cells in POA of ovary-intact ewes. Recent findings of photoperiod-induced changes in numbers of ER-ir cells in hamsters (Kramer et al., 2008) and deer mice (Trainor et al., 2007) are consistent with this earlier work. In the latter study, ERα expression was increased, and ERβ expression decreased, by winter day lengths. The loci of these effects led these authors to interpret them as relevant to seasonal changes in sociosexual behaviors.

It is important to note that most of these studies were performed in gonad-intact animals. Expression of the receptors for gonadal steroid hormones may be regulated by their endogenous ligands (Schreihofer et al., 2000). Thus, it is often not clear whether effects of photoperiod upon steroid hormone binding or receptor expression are secondary to seasonal changes in secretion of endogenous gonadal steroids, or whether they are primary causes of fluctuations in the sensitivity or response to these hormones. A few experiments have addressed this issue. In male Siberian hamsters that were castrated and clamped at a physiological range of testosterone levels, a relatively modest photoperiodic modulation of androgen receptor expression occurred in the arcuate nucleus and the BNST (Bittman et al., 2003). In studies of ovariectomized female Syrian hamsters, short days and MEL injections reduce the number of ERα-ir cells in the ventromedial hypothalamus and POA (Lawson et al., 1992; Mangels et al., 1998). An important function of estrogen is to "prime" neuroendocrine targets for responsiveness to peri- and postovulatory progestins. Thus, it is significant that exposure of female hamsters to short days reduced progestin receptor induction by estradiol in the ventromedial nucleus, amygdala, and POA (Mangels et al., 1998).

Even within particular brain regions in which gonadal steroid hormones exert photoperiodically modulated behavioral or feedback effects, the population of functionally important cells could be heterogeneous. Thus, it may prove to be necessary to examine seasonal changes in gonadal steroid receptor expression within particular cell phenotypes using double-label techniques. However, this strategy may not succeed if the photoperiodic or steroid influences on different cells are conveyed transsynaptically to a final common pathway for expression of seasonal changes. Furthermore, there is a certain amount of guesswork involved in the decision not only of *where*, but of *when*, to look for steroid-photoperiod interactions: changes in reproductive function with season often occur gradually, such that the initiating neurochemical events may precede changes in behavior or gonadotropin secretion by many weeks. Similarly, the onset of refractoriness to initially effective photoperiods may arise from neural events that begin while the gonads and gonadotropins are still showing the most dramatic response to day length. Furthermore, posttranslational events including phosphorylation, dimerization, and translocation of steroid receptor proteins may be critically affected by photoperiod. These functionally important changes would not be detected by studies that measure mRNA or immunoreactive peptide.

To regulate expression of specific genes, steroid hormone receptors must recruit nuclear receptor coactivators to activate the basal transcription machinery and remodel chromatin structure. Thus, changes in steroid efficacy could result from changes in interactions between steroid hormone receptors and these cofactors. In support of this idea, photoperiod regulates steroid receptor cofactor-1 (SRC-1) immunoreactivity in the BNST and medial amygdala of Siberian hamsters, and such effects may be androgen dependent (Tetel et al., 2004). Though statistically significant, the effect of photoperiod is not substantial. Nevertheless, regulation of steroid receptor cofactors by day length may contribute to seasonal changes in feedback or behavioral actions of gonadal hormones and these effects may be additive or multiplicative with other levels of photoperiodic control. The possibility that seasonal fluctuations in neuroendocrine function depend on epigenetic changes, including histone acetylation or methylation, merits investigation (Kouzarides, 2007). Furthermore, the possibility that photoperiod or MEL regulates nonclassical pathways by which steroid receptors in the cell membrane exert rapid cellular responses appears not to have been explored (Boulware and Mermelstein, 2009).

Ultimately day length–dependent signals that regulate reproduction must be relayed to the GnRH cells that regulate the release of gonadotropins from the pars distalis. Although GnRH peptide levels, as indicated by immunostaining, change gradually upon transfer to short photoperiods (Yellon, 1994), exposure of photoinhibited hamsters to long days acutely stimulates transcription of GnRH in the POA (Ronchi et al., 1992; Porkka-Heiskanen et al., 1997). In *Peromyscus maniculatus*, depolarization-elicited GnRH release from isolated hypothalamus (presumably an estimate of the releasable pool) is reduced by short photoperiod (Mintz et al., 2007). There is no evidence to indicate that such effects of day length reflect an influence of pineal or gonadal hormones directly upon the GnRH cells. Instead, it is most likely that photoperiodic and gonadal signals exert their impact through at least some of the many synaptic inputs to the GnRH cells, or to their terminals in the median eminence.

Pharmacological manipulations in ewes indicate that the ability of estradiol to slow LH pulses in anestrus animals depends upon dopaminergic (D_2) transmission, and lesion studies implicate the retrochiasmatic A15 dopaminergic cell group (Anderson et al., 2003; Hardy et al., 2003). In light of the finding that A15 neurons lack ERs, it is proposed that these cells relay input from estrogen-sensitive cells to the GnRH neurons. Estrogen may modulate $GABA_A$ and $GABA_B$ inhibitory input to the A15 dopaminergic cells, thus regulating their inhibitory output to GnRH neurons (Goodman et al., 2000; Bogusz et al., 2008).

Photoperiodic effects on reproduction may reflect environmental regulation of brain plasticity. The thyroid gland, which plays an important role in brain development, also participates in regulation of annual reproductive cycles in adult birds and mammals. Photoperiod and MEL regulate *Dio2* expression in sheep and hamster brain and thus control the conversion of tetraiodothyronine (T4) to the more active triiodothyronine (T3; Yasuo et al. 2007; Hannon et al., 2008; see Chapter 21). Thyroidectomy of ewes blocks the onset of photorefractoriness to short days which terminates the breeding season, and local implants of thyroid hormone into the ventromedial POA reverse this effect (Anderson et al., 2003). The finding that synaptic input to dopaminergic cells in the A15 group changes seasonally suggests that structural remodeling of the POA contributes to seasonal changes in estrogen negative feedback in sheep (Bogusz et al., 2008). Seasonal changes in numbers of specific synaptic inputs to GnRH neurons, including GABA-ir, neuropeptide Y–ir, and β-endorphin–ir contacts, have also been reported in ewes (Jansen et al., 2003). The extent of glial ensheathment of GnRH cells also varies with season. Photoperiod regulates expression of the polysialated form of neuronal cell adhesion molecule in the preoptic/hypothalamic region in both sheep and hamsters (Lee et al., 1995; Viguie et al., 2001). Photoperiodic regulation of nestin expression in the medial basal hypothalamus of Siberian hamsters also supports the idea that seasonal changes are associated with remodeling of neural networks (Barrett et al., 2006b). Cooke et al. (2002) reported that androgen increases the volume of the anterior medial amygdala of Siberian hamsters only in long days. Furthermore, photoperiodic regulation of new cell incorporation and dendritic morphology in hippocampus and the main and accessory olfactory bulbs has been reported in Syrian hamsters, voles, and deer mice (Huang et al., 1998; Galea and McEwen, 1999; Pyter et al., 2005).

NEUROPEPTIDERGIC EFFECTORS OF SEASONAL REPRODUCTION

GnRH secretion is regulated not only by GABA, glutamate, and monoaminergic transmitters, but also by numerous neuromodulatory peptides of the hypothalamus and POA. Research into this subject is complicated by overlap of the negative and positive feedback axes and the potential for the same peptides, produced in the same or different cell groups, to participate in both pulsatile and surge modes of GnRH secretion. Photoperiod may regulate expression of several of these peptides and their receptors, as well as extrahypothalamic signals that impinge upon them. In addition, hypothalamic mechanisms that regulate the release of prolactin releasing and inhibitory factors into the hypothalamo-hypophyseal portal system most likely contribute to seasonal changes in reproductive function. Photoperiodic regulation of prolactin release is particularly complex, not only because of the multifactorial nature of hypothalamic control, but also because of the importance of the pars tuberalis in regulation of the lactotropes. Indeed, photoperiod and melatonin continue to regulate prolactin release in rams in which the pituitary complex has been disconnected from the hypothalamus (Lincoln et al., 2006). This finding suggests that the principal seasonal control of prolactin secretion is exerted at the level of the pars tuberalis.

Photoperiod also regulates several hypothalamic peptides that control body weight and metabolism. The expression of many of these modulators changes seasonally, and such fluctuations may contribute to or mediate annual rhythms (Ellis et al., 2008). Neuropeptide Y, which is produced in cells of the arcuate nucleus and is likely to play an important role in seasonal changes in appetite, is under photoperiodic control in the ewe (Skinner and Herbison, 1997). In both sheep and hamsters, photoperiod regulates the production of opiates and their receptors (Lincoln et al., 1987; Roberts et al., 1987; Tubbiola et al., 1989). Day length and testosterone interact to regulate the expression of *pro-opiomelanocortin* in the arcuate nucleus of Syrian hamsters (Bittman et al., 1999). β-Endorphin staining is also reduced by short-day exposure in ewes (Skinner and Herbison, 1997). Dynorphin cells in the arcuate, which likely coexpress progestin receptors and kisspeptin, constitute another opiatergic system that may participate in responses to photoperiod (Goodman et al., 2007). Not only do endogenous opiates participate in seasonal changes in gonadotropin secretion and behavior, but they also interact with other photoperiodically regulated, medial basal hypothalamic peptides that control appetite, body weight, and metabolism (Mercer et al., 2003; Schuhler et al., 2003; Anukulkitch et al., 2007; Smith et al., 2008b).

RFamide Peptide Control of Seasonal Reproduction

The identification of the cardioexcitatory neuropeptide containing the C-terminal Phe-Met-Arg-Phe-NH$_2$ (FMRFamide) in the ganglia of the clam, *Macrocallista nimbosa* (Price and Greenberg, 1977) eventually led to the discovery of a family of peptides that markedly regulate the reproductive neuroendocrine axes of vertebrates. Members of a class of neuropeptides called RFamides (sharing the Arg-Phe-NH$_z$ motif on their C-terminus) have begun to emerge as key mediators of seasonal reproduction (reviewed in Greives et al., 2008). One such peptide, kisspeptin, acts to stimulate the GnRH system (reviewed in Kauffman et al., 2007), whereas another, gonadotropin inhibitory hormone (GnIH), suppresses GnRH and gonadotropin secretion (Tsutsui et al., 2000; Bentley et al., 2003; Kriegsfeld et al., 2006). Given the localization of these peptides in the brains of rodents and sheep, both are in a position to relay circadian and/or photoperiodic information to the reproductive axis and perhaps to act synergistically to exercise precise control over seasonal reproduction (figure 20.4).

Gonadotropin-Inhibitory Hormone

Using enzyme-linked immunosorbant assays with antibodies directed against the RF motif, a novel RFamide peptide was identified that rapidly and dose-dependently inhibits gonadotropin release from cultured quail pituitaries (Tsutsui et al., 2000). Because this peptide acted, at least in part, at the level of the pituitary to inhibit gonadotropin secretion, Tsutsui and colleagues named this peptide GnIH. More recent evidence in birds and mammals indicates that GnIH also acts directly on the GnRH system. In birds, GnIH-ir cell bodies are found in the PVN with fibers targeting both the pituitary and GnRH cells (Bentley et al., 2003). In rodents (including rats, mice, and hamsters), GnIH cells are situated in the DMH, with widespread projections to both GnRH-ir cells and the external layer of the median eminence (Kriegsfeld, 2006; Kriegsfeld et al., 2006). As in birds, injections of GnIH lead to rapid suppression of LH in rats, mice, and Syrian hamsters. Importantly, GnIH cells express ERs and may mediate steroid negative feedback on the reproductive axis (Kriegsfeld, 2006), placing these cells in a position to mediate steroid-dependent effects of photoperiod (described previously).

Kisspeptin

Although the product of the *Kiss-1* gene is now best known for its stimulatory role in reproductive function, it was originally discovered in a screen for tumor metastasis suppressors, and its protein product was initially named metastin (Dungan et al., 2006). At approximately the same time, this RFamide peptide was shown to be the endogenous ligand for the orphan G-protein–coupled receptor, GPR54 (reviewed in Kriegsfeld, 2006; Kauffman et al., 2007). In 2003, a population of individuals with hypogonadotropic hypogonadism was shown to have a mutation in *GPR54* (de Roux et al., 2003), and this peptide was renamed kisspeptin based on this discovery out of Hershey, Pennsylvania. Administration of kisspeptin leads to immediate-early gene expression (i.e., FOS) in GnRH cells (Irwig et al., 2004; Matsui et al., 2004), suggesting direct actions on this neurosecretory population. Kisspeptin does not appear to act at the level of the pituitary; kisspeptin-induced gonadotropin secretion is blocked across species by the GnRH receptor antagonist, acyline (Gottsch et al., 2004; Irwig et al., 2004; Shahab et al., 2005). A majority of GnRH cells express *GPR54* mRNA across species (Irwig et al., 2004; Han et al., 2005;

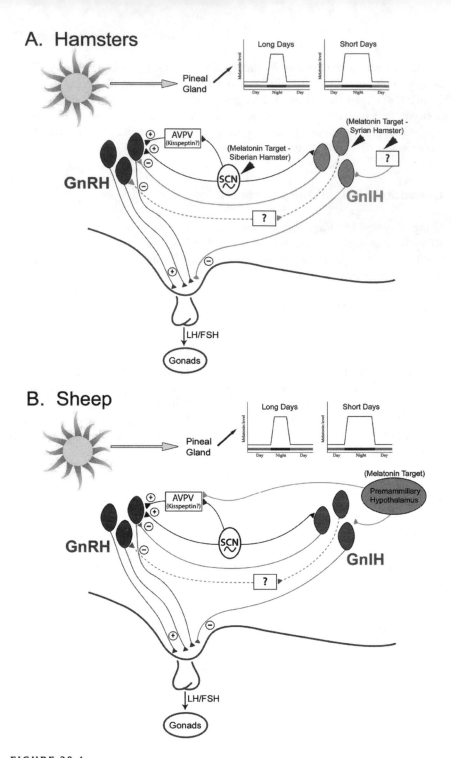

FIGURE 20.4.
A model depicting potential pathways mediating RFamide peptide control of seasonal
reproduction in hamsters (A) and sheep (B). In both hamsters and sheep, light is

Messager et al., 2005), further indicating direct actions of kisspeptin on the GnRH system. In rodents, kisspeptin-ir cell bodies are concentrated in the AVPV and arcuate nuclei (Kauffman et al., 2007), whereas sheep express kisspeptin in both the POA and the arcuate (Estrada et al., 2006; Caraty et al., 2007). As with the GnIH system in rodents, kisspeptin cells contain sex-steroid receptors, suggesting they are a direct target of gonadal steroid hormones (Smith, 2008).

RFamide Peptides and Photoperiodic Regulation

The first evidence for a role of RFamide peptides in seasonal breeding came from studies in quail in which the eyes and pineal gland were removed In order to eliminate endogenous MEL. These manipulations led to a marked reduction in GnIH mRNA and peptide expression (Ubuka et al., 2005). MEL replacement reversed these effects. This finding set the stage for investigating the role of GnIH and other RFamide peptides across species. As discussed above, several hypothalamic sites have been identified as MEL targets and their relative importance differs between species. Across these well-studied species, evidence is mounting that GnIH and kisspeptin play a key role in seasonal breeding.

As described previously, the DMH is a key site for MEL action in Syrian hamsters (Maywood and Hastings, 1995; Maywood et al., 1996; Bae et al., 1999; Lewis et al., 2002). The DMH also expresses androgen receptors, suggesting it may be a site at which photoperiod/MEL modulates sensitivity to steroid negative feedback. Given the location of GnIH in the DMH across rodent species (Kriegsfeld et al., 2006) and the critical role for the DMH in seasonal breeding in Syrian hamsters, GnIH has attracted the attention of researchers in this area. Furthermore, the SCN projects to the DMH in this species (Kriegsfeld et al., 2004) and more than 60% of GnIH cells in this area are contacted by SCN terminal fibers (Gibson et al., 2008). Although the SCN is not required for MEL-induced reproductive regression in this species, the possibility for circadian interactions with a MEL target site further suggests that GnIH cells play a role in seasonal breeding.

Two studies investigating the effects of photoperiod on GnIH expression in Syrian hamsters have uncovered similar, but unanticipated, findings. Contrary to expectation, extended exposure to inhibitory day lengths leads to *suppression* of

transduced into an MEL signal of short duration during summer and long duration during winter. The MEL message is decoded in the DMH in Syrian hamsters, potentially increasing negative feedback actions of sex steroids via enhanced GnIH inhibition of the GnRH system. In Siberian hamsters, MEL acts on the circadian clock in the SCN which, in turn, may communicate day-length information to both kisspeptin and GnIH cells. Other sites containing MEL receptors (nucleus reuniens, paraventricular nucleus of the thalamus) also have access to the GnRH system (not shown). In sheep, MEL acts in the premammillary hypothalamus which presumably communicates with both the kisspeptin and GnIH networks to control positive and negative balance of inputs to the GnRH system. Participation of the dopaminergic A15 cell group is also likely (not shown).

GnIH immunostaining and mRNA (Revel et al., 2008). Similar results are seen following 60 days of melatonin administration to long-day animals (Revel et al., 2008). This is difficult to reconcile with a role for GnIH in reproductive suppression, as its expression is lowest when the hypothalamo-hypophyseal-gonadal axis is maximally inhibited. One possibility is that Syrian hamsters require enhanced GnIH expression to suppress GnRH during the initial period of regression, but that this level of inhibition is not necessary to maintain a fully regressed reproductive axis and low testosterone concentrations. Further examination of the pattern of GnIH expression throughout the development of reproductive quiescence are necessary to examine this possibility. Less likely, GnIH may not participate in seasonal changes in reproductive function, and another DMH peptide may serve this function. Future studies using molecular approaches to directly target GnIH function/expression are necessary to understand the role of this peptide in seasonal breeding.

In contrast to Syrian hamsters, the SCN is a critical MEL target tissue for short-day–induced regression of the reproductive system in Siberian hamsters (Bartness et al., 1991; Bittman et al., 1991). The possibility that the DMH is important for photoperiodic responses in Siberian hamsters has not been directly examined. According to one model (figure 20.4), the SCN communicates information about MEL duration directly to GnRH neurons or to intermediate sites upstream of GnRH, such as kisspeptin and GnIH cells (reviewed in Smith and Clarke, 2007). As in Syrian hamsters, GnIH is suppressed by short-day exposure in Siberian hamsters (Revel et al., 2006; Mason et al., 2007a). In male and female Siberian hamsters, photoperiod-driven changes in reproduction are also associated with marked changes in kisspeptin -ir (Greives et al., 2007, 2008; Mason et al., 2007a). Kisspepti-ir is reduced in the AVPV of animals held in short days, as would be expected in light of its stimulatory input to GnRH cells. Patterns in the arcuate nucleus are reversed, however: kisspeptin-ir is significantly elevated in short days (Greives et al., 2007; Mason et al., 2007b). It is unclear whether kisspeptin neurons in the arcuate contribute to the reproductive response to photoperiod in Siberian hamsters. In Syrian hamsters, arcuate *Kiss-1* mRNA is increased in long days and decreased in short days. Whether or not kisspeptin is present in the AVPV of Syrian hamsters remains to be definitively determined, although recent work indicates kisspeptin peptide expression in this region (Kriegsfeld and Williams, unpublished data). It is possible that posttranscriptional/translational events account for the differences seen among these studies, or that species differ in the direction of kisspeptin change. Future studies combining mRNA and peptide quantification are necessary.

In sheep, GnIH neurons are prominent in the premammillary hypothalamus (Smith et al., 2008a), whereas kisspeptin cells are concentrated in the POA and arcuate (Estrada et al., 2006; Smith et al., 2007). Kisspeptin can stimulate ovulation in seasonally acyclic ewes (Caraty et al., 2007), suggesting a potent role for this peptide in sheep reproduction. In the breeding season, kisspeptin (protein and mRNA) is increased in the arcuate nucleus compared to reproductively inhibited ewes (Smith, 2008; Wagner et al., 2008). GnIH protein shows the opposite pattern, with decreased expression of the peptide in the PVN and DMH during the breeding season (Smith et al., 2008a). An analogous inverse relationship is seen in the number of kisspeptin- and GnIH-containing contacts onto GnRH neurons: kisspeptin contacts onto GnRH neurons are increased in the mediobasal hypothalamus and GnIH-GnRH contacts are decreased in the POA during the breeding season (Smith et al., 2008a). Together, these findings suggest that kisspeptin and GnIH act in a coordinated fashion to provide positive and negative drive to the reproductive axis and control seasonal breeding. Experimental manipulation of the relative concentrations of these peptides across the reproductive cycle is required to evaluate this hypothesis in seasonal breeders.

CONCLUSIONS

Among animals that inhabit temperate environments, natural selection has favored those individuals that curtail reproduction when available energy must be reserved for behavioral and physiological processes essential for survival. Instead, mammals with short gestation periods, such as small rodents, mate and breed during times of year when conditions are less challenging. Larger species such as sheep have longer gestation periods and thus mate during winter in order to give birth in the spring. Diverse mechanisms regulate the timing of fertility, but both long- and short-day breeders make use of the circadian system in order to respond to photoperiodic cues. In some species this information entrains a circannual oscillator, whereas in others it induces and breaks refractory phases. In each case, day length information is relayed from the circadian pacemaker to the pineal gland in order to generate a seasonally appropriate pattern of MEL secretion. The duration of MEL regulates gonadotropin secretion, metabolic and behavioral adjustments through actions on brain targets including the DMH and SCN of hamsters and the premammillary hypothalamus of sheep. MEL also regulates prolactin secretion through actions on the pars tuberalis. Critical to seasonal changes in reproduction are changes in the sensitivity of the brain to behavioral and negative feedback actions of gonadal hormones, but photoperiod also triggers changes in patterns of GnRH release that are independent of sex steroids. Photoperiod regulates expression of many brain peptides; neurons that express the RFamides, kisspeptin and GnIH, may serve as relatively direct targets of MEL in order to provide a precise balance of positive and negative inputs to the GnRH system throughout the year. Despite considerable recent progress, the precise pathways and mechanisms that underlie seasonal regulation of mammalian reproduction remain to be fully understood.

References

Almeida OF and Lincoln GA. (1984). Reproductive photorefractoriness in rams and accompanying changes in the patterns of melatonin and prolactin secretion. Biol Reprod 30(1): 143–158.

Anderson GM, Hardy SL, Valent M, Billings HJ, Connors JM, and Goodman RL. (2003). Evidence that thyroid hormones act in the ventromedial preoptic area and the premammillary region of the brain to allow the termination of the breeding season in the ewe. Endocrinology 144(7): 2892–2901.

Andersson H, Johnston JD, Messager S, Hazlerigg D, and Lincoln G. (2005). Photoperiod regulates clock gene rhythms in the ovine liver. Gen Comp Endocrinol 142(3): 357–363.

Anukulkitch C, Rao A, Dunshea FR, Blache D, Lincoln GA, and Clarke IJ. (2007). Influence of photoperiod and gonadal status on food intake, adiposity, and gene expression of hypothalamic appetite regulators in a seasonal mammal. Am J Physiol Regul Integr Comp Physiol 292(1): R242–R252.

Avigdor M, Sullivan SD, and Heideman PD. (2005). Response to selection for photoperiod responsiveness on the density and location of mature GnRH-releasing neurons. Am J Physiol Regul Integr Comp Physiol 288(5): R1226–R1236.

Badura LL and Goldman BD. (1992). Central sites mediating reproductive responses to melatonin in juvenile male Siberian hamsters. Brain Res 598(1–2): 98–106.

Badura LL, Yant WR, and Nunez AA. (1987). Photoperiodic modulation of steroid-induced lordosis in golden hamsters. Physiol Behav 40(4): 551–554.

Bae HH, Mangels RA, Cho BS, Dark J, Yellon SM, and Zucker I. (1999). Ventromedial hypothalamic mediation of photoperiodic gonadal responses in male Syrian hamsters. J Biol Rhythms 14(5): 391–401.

Barassin S, Kalsbeek A, Saboureau M, Bothorel B, Vivien-Roels B, Malan A, et al. (2000). Potentiation effect of vasopressin on melatonin secretion as determined by trans-pineal microdialysis in the rat. J Neuroendocrinol 12(1): 61–68.

Barrett P, Ivanova E, Graham ES, Ross AW, Wilson D, Ple H, et al. (2006a). Photoperiodic regulation of cellular retinoic acid-binding protein 1, GPR50 and nestin in tanycytes of the third ventricle ependymal layer of the Siberian hamster. J Endocrinol 191(3): 687–698.

Barrett P, Ivanova E, Graham ES, Ross AW, Wilson D, Ple H, et al. (2006b). Photoperiodic regulation of cellular retinoic acid-binding protein 1, GPR50 and nestin in tanycytes of the third ventricle ependymal layer of the Siberian hamster. J Endocrinol 191(3): 687–698.

Bartlewski PM, Beard AP, and Rawlings NC. (1999a). Ovarian function in ewes at the onset of the breeding season. Anim Reprod Sci 57(1–2): 67–88.

Bartlewski PM, Beard AP, and Rawlings NC. (1999b). Ovarian function in ewes during the transition from breeding season to anoestrus. Anim Reprod Sci 57(1–2): 51–66.

Bartness TJ, Goldman BD, and Bittman EL. (1991). SCN lesions block responses to systemic melatonin infusions in Siberian hamsters. Am J Physiol 260: R102–R112.

Bartness TJ, Powers JB, Hastings MH, Bittman EL, and Goldman BD. (1993). The timed infusion paradigm for melatonin delivery: What has it taught us about the melatonin signal, its reception, and the photoperiodic control of seasonal responses? J Pineal Res 15(4): 161–190.

Baulieu EE and Robel P. (1990). Neurosteroids: A new brain function? J Steroid Biochem Mol Biol 37(3): 395–403.

Beccavin C, Malpaux B, and Tillet Y. (1998). Effect of oestradiol and photoperiod on TH mRNA concentrations in A15 and A12 dopamine cell groups in the ewe. J Neuroendocrinol 10(1): 59–66.

Bentley GE, Perfito N, Ukena K, Tsutsui K, and Wingfield JC. (2003). Gonadotropin-inhibitory peptide in song sparrows (*Melospiza melodia*) in different reproductive conditions, and in house sparrows (*Passer domesticus*) relative to chicken-gonadotropin-releasing hormone. J Neuroendocrinol 15(8): 794–802.

Bittman EL, Bartness TJ, Goldman BD, and DeVries GJ. (1991). Suprachiasmatic and paraventricular control of photoperiodism in Siberian hamsters. Am J Physiol 260(1 pt 2): R90–R101.

Bittman EL and Blaustein JD. (1990). Effects of day length on sheep neuroendocrine estrogen and progestin receptors. Am J Physiol 258(1 pt 2): R135–R142.

Bittman EL, Dempsey RJ, and Karsch FJ. (1983). Pineal melatonin secretion drives the reproductive response to daylength in the ewe. Endocrinology 113(6): 2276–2283.

Bittman EL, Ehrlich DA, Ogdahl JL, and Jetton AE. (2003). Photoperiod and testosterone regulate androgen receptor immunostaining in the Siberian hamster brain. Biol Reprod 69(3): 876–884.

Bittman EL, Goldman BD, and Zucker I. (1979). Testicular responses to melatonin are altered by lesions of the suprachiasmatic nuclei in golden hamsters. Biol Reprod 21(3): 647–656.

Bittman EL, Hegarty CM, Layden MQ, and Jonassen JA. (1990). Influences of photoperiod on sexual behaviour, neuroendocrine steroid receptors and adenohypophysial hormone secretion and gene expression in female golden hamsters. J Mol Endocrinol 5(1): 15–25.

Bittman EL, Jonassen JA, and Hegarty CM. (1992). Photoperiodic regulation of pulsatile luteinizing hormone secretion and adenohypophyseal gene expression in female golden hamsters. Biol Reprod 47(1): 66–71.

Bittman EL and Karsch FJ. (1984). Nightly duration of pineal melatonin secretion determines the reproductive response to inhibitory day length in the ewe. Biol Reprod 30(3): 585–593.

Bittman EL, Kaynard AH, Olster DH, Robinson JE, Yellon SM, and Karsch FJ. (1985). Pineal melatonin mediates photoperiodic control of pulsatile luteinizing hormone secretion in the ewe. Neuroendocrinology 40(5): 409–418

Bittman EL and Kley LC. (1988). Influences of photoperiod on nuclear androgen receptor occupancy in neuroendocrine tissues of the golden hamster. Neuroendocrinology 47(1): 61–67.

Bittman EL, Tubbiola ML, Foltz G, and Hegarty CM. (1999). Effects of photoperiod and androgen on proopiomelanocortin gene expression in the arcuate nucleus of golden hamsters. Endocrinology 140(1): 197–206.

Bittman EL and Weaver DR. (1990). The distribution of melatonin binding sites in neuroendocrine tissues of the ewe. Biol Reprod 43(6): 986–993.

Bittman EL and Zucker I. (1977). Influences of the adrenal gland and photoperiod on the hamster oestrus cycle. J Reprod Fertil 50(2): 331–333.

Blank JL and Desjardins C. (1986). Photic cues induce multiple neuroendocrine adjustments in testicular function. Am J Physiol 250(2 pt 2): R199–R206.

Blank JL and Freeman DA. (1991). Differential reproductive response to short photoperiod in deer mice: Role of melatonin. J Comp Physiol A 169(4): 501–506.

Blank JL, Nelson RJ, Vaughan MK, and Reiter RJ. (1988). Pineal melatonin content in photoperiodically responsive and non-responsive phenotypes of deer mice. Comp Biochem Physiol A 91(3): 535–537.

Bogusz AL, Hardy SL, Lehman MN, Connors JM, Hileman SM, Sliwowska JH, et al. (2008). Evidence that gamma-aminobutyric acid is part of the neural circuit mediating estradiol negative feedback in anestrous ewes. Endocrinology 149(6): 2762–2772.

Boulware MI and Mermelstein PG. (2009). Membrane estrogen receptors activate metabotropic glutamate receptors to influence nervous system physiology. Steroids 74(7): 608–613.

Bridges RS and Goldman BD. (1975). Diurnal rhythms in gonadotropins and progesterone in lactating and photoperiod induced acyclic hamsters. Biol Reprod 13(5): 617–622.

Brock MA. (1974). Circannual rhythms in insects. In *Circannual Clocks: Annual Biological Rhythms* (ET Pengelley, ed). New York: Academic Press, pp. 11–53.

Bünning E. (1936). Die endogene tagesrhythmik als grundlage der photoperiodischen reaktion. Ber Dtsch Bot Ges 54: 590–607.

Canguilhem B and Marx C. (1973). Regulation of the body weight of the european hamster during the annual cycle. Pflugers Arch 338(2): 169–175.

Caraty A, Smith JT, Lomet D, Ben Said S, Morrissey A, Cognie J, et al. (2007). Kisspeptin synchronizes preovulatory surges in cyclical ewes and causes ovulation in seasonally acyclic ewes. Endocrinology 148(11): 5258–5267.

Carlson LL, Zimmermann A, and Lynch GR. (1989). Geographic differences for delay of sexual maturation in *Peromyscus leucopus*: Effects of photoperiod, pinealectomy, and melatonin. Biol Reprod 41(6): 1004–1013.

Carr AJ, Johnston JD, Semikhodskii AG, Nolan T, Cagampang FR, Stirland JA, et al. (2003). Photoperiod differentially regulates circadian oscillators in central and peripheral tissues of the Syrian hamster. Curr Biol 13(17): 1543–1548.

Carter DS and Goldman BD. (1983). Antigonadal effects of timed melatonin infusion in pinealectomized male djungarian hamsters (*Phodopus sungorus* sungorus): Duration is the critical parameter. Endocrinology 113(4): 1261–1267.

Carter DS, Herbert J, and Stacey PM. (1982). Modulation of gonadal activity by timed injections of melatonin in pinealectomized or intact ferrets kept under two photoperiods. J Endocrinol 93(2): 211–222.

Cooke BM, Hegstrom CD, and Breedlove SM. (2002). Photoperiod-dependent response to androgen in the medial amygdala of the Siberian hamster, *Phodopus sungorus*. J Biol Rhythms 17(2): 147–154.

Dark J, Johnston PG, Healy M, and Zucker I. (1983). Latitude of origin influences photoperiodic control of reproduction of deer mice (*Peromyscus maniculatus*). Biol Reprod 28(1): 213–220.

de la Iglesia HO, Meyer J, and Schwartz WJ. (2004). Using Per gene expression to search for photoperiodic oscillators in the hamster suprachiasmatic nucleus. Brain Res Mol Brain Res 127(1-2): 121–127.

de Roux N, Genin E, Carel JC, Matsuda F, Chaussain JL, and Milgrom E. (2003). Hypogonadotropic hypogonadism due to loss of function of the KiSS1-derived peptide receptor GPR54. Proc Natl Acad Sci USA 100(19): 10972–10976.

Desjardins C, Bronson FH, and Blank JL. (1986). Genetic selection for reproductive photoresponsiveness in deer mice. Nature 322(6075): 172–173.

Dubocovich ML, Yun K, Al-Ghoul WM, Benloucif S, and Masana MI. (1998). Selective MT2 melatonin receptor antagonists block melatonin-mediated phase advances of circadian rhythms. FASEB J 12(12): 1211–1220.

Ducker MJ, Bowman JC, and Temple A. (1973). The effect of constant photoperiod on the expression of oestrus in the ewe. J Reprod Fertil 19(suppl): 143–150.

Dufourny L, Levasseur A, Migaud M, Callebaut I, Pontarotti P, Malpaux B, et al. (2008). GPR50 is the mammalian ortholog of Mel1c: Evidence of rapid evolution in mammals. BMC Evol Biol 8: 105.

Duncan MJ, Takahashi JS, and Dubocovich ML. (1989). Characteristics and autoradiographic localization of 2-[^{125}I]iodomelatonin binding sites in djungarian hamster brain. Endocrinology 125(2): 1011–1018.

Dungan HM, Clifton DK, and Steiner RA. (2006). Minireview: Kisspeptin neurons as central processors in the regulation of gonadotropin-releasing hormone secretion. Endocrinology 147(3): 1154–1158.

Elliott JA, Stetson MH, and Menaker M. (1972). Regulation of testis function in golden hamsters: A circadian clock measures photoperiodic time. Science 178(62): 771–773.

Elliott JA and Tamarkin L. (1994). Complex circadian regulation of pineal melatonin and wheel running in Syrian hamsters. J Comp Physiol A 174(4): 469–484.

Ellis C, Moar KM, Logie TJ, Ross AW, Morgan PJ, and Mercer JG. (2008). Diurnal profiles of hypothalamic energy balance gene expression with photoperiod manipulation in the Siberian hamster, *Phodopus sungorus*. Am J Physiol Regul Integr Comp Physiol 294(4): R1148–R1153.

Estrada KM, Clay CM, Pompolo S, Smith JT, and Clarke IJ. (2006). Elevated KiSS-1 expression in the arcuate nucleus prior to the cyclic preovulatory gonadotrophin-releasing hormone/lutenising hormone surge in the ewe suggests a stimulatory role for kisspeptin in oestrogen-positive feedback. J Neuroendocrinol 18(10): 806–809.

Farner DS. (1975). Photoperiodic controls in the secretion of gonadotropins in birds. Amer Zool 15(suppl 1): 117–135.

Freeman DA and Goldman BD. (1997). Evidence that the circadian system mediates photoperiodic nonresponsiveness in Siberian hamsters: The effect of running wheel access on photoperiodic responsiveness. J Biol Rhythms 12(2): 100–109.

Freeman DA and Zucker I. (2001). Refractoriness to melatonin occurs independently at multiple brain sites in Siberian hamsters. Proc Natl Acad Sci USA 98(11): 6447–6452.

Galea LA and McEwen BS. (1999). Sex and seasonal differences in the rate of cell proliferation in the dentate gyrus of adult wild meadow voles. Neuroscience 89(3): 955–964.

Gallegos-Sanchez J, Delaleu B, Caraty A, Malpaux B, and Thiery JC. (1997). Estradiol acts locally within the retrochiasmatic area to inhibit pulsatile luteinizing-hormone release in the female sheep during anestrus. Biol Reprod 56(6): 1544–1549.

Gaston S and Menaker M. (1967). Photoperiodic control of hamster testis. Science 158(803): 925–928.

Gibson EM, Humber SA, Jain S, Williams WP, Zhao S, Bentley GE, et al. (2008). Alterations in RFamide-related peptide expression are coordinated with the preovulatory luteinizing hormone surge. Endocrinology 149(10): 4958–4969. .

Goldman BD. (2001). Mammalian photoperiodic system: Formal properties and neuroendocrine mechanisms of photoperiodic time measurement. J Biol Rhythms 16(4): 283–301.

Goodman RL and Karsch FJ. (1980). Pulsatile secretion of luteinizing hormone: Differential suppression by ovarian steroids. Endocrinology 107(5): 1286–1290.

Goodman RL and Karsch FJ. (1981). A critique of the evidence on the importance of steroid feedback to seasonal changes in gonadotrophin secretion. J Reprod Fertil 30(suppl). 1–13.

Goodman RL, Legan SJ, Ryan KD, Foster DL, and Karsch FJ. (1981). Importance of variations in behavioural and feedback actions of oestradiol to the control of seasonal breeding in the ewe. J Endocrinol 89(2): 229–240.

Goodman RL, Lehman MN, Smith JT, Coolen LM, de Oliveira CV, Jafarzadehshirazi MR, et al. (2007). Kisspeptin neurons in the arcuate nucleus of the ewe express both dynorphin A and neurokinin B. Endocrinology 148(12): 5752–5760.

Goodman RL, Thiery JC, Delaleu B, and Malpaux B. (2000). Estradiol increases multiunit electrical activity in the A15 area of ewes exposed to inhibitory photoperiods. Biol Reprod 63(5): 1352–1357.

Gorman MR, Freeman DA, and Zucker I. (1997). Photoperiodism in hamsters: Abrupt versus gradual changes in day length differentially entrain morning and evening circadian oscillators. J Biol Rhythms 12(2): 122–135.

Gorman MR and Zucker I. (1995a). Seasonal adaptations of Siberian hamsters. II. pattern of change in daylength controls annual testicular and body weight rhythms. Biol Reprod 53(1): 116–125.

Gorman MR and Zucker I. (1995b). Testicular regression and recrudescence without subsequent photorefractoriness in Siberian hamsters. Am J Physiol 269(4 pt 2): R800–R806.

Gorman MR and Zucker I. (1997). Environmental induction of photononresponsiveness in the Siberian hamster, *Phodopus sungorus*. Am J Physiol 272(3 pt 2): R887–R895.

Goss RJ. (1984). Photoperiodic control of antler cycles in deer. VI. Circannual rhythms on altered day lengths. J Exp Zool 230(2): 265–271.

Gottsch ML, Cunningham MJ, Smith JT, Popa SM, Acohido B, Crowley WF, et al. (2004). A role for kisspeptins in the regulation of gonadotropin secretion in the mouse. Endocrinology 145(9): 4073–4077.

Gram WD, Heath HW, Wichman HA, and Lynch GR. (1982). Geographic variation in *Peromyscus leucopus*: Short-day induced reproductive regression and spontaneous recrudescence. Biol Reprod 27(2): 369–373.

Greives TJ, Kriegsfeld LJ, Bentley GE, Tsutsui K, and Demas GE. (2008). Recent advances in reproductive neuroendocrinology: A role for RFamide peptides in seasonal reproduction? Proc Biol Sci 275(1646): 1943–1951.

Greives TJ, Mason AO, Scotti MA, Levine J, Ketterson ED, Kriegsfeld LJ, and Demas GE. (2007). Environmental control of kisspeptin: Implications for seasonal reproduction. Endocrinology 148(3): 1158–1166.

Gwinner E. (2003). Circannual rhythms in birds. Curr Opin Neurobiol 13(6): 770–778.

Han SK, Gottsch ML, Lee KJ, Popa SM, Smith JT, Jakawich SK, et al. (2005). Activation of gonadotropin-releasing hormone neurons by kisspeptin as a neuroendocrine switch for the onset of puberty. J Neurosci 25(49): 11349–11356.

Hanon EA, Lincoln GA, Fustin JM, Dardente H, Masson-Pevet M, Morgan PJ, and Hazlerigg DG (2008). Ancestral TSH mechanism signals summer in a photoperiodic mammal. Curr Biol 18(15): 1147–1152.

Hardy SL, Anderson GM, Valent M, Connors JM, and Goodman RL. (2003). Evidence that estrogen receptor alpha, but not beta, mediates seasonal changes in the response of the ovine retrochiasmatic area to estradiol. Biol Reprod 68(3): 846–852.

Heideman PD, Kane SL, and Goodnight AL. (1999). Differences in hypothalamic 2-[^{125}I]iodomelatonin binding in photoresponsive and non-photoresponsive white-footed mice, *Peromyscus leucopus*. Brain Res 840(1–2): 56–64.

Horton TH. (1984). Growth and reproductive development of male *Microtus montanus* is affected by the prenatal photoperiod. Biol Reprod 31(3): 499–504.

Horton TH. (1985). Cross-fostering of voles demonstrates in utero effect of photoperiod. Biol Reprod 33(4): 934–939.

Huang L, DeVries GJ, and Bittman EL. (1998). Photoperiod regulates neuronal bromodeoxyuridine labeling in the brain of a seasonally breeding mammal. J Neurobiol 36(3): 410–420.

Hyde LL and Underwood H. (1993). Effects of nightbreak, T-cycle, and resonance lighting schedules on the pineal melatonin rhythm of the lizard *Anolis carolinensis*: Correlations with the reproductive response. J Pineal Res 15(2): 70–80.

Imaizumi T and Kay SA. (2006). Photoperiodic control of flowering: Not only by coincidence. Trends Plant Sci 11(11): 550–558.

Inagaki N, Honma S, Ono D, Tanahashi Y, and Honma K. (2007). Separate oscillating cell groups in mouse suprachiasmatic nucleus couple photoperiodically to the onset and end of daily activity. Proc Natl Acad Sci USA 104(18): 7664–7669.

Inai Y, Nagai K, Ukena K, Oishi T, and Tsutsui K. (2003). Seasonal changes in neurosteroid concentrations in the amphibian brain and environmental factors regulating their changes. Brain Res 959(2): 214–225.

Irwig MS, Fraley GS, Smith JT, Acohido BV, Popa SM, Cunningham J, et al. (2004). Kisspeptin activation of gonadotropin releasing hormone neurons and regulation of KiSS-1 mRNA in the male rat. Neuroendocrinology 80(4): 264–267.

Jansen HT, Cutter C, Hardy S, Lehman MN, and Goodman RL. (2003). Seasonal plasticity within the gonadotropin-releasing hormone (GnRH) system of the ewe: Changes in identified GnRH inputs and glial association. Endocrinology 144(8): 3663–3676.

Johnston JD, Tournier BB, Andersson H, Masson-Pevet M, Lincoln GA, and Hazlerigg DG. (2006). Multiple effects of melatonin on rhythmic clock gene expression in the mammalian pars tuberalis. Endocrinology 147(2): 959–965.

Jorgenson KL and Schwartz NB. (1984). Effect of steroid treatment on tonic and surge secretion of LH and FSH in the female golden hamster: Effect of photoperiod. Neuroendocrinology 39(6): 549–554.

Kalsbeek A, Garidou ML, Palm IF, Van Der Vliet J, Simonneaux V, Pevet P, et al. (2000). Melatonin sees the light: Blocking GABA-ergic transmission in the paraventricular nucleus induces daytime secretion of melatonin. Eur J Neurosci 12(9): 3146–3154.

Karsch FJ, Bittman EL, Foster DL, Goodman RL, Legan SJ, and Robinson JE. (1984). Neuroendocrine basis of seasonal reproduction. Recent Prog Horm Res 40: 185–232.

Karsch FJ, Goodman RL, and Legan SJ. (1980). Feedback basis of seasonal breeding: Test of an hypothesis. J Reprod Fertil 58(2): 521–535.Kauffman AS, Clifton DK, and Steiner RA. (2007). Emerging ideas about kisspeptin- GPR54 signaling in the neuroendocrine regulation of reproduction. Trends Neurosci 30(10): 504–511.

Klein DC. (2004). The 2004 Aschoff/Pittendrigh lecture: Theory of the origin of the pineal gland—a tale of conflict and resolution. J Biol Rhythms 19(4): 264–279.

Kliman RM and Lynch GR. (1992). Evidence for genetic variation in the occurrence of the photoresponse of the Djungarian hamster, *Phodopus sungorus*. J Biol Rhythms 7(2),161–173.

Ko CH and Takahashi JS. (2006). Molecular components of the mammalian circadian clock. Hum Mol Genet 15(spec no 2): R271–R277.

Kobayashi Y and Weigel D. (2007). Move on up, it's time for change—mobile signals controlling photoperiod-dependent flowering. Genes Dev 21(19): 2371–2384.

Korf HW and von Gall C. (2006). Mice, melatonin and the circadian system. Mol Cell Endocrinol 252(1–2): 57–68.

Kouzarides T (2007). Chromatin modifications and their function. Cell 128: 693–705.

Kramer KM, Simmons JL, and Freeman DA. (2008). Photoperiod alters central distribution of estrogen receptor alpha in brain regions that regulate aggression. Horm Behav 53(2): 358–365.

Krey LC, Ronchi E, and Bittman EL. (1989). Effects of daylength on androgen metabolism and pulsatile luteinizing hormone secretion in male golden hamsters. Neuroendocrinology 50(5): 533–542.

Kriegsfeld LJ. (2006). Driving reproduction: RFamide peptides behind the wheel. Horm Behav 50(5): 655–666.

Kriegsfeld LJ, Leak RK, Yackulic CB, LeSauter J, and Silver R. (2004). Organization of suprachiasmatic nucleus projections in Syrian hamsters (*mesocricetus auratus*): An anterograde and retrograde analysis. J Comp Neurol 468(3): 361–379.

Kriegsfeld LJ, Mei DF, Bentley GE, Ubuka T, Mason AO, Inoue K, et al. (2006). Identification and characterization of a gonadotropin-inhibitory system in the brains of mammals. Proc Natl Acad Sci USA 103(7): 2410–2415.

Lawson NO, Wee BE, Blask DE, Castles CG, Spriggs LL, and Hill SM. (1992). Melatonin decreases estrogen receptor expression in the medial preoptic area of inbred (LSH/SsLak) golden hamsters. Biol Reprod 47(6): 1082–1090.

Lee TM and Zucker I. (1988). Vole infant development is influenced perinatally by maternal photoperiodic history. Am J Physiol 255(5 pt 2): R831–R838.

Lee TM and Zucker I. (1991). Suprachiasmatic nucleus and photic entrainment of circannual rhythms in ground squirrels. J Biol Rhythms 6: 315–330.

Lee W, Watanabe M, and Glass JD. (1995). Photoperiod affects the expression of neural cell adhesion molecule and polysialic acid in the hypothalamus of the Siberian hamster. Brain Res 690(1): 64–72.

Legan SJ, Goodman RL, Ryan KD, Foster DL, and Karsch FJ. (1985). Can the transition into anoestrus in the ewe be accounted for solely by insufficient tonic LH secretion? J Endocrinol 106(1): 55–60.

Legan SJ and Karsch FJ. (1980). Photoperiodic control of seasonal breeding in ewes: Modulation of the negative feedback action of estradiol. Biol Reprod 23: 1061–1068.

Legan SJ, Karsch FJ, and Foster DL. (1977). The endocrin control of seasonal reproductive function in the ewe: A marked change in response to the negative feedback action of estradiol on luteinizing hormone secretion. Endocrinology 101(3): 818–824.

Lehman MN, Bittman EL, and Newman SW. (1984). Role of the hypothalamic paraventricular nucleus in neuroendocrine responses to daylength in the golden hamster. Brain Res 308(1): 25–32.

Levoye A, Dam J, Ayoub MA, Guillaume JL, Couturier C, Delagrange P, et al. (2006). The orphan GPR50 receptor specifically inhibits MT1 melatonin receptor function through heterodimerization. EMBO J 25(13): 3012–3023.

Lewis D, Freeman DA, Dark J, Wynne-Edwards KE, and Zucker I. (2002). Photoperiodic control of oestrous cycles in Syrian hamsters: Mediation by the mediobasal hypothalamus. J Neuroendocrinol 14(4): 294–299.

Lincoln GA. (1992). Administration of melatonin into the mediobasal hypothalamus as a continuous or intermittent signal affects the secretion of follicle stimulating hormone and prolactin in the ram. J Pineal Res 12(3): 135–144.

Lincoln GA. (2006). Decoding the nightly melatonin signal through circadian clockwork. Mol Cell Endocrinol 252(1–2): 69–73.

Lincoln GA, Clarke IJ, Hut RA, and Hazelrigg DG. (2006). Characterizing a mammalian circannual pacemaker. Science 314(5807): 1941–1944.

Lincoln GA, Ebling FJ, and Martin GB. (1987). Endogenous opioid control of pulsatile LH secretion in rams: Modulation by photoperiod and gonadal steroids. J Endocrinol 115(3): 425–438.

Lincoln GA and Maeda KI. (1992). Reproductive effects of placing micro-implants of melatonin in the mediobasal hypothalamus and preoptic area in rams. J Endocrinol 132(2): 201–215.

Liu T and Borjigin J. (2005). *N*-Acetyltransferase is not the rate-limiting enzyme of melatonin synthesis at night. J Pineal Res 39(1): 91–96.

Malpaux B, Daveau A, Maurice-Mandon F, Duarte G, and Chemineau P. (1998). Evidence that melatonin acts in the premammillary hypothalamic area to control reproduction in the ewe: Presence of binding sites and stimulation of luteinizing hormone secretion by in situ microimplant delivery. Endocrinology 139(4): 1508–1516.

Malpaux B, Daveau A, Maurice F, Gayrard V, and Thiery JC. (1993). Short-day effects of melatonin on luteinizing hormone secretion in the ewe: Evidence for central sites of action in the mediobasal hypothalamus. Biol Reprod 48(4): 752–760.

Mangels RA, Powers JB, and Blaustein JD. (1998). Effect of photoperiod on neural estrogen and progestin receptor immunoreactivity in female Syrian hamsters. Brain Res 796(1–2): 63–74.

Mani SK, Allen JM, Lydon JP, Mulac-Jericevic B, Blaustein JD, DeMayo FJ, et al. (1996). Dopamine requires the unoccupied progesterone receptor to induce sexual behavior in mice. Mol Endocrinol 10(12): 1728–1737.

Margraf RR, Zlomanczuk P, Liskin LA, and Lynch GR. (1991). Circadian differences in neuronal activity of the suprachiasmatic nucleus in brain slices prepared from photo-responsive and photo-non-responsive Djungarian hamsters. Brain Res 544(1): 42–48.

Martinet L, Mondain-Monval M, and Monnerie R. (1992). Endogenous circannual rhythms and photorefractoriness of testis activity, moult and prolactin concentrations in mink (*Mustela vison*). J Reprod Fertil 95(2): 325–338.

Mason AO, Duffy SP, Bentley GE, Ubuka T, Tsutsui K, Silver R, and Kriegsfeld LJ. (2007a). *Effects of Photoperiod and Reproductive Condition on the RFRP System in Syrian Hamsters.* San Diego, CA: Society for Neuroscience.

Mason AO, Greives TJ, Scotti MA, Levine J, Frommeyer S, Ketterson ED, et al. (2007b). Suppression of kisspeptin expression and gonadotropic axis sensitivity following exposure to inhibitory day lengths in female Siberian hamsters. Horm Behav 52(4): 492–498.

Masson-Pevet M, Naimi F, Canguilhem B, Saboureau M, Bonn D, and Pevet P. (1994). Are the annual reproductive and body weight rhythms in the male european hamster (*Cricetus cricetus*) dependent upon a photoperiodically entrained circannual clock? J Pineal Res 17(4): 151–163.

Matsui H, Takatsu Y, Kumano S, Matsumoto H, and Ohtaki T. (2004). Peripheral administration of metastin induces marked gonadotropin release and ovulation in the rat. Biochem Biophys Res Commun 320(2): 383–388.

Maywood ES, Bittman EL, and Hastings MH. (1996). Lesions of the melatonin- and androgen-responsive tissue of the dorsomedial nucleus of the hypothalamus block the gonadal response of male Syrian hamsters to programmed infusions of melatonin. Biol Reprod 54(2): 470–477.

Maywood ES, Buttery RC, Vance GH, Herbert J, and Hastings MH. (1990). Gonadal responses of the male Syrian hamster to programmed infusions of melatonin are sensitive to signal duration and frequency but not to signal phase nor to lesions of the suprachiasmatic nuclei. Biol Reprod 43(2): 174–182.

Maywood ES and Hastings MH. (1995). Lesions of the iodomelatonin-binding sites of the mediobasal hypothalamus spare the lactotropic, but block the gonadotropic response of male Syrian hamsters to short photoperiod and to melatonin. Endocrinology 136(1): 144–153.

McNulty S, Ross AW, Barrett P, Hastings MH, and Morgan PJ. (1994). Melatonin regulates the phosphorylation of CREB in ovine pars tuberalis. J Neuroendocrinol 6(5): 523–532.

Mercer JG, Ellis C, Moar KM, Logie TJ, Morgan PJ, and Adam CL. (2003). Early regulation of hypothalamic arcuate nucleus CART gene expression by short photoperiod in the Siberian hamster. Regul Pept 111(1–3): 129–136.

Meredith JM, Turek FW, and Levine JE. (1998). Effects of gonadotropin-releasing hormone pulse frequency modulation on the reproductive axis of photoinhibited male Siberian hamsters. Biol Reprod 59(4): 813–819.

Messager S, Chatzidaki EE, Ma D, Hendrick AG, Zahn D, Dixon J, et al. (2005). Kisspeptin directly stimulates gonadotropin-releasing hormone release via G protein-coupled receptor 54. Proc Natl Acad Sci USA 102(5): 1761–1766.

Messager S, Hazlerigg DG, Mercer JG, and Morgan PJ. (2000). Photoperiod differentially regulates the expression of Per1 and ICER in the pars tuberalis and the suprachiasmatic nucleus of the Siberian hamster. Eur J Neurosci 12(8): 2865–2870.

Meyer SL and Goodman RL. (1986). Separate neural systems mediate the steroid-dependent and steroid-independent suppression of tonic luteinizing hormone secretion in the anestrous ewe. Biol Reprod 35(3): 562–571.

Meyer-Bernstein EL, Jetton AE, Matsumoto SI, Markuns JF, Lehman MN, and Bittman EL. (1999). Effects of suprachiasmatic transplants on circadian rhythms of neuroendocrine function in golden hamsters. Endocrinology 140(1): 207–218.

Miernicki M, Pospichal MW, and Powers JB. (1990). Short photoperiods affect male hamster sociosexual behaviors in the presence and absence of testosterone. Physiol Behav 47(1): 95–106.

Mintz EM, Lavenburg KR, and Blank JL. (2007). Short photoperiod and testosterone-induced modification of GnRH release from the hypothalamus of *peromyscus maniculatus*. Brain Res 1180: 20–28.

Mittag M, Kiaulehn S, and Johnson CH. (2005). The circadian clock in *Chlamydomonas reinhardtii*. What is it for? What is it similar to? Plant Physiol 137(2): 399–409.

Morgan PJ, Barrett P, Howell HE, and Helliwell R. (1994). Melatonin receptors: Localization, molecular pharmacology and physiological significance. Neurochem Int 24(2): 101–146.

Morgan PJ and Mercer JG. (2001). The regulation of body weight: Lessons from the seasonal animal. Proc Nutr Soc 60(1): 127–134.

Morgan PJ, Webster CA, Mercer JG, Ross AW, Hazlerigg DG, MacLean A, et al. (1996a). The ovine pars tuberalis secretes a factor(s) that regulates gene expression in both lactotropic and nonlactotropic pituitary cells. Endocrinology 137(9): 4018–4026.

Morgan PJ, Williams LM, Barrett P, Lawson W, Davidson G, Hannah L, et al. (1996b). Differential regulation of melatonin receptors in sheep, chicken and lizard brains by cholera and pertussis toxins and guanine nucleotides. Neurochem Int 28(3): 259–269.

Mori Y, Tanaka M, Maeda K, Hoshino K, and Kano Y. (1987). Photoperiodic modification of negative and positive feedback effects of oestradiol on LH secretion in ovariectomized goats. J Reprod Fertil 80(2): 523–529.

Morin LP, Fitzgerald KM, Rusak B, and Zucker I. (1977). Circadian organization and neural mediation of hamster reproductive rhythms. Psychoneuroendocrinology 2(1): 73–98.

Nakao N, Ono H, Yamamura T, Anraku T, Takagi T, Higashi K, et al. (2008). Thyrotrophin in the pars tuberalis triggers photoperiodic response. Nature 452(7185): 317–322.

Nanda KK and Hamner KC. (1958). Photoperiodic time measurement in the Biloxi soybean. Bot Gaz 120: 14–25.

Nelson RJ. (1987). Photoperiod-nonresponsive morphs: A possible variable in microtine population density fluctuations. Am Nat 130: 350–369.

Niklowitz P, Lerchl A, and Nieschlag E. (1994). Photoperiodic responses in Djungarian hamsters (*Phodopus sungorus*): Importance of light history for pineal and serum melatonin profiles. Biol Reprod 51(4): 714–724.

Nuesslein-Hildesheim B, O'Brien JA, Ebling FJ, Maywood ES, and Hastings MH. (2000). The circadian cycle of mPER clock gene products in the suprachiasmatic nucleus of the Siberian hamster encodes both daily and seasonal time. Eur J Neurosci 12(8): 2856–2864.

Ono H, Hoshino Y, Yasuo S, Watanabe M, Nakane Y, Murai A, Ebihara S, Korf H-W, and Yoshimura T. (2008). Involvement of thyrotropin in photoperiodic signal transduction in mice. Proc Natl Acad Sci USA 105(47): 18238–18242.

Ortavant R, Bocquier F, Pelletier J, Ravault JP, Thimonier J, and Volland-Nail P. (1988). Seasonality of reproduction in sheep and its control by photoperiod. Austral J Biol Sci 41(1): 69–85.

Paul MJ, Zucker I, and Schwartz WJ. (2008). Tracking the seasons: The internal calendars of vertebrates. Philos Trans R Soc Lond B Biol Sci 363(1490): 341–346.

Pelletier J, Bodin L, Hanocq E, Malpaux B, Teyssier J, Thimonier J, et al. (2000). Association between expression of reproductive seasonality and alleles of the gene for mel(1a) receptor in the ewe. Biol Reprod 62(4): 1096–1101.

Pengelley ET and Asmundson SJ. (1971). Annual biological clocks. Sci Am 224: 72–79.

Perreau-Lenz S, Kalsbeek A, Garidou ML, Wortel J, van der Vliet J, van Heijningen C, et al. (2003). Suprachiasmatic control of melatonin synthesis in rats: Inhibitory and stimulatory mechanisms. Eur J Neurosci 17(2): 221–228.

Perreau-Lenz S, Kalsbeek A, Pevet P, and Buijs RM. (2004). Glutamatergic clock output stimulates melatonin synthesis at night. Eur J Neurosci 19(2): 318–324.

Pickard GE and Turek FW. (1983). The hypothalamic paraventricular nucleus mediates the photoperiodic control of reproduction but not the effects of light on the circadian rhythm of activity. Neurosci Lett 43(1): 67–72.

Pittendrigh CS. (1981). Circadian organization and the photoperiodic phenomena. In *Colston Symposium, Biological Clocks and Seasonal Reproductive Cycles*, vol 32(BK Follett and DE Follett eds). Bristol, UK: Butterworth-Heinemann, pp. 1–35.

Plautz JD, Kaneko M, Hall JC, and Kay SA. (1997). Independent photoreceptive circadian clocks throughout *Drosophila*. Science 278(5343): 1632–1635.

Porkka-Heiskanen T, Khoshaba N, Scarbrough K, Urban JH, Vitaterna MH, Levine JE, et al. (1997). Rapid photoperiod-induced increase in detectable GnRH mRNA-containing cells in Siberian hamster. Am J Physiol 273(6 pt 2), R2032–9.

Pospichal MW, Karp JD, and Powers JB. (1991). Influence of day length on male hamster sexual behavior: Masking effects of testosterone. Physiol Behav 49(3): 417–422.

Prendergast BJ, Flynn AK, and Zucker I. (2000a). Triggering of neuroendocrine refractoriness to short-day patterns of melatonin in Siberian hamsters. J Neuroendocrinol 12(4): 303–310.

Prendergast BJ and Freeman DA. (1999). Pineal-independent regulation of photo-nonresponsiveness in the Siberian hamster (*Phodopus sungorus*). J Biol Rhythms 14(1): 62–71.

Prendergast BJ, Gorman MR, and Zucker I. (2000b). Establishment and persistence of photoperiodic memory in hamsters. Proc Natl Acad Sci USA 97(10): 5586–5591.

Prendergast BJ, Kriegsfeld LJ, and Nelson RJ. (2001). Photoperiodic polyphenisms in rodents: Neuroendocrine mechanisms, costs, and functions. Q Rev Biol 76(3): 293–325.

Price DA and Greenberg MJ. (1977). Purification and characterization of a cardioexcitatory neuropeptide from the central ganglia of a bivalve mollusc. Prep Biochem 7(3–4): 261–281.

Puchalski W, Kliman R, and Lynch GR. (1988). Differential effects of short day pretreatment on melatonin-induced adjustments in Djungarian hamsters. Life Sci 43(12): 1005–1012.

Puchalski W and Lynch GR. (1986). Evidence for differences in the circadian organization of hamsters exposed to short day photoperiod. J Comp Physiol A 159(1): 7–11.

Pyter LM, Reader BF, and Nelson RJ. (2005). Short photoperiods impair spatial learning and alter hippocampal dendritic morphology in adult male white-footed mice (*Peromyscus leucopus*). J Neurosci 25(18): 4521–4526.

Raitiere MN, Garyfallou VT, and Urbanski HF. (1997). Lesions in the anterior bed nucleus of the stria terminalis in Syrian hamsters block short-photoperiod-induced testicular regression. Biol Reprod 57(4): 796–806.

Reiter RJ. (1973). Pineal control of a seasonal reproductive rhythm in male golden hamsters exposed to natural daylight and temperature. Endocrinology 92(2): 423–430.

Revel FG, Saboureau M, Masson-Pevet M, Pevet P, Mikkelsen JD, and Simonneaux V. (2006). Kisspeptin mediates the photoperiodic control of reproduction in hamsters. Curr Biol 16(17): 1730–1735.

Revel FG, Saboureau M, Pevet P V Simonneaux V, and Mikkelsen JD. (2008). RFamide-related peptide gene is a melatonin-driven photoperiodic gene. Endocrinology 149(3): 902–912.

Roberts AC, Martensz ND, Hastings MH, and Herbert J. (1987). The effects of castration, testosterone replacement and photoperiod upon hypothalamic beta-endorphin levels in the male Syrian hamster. Neuroscience 23(3): 1075–1082.

Robinson JE and Karsch FJ. (1984). Refractoriness to inductive day lengths terminates the breeding season of the suffolk ewe. Biol Reprod 31: 656–663.

Robinson JE, Wayne NL, and Karsch FJ. (1985). Refractoriness to inhibitory day lengths initiates the breeding season of the suffolk ewe. Biol Reprod 32(5): 1024–1030.

Ronchi E, Krey LC, and Pfaff DW. (1992). Steady state analysis of hypothalamic GnRH mRNA levels in male Syrian hamsters: Influences of photoperiod and androgen. Neuroendocrinology 55(2): 146–155.

Ruf T, Korytko AI, Stieglitz A, Lavenburg KR, and Blank JL. (1997). Phenotypic variation in seasonal adjustments of testis size, body weight, and food intake in deer mice: Role of pineal function and ambient temperature. J Comp Physiol B 167(3): 185–192.

Saunders DS. (1997). Insect circadian rhythms and photoperiodism. Invert Neurosci 3(2–3): 155–164.

Saunders DS. (2005). Erwin Bünning and Tony Lees, two giants of chronobiology, and the problem of time measurement in insect photoperiodism. J Insect Physiol 51(6): 599–608.

Schaap J, Albus H, VanderLeest HT, Eilers PH, Detari L, and Meijer JH. (2003). Heterogeneity of rhythmic suprachiasmatic nucleus neurons: Implications for circadian waveform and photoperiodic encoding. Proc Natl Acad Sci USA 100(26): 15994–15999.

Schmidt KL, Pradhan DS, Shah AH, Charlier TD, Chin EH, and Soma KK. (2008). Neurosteroids, immunosteroids, and the balkanization of endocrinology. Gen Comp Endocrinol 157(3): 266–274.

Schreihofer DA, Stoler MH, and Shupnik MA. (2000). Differential expression and regulation of estrogen receptors (ERs) in rat pituitary and cell lines: Estrogen decreases ERalpha protein and estrogen responsiveness. Endocrinology 141(6): 2174–2184.

Schuhler S, Horan TL, Hastings MH, Mercer JG, Morgan PJ, and Ebling FJ. (2003). Decrease of food intake by MC4-R agonist MTII in Siberian hamsters in long and short photoperiods. Am J Physiol Regul Integr Comp Physiol 284(1): R227–R232.

Seegal RF and Goldman BD. (1975). Effects of photoperiod on cyclicity and serum gonadotropins in the Syrian hamster. Biol Reprod 12(2): 223–231.

Shahab M, Mastronardi C, Seminara SB, Crowley WF, Ojeda SR, and Plant TM. (2005). Increased hypothalamic GPR54 signaling: A potential mechanism for initiation of puberty in primates. Proc Natl Acad Sci USA 102(6): 2129–2134.

Silver R, LeSauter J, Tresco PA, and Lehman MN. (1996). A diffusible coupling signal from the transplanted suprachiasmatic nucleus controlling circadian locomotor rhythms. Nature 382(6594): 810–813.

Sim JA, Skynner MJ, and Herbison AE. (2001). Direct regulation of postnatal GnRH neurons by the progesterone derivative allopregnanolone in the mouse. Endocrinology 142(10): 4448–4453.

Simpson SM, Follett BK, and Ellis DH. (1982). Modulation by photoperiod of gonadotrophin secretion in intact and castrated Djungarian hamsters. J Reprod Fertil 66(1): 243–250.

Skinner DC and Herbison AE. (1997). Effects of photoperiod on estrogen receptor, tyrosine hydroxylase, neuropeptide Y, and beta-endorphin immunoreactivity in the ewe hypothalamus. Endocrinology 138(6): 2585–2595.

Skinner DC and Malpaux B. (1999). High melatonin concentrations in third ventricular cerebrospinal fluid are not due to Galen vein blood recirculating through the choroid plexus. Endocrinology 140(10): 4399–4405.

Sliwowska JH, Billings HJ, Goodman RL, Coolen LM, and Lehman MN. (2004). The premammillary hypothalamic area of the ewe: Anatomical characterization of a melatonin target area mediating seasonal reproduction. Biol Reprod 70(6): 1768–1775.

Smith JT. (2008). Kisspeptin signalling in the brain: Steroid regulation in the rodent and ewe. Brain Res Brain Res Rev 57(2): 288–298.

Smith JT and Clarke IJ. (2007). Kisspeptin expression in the brain: Catalyst for the initiation of puberty. Rev Endocr Metab Disord 8(1): 1–9.

Smith JT, Clay CM, Caraty A, and Clarke IJ. (2007). Kiss-1 messenger ribonucleic acid expression in the hypothalamus of the ewe is regulated by sex steroids and season. Endocrinology 148(3): 1150–1157.

Smith JT, Coolen LM, Kriegsfeld LJ, Sari IP, Jaafarzadehshirazi MR, Maltby M, et al. (2008a). Variation in kisspeptin and gonadotropin-inhibitory hormone expression and terminal connections to GnRH neurons in the brain: A novel medium for seasonal breeding in the sheep. Endocrinology 149(11):5770–5782.

Smith JT, Rao A, Pereira A, Caraty A, Millar RP, and Clarke IJ. (2008b). Kisspeptin is present in ovine hypophysial portal blood but does not increase during the preovulatory luteinizing hormone surge: Evidence that gonadotropes are not direct targets of kisspeptin in vivo. Endocrinology 149(4): 1951–1959.

Song CK, Bartness TJ, Petersen SL, and Bittman EL. (2000). Co-expression of melatonin (MEL1a) receptor and arginine vasopressin mRNAs in the Siberian hamster suprachiasmatic nucleus. J Neuroendocrinol 12(7): 627–634.

Stetson MH, Ray SL, Creyaufmiller N, and Horton TH. (1989). Maternal transfer of photoperiodic information in Siberian hamsters. II. The nature of the maternal signal, time of signal transfer, and the effect of the maternal signal on peripubertal reproductive development in the absence of photoperiodic input. Biol Reprod 40(3): 458–465.

Stetson MH and Watson-Whitmyre M. (1976). Nucleus suprachiasmaticus: The biological clock in the hamster? Science 191(4223): 197–199.

Stetson MH, Watson-Whitmyre M, and Matt KS. (1977). Termination of photorefractoriness in golden hamsters-photoperiodic requirements. J Exp Zool 202(1): 81–88.

Stoleru D, Nawathean P, Fernandez MP, Menet JS, Ceriani MF, and Rosbash M. (2007). The *Drosophila* circadian network is a seasonal timer. Cell 129(1): 207–219.

Tamarkin L, Hutchison JS, and Goldman BD. (1976). Regulation of serum gonadotropins by photoperiod and testicular hormone in the Syrian hamster. Endocrinology 99(6): 1528–1533.

Tetel MJ, Ungar TC, Hassan B, and Bittman EL. (2004). Photoperiodic regulation of androgen receptor and steroid receptor coactivator-1 in Siberian hamster brain. Brain Res Mol Brain Res 131(1–2): 79–87.

Teubner BJ and Freeman DA. (2007). Different neural melatonin-target tissues are critical for encoding and retrieving day length information in Siberian hamsters. J Neuroendocrinol 19(2): 102–108.

Tournier BB, Menet JS, Dardente H, Poirel VJ, Malan A, Masson-Pevet M, et al. (2003). Photoperiod differentially regulates clock genes' expression in the suprachiasmatic nucleus of Syrian hamster. Neuroscience 118(2): 317–322.

Trainor BC, Rowland MR, and Nelson RJ. (2007). Photoperiod affects estrogen receptor alpha, estrogen receptor beta and aggressive behavior. Eur J Neurosci 26(1): 207–218.

Travnickova Z, Sumova A, Peters R, Schwartz WJ, and Illnerova H. (1996). Photoperiod-dependent correlation between light-induced SCN c-*fos* expression and resetting of circadian phase. Am J Physiol 271(4 pt 2): R825–R831.

Trillmich, F. (2000). Effects of low temperature and photoperiod on reproduction in the female wild guinea pig (*Cavia aperea*). J. Mammol 81(2): 586–594.

Tsutsui K, Saigoh E, Ukena K, Teranishi H, Fujisawa Y, Kikuchi M, et al. (2000). A novel avian hypothalamic peptide inhibiting gonadotropin release. Biochem Biophys Res Commun 275(2): 661–667.

Tubbiola ML, Nock B, and Bittman EL. (1989). Photoperiodic changes in opiate binding and their functional implications in golden hamsters. Brain Res 503(1): 91–99.

Turek FW. (1977). The interaction of the photoperiod and testosterone in regulating serum gonadotropin levels in castrated male hamsters. Endocrinology 101(4): 1210–1215.

Ubuka T, Bentley GE, Ukena K, Wingfield JC, and Tsutsui K. (2005). Melatonin induces the expression of gonadotropin-inhibitory hormone in the avian brain. Proc Natl Acad Sci USA 102(8): 3052–3057.

Urbanski HF, Simpson SM, Ellis DH, and Follett BK. (1983). Secretion of follicle-stimulating hormone and luteinizing hormone in castrated golden hamsters during exposure to various photoperiods and to natural daylengths. J Endocrinol 99(3): 379–386.

VanderLeest HT, Houben T, Michel S, Deboer T, Albus H, Vansteensel MJ, et al. (2007). Seasonal encoding by the circadian pacemaker of the SCN. Curr Biol 17(5): 468–473.

Viguie C, Jansen HT, Glass JD, Watanabe M, Billings HJ, Coolen L, et al. (2001). Potential for polysialylated form of neural cell adhesion molecule-mediated neuroplasticity within the gonadotropin-releasing hormone neurosecretory system of the ewe. Endocrinology 142(3): 1317–1324.

von Gall C, Weaver DR, Moek J, Jilg A, Stehle JH, and Korf HW. (2005). Melatonin plays a crucial role in the regulation of rhythmic clock gene expression in the mouse pars tuberalis. Ann NY Acad Sci 1040; 508–511.

Wagner GC, Johnston JD, Clarke IJ, Lincoln GA, and Hazlerigg DG. (2008). Redefining the limits of day length responsiveness in a seasonal mammal. Endocrinology 149(1): 32–39.

Wallace JM, McNeilly AS, and Baird DT. (1986). Induction of ovulation during anoestrus in two breeds of sheep with multiple injections of LH alone or in combination with FSH. J Endocrinol 111(1): 181–190.

Wang LM, Suthana NA, Chaudhury D, Weaver DR, and Colwell CS. (2005). Melatonin inhibits hippocampal long-term potentiation. Eur J Neurosci 22(9): 2231–2237.

Weaver DR, Carlson LL, and Reppert SM. (1990). Melatonin receptors and signal transduction in melatonin-sensitive and melatonin-insensitive populations of white-footed mice (*Peromyscus leucopus*). Brain Res 506(2): 353–357.

Weaver DR, Liu C, and Reppert SM. (1996). Nature's knockout: The Mel1b receptor is not necessary for reproductive and circadian responses to melatonin in Siberian hamsters. Mol Endocrinol 10(11): 1478–1487.

Weaver DR and Reppert SM. (1990). Melatonin receptors are present in the ferret pars tuberalis and pars distalis, but not in brain. Endocrinology 127(5): 2607–2609.

Weaver DR, Rivkees SA, and Reppert SM. (1989). Localization and characterization of melatonin receptors in rodent brain by in vitro autoradiography. J Neurosci 9(7): 2581–2590.

Whisnant CS and Goodman RL. (1990). Further evidence that serotonin mediates the steroid-independent inhibition of luteinizing hormone secretion in anestrous ewes. Biol Reprod 42(4): 656–661.

Whitmore D, Foulkes NS, and Sassone-Corsi P. (2000). Light acts directly on organs and cells in culture to set the vertebrate circadian clock. Nature 404(6773): 87–91.

Woodfill CJ, Wayne NL, Moenter SM, and Karsch FJ. (1994). Photoperiodic synchronization of a circannual reproductive rhythm in sheep: Identification of season-specific time cues. Biol Reprod 50(4): 965–976.

Yan L and Silver R (2008). Day-length encoding through tonic photic effects in the retinorecipient SCN region. Eur J Neurosci 28(1): 2108–2115.

Yasuo S, Yoshimura T, Ebihara S, and Korf HW (2007). Temporal dynamics of type 2 deiodinase expression after melatonin injections in Syrian hamsters. Endocrinology 148(9): 4385-4392.

Yasuo S, von Gall C, Weaver DR, and Korf HW (2008). Rhythmic expression of clock genes iin the ependymal cell layer of the third ventricle of rodents is independent of melatonin signaling. Eur J Neurosci 28(12): 2443–2450.

Yeates NTM. (1949). The effect of light on the breeding season, gestation, and birth weight of Merino sheep. Aust J Agricul Res 7(5): 440–446.

Yellon SM. (1994). Effects of photoperiod on reproduction and the gonadotropin-releasing hormone-immunoreactive neuron system in the postpubertal male Djungarian hamster. Biol Reprod 50(2): 368–372.

Yellon SM, Bittman EL, Lehman MN, Olster DH, Robinson JE, and Karsch FJ. (1985). Importance of duration of nocturnal melatonin secretion in determining the reproductive response to inductive photoperiod in the ewe. Biol Reprod 32(3): 523–529.

Zucker I, Lee TM, and Dark J. (1991). The suprachiasmatic nucleus and the annual rhythm of mammals. In *Suprachiasmatic Nucleus: The Mind's Clock* (DC Klein and SM Reppert, eds). New York: Academic Press, pp. 246–260.

Zucker I and Licht P. (1983). Circannual and seasonal variations in plasma luteinizing hormone levels of ovariectomized ground squirrels (*Spermophilus lateralis*). Biol Reprod 28(1): 178–185.

21

Genetic and Molecular Mechanisms of Mammalian Photoperiodism

David Hazlerigg

SEASONAL PHOTOPERIODIC RESPONSES IN MAMMALS

In using the annual cycle of photoperiodic change to gate seasonal transitions in life history state, mammals achieve two feats of information processing that present challenges to mechanistic explanation. First, whereas day length changes progressively according to a sinusoidal function over the year, seasonal transitions in physiology are generally much more discontinuous in nature, switched on or off over narrow time windows, as is the case for the autumn rut in deer, spring termination of embryonic diapause in mustelids, or seasonal molting and pelage growth in Siberian hamsters. Second, mammalian photoperiodic responses show a pronounced "history dependence." In the lab this is seen in the differential responses of hamsters to exposure to static as opposed to increasing or decreasing photoperiods, independent of absolute photoperiod duration (Gorman and Zucker, 1995). This sensitivity to photoperiodic history is one manifestation of the existence of endogenous long-term timing mechanisms, which interact with the photoperiodic response. These long-term timing processes are particularly strong in hibernating mammals, which may stay underground for many months without exposure to photoperiodic information, and yet still time emergence precisely the following spring (reviewed in Kortner and Geiser, 2000). Under laboratory conditions, the hibernation timer of chipmunks has been shown to "free run" in constant darkness over many years, implying the existence of a long-term oscillator underlying the seasonal timing process in this species (Kondo et al., 2006). Similar circannual timing mechanisms are also evident in the seasonal cycles of breeding and prolactin secretion in sheep (Woodfill et al., 1991; Gomez-Brunet et al., 2008). While short-lived rodent species do not express full circannual rhythms, spontaneous reversion to a summer type physiological state despite continuing exposure to winter like photoperiod is a common feature in this group. It is unclear whether this so-called "photorefractoriness" stems from similar mechanisms as circannual rhythm generation in longer lived seasonal species.

The following discussion considers the challenge of accounting for the mechanisms behind these features of seasonal photoperiodic timing in mammals for two well-defined seasonal responses: reproductive activation, and the molting cycle.

NEUROANATOMICAL BASIS TO THE SEASONAL CONTROL OF BREEDING AND THE MOLT

Both seasonal breeding and molting ultimately depend on changes in secretion of anterior pituitary hormones, the gonadotropins and prolactin, respectively. These hormones differ qualitatively in the ways in which they are controlled.

Gonadotropin secretion is governed by changes in the pattern of pulsatile release of the hypothalamic releasing hormone gonadotropin-releasing hormone (GnRH) into the pituitary portal vessels, which then arrives as pulses in the pars distalis (PD). Analysis of this so-called "GnRH pulse generator" in seasonal sheep indicates that the intrinsic pulse frequency is suppressed during the anoestrus season, and sensitivity to negative feedback inhibition by gonadal steroid hormones is increased (reviewed in Karsch et al., 1984). Seasonal reproductive activation depends on a reversal of these negative influences: heightened intrinsic pulse generator activity, and attenuated steroidal negative feedback effects. Hence, efforts to understand seasonal reproductive control have focused on hypothalamic effects leading to changes in GnRH pulse generator function. In this regard, a major recent focus has been the possible role of the RFamide neuropeptide kisspeptin in the seasonal control of GnRH release. Kisspeptin came to prominence with the demonstration that pubertal reproductive activation in mice and humans depended critically on increased kisspeptin activation of GnRH neurons (Seminara et al., 2003). More recently, it has been shown that seasonal reproductive inhibition in golden hamsters and in sheep involves reduced expression of kisspeptin in the arcuate region of the hypothalamus (Revel et al., 2006a; Wagner et al., 2008), and infusion of kisspeptin into the cerebroventricular system can override seasonal reproductive inhibition in this hamsters (Revel et al., 2006a).

Prolactin secretion differs from that of the gonadotropins, or indeed, other anterior pituitary hormones, in that the principal mechanism of control is through release inhibition. In culture, lactotrophic cells exhibit constitutive secretion of prolactin, whereas other anterior pituitary hormonal cell types require stimulation. Dopamine is well documented as a prolactin-release–inhibiting factor (Ben-Jonathan et al., 1989), and some studies have reported evidence that dopamine turnover changes as a function of seasonal status (Curlewis, 1992), but this appears unable to account for seasonal changes in prolactin secretion (Curlewis, 1992; Lincoln, 2002). Instead, attention has focused on extrahypothalamic, intrapituitary regulation of prolactin secretion in seasonal mammals, and these processes are discussed further below.

ORGANIZATION OF THE MAMMALIAN "PHOTOPERIODIC AXIS"

To understand how seasonal neuroendocrine outputs are regulated by photoperiod, it is necessary to consider the axis through which photoperiodic information reaches the hypothalamus and the anterior pituitary. In mammals, this is effected through what can be described as a "photoperiodic axis" arranged linearly from light

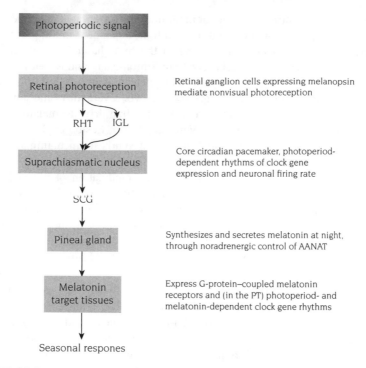

FIGURE 21.1.
The "photoperiodic axis" in mammals. This axis is defined anatomically by those functionally connected structures in which rhythmical physiology reflects photoperiod. Organization is shown schematically on the left, with comments on features at each level of the axis on the right. Abbreviations: RHT, retinohypothalamic tract; IGL, intergeniculate leaflet; SCG, superior cervical ganglion; AANAT, aryl amine *N*-acetyltransferase; PT, pars tuberalis.

reception in the retina, through a core circadian pacemaker in the hypothalamus and its effects on pineal melatonin production and culminating with melatonin receptor expressing cells (Hazlerigg and Wagner, 2006) (figure 21.1). From the perspective of seasonal photoperiodism, the key property of this axis is that it produces a light modulated circadian rhythm of melatonin release. Melatonin secretion is restricted to the night, continuing from after dark onset, through to dawn—and changes in the duration of this nocturnal melatonin signal are both necessary and sufficient to drive seasonal endocrine responses in mammals (Bartness et al., 1993).

THE CONTROL OF MELATONIN SYNTHESIS: ENCODING THE PHOTOPERIODIC MESSAGE

Melatonin is an indole amine neurohormone derived from the aromatic amino acid tryptophan and is the principal secretory product of the pineal gland.

In mammals, unlike birds and other nonmammalian vertebrates, the pineal is not directly light responsive, but depends upon a light-dependent input, processed by the master circadian pacemaker residing in the hypothalamic suprachiasmatic nucleus (SCN) (reviewed comprehensively in Simonneaux and Ribelayga, 2003). The circadian basis to melatonin synthesis is clearly demonstrated by studies in which animals are kept under constant darkness over several days and sampled for melatonin synthesis—demonstrating a clear circadian drive to melatonin secretion entrained by light (Illnerova and Vanecek, 1982). Lesion of the SCN circadian pacemaker abolishes rhythmical melatonin synthesis in mammals (Klein, 1979).

According to this framework, photoperiodic modulation of pineal function derives primarily from the effects of the changing light-dark cycle on the function of the SCN. Light reaches the SCN via the retinohypothalamic tract and also via the intergeniculate leaflet (Card and Moore, 1991), and it has been shown that specialized nonvisual photoreceptive cells within the retina—the melanopsin-expressing retinal ganglion cells—mediate the effects of light on SCN function (Hattar et al., 2002).

Within the SCN, changes in day length lead to altered profiles of circadian gene expression and of electrical activity, with spatiotemporal effects on these parameters having been reported in some recent studies (Jagota et al., 2000; Hazlerigg et al., 2005; Inagaki et al., 2007; Naito et al., 2008). These effects are presumed to lead to quantitative changes in SCN outputs, including signals reaching the pineal gland via the superior cervical ganglion. Studies by Kalsbeek and colleagues indicate that the SCN governs pineal activity by reciprocal switching between inhibitory day time release of GABA and stimulatory nighttime release of glutamate, with both these signals travelling through relays in the paraventricular nucleus and the superior cervical ganglion (Perreau-Lenz et al., 2003).

The key input to the pineal gland is the neurotransmitter noradrenalin (NA), which acts primarily through β-adrenergic receptors to increase melatonin synthesis at night. This is achieved through increases in the expression and/or activity of the rate limiting enzyme for melatonin biosynthesis, aryl amine *N*-acetyltransferase (AANAT). In rodents, this control occurs at a transcriptional level, through cyclic adenosine monophosphate (cAMP)-responsive elements (CREs) present in the AANAT gene; in ungulates, control is post-translational, through cAMP-dependent effects on the activity and stability of AANAT protein. This difference may account for generally longer delay between dark and melatonin onset in rodents compared to ungulates (Simonneaux and Ribelayga, 2003).

In rodents, control of AANAT RNA expression also involves the induction of transcriptional repressors, in particular inducible cAMP-element repressor (ICER), which may act as a delayed mechanism to shut down AANAT transcription during the latter part of the night (Stehle et al., 1993). This induction of repressive transcription factors may also provide a mechanism ensuring that melatonin synthesis does not take place during the day in rodents.

DECODING THE MELATONIN SIGNAL: DURATION AS THE CRITICAL FEATURE

Melatonin injections given daily in the late afternoon or before dawn are sufficient to induce a short-day typical gonadal response in Syrian hamsters (*Mesocricetus auratus*) acclimated to long days; other injection times are ineffective (Tamarkin et al., 1976). Although this observation appeared at first glance to suggest the existence of melatonin-sensitive phases in the late afternoon and around dawn, this is not now generally accepted. Instead, based largely on experiments in pinealectomized hamsters given timed subcutaneous infusions of melatonin, short-day responses are believed to occur when the melatonin signal exceeds a certain critical continuous duration (Bartness et al., 1993). Thus, morning or evening injections to pineal-intact hamsters are believed to act by augmenting the duration of the endogenous nocturnal melatonin signal. Nevertheless, some authors have argued that seasonal photoperiodic responses involve a melatonin-driven rhythm of sensitivity to melatonin in relevant target tissues (Pitrosky et al., 1995). As described below, a molecular analysis of melatonin action suggests that this interpretation of in vivo results may have a basis in underlying molecular mechanism.

MELATONIN TARGETS RELEVANT TO THE SEASONAL PHOTOPERIODIC RESPONSE: EVIDENCE FOR THE CLASSICAL TWO-SITE MODEL

From the foregoing discussion of the levels at which seasonal control of reproduction and molting takes place, we would predict that melatonin might act on targets within the hypothalamus and/or the anterior pituitary. Confirming this, in vitro autoradiography using a radioanalog of melatonin, $[^{125}I]2$-iodomelatonin (IMEL) revealed the presence of melatonin binding sites within the hypothalamus and pituitary across a range of seasonal mammals (reviewed in Morgan et al., 1994). Intriguingly a survey across different seasonal mammals reveals considerable inconsistencies in IMEL binding seen within the hypothalamus, whereas within the pituitary the pars tuberalis (PT) is consistently recorded as the site of most intense IMEL binding (Morgan et al., 1994).

Building on the IMEL data, several groups performed functional studies aimed at determining which sites within the hypothalamus or pituitary mediated the seasonal photoperiodic effects of melatonin. These were undertaken in seasonal hamsters and in sheep, and collectively led a "two-site" model for melatonin action.

Firstly, based upon the effects of neuroanatomical lesions and also micro-implantation of melatonin, it was concluded that sites in the mediobasal hypothalamus, in particular the arcuate, dorsomedial, or ventromedial hypothalamic nuclei mediated melatonin's effects on reproductive activation (Lincoln and Maeda, 1992; Malpaux et al., 1993, 1998; Maywood et al., 1995, 1996).

Secondly, it was concluded that the PT was crucially involved in the seasonal control of prolactin secretion, operating through a hypothalamus-independent mechanism. This surprising inference was drawn following the observation that Soay sheep that had been subjected to surgical isolation of the pituitary from direct hypothalamic input continued to show seasonal cycles of prolactin secretion, despite gross disruption of other descending neuroendocrine control (Lincoln and Clarke, 1994, 1995). Prolactin-secreting cells are not directly melatonin sensitive, suggesting that the PT acts as an intrapituitary relay to maintain only the seasonal prolactin regulation in the hypothalamopituitary-disconnected animal. *In vitro* coculture of PT and PD cells from sheep and hamsters, suggests that the PT produces a prolactin-releasing factor, but the identity of this has so far not been established (Hazlerigg et al., 1996; Stirland et al., 2001)

This two-site model has stood largely unchallenged for over a decade, but recent insights suggest that it may now be due for revision. This is returned to later in this chapter.

THE MOLECULAR NATURE OF IMEL BINDING SITES: G-PROTEIN–COUPLED MELATONIN RECEPTORS

In 1996, Reppert and colleagues succeeded in cloning the melatonin receptor, first in the clawed frog, *Xenopus laevis* and subsequently in mammals (Ebisawa et al., 1994; Reppert et al., 1994). In mammals, two melatonin receptor subtypes have been isolated, termed subtype I (MT_1) and subtype II (MT_2). Both MT_1 and MT_2 are G-protein–coupled receptors, with an affinity in the order of 10 pmol/liter (pM) (Reppert et al., 1996). This corresponds well to the pattern of melatonin secretion that produces a daily variation in plasma titers in the range of 0–100 pM over the daily cycle. More recently, an additional IMEL-binding moiety, termed MT_3 was cloned and characterized as a quinone reductase. This has a 1,000-fold lower affinity for melatonin (Nosjean et al., 2000), and probably does not contribute to physiological melatonin signaling mechanisms.

In situ hybridization analysis of MT_1/MT_2 expression reveals that MT_1 is the dominantly expressed subtype, with MT_2 expression being undetectable in the brain or pituitary in many species. Moreover, in Siberian hamsters (*Phodopus sungorus*), a highly photoperiodic species, there is an in-frame stop codon within the predicted coding region of the MT_2 gene, which appears to render it a pseudogene, giving no functional contribution (Weaver et al., 1996). This and the demonstration that transgenic knockout of MT_1 in mice renders the PT unresponsive to melatonin (von Gall et al., 2002), argue strongly that the neuroendocrine effects of melatonin derive largely from effects mediated through MT_1 receptors.

A number of studies have reported daily, photoperiod- or pineal-dependent changes in levels of melatonin binding site expression in the PT (Gauer et al., 1992, 1993). These studies show a negative effect of melatonin on levels of melatonin receptor expression in this tissue, which might be suggestive of a mechanism through which photoperiodic history could be relayed through long-term changes in melatonin sensitivity. Against this, there is no evidence that these reported changes in binding site density lead to altered melatonin signal transduction in the PT.

DURATIONAL INTERPRETATION OF THE MELATONIN SIGNAL THROUGH MT$_1$-EXPRESSING CELLS IN THE PT

cAMP Signal Transduction

Much of our knowledge of melatonin signal processing in a photoperiodic context comes from studies of the PT. Studies of PT cells in vitro indicate that melatonin's principal mode of action is through acute inhibition and chronic sensitization of cAMP-mediated signal transduction (Morgan et al., 1989; Hazlerigg et al., 1993). These actions of melatonin are presumed to interact with endogenous adenylate cyclase (AC)-activating signals, including adenosine, to induce CRE-regulated gene expression (von Gall et al., 2002). The chronic AC-sensitizing effect of melatonin may account for the importance of melatonin signal continuity noted in timed infusion of melatonin to pinealectomized animals (Bartness et al., 1993), but is not in itself sufficient as a mechanism for measuring changes in melatonin signal duration. This is because similar levels of AC sensitization are seen in PT cells exposed to melatonin for 8 or 16 h, even though signals of these durations have quite different seasonal photoperiodic effects (Hazlerigg et al., 1993).

Effects on Clock Genes

Great progress has been made in hammering out the details of how circadian rhythms are generated at a cellular level (Bell-Pedersen et al., 2005). It is believed that the core of this process depends upon the transcriptional activating effects of a complex containing the transcription factors CLOCK and BMAL1 (figure 21.2A). This acts at E-box (consensus CACGTG) elements in the promoter regions of the period (*Per*) and cryptochrome (*Cry*) genes, and in turn the protein products of these genes (PER and CRY) form a repressive complex that inhibits the actions of CLOCK:BMAL1-containing complexes. Additional feedback control of the expression of the *Bmal1* gene depends on the product of the gene *Rev-erbα*, which is also an E-box regulated gene (Preitner et al., 2002). These interconnected feedback loops, coupled with positive and negative effects on the expression of clock controlled genes generate circadian physiology.

Intriguingly, it has recently become apparent that different aspects of the nocturnal melatonin signal are associated with distinctive effects on the expression of different clock genes in the PT (Messager et al., 1999; Lincoln et al., 2002). First, the onset of melatonin secretion with nightfall induces the expression of the clock gene *Cry1* (Lincoln et al., 2002), and second the termination of the melatonin signal

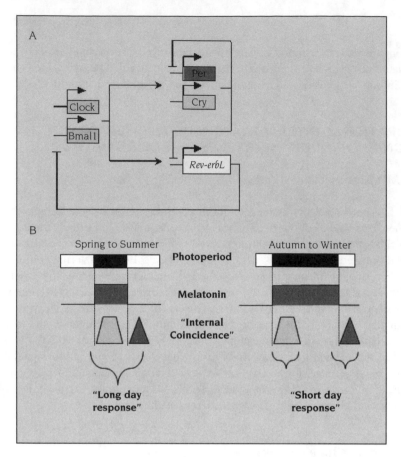

FIGURE 21.2.

Clock gene–based "internal coincidence" hypothesis for melatonin action in the PT. (A) A simplified scheme showing principal molecular components driving circadian oscillations in mammalian cells. This "two-loop" model proposes that oscillations derive from positive elements, CLOCK and BMAL1 driving transcription of negative elements *Per* and *Cry*, which form a transcriptionally repressive complex that feeds back negatively on the effects of CLOCK and BMAL1. These effects are mediated through E-box (CACGTG)- or E′-box (CANNTG)-containing regions in the *Per* and *Cry* genes. Additionally, transcription of the *Bmal*1 gene is rhythmically controlled through retinoid related responses elements (ROREs), which are sensitive to negative feedback control by the transcription factor REV-ERBL . Since Rev-erbα is itself an E-box–regulated gene, this forms a second negative feedback loop in the circadian clock. The interconnection between E-box based and RORE-based circadian oscillations allows complex circadian programs of clock controlled gene expression, with peak phases of gene expression matched to particular cellular requirements. (B) Derived function in clock gene networks in PT cells, leading to "Internal coincidence" timing. Here, melatonin exerts direct transcriptional control on the *Per*1 and *Cry*1 genes through separate mechanisms, cAMP-mediated transactivation of *Per*1 at dawn, and evening transactivation of *Cry*1 by an as yet undefined pathway. By varying the relative timing of these two events, it is postulated that photoperiodic changes in melatonin signal duration lead to differing PER1/CRY1 transcriptional effects in PT cells. According to this model, spring to summer photoperiods result in a short CRY1-PER1 interval, and hence more PER1-CRY1 complex formation, than is the case on autumn to winter photoperiods, and this distinguishes long- from short-day responses.

at dawn induces the expression of *Per1* (Messager et al., 1999, Lincoln et al., 2002). Moreover, in pinealectomized Syrian hamsters or transgenic mice (*Mus musculus*) lacking the MT_1 melatonin receptor, no circadian gene expression is observable in the PT (Messager et al., 2001, von Gall et al., 2002). The dawn–*Per1* relationship can be understood in terms of the known cAMP sensitivity of this gene, and AC regulatory effects of melatonin (Hazlerigg et al., 1993; von Gall et al., 2002), but the evening effect of rising melatonin levels on *Cry1* expression has so far not been dissected at the signal transduction level.

Because onset and offset of the melatonin signal independently regulate *Cry1* and *Per1*, changing the photoperiod—and hence the melatonin signal duration—leads to changing "coincidence" between the times of peak expression of these two genes (Lincoln et al., 2002). This led to a novel hypothesis to explain seasonal decoding of the melatonin signal that invoked the concept of "internal coincidence" (Lincoln et al., 2002) (figure 21.2B). This model is based upon the assumption that PER and CRY proteins interact with one another to control the expression genes containing appropriate response elements in their promoter—a concept derived from the molecular clock model outlined above, and having strong biochemical support (Bell-Pedersen et al., 2005). Hence the model predicts that depending on the interval between dusk and dawn—relayed via the melatonin signal—the extent of daily formation of functional PER:CRY complexes will vary over the course of the year. This model hinges upon the idea that protein levels undergo similar variation to that observed for RNA expression; in the case of PER1, published data support this view (Nuesslein-Hildesheim et al., 2000), and recent studies in the mouse PT suggest that nuclear PER1 and CRY1 protein levels peak synchronously (Jilg et al., 2005).

Studies in the Soay sheep have further investigated how PT clock gene expression changes in response to acute changes in photoperiod and to prolonged exposure to constant photoperiod (Hazlerigg et al., 2004; Lincoln et al., 2005). An acute switch from 8- to 16-h light per day causes immediate establishment of a long-day–typical pattern of *Per1* and *Cry1* gene expression, but for the clock controlled gene *Rev-erbL* a transition state distinct from either the short- or long-day acclimated states was observed (Hazlerigg et al., 2004). This result indicates that the molecular clockwork in PT cells may carry a form of photoperiodic history dependence. Contrastingly, in animals kept under long photoperiods for 30 weeks, showing a photorefractory decline in prolactin secretion, PT clock gene expression profiles were indistinguishable from those in animals kept under long photoperiods for 6 weeks (Lincoln et al., 2005). This result, and similar data in Syrian hamsters kept under short photoperiods (Johnston et al., 2003), implies that photorefractoriness does not develop from insensitivity of clock genes to melatonin.

Recently this clock gene based "internal coincidence" model has been explored in a study of the photoperiodic range limits over which *Per1* and *Cry1* RNA expression can be seen in the PT. Here it was shown that when photoperiod limits the duration of melatonin signal, *Per1* expression ceases to be detectable, consistent with the concept that AC sensitization is necessary for this molecular response to melatonin (Wagner et al., 2008). Under these conditions, animals cease to express summer phenotypic attributes, consistent with the hypothesis that clock gene effects may be important for decoding the melatonin signal.

At the present time, the development of techniques to manipulate *clock* gene expression in the PT of seasonal animals, as a means to testing this hypothesis is an outstanding research objective.

PHOTOPERIODIC EFFECTS ON INTRAHYPOTHALAMIC GENE EXPRESSION: A NOVEL ROLE FOR LOCAL THYROID HORMONE METABOLISM

Although the PT is relatively tractable to cellular and molecular analysis of mela-tonin action, this does not apply for candidate intrahypothalamic melatonin target sites. This is because melatonin receptor expressing cells in these areas appear at much lower density amongst complex mixed cell populations. For this reason, anal-yses of photoperiodic effects within the hypothalamus have focused on changes in gene expression, leaving the issue of melatonin-dependent control as a secondary question.

The most striking recent development in this area has been the discovery that, in mammals and birds, intrahypothalamic bioavailability of triiodothyronine (T_3), the most biologically active form of thyroid hormone (TH), is governed through photoperiod-dependent changes in deiodinase gene expression (Yoshimura et al., 2003; Yasuo et al., 2005, 2006, 2007a; Revel et al., 2006b; Barrett et al., 2007). These enzymes govern the metabolism of TH, with deiodinase II (DIO2) promoting the conversion of thyroxine (T_4) into active T_3, whereas deiodinase III (DIO3) con-verts T_4 into inactive reverse T_3, and inactivates T_3 by conversion to diiodothyronine (T_2) (figure 21.3).

Long photoperiods are associated with heightened expression of DIO2 in rodents and sheep, while short photoperiods produce the reciprocal effect, and promote heightened expression of DIO3. These effects on deiodinase gene expres-sion in mammals have been shown to be melatonin-dependent (Revel et al., 2006b; Yasuo et al., 2007b; Barrett et al., 2007) This pattern of reciprocal DIO2 and DIO3 regulation predicts that hypothalamic T_3 levels will tend to be increased in spring and summer and will decline in autumn and winter. The functional importance of this regulation in mammals is supported by the recent finding that in Siberian ham-sters, short-photoperiod–induced reproductive inhibition is overcome by placement of T_3-containing microimplants into the hypothalamus—presumably bypassing DIO control of T_3 availability in this region (Barrett et al., 2007).

These recent data provide a mechanistic explanation of the established link between thyroid function and the termination of the breeding season in mammals (Nicholls et al., 1988). Furthermore, they explain the finding that local implants of T_4 into the basal hypothalamus of ewes were sufficient to overcome this effect of thyroidectomy (Anderson et al., 2003).

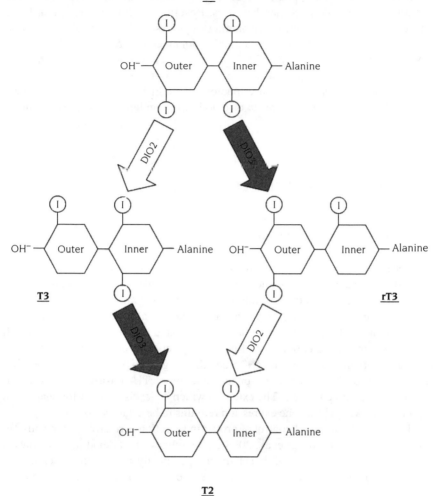

FIGURE 21.3.
Processing of thyroid hormone by deiodinases. Thyroxine (T_4) is the principal circulating form of thyroid hormone but has relatively low biological potency for activation of nuclear thyroid hormone receptors. This potency is increased markedly by conversion to triiodothyronine (T_3) through outer ring deiodination. This is mediated by type II deiodinase (DIO2). Conversely, T_4 can be converted to an inactive form, reverse T_3 (rT_3) by inner ring deiodination mediated by type III deiodinase (DIO3). Both T_3 and rT_3 can be further metabolized by DIO3 or DIO2, respectively, leading to diiodothyronine (T_2) formation.

It is unclear whether photoperiodic history effects emanate from changes in deiodinase expression. In Syrian hamsters, Revel et al. (2006b) found that DIO2 faithfully reflects photoperiod, remaining suppressed during prolonged short-photoperiod exposure, even when animals spontaneously reactivate the gonadal axis. In contrast, Barrett et al. (2007) have reported a gradual decline in DIO3 expression during prolonged SP exposure in Siberian hamsters, preceding reactivation of the gonadal axis. Hence, photoperiodic history dependence may be a feature of some but not all aspects of deiodinase regulation of thyroid hormone bioavailability.

THE LOCATION AND NATURE OF DEIODINASE-EXPRESSING CELLS

Hypothalamic deiodinase gene expression is concentrated to the ependymal cell layer surrounding the lower portion of the third ventricle, and in particular to a specialized population of cells known as tanycytes (Rodriguez et al., 2005). These cells form projections into the external zone of the median eminence and have been variously implicated in uptake and transport of molecules into or out of the blood–brain barrier, and in controlling the access of hypothalamic neurosecretory cell endings to capillaries in the primary plexus of the portal system (Rodriguez et al., 2005). Additionally, the subventricular zone inhabited by these cells is a site of cell proliferation in the vertebrate central nervous system (Xu et al., 2005). Effects on TH availability mediated by changes in deiodinase gene expression might therefore influence any of these processes—or the function of TH-sensitive cells in neighboring hypothalamic regions. The extent to which T_3 levels account for changes in hypothalamic kisspeptin gene expression remains to be established.

Interestingly, seasonal control of prolactin secretion is relatively insensitive to systemic TH status (Billings et al., 2002). This suggests that deiodinase regulation may constitute a hypothalamic mechanism specifically related to seasonal control of reproduction, and possibly energy metabolism, separate from the prolactin axis.

LINK BETWEEN MELATONIN SIGNAL TRANSDUCTION AND DEIODINASE-EXPRESSING CELLS

An outstanding question regarding the photoperiodic control of hypothalamic deiodinase gene expression in mammals, concerns how these events are linked to melatonin receptor pathways. Analysis of melatonin receptor distribution by in situ hybridization fails to detect expression in the ependymal layer, suggesting that neighboring sites, serve a relay function. Within the hypothalamus itself, it is not clear which tissue would serve this purpose, and although DIO regulation is conserved across species, IMEL binding patterns vary, while MT_1 expression is not consistently detectable in the hypothalamus at all. In contrast, the PT lies immediately adjacent, in contact with mediobasal tanycytes in all mammals and consistently expresses high levels of MT_1. For these reasons, the possibility that the PT serves as a relay between melatonin and DIO expression control holds considerable attraction.

In quail, it has recently been proposed that this relay function might be served by secretion of thyrotropin (TSH) by PT cells (Nakao et al., 2008). This is based on screenings of gene expression changes following induction of a long-day response, and on the finding that TSH receptors are present within the mediobasal hypothalamus of quail. These basic ingredients for a TSH-dependent link between the PT and DIO regulation are also present in mammals: there is a well documented control of TSH expression in the PT of mammals, and TSH receptor expression is present in sites where *DIO2* gene expression is photoperiodically regulated (figure 21.4). Strong support for this hypothesis is provided by the recent demonstrations, first in Soay sheep (Hanon et al., 2008) and then in mice (Ono et al., 2008), that TSH controls *DIO2* gene expression in the mediobasal hypothalamus. Thus it appears that, in mammals, PT melatonin receptor expressing cells govern seasonal *DIO* expression through local TSH-mediated effects.

PERSPECTIVES AND PROSPECTS

The proposal that the PT acts as a melatonin-dependent relay by influencing hypothalamic deiodinase gene expression, appears to undermine the classical "two-site" model for melatonin action outlined in this chapter. The lesioning experiments that led investigators to assume an intrahypothalamic melatonin target tissue governed the reproductive response appear now to need reinterpretation. Since these presumptive sites lie very close to the PT itself—as well as to the periventricular tanycytes—it is possible that lesions of the targets of the PT account for the observed results.

More difficult for a revised PT–TSH–tanycyte model for melatonin action is the issue of intrapituitary effects of melatonin on the prolactin axis. Unlike reproduction, thyroidectomy does not appear to influence springtime changes in prolactin secretion (Billings et al., 2002), suggesting that DIO2/DIO3 regulation has little to do with this seasonal response. Potentially some tanycytic cells would form part of the isolated PT-PD structure following the pituitary disconnection protocol of Lincoln and Clarke (1994), raising the possibility that some additional DIO-independent function of these cells might lead to cycles of prolactin secretion. Alternatively, it remains quite possible that a tanycyte-independent output of PT cells directly interacts with PD lactotrophs. It is highly unlikely that this role would be played by TSH, however, because there is no evidence for TSH-receptor expression in the PD. Resolving these issues, and the question of how melatonin signal duration is transduced into changes in PT output, be it through TSH or additional unidentified factor(s) are the key research priorities.

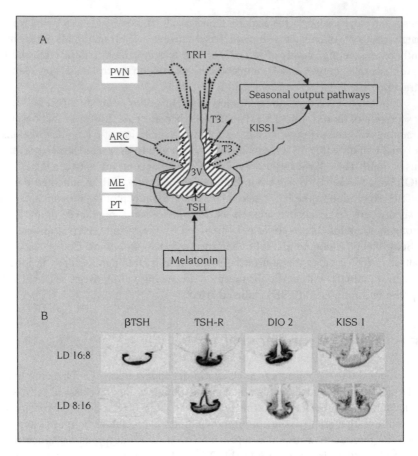

FIGURE 21.4.
Model for PT-derived TSH control hypothalamic DIO2 expression and seasonal output.
(A) Anatomical scheme showing the elements of the model. Melatonin acts on the PT to
cause photoperiodic changes in the expression of βTSH RNA expression, which leads in
turn to altered secretion of TSH. This acts locally on TSH receptor (TSH-R)-expressing
tanycytes in the median eminence (ME), and extending up into the periventricular basal
hypothalamus. These cells respond to changes in TSH by altering levels of expression of
DIO2 and/or DIO3, leading to changes in T₃ levels in the hypothalamus, permeating to
adjacent structures such as the arcuate nucleus (ARC) and the paraventricular nucleus
(PVN). This results in changes in neural activity in these areas, ultimately leading to
changes in seasonal output. 3V, third ventricle. (B) Representative autoradiographic
images of radioactive in situ hybridizations with antisense probes to βTSH, TSH receptor
(TSH-R), DIO2, and the kisspeptin encoding gene KISS 1, performed on tissue from
Soay sheep acclimated to the indicated photoperiods for sox weeks. Note the strong
photoperiodic effects on PT βTSH expression and on DIO2 expression in the TSH-
R–expressing region of the median eminence. KISS-1 expression in the adjacent arcuate
nucleus is also photoperiod dependent, but the relationship between this and DIO2
regulation has not been established.

Acknowledgments   The author thanks the U.K. Biotechnology and Biological Sciences Research Council for continuing financial support.

References

Anderson GM, Hardy SL, Valent M, Billings HJ, Connors JM, and Goodman RL. (2003). Evidence that thyroid hormones act in the ventromedial preoptic area and the premammillary region of the brain to allow the termination of the breeding season in the ewe. Endocrinology 144: 2892–2901.

Barrett P, Ebling FJ, Schuhler S, Wilson D, Ross AW, Warner A, Jethwa P, Boelen A, Visser TJ, Ozanne DM, Archer ZA, Mercer JG, and Morgan PJ. (2007). Hypothalamic thyroid hormone catabolism acts as a gatekeeper for the seasonal control of body weight and reproduction. Endocrinology 148: 3608–3617.

Bartness TJ, Powers JB, Hastings MH, Bittman EL, and Goldman BD. (1993). The timed infusion paradigm for melatonin delivery: What has it taught us about the melatonin signal, its reception, and the photoperiodic control of seasonal responses? J Pineal Res 15: 161–190.

Bell-Pedersen D, Cassone VM, Earnest DJ, Golden SS, Hardin PE, Thomas TL, and Zoran MJ. (2005). Circadian rhythms from multiple oscillators: Lessons from diverse organisms. Nature Rev Genet 6: 544–556.

Ben-Jonathan N, Arbogast LA, and Hyde, JF. (1989). Neuroendocrine regulation of prolactin-release. Prog Neurobiol 33: 399–447.

Billings HJ, Viguie C, Karsch FJ, Goodman RL, Connors JM, and Anderson GM. (2002). Temporal requirements of thyroid hormones for seasonal changes in LH secretion. Endocrinology 143: 2618–2625.

Card JP and Moore RY. (1991). The organisation of visual circuits influencing the circadian activity of the suprachiasmatic nucleus. In *Suprachiasmatic Nucleus: The Mind's Clock* (DC Klein and SM Reppert, eds). New York: Academic Press, pp. 51–76.

Curlewis JD. (1992). Seasonal prolactin secretion and its role in seasonal reproduction—a review. Reprod Fertil Dev 4: 1.

Ebisawa T, Karne S, Lerner MR, and Reppert SM. (1994). Expression cloning of a high-affinity melatonin receptor from *Xenopus* dermal melanophores. Proc Natl Acad Sci USA 91: 6133–6137.

Gauer F, Massonpevet M, and Pevet P. (1992). Pinealectomy and constant illumination increase the density of melatonin binding-sites in the pars-tuberalis of rodents. Brain Res 575: 32–38.

Gauer F, Massonpevet M, and Pevet P. (1993). Melatonin receptor density is regulated in rat pars tuberalis and suprachiasmatic nuclei by melatonin itself. Brain Res 602: 153–156.

Gomez-Brunet A, Santiago-Moreno J, del Campo A, Malpaux B, Chemineau P, Tortonese DJ, Gonzalez-Bulnes A, and Lopez-Sebastian A. (2008). Endogenous circannual cycles of ovarian activity and changes in prolactin and melatonin secretion in wild and domestic female sheep maintained under a long-day photoperiod. Biol Reprod 78: 552–562.

Gorman MR and Zucker I. (1995). Seasonal adaptations of Siberian hamsters. 2. Pattern of change in day length controls annual testicular and body-weight rhythms. Biol Reprod 53: 116–125.

Hanon E, Lincoln GA, Fustin J-M, Dardente H, Masson-Pevet M, Morgan PJ, and Hazlerigg DG. (2008). Ancestral TSH mechanism signals summer in a seasonal mammal. Curr Biol 18: 1147–1152.

Hattar S, Liao HW, Takao M, Berson DM, and Yau KW. (2002). Melanopsin-containing retinal ganglion cells: Architecture, projections, and intrinsic photosensitivity. Science 295: 1065–1070.

Hazlerigg DG, Andersson H, Johnston JD, and Lincoln G. (2004). Molecular characterization of the long-day response in the Soay sheep, a seasonal mammal. Curr Biol 14: 334–339.

Hazlerigg DG, Ebling FJP, and Johnston JD. (2005). Photoperiod differentially regulates gene expression rhythms in the rostral and caudal SCN. Curr Biol 15: R449–R450.

Hazlerigg DG, Gonzalez-Brito A, Lawson W, Hastings MH, and Morgan PJ. (1993). Prolonged exposure to melatonin leads to time-dependent sensitization of adenylate cyclase and down-regulates melatonin receptors in pars tuberalis cells from ovine pituitary. Endocrinology 132: 285–292.

Hazlerigg DG, Hastings MH, and Morgan PJ. (1996). Production of a prolactin releasing factor by the ovine pars tuberalis. J Neuroendocrinol 8: 489–492.

Hazlerigg DG and Wagner GC. (2006). Seasonal photoperiodism in vertebrates: from coincidence to amplitude. Trends Endocrinol Metab 17: 83–91.

Illnerova H and Vanecek J. (1982). 2-Oscillator structure of the pacemaker controlling the circadian-rhythm of N-acetyltransferase in the rat pineal-gland. J Comp Physiol 145: 539–548.

Inagaki N, Honma S, Ono D, Tanahashi Y, and Honma K. (2007). Separate oscillating cell groups in mouse suprachiasmatic nucleus couple photoperiodically to the onset and end of daily activity. Proc Natl Acad Sci USA 104: 7664–7669.

Jagota A, de la Iglesia HO, and Schwartz WJ. (2000). Morning and evening circadian oscillations in the suprachiasmatic nucleus in vitro. Nat Neurosci 3: 372–376.

Jilg A, Moek J, Weaver DR, Korf HW, Stehle JH, and von Gall C. (2005). Rhythms in clock proteins in the mouse pars tuberalis depend on MT_1 melatonin receptor signalling. Eur J Neurosci 22: 2845–2854.

Johnston JD, Cagampang FR, Stirland JA, Carr AJ, White MR, Davis JR, and Loudon AS. (2003). Evidence for an endogenous per1- and ICER-independent seasonal timer in the hamster pituitary gland. FASEB J 17: 810–815.

Karsch FJ, Bittman EL, Foster DL, Goodman RL, Legan SJ, and Robinson JE. (1984). Neuroendocrine basis of seasonal reproduction. Rec Prog Horm Res 40: 185–232.

Klein DC. (1979). Pineal N-acetyltransferase and hydroxyindole-O-methyltransferase—control by the retinohypothalamic tract and the suprachiasmatic nucleus. Brain Res 174: 245.

Kondo N, Sekijima T, Kondo J, Takamatsu N, Tohya K, and Ohtsu T. (2006). Circannual control of hibernation by HP complex in the brain. Cell 125: 161–172.

Kortner G and Geiser F. (2000). The temporal organization of daily torpor and hibernation: Circadian and circannual rhythms. Chronobiol Int 17: 103–128.

Lincoln G. (2002). Noradrenaline and dopamine regulation of prolactin secretion in sheep: Role in prolactin homeostasis but not photoperiodism. J Neuroendocrinol 14: 36–44.

Lincoln GA and Clarke IJ. (1994). Photoperiodically-induced cycles in the secretion of prolactin in hypothalamo-pituitary disconnected rams: Evidence for translation of the melatonin signal in the pituitary gland. J Neuroendocrinol 6: 251–260.

Lincoln GA and Clarke IJ. (1995). Evidence that melatonin acts in the pituitary gland through a dopamine-independent mechanism to mediate effects of daylength on the secretion of prolactin in the ram. J Neuroendocrinol 7: 637–643.

Lincoln GA, Johnston JD, Andersson H, Wagner G, and Hazlerigg DG. (2005). Photorefractoriness in mammals: Dissociating a seasonal timer from the circadian-based photoperiod response. Endocrinology 146: 3782–3790.

Lincoln GA and Maeda KI. (1992). Reproductive effects of placing micro-implants of melatonin in the mediobasal hypothalamus and preoptic area in rams. J Endocrinol. 132: 201–215.

Lincoln G, Messager S, Andersson H, and Hazlerigg D. (2002). Temporal expression of seven clock genes in the suprachiasmatic nucleus and the pars tuberalis of the sheep: Evidence for an internal coincidence timer. Proc Natl Acad Sci USA 99: 13890–13895.

Malpaux B, Daveau A, Maurice-Mandon F, Duarte G, and Chemineau P. (1998). Evidence that melatonin acts in the premammillary hypothalamic area to control reproduction in the ewe: Presence of binding sites and stimulation of luteinizing hormone secretion by in situ microimplant delivery. Endocrinology 139: 1508–1516.

Malpaux B, Daveau A, Maurice F, Gayrard V, and Thiery JC. (1993). Short-day effects of melatonin on luteinizing hormone secretion in the ewe: Evidence for central sites of action in the mediobasal hypothalamus. Biol Reprod 48: 752–760.

Maywood ES, Bittman EL, and Hastings MH. (1996). Lesions of the melatonin- and androgen-responsive tissue of the dorsomedial nucleus of the hypothalamus block the gonadal response of male Syrian hamsters to programmed infusions of melatonin. Biol Reprod 54: 470–477.

Maywood ES and Hastings MH. (1995). Lesions of the iodomelatonin-binding sites of the mediobasal hypothalamus spare the lactotropic, but block the gonadotropic response of male Syrian hamsters to short photoperiod and to melatonin. Endocrinology 136: 144–153.

Messager S, Garabette ML, Hastings MH, and Hazlerigg DG. (2001). Tissue-specific abolition of Per1 expression in the pars tuberalis by pinealectomy in the Syrian hamster. Neuroreport 12: 579–582.

Messager S, Ross AW, Barrett P, and Morgan PJ. (1999). Decoding photoperiodic time through Per1 and ICER gene amplitude. Proc Natl Acad Sci USA 96: 9938–9943.

Morgan PJ, Barrett P, Howell HE, and Helliwell R. (1994). Melatonin receptors: Localization, molecular pharmacology and physiological significance. Neurochem Int 24: 101–146.

Morgan PJ, Williams LM, Davidson G, Lawson W, and Howell H. (1989). Melatonin receptors on ovine pars tuberalis: Characterisation and autoradiographical localization. J Neuroendorcinol 1. 1–4.

Naito E, Watanabe T, Tei H, Yoshimura T, and Ebihara S. (2008). Reorganization of the suprachiasmatic nucleus coding for day length. J Biol Rhythms 23: 140–149.

Nakao N, Ono H, Yamamura T, Anraku T, Takagi T, Higashi K, Yasuo S, Katou Y, Kageyama S, Uno Y, Kasukawa T, Iigo M, Sharp PJ, Iwasawa A, Suzuki Y, et al. (2008). Thyrotrophin in the pars tuberalis triggers photoperiodic response. Nature 452: 317–323.

Nicholls TJ, Follett BK, Goldsmith AR, and Pearson H. (1988). Possible homologies between photorefractoriness in sheep and birds: The effect of thyroidectomy on the length of the ewe's breeding season. Reprod Nutr Dev 28: 375–385.

Nosjean O, Ferro M, Coge F, Beauverger P, Henlin JM, Lefoulon F, Fauchere JL, Delagrange P, Canet E, and Boutin JA. (2000). Identification of the melatonin-binding site MT_3 as the quinone reductase 2. J Biol Chem 275: 31311–31317.

Nuesslein-Hildesheim B, O'Brien JA, Ebling JP, Maywood ES, and Hastings MH. (2000). The circadian cycle of mPER clock gene products in the suprachiasmatic nucleus of the Siberian hamster encodes both daily and seasonal time. Eur J Neurosci 12: 2856–2864.

Ono H, Hoshino Y, Yasuo S, Watanabe M, Nakane Y, Murai A, Ebihara S, Korf HW, and Yoshimura T. (2008). Involvement of thyrotropin in photoperiodic signal transduction in mice. Proc Natl Acad Sci USA 47: 18238–18242.

Perreau-Lenz S, Kalsbeek A, Garidou A, Wortel J, van der Vliet J, van Heijningen C, Simonneaux V, Pevet P, and Buijs RM. (2003). Suprachiasmatic control of melatonin synthesis in rats: Inhibitory and stimulatory mechanisms. Eur J Neurosci 17: 221–228.

Pitrosky B, Kirsch R, Vivienroels B, Georgbentz I, Canguilhem B, and Pevet P. (1995). The photoperiodic response in Syrian-hamster depends upon a melatonin-driven circadian-rhythm of sensitivity to melatonin. J Neuroendocrinol 7: 889–895.

Preitner N, Damiola F, Molina LL, Zakany J, Duboule D, Albrecht U, and Schibler U. (2002). The orphan nuclear receptor REV-ERB alpha controls circadian transcription within the positive limb of the mammalian circadian oscillator. Cell 110: 251–260.

Reppert SM, Weaver Dr, and Ebisawa T. (1994). Cloning and characterization of a mammalian melatonin receptor that mediates reproductive and circadian responses. Neuron 13: 1177–1185.

Reppert SM, Weaver DR, and Godson C. (1996). Melatonin receptors step into the light: Cloning and classification of subtypes. Trends Pharmacol Sci 17: 100–102.

Revel FG, Saboureau M, Masson-Pevet M, Pevet P, Mikkelsen JD, and Simonneaux V. (2006a). Kisspeptin mediates the photoperiodic control of reproduction in hamsters. Curr Biol 16: 1730–1735.

Revel FG, Saboureau M, Pevet P, Mikkelsen JD, and Simonneaux V. (2006b). Melatonin regulates type 2 deiodinase gene expression in the Syrian hamster. Endocrinology 147: 4680–4687.

Rodriguez EM, Blazquez JL, Pastor FE, Pelaez B, Pena P, Peruzzo B, and Amat P. (2005). Hypothalamic tanycytes: A key component of brain-endocrine interaction. Int Rev Cytol 247: 87–164.

Seminara SB, Messager S, Chatzidaki EE, Thresher RR, Acierno JS, Shagoury JK, Bo-Abbas Y, Kuohung W, Schwinof KM, Hendrick AG, Zahn D, Dixon J, Kaiser UB, Slaugenhaupt SA, Gusella JF, et al. (2003). The GPR54 gene as a regulator of puberty. New Engl J Med 349: 1614–1618.

Simonneaux V and Ribelayga C. (2003). Generation of the melatonin endocrine message in mammals: A review of the complex regulation of melatonin synthesis by norepinephrine, peptides, and other pineal transmitters. Pharmacol Rev 55: 325–395.

Stehle JH, Foulkes NS, Molina CA, Simonneaux V, Pevet P, and Sassonecorsi P. (1993). Adrenergic signals direct rhythmic expression of transcriptional repressor CREM in the pineal-gland. Nature 365: 314–320.

Stirland JA, Johnston JD, Cagampang FRA, Morgan PJ, Castro MG, White MRH, Davis JRE, and Loudon ASI. (2001). Photoperiodic regulation of prolactin gene expression in the Syrian hamster by a pars tuberalis-derived factor. J Neuroendocrinol 13: 147–157.

Tamarkin L, Westrom WK, Hamill AI, and Goldman BD. (1976). Effect of melatonin on reproductive systems of male and female Syrian-hamsters—diurnal rhythm in sensitivity to melatonin. Endocrinology 99: 1534–1541.

von Gall C, Garabette ML, Kell CA, Frenzel S, Dehghani F, Schumm-Draeger PM, Weaver DR, Korf HW, Hastings MH, and Stehle JH. (2002). Rhythmic gene expression in pituitary depends on heterologous sensitization by the neurohormone melatonin. Nat Neurosci 5: 234–238.

Wagner GC, Johnston JD, Clarke IJ, Lincoln GA, and Hazlerigg DG. (2008). Redefining the limits of day length responsiveness in a seasonal mammal. Endocrinology 149: 32–39.

Weaver DR, Liu C, and Reppert SM. (1996). Nature's knockout: The Mel(1b) receptor is not necessary for reproductive and circadian responses to melatonin in Siberian hamsters. Mol Endocrinol 10: 1478–1487.

Woodfill CJI, Robinson JE, Malpaux B, and Karsch FJ. (1991). Synchronization of the circannual reproductive rhythm of the ewe by discrete photoperiodic signals. Biol Reprod 45: 110–121.

Xu Y, Tamamaki N, Noda T, Kimura K, Itokazu Y, Matsumoto N, Dezawa M, and Ide C. (2005). Neurogenesis in the ependymal layer of the adult rat 3rd ventricle. Exp Neurol 192: 251–264.

Yasuo S, Nakao N, Ohkura S, Iigo M, Hagiwara S, Goto A, Ando H, Yamamura T, Watanabe M, Watanabe T, Oda S, Maeda K, Lincoln GA, Okamura H, Ebihara S, and Yoshimura T. (2006). Long-day suppressed expression of type 2 deiodinase gene in the mediobasal hypothalamus of the Saanen goat, a short-day breeder: Implication for seasonal window of thyroid hormone action on reproductive neuroendocrine axis. Endocrinology 147: 432–440.

Yasuo S, Watanabe M, Iigo M, Nakamura TJ, Watanabe T, Takagi T, Ono H, Ebihara S, and Yoshimura T. (2007a). Differential response of type 2 deiodinase gene expression to photoperiod between photoperiodic Fischer 344 and nonphotoperiodic Wistar rats. Am J Physiol Regul Integr Comp Physiol 292: R1315–R1319.

Yasuo S, Watanabe M, Nakao N, Takagi T, Follett BK, Ebihara S, and Yoshimura T. (2005). The reciprocal switching of two thyroid hormone-activating and -inactivating enzyme genes is involved in the photoperiodic gonadal response of Japanese quail. Endocrinology 146: 2551–2554.

Yasuo S, Yoshimura T, Ebihara S, and Korf HW. (2007b). Temporal dynamics of type 2 deiodinase expression after melatonin injections in Syrian hamsters. Endocrinology 148: 4385–4392.

Yoshimura T, Yasuo S, Watanabe M, Iigo M, Yamamura T, Hirunagi K, and Ebihara S. (2003). Light-induced hormone conversion of T_4 to T_3 regulates photoperiodic response of gonads in birds. Nature 426: 178–181.

Epilogue: Future Directions

David L. Denlinger, David E. Somers, and Randy J. Nelson

Because of the tilt and elliptical orbit of our planet around the sun, ambient conditions change dramatically over the course of the year. Seasonal fluctuations in energy availability create thermoregulatory and other physiological demands that serve as potent selective forces in the evolution of life. At high latitudes and altitudes, the seasonal fluctuations in temperature between winter and summer are substantial, and individuals that inhabit these challenging conditions have evolved mechanisms to cope with wide alterations in ambient conditions. Animals that withstand severe seasonal fluctuations undergo such striking seasonal adaptations that summer- and winter-captured individuals of the same species can be mistaken for different species. Development of these seasonal adaptations permits plants and animals to exploit specific spatial niches that vary in quality over time.

Appropriate timing of reproduction is perhaps the central seasonal adjustment made by living organisms. To prepare in advance for seasonal changes in ambient conditions, plants and animals must be able to determine the season of the year. Seasonal adaptations often require significant time to develop; thus, the ability to ascertain the time of year in order to *anticipate* conditions is critical for survival. Several mechanisms have evolved that provide seasonal information. In some animal species, an annual clock, analogous to the circadian clock, has evolved; in other species, especially in plants, circadian clocks are used to determine day lengths (i.e., photoperiod), and day length information is transduced into seasonally appropriate responses.

Because breeding cycles are perhaps the most salient of seasonal phenomena, it is not surprising that most studies of photoperiodism have focused on reproductive functions. Production of meat animals, eggs, or flowers and control of insect breeding have driven much research on photoperiodism. However, studies on animal growth or fur development have waned over the past two decades. This may reflect the dwindling number of photoperiodic traits and species under study.

One trend that is evident by reviewing the literature from the past 25 years is a reduction in the number of taxa that currently are being studied. For example, among invertebrates, there has been a conspicuous decline in the use of some invertebrates that were once popular study organisms, for example, mollusks and copepods, as researchers shift their attention more to model organisms. The power of model organisms cannot be denied, but in some cases, model organisms (e.g., *Drosophila melanogaster*) have a photoperiodic response that is much less robust than in other species. It is also evident that important differences exist between species. For example, the clock proteins PERIOD and TIMELESS form a heterodimer that enters the nucleus in *D. melanogaster* but remains in the cytoplasm in Lepidoptera. The route of reception for light signals also varies among species. Sometimes the light signals used for photoperiodism enter through the eyes, whereas in other cases the reception is extraretinal. Such differences suggest it is not possible to develop an adequate model that will be universally applicable from information derived from a single species, and thus we argue that a broad comparative approach is still important for developing a comprehensive understanding of photoperiodism.

Features that are unique to certain species can also be exploited for addressing specific questions in this field. For example, as pointed out in chapter 8 by Numata and Udaka, the large but relatively few neurons in snails may uniquely facilitate progress in deciphering function of the neural network regulating photoperiodism. Among mammals, studies of photoperiod have become mainly focused on two model systems, hamsters (long-day breeders) and sheep (short-day breeders). This focus on model species may yield general rules and mechanisms underlying photoperiod time measurement, but this approach neglects the enormous diversity in seasonal traits and potentially ignores broad solutions to common seasonal challenges. The trends in plant research are very promising, however, as the initial molecular studies in the annual dicotyledonous long-day plant *Arabidopsis* have set the stage for very recent parallel work in rice (an annual short-day monocot), Japanese morning glory (the highly photoperiodically responsive short-day dicot, *Pharbitis*, = *Ipomoea*), and the small, aquatic dicot *Lemna*. Additionally, perennial woody plants, such as poplar, have also come under study very recently. As even more divergent plant species (e.g., the moss *Physcomitrella*) are examined, the commonality of the underlying mechanism of photoperiodism will become much clearer. The emerging themes so far suggest a great degree of similarity, fundamentally grounded in a circadian-clock based timing scheme. A return to some the species used in the early, classic studies of photoperiodism, armed with the new molecular findings from model plants, could even further elucidate photoperiodic processes.

Another intriguing but open question is how photoperiodism links with other signaling systems to exert seasonal effects. Clearly, diverse forms of environmental input can feed into a common decision-making site. For example, in many aquatic organisms, including snails and copepods, temperature cycles may be just as important as photoperiod in dictating seasonal events, and temperature signals can replace photoperiodism in insects living in tropical regions. The common pattern for programming an overwintering diapause in temperate region insect species is for low temperature to somehow enhance the effect of short day lengths. A similar pattern is observed among small mammals as low temperatures enhance the reproductive inhibitory signals provided by short days. Although some attractive models have been proposed, how temperature modifies or replaces the photoperiodic signals remains poorly understood. In plants, this synthesis has already begun (chapter 6), instigated by the observation that low-temperature treatment can potentiate responsiveness to day-length signals. A molecular framework is in place, but there is still much to be done to fill the gaps. Additionally, the exciting finding of a common molecular link between photoperiodically induced flower formation and dormancy onset in poplar (chapter 5) provides another system where the role of temperature in bud break and dormancy/cold hardiness now can be examined in a new context.

Our understanding of the molecular mechanisms of photoperiodism lags behind our understanding of the molecular control of circadian rhythmicity. The power of *Drosophila* genetics has yielded an impressive array of genes linked to the regulation of circadian rhythms. Recent progress on photoperiodism in *Arabidopsis* is beginning to close that gap for plants, but there are still many missing links in our knowledge of photoperiodism in animal systems. Though most of the genes first identified in *Drosophila* appear to be present also in other animal systems, in only a few cases do we have precise information concerning the functions of these genes in photoperiodism. Photoperiodism also raises some unique questions that are not quite as germane to the circadian literature. Commonly, the reception of short day lengths involved in programming a photoperiodic event (e.g., insect diapause) occurs far in advance of the manifestation of that event, whereas most circadian events occur in close temporal proximity to reception of the effective light signal. For example, the short day lengths that are critical for programming a pupal diapause may be received by the brain during embryonic development or in some cases a full generation earlier than the stage that enters diapause. How is such information stored by the brain and then later retrieved to evoke a developmental response at the correct time? Though it is clear that the brain is the storage site of such information, we have few insights as to how such information is stored. The termination of a developmental arrest such as diapause requires a precisely timed release of neurohormones that is obviously dictated by the brain, but how the brain knows when to release these hormonal signals is unclear. In most cases, photoperiodic signals are not involved in the termination of diapause, only in the inception, but there are some well-known cases (e.g., the silk moth *Antheraea pernyi*), where day length provides the trigger for reinitiating development, as well.

Often the photoperiodic response is a binary decision, for example, undergo diapause or puberty, change color, flower, or grow. How does the annual cycle of gradually changing day length get translated to a yes–no decision? Melatonin appears to be a physiological signal for night length among many vertebrates. Because melatonin secretion is limited to night, the duration of melatonin secretion appears to encode night length. Thus, a relatively long duration of melatonin secretion depicts a long night (or short day), whereas a relatively short duration of melatonin secretion encodes a short night (or long day), resulting in a winter or summer phenotype, respectively. The concept of "critical day length" was initially proposed as a decision point above and below which long-day breeders exhibit the summer and winter phenotypes, respectively. Critical day lengths were originally established with a protocol that transferred individuals between invariant long or short photoperiods (e.g., LD 16:8 to LD 8:16) in a single day, an event that fails to resemble the gradual changes in day length experienced in nature (<6 min/day). Generally, the critical day length is determined under conditions of thermoneutrality, access to ad libitum food and water, social isolation, and other conditions that seldom occur in nature. Effects of photoperiod history and simulated natural photoperiods require adjustment of the critical day length concept. The pattern of change in day length codes important information, and in some instances, changes in day length are more salient than absolute length of the day. How these subtle patterns in day length change are translated into the molecular mechanisms underlying photoperiodism remains unspecified in many animal systems. In plants, the phasing of the expression of the *CONSTANS* gene and the photosensitivity of its protein product provide good insights into this mechanism. However, there are numerous *CONSTANS-LIKE* genes for which a role in photoperiodism has not yet been discerned. Plants often have gene families where apparent redundancy can obscure subtle effects of various members, or lead to errant presumptions of similar function. More work is needed to determine how the *CONSTANS-LIKE* genes fit into the larger picture.

What common molecular themes are emerging? Cyclic expression of clock genes in the suprachiasmatic nucleus (SCN) of the mammalian hypothalamus and in extra-SCN tissues is a key component of an autoregulatory feedback loop that confers circadian periodicity on these cells. Because several of these genes are phase-locked to dusk and dawn, their expression provides a molecular representation of ambient photoperiod, which may complement the neuroendocrine representation of day length contained in the nightly melatonin signal. Expression of at least some clock genes is driven by melatonin treatment; for example, in hamsters and sheep, melatonin suppresses *Per1* expression in the pars tuberalis, and the dawn rise in *Per1* expression reflects the withdrawal of circulating melatonin. In contrast, expression of other clock genes (e.g., *Cry1* and *Cry2*) is elevated by melatonin and peaks after dusk. Which features of the rhythmic production of melatonin-evoked and melatonin-suppressed clock gene expression provide the internal representation of ambient day length? Obvious parameters include (1) the duration of elevated production of a given clock gene product, (2) phase relations between distinct clock gene products, and (3) the simple existence of rhythmic production of a given clock gene product. Although the phasing of clock gene expression may recapitulate the phasing of the *CONSTANS* gene expression in plants, it remains to be determined how these changes in gene expression are translated into cellular signals that provoke seasonal adaptations.

Common molecular themes will also surely emerge as we continue to expand research on dormancy mechanisms to additional species. One theme that is already evident from work on insects, brine shrimp, mammals, and plants is the common expression of stress resistant genes during dormancy. Insulin signaling is also a theme common to nematodes and insects, although relatively few species have actually been examined thus far. Cell cycle arrest is another feature likely to be shared by diverse species during dormancy. But at this point, information remains fragmentary. Few upstream and downstream connections have yet been made, so we have only pieces of the puzzle that we hope will someday fit into a comprehensive network leading from light perception to execution of the photoperiodic response. Diapause is the photoperiodic response among invertebrates that is best known, but equally interesting will be deciphering such networks for seasonal polyphenisms and complex seasonal patterns of development as seen in aphids. Good model systems for probing photoperiodically cued morphological changes are needed. Progress toward deciphering links between photoperiod and the hormonal action that regulates seasonal polyphenisms is currently hampered by our poor current understanding of the neurohormones and molecular events that direct the morphological responses. We anticipate that some of the sequencing initiatives currently under way will be helpful in identifying key elements involved in photoperiodism in new nonmodel systems. Upcoming sequencing information, coupled with the success of RNA interference, portends a bright future for probing photoperiodism in nonmodel organisms that display a robust photoperiodic response.

Recently, the role of thyroid hormones as important mediators of photoperiodism in animals has emerged. Although the importance of thyroid hormones in seasonal adaptations among fishes has been known for some time, their importance in avian and mammalian photoperiod responsiveness is only now becoming appreciated. As noted in the chapters by Yoshimura and Sharp (chapter 18) and Hazlerigg (chapter 21), photoperiod has relatively small, if any, influences on circulating triiodothyronine (T_3) or thyroxine (T_4). Thyroidectomy does not block short-day–induced gonadal regression. However, short days or melatonin treatment decreases expression of hypothalamic iodothyronine deiodinase type 2 (DIO2; converts T_4 to T_3) but increases DIO3 (inactivates T_3 and T_4). Thus, local hypothalamic conversion of T_4 into the biologically active T_3 may be an evolutionary conserved mechanism to modulate photoperiodic responsiveness among vertebrates.

Two recent, human-driven events raise new concerns relevant to photoperiodism. It is evident that photoperiodism plays a critical role in coordinating plant and animal development with the seasonal environment, fine-tuning periods of activity and reproduction with periods of food supply. The rapid temperature change that the earth is now experiencing implies that current photoperiodic signals will soon not be appropriate for identification of the most suitable seasons. Hopefully, there is sufficient genetic variability in photoperiodic responses to allow organisms to adapt to new climatic realities by altering their photoperiodic responses and/or by migrating to newly appropriate seasonal environments, but clearly these sorts of changes have the potential for massive disruption of current population structures, species interactions, and geographic distributions. Plants may be particularly sensitive to changes in ambient environmental temperature because bud break—the initiation of new growth in the spring—is largely temperature driven. This means that at higher temperatures there will be earlier growth initiation, which could put the plant developmentally out of synchrony with the photoperiod that would normally be optimal for floral initiation, or in competition with other plants that might not normally be present in high abundance at that time of year. A very recent report illustrates that a warming climate may already be changing distributions of plant species, to the detriment of some (Pennisi, 2008). Interestingly, day-neutral species, where flowering is not based on day length but on temperature or developmental cues, may be the most plastic and therefore better able to adapt to these changes.

A less appreciated change is the dramatic and accelerating increase in light pollution, as pointed out in chapter 9 by Marcus and Scheef. Although we are not aware of any documented cases of disruption of photoperiodism as a result of light pollution, this is an issue deserving of attention. Because the threshold for light perception in photoperiodism is especially low, the potential for disrupting seasonal timing of development in plants and animals through this human activity is a distinct possibility.

Finally, the extent to which humans are influenced by photoperiod deserves attention. It is certainly the case that humans display strong seasonal adjustments in physiology and behavior, including depression, sleep, cardiovascular disease, homicides, multiple sclerosis, autism, food intake, and energy expenditure. The extent to which human seasonality reflects photoperiodism, however, remains largely unspecified. Because humans have developed a number of ways to buffer themselves from the elements, it is surprising that seasonal traits remain extant. In common with other mammals, humans release melatonin from the pineal gland during the night. The duration of human melatonin secretion reflects night length and thus provides a convenient mechanism by which photoperiod could be measured. Also, in common with other mammals examined, the nightly melatonin secretion can be truncated by light exposure, and the role of light at night on seasonal adaptations among industrialized humans remains controversial. There remains no convincing evidence that melatonin is linked to annual fluctuations in human breeding. However, other annual rhythms may be significantly affected by photoperiod and its underlying transduction mechanisms, including the duration of melatonin secretion. Thus, understanding and manipulation of human seasonal rhythms as part of the treatment for seasonal affective disorder, sleep disorders, infectious diseases, bipolar disorders, cancers, or inflammatory diseases, including cardiovascular disease, is important. An appreciation of the interaction of seasonal rhythms in susceptibility to disease in host species and the seasonal rhythm of vector prevalence could be important in understanding important diseases such as malaria. An appreciation of such photoperiod-driven fluctuations could also provide insight into emerging diseases. For example, the incidence of sudden acute respiratory syndrome (SARS) might reflect the seasonal proximity between humans and masked palm civets in rural China. Thus, an understanding of the mechanisms underlying photoperiodism among different flora and fauna remains an important goal in the integration of our knowledge of complex interactions in biology.

Reference

Pennisi E. (2008). Evolution: Where have all Thoreau's flowers gone? Science 321: 24–25.

Index